Advances in Experimental Medicine and Biology

Volume 982

More information about this series at http://www.springer.com/series/5584

Gaetano Santulli
Editor

Mitochondrial Dynamics in Cardiovascular Medicine

Foreword by Dr. Wilson S. Colucci

Springer

Editor
Gaetano Santulli
Dept. of Biomedical Advanced Sciences
Federico II University
Naples, Italy

Columbia University Medical Center
New York Presbyterian Hospital—Manhattan
New York, NY, USA

ISSN 0065-2598 ISSN 2214-8019 (electronic)
Advances in Experimental Medicine and Biology
ISBN 978-3-319-55329-0 ISBN 978-3-319-55330-6 (eBook)
DOI 10.1007/978-3-319-55330-6

Library of Congress Control Number: 2017937863

Printed on acid-free paper

This Springer imprint is published by Springer Nature
The registered company is Springer International Publishing AG
The registered company address is: Gewerbestrasse 11, 6330 Cham, Switzerland

Foreword

The field of mitochondrial biology is rapidly changing. Mitochondria certainly play a fundamental role in providing energy for all cell types, and are of particular importance in meeting the finely tuned energetic demands of the cardiovascular system. In addition, it is increasingly recognized that mitochondria are also key participants in a large number of nonenergetic biological pathways that, on the one hand, are essential to normal cellular function, and on the other hand, are central to a plethora of pathological processes that lead to cardiovascular disease.

The goal of this compendium is to provide a single place where the reader can obtain timely and authoritative information in this evolving field. Thanks to the informed guidance of the editor, Dr. Gaetano Santulli, *Mitochondrial Dynamics in Cardiovascular Medicine* brings together articles from prominent experts in the areas of mitochondrial biology, physiology, and disease – all with a focus on the cardiovascular system.

Basic areas of physiology are summarized and provide the background for detailed, cutting-edge reviews of the role of mitochondria in a wide array of cardiovascular processes from cardiac hypertrophy and heart failure to arrhythmias and ischemia. This book concludes with chapters on potential new therapeutic approaches – mitochondrial transplantation and mitochondria-targeted antioxidants. The liberal use of color figures, diagrams, and tables add to the value of this resource.

This is a book that will appeal to readers of all backgrounds, from student to scholar. Individually, the chapters in this book are a valuable resource. Together, they provide a comprehensive state-of-the-art treatment of the role of this dynamic and exciting organelle in the cardiovascular system.

Boston University School of Medicine
Boston, MA, USA

Wilson S. Colucci M.D.

Contents

1 **Functional Implications of Cardiac Mitochondria Clustering** 1
Felix T. Kurz, Miguel A. Aon, Brian O'Rourke, and Antonis A. Armoundas

2 **Mitochondrial Calcium Handling in Physiology and Disease** 25
Veronica Granatiero, Diego De Stefani, and Rosario Rizzuto

3 **The In Vivo Biology of the Mitochondrial Calcium Uniporter** 49
Julia C. Liu, Randi J. Parks, Jie Liu, Justin Stares, Ilsa I. Rovira, Elizabeth Murphy, and Toren Finkel

4 **Mitochondrial Bioenergetics and Dysfunction in Failing Heart** 65
Freya L. Sheeran and Salvatore Pepe

5 **Mitochondrial Mutations in Cardiac Disorders** 81
Sung Ryul Lee and Jin Han

6 **Mitochondrial Function in Non-ischemic Heart Failure** 113
Anisha A. Gupte and Dale J. Hamilton

7 **Mitochondria in Ischemic Heart Disease** 127
L. Maximilian Buja

8 **Mitochondrial Bioenergetics During Ischemia and Reperfusion** 141
Alicia E. Consolini, María I. Ragone, Patricia Bonazzola, and Germán A. Colareda

9 **Mechanistic Role of mPTP in Ischemia-Reperfusion Injury** 169
Giampaolo Morciano, Massimo Bonora, Gianluca Campo, Giorgio Aquila, Paola Rizzo, Carlotta Giorgi, Mariusz R. Wieckowski, and Paolo Pinton

10 Functional Role of Mitochondria in Arrhythmogenesis 191
Jessica Gambardella, Daniela Sorriento, Michele Ciccarelli, Carmine Del Giudice, Antonella Fiordelisi, Luigi Napolitano, Bruno Trimarco, Guido Iaccarino, and Gaetano Santulli

11 Mitochondria and Cardiac Hypertrophy 203
Heberty di Tarso Fernandes Facundo, Robert Eli Brainard, Francisco Rodrigo de Lemos Caldas, and Aline Maria Brito Lucas

12 Connexin 43 and Mitochondria in Cardiovascular Health and Disease 227
Kerstin Boengler and Rainer Schulz

13 Mitochondrial Mechanosensor Microdomains in Cardiovascular Disorders 247
Michele Miragoli and Aderville Cabassi

14 Mechanistic Role of Thioredoxin 2 in Heart Failure 265
Chaofei Chen, Haixuan Chen, Huanjiao Jenny Zhou, Weidong Ji, and Wang Min

15 Mitochondria in Structural and Functional Cardiac Remodeling 277
Natalia Torrealba, Pablo Aranguiz, Camila Alonso, Beverly A. Rothermel, and Sergio Lavandero

16 Functional Role of Nox4 in Autophagy 307
Maurizio Forte, Silvia Palmerio, Derek Yee, Giacomo Frati, and Sebastiano Sciarretta

17 Mitochondrial Ubiquitin Ligase in Cardiovascular Disorders 327
Tao Yu, Yinfeng Zhang, and Pei-feng Li

18 Nitrite-Nitric Oxide Signaling and Cardioprotection 335
Matthias Totzeck, Ulrike B. Hendgen-Cotta, and Tienush Rassaf

19 Peripheral Blood Mitochondrial DNA and Myocardial Function 347
Tatiana Kuznetsova and Judita Knez

20 Mitochondrial Proton Leak Plays a Critical Role in Pathogenesis of Cardiovascular Diseases 359
Jiali Cheng, Gayani Nanayakkara, Ying Shao, Ramon Cueto, Luqiao Wang, William Y. Yang, Ye Tian, Hong Wang, and Xiaofeng Yang

21 Mitochondria and Angiogenesis 371
Raluca Marcu, Ying Zheng, and Brian J. Hawkins

22 High-Density Lipoprotein Regulation of Mitochondrial Function 407
C. Roger White, Geeta Datta, and Samantha Giordano

23 MitomiRs Keep the Heart Beating 431
Samarjit Das, Hannah R. Vasanthi, and Ramesh Parjapath

24 Mitochondrial Dysfunction in Cardiovascular Aging 451
Charles L. Hoppel, Edward J. Lesnefsky, Qun Chen, and Bernard Tandler

25 Insulin Resistance and Mitochondrial Dysfunction 465
Alba Gonzalez-Franquesa and Mary-Elizabeth Patti

26 Mechanistic Role of Kinases in the Regulation of Mitochondrial Fitness 521
Daniela Sorriento, Jessica Gambardella, Antonella Fiordelisi, Bruno Trimarco, Michele Ciccarelli, Guido Iaccarino, and Gaetano Santulli

27 Mitochondria Damage and Kidney Disease 529
Pu Duann and Pei-Hui Lin

28 Mitochondrial Dysfunction in the Diabetic Kidney 553
Kumar Sharma

29 Prohibitin Signaling at the Kidney Filtration Barrier 563
Christina Ising and Paul T. Brinkkoetter

30 Mitochondrial Heteroplasmy 577
George B. Stefano, Christina Bjenning, Fuzhou Wang, Nan Wang, and Richard M. Kream

31 Mitochondrial Transplantation in Myocardial Ischemia and Reperfusion Injury 595
Borami Shin, Douglas B. Cowan, Sitaram M. Emani, Pedro J. del Nido, and James D. McCully

32 Mitochondria-Targeted Antioxidants for the Treatment of Cardiovascular Disorders 621
Hyoung Kyu Kim and Jin Han

Chapter 1
Functional Implications of Cardiac Mitochondria Clustering

Felix T. Kurz, Miguel A. Aon, Brian O'Rourke, and Antonis A. Armoundas

Mitochondria in cardiac myocytes are densely packed to form a cell-wide network of communicating organelles that accounts for about 35% of the myocyte's volume [1]. The lattice-like arrangement of mitochondria, mostly in long and dense rows parallel to the cardiac myocyte myofilaments, has the structural features of a highly ordered network [2]. This specific mitochondrial network architecture ensures that the large demand of adenosine triphosphate (ATP) of cardiac myocytes is met and appropriately distributed. Mitochondria synthesize approximately 30 kg of ATP each day to both provide energy for the basic cellular metabolism and to secure basic physiological functions of the cardiovascular system such as the maintenance of pulmonary and systemic blood pressure during heart contractions [3]. Their intracellular position is closely associated with the sarcoplasmic reticulum and the myofilaments to facilitate cellular distribution of ATP [4]. However, the role of mitochondrial structure and function is not only limited to ATP generation;

F.T. Kurz (✉)
Department of Neuroradiology, Heidelberg University Hospital, Heidelberg, Germany

Massachusetts General Hospital, Cardiovascular Research Center, Harvard Medical School, Charlestown, MA, USA
e-mail: felix.kurz@med.uni-heidelberg.de

M.A. Aon
Laboratory of Cardiovascular Science, National Institute on Aging, National Institutes of Health, Baltimore, MD, USA

B. O'Rourke
Division of Cardiology, Department of Medicine, Johns Hopkins University, Baltimore, MD, USA

A.A. Armoundas (✉)
Massachusetts General Hospital, Cardiovascular Research Center, Harvard Medical School, Charlestown, MA, USA
e-mail: aarmoundas@partners.org

G. Santulli (ed.), *Mitochondrial Dynamics in Cardiovascular Medicine*, Advances in Experimental Medicine and Biology 982,
DOI 10.1007/978-3-319-55330-6_1

mitochondria participate in and control numerous metabolic pathways and signaling cascades such as the calcium signaling, redox oxidation, β-oxidation of fatty acids, oxidative phosphorylation, the synthesis of aminoacids, heme and steroids, and cellular apoptosis [5]. Their structural and functional diversity thus surpasses any other cellular organelle.

Specifically, the control of mitochondrial oxidative phosphorylation is dependent on the arrangement, interplay and interdependence of mitochondrial ion channels, transporters, and pumps that are located in an inner membrane that is selectively permeable to enable the adjustment of mitochondrial volume, inner membrane potential $\Delta\Psi_m$ and mitochondrial redox potential [6]. Thus, the mitochondrial membranes facilitate and balance cellular energy supply and demand.

The Mitochondrial Network in Cardiac Myocytes

Mitochondrial Inner Membrane Oscillations

When subjected to critical amounts of oxidative stress or substrate deprivation, the mitochondrial network in cardiac myocytes may pathophysiologically transition into a state where the inner membrane potential $\Delta\Psi_m$ of a large amount of mitochondria depolarizes and oscillates [6–9]. Oscillations of individual mitochondria in intact myocytes were first documented in the early 1980s when Berns et al. excited quiescent cardiac myocytes with a focal laser beam to induce transient cycles of alternating $\Delta\Psi_m$ depolarization and repolarization [10]. Mitochondrial inner membrane depolarizations were also detected in smooth muscle cells [11], cultured neurons [12], and in isolated mitochondria [13]. A functional link between metabolic oscillations and cardiac function was demonstrated when O'Rourke et al. [14] showed that cyclical activation of ATP-sensitive potassium currents in guinea pig cardiac myocytes under substrate deprivation was associated with low frequency oscillations in the myocyte's action potential duration and excitation-contraction coupling. These oscillations were accompanied by a synchronous oxidation and reduction of the intra-cellular nicotinamide adenine dinucleotide (NADH) concentration.

While it was initially believed that this mechanism was due to changes in glucose metabolism, subsequent studies detected an association of the oscillations with redox transients of mitochondrial flavoprotein and waves of mitochondrial $\Delta\Psi_m$ depolarization [15], therefore implying a mitochondrial origin of the phenomenon. Network-wide mitochondrial membrane potential oscillations could be initiated reliably by the application of a localized laser-flash that induced a depolarization of $\Delta\Psi_m$ in just a few mitochondria [16]. The association of $\Delta\Psi_m$ oscillations with reactive oxygen species (ROS) was discovered by Zorov and Sollot who showed that a localized laser-flash produced a high number of free radicals in cardiac mitochondria in an autocatalytic manner, concomitant with depolarization, referred to as ROS-induced ROS release [17, 18]. Experimentally, synchronous $\Delta\Psi_m$ oscillations were observed in small contiguous mitochondrial clusters as well as in large clusters that span the whole myocyte [16], or as mitochondrial $\Delta\Psi_m$ depolarization waves that could pass between neighboring cardiac myocytes through intercalated discs [15]. Mitochondrial communication may also be facilitated through the formation of intermitochondrial junctions [19], which could facilitate wave propagation.

Generally, mitochondrial network coordination is heavily influenced by the intricate balance of ROS generation and ROS scavenging within the myocyte: when a critical amount of network mitochondria exhibits a destabilized inner membrane potential, the increased presence of ROS can lead the network into a state of ROS-induced ROS release [16, 20]. It is likely that the antioxidant flux is lowered to a point at which ROS production cannot be scavenged fast enough – this causes a flux mismatch that rapidly goes out of control such that multiple mitochondria form a myocyte-wide cluster of synchronously oscillating mitochondria, also called spanning cluster; these mitochondria lock into an oscillatory mode with low frequencies and high amplitudes [6, 8, 20].

The cardiac mitochondrial network can also behave in an oscillatory mode in the absence of metabolic or oxidative stress. A computational model of the mitochondrial oscillator has predicted mitochondrial behavior that was characterized by low-amplitude, high-frequency oscillations [21], which was later demonstrated experimentally [8]. Since, the collective mitochondrial behavior happened (i) in the absence of metabolic/oxidative stress, (ii) did not involve collapse or large excursions of the membrane potential (i.e., microvolts to a few millivolts), and (iii) could be clearly distinguished from the high-amplitude, low-frequency oscillations involving strongly coupled mitochondria through elevated ROS levels, this regimen was dubbed "physiological" [6, 8]. The transition between normal and pathological behavior in mitochondrial function can be determined by the abundance and interplay between cytoplasmic (Cu, Zn SOD) and matrix (Mn SOD) superoxide dismutases with ROS generation in the respiratory chain, as recently demonstrated [22].

In intact cardiac myocytes, ROS generation from mitochondrial respiration mainly originates in complex III and complex I, and is favored by highly reduced redox potentials and high $\Delta\Psi_m$, typically corresponding to either inhibition or low respiratory flux [6, 16]. The occurrence of $\Delta\Psi_m$ limit-cycle oscillations could be suppressed by adding inhibitors of mitochondrial respiration or ligands of the mitochondrial benzodiazepine receptor, but not adding inhibitors of the mitochondrial permeability transition pore (mPTP) [23]. This class of $\Delta\Psi_m$ oscillations was consistent with a ROS-induced ROS release mechanism involving the activation of mitochondrial inner membrane anion channels (IMAC), see also below and [6, 16, 24].

Mitochondrial Morphology, Dynamics and Morphodynamics

Changes in the morphology of mitochondria or their dynamic characteristics can influence their collective behavior. In cardiac myocytes, mitochondria can be perinuclear, intermyofibrilar and subsarcolemmal, and possess different morphologies as well as proteomic and physiological status [25–29]. This (static) morphological heterogeneity is associated with different mitochondrial functions [30, 31]; for instance, mitochondria in intermyofibrillar locations show a reduced content of flavoproteins compared to those beneath the sarcolemma, thus resulting in different oxidative capacities for NADH- and FADH-generating substrates [30, 32].

The role of dynamic heterogeneity becomes important for pertubations of the mitochondrial network during oxidative or metabolic stress: the mitochondrial network's response to stress can affect not only intracellularly clustered mitochondria, but scale to the majority of cellular mitochondria and even to the whole-organ level, thus leading to emergent phenomena, such as ventricular arrhythmias [20, 33–35]. In many non-cardiac myocytes, mitochondrial mobility and morphological plasticity or morphodynamics is affected by proteins that facilitate fusion and fission processes, by cytoskeletal proteins (through disruptive structural changes of the cytoskeleton) [36], and changes in myocyte energy metabolism or calcium homeostasis [37, 38]. Recently, it was shown that redox signaling also occurs in clusters of morphologically changing axonal mitochondria, suggesting a morphological and functional inter-mitochondrial coupling in neurons [39, 40]. However, apart from "pre-fusion" events at inter-mitochondrial junctions [19], fusion and fission events as well as mitochondrial mobility have not been observed in adult cardiac myocytes so far, although mitochondrial fusion has been observed in skeletal muscle [41]. However, there is growing evidence from mitochondrial protein gene studies that mitochondrial morphodynamics may play a role in the maintenance of mitochondrial fitness [42].

Spatio-Temporal Behavior of Individual Mitochondrial Oscillators

While many studies have investigated the response of the whole mitochondrial network in external stress stimuli (i.e. ischemia-reperfusion injury or oxidative stress) at the inner mitochondrial membrane [8, 16], more recent ones have focused on the role of individual mitochondria and their spatio-temporal behavior [9, 35, 40, 43, 44].

Alterations of the mitochondrial membrane potential $\Delta\Psi_m$ can be monitored with the positively charged lipophilic fluorescent dyes tetramethylrhodamine methyl or ethyl ester (TMRM or TMRE), that redistribute from the matrix to the cytoplasm with a depolarization of $\Delta\Psi_m$ [45] (see Fig. 1.1). Myocyte-wide $\Delta\Psi_m$ oscillations can then be triggered by a perturbation of the mitochondrial network through thiol-oxidizing substances such as diamide, or by a highly localized laser flash [16, 46]: the ensuing sequence of locally confined ROS generation and induced autocatalytic ROS release [16, 17, 47] can lead to a critical accumulation of ROS that initiates a propagating wave of $\Delta\Psi_m$ depolarization [7]. However, the reaction of each individual mitochondrion within the myocyte is highly variable: while a major-

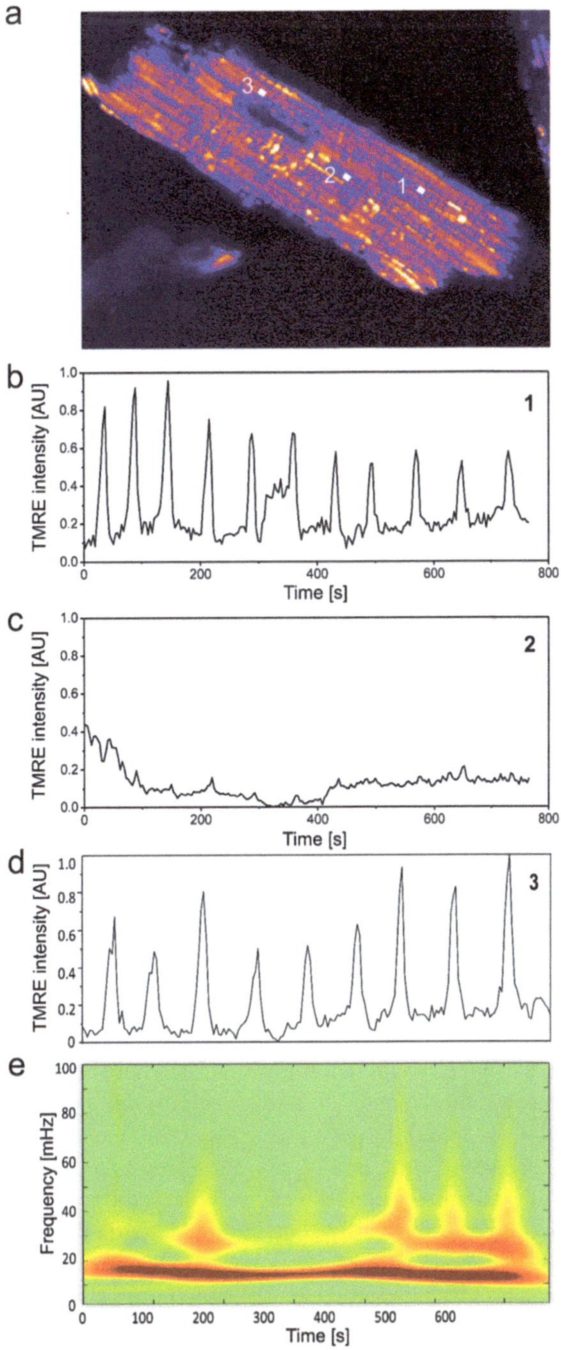

Fig. 1.1 Dynamic mitochondrial oscillations and wavelet transform analysis. (**a**) Densely packed mitochondria in an adult cardiac myocyte, monitored with the fluorescence dye TMRE. (**b–d**) Perturbations of the mitochondrial network can result in oscillatory patterns of individual mitochondria that differ in occurrence and frequency content. (**e**) Dynamic frequency content of oscillating mitochondria can be analyzed with the absolute wavelet transform: the major frequency component of mitochondrion 3 (**d**) varies between 15 and 20 mHz (Reprinted with permission from Ref. [81]. Copyright 2014)

ity of mitochondria synchronize to cell-wide $\Delta\Psi_m$ oscillations, some mitochondria do not participate whereas others participate only temporarily (see Fig. 1.1b–d) [9, 43]. This diversified mitochondrial behavior (in frequency and amplitude) is directly associated with the scavenging capacity of intracellular ROS on a local level within the myocyte, as well as the inner membrane efflux of superoxide ions.

A large cluster of mitochondria with synchronized $\Delta\Psi_m$ oscillations can recruit additional mitochondria in analogy to a network of weakly coupled oscillators (see below), whereas some mitochondria may exit the cluster, i.e. show $\Delta\Psi_m$ oscillations that are not synchronized with the cluster, because their local ROS defense mechanisms are exhausted [9, 43]. Thus, this complex interplay of ROS generation and scavenging determines the individual mitochondrial frequency and amplitude.

Since the frequencies of these oscillations change in time, they are best described by methodologies that do not assume an a priori stationarity of $\Delta\Psi_m$ signals, such as wavelet transform analysis. Unlike standard methods of statistical analysis, that can only be applied if the nonstationarity of the signals is associated with the low-frequency part of the observed power spectrum, wavelet analysis allows analyzing the full range of frequencies of the nonstationary mitochondrial network. The wavelet analysis provides a time-resolved picture of the frequency content of each mitochondrial signal, see Fig. 1.1e. Earlier methodologies, e.g. relative dispersion analysis or power spectral analysis, although they provided a glimpse of the functional mitochondrial network properties such as the long-term memory of mitochondrial oscillatory dynamics [7], probed the average behavior of the whole myocyte (single signal trace) [8], without discriminating individual mitochondria. Therefore, application of wavelet analysis on individual mitochondria to characterize the dynamic behavior of the mitochondrial network across the myocyte, was a significant step forward in the systematic analysis of the whole population of mitochondrial oscillators, accounting for both their frequency dynamics and spatio-temporal correlations [9, 43].

Mitochondrial Clusters with Similar Oscillatory Dynamics

It has been shown that mitochondria whose signals are highly correlated during a specific time range, i.e. mitochondria with a similar frequency content, could be grouped into different clusters [9]. Then, the largest cluster of similarly oscillating mitochondria forms a highly correlated and coherent $\Delta\Psi_m$ signal that drives the cell-wide oscillations which are observed during the recording. These major clusters do not necessarily contain mitochondria that demonstrate strong spatial contiguity (albeit the data is showing only a cross section of the network, therefore synchronized clusters that are "non-contiguous" could be connected as a 3D structure through other planes), nor do they have to coincide with local ROS levels as in the case of a "spanning cluster" as proposed in [8]. In fact, the spatial contiguity of major clusters is substrate dependent (such as glucose, pyruvate, lactate and β-hydroxybutyrate) through their influence on the cellular redox potential [35].

The size of mitochondrial clusters is related to the cluster frequency and phase: mitochondria within the cluster adjust their frequencies and phases to a common one, while large clusters comprising many (coupled) mitochondria need a longer time to synchronize [48, 49]. Similarly, clusters with large separation distances among their components or a low contiguity, take longer to synchronize compared to clusters with smaller separation distances among their components. In both cases, the common cluster frequency remains low and is characterized by an inverse relation between common cluster frequency and cluster size [9, 35].

For cardiac myocytes superfused with glucose, pyruvate, lactate or β-hydroxybutyrate, local laser-flashing still leads to synchronized cell-wide oscillations of mitochondrial clusters. However, the network's adaptation to the specific metabolic conditions leads to differences in mitochondrial cluster synchronization and dynamics. These differences can be attributed to the altered cardiac myocyte catabolism and ROS/redox balance [35]. For substrates such as pyruvate and β-hydroxybutyrate that generally lead to strongly reduced redox potential (in comparison with glucose or lactate), the rate of change of cluster size versus averaged cluster frequency is higher than for perfusion with glucose or lactate, see Fig. 4 in [35]; yet, typical cluster frequencies for β -hydroxybutyrate were found in a much smaller range (5–20 mHz) than those for pyruvate (10–70 mHz) [35]. A large range of mitochondrial frequencies is a strong indicator of desynchronization processes that are based on dynamic heterogeneity. Likewise, an increased dispersion in cluster size, as observed for perfusion with pyruvate and β-hydroxybutyrate, is a signature of an increasingly fragmented mitochondrial network with several smaller clusters of similar size that have different frequencies. In contrast, a low mitochondrial frequency dispersion is an indicator of a rapid increase in the formation of mitochondrial network clusters during ROS-induced ROS release, such that an increasingly rapid recruitment of mitochondrial oscillators would result in a decrease of clusters with different frequencies. Since in a pathophysiological state, a slow growth in cluster size after the rapid onset of a synchronized cell-wide cluster oscillation is a sign of exhausted ROS scavenging capacity (reflected by a decrease of $\Delta\Psi_m$ amplitude), quantification of cluster synchronization and contiguity properties provide important information about the myocyte's propensity to death [9]. These findings help provide a better understanding of the altered cellular metabolism observed in diabetic cardiomyopathy or heart failure, where the myocyte's response to the switch in substrate utilization, e.g. through an increased utilization of glucose, can be protective or maladaptive [49–51].

Apart from isolated cardiac myocytes, such a relation could also be observed in mitochondria in intact hearts preparations, where neighboring cardiac myocytes are linked electrically and chemically via gap junctions [35, 52]. In those experiments, mitochondrial populations of several myocytes show cell-wide $\Delta\Psi_m$ signal oscillations following a laser flash, similar to those observed in isolated or pairs of cardiac myocytes [15, 52]. Such behavior could also be observed in rat salivary glands in vivo [53]. Generally, mitochondrial oscillations in intact tissue are faster than in isolated myocytes with an approximate frequency range of 10–80 mHz in cardiac tissue and 50–200 mHz in salivary glands [35, 53], and their frequency is not as strongly influenced by the size of the cluster as in isolated myocytes [35].

The Mitochondrial Network as a Network of Coupled Oscillators

The intricate spatio-temporal dynamics of mitochondrial networks results from many forms of inter-mitochondrial interactions that involve several metabolic pathways. Such pathways include the modification of gene expression, electrochemical interactions that are mediated by ROS through ROS-induced ROS release, other (e.g. exogenous) forms of control exerted on the local redox environment [54], and morphological changes in neighboring mitochondria [19]. In general terms, signaling or communication between (intra-cellular) organelles modulates the relation between cellular networks that involve the transformation (generation) of mass (energy), as represented by the metabolome, and those networks that carry information as represented by the proteome, genome, and transcriptome. A unique phenotypical signature of a myocyte is then given by the fluxome which represents the network of all cellular reaction-fluxes: it interacts with all other networks and provides a description of the dynamic content of all cellular processes [55]. Emergent properties of the cardiac mitochondrial network, i.e. properties that cannot be derived from the simple sum of behavior of the network's individual components represent a distinguishing trait of biological complexity and provide the opportunity to study the dynamical coordination of functional intra-cellular relations.

Emergent Properties of the Mitochondrial Network: Mitochondrial Criticality

Biological networks exhibit mostly scale-free dynamics and topologies: the structure of the network is represented by several network hubs, i.e. nodes with a large number of links to other nodes, and a large number of nodes with low connectivity [56]. Analysis of emergent properties of these networks, e.g. large-scale responses to external or internal stimuli (perturbations), synchronization, general collective behavior, vulnerability to damage and ability to self-repair, has been the subject of intensive research endeavors [57–59]. In those studies, the scale-free character of the networks is revealed by the presence of network constituents with different frequencies on highly dependent temporal scales [60]. However, as pointed out by Rosenfeld et al. [61], the network's robustness to perturbations and modularity cannot be determined using a reductionist approach from knowledge of all elementary biochemical reactions at play, but instead requires network conceptualizations known from systems biology.

In the cardiac network of mitochondrial oscillators, the synchronized, myocyte-wide, low-frequency, high-amplitude depolarizations of $\Delta\Psi_m$ represent an emerging self-organized spatio-temporal process in response to a perturbation of individual mitochondrial oscillators that are weakly or strongly coupled in the physiological or pathophysiological state, respectively [8, 62–64]. The oscillation period of an individual mitochondrial oscillator is therefore not static but can be modulated by long-term temporal correlations within the mitochondrial network [8, 21]. Such synchronization processes in the form of biochemical rhythms, as observed in many systems of chemically, physically or chemico-physically coupled oscillators, serve the purpose of generating and sustaining spatio-temporal organization through numerous regulatory mechanisms: the resulting dynamically changing oscillations are used by cardiac myocytes to achieve and control internal organization [62–65]. During oxidative or metabolic stress, the mitochondrial network's response can affect not only intracellularly clustered mitochondria, but scale-up to affect the majority of cellular mitochondria and even the whole-organ, thus leading to myocardial malfunction, and ultimately to ventricular arrhythmias [24, 33, 34]. For example, during reperfusion following ischemic injury, cardiac myocytes become oxidatively stressed due to substrate deprivation (e.g., oxygen, glucose), leading to mitochondrial membrane potential collapse at the onset of oscillations [16, 46], transient inexcitability [16] and, eventually, ventricular fibrillation [24, 34, 66]. Experimental measures to prevent these effects include the application of the mBzR antagonist 4′-chlorodiazepam that modulates IMAC activity [34]. The prominent role of IMAC opening in ROS-induced ROS release and $\Delta\Psi_m$ depolarization was further supported by blocking mPTP via cyclosporin A, which had no effect on ensuing post-ischemic arrhythmias [24, 34] (see also below and Fig. 1.2).

As was previously demonstrated [9, 20], synchronization of mitochondrial oscillations can be achieved by a critical amount of spontaneously synchronized oscillators, a concept termed ‘mitochondrial criticality’ [7, 20]: a number of synchronously oscillating mitochondria recruits more mitochondria to eventually lead the majority of network mitochondria into myocyte-wide synchronized $\Delta\Psi_m$ oscillations. A transition of this critical threshold into a state of collective, synchronized mitochondrial behavior is equivalent to that of a phase transition between states of matter in thermodynamics. However, in the mitochondrial network it describes a transition from a physiological state into a pathophysiological state. Such transitions among large populations of locally connected mitochondria can be described with the percolation theory which takes into consideration the long-range connectivity of (random) systems and employs a probabilistic model to assess the collective dynamics of the inter-connected network of mitochondria [67]. For ideal lattice-like networks, which is a convenient description of the dense, quasi-uniform and spatially arranged cardiac mitochondria, the theoretical percolation threshold can be identified as pc = 0.59; this threshold is close to experimentally obtained values of mitochondrial criticality where a transition into myocyte-wide $\Delta\Psi_m$ oscillations was observed in about 60% of all mitochondria within the mitochondrial network with significantly

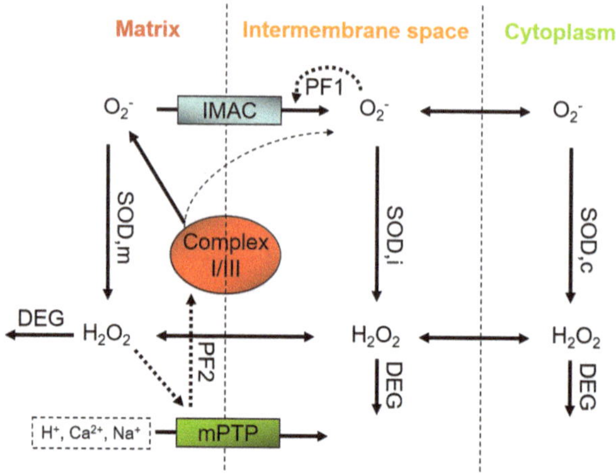

Fig. 1.2 Schematic description of mitochondrial ROS-induced ROS release. The mitochondrial respiration complexes I and III produce superoxide anions (O_2^-) and release them in the mitochondrial matrix. The anions can then reach the intermembrane space (IS) through IMAC, although small quantities directly exit complex III (dashed arrow that originates from complex III). Superoxide dismutases (SOD) in matrix, IS and cytoplasm convert superoxide anions into hydrogen peroxide that can diffuse freely between all compartments. Its degradation is effected by peroxidases (DEG). One positive feedback loop (PF1) is initiated by the activation of IMAC through superoxide anions in the IS to eventually depolarize the mitochondrial inner membrane potential [16, 21]. The resulting stimulation of SODs can, if passing a critical threshold, deactivate IMAC to allow electron transport and repolarization of the membrane potential. Byproducts of hydrogen peroxide-initiated redox processes can trigger mPTP opening (e.g. oxidized lipids or hydroxyl radicals), thus leading to (temporarily) increased production of superoxide anions in the respiratory chain to further open mPTPs in a positive feedback loop (PF2). (This diagram has been adapted from its original version in Yang et al. [71]).

increased ROS levels [20]. For mitochondria in close vicinity to the percolation threshold, small perturbations can thus provoke a transition into global limit cycle oscillations.

Stochasticity and Mitochondrial Network Modeling

Computational models may help obtain a deeper understanding of the intricate interdependencies of mitochondrial networks that lead to emergent behavior. Today, there are several computational models that describe mitochondrial network dynamics, see e.g. [68] for a detailed review.

A mathematical model that has employed ordinary differential equations to represent matrix concentration time rates of ADP, NADH, calcium, Krebs cycle intermediates and $\Delta\Psi_m$ has been recently reported by Cortassa et al. [69]. In this model, the role of IMAC opening in ROS-induced ROS release and ROS diffusion dynamics were incorporated to correctly reproduce mitochondrial respiration, calcium dynamics, bioenergetics, and the generation of $\Delta\Psi_m$ limit cycle oscillations [21]. The model provided evidence for the existence of synchronized clusters of mitochondria that entrain more mitochondria to participate in synchronous $\Delta\Psi_m$ oscillations [47]. Other models have been developed based on the hypothesis of mPTP-mediated ROS-induced ROS release [18, 70] or a combination of mPTP- and IMAC-mediated ROS-induced ROS release [71] (see also Fig. 1.2 for physiological details). Another probabilistic, agent-based model has considered the switch of superoxide anions to hydrogen peroxide, to model ROS signaling for increasingly distant neighboring mitochondria, thus facilitating the inhibition of ROS-induced ROS release through gluthathione peroxidase 1 in the cytosol [72].

Dynamic tubular reticulum models of mitochondrial networks that involve the modeling of molecular fission and fusion events have been recently proposed [73, 74]. Some use graph-based models of the mitochondrial network with a set of simple rules of network connectivity, see e.g. [75], which predict topological properties of the mitochondrial network. However, this approach does not take into account spatial information about the proximity and positioning of neighboring mitochondria.

Overall, the dynamic response of mitochondrial networks to stimuli is due to the stochastic nature of the underlying processes. These processes can be modeled by the addition of intrinsic noise (that emerges from the stochastic processes at small timescales), into system-defining differential equations [61, 76, 77]. However, although in low-dimensional non-linear systems the introduction of random noise may lead to deterministic chaos [78], in high-dimensional non-linear systems with strong links among their components stochasticity may exhibit synchronized behavior [79], as is the case for cardiac mitochondria.

In this case, criteria for stochasticity of high-dimensional non-linear systems are easily fulfilled for strong links between components [79]. This is especially the case for mitochondria in cardiac myocytes that are morphologically and functionally coupled to reveal complex spatio-temporal dynamics [19, 80, 81], see also below.

Mitochondrial Coupling in a Stochastic Phase Model

Synchronization phenomena in networks of coupled nonlinear oscillators were first studied mathematically a few decades ago by Winfree and Kuramoto [82, 83]. The Kuramoto model assumes a (large) group of identical or nearly identical limit-cycle oscillators that are weakly coupled and whose oscillations are driven by intrinsic natural frequencies which are normally distributed around a mean frequency [84]. It provides an exact solution for the critical strength of coupling that allows a transition into a fully synchronized state [85]. The simplicity of the model and its analytical power have made it a frequently applied model in the context of circadian biology, but also in many other systems that contain biologically, physically or chemically coupled oscillators (reviewed in detail in [84, 85]).

The coupling constant in Kuramoto models is time-independent and identical for every pair of coupled oscillators [84]. In cardiac myocytes, however, the dynamic behavior of mitochondrial oscillating frequencies, as well as the presence of structurally varying areas of mitochondrial density or arrangement within the myocyte suggest a position-dependent and time-varying mitochondrial coupling.

We have recently studied the cardiac inter-mitochondrial coupling by adopting an extended Kuramoto-type model, to determine the time-dependent coupling constants of individual mitochondrial oscillators [35]. In this model, we combined wavelet analysis results with a set of stochastic differential equations that describe drifting frequencies with the intention to develop a minimum-order model that uses only two parameters to characterize inter-mitochondrial coupling: the intrinsic oscillatory frequency and the mitochondrial coupling constant [35]. By mapping and identifying each mitochondrion within the mitochondrial network, the exact nearest neighbors of each mitochondrion were used to describe the spatial variance of mitochondrial coupling, i.e. inter-mitochondrial coupling constants were considered for the local nearest-neighbor environment. The concept of local coupling acknowledges inter-mitochondrial coupling through physical connections, i.e. coupling through diffusion of chemical coupling agents such as ROS and/or forms of inter-mitochondrial communication at inter-mitochondrial junctions [19]. It also emphasizes the role of local ROS scavenging mechanisms that prevent long-range diffusion. Using this approach, we were able to determine the averaged dynamic coupling constant (of local mean field type) of each mitochondrion to all of its nearest neighbors, see Fig. 1.3. Quantitative understanding of local myocardial coupling has important implications in the study of the myocyte's response to metabolic and oxidative challenges during its transition to a pathophysiological state.

Coupling Dynamics in Mitochondrial Frequency Clusters

Inter-mitochondrial coupling within the cardiac myocyte can be studied by comparing the spatio-temporal distribution of inter-mitochondrial coupling constants of mitochondria that belong to the (largest) mitochondrial cluster with similar frequencies, against those of the remaining ("non-cluster") mitochondria. Cellular conditions that promote low coupling constant values for the cluster mitochondria versus the remaining oscillating mitochondria (that may also be organized in clusters), indicate similar structural organization principles [35].

By relating the inter-mitochondrial coupling within a cluster with the coherence (a measure of the similarity of the mitochondrial $\Delta\Psi m$ oscillations) of the cluster (Fig. 1.3d), we have shown that larger clusters show stronger cluster coupling constants rendering more coherent cluster oscillations. However, for small cluster areas that occur when a myocyte is superfused with lactate (see also Fig. 4 in [35]), an increase in cluster coherence is not accompanied by an increase in cluster coupling. Such behavior is indicative of a greater influence of spatial rather than temporal coupling on the network of mitochondrial oscillators. During superfusion with

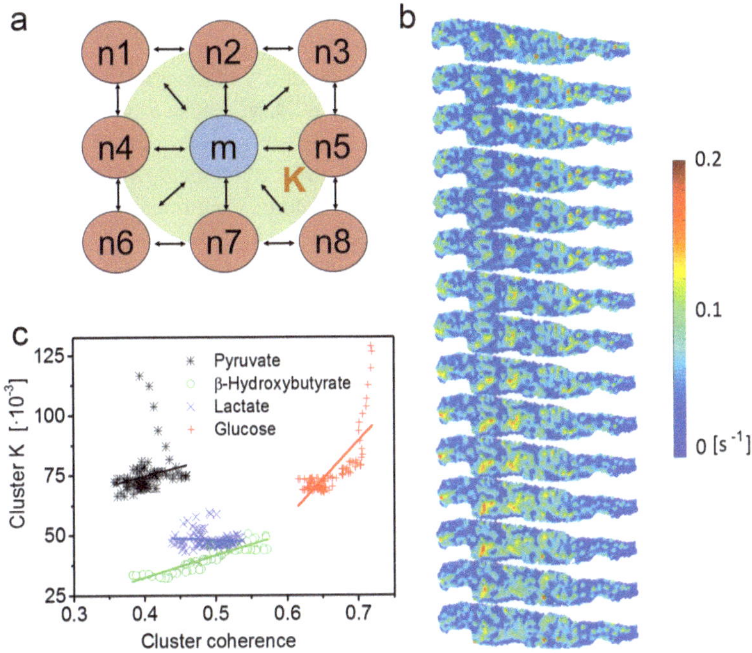

Fig. 1.3 Dynamic inter-mitochondrial coupling and relation to frequency cluster coherence. (**a**) Dynamically changing inter-mitochondrial coupling can be described with a stochastic phase model that provides local coupling constants K of mean-field form (*green cloud*) for each mitochondrion *m* to its nearest neighbors *n*. Inter-mitochondrial coupling is mediated by diffusive coupling agents that interact with the inner mitochondrial membrane, see also Fig. 1.2. (**b**) Maps of coupling constants at different time-points (from *top* to *bottom* with time intervals of 17.5 s): strong coupling is indicated in yellow and red colors, weak coupling in green and blue colors. Clustered nests of highly coupled mitochondria indicate synchronously oscillating mitochondria, whereas regions with low coupling may be associated with a local exhaustion of antioxidant defense systems. (**c**) Coupling constant K versus coherence for mitochondria in a large frequency cluster (details see main text) for cell-superperfusion with different substrates. Generally, cluster coupling increases with coherence, whereby the rate of increase is associated with the size of the cluster (see also Fig. 4 in [35]). Lactate-perfusion is associated with small cluster areas such that cluster coupling is not increasing with cluster coherence. Linear fit curves are shown in their respective substrate color (Reprinted with permission from Ref. [35]. Copyright 2015)

pyruvate an early decrease of coupling strength (Fig. 1.3d) can be attributed to a strong drift towards the common cluster frequency of those mitochondria with the initially largest difference from the common cluster frequency. When cluster coherence levels at this point are low, which is likely is for small clusters (as experimentally verified for pyruvate perfusion (Fig. S5 in [35]), mitochondria might rapidly drop in and out of the cluster. In addition, the distribution of mitochondrial oscillator frequencies is narrowed during their rapid recruitment into the major frequency cluster, leaving fewer clusters with different frequencies. In contrast, high frequency distributions and high rates of change of cluster size compared to the common cluster frequency are a sign of desynchronization through dynamic heterogeneity.

In this system, the observed strong drift of oscillatory frequencies relates to the fundamental concept of synchronization within the mitochondrial oscillator network: its mechanism can be explained by the dynamics of local ROS generation and scavenging mechanisms that involve the mitochondrial matrix as well as extramitochondrial compartments [86]. Then, the slow drift of each mitochondrial oscillator towards the actual mitochondrial frequency is in agreement with the notion of diffusion-mediated signaling, with ROS as the primary coupling agent that is locally restricted and mainly dependent on local ROS density changes.

For small areas of structurally dense, phase-coupled mitochondrial oscillators, local ROS concentration is strongly increased through ROS-induced ROS release, thus leading to the recruitment of neighboring mitochondria. Consequently, these growing clusters lose spatial contiguity and, therefore, show a decrease in their common frequency which, in turn, leads to a decrease of the frequency of local ROS release and to declining local ROS levels, especially along the cluster perimeter. This effect naturally decreases inter-mitochondrial coupling. The formation of spanning clusters, however, provides for a slight increase in the overall basal ROS level to coincidentally increase local inter-mitochondrial coupling. Generally, these observations indicate a preference of spatial coupling over temporal coupling that affect synchronized mitochondrial oscillator networks: early cluster formation mainly drives the correlation between common cluster frequency and local mitochondrial coupling. During this time, the mean cluster coupling constant is large (see Fig. 5 in [15]). Cluster size and local coupling, however, mainly show correlative effects when the cluster of synchronized mitochondrial oscillators involves a majority of the mitochondrial network. At this point, a spanning cluster has already been formed and most mitochondria participate in cell-wide $\Delta\Psi_m$ oscillations.

ROS as primary inter-mitochondrial coupling agents and products of mitochondrial respiration take a central role in cell homeostasis and its response to oxidative or metabolic stress. Under physiological conditions, mitochondria minimize their ROS emission as they maximize their respiratory rate and ATP synthesis according to the redox-optimized ROS balance hypothesis [87]. Stress alters this relationship through increased ROS release and a diminished energetic performance [88]. Thereby, the compartmentation of the redox environment within the cellular compartments such as the cytoplasm and mitochondria, is important in the control of ROS dynamics and their target processes [86]. Within physiological limits, increased mitochondrial metabolism resulting in enhanced ROS generation can function as an adaptive mechanism leading to increased antioxidant capacity and stress resistance, according to the concept of (mito)hormesis [89, 90]. Generally speaking, controlled ROS emission may serve as a robust feedback regulatory mechanism of both ROS generation and scavenging [88]. Dysfunctional mitochondria alter their redox state to elicit pathological oxidative stress, which has been implicated in ageing [91, 92], metabolic disorders like diabetes [93], as well as in cardiovascular [34, 94], neoplastic [95], and neurological diseases [40, 96]. Chronic exposure to increased ROS levels can affect excitation-contraction coupling and maladaptive cardiac remodeling that may lead or increase the propensity to cardiac arrhythmias [34, 97, 98].

Mitochondrial Scale-Free Network Topology and Functional Clustering

Self-Similar Dynamics and Fractal Behavior in Mitochondrial Networks

The defining feature of time series signals with scale-free dynamics is the similarity of frequency-amplitude relations on small, intermediate, and large time scales. In a network of coupled oscillators this translates into new links between oscillators that form in higher numbers depending upon the number of already existing links. In scale-free networks, the occurrence of hubs (i.e., nodes with multiple input and output links) increases the probability of new links being formed [99], an effect known as *preferential attachment* [58] or the "Matthew effect" in the sociological domain [100].

Such fractal or self-similar dynamics develop because of a multiplicative interdependence among these processes and has been observed in cardiac myocytes subjected to oxidative stress [20, 34, 101], in self-organized chemostat yeast culture [62, 102, 103] and in continuous single-layered films of yeast cells exhibiting large-scale synchronized oscillations of NAD(P)H and $\Delta\Psi_m$ [62, 104]. Using wavelet analysis, and as predicted by fractal dynamics, one observes an intricate scale-free frequency content with similar features on different time scales (Fig. 1.4) [60].

For cardiac mitochondrial networks in the physiological regime, it could be demonstrated that the failure or disruption of a single element beyond a critical threshold can result in a failure of the entire network; such scaling towards higher levels of organization in the mitochondrial network obeys an inverse power law behavior that implies an inherent correlation of cellular processes operating at different timescales [8, 58]. The observation of scale-free dynamics in mitochondrial oscillator networks is in agreement with their representation as weakly coupled oscillators in the physiological domain (in the presence of low ROS levels), and their role as energetic, redox and signaling hubs [105].

Functional Connectivity and Mitochondrial Topological Clustering

Dynamic changes in the network's topology during synchronized myocyte-wide oscillations in the pathophysiological state reveal important information about functional relations between distant mitochondria displaying similar signal dynamics. These relations cannot be explained alone by local coupling through diffusion or structural connections. As in networks of communicating neurons, the spontaneous occurrence of changes in the clusters' oscillatory patterns observed at markedly different locations, indicates strong and robust functional correlations; interestingly,

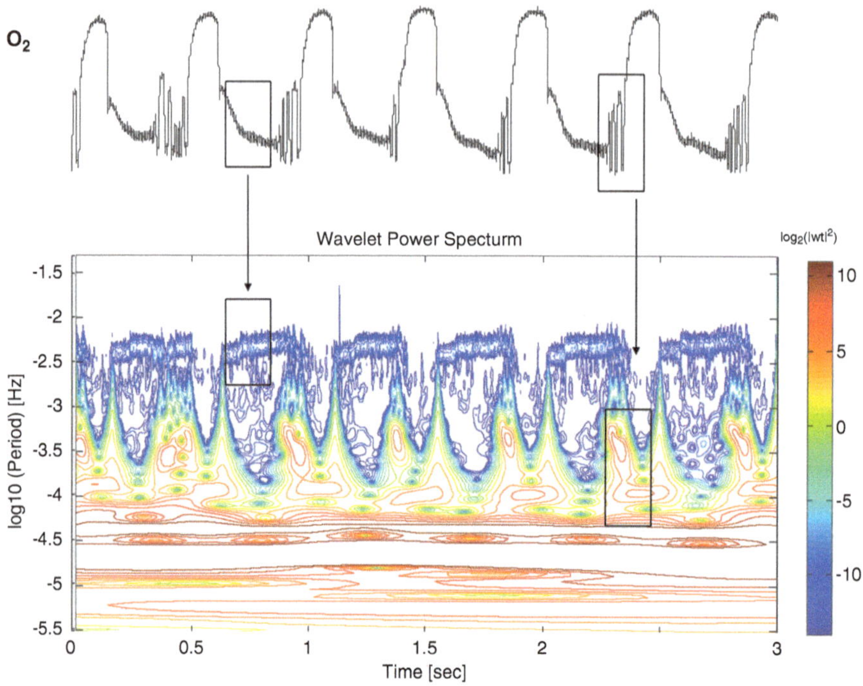

Fig. 1.4 Wavelet analysis of the O_2 signal time series obtained from the self-organized multioscillatory continuous culture of S. cerevisiae. (**a**) Relative membrane inlet mass spectrometric signal of O_2 versus time for S. cerevisiae. The time scale is not shown, but corresponds to hours after the start of continuous fermentor operation. Large-amplitude oscillations show substantial cycle-to-cycle variability with 13.6±1.3 h (N = 8). Other evident oscillation periods are ~40 min and ~4 min. This figure was provided courtesy of Prof. David Lloyd, affiliated with Cardiff University, United Kingdom. (**b**) Logarithmic absolute squared wavelet transform over logarithmic frequency and time. At any time, the wavelet transform uncovers the predominant frequencies and reveals a complex and fine dynamic structure that is associated with mitochondrial fractal dynamics. *Boxes* and *arrows* show the correspondence between the time series and the wavelet plot

the time-scales of these changes are similar in mitochondrial and neuronal networks [106, 107].

Mitochondrial network functionality can be assessed by measuring the *functional connectedness* between mitochondrial network nodes, i.e. by obtaining a quantitative measure of the functional relations that span the dynamically changing mitochondrial network topology [108].

Such a measure of mitochondrial functional connectedness can be the clustering coefficient, similar to that employed by Eguiluz et al. for scale-free functional brain networks [109]. In that case, for a specific network constituent, the clustering coefficient provides a measure of connectedness between neighbors [110]. For instance, clustering plays an important role in neuronal phase synchronization [111]. In the case of mitochondrial networks, dynamic functional clustering significantly differs from random Erdös-Rényi networks [112], even in the presence of similar number

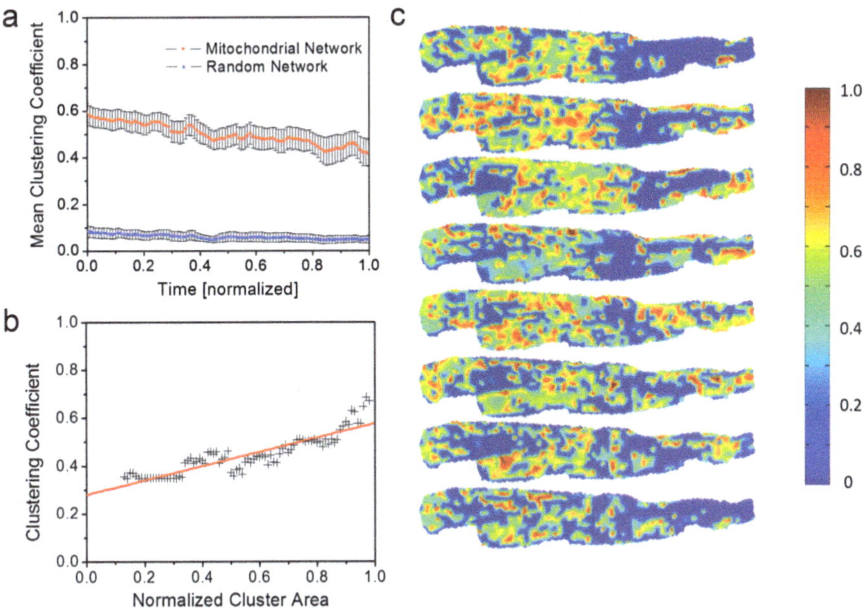

Fig. 1.5 Dynamic functional clustering in mitochondrial networks. (**a**) Comparison of dynamic functional clustering for mitochondrial networks (*red line*) and random networks (*blue line*) with the same number of links on the same set of vertices. Mitochondrial networks are scale-free networks whose functional organization is significantly different from random networks. (**b**) Functional clustering for a large cluster of similarly oscillating mitochondria versus the cluster's area. Functional clustering is increasing with the number of synchronously oscillating mitochondria. (**c**) Consecutive maps of distributions of functional clustering coefficients in a cardiac myocyte. Some regions of the cell maintain a high functional clustering during the recording (*middle left* area of the cell) whereas other regions only temporarily display functional clustering (*far right area* of the cell) (**a** and **b** reprinted with permission from Ref. [81] . Copyright 2014)

of links [81] (Fig. 1.5). This result supports the concept that mitochondrial networks exhibit structural-functional unity [113]. Therefore, dynamic mitochondrial clustering is consistent with its functional properties that are associated with mitochondrial positioning within the network, as previously described [30, 31]. In Fig. 1.5c, the dynamic nature of topological clustering is shown for a glucose-perfused cardiac myocyte during synchronized membrane potential oscillations. While some regions on the left part of the myocyte maintain high clustering, mitochondria on the right side show higher clustering only during the middle portion of the recording, as the spatio-temporal topology evolves.

To further investigate the mitochondrial network dynamics, future studies will need to identify and quantify functional ensembles of strongly interconnected mitochondria [80, 114]. These studies may shed new light into populations of coexistent subsarcolemmal and intermyofibrillar mitochondria that exhibit different roles in the myocyte (patho)physiology [28, 29, 115].

Concluding Remarks

Biological networks within one cell, organ or organism are not isolated entities but exhibit common nodes. Introducing higher-dimensional node degrees generates multi-layer, interdependent, networks. So far, our understanding of such networks is still in its infancy (see [116, 117] for detailed reviews). For example, within the cardiac mitochondrial network, at least two layers can be distinguished, corresponding to structural connections via inter-mitochondrial junctions, and communication through mechanisms that control local ROS balance [19, 35, 80, 118].

Stressful events can trigger synchronized mitochondrial oscillations in which mitochondria functionally self-recruit in frequency clusters [9, 16, 35, 81]. In functional terms, mitochondria can be seen as alternating between states of excitability and quiescence modulated by ROS signaling, thus modulating myocyte's function and energy metabolism [119]. In cardiac myocytes under critical situations, e.g. at myocyte-wide ROS accumulation close to the percolation threshold, inter-mitochondrial functional, structural and dynamic coupling can lead to electrical and contractile dysfunction, and ultimately to myocyte death and whole heart dysfunction. Eventually, under criticality conditions, perturbation of a single mitochondrion could propagate to the remaining mitochondria, leading to myocyte death.

A mitochondrion's gatekeeper role, at the brink of cardiac myocyte death, can only be understood when its function in the mitochondrial network can be predicted either experimentally or through quantitative modeling in which case the mitochondrial parameters are inferred from the available experimental data. Although an increase in computational efficiency has helped modeling make impressive progress in recent years [68], a complete description of the mitochondrial network is still evolving. Stochastic models have been well suited to provide system-wide mitochondrial parameters, such as dynamic coupling constants and functional clustering coefficients in large networks of coupled mitochondrial oscillators.

Synergistic, experimental and computational investigations have unveiled emergent properties of the complex mitochondrial oscillatory networks that describe cluster partitioning of synchronized oscillators with different oscillatory frequencies [120, 121], as well as the chimeric co-existence of synchronized and non-synchronized oscillators [122, 123]. These results coincide with our observations of mitochondrial oscillations in cardiac myocytes, thus supporting the view of network-coupled mitochondrial oscillators. The computational approach promises further discernment of the dynamic spatio-temporal relations within biological networks, while aiding our understanding of its functional implications for cell survival. Therefore, the quantitative description of inter-mitochondrial coupling and functional connectedness provides parameters whose dynamic behavior under pathological conditions may help evaluate the impact of, e.g., metabolic disorders, ischemia-reperfusion injury, heart failure, on mitochondrial networks in cardiac, cancer and neurodegenerative diseases.

To further investigate the mitochondrial network dynamics, future studies will need to identify and quantitate functional ensembles of strongly interconnected mitochondria [80, 114]. These studies may shed new light into populations of coexistent subsarcolemmal and intermyofibrillar mitochondria that exhibit different functional outputs [28, 29, 115].

Acknowledgements The work was supported by a Grant-in-Aid (#15GRNT23070001) from the American Heart Association (AHA) and NIH grant 1 R01 HL135335-01.

References

1. Page E, McCallister LP. Quantitative electron microscopic description of heart muscle cells; application to normal, hypertrophied and thyroxin-stimulated hearts. Am J Cardiol. 1973;31:172–81.
2. Covian R, Balaban RS. Cardiac mitochondrial matrix and respiratory complex protein phosphorylation. Am J Physiol Heart Circ Physiol. 2012;303:H940–66.
3. Ferrari R, Censi S, Mastrorilli F, Boraso A. Prognostic benefits of heart rate reduction in cardiovascular disease. Eur Hear J Suppl. 2003;5:G10–4.
4. Ruiz-Meana M, Fernandez-Sanz C, Garcia-Dorado D. The SR-mitochondria interaction: a new player in cardiac pathophysiology. Cardiovasc Res. 2010;88:30–9.
5. Alberts B, Bray D, Lewis J, Raff M, Roberts K, Watson JD. Molecular biology of the cell. New York: Garland Science; 1995.
6. Aon MA, Cortassa S, O'Rourke B. Mitochondrial oscillations in physiology and pathophysiology. Adv Exp Med Biol. 2008;641:98–117.
7. Aon MA, Cortassa S, Akar FG, O'Rourke B. Mitochondrial criticality: a new concept at the turning point of life or death. Biochim Biophys Acta. 2006;1762(2):232–40.
8. Aon MA, Cortassa S, O'Rourke B. The fundamental organization of cardiac mitochondria as a network of coupled oscillators. Biophys J. 2006;91(11):4317–27.
9. Kurz FT, Aon MA, O'Rourke B, Armoundas AA. Spatio-temporal oscillations of individual mitochondria in cardiac myocytes reveal modulation of synchronized mitochondrial clusters. Proc Natl Acad Sci U S A. 2010;107(32):14315–20.
10. Berns MW, Siemens AE, Walter RJ. Mitochondrial fluorescence patterns in rhodamine 6G-stained myocardial cells in vitro: analysis by real-time computer video microscopy and laser microspot excitation. Cell Biophys. 1984;6(4):263–77.
11. O'Reilly CM, Fogarty KE, Drummond RM, Tuft RA, Walsh JVJr. Spontaneous mitochondrial depolarizations are independent of SR Ca2+ release. Biophys J. 2003;85(5):3350–7.
12. Loew LM, Tuft RA, Carrington W, Fay FS. Imaging in five dimensions: time-dependent membrane potentials in individual mitochondria. Biophys J. 1993;65(6):2396–407.
13. Huser J, Rechenmacher CE, Blatter LA. Imaging the permeability pore transition in single mitochondria. Biophys J. 1998;74(4):2129–37.
14. O'Rourke B, Ramza B, Marban E. Oscillations of membrane current and excitability driven by metabolic oscillations in heart cells. Science. 80-1994;265:962–6.
15. Romashko DN, Marban E, O'Rourke B. Subcellular metabolic transients and mitochondrial redox waves in heart cells. Proc Natl Acad Sci U S A. 1998;95(4):1618–23.
16. Aon MA, Cortassa S, Marban E, O'Rourke B. Synchronized whole cell oscillations in mitochondrial metabolism triggered by a local release of reactive oxygen species in cardiac myocytes. J Biol Chem. 2003;278(45):44735–44.
17. Zorov DB, Filburn CR, Klotz LO, Zweier JL, Sollot SJ. Reactive oxygen species (ROS)-induced ROS release: a new phenomenon accompanying induction of the mitochondrial permeability transition in cardiac myocytes. J Exp Med. 2000;192:1001–14.
18. Zorov DB, Juhaszova M, Sollot SJ. Mitochondrial ROS-induced ROS release: an update and review. Biochim Biophys Acta. 2006;1757:509–17.

19. Picard M. Mitochondrial synapses: intracellular communication and signal integration. Trends Neurosci. 2015;38:468–74.
20. Aon MA, Cortassa S, O'Rourke B. Percolation and criticality in a mitochondrial network. Proc Natl Acad Sci U S A. 2004;101(13):4447–52.
21. Cortassa S, Aon MA, Winslow RL, O'Rourke B. A mitochondrial oscillator dependent on reactive oxygen species. Biophys J. 2004;87(3):2060–73.
22. Kembro JM, Cortassa S, Aon MA. Complex oscillatory redox dynamics with signaling potential at the edge between normal and pathological mitochondrial function. Front Physiol. 2014;5:257.
23. O'Rourke B. Pathophysiological and protective roles of mitochondrial ion channels. J Physiol. 2000;529(1):23–36.
24. Aon MA, Cortassa S, Akar FG, Brown DA, Zhou L, O'Rourke B. From mitochondrial dynamics to arrhythmias. Int J Biochem Cell Biol. 2009;41(10):1940–8.
25. Kuznetsov AV, Usson Y, Leverve X, Margreiter R. Subcellular heterogeneity of mitochondrial function and dysfunction: evidence obtained by confocal imaging. Mol Cell Biochem. 2004;256–257(1–2):359–65.
26. Manneschi L, Federico A. Polarographic analyses of subsarcolemmal and intermyofibrillar mitochondria from rat skeletal and cardiac muscle. J Neurol Sci. 1995;128(2):151–6.
27. Croston TL, Thapa D, Holden AA, Tveter KJ, Lewis SE, Shepherd DL, et al. Functional deficiencies of subsarcolemmal mitochondria in the type 2 diabetic human heart. Am J Physiol Heart Circ Physiol. 2014;307(1):H54–65.
28. Dabkowski ER, Williamson CL, Bukowski VC, Chapman RS, Leonard SS, Peer CJ, et al. Diabetic cardiomyopathy-associated dysfunction in spatially distinct mitochondrial subpopulations. Am J Physiol Heart Circ Physiol. 2009;296(2):H359–69.
29. Lesnefsky EJ, Chen Q, Hoppel CL. Mitochondrial metabolism in aging heart. Circ Res. 2016;118(10):1593–611.
30. Kuznetsov AV, Mayboroda O, Kunz D, Winkler K, Schubert W, Kunz WS. Functional imaging of mitochondria in saponin-permeabilized mice muscle fibers. J Cell Biol. 1998;140(5):1091–9.
31. Lesnefsky EJ, Tandler B, Ye J, Slabe TJ, Turkaly J, Hoppel CL. Myocardial ischemia decreases oxidative phosphorylation through cytochrome oxidase in subsarcolemmal mitochondria. Am J Phys. 1997;273(3 Pt 2):H1544–54.
32. Philippi M, Sillau AH. Oxidative capacity distribution in skeletal muscle fibers of the rat. J Exp Biol. 1994;189:1–11.
33. O'Rourke B, Cortassa S, Aon MA. Mitochondrial ion channels: gatekeepers of life and death. Physiology. 2005;20:303–15.
34. Akar FG, Aon MA, Tomaselli GF, O'Rourke B. The mitochondrial origin of postischemic arrhythmias. J Clin Invest. 2005;115(12):3527–35.
35. Kurz FT, Derungs T, Aon MA, O'Rourke B, Armoundas AA. Mitochondrial networks in cardiac myocytes reveal dynamic coupling behavior. Biophys J. 2015;108(8):1922–33.
36. Anesti V, Scorrano L. The relationship between mitochondrial shape and function and the cytoskeleton. Biochim Biophys Acta. 2006;1757:692–9.
37. Rossignol R, Gilkerson R, Aggeler R, Yamagata K, Remington SJ, Capaldi RA. Energy substrate modulates mitochondrial structure and oxidative capacity in cancer cells. Cancer Res. 2004;64:985–93.
38. Yi M, Weaver D, Hajnoczky G. Control of mitochondrial motility and distribution by the calcium signal: a homeostatic circuit. J Cell Biol. 2004;167:661–72.
39. Breckwoldt MO, Pfister FMJ, Bradley PM, Marinković P, Williams PR, Brill MS, et al. Multiparametric optical analysis of mitochondrial redox signals during neuronal physiology and pathology in vivo. Nat Med. 2014;20(5):555–60.
40. Breckwoldt MO, Armoundas AA, Aon MA, Bendszus M, O'Rourke B, Schwarzländer M, et al. Mitochondrial redox and pH signaling occurs in axonal and synaptic organelle clusters. Sci Rep. 2016;6:23251.

41. Eisner V, Lenaers G, Hajnóczky G. Mitochondrial fusion is frequent in skeletal muscle and supports excitation-contraction coupling. J Cell Biol. 2014;205(2):179–95.
42. Lee S, Park YY, Kim SH, Nguyen OT, Yoo YS, Chan GK, Sun X, Cho H. Human mitochondrial Fis1 links to cell cycle regulators at G2/M transition. Cell Mol Life Sci. 2014;71(4):711–25.
43. Kurz FT, Aon MA, O'Rourke B, Armoundas AA. Wavelet analysis reveals heterogeneous time-dependent oscillations of individual mitochondria. Am J Physiol Heart Circ Physiol. 2010;299(5):H1736–40.
44. Nivala M, Korge P, Nivala M, Weiss JN, Qu Z. Linking flickering to waves and whole-cell oscillations in a mitochondrial network model. Biophys J. 2011;101(9):2102–11.
45. Scaduto RC, Grotyohann LW. Measurement of mitochondrial membrane potential using fluorescent rhodamine derivatives. Biophys J. 1999;76(1):469–77.
46. Aon MA, Cortassa S, Maack C, O'Rourke B. Sequential opening of mitochondrial ion channels as a function of glutathione redox thiol status. J Biol Chem. 2007;282(30):21889–900.
47. Zhou L, Aon MA, Almas T, Cortassa S, Winslow RL, O'Rourke B. A reaction-diffusion model of ROS-induced ROS release in a mitochondrial network. PLoS Comput Biol. 2010;6(1):e1000657.
48. Li P, Yi Z. Synchronization of Kuramoto oscillators in random complex networks. Phys A Stat Mech Appl. 2008;387(7):1669–74.
49. Lopaschuk GD, Ussher JR, Folmes CDL, Jaswal JS, Stanley WC. Myocardial fatty acid metabolism in health and disease. Physiol Rev. 2010;90:207–58.
50. Stanley WC, Recchia FA, Lopaschuk GD. Myocardial substrate metabolism in the normal and failing heart. Physiol Rev. 2005;85(3):1093–129.
51. Aon MA, Cortassa S (n.d.). Mitochondrial network energetics in the heart. Wiley Interdiscip Rev Syst Biol Med. 4(6):599–613.
52. Slodzinski MK, Aon MA, O'Rourke B. Glutathione oxidation as a trigger of mitochondrial depolarization and oscillation in intact hearts. J Mol Cell Cardiol. 2008;45(5):650–60.
53. Porat-Shliom N, Chen Y, Tora M, Shitara A, Masedunskas A, Weigert R, et al. In vivo tissue-wide synchronization of mitochondrial metabolic oscillations. Cell Rep. 2014;9(2):514–21.
54. Cortassa S, Aon MA. Dynamics of mitochondrial redox and energy networks: insights from an experimental-computational synergy. In: Aon MA, Saks V, Schlattner U, editors. Systems biology of metabolic and signaling networks. energy, mass and information transfer. 1st ed. Berlin: Springer; 2013. p. 115–44.
55. Cascante M, Marin S. Metabolomics and fluxomics approaches. Essays Biochem. 2008;45:67–81.
56. Barrat A, Barthélemy M, Vespignani A. Dynamical processes on complex networks. Cambridge, UK: Cambridge University Press; 2008.
57. Albert R, Jeong H, Barabasi A. Error and attack tolerance of complex networks. Nature. 2000;406(6794):378–82.
58. Barabasi AL, Oltvai ZN. Network biology: understanding the cell's functional organization. Nat Rev Genet. 2004;5(2):101–13.
59. Alon U. Network motifs: theory and experimental approaches. Nat Rev Genet. 2007;8(6):450–61.
60. Aon MA, Cortassa S, Lloyd D. Chaos in biochemistry and physiology. Encyclopedia of molecular cell biology and molecular medicine. Weinheim: Wiley-VCH Verlag GmbH & Co. KGaA; 2011.
61. Rosenfeld S. Mathematical descriptions of biochemical networks: stability, stochasticity, evolution. Prog Biophys Mol Biol. 2011;106(2):400–9.
62. Aon MA, Roussel MR, Cortassa S, O'Rourke B, Murray DB, Beckmann M, et al. The scale-free dynamics of eukaryotic cells. PLoS One. 2008;3(11):e3624.
63. Yates FE. Self-organizing systems. In: Boyd CAR, Noble D, editors. The logic of life. Oxford: Oxford University Press; 1993. p. 189–218.

64. Mazloom AR, et al. Modeling a complex biological network with temporal heterogeneity: cardiac myocyte plasticity as a case study. Complex. 2009;1(LNICST 4):467–86.
65. Tsiairis CD, Aulehla A, Aulehla A, Wiegraebe W, Baubet V, Wahl MB, et al. Self-organization of embryonic genetic oscillators into spatiotemporal wave patterns. Cell. 2016;164(4):656–67.
66. Brown DA, Aon MA, Frasier CR, Sloan RC, Maloney AH, Anderson EJ, et al. Cardiac arrhythmias induced by glutathione oxidation can be inhibited by preventing mitochondrial depolarization. J Mol Cell Cardiol. 2010;48(4):673–9.
67. Kesten H. What is ... Percolation? Not Am Math Soc. 2006;53(5):572–3.
68. Zhou L, O'Rourke B. Cardiac mitochondrial network excitability: insights from computational analysis. Am J Physiol Heart Circ Physiol. 2012;302(11):H2178–89.
69. Cortassa S, Aon MA, Marban E, Winslow RL, O'Rourke B. An integrated model of cardiac mitochondrial energy metabolism and calcium dynamics. Biophys J. 2003;84(4):2734–55.
70. Brady NR, Hamacher-Brady A, Westerhoff HV, Gottlieb RA. A wave of reactive oxygen species (ROS)-induced ROS release in a sea of excitable mitochondria. Antioxid Redox Signal. 2006;8(9–10):1651–65.
71. Yang L, Paavo K, Weiss JN, Qu Z. Mitochondrial oscillations and waves in cardiac myocytes: insights from computational models. Biophys J. 2010;98:1428–38.
72. Park J, Lee J, Choi C. Mitochondrial network determines intracellular ROS dynamics and sensitivity to oxidative stress through switching inter-mitochondrial messengers. PLoS ONE. 2011;6:e23211.
73. Liu X, Weaver D, Shirihai O, Hajnóczky G. Mitochondrial "kiss-and-run": interplay between mitochondrial motility and fusion-fission dynamics. EMBO J. 2009;28(20):3074–89.
74. Twig G, Liu X, Liesa M, Wikstrom JD, Molina AJA, Las G, et al. Biophysical properties of mitochondrial fusion events in pancreatic beta-cells and cardiac cells unravel potential control mechanisms of its selectivity. Am J Phys Cell Physiol. 2010;299(2):C477–87.
75. Sukhorukov VM, Dikov D, Reichert AS, Meyer-Hermann M, Scheckhuber C, Erjavec N, et al. Emergence of the mitochondrial reticulum from fission and fusion dynamics. Beard DA, editorPLoS Comput Biol. 2012;8(10):e1002745.
76. Blake WJ, Kaern M, Cantor CR, Collins JJ. Noise in eukaryotic gene expression. Nature. 2003;422(6932):633–7.
77. Kepler TB, Elston TC. Stochasticity in transcriptional regulation: origins, consequences, and mathematical representations. Biophys J. 2001;81(6):3116–36.
78. Lorenz EN. Deterministic nonperiodic flow. J Atmos Sci. 1963;20(2):130–41.
79. Rosenfeld S, Kapetanovic I. Systems biology and cancer prevention: all options on the table. Gene Regul Syst Biol. 2008;2:307–19.
80. Aon MA, Camara AKS. Mitochondria: hubs of cellular signaling, energetics and redox balance. A rich, vibrant, and diverse landscape of mitochondrial research. Front Physiol. 2015;6:94.
81. Kurz FT, Aon MA, O'Rourke B, Armoundas AA. Cardiac mitochondria exhibit dynamic functional clustering. Front Physiol. 2014;5:329.
82. Winfree AT. Biological rhythms and the behavior of populations of coupled oscillators. J Theor Biol. 1967;16:15–42.
83. Kuramoto Y. Chemical oscillations, waves, and turbulence. Berlin: Springer; 1984.
84. Acebrón JA, Bonilla LL, Vicente P, Conrad J, Ritort F, Spigler R. The Kuramoto model: a simple paradigm for synchronization phenomena. Rev Mod Phys. 2005;77:137–85.
85. Strogatz SH. From Kuramoto to Crawford: exploring the onset of synchronization in population of coupled oscillators. Phys D. 2000;143:1–20.
86. Kembro JM, Aon MA, Winslow RL, O'Rourke B, Cortassa S. (n.d.) Integrating mitochondrial energetics, redox and ROS metabolic networks: a two-compartment model. Biophys J. 104(2):332–43.
87. Aon MA, Cortassa S, O'Rourke B. Redox-optimized ROS balance: a unifying hypothesis. Biochim Biophys Acta. 1797;6–7:865–77.
88. Cortassa S, O'Rourke B, Aon MA. Redox-optimized ROS balance and the relationship between mitochondrial respiration and ROS. Biochim Biophys Acta. 2014;1837(2):287–95.

89. Calabrese EJ. Hormesis and medicine. Br J Clin Pharmacol. 2008;66(5):594–617.
90. Ristow M, Zarse K. How increased oxidative stress promotes longevity and metabolic health: the concept of mitochondrial hormesis (mitohormesis). Exp Gerontol. 2010;45(6):410–8.
91. Bratic I, Trifunovic A. Mitochondrial energy metabolism and ageing. Biochim Biophys Acta. 1797;6–7:961–7.
92. Aon MA, Cortassa S, Juhaszova M, Sollott SJ. Mitochondrial health, the epigenome and healthspan. Clin Sci (Lond). 2016;130(15):1285–305.
93. Tocchetti CG, Caceres V, Stanley BA, Xie C, Shi S, Watson WH, et al. GSH or palmitate preserves mitochondrial energetic/redox balance, preventing mechanical dysfunction in metabolically challenged myocytes/hearts from type 2 diabetic mice. Diabetes. 2012;61(12):3094–105.
94. Stary V, Puppala D, Scherrer-Crosbie M, Dillmann WH, Armoundas AA. SERCA2a upregulation ameliorates cellular alternans induced by metabolic inhibition. J Appl Physiol. 2016;120(8):865–75.
95. Wallace DC. Mitochondria and cancer. Nat Rev Cancer. 2012;12(10):685–98.
96. Nunnari J, Suomalainen A. Mitochondria: in sickness and in health. Cell. 2012;148(6):1145–59.
97. Giordano FJ. Oxygen, oxidative stress, hypoxia, and heart failure. J Clin Invest. 2005; 115(3):500–8.
98. Maack C, Dabew ER, Hohl M, Schäfers H-J, Böhm M. Endogenous activation of mitochondrial KATP channels protects human failing myocardium from hydroxyl radical-induced stunning. Circ Res. 2009;105(8):811–7.
99. Keller EF. Revisiting "scale-free" networks. Bioessays. Wiley Periodicals, Inc. 2005;27(10): 1060–68.
100. Merton RK. The Matthew effect in science. Science. (80-)1968;159:56–63.
101. West BJ. Fractal physiology and the fractional calculus: a perspective. Front Physiol. 2010;1:12.
102. Roussel MR, Lloyd D. Observation of a chaotic multioscillatory metabolic attractor by real-time monitoring of a yeast continuous culture. FEBS J. 2007;274(4):1011–8.
103. Murray AJ. Mitochondria at the extremes: pioneers, protectorates, protagonists. Extrem Physiol Med. 2014;3:10.
104. Aon MA, Cortassa S, Lemar KM, Hayes AJ, Lloyd D. Single and cell population respiratory oscillations in yeast: a 2-photon scanning laser microscopy study. FEBS Lett. 2007;581(1):8–14.
105. Aon M, Cortassa S, O'Rourke B. Chapter 4: On the network properties of mitochondria. In: Saks V, editor. Molecular system bioenergetics. Weinheim: Wiley-VCH Verlag GmbH & Co. KG; 2007.
106. Fox MD, Snyder AZ, Vincent JL, Corbetta M, Van Essen DC, Raichle ME. The human brain is intrinsically organized into dynamic, anticorrelated functional networks. Proc Natl Acad Sci U S A. 2005;102(27):9673–8.
107. Greicius MD, Krasnow B, Reiss AL, Menon V. Functional connectivity in the resting brain: a network analysis of the default mode hypothesis. Proc Natl Acad Sci U S A. 2003;100(1):253–8.
108. Passingham RE, Stephan KE, Kötter R. The anatomical basis of functional localization in the cortex. Nat Rev Neurosci. 2002;3(8):606–16.
109. Eguíluz VM, Chialvo DR, Cecchi GA, Baliki M, Apkarian AV. Scale-free brain functional networks. Phys Rev Lett. 2005;94(1):18102.
110. Watts DJ, Strogatz SH. Collective dynamics of "small-world" networks. Nature. 1998;393:440–2.
111. Percha B, Dzakpasu R, Zochowski M, Parent J. Transition from local to global phase synchrony in small world neural network and its possible implications for epilepsy. Phys Rev E Stat Nonlinear Soft Matter Phys. 2005;72(3 Pt 1):31909.
112. Erdős P, Rényi A. On the evolution of random graphs. Publ Math Inst Hung Acad Sci. 1960;5:17–61.
113. Viola HM, Arthur PG, Hool LC. Evidence for regulation of mitochondrial function by the L-type Ca2+ channel in ventricular myocytes. J Mol Cell Cardiol. 2009;46(6):1016–26.

114. Harenberg S, Bello G, Gjeltema L, Ranshous S, Harlalka J, Seay R, et al. Community detection in large-scale networks: a survey and empirical evaluation. Wiley Interdiscip Rev Comput Stat. 2014;6(6):426–39.
115. Dabkowski ER, Baseler WA, Williamson CL, Powell M, Razunguzwa TT, Frisbee JC, et al. Mitochondrial dysfunction in the type 2 diabetic heart is associated with alterations in spatially distinct mitochondrial proteomes. Am J Physiol Heart Circ Physiol. 2010;299(2): H529–40.
116. Kivela M, Arenas A, Barthelemy M, Gleeson JP, Moreno Y, Porter MA. Multilayer networks. J Complex Networks. Oxford University Press2014;2(3):203–71.
117. Boccaletti S, Bianconi G, Criado R, del Genio CI, Gómez-Gardeñes J, Romance M, et al. The structure and dynamics of multilayer networks. Phys Rep. 2014;544(1):1–122.
118. Kurz FT, Kembro JM, Flesia AG, Armoundas AA, Cortassa S, Aon MA, et al. Network dynamics: quantitative analysis of complex behavior in metabolism, organelles, and cells, from experiments to models and back. Wiley Interdiscip Rev Syst Biol Med. 2017;9(1):e1352.
119. Wang W, Gong G, Wang X, Wei-LaPierre L, Cheng H, Dirksen R, et al. Mitochondrial flash: integrative reactive oxygen species and pH signals in cell and organelle biology. Antioxid Redox Signal. 2016;25(9):534–49.
120. Belykh VN, Osipov GV, Petrov VS, Suykens JAK, Vandewalle J. Cluster synchronization in oscillatory networks. Chaos. 2008;18:37106.
121. Wang X, Zhou C, Lai CH. Multiple effects of gradient coupling on network synchronization. Phys Rev E. 2008;77(5):056208.1–5.
122. Tinsley MR, Nkomo S, Showalter K. Chimera and phase-cluster states in populations of coupled chemical oscillators. Nat Phys. 2012;8(9):662–5.
123. Nkomo S, Tinsley MR, Showalter K. Chimera states in populations of nonlocally coupled chemical oscillators. Phys Rev Lett. 2013;110(24):244102.

Chapter 2
Mitochondrial Calcium Handling in Physiology and Disease

Veronica Granatiero, Diego De Stefani, and Rosario Rizzuto

Calcium as Important Second Messenger

In all multicellular organisms, a precise and efficient cell-to-cell communication is required for coordinating the vast majority of physiological processes. Hundreds of diverse molecules (e.g. ions, metabolites, peptides and proteins) can carry information from one cell to another, thus acting as signaling entities that interact with specific receptors generally located on the plasma membrane. Despite this huge complexity, only a small number of intracellular transduction pathways (e.g. Ca^{2+}, cAMP or phosphorylation cascades) can ensure the proper decoding of all these diverse signals. One of the most important intracellular messengers is Ca^{2+}. Indeed, between cytosol and the extracellular environment there are both chemical and electrical gradients that represent the main driving force for changes of intracellular Ca^{2+} concentration ($[Ca^{2+}]$). Cells invest a substantial quota of their energy to guarantee this electrochemical gradient to allow rapid and transient changes in $[Ca^{2+}]$ that can largely differ through space and time [1], thus generating a plethora of signaling patterns that can be differentially decoded. Ca^{2+}-dependent signal transduction requires the binding to buffering proteins, the compartmentalization into intracellular compartments and the extrusion outside the cells [2]. Ca^{2+} binding triggers changes in protein shape and charge and consequently activates or inhibits protein functions. The best known Ca^{2+}-binding protein is calmodulin that, in concert with other Ca^{2+} buffers and intracellular compartments, controls the spatiotemporal patterning of Ca^{2+} signals [3]. These dynamic changes of $[Ca^{2+}]$ thus trigger a number of cellular events, including muscle contraction, hormone secretion, synaptic

V. Granatiero • D. De Stefani • R. Rizzuto (✉)
Department of Biomedical Sciences, University of Padova, Padua, Italy
e-mail: diego.destefani@gmail.com; rosario.rizzuto@unipd.it

G. Santulli (ed.), *Mitochondrial Dynamics in Cardiovascular Medicine*, Advances in Experimental Medicine and Biology 982,
DOI 10.1007/978-3-319-55330-6_2

transmission, cellular proliferation, apoptosis and others [4–7]. The universality and versatility of Ca^{2+} signaling is thus guaranteed by a sophisticated machinery of ion channels, pumps, and exchangers that drive the flux of Ca^{2+} ions across the plasma membrane and across the membrane of intracellular organelles, such as the Endoplasmic Reticulum (ER) and mitochondria (see Chap. 10). During last decades, an increasing number of researchers focused their efforts in the identification and characterization of all the import/export mechanisms of Ca^{2+} homeostasis, but the whole scenario is still not complete. Ca^{2+} signaling can be activated through two main mechanisms. The first is the Ca^{2+} entry from the plasma membrane channels, which can be classified into four groups, Store Operated Ca^{2+} Channels (SOCs), which open following intracellular store depletion [8], Receptor Operated Ca^{2+} Channels (ROCs), which open after an external ligand binds the receptor [9], Voltage Operated Ca^{2+} Channel (VOCs), which are sensitive to decreases of membrane potential and Second Messenger Operated Ca^{2+} Channels (SMOCs) which open after binding of second messenger on the inner side of the membrane [10, 11]. The second mechanism implies Ca^{2+} release from intracellular depots. The most important store is the ER, but recent works demonstrated that also other organelles, such as Golgi apparatus, endosome and lysosome are able to participate in Ca^{2+} signaling [12, 13]. In this scenario, the best described pathway involves the release of IP_3 after activation of a G-protein coupled receptor, and the consequent Ca^{2+} release from the ER through the binding to the IP_3R. Other intracellular Ca^{2+}-releasing channels are the Ryanodine Receptor (RyR) [14] or the Two-pore channels (TPCs) activated by NAADP [15]. Termination of Ca^{2+} signals is finally completed through Ca^{2+} extrusion from the cytoplasm by various pumps and exchangers, e.g. SERCA (sarco-endoplasmic reticulum ATPase) or PMCA [16].

This chapter is primarily focused on the role of mitochondria in cellular Ca^{2+} handling. Indeed, many subcellular compartments, including mitochondria, are key regulators of Ca^{2+} signaling, by actively participating to shape global Ca^{2+} waves and determining specific cellular functions.

Biophysical and Structural Basis of Mitochondrial Ca^{2+} Handling

The relationship between mitochondria and Ca^{2+} dates back to more than 50 years ago, when Engstrom's group, in the 1961, demonstrated for the first time that energized mitochondria, when exposed to Ca^{2+} pulses, were able to rapidly and efficiently take up large quantity of this cation [17]. These organelles are characterized by a distinctive structure. They are double membrane-bounded organelles thought to be derived from an proteobacterium-like ancestor, presumably due to a single ancient invasion occurred more than 1.5 billion years ago [18, 19]. The basic evidence of this endosymbiont theory is the existence of the mitochondrial DNA (mtDNA), with structural and functional analogies to bacterial genomes. Mitochondria are defined by two structurally and functionally different membranes, the outer membrane (OMM) and the inner membrane (IMM), characterized by invaginations called "cristae", which enclose the mitochondrial matrix. The space between these two structures is traditionally called intermembrane space (IMS). Cristae are not random folds, but rather internal compartments formed by profound invaginations, originating from very tiny "point-like structures" in the inner membrane [20]. These narrow tubular structures, called cristae junctions, can limit the diffusion of molecules from the intra-cristae space towards the IMS, thus creating a micro-environment where respiratory chain complexes, and also other proteins, are hosted and protected from random diffusion. The OMM contains high amount of a specific transport protein, VDAC (Voltage-Dependent Anion Channel), which is able to form pores on the membrane, becoming permeable to ions and metabolites up to 5 kDa [21]. Conversely, the IMM is a highly selective membrane that tightly controls the exchange of ions and metabolites through specialized molecular machineries. The chemiosmotic theory of energy transfer was first proposed by Peter Mitchell [22], who suggested that the electrochemical gradient across the IMM is utilized by the F_1/F_0 ATPase to convert the energy of NADH and $FADH_2$, generated by the breaking down of energy rich molecules, such as glucose or lipids, into ATP. This gradient ($\Delta\mu$) is composed by both electrical ($\Delta\Psi$) and chemical components (ΔpH). This potential represents a huge driving force that allows the accumulation of cations into the matrix, and is normally estimated to be approximately −180 mV [23]. Another key feature of mitochondria is their spatial organization in the cells. They are not solitary organelles, but they make contacts with several other structures, especially with the ER. Indeed, the physical and functional coupling of these two organelles in living cells was originally found to determine the Ca^{2+} and lipid transfer between the two organelles. Several works indeed underlined the presence of overlapping regions of two organelles and allowed to estimate the area of the contact sites as 5–20% of total mitochondrial surface [24, 25]. More recently, electron tomography techniques allowed to estimate an even smaller distance (10–25 nm), as well as the presence of trypsin-sensitive tethers between the two membranes [26].

In mammals, many proteins have been identified to be indirectly involved in the regulation of ER–mitochondria functional interaction, such as some chaperones, PACS-2, BAP31 and NOGO-A. Searching for the long-sought direct tether, de Brito and Scorrano and co-workers have recently identified MFN2 as important mammalian protein directly bridging the two organelles. MFN2 is enriched in the ER–mitochondria interface and connects ER with mitochondria via direct interactions between the protein localized in the ER and MFN1 or MFN2 present in the OMM. They also showed that genetic ablation of MFN2 causes an increase in the distance between the two organelles with a consequent impairment of mitochondrial Ca^{2+} uptake [27]. The mechanistic role of MFN2 in tethering the two organelles was confirmed in different systems [28–30], but this notion was also recently challenged [31, 32]. Nevertheless, there is a broad agreement in considering ER/mitochondria contact sites as critical determinant of the rapid transport of Ca^{2+} into the matrix. Indeed, some mitochondria are strategically positioned near the ER Ca^{2+} releasing sites, and are thus exposed to a microdomain where the local [Ca^{2+}] is significantly higher compared to the bulk cytosol [24, 25, 33–37]. In addition, a specific direct coupling between the IP_3R and Ca^{2+} channels of the OMM has been described [38], thus further underlining the pivotal role that organelle tethering play in cellular Ca^{2+} homeostasis.

Mitochondrial Ca^{2+} Homeostasis

Mitochondrial matrix $[Ca^{2+}]$ ($[Ca^{2+}]_{mt}$) is the result of the dynamic equilibrium between two continuous and opposite processes, i.e. mitochondrial Ca^{2+} influx vs efflux. The basic functional characteristics of mitochondrial Ca^{2+} uptake and release have been firmly established in the 1960s and 1970s. Ca^{2+} influx requires an intact mitochondrial membrane potential (negative inside) and a so called electrogenic Ca^{2+} uniporter (i.e. resulting in the net transport of two positive charges the matrix) [17, 39, 40]. Conversely, the efflux from the matrix depends on two distinct mechanisms, (i) an ubiquitous H^+/Ca^{2+} exchange (likely to be electroneutral) [41] and (ii) a Na^+/Ca^{2+} exchange (likely to be electrogenic, with 3 or 4 Na^+ ions per Ca^{2+}) [42, 43].

Mitochondrial Ca^{2+} Uptake

3.1.1.The Mitochondrial Ca^{2+} Uniporter (MCU)

In the last 50 years, several efforts have been spent to identify the channel responsible for ruthenium red (RuR)-sensitive mitochondrial Ca^{2+} entry. However, only in the last 10 years significant successes have been accomplished. The first fundamental step was indeed obtained in Clapham's lab in the 2004, when the Ca^{2+} channel of the inner mitochondrial membrane was finally directly measured. In this work, single mitoplasts (i.e. mitochondria lacking the outer mitochondrial membrane) were isolated from COS7 cells and patch clamp was performed. Accordingly, it was unambiguously shown that organelle Ca^{2+} entry was mediated by a Mitochondrial Ca^{2+} Uniporter (MCU), here also named MiCa. This channel is inward rectifying, highly Ca^{2+}-selective, sensitive to RuR and characterized by an enormous Ca^{2+} carrying capacity (half-saturation at 19 mM) with no Ca^{2+}-dependent inactivation [44]. Nonetheless, despite the detailed electrophysiological characterization of MCU, its molecular identity remained a mystery. The first step in the right direction came in 2010, when the group of Vamsi Mootha identified a mitochondrial protein, named MICU1, specifically affecting mitochondrial Ca^{2+} uptake [45]. However, it was immediately evident that MICU1 was a regulator rather than the channel per se. Only 1 year later, our group and Mootha's lab independently identified another previously uncharacterized mitochondrial protein acting as the bona fide mitochondrial Ca^{2+} uniporter [46, 47]. The MCU gene (previously known as CCDC109A) is broadly conserved among metazoans and plants, but absent in yeasts [48, 49], and encodes for a 40 kDa protein (running at 35 kDa due to the cleavage of the organelle localization peptide) with two coiled-coil domains and two transmembrane domains separated by a short loop (EYSWDIMEP). The loop is enriched in acidic residues and exposed to the intermembrane space, while the vast majority of the protein is located inside the matrix [50]. Residues E256, D260 and E263 are critical for Ca^{2+} channeling function, since their replacement with both uncharged and positively charged amino acids result in the loss of Ca^{2+} permeation [47, 51]. Moreover, substitution of Ser^{258} leads to decreased ruthenium red sensitivity [46, 52]. The recent elucidation of the 3D structure revealed that the functional channel is a homopentamer stabilized by the coiled-coil motif protruding into the mitochondrial matrix [53]. However, it is now clear that MCU is the key component of a higher order macromolecular complex, named the MCU complex, that represents the whole molecular machinery mediating the electrophoretic transport of Ca^{2+} cations inside the organelle matrix [54, 55]. MCU alone is necessary and sufficient (at least *in vitro*) for mitochondrial Ca^{2+} uptake, thus indicating that it represents the pore forming subunit [47], but what happens in vivo is still under intense investigation.

The MCU Complex

The first indication that MCU is part of a bigger complex was revealed even before the discovery of the channel itself, with the identification of MICU1 [45]. In the last 5 years, a growing amount of proteins have been shown to interact and modify MCU activity. At the moment, the general consensus is that the MCU complex can be composed by three different membrane components, MCU [46, 47], its isoform MCUb [51] and the recently identified EMRE (essential MCU regulator, previously known as C22orf32) [56], and some associated regulators (see later). The MCUb protein has a high similarity in sequence with MCU, but at least a couple of amino acid substitutions in the loop domain are predicted to be critical for ion permeation and suggest that this protein has a lower efficiency in Ca^{2+} transport. Accordingly, MCUb overexpression drastically reduces the mitochondrial Ca^{2+} uptake, while its silencing exerts the opposite effect, thus indicating that MCUb is the endogenous dominant-negative isoform of MCU [51]. Intriguingly, the MCU/MCUb expression ratio widely varies among tissues, suggesting that this mechanism could account at least in part for the differences in MCU activity displayed by different tissues [57]. On the other hand, EMRE is a small protein of about 10 kDa, with a predicted mitochondrial targeted sequence containing a single transmembrane domain and a highly conserved C-terminus rich in aspartate residues. This protein was identified in the 2013 using a quantitative mass spectrometry approach in cell culture (SILAC) [56]. As to its function, EMRE appears to be required for Ca^{2+} channeling activity in vivo and to mediate the binding of MICU1 to the channel. In the contrast to its essential role, EMRE homologs are not present in plants, fungi or protozoa in with MCU and MICU1 are highly conserved. However, knockout of EMRE is able to abrogate the mitochondrial Ca^{2+} uptake and that the overexpression of MCU in these conditions was not able per se to restore the normal mitochondrial Ca^{2+} uptake. More recently, Foskett and colleagues proposed that the C-terminal region of EMRE acts as Ca^{2+} sensor and ensures the inhibition of uniporter activity under high matrix [Ca^{2+}] [58]. However, the membrane topology of EMRE is still debated and additional studies will be necessary to carefully dissect its contribution to the whole MCU complex. In addition to these core membrane components, the MCU complex also includes soluble components localized in the intermembrane space that accounts for the so-called sigmoidicity of the mitochondrial Ca^{2+} uptake system.

The Sigmoidicity of Mitochondrial Ca^{2+} Uptake

One well established feature of organelle Ca^{2+} handling is the sigmoidal relationship between extramitochondrial [Ca^{2+}] and Ca^{2+} uptake into the matrix. In resting cells, [Ca^{2+}] in the cytosol ([Ca^{2+}]$_{cyt}$) is approximately 100 nM, thanks to the activity of Ca^{2+}-ATPases (e.g. PMCA, SERCA, SPCA). In this condition, MCU activity is very low even in front of the huge driving force represented by the mitochondrial membrane potential, in order to prevent vicious cycling of the cation and consequent energy drain. As soon as Ca^{2+}-signaling is activated, [Ca^{2+}]$_{cyt}$ rises and MCU gets activated and leads to a very large cation accumulation inside the matrix, thus ensuring rapid mitochondrial Ca^{2+} uptake. This behavior requires both a gatekeeper (i.e. a MCU inhibitor working at resting [Ca^{2+}]$_{cyt}$) as well as cooperative activator (i.e. a MCU enhancer that becomes active after Ca^{2+}-signaling activation). Importantly, these features are not part of the MCU channel per se but they are rather provided by the MICU protein family, that includes three distinct members (MICU1, MICU2 and MICU3). Originally, MICU1 was shown to possess EF-hand domains and be associated with the inner mitochondrial membrane. Its silencing compromises mitochondrial Ca^{2+} uptake in intact cells and isolated organelle, without affecting mitochondrial membrane potential or respiration [45]. Few years later, Madesh and coworkers demonstrated that MICU1 inhibits channel function in unstimulated cells, thus proposing MICU1 as the essential gatekeeper of MCU-mediated mitochondrial Ca^{2+} uptake. Indeed, MICU1 silencing causes an increase of mitochondrial Ca^{2+} levels in basal conditions, sensitizing the cells to excessive ROS production and cell death [59]. In parallel, Hajnoczky's group confirmed the role of MICU1 as MCU gatekeeper, and further demonstrated that MICU1 also act as a cooperative activator of MCU activity at high cytosolic [Ca^{2+}] [60]. Finally, we recently added another level of complexity to this picture. Indeed, we found that MICU1 forms dimers with its isoform MICU2 through a disulfide bridge. Thus, according to our hypothesis, MICU1 and MICU2 represents a unique entity, widely expressed in all tissues, and represent the molecular machinery underlying the sigmoidicity of mitochondrial Ca^{2+} uptake. In this model, MICU1 acts predominantly as an MCU activator, although it can still retain some residual gatekeeping function in some situation. Conversely, MICU2 act as a genuine MCU inhibitor, thus guaranteeing the normal physiological activation threshold. Accordingly, at resting [Ca^{2+}], the prevailing inhibitory effect of MICU2 ensures minimal Ca^{2+} accumulation in the presence of a very large driving force for cation accumulation, thus preventing the deleterious effects of Ca^{2+} cycling and matrix overload. As soon as extramitochondrial [Ca^{2+}] increases, Ca^{2+}-dependent inactivation of MICU2 and the concomitant activation of MICU1 guarantee the prompt initiation of rapid mitochondrial Ca^{2+} accumulation, thus stimulating aerobic metabolism and increasing ATP production [61].

Finally, other MCU regulators have been proposed, including MCUR1 (also known as CCDC90A) and SLC25A23 (also known as SCaMC-3). They are both are membrane proteins shown to interact with MCU and modulate its activity [62, 63]. However, these two protein are also involved in other processes, such as the assembly of complex IV [64] or ATP/Pi exchange [65–67] respectively. This opens the possibility that the functional effect of these proteins on mitochondrial Ca^{2+} homeostasis could be an indirect secondary effect, as better discussed elsewhere [68].

Mitochondrial Ca^{2+} Efflux

As to the Ca^{2+}-release pathways from mitochondria, their functional existence has been characterized already in the 1970s [41, 43]. Ca^{2+} efflux from mitochondria depends on two different mechanisms, one mediated by a $2H^+/Ca^{2+}$ antiporter (mHCX) and expressed in most of the cells and the other by a 3 (or 4) Na^+/Ca^{2+} exchanger (mNCX), predominant in excitable tissues. The molecular identity of the $2H^+/Ca^{2+}$ exchanger is still a matter of debate, but the recent works by Clapham and coworkers suggests that Letm1 is a strong candidate [69, 70], although this notion is not yet widely accepted [71, 72]. Conversely, a universal consensus exists on the identity of the mitochondrial Na^+/Ca^{2+} exchanger. Indeed, in 2010 Sekler and coworkers demonstrated that the gene NCLX (also known as NCKX6, SLC24A6 or SLC8B1) encodes for a mitochondrial protein necessary for Na^+-dependent Ca^{2+} clearance from the matrix [73]. Interestingly, many years before, an unidentified mitochondrial protein, but similar in molecular weight to NCLX, was reported to exhibit Na^+/Ca^{2+} exchange activity when purified and reconstituted in vitro [74, 75]. The NCLX protein shares only low sequence homology (approximately 20%) to Na^+/Ca^{2+} exchangers of the plasma membrane and the exact stoichiometry of Na^+/Ca^{2+} exchange is still unknown. Based on thermodynamic considerations, it should catalyze the exchange of 3–4 Na^+ every Ca^{2+}, thus causing the net import of one or two positive charges into the matrix. In line with this, molecular dynamics simulation based on the crystal structure of a bacterial Na^+/Ca^{2+} exchanger [76] suggests a 3:1 stoichiometry [77], that should apply also to mNCX. Up to now, the contribution of NCLX to mitochondrial Ca^{2+} extrusion has been confirmed in a variety of cellular models, including B lymphocytes, pancreatic beta cells, cardiomyocyte-like cell lines and astrocytes. Unfortunately, animal models lacking the NCLX gene are still missing at this moment, but in the future they will provide essential information on the physiopathological role of mitochondrial Ca^{2+} homeostasis.

The Pathophysiological Role of Mitochondrial Ca^{2+} Signaling

The molecular identification of the components of the MCU complex finally opened the possibility to address the physiological role of mitochondrial Ca^{2+} uptake by genetic approaches, thus overcoming the limits of classical pharmacological approaches, in particular considering the lack (or, at least, the poor performance) of MCU-targeting drugs suitable for in vivo applications. However, in the last three decades a huge amount of experimental work has been carried out using either pharmacological or indirect genetic tools to target mitochondrial Ca^{2+} homeostasis and dissect its patho-physiological role. According to these studies, the general consensus is that Ca^{2+} inside mitochondria acts a pleiotropic signal, with different cellular outcomes that depend on the investigated cell type, the metabolic state and the concomitant presence of other stress signals. On one hand, Ca^{2+} plays a regulatory role within the organelle itself ranging from the regulation of ATP production to the release of caspases cofactor with consequent cell death. On the other hand, mitochondrial Ca^{2+} can exert its function at the whole cell level, e.g. by regulating cation homeostasis both locally and globally [78]. However, in all the studies that preceded the molecular identification of MCU, the experimental approaches used to address the physiological relevance of mitochondrial Ca^{2+} homeostasis was (i) the use of drugs interfering, more or less specifically, with the Ca^{2+} uptake (e.g. uncouplers, ruthenium red, Ru360) or release (e.g. CGP37157) processes and (ii) genetic manipulations indirectly modifying the amount of Ca^{2+} that mitochondria can accumulate (e.g. regulation of Bcl2 family proteins or MAM-localized proteins). However, none of these approaches is specific, since they either modify other organelle functions (e.g. ATP synthesis), or affect extramitochondrial processes (e.g. Ca^{2+}-depletion of intracellular stores, IP_3R activity).

Even the data obtained with the most specific MCU inhibitors, i.e. ruthenium red and Ru360, should be interpreted with caution. Indeed, it has been shown that in several cellular models, the functional effects of these two compounds are not dependent on their inhibition of MCU [79], in line with the predicted poor membrane solubility of these drugs. Fortunately, the recent identification and elucidation of the MCU complex, although not yet definitive, has opened the field to genetic loss- or gain-of function approaches to unambiguously determine the role of mitochondrial Ca^{2+} uptake in organism pathophysiology. Due to the intrinsic novelty of these discoveries, the picture is still largely incomplete, and apparently contrasting data are being published. While the studies carried out in cultured cells concern both the overexpression and downregulation of all the proteins in the complex, in vivo models (at the moment this chapter was written) are available only for MCU KO and overexpression. Here, we will critically synthetize the main findings on the role of mitochondrial Ca^{2+} transport in both physiological and pathological conditions.

Regulation of Cellular Bioenergetics and Autophagy

In the last 5 years, the genetic manipulation of MCU complex components has been carried out in several cultured cells. Overall, the relationship between mitochondrial Ca^{2+} uptake and the regulation of energy production has been confirmed in most of them. For example, in pancreatic β-cells, glucose-induced insulin release is dependent on the stimulation of mitochondrial oxidative metabolism, and it involves both K_{ATP}-dependent [80] and K_{ATP}-independent [81] regulation of exocytosis. Stimulation of mitochondrial dehydrogenases by organelle matrix $[Ca^{2+}]$ is a critical step critical in this process [82]. Indeed, MCU silencing impairs the Ca^{2+}-dependent phase of glucose-induced ATP increase [83, 84], and hence insulin secretion [85, 86]. In line with this, NCLX silencing temporally enhances the glucose-induced increase in oxidative phosphorylation [87, 88] and cellular ATP levels [84]. Interestingly, also silencing of MICU1 (that increases basal $[Ca^{2+}]_{mt}$ but reduces agonists-induced transients) has a similar effect in this cellular model [85], thus suggesting that tonic rises in $[Ca^{2+}]_{mt}$, and not constitutively high resting $[Ca^{2+}]_{mt}$, are decoded and translated into a boost of ATP production. This is further supported by the observation that, at least in HeLa cells, silencing of MICU1 do not modify basal respiration levels, but it rather dampens agonists-induced increase in oxygen consumption [59]. Overall, these data demonstrated that mitochondrial Ca^{2+} accumulation, mediated by MCU and modulated by NCLX, is thus required for normal glucose sensing by pancreatic β-cells. Moreover, direct coupling of mitochondrial Ca^{2+} transients and oxidative metabolism has also been confirmed in other cellular model (as well as in vivo, see below). Overexpression of MCU in human fibroblasts derived from both healthy subjects and patients affected by mitochondrial disorders can efficiently decrease AMPK phosphorylation levels [89], i.e. the major cellular metabolic sensor of AMP/ATP ratio. Similarly, MCU silencing enhances starvation-induced increase of phospho-AMPK levels in MDA-MB231, a breast cancer derived human cell line [90].

Autophagy is a broadly conserved adaptive response to energetic defects [91], as well as the leading mechanism for eliminating damaged organelles [92, 93]. Recently, a seminal work by Foskett and colleagues clearly demonstrated that mitochondrial Ca^{2+} signaling has a fundamental role in the regulation of autophagy [94]. In addition, the removal of damaged or non-functional mitochondria is regulated by a specific autophagic process named "mitophagy", that acts as a critical organelle quality control pathway [95]. In this context, mitochondrial Ca^{2+} has been recently shown to be a potential specific signal in the control of mitophagy. Indeed, downregulation of ER to mitochondrial Ca^{2+} transfer can effectively decrease parkin-mediated organelle removal [96]. In addition, we also demonstrated a causal link between mitochondrial Ca^{2+} signaling and autophagy/mitophagy in a cellular model of mitochondrial disease [89]. Mitochondrial disorders are indeed a wide range of clinical syndromes caused by inherited mutations in the mitochondrial DNA (mtDNA). Given the complexity of mitochondrial genetics and biochemistry, the clinical manifestations of mitochondrial disorders are extremely heterogeneous. They range from lesions of single tissues or structures, such as the optic nerve in Leber's hereditary optic neuropathy (LHON), to more widespread lesions including myopathies, encephalomyopathies, cardiomyopathies, or complex multisystem syndromes with onset ranging from neonatal to adult life, such as MELAS (mitochondrial encephalomyopathy, lactic acidosis, and stroke-like episodes [97, 98]. In this pathological context, the direct correlation between mitochondrial Ca^{2+} and autophagy is supported by the demonstration that genetically or pharmacological approaches that enhance mitochondrial Ca^{2+} uptake can efficiently slow down the hyperactivated autophagic flux, through an AMPK-dependent mechanism [89]. Overall, the emerging picture consistently shows that the regulation of the autophagy is dependent on the modulation of oxidative metabolism by $[Ca^{2+}]_{mt}$. Accordingly, the silencing of MCU, MICU1 and MCUR1, all conditions that decrease organelle Ca^{2+} accumulation after IP_3-coupled stimuli, increase the formation of autophagosomes and are accompanied by higher AMPK phosphorylation in both HeLa and HEK293 cells [59, 63]. In conclusion, the positive role that $[Ca^{2+}]_{mt}$ exerts on energy production has been well confirmed by genetic approaches targeting the MCU complex in cultured cells.

Regulation of Cell Death

As to the Ca^{2+}-dependent regulation of cell death, the situation is more complex, with apparently slightly contrasting results. As reported above, all conditions that decreases $[Ca^{2+}]$ should in principle protect cells from toxic stimuli. In line with this view, MCU overexpression enhances cell death triggered by a variety of stimuli (including H_2O_2, C2-ceramide and NMDA) in HeLa cells [47, 99], cerebellar granule neurons (CGNs) [99] and primary cortical neurons [100]. Notably, in cortical neurons overexpression of MCU enhances cell death per se, even in the absence of excitotoxic agents [100], thus confirming the idea that different cell types have different susceptibility to mitochondrial Ca^{2+} overload. Accordingly, in the same experimental settings, MCU silencing exerted the opposite effect, i.e. protected cells from apoptosis [99, 100]. However, in one report, primary human epithelial cells (HMEC), but not HeLa or MDA-MB-231 (a breast cancer-derived line), displayed reduced survival after C2-ceramide treatment when MCU was downregulated. MCU was also dispensable for apoptosis triggered by paclitaxel (a cytoskeletal drug) and ionizing radiations in MDA-MB-231 [90]. In the same cell line, MCU silencing was even shown to enhance caspase-independent cell death when moderate dose of ionomycin (3 μM) was used as stimulus, and ineffective at higher doses (10 μM) or after Bcl-2 inhibition with ABT-263 [101]. Conversely, in a similar cell line, the MDA-MB-435S, silencing of MCU slightly protected from a different form of cell death, named paraptosis, induced by celastrol [102]. However, it must be stressed that (i) not all death stimuli depend on organelle Ca^{2+} overload and (ii) mitochondrial Ca^{2+} uptake is not an proapoptotic signal per se, but it rather requires the concomitant activation of other signaling pathways that can be cell-type specific. It was recently shown that MCU mRNA is one of the targets of miR-25 [103]. Therefore, overexpression of miR-25 leads to a decrease of MCU at protein level and of mitochondrial Ca^{2+} transients. Accordingly, upregulation of miR-25 protects both HeLa [103] and H9c2 (derived from rat myocardium) [104] cells from death induced by H_2O_2 and C2-ceramide, but not by staurosporine (a stimulus that does not strictly rely on mitochondrial Ca^{2+} overload). In addition, treatment of PC3 (derived from prostate cancer) cells with a specific miR-25 antagonist, efficiently enhances cell death [103], thus further reinforcing the notion that mitochondrial Ca^{2+} is a sensitizing element for apoptosis.

Some additional insights into the relationship between mitochondrial Ca^{2+} homeostasis and apoptosis can be obtained by looking at the other components of the MCU complex. Indeed, it has been demonstrated that in HeLa [59, 60, 90], HMEC and primary hepatocytes [60], but not in MDA-MB-231 [90], silencing of MICU1 enhances cell death triggered by C2-ceramide or H_2O_2. Comparable results have been obtained also in primary endothelial cells exposed to lipopolysaccharide and cycloheximide [59]. It is here worth noting that loss of MICU1 leads on one hand to higher mitochondrial Ca^{2+} uptake at low $[Ca^{2+}]_{cyt}$, but on the other it also lowers organelle cation accumulation when $[Ca^{2+}]_{cyt}$ rises [60, 61]. As a consequence, in MICU1 silenced cells, the transfer of low amplitude Ca^{2+} signals should be favored, while large and rapid increases of $[Ca^{2+}]_{mt}$ triggered by temporally restricted Ca^{2+} signals should be impaired. In line with this, Ca^{2+}-dependent apoptotic stimuli generates increases in $[Ca^{2+}]_{cyt}$ that are small in amplitude but prolonged in time [105–107], as opposed to normal physiological agonists (e.g. IP_3-mobilizing agents), which triggers large but transient rises in $[Ca^{2+}]_{cyt}$. Indeed, silencing of MICU1 potentiates the transfer of apoptotic stimuli (i.e. small and prolonged signals) and dampens the burst on oxidative metabolism triggered by IP3-coupled stimuli (i.e. large and fast signals) [59]. Finally, also the knockdown of another putative MCU modulator, SLC25A23, a condition that inhibit mitochondrial Ca^{2+} accumulation, protect from cell death induced by oxidative stress [62]. Regarding the exact cell death mechanism downstream to $[Ca^{2+}]_{mt}$ rises, generation of reactive oxygen species (ROS) is one of obvious candidate, although different results have been obtained. According to Madesh and coworkers, silencing of MICU1 leads to an increase of basal mitochondrial ROS levels that in turn enhances the susceptibility to apoptotic stimuli [59]. Conversely, Hajnoczky and coworkers showed that the basal ROS levels are not affected by MICU1 deficiency. However, during prolonged store operated Ca^{2+} entry, a gradual increase in mitochondrial H_2O_2 production was detected when MICU1 was downregulated, maybe contributing to enhanced cell death [60]. Finally, no differences in ROS levels were detected after regulation of either MICU1 or MCU in MDA-MB-231 cells [90], a model were MCU appears to be dispensable for cell death. However, we also recently reported that the genetic inhibition of MCU in this cell line limits tumor progression both in vitro an in vivo through a ROS/HIF1α-dependent mechanism [108]. Overall, the available data obtained in cultured cells support the idea that mitochondrial Ca^{2+} overload can be a trigger of cell death, although its contribution can largely vary depending on the cell type.

Targeting Mitochondrial Ca^{2+} Uptake In Vivo

The vast majority of studies cited so far agrees on the fact that mitochondrial Ca^{2+} is a pleiotropic signal that regulates many essential aspects of cellular physiology. Starting from these observation, one could easily expect that the genetic ablation of the MCU gene would be detrimental for so many functions at whole organism level to be incompatible with life (i.e. leading to embryonic lethality). However, the first attempt to target MCU in *Mus musculus* showed a very surprising result that puzzled most of the scientists in the field. Indeed, in 2013 Finkel and coworkers published a paper showing a viable MCU knockout (KO) mouse, characterized by the lack of any significant phenotype except for a slight decrease of skeletal muscle peak performance [109]. MCU-KO mice showed no histological aberrations or evident dysfunctions in any organ, neither in high energy demanding tissues such as heart, skeletal muscle and brain. In this model, differences are lacking not only in resting conditions, but also under both physiological (e.g. isoprenaline) or pathological (e.g. ischemia/reperfusion injury in the heart) stimuli [109, 110]. However, it became evident that ablation of MCU in the standard C57BL/6 strain actually leads to embryonic lethality, and this notion has been confirmed in different and independent mouse models (see below). Viable mice could indeed be obtained only in a mixed C57BL/6×CD1 background, and even in this strain the birth ratio of the homozygous MCU null mice is half of the expected [111], thus indicating the requirement for mitochondrial Ca^{2+} uptake during embryo development. How can the genetic background have such a profound impact remains an enigma, although similar results are not new [112]. For example, knockout of the transforming growth factor-β1 (TGF-β1) leads to different outcomes that largely depend on the mouse strain. Embryonic lethality is complete when backcrossed in C57/BL6, 50% in a mixed 129×CF1 and just 20% in BALB/c [113]. Similarly, endothelial growth factor (EGF)-receptor-deficient mice show large variation in survival, ranging from peri-implantation to postnatal lethality, according to the strain of the developing embryos [114, 115]. In any case, the lack of any phenotype of the viable MCU-KO mice is still puzzling. The obvious explanation is that some kind of compensation could take place, for example due to the activation of alternative routes for mitochondrial Ca^{2+} uptake. However, the evidences provided by Finkel and coworker are compelling and demonstrate that MCU is the only mechanism leading to the fast, energy-driven organelle Ca^{2+} uptake in a variety of experimental setups (see Chap. 3). The only concern lies in the observation that in MCU-KO mitochondria, in steady state, matrix Ca^{2+} content, although greatly reduced, is still measurable, maybe suggesting that alternative pathways for slow Ca^{2+} uptake could exist. Again, also the absence of the "expected" phenotype is not new when dealing with loss-of-function models [116]. For example, knockout of creatine kinase, an enzyme that control a critical parameter such as the intracellular ATP levels, leads to no evident phenotype [117]. Even more strikingly, genetic ablation of myoglobin, the oxygen-binding protein of striated skeletal muscle, resulted in no consequences [118].

Interestingly, in myoglobin-KO mice non-obvious multiple compensatory mechanisms take place, all converging to steepen the oxygen pressure gradient to the mitochondria, including higher capillary density, smaller cell width, elevated hematocrit and increased coronary flow and coronary flow reserve [119–121]. Other examples of unpredictable compensations include cGKII- [122] and GDNF-KO [123] mice. More generally, biological robustness should be taken into account when evaluating these models. For example, in yeasts only 13% of genes are essential when mutations in single genes are analyzed [124, 125], but most of the genes are only apparently dispensable, as demonstrated when the functional buffering of biochemical pathway is taken into account [126]. In our biased opinion, a compensation in the downstream signals activated by mitochondrial Ca^{2+} transient (that are still largely unknown) must be present in these mice, a concept that is supported also by results obtained in other related genetic models. Indeed, Anderson and coworkers generated a heart-specific transgenic mouse model expressing a dominant negative MCU isoform, $MCU^{D260Q,E263Q}$ (DN-MCU), in the same mixed C57BL/6×CD1 background of the constitutive MCU-KO model [127]. In cultures cells this mutant is not able to completely abolish organelle Ca^{2+} accumulation [47], but mitochondria from DN-MCU expressing hearts show no measurable Ca^{2+} uptake. In any case, this genetic model shows a clear phenotype, i.e. the lack of isoprenaline-induced fight or flight response. In particular, these mice are incapable of heart rate acceleration induced by β-adrenergic stimulation, an effect completely absent in the constitutive MCU-KO model [110]. The mechanism behind this effect depends on the enhancement of oxidative phosphorylation triggered by $[Ca^{2+}]_{mt}$ increase. Indeed, while MCU appears to be dispensable for normal heart function, under physiological stress, the MCU-dependent boost of ATP production is necessary to sustain the increased demand of SERCA pumps and the proper Ca^{2+} load of the sarcoplasmic reticulum in pacemaker (SAN) cells. In line with this, DN-MCU expressing hearts exhibit impaired performance at increasing workloads (lower ventricular developed pressure at >600 bpm) and a higher oxygen consumption rate (OCR) at any heart rate [127]. Moreover, DN-MCU expressing cardiomyocytes display a clear extramitochondrial adaptation (i.e. higher diastolic $[Ca^{2+}]_{cyt}$) that depends on the reduced ATP availability. However, despite this phenotype, also in this model the susceptibility to ischemia reperfusion injury is comparable to control animals [128]. Conversely, this notion has been recently challenged by a new mouse model developed by Molkentin and coworkers. They indeed generated a conditional MCU knockout model with two LoxP sites between exons 5 and 6 ($MCU^{fl/fl}$) that were crossed with mice expressing a tamoxifen-inducible Cre recombinase (MerCreMer [MCM]) driven by the cardiomyocyte-specific α-myosin heavy chain promoter. MCU gene deletion was induced in adult (i.e. 8 weeks) mice and the cardiac function was evaluated [129]. Interestingly, in a parallel study, Elrod and colleagues crossed the $MCU^{fl/fl}$ animals with mice constitutively expressing Cre recombinase (B6.CMV-Cre). No homozygous $MCU^{-/-}$ mice could be obtained due to complete embryo lethality, thus confirming the requirement for MCU during development [130]. Ablation of MCU in adult heart leads to an 80% decrease of MCU at the protein level, most likely due to the mosaicism of the MCM strain [131,

132]. Therefore, this lead to a drastic, but not complete, reduction of mitochondrial Ca^{2+} uptake has been observed, as can be easily appreciated when looking to intra-mitochondrial Ca^{2+} dynamics (as opposed to indirect extramitochondrial $[Ca^{2+}]$ measurement in isolated organelle, which is a less sensitive method especially when dealing low mitochondrial Ca^{2+} uptake rates). Cardiac-specific MCU-KO mice show no major defects, even 40 weeks after gene ablation, in both normal conditions and after cardiac pressure overload induced by transverse aortic constriction (TAC) [129]. However, this genetic model not only show the lack of any fight or flight response, as previously reported [127], but it also display a clear additional phenotype. Indeed, MCM-MCU-KO mice are strongly protected from I/R injury [129, 130], in line with the predicted role that mitochondrial Ca^{2+} plays in cell death, but in sharp contrast to other MCU-KO models [109, 128]. In support of this notion, mitochondrial Ca^{2+} overload has been shown to be a key determinant of cardiac cell death also in different models of heart failure [133]. However, these discrepancies are difficult to solve, in particular considering the lack of standardization among these models. On one hand, the mice were generated in two different strains (pure C57BL/6 and mixed C57BL/6×CD1), and on the other inhibition of MCU during development is likely to activate compensatory networks. In order to solve some of these issue, it is worth looking to data produced by the International Mouse Phenotype Consortium (IMPC) with the aim to generate standardized procedures for phenotyping and provide reliable insights into the function of the whole mammalian genome [134–137]. The consortium already generated KO models for MCU (http://www.mousephenotype.org/phenotype-archive/genes/MGI%3A3026965), MCUb (http://www.mousephenotype.org/phenotype-archive/genes/MGI%3A1914065) and EMRE (http://www.mousephenotype.org/phenotype-archive/genes/MGI%3A1916279). Some phenotypic data are already available and will be discussed here, although it must be stressed that overall the picture is not yet complete and some of the data must still undergo to quality control processes, and should be thus taken with caution. According to these data, genetic ablation of MCU or EMRE (both conditions that completely block mitochondrial Ca^{2+} uptake) leads to embryonic lethality, thus further reinforcing the requirement of mitochondrial Ca^{2+} signaling during embryogenesis. Unexpectedly, $MCU^{+/-}$ heterozygous mice show several impaired physiological parameters, including a decreased cardiac stroke volume, abnormal fasting glucose level and fat amount (although for these last two parameters, males and females behave differently, thus indicating sexual dimorphism). $EMRE^{+/-}$ heterozygous mice display instead neurological defects (e.g. lower center distance travelled) and impairment of blood glucose homeostasis near statistical significance. Finally, $MCUb^{-/-}$ homozygous mice are viable (as one would expect) but show vestibular impairment (low contact righting score) and electrocardiogram defects. Interestingly, considering the high expression of MCUb in the immune system, a tendency (although not significant) to higher lymphocytic count is also present. Overall, these data suggest that the modulation of MCU complex components has a broad potential impact on physiology at whole organism level.

In line with this view, mitochondrial Ca^{2+} uptake has been shown to play a central role in skeletal muscle pathophysiology. In this case, in order to avoid potential compensatory effects activated during embryonic development, adeno-associated viral (AAV) particles for both the overexpression and silencing of MCU were used in either newborn or adult mice. MCU overexpression and downregulation causes muscular hypertrophy and atrophy, respectively. Moreover, MCU overexpression also protects from denervation-induced muscle atrophy triggered by sciatic nerve resection. Regarding the mechanisms behind this phenotype, they appear to be independent of the control of aerobic metabolism, since (i) PDH activity, although defective in MCU-silenced muscles was unaffected in MCU-overexpressing muscles, (ii) the hypertrophy was comparable in both oxidative and glycolytic muscles (in the latter the contribution of mitochondrial metabolism should play a relatively marginal role) and (iii) analysis of aerobic metabolism revealed no major alterations. Conversely, this effect rather relies on two major hypertrophic pathways of skeletal muscle, PGC-1α4 and IGF1-Akt/PKB, leading to marked increase in protein synthesis. These results indicate the existence of a Ca^{2+}-dependent mitochondria-to-nucleus signaling route that links organelle physiology to the control of muscle mass. It also should be underlined here that the apparent differences in phenotype between gene knockout and gene silencing approaches are not new [138–140] and are thought to be due to the activation of a compensatory network to buffer against deleterious mutations (i.e. gene knockout), which are not observed after gene knockdown [141].

Genetic manipulation of MCU has also been carried out in other organisms. Indeed, both silencing and conditional knockout of MCU in *Trypanosome brucei* impairs energy production, enhances cellular recycling through autophagy and produces marked defects of growth *in vitro* and infectivity in mice [142]. Similarly, developmental defects are also evident in *Danio rerio* injected with anti-MCU morpholinos [143], underlining the physiological requirement of an efficient organelle Ca^{2+} uptake machinery in simple vertebrates. Conversely, knockout of MCU in *Caenorhabditis elegans* leads to apparently viable and fertile worms, although the productions of ROS and wound healing are impaired [144].

Also in the case of the MCU regulators, the situation is still controversial. Very recently, Hajnoczky and coworkers described the first attempt to target MICU1 in vivo [145]. Embryonic ablation of MICU1 causes perinatal lethality in mice without causing gross anatomical defects. Most importantly, liver-specific MCU deletion determines the striking failure in organ regeneration after partial hepatectomy. MICU1-deficient hepatocytes show exaggerated Ca^{2+} overload-induced mitochondrial permeability transition pore (PTP) opening, and indeed the phenotype is completely reversed in vivo by using the PTP inhibitor NIM811 [145]. Conceptually similar results have been obtained in a different model of MICU1-KO mice generated by Finkel and coworkers (see Chap. 3). However, in this model the perinatal mortality is only partial, and the phenotype of surviving MICU1-KO animals spontaneous improves during aging, most likely due to a compensation through a decreased expression level of EMRE [146].

Finally, the pleiotropic role of mitochondrial Ca^{2+} uptake has also been confirmed in human pathophysiology. Few families carrying a loss-of-function mutation of MICU1 have been identified. Homozygous individuals for this mutation are characterized by early-onset proximal muscle weakness with a static course, moderately or grossly elevated serum creatine kinase levels accompanied by learning difficulties and a progressive extrapyramidal movement disorder. Fibroblasts derived from these patients show the expected impairment of MCU gating accompanied by mitochondrial fragmentation and decreased oxidative metabolism [147, 148]. Although the pathogenesis lying behind this disease is impossible to dissect at the moment, it needs stressing that this finding underlines the importance of mitochondrial Ca^{2+} transport in humans, as a change in one of the regulatory proteins in the complex results in a genetic disease with devastating symptoms.

Overall, it is still difficult at the moment drawing any definitive conclusion on the role of mitochondrial Ca^{2+} handling in pathophysiology. Still, despite the initial setback caused by the lack of significant phenotype of the MCU-KO, the new available models are starting to uncover the real contribution of mitochondrial Ca^{2+} signals to physiological processes. New questions need to be addressed and new cellular pathways are waiting to be discovered, in an unexpectedly exciting exploding field.

References

1. Clapham DE. Calcium signaling. Cell. 2007;131:1047–58. doi:10.1016/j.cell.2007.11.028.
2. Berridge MJ. Inositol trisphosphate and calcium signalling mechanisms. Biochim Biophys Acta. 2009;1793:933–40. doi:10.1016/j.bbamcr.2008.10.005.
3. Hoeflich KP, Ikura M. Calmodulin in action: diversity in target recognition and activation mechanisms. Cell. 2002;108:739–42.
4. Berridge MJ, Bootman MD, Roderick HL. Calcium signalling: dynamics, homeostasis and remodelling. Nat Rev Mol Cell Biol. 2003;4:517–29. doi:10.1038/nrm1155.
5. Hajnoczky G, Robb-Gaspers LD, Seitz MB, et al. Decoding of cytosolic calcium oscillations in the mitochondria. Cell. 1995;82:415–24. doi: 0092-8674(95)90430-1 [pii]
6. Rizzuto R. Calcium mobilization from mitochondria in synaptic transmitter release. J Cell Biol. 2003;163:441–3. doi:10.1083/jcb.200309111jcb.200309111. [pii]
7. Santulli G, Marks AR. Essential roles of intracellular calcium release channels in muscle, brain, metabolism, and aging. Curr Mol Pharmacol. 2015;8:206–22. doi:10.2174/187446720 8666150507105105.
8. Feske S. ORAI1 and STIM1 deficiency in human and mice: roles of store-operated Ca2+ entry in the immune system and beyond. Immunol Rev. 2009;231:189–209. doi:10.1111/j.1600-065X.2009.00818.x.
9. McFadzean I, Gibson A. The developing relationship between receptor-operated and store-operated calcium channels in smooth muscle. Br J Pharmacol. 2002;135:1–13. doi:10.1038/sj.bjp.0704468.
10. Doering CJ, Zamponi GW. Molecular pharmacology of high voltage-activated calcium channels. J Bioenerg Biomembr. 2003;35:491–505. doi:10.1023/B:JOBB.0000008022.50702.1a.
11. Meldolesi J, Pozzan T. Pathways of Ca2+ influx at the plasma membrane: voltage-, receptor-, and second messenger-operated channels. Exp Cell Res. 1987;171:271–83.
12. Calcraft PJ, Ruas M, Pan Z, et al. NAADP mobilizes calcium from acidic organelles through two-pore channels. Nature. 2009;459:596–600. doi:10.1038/nature08030.

13. Pinton P, Pozzan T, Rizzuto R. The Golgi apparatus is an inositol 1,4,5-trisphosphate-sensitive Ca2+ store, with functional properties distinct from those of the endoplasmic reticulum. EMBO J. 1998;17:5298–308. doi:10.1093/emboj/17.18.5298.
14. Ryu SY, Beutner G, Kinnally KW, et al. Single channel characterization of the mitochondrial ryanodine receptor in heart mitoplasts. J Biol Chem. 2011;286:21324–9. doi: C111.245597 [pii]10.1074/jbc.C111.245597
15. Marchant JS, Patel S. Two-pore channels at the intersection of endolysosomal membrane traffic. Biochem Soc Trans. 2015;43:434–41. doi:10.1042/BST20140303.
16. Brini M, Carafoli E. Calcium pumps in health and disease. Physiol Rev. 2009;89(4):1341–78. doi:10.1152/physrev.00032.2008.
17. Deluca HF, Engstrom GW. Calcium uptake by rat kidney mitochondria. Proc Natl Acad Sci U S A. 1961;47:1744–50.
18. Carvalho DS, Andrade RFS, Pinho STR, et al. What are the evolutionary origins of mitochondria? A complex network approach. PLoS One. 2015;10:1–21. doi:10.1371/journal.pone.0134988.
19. Dyall SD, Brown MT, Johnson PJ. Ancient invasions: from endosymbionts to organelles. Science. 2004;304:253–7. doi:10.1126/science.1094884.
20. Mannella CA. Structure and dynamics of the mitochondrial inner membrane cristae. Biochim Biophys Acta. 2006;1763:542–8. doi:10.1016/j.bbamcr.2006.04.006.
21. Shoshan-Barmatz V, De Pinto V, Zweckstetter M, et al. VDAC, a multi-functional mitochondrial protein regulating cell life and death. Mol Asp Med. 2010;31:227–85. doi: S0098-2997(10)00021-X [pii]10.1016/j.mam.2010.03.002
22. Mitchell P, Moyle J. Chemiosmotic hypothesis of oxidative phosphorylation. Nature. 1967;213:137–9.
23. Bernardi P. Mitochondrial transport of cations: channels, exchangers, and permeability transition. Physiol Rev. 1999;79:1127–55.
24. Rizzuto R, Brini M, Murgia M, Pozzan T. Microdomains with high Ca2+ close to IP3-sensitive channels that are sensed by neighboring mitochondria. Science. (80-. 1993;262:744–7.
25. Rizzuto R, Pinton P, Carrington W, et al. Close contacts with the endoplasmic reticulum as determinants of mitochondrial Ca2+ responses. Science. (80-. 1998;280:1763–6.
26. Csordas G, Renken C, Varnai P, et al. Structural and functional features and significance of the physical linkage between ER and mitochondria. J Cell Biol. 2006;174:915–21. doi: jcb.200604016 [pii]10.1083/jcb.200604016
27. de Brito OM, Scorrano L. Mitofusin 2 tethers endoplasmic reticulum to mitochondria. Nature. 2008;456:605–10. doi: nature07534 [pii]10.1038/nature07534
28. Area-Gomez E, Del Carmen Lara Castillo M, Tambini MD, et al. Upregulated function of mitochondria-associated ER membranes in Alzheimer disease. EMBO J. 2012;31:4106–23. doi:10.1038/emboj.2012.202.
29. Chen Y, Csordás G, Jowdy C, et al. Mitofusin 2-containing mitochondrial-reticular microdomains direct rapid cardiomyocyte bioenergetic responses via interorganelle Ca(2+) crosstalk. Circ Res. 2012;111:863–75. doi:10.1161/CIRCRESAHA.112.266585.
30. Sebastián D, Hernández-Alvarez MI, Segalés J, et al. Mitofusin 2 (Mfn2) links mitochondrial and endoplasmic reticulum function with insulin signaling and is essential for normal glucose homeostasis. Proc Natl Acad Sci U S A. 2012;109:5523–8. doi:10.1073/pnas.1108220109.
31. Cosson P, Marchetti A, Ravazzola M, Orci L. Mitofusin-2 independent juxtaposition of endoplasmic reticulum and mitochondria: an ultrastructural study. PLoS One. 2012;7:e46293. doi:10.1371/journal.pone.0046293.
32. Filadi R, Greotti E, Turacchio G, et al. Mitofusin 2 ablation increases endoplasmic reticulum–mitochondria coupling. Proc Natl Acad Sci. 2015;112:E2174–81. doi:10.1073/pnas.1504880112.
33. Csordás G, Thomas AP, Hajnoczky G, et al. Quasi-synaptic calcium signal transmission between endoplasmic reticulum and mitochondria. EMBO J. 1999;18:96–108. doi:10.1093/emboj/18.1.96.

34. Csordas G, Varnai P, Golenar T, et al. Imaging interorganelle contacts and local calcium dynamics at the ER-mitochondrial interface. Mol Cell. 2010;39:121–32. doi: S1097-2765(10)00496-X [pii]10.1016/j.molcel.2010.06.029
35. Giacomello M, Drago I, Bortolozzi M, et al. Ca2+ hot spots on the mitochondrial surface are generated by Ca2+ mobilization from stores, but not by activation of store-operated Ca2+ channels. Mol Cell. 2010;38:280–90. doi: S1097–2765(10)00280–7 [pii]10.1016/j.molcel.2010.04.003
36. Hajnoczky G, Csordas G, Madesh M, Pacher P. Control of apoptosis by IP(3) and ryanodine receptor driven calcium signals. Cell Calcium. 2000;28:349–63. doi:10.1054/ceca.2000.0169S0143-4160(00)90169-6. [pii]
37. Rizzuto R, Simpson AW, Brini M, Pozzan T. Rapid changes of mitochondrial Ca2+ revealed by specifically targeted recombinant aequorin. Nature. 1992;358:325–7. doi:10.1038/358325a0.
38. Szabadkai G, Bianchi K, Várnai P, et al. Chaperone-mediated coupling of endoplasmic reticulum and mitochondrial Ca2+ channels. J Cell Biol. 2006;175:901–11. doi:10.1083/jcb.200608073.
39. Lehninger AL, Rossi CS, Greenawalt JW. Respiration-dependent accumulation of inorganic phosphate and Ca ions by rat liver mitochondria. Biochem Biophys Res Commun. 1963;10:444–8.
40. Vasington FD, Murphy JV. Ca2+ ion uptake by rat kidney mitochondria and its dependence on respiration and phosphorylation. J Biol Chem. 1962;237:2670–7.
41. Pozzan T, Bragadin M, Azzone GF. Disequilibrium between steady-state Ca2+ accumulation ratio and membrane potential in mitochondria. Pathway and role of Ca2+ efflux. Biochemistry. 1977;16:5618–25.
42. Carafoli E, Tiozzo R, Lugli G, et al. The release of calcium from heart mitochondria by sodium. J Mol Cell Cardiol. 1974;6:361–71.
43. Crompton M, Kunzi M, Carafoli E. The calcium-induced and sodium-induced effluxes of calcium from heart mitochondria. Evidence for a sodium-calcium carrier. Eur J Biochem. 1977;79:549–58.
44. Kirichok Y, Krapivinsky G, Clapham DE. The mitochondrial calcium uniporter is a highly selective ion channel. Nature. 2004;427:360–4. doi:10.1038/nature02246nature02246. [pii]
45. Perocchi F, Gohil VM, Girgis HS, et al. MICU1 encodes a mitochondrial EF hand protein required for Ca(2+) uptake. Nature. 2010;467:291–6. doi: nature09358 [pii]10.1038/nature09358
46. Baughman JM, Perocchi F, Girgis HS, et al. Integrative genomics identifies MCU as an essential component of the mitochondrial calcium uniporter. Nature. 2011;476:341–5. doi: nature10234 [pii]10.1038/nature10234
47. De Stefani D, Raffaello A, Teardo E, et al. A forty-kilodalton protein of the inner membrane is the mitochondrial calcium uniporter. Nature. 2011;476:336–40. doi: nature10230 [pii]10.1038/nature10230
48. Bick AG, Calvo SE, Mootha VK. Evolutionary diversity of the mitochondrial calcium uniporter. Science. (80-. 2012;336:886. doi:10.1126/science.1214977.
49. Cheng Y, Perocchi F. ProtPhylo: identification of protein-phenotype and protein-protein functional associations via phylogenetic profiling. Nucleic Acids Res. 2015. gkv455-; doi:10.1093/nar/gkv455.
50. Martell JD, Deerinck TJ, Sancak Y, et al. Engineered ascorbate peroxidase as a genetically encoded reporter for electron microscopy. Nat Biotechnol. 2012;30:1143–8. doi:10.1038/nbt.2375.
51. Raffaello A, De Stefani D, Sabbadin D, et al. The mitochondrial calcium uniporter is a multimer that can include a dominant-negative pore-forming subunit. EMBO J. 2013;32:2362–76. doi:10.1038/emboj.2013.157.
52. Chaudhuri D, Sancak Y, Mootha VK, Clapham DE. MCU encodes the pore conducting mitochondrial calcium currents. elife. 2013;2:e00704. doi:10.7554/eLife.00704.
53. Oxenoid K, Dong Y, Cao C, et al. Architecture of the mitochondrial calcium uniporter. Nature. 2016;533:269–73. doi:10.1038/nature17656.

54. De Stefani D, Patron M, Rizzuto R. Structure and function of the mitochondrial calcium uniporter complex. Biochim Biophys Acta. 2015;1853:2006–11. doi:10.1016/j.bbamcr.2015.04.008.
55. De Stefani D, Rizzuto R. Molecular control of mitochondrial calcium uptake. Biochem Biophys Res Commun. 2014;449:373–6. doi:10.1016/j.bbrc.2014.04.142.
56. Sancak Y, Markhard AL, Kitami T, et al. EMRE is an essential component of the mitochondrial calcium uniporter complex. Science. 2013;342:1379–82. doi:10.1126/science.1242993.
57. Fieni F, Lee SB, Jan YN, Kirichok Y. Activity of the mitochondrial calcium uniporter varies greatly between tissues. Nat Commun. 2012;3:1317. doi:10.1038/ncomms2325.
58. Vais H, Mallilankaraman K, Mak D-OD, et al. EMRE is a matrix Ca(2+) sensor that governs gatekeeping of the mitochondrial Ca(2+) uniporter. Cell Rep. 2016;14:403–10. doi:10.1016/j.celrep.2015.12.054.
59. Mallilankaraman K, Doonan P, Cardenas C, et al. MICU1 is an essential gatekeeper for MCU-mediated mitochondrial Ca(2+) uptake that regulates cell survival. Cell. 2012;151:630–44. doi:10.1016/j.cell.2012.10.011.
60. Csordas G, Golenar T, Seifert EL, et al. MICU1 controls both the threshold and cooperative activation of the mitochondrial Ca(2)(+) uniporter. Cell Metab. 2013;17:976–87. doi:10.1016/j.cmet.2013.04.020.
61. Patron M, Checchetto V, Raffaello A, et al. MICU1 and MICU2 finely tune the mitochondrial Ca2+ uniporter by exerting opposite effects on MCU activity. Mol Cell. 2014;53:726–37. doi:10.1016/j.molcel.2014.01.013.
62. Hoffman NE, Chandramoorthy HC, Shanmughapriya S, et al. SLC25A23 augments mitochondrial Ca2+ uptake, interacts with MCU, and induces oxidative stress-mediated cell death. Mol Biol Cell. 2014;25:936–47. doi:10.1091/mbc.E13-08-0502.
63. Mallilankaraman K, Cardenas C, Doonan PJ, et al. MCUR1 is an essential component of mitochondrial Ca(2+) uptake that regulates cellular metabolism. Nat Cell Biol. 2012;14:1336–43. doi:10.1038/ncb2622.
64. Paupe V, Prudent J, Dassa EP, et al. CCDC90A (MCUR1) is a cytochrome c oxidase assembly factor and not a regulator of the mitochondrial calcium uniporter. Cell Metab. 2015;21:109–16. doi:10.1016/j.cmet.2014.12.004.
65. Amigo I, Traba J, González-Barroso MM, et al. Glucagon regulation of oxidative phosphorylation requires an increase in matrix adenine nucleotide content through Ca2+ activation of the mitochondrial ATP-Mg/Pi carrier SCaMC-3. J Biol Chem. 2013;288:7791–802. doi:10.1074/jbc.M112.409144.
66. Rueda CB, Llorente-Folch I, Amigo I, et al. Ca(2+) regulation of mitochondrial function in neurons. Biochim Biophys Acta. 2014;1837:1617–24. doi:10.1016/j.bbabio.2014.04.010.
67. Rueda CB, Traba J, Amigo I, et al. Mitochondrial ATP-Mg/Pi carrier SCaMC-3/Slc25a23 counteracts PARP-1-dependent fall in mitochondrial ATP caused by excitotoxic insults in neurons. J Neurosci. 2015;35:3566–81. doi:10.1523/JNEUROSCI.2702-14.2015.
68. De Stefani D, Rizzuto R, Pozzan T. Enjoy the trip: calcium in mitochondria back and forth. Annu Rev Biochem. 2016;85:161–92. doi:10.1146/annurev-biochem-060614-034216.
69. Jiang D, Zhao L, Clapham DE. Genome-wide RNAi screen identifies Letm1 as a mitochondrial Ca2+/H+ antiporter. Science. (80-. 2009;326:144–7. doi:10.1126/science.1175145.
70. Tsai MF, Jiang D, Zhao L, et al. Functional reconstitution of the mitochondrial Ca2+/H+ antiporter Letm1. J Gen Physiol. 2014;143:67–73. doi:10.1085/jgp.201311096.
71. De Marchi U, Santo-Domingo J, Castelbou C, et al. NCLX protein, but not LETM1, mediates mitochondrial Ca2+ extrusion, thereby limiting Ca2+-induced NAD(P)H production and modulating matrix redox state. J Biol Chem. 2014;289:20377–85. doi:10.1074/jbc.M113.540898.
72. Nowikovsky K, Pozzan T, Rizzuto R, et al. Perspectives on: SGP symposium on mitochondrial physiology and medicine: the pathophysiology of LETM1. J Gen Physiol. 2012;139:445–54. doi:10.1085/jgp.201110757.

73. Palty R, Silverman WF, Hershfinkel M, et al. NCLX is an essential component of mitochondrial Na+/Ca2+ exchange. Proc Natl Acad Sci U S A. 2010;107:436–41. doi: 0908099107 [pii]10.1073/pnas.0908099107
74. Li W, Shariat-Madar Z, Powers M, et al. Reconstitution, identification, purification, and immunological characterization of the 110-kDa Na+/Ca2+ antiporter from beef heart mitochondria. J Biol Chem. 1992;267:17983–9.
75. Paucek P, Jabůrek M, Jaburek M. Kinetics and ion specificity of Na(+)/Ca(2+) exchange mediated by the reconstituted beef heart mitochondrial Na(+)/Ca(2+) antiporter. Biochim Biophys Acta. 2004;1659:83–91. doi:10.1016/j.bbabio.2004.03.019.
76. Liao J, Li H, Zeng W, et al. Structural insight into the ion-exchange mechanism of the sodium/calcium exchanger. Science. (80-. 2012;335:686–90. doi:10.1126/science.1215759.
77. Marinelli F, Almagor L, Hiller R, et al. Sodium recognition by the Na+/Ca2+ exchanger in the outward-facing conformation. Proc Natl Acad Sci U S A. 2014;111:E5354–62. doi:10.1073/pnas.1415751111.
78. Rizzuto R, De Stefani D, Raffaello A, Mammucari C. Mitochondria as sensors and regulators of calcium signalling. Nat Rev Mol Cell Biol. 2012;13:566–78. doi:10.1038/nrm3412.
79. Hajnoczky G, Csordas G, Das S, et al. Mitochondrial calcium signalling and cell death: approaches for assessing the role of mitochondrial Ca2+ uptake in apoptosis. Cell Calcium. 2006;40:553–60. doi:10.1016/j.ceca.2006.08.016.
80. Eliasson L, Renström E, Ding W-G, et al. Rapid ATP-dependent priming of secretory granules precedes Ca 2+ -induced exocytosis in mouse pancreatic B-cells. J Physiol. 1997;503:399–412. doi:10.1111/j.1469-7793.1997.399bh.x.
81. Takahashi N, Kadowaki T, Yazaki Y, et al. Post-priming actions of ATP on Ca2+-dependent exocytosis in pancreatic beta cells. Proc Natl Acad Sci. 1999;96:760–5. doi:10.1073/pnas.96.2.760.
82. Wiederkehr A, Szanda G, Akhmedov D, et al. Mitochondrial matrix calcium is an activating signal for hormone secretion. Cell Metab. 2011;13:601–11. doi:10.1016/j.cmet.2011.03.015.
83. Tarasov AI, Semplici F, Li D, et al. Frequency-dependent mitochondrial Ca(2+) accumulation regulates ATP synthesis in pancreatic beta cells. Pflugers Arch. 2013;465:543–54. doi:10.1007/s00424-012-1177-9.
84. Tarasov AI, Semplici F, Ravier MA, et al. The mitochondrial Ca2+ uniporter MCU is essential for glucose-induced ATP increases in pancreatic beta-cells. PLoS One. 2012;7:e39722. doi:10.1371/journal.pone.0039722.
85. Alam MR, Groschner LN, Parichatikanond W, et al. Mitochondrial Ca2+ uptake 1 (MICU1) and mitochondrial ca2+ uniporter (MCU) contribute to metabolism-secretion coupling in clonal pancreatic beta-cells. J Biol Chem. 2012;287:34445–54. doi:10.1074/jbc.M112.392084.
86. Quan X, Nguyen TT, Choi S-K, et al. Essential role of mitochondrial Ca2+ uniporter in the generation of mitochondrial pH gradient and metabolism-secretion coupling in insulin-releasing cells. J Biol Chem. 2015;290:4086–96. doi:10.1074/jbc.M114.632547.
87. Nita II, Hershfinkel M, Kantor C, et al. Pancreatic beta-cell Na+ channels control global Ca2+ signaling and oxidative metabolism by inducing Na+ and Ca2+ responses that are propagated into mitochondria. FASEB J. 2014;28:3301–12. doi:10.1096/fj.13-248161.
88. Nita II, Hershfinkel M, Lewis EC, Sekler I. A crosstalk between Na(+) channels, Na(+)/K(+) pump and mitochondrial Na(+) transporters controls glucose-dependent cytosolic and mitochondrial Na(+) signals. Cell Calcium. 2015;57:69–75. doi:10.1016/j.ceca.2014.12.007.
89. Granatiero V, Giorgio V, Calì T, et al. Reduced mitochondrial Ca2+ transients stimulate autophagy in human fibroblasts carrying the 13514A>G mutation of the ND5 subunit of NADH dehydrogenase. Cell Death Differ. 2016;23:231–41. doi:10.1038/cdd.2015.84.
90. Hall DD, Wu Y, Domann FE, et al. Mitochondrial calcium uniporter activity is dispensable for MDA-MB-231 breast carcinoma cell survival. PLoS One. 2014;9:e96866. doi:10.1371/journal.pone.0096866.
91. Galluzzi L, Pietrocola F, Levine B, Kroemer G. Metabolic control of autophagy. Cell. 2014;159:1263–76. doi:10.1016/j.cell.2014.11.006.

92. Ashrafi G, Schwarz TL. The pathways of mitophagy for quality control and clearance of mitochondria. Cell Death Differ. 2013;20:31–42. doi:10.1038/cdd.2012.81.
93. Youle RJ, Narendra DP. Mechanisms of mitophagy. Nat Rev Mol Cell Biol. 2011;12:9–14. doi:10.1038/nrm3028.
94. Cárdenas C, Miller RA, Smith I, et al. Essential regulation of cell bioenergetics by constitutive InsP3 receptor Ca2+ transfer to mitochondria. Cell. 2010;142:270–83. doi: S0092-8674(10)00629-X [pii]10.1016/j.cell.2010.06.007
95. Kim I, Rodriguez-Enriquez S, Lemasters JJ. Selective degradation of mitochondria by mitophagy. Arch Biochem Biophys. 2007;462:245–53. doi:10.1016/j.abb.2007.03.034.
96. MacVicar TDB, Mannack LVJC, Lees RM, Lane JD. Targeted siRNA screens identify ER-to-mitochondrial calcium exchange in autophagy and mitophagy responses in RPE1 cells. Int J Mol Sci. 2015;16:13356–80. doi:10.3390/ijms160613356.
97. Viscomi C, Bottani E, Zeviani M. Emerging concepts in the therapy of mitochondrial disease. Biochim Biophys Acta. 2015;1847(6–7):544–57. doi:10.1016/j.bbabio.2015.03.001.
98. Zeviani M, Carelli V. Mitochondrial disorders. Curr Opin Neurol. 2007;20:564–71. doi:10.1097/WCO.0b013e3282ef58cd.
99. Liao Y, Hao Y, Chen H, et al. Mitochondrial calcium uniporter protein MCU is involved in oxidative stress-induced cell death. Protein Cell. 2015;6:434–42. doi:10.1007/s13238-015-0144-6.
100. Qiu J, Tan YW, Hagenston AM, et al. Mitochondrial calcium uniporter Mcu controls excitotoxicity and is transcriptionally repressed by neuroprotective nuclear calcium signals. Nat Commun. 2013;4:2034. doi:10.1038/ncomms3034.
101. Curry MC, Peters AA, Kenny PA, et al. Mitochondrial calcium uniporter silencing potentiates caspase-independent cell death in MDA-MB-231 breast cancer cells. Biochem Biophys Res Commun. 2013;434:695–700. doi:10.1016/j.bbrc.2013.04.015.
102. Yoon MJ, Lee AR, Jeong SA, et al. Release of Ca 2+ from the endoplasmic reticulum and its subsequent influx into mitochondria trigger celastrol-induced paraptosis in cancer cells. Oncotarget. 2014;5:6816–31.
103. Marchi S, Lupini L, Patergnani S, et al. Downregulation of the mitochondrial calcium uniporter by cancer-related miR-25. Curr Biol. 2013;23:58–63. doi:10.1016/j.cub.2012.11.026.
104. Pan L, Huang B-J, Ma X-E, et al. MiR-25 protects cardiomyocytes against oxidative damage by targeting the mitochondrial calcium uniporter. Int J Mol Sci. 2015;16:5420–33. doi:10.3390/ijms16035420.
105. Baumgartner HK, Gerasimenko JV, Thorne C, et al. Calcium elevation in mitochondria is the main Ca2+ requirement for mitochondrial permeability transition pore (mPTP) opening. J Biol Chem. 2009;284:20796–803. doi: M109.025353 [pii]10.1074/jbc.M109.025353
106. De Stefani D, Bononi A, Romagnoli A, et al. VDAC1 selectively transfers apoptotic Ca(2+) signals to mitochondria. Cell Death Differ. 2012;19:267–73. doi: cdd201192 [pii]10.1038/cdd.2011.92
107. Pinton P, Ferrari D, Rapizzi E, et al. The Ca2+ concentration of the endoplasmic reticulum is a key determinant of ceramide-induced apoptosis: significance for the molecular mechanism of Bcl-2 action. EMBO J. 2001;20:2690–701. doi:10.1093/emboj/20.11.2690.
108. Tosatto A, Sommaggio R, Kummerow C, et al. The mitochondrial calcium uniporter regulates breast cancer progression via HIF-1α. EMBO Mol Med. 2016;8:569–85. doi: 10.15252/emmm.201606255
109. Pan X, Liu J, Nguyen T, et al. The physiological role of mitochondrial calcium revealed by mice lacking the mitochondrial calcium uniporter. Nat Cell Biol. 2013;15:1464–72. doi:10.1038/ncb2868.
110. Holmstrom KM, Pan X, Liu JC, et al. Assessment of cardiac function in mice lacking the mitochondrial calcium uniporter. J Mol Cell Cardiol. 2015;85:178–82. doi:10.1016/j.yjmcc.2015.05.022.
111. Murphy E, Pan X, Nguyen T, et al. Unresolved questions from the analysis of mice lacking MCU expression. Biochem Biophys Res Commun. 2014;449:384–5. doi:10.1016/j.bbrc.2014.04.144.

112. Rivera J, Tessarollo L. Genetic background and the dilemma of translating mouse studies to humans. Immunity. 2008;28:1–4. doi:10.1016/j.immuni.2007.12.008.
113. Kallapur S, Ormsby I, Doetschman T. Strain dependency of TGFbeta1 function during embryogenesis. Mol Reprod Dev. 1999;52:341–9. doi:10.1002/(SICI)1098-2795(199904)52:4<341::AID-MRD2>3.0.CO;2-N.
114. Sibilia M, Wagner EF. Strain-dependent epithelial defects in mice lacking the EGF receptor. Science. 1995;269:234–8.
115. Threadgill DW, Dlugosz AA, Hansen LA, et al. Targeted disruption of mouse EGF receptor: effect of genetic background on mutant phenotype. Science. 1995;269:230–4.
116. Barbaric I, Miller G, Dear TN. Appearances can be deceiving: phenotypes of knockout mice. Brief Funct Genomic Proteomic. 2007;6:91–103. doi:10.1093/bfgp/elm008.
117. Steeghs K, Heerschap A, de Haan A, et al. Use of gene targeting for compromising energy homeostasis in neuro-muscular tissues: the role of sarcomeric mitochondrial creatine kinase. J Neurosci Methods. 1997;71:29–41. doi:10.1016/S0165-0270(96)00124-0.
118. Garry DJ, Ordway GA, Lorenz JN, et al. Mice without myoglobin. Nature. 1998;395:905–8. doi:10.1038/27681.
119. Godecke A, Flogel U, Zanger K, et al. Disruption of myoglobin in mice induces multiple compensatory mechanisms. Proc Natl Acad Sci. 1999;96:10495–500. doi:10.1073/pnas.96.18.10495.
120. Ordway GA, Garry DJ. Myoglobin: an essential hemoprotein in striated muscle. J Exp Biol. 2004;207:3441–6. doi:10.1242/jeb.01172.
121. Wittenberg JB. Myoglobin function reassessed. J Exp Biol. 2003;206:2011–20. doi:10.1242/jeb.00243.
122. Kim S, Titcombe RF, Zhang H, et al. Network compensation of cyclic GMP-dependent protein kinase II knockout in the hippocampus by Ca2+-permeable AMPA receptors. Proc Natl Acad Sci U S A. 2015;112:3122–7. doi:10.1073/pnas.1417498112.
123. Pascual A, Hidalgo-Figueroa M, Piruat JI, et al. Absolute requirement of GDNF for adult catecholaminergic neuron survival. Nat Neurosci. 2008;11:755–61. doi:10.1038/nn.2136.
124. Giaever G, Chu AM, Ni L, et al. Functional profiling of the *Saccharomyces cerevisiae* genome. Nature. 2002;418:387–91. doi:10.1038/nature00935.
125. Wagner A. Robustness against mutations in genetic networks of yeast. Nat Genet. 2000;24:355–61. doi:10.1038/74174.
126. Deutscher D, Meilijson I, Kupiec M, Ruppin E. Multiple knockout analysis of genetic robustness in the yeast metabolic network. Nat Genet. 2006;38:993–8. doi:10.1038/ng1856.
127. Wu Y, Rasmussen TP, Koval OM, et al. The mitochondrial uniporter controls fight or flight heart rate increases. Nat Commun. 2015;6:6081. doi:10.1038/ncomms7081.
128. Rasmussen TP, Wu Y, Joiner MA, et al. Inhibition of MCU forces extramitochondrial adaptations governing physiological and pathological stress responses in heart. Proc Natl Acad Sci U S A. 2015; doi:10.1073/pnas.1504705112.
129. Kwong JQ, Lu X, Correll RN, et al. The mitochondrial calcium uniporter selectively matches metabolic output to acute contractile stress in the heart. Cell Rep. 2015;12:15–22. doi:10.1016/j.celrep.2015.06.002.
130. Luongo TS, Lambert JP, Yuan A, et al. The mitochondrial calcium uniporter matches energetic supply with cardiac workload during stress and modulates permeability transition. Cell Rep. 2015;12:23–34. doi:10.1016/j.celrep.2015.06.017.
131. Oka T, Maillet M, Watt AJ, et al. Cardiac-specific deletion of Gata4 reveals its requirement for hypertrophy, compensation, and myocyte viability. Circ Res. 2006;98:837–45. doi:10.1161/01.RES.0000215985.18538.c4.
132. Sohal DS, Nghiem M, Crackower MA, et al. Temporally regulated and tissue-specific gene manipulations in the adult and embryonic heart using a tamoxifen-inducible Cre protein. Circ Res. 2001;89:20–5.
133. Santulli G, Xie W, Reiken SR, Marks AR. Mitochondrial calcium overload is a key determinant in heart failure. Proc Natl Acad Sci U S A. 2015;112:11389–94. doi:10.1073/pnas.1513047112.

134. Brown SDM, Moore MW. Towards an encyclopaedia of mammalian gene function: the International Mouse Phenotyping Consortium. Dis Model Mech. 2012;5:289–92. doi:10.1242/dmm.009878.
135. Brown SDM, Moore MW. The international mouse phenotyping consortium: past and future perspectives on mouse phenotyping. Mamm Genome. 2012;23:632–40. doi:10.1007/s00335-012-9427-x.
136. Dickinson ME, Flenniken AM, Ji X, et al. High-throughput discovery of novel developmental phenotypes. Nat Publ Gr. 2016; doi:10.1038/nature19356.
137. Morgan H, Simon M, Mallon A-M. Accessing and mining data from large-scale mouse phenotyping projects. Int Rev Neurobiol. 2012;104:47–70. doi:10.1016/B978-0-12-398323-7.00003-3.
138. De Souza AT, Dai X, Spencer AG, et al. Transcriptional and phenotypic comparisons of Ppara knockout and siRNA knockdown mice. Nucleic Acids Res. 2006;34:4486–94. doi:10.1093/nar/gkl609.
139. Kok FO, Shin M, Ni C-W, et al. Reverse genetic screening reveals poor correlation between morpholino-induced and mutant phenotypes in Zebrafish. Dev Cell. 2014;32:97–108. doi:10.1016/j.devcel.2014.11.018.
140. Stainier DYR, Kontarakis Z, Rossi A. Making sense of anti-sense data. Dev Cell. 2015;32:7–8. doi:10.1016/j.devcel.2014.12.012.
141. Rossi A, Kontarakis Z, Gerri C, et al. Genetic compensation induced by deleterious mutations but not gene knockdowns. Nature. 2015;524:230–3. doi:10.1038/nature14580.
142. Huang G, Vercesi AE, Docampo R. Essential regulation of cell bioenergetics in Trypanosoma brucei by the mitochondrial calcium uniporter. Nat Commun. 2013;4:2865. doi:10.1038/ncomms3865.
143. Prudent J, Popgeorgiev N, Bonneau B, et al. Bcl-wav and the mitochondrial calcium uniporter drive gastrula morphogenesis in zebrafish. Nat Commun. 2013;4:2330. doi:10.1038/ncomms3330.
144. Xu S, Chisholm AD. *C. elegans* epidermal wounding induces a mitochondrial ROS burst that promotes wound repair. Dev Cell. 2014;31:48–60. doi:10.1016/j.devcel.2014.08.002.
145. Antony AN, Paillard M, Moffat C, et al. MICU1 regulation of mitochondrial Ca(2+) uptake dictates survival and tissue regeneration. Nat Commun. 2016;7:10955. doi:10.1038/ncomms10955.
146. Liu JC, Liu J, Holmström KM, et al. MICU1 serves as a molecular gatekeeper to prevent in vivo mitochondrial calcium overload. Cell Rep. 2016;16:1561–73. doi:10.1016/j.celrep.2016.07.011.
147. Lewis-Smith D, Kamer KJ, Griffin H, et al. Homozygous deletion in MICU1 presenting with fatigue and lethargy in childhood. Neurol Genet. 2016;2:e59. doi:10.1212/NXG.0000000000000059.
148. Logan CV, Szabadkai G, Sharpe JA, et al. Loss-of-function mutations in MICU1 cause a brain and muscle disorder linked to primary alterations in mitochondrial calcium signaling. Nat Genet. 2014;46:188–93. doi:10.1038/ng.2851.

Chapter 3
The In Vivo Biology of the Mitochondrial Calcium Uniporter

Julia C. Liu, Randi J. Parks, Jie Liu, Justin Stares, Ilsa I. Rovira, Elizabeth Murphy, and Toren Finkel

Decades of research have established the importance of regulating mitochondrial calcium both for stimulating ATP production [1], as well as modulating cell death through the calcium-dependent opening of the mitochondrial permeability transition pore (mPTP) [2]. How calcium passes through the inner mitochondrial membrane remained elusive until 2011, when two groups simultaneously identified CCDC109A as the gene encoding the bona fide mitochondrial calcium uniporter, renaming this previously uncharacterized gene product MCU [3, 4]. In initial cell culture experiments, MCU silencing severely attenuated rapid mitochondrial calcium uptake [3, 4]. Additional analysis revealed that in mammalian cells the calcium uniporter complex comprises not only MCU but several additional proteins that are thought to regulate MCU-mediated calcium entry into the mitochondria [5]. The first member of the uniporter complex to be described was MICU1 [6]. MICU1 and its paralogues MICU2 and MICU3 are thought to be localized in the intermembrane space and involved in the gatekeeping of the uniporter [7, 8]. Based on work done in cell culture models, MICU1 appears to inhibit MCU opening at low cytoplasmic calcium levels [7–10] and may also contribute to channel activation at high

J.C. Liu • J. Liu • J. Stares • I.I. Rovira
Center for Molecular Medicine, National Heart Lung and Blood Institute, Bethesda, MD 20892, USA

R.J. Parks • E. Murphy
Systems Biology Center, National Heart Lung and Blood Institute, Bethesda, MD 20892, USA

T. Finkel, MD, PhD (✉)
Center for Molecular Medicine, National Heart Lung and Blood Institute, Bethesda, MD 20892, USA

NIH, Bldg 10/CRC 5-3330, Bethesda, MD 20892, USA
e-mail: finkelt@nih.gov

G. Santulli (ed.), *Mitochondrial Dynamics in Cardiovascular Medicine*, Advances in Experimental Medicine and Biology 982,
DOI 10.1007/978-3-319-55330-6_3

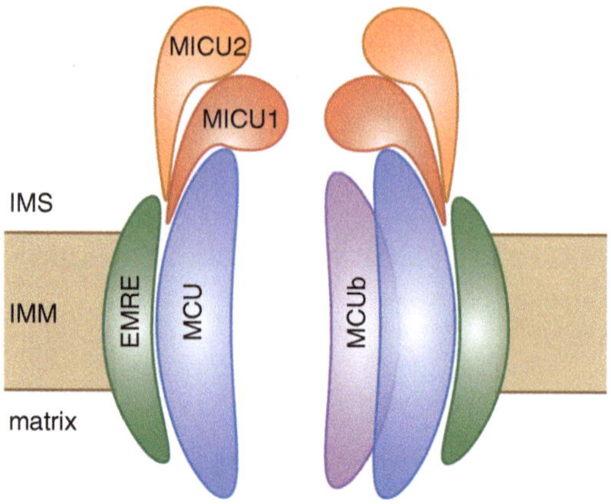

Fig. 3.1 The uniporter is a multiprotein complex. Representative diagram of some of the known components of the uniporter. The predominance of evidence suggests that MICU1 and MICU2 are located in the inner mitochondrial space (*IMS*), while MCU, MCUb, and EMRE span the inner mitochondrial membrane (*IMM*). See text for details.

calcium levels [8]. Recently, a single pass membrane protein named Essential MCU Regulator (EMRE) was described to be necessary for mitochondrial calcium uptake [11]. EMRE is proposed to act as a scaffold, mediating the interaction between MCU and other proteins belonging to the complex, including MICU1 and MICU2. Furthermore, the pore is formed by multimers of MCU as well as MCUb, a protein with high sequence homology with MCU that appears to negatively regulate calcium uptake. The relative stoichiometry of MCU and MCUb potentially contributes to variable uniporter activity among tissues [12]. Together, these genetic, biochemical, and structural studies have provided a relatively complete picture of the organizational structure of the uniporter complex (Fig. 3.1). However, the limited energetic demand of cells in culture suggested that to better understand the physiological role of mitochondrial calcium regulation, in vivo models would be needed. Here, we review the initial in vivo observations characterizing the effects of disrupting uniporter components in non-mammalian and mammalian model organisms.

Analysis of the Uniporter in Model Organisms

Homologues of MCU are present in a wide array of organisms including most plants and metazoans (Fig. 3.2). In contrast, many fungi and protozoans lack evidence for either MCU-like or MICU1-like proteins [13]. Similarly, there is no apparent uniporter activity evident in yeast [14]. Interestingly, the filamentous fungus *Neurospora crassa* has an MCU homologue [13], yet does not appear to have canonical mitochondrial calcium uptake activity [14], raising questions regarding what function this MCU homologue actually performs. Most organisms that have MCU-like proteins also have MICU1-like proteins, with the exception of the fungi possessing MCU homologues, including *N. crassa*, which typically appear to lack MICU1 homologues [13]. The wide diversity in domain structure of the MCU homologues in these fungi [13], coincident with the absence of MICU1, suggests that there are

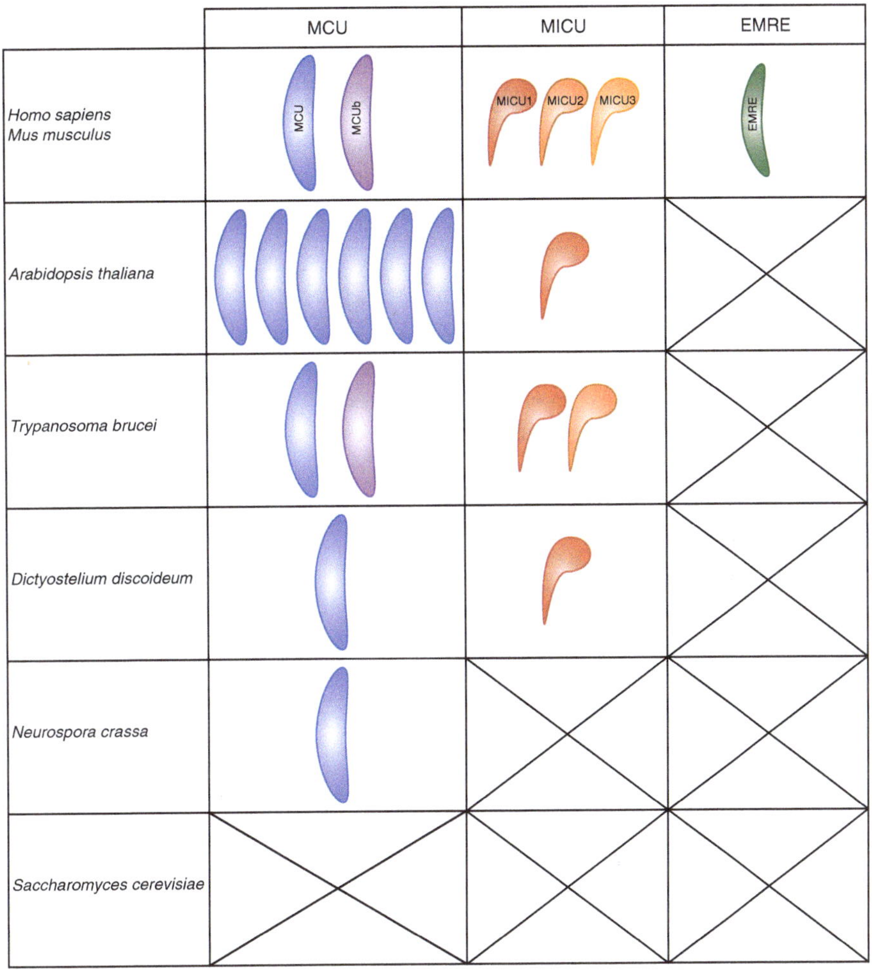

Fig. 3.2 Evolutionary diversity of the uniporter components. In both humans and mice there are two forms of MCU (MCU and MCUb), three MICU family members, and one EMRE. In other species, the quantity of these elements can vary considerably, including the occurrence of EMRE only in metazoans and the complete absence of a uniporter in yeast.

multiple evolutionary approaches to regulate mitochondrial calcium entry. In contrast to the broad evolutionary range of MCU and MICU1, EMRE is specific for metazoans [11]. These evolutionary differences have allowed for heterologous systems to explore functional requirements of the uniporter. For instance, the yeast *S. cerevisiae,* which lacks any uniporter, has been used as a platform for in vivo reconstitution [15]. Interestingly, this analysis has demonstrated that human MCU requires EMRE for calcium uptake, while the structurally distinct MCU found in *Dictyostelium discoideum* (DdMCU) does not [15]. Indeed, DdMCU restores calcium uptake in human cells lacking either MCU or EMRE. As such, these results establish two important principles. First, lessons gleaned from organisms such as yeast that lack any uniporter tell us that there is no absolute requirement for uniporter activity to maintain mitochondrial integrity or function. Second, in organisms that contain uniporter activity, there is a strong predilection for the occurrence of both MCU and MICU1 homologues to be present although evolutionary-derived variations in the structure of MCU can obviate the requirement for either MICU1-like or EMRE-like proteins.

Analysis of model systems has revealed significant information regarding the physiological function of the uniporter. One early example involved the role of the uniporter in the biology of *Trypanosoma brucei*, the parasite responsible for African sleeping sickness [16]. This parasite can exist in several distinct life cycle stages, including the procyclic form (PCF), in which the parasite resides within its normal vector the tse tse fly, and the bloodstream form (BSF), in which it circulates in the blood of its mammalian host. Remarkably, both the PCF and BSF forms contain a single mitochondrion. Moreover, the PCF mitochondrion has intact and fully functional oxidative phosphorylation capacity, while the BSF form does not. This system therefore is unique in that the two different forms of the parasite each have a single mitochondrion, yet the activity of that single mitochondrion is completely different depending on the lifecycle of the organism. When these authors ablated the *Trypanosoma brucei* uniporter (TbMCU) they were able, as expected, to reduce calcium uptake into the mitochondria. In addition, growth of both the PCF and BSF forms was inhibited in the absence of TbMCU. Similarly, mice injected with *Trypanosoma* containing a conditional knockout of TbMCU had decreased parasitemia and increased survival [16]. Given that the BSF form of *Trypanosoma brucei* does not rely on its mitochondrion for ATP (this form is entirely dependent on glycolysis) [17], this would suggest that mitochondrial calcium is required for some other function than simple bioenergetics in order to maintain viability of the parasite. The authors provide at least supportive evidence that in the absence of the uniporter there is an inability to generate acetyl-CoA due to a calcium-mediated reduction in pyruvate dehydrogenase (PDH) activity [16]. Interestingly, this impairment in PDH is also seen in mammalian models where MCU is deleted [18].

There has been some limited analysis of the requirement of the uniporter and its components in plants. In *Arabidopsis*, neither EMRE or MCUb exist, while there appear to be at least six MCU homologues present [19]. One of these homologues (At5g66650) appears to be targeted to the chloroplast, consistent with past observations that light-dependent calcium influx occurs in chloroplasts and this influx can be blocked by ruthenium red, a pharmacological inhibitor of the uniporter [20].There also appears to be differential tissue expression of these plant MCU homologues with some expressed highly in the root, while others are more abundant in the flower or guard cells [19]. This would imply that the diversification of MCU homologues might provide specialized calcium-dependent roles for individual tissues within the plant. Interestingly, while most plant species contain two MICU1 homologues (similar to MICU1 and MICU2), *Arabidopsis* contains only one such family member [21]. This single *Arabidopsis* MICU1 family member (At-MICU) appears to contain three EF hands, rather than the two EF hands found in both MICU1 and MICU2. As such, it would appear that with At-MICU, *Arabidopsis* has incorporated MICU1 and MICU2 function within a single molecule. Plants lacking At-MICU appear to develop normally but have an increase in steady state mitochondrial calcium levels [21]. This same calcium-overloaded phenotype is seen in tissues or mitochondria derived from mice lacking MICU1 [22, 23].

Analysis of model organisms has also revealed an interesting relationship between the uniporter and mitochondrial reactive oxygen species (mtROS) production. This was evident in the previously discussed *Trypanosoma* model where overexpression of TbMCU was demonstrated to increase mtROS production [16]. In *C. elegans*, skin wounding causes local superoxide production that appears to be required for actin-based wound closure [24]. Previous work had shown that a sustained rise in cytosolic calcium was also required for wound repair in this model [25]. The mechanistic link between these two phenomena came from the observation that MCU-null worms show reduced mitochondrial calcium uptake after wounding, as well as reduced mtROS production. Thus, a pathway induced by skin wounding leads to increased mitochondrial calcium uptake, increased mtROS production and the subsequent redox-dependent activation of the actin cytoskeleton [24]. While in this case the uniporter-dependent production of mtROS was beneficial, in other cases it appears to confer toxicity. For instance, a recent study sought to examine the toxicity of aminoglycoside antibiotics using a zebrafish model. Just as in human patients, in zebrafish, aminoglycosides can induce inner ear damage by permanently damaging the mechanosensory hair cells within the ear. Again, previous evidence suggested that this damage was mediated, at least in part, by ROS [26, 27]. However, in this case, it appears that in the dying hair cell, ROS production is dependent on mitochondrial calcium uptake through the uniporter [28]. As such, while calcium uptake through the uniporter appears to generate increased mtROS production in both worms and zebrafish, these two examples suggest that this can either be beneficial or harmful depending on the context.

Finally, one recent, fascinating report analyzed the role of the uniporter in *Drosophila*. This work was stimulated by a previous screen looking for genes that affected olfactory memory in the fly. Interestingly, this analysis suggested that silencing the *Drosophila* form of MCU or MICU1 impaired memory formation [29]. Following up on this observation, a recent study demonstrated that knockdown of *Drosophila* MCU (dMCU) within mushroom body neurons, a region essential for olfactory memory, resulted in impaired memory formation [30]. Interestingly, this defect was selective, as it did not affect the capacity to learn; instead, it only altered the capacity to remember at one or three hours after the conditioning stimulus [30]. Moreover, knockdown of dMCU only altered adult memory if performed during pupation, as a similar knockdown in adult flies had no effect [30].

Mouse Models

The generation of the first mouse model of MCU deletion was described in 2013 and represented an important step toward better defining the physiological impact of mitochondrial calcium [18]. While embryonically lethal in a pure genetic background such as C57BL/6, MCU-deficient mice could be generated in a mixed genetic background (e.g. CD1). Mitochondria derived from these $MCU^{-/-}$ mice were unable to rapidly uptake calcium, confirming results from cells lacking MCU. Unexpectedly, these mice displayed a seemingly modest phenotype and were only slightly smaller than wild-type littermates [18]. Despite reduced calcium in the mitochondrial matrix of $MCU^{-/-}$ skeletal muscle, there was no significant change in basal metabolism measured in either mouse embryonic fibroblasts (MEFs) or isolated mitochondria. When challenged with strenuous exercise requiring a rapid increase in skeletal muscle work load, however, $MCU^{-/-}$ mice were measurably impaired in exercise capacity (Fig. 3.3). Furthermore, addition of extra-mitochondrial calcium to wild-type cardiac mitochondria induced mPTP opening as evidenced by mitochondrial swelling, and this response was, as expected, inhibited by cyclosporine A (CsA), an mPTP inhibitor, or by Ru360, a ruthenium red analog and uniporter inhibitor. In contrast, $MCU^{-/-}$ mitochondria were completely resistant to calcium-induced mPTP opening, even at high extra-mitochondrial calcium concentrations (e.g. 0.5 mM). On the other hand, $MCU^{-/-}$ MEFs were not more sensitive than wild-type cells to death-inducing agents including hydrogen peroxide, tunicamycin, doxorubicin, and thapsigargin [18]. Since calcium overload is reported to play a role in injury and necrosis in the setting of ischemia-reperfusion (I/R), hearts from wild-type and $MCU^{-/-}$ mice were subjected to an ex vivo Langendorff model of I/R injury. Surprisingly, overall $MCU^{-/-}$ hearts were not protected. However, while CsA conferred significant protection to wild-type hearts, it did not protect $MCU^{-/-}$ hearts. These results suggest that cell death in the absence of MCU may be mediated through a mPTP-independent pathway, begging the questions of whether and how cell death signaling may be rewired in the setting of chronic MCU inhibition.

A subsequent study analyzed cardiac function in the same strain of $MCU^{-/-}$ mice [31]. This analysis revealed that despite the fact that $MCU^{-/-}$ heart mitochondria had attenuated calcium uptake and reduced matrix calcium levels, basal cardiac function was normal at both 12 and 20 months of age. Moreover, the ability of $MCU^{-/-}$ hearts to respond to certain stresses – either pharmacological (isoproterenol stimulation) or physiological (surgical transverse aortic constriction) – appeared indistinguishable from wild-type littermates. This lack of a basal or stress-induced phenotype was quite unexpected given the numerous important roles mitochondrial calcium was believed to play in cardiac physiology [32].

To further investigate the role of MCU in the heart, other groups have generated mice with transgenic expression of a dominant-negative (DN) MCU [33]. These mice were used to focus on the involvement of MCU in increasing ATP production in order to fuel demand-dependent heart rate increases. To inhibit MCU-mediated

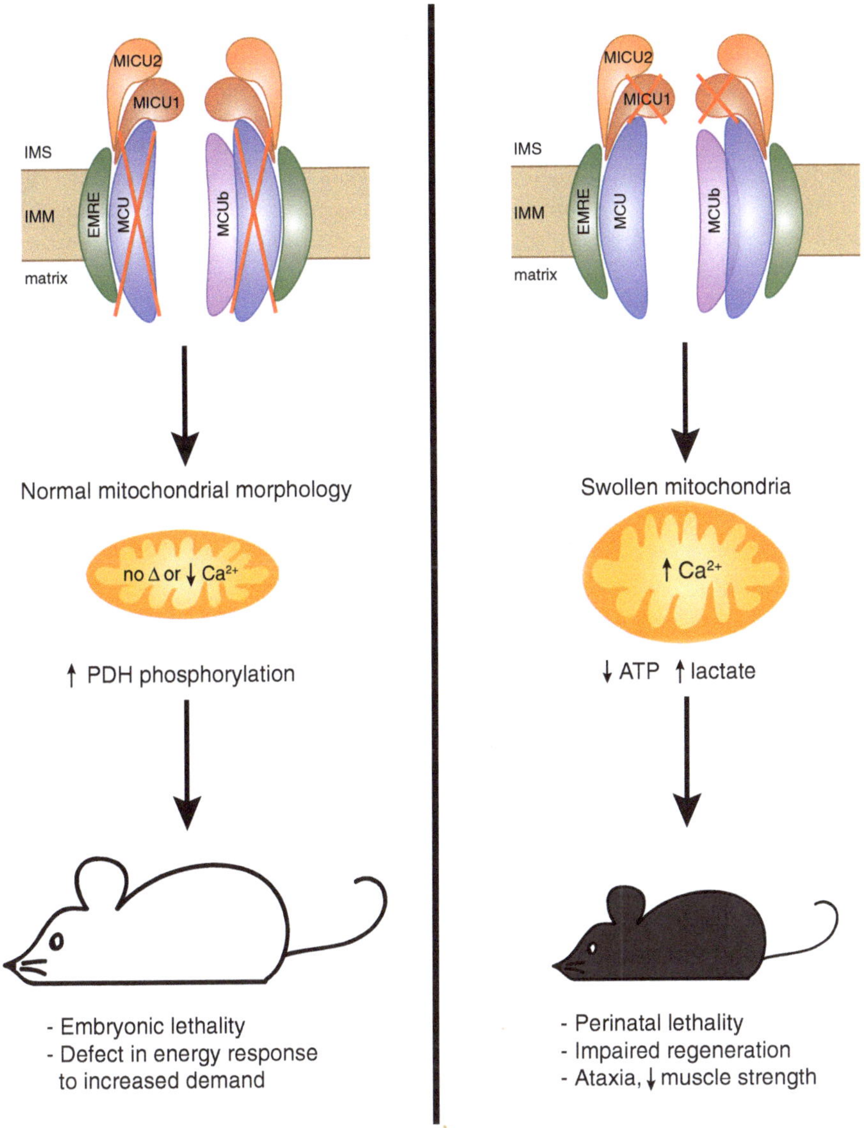

Fig. 3.3 Insights from genetic manipulations of MCU and MICU1 in mouse models. Summary of some of the findings from reports in which mice lacking either MCU (*left panel*) or MICU1 (*right panel*) have been characterized. These include effects on resting mitochondrial calcium levels and mitochondrial morphology, as well as alterations at the tissue and organ level. See text for additional details.

mitochondrial calcium uptake in the myocardium, a mouse containing a DN-MCU construct driven by the α-myosin heavy chain promoter was generated. This dominant negative mutation is in the pore-forming region and takes advantage of the observation that within the inner mitochondrial membrane, MCU forms oligomers [34]. In vivo expression of DN-MCU prevented rapid, MCU-mediated, mitochondrial calcium uptake. Nonetheless, under basal conditions, DN-MCU mice had hearts with normal chamber size and function. These animals did exhibit a reduction in atrial ATP levels when compared with littermate controls. While basal heart rates were only mildly slower in DN-MCU hearts, in unrestrained and un-sedated mice, the heart rate increase following isoproterenol stimulation was less pronounced in the DN-MCU mice. Cardiac sinoatrial nodal (SAN) pacemaker cells isolated from DN-MCU mice also showed reduced sarcoplasmic reticulum (SR) calcium content after physiological stress (isoproterenol stimulation) compared with wild-type SAN cells. This defect could be rescued by exogenous ATP. These data suggested that in this model, mitochondrial calcium uptake through MCU plays a role in ATP synthesis needed for maintaining overall calcium homeostasis. As such, the study concluded that MCU is required for the physiologically important 'fight or flight' stress response.

In a second study, the same mouse model of dominant-negative MCU was used to examine certain stress responses in the absence of mitochondrial calcium uptake [35]. Unexpectedly, isolated Langendorff-perfused DN-MCU hearts exhibited higher oxygen consumption rates (OCR) than WT. As mitochondrial morphology, protein and DNA content seemed unaltered; a potential explanation would be that DN-MCU hearts were energetically less efficient than WT. In contrast, permeabilized myocardial fibers or isolated mitochondria from DN-MCU mice exhibited only a modest reduction, if any, in oxygen consumption. Similar to what was reported in the whole-body MCU knockout [18], phosphorylation of PDH was significantly increased, and hence PDH was less active in DN-MCU hearts compared with WT. Monitoring membrane potential and changes in reactive oxygen species (ROS) during I/R revealed significantly reduced ROS in DN-MCU hearts, but no difference in infarct area following I/R was found [35]. As such, this lack of protection was consistent with that seen in total body MCU knockout mice [18].

In contrast to these models of constitutive MCU deletion or inhibition, two collaborative groups recently developed a conditional knockout mouse model in which MCU can be deleted in specific tissues by Cre-mediated methods [36, 37]. When $MCU^{fl/fl}$ mice were crossed to a transgenic germline-Cre model, no $MCU^{-/-}$ pups were obtained, implying that as previously discussed homozygous MCU deletion results in embryonic lethality in inbred mice. Crossing $MCU^{fl/fl}$ to αMHC-Cre to specifically delete MCU in cardiomyocytes resulted in about 80 % reduction of MCU protein in adult cardiomyocytes. Unlike in the global MCU knockout [31], matrix calcium in adult cardiomyocytes did not decrease with MCU deletion, although mitochondrial calcium uptake was eliminated [36, 37]. Using a tamoxifen-inducible heart-specific Cre, the authors next assessed I/R injury using a left coronary artery (LCA) ligation model. As would be expected if calcium uptake caused mPTP opening, MCU conditional knockout (cKO) mice were protected from injury, exhibiting reduced TUNEL staining and infarct sizes that measured roughly half of WT littermates. The discrepancy between these positive results and the lack of protection seen in MCU knockout or DN-MCU mice is currently unexplained. Possibilities include technical differences such as differences in the model of I/R injury (e.g. in vivo LCA ligation as opposed to an *ex vivo* Langendorff model).

More likely, these disparities reflect differences in the timing and duration of MCU deletion. In the MCU cKO, tamoxifen was administered to adult mice, several weeks before the response to I/R was assayed. In contrast, in the $MCU^{-/-}$ and DN-MCU mice, MCU activity was absent throughout development. This suggests that there might be important physiological differences between acute and chronic deletion of MCU.

Further analysis of the MCU cKO mice revealed a number of interesting phenotypes. Similar to what has been observed in other models, even 1 year after MCU deletion, no pathological effects on mitochondrial, cellular, or cardiac structure were observed [36]. When acute cardiac stress was induced by intravenous infusion of isoproterenol, PDH phosphorylation was higher in the MCU cKO versus WT, and correspondingly isoproterenol-stimulated PDH enzymatic activity in MCU cKO was decreased [37]. This finding was consistent with observations in both non-mammalian model organisms [16] and other MCU deletion models [18]. Regarding metabolism, no differences in basal respiration were found in adult cardiomyocytes, although the MCU cKO cells were refractory to isoproterenol-induced maximal oxygen consumption rates [37]. In another cardiac stress test, transverse aortic constriction was performed on MCU cKO and control mice, but led to no differences in cardiac pathology or function [36], consistent with the data from the $MCU^{-/-}$ mice [31]. When mice were challenged to sprint on a treadmill, when only two minutes of warm-up time was allowed, MCU cKO mice displayed significantly reduced running capacity compared with controls. This difference disappeared when 30 min of warm-up was allowed. Hence, while subtle, it would appear that adult cardiac-specific deletion of MCU impairs the ability of the heart to adapt to certain acute stresses that require acute increases in mitochondrial metabolism.

Studies of the mitochondrial calcium uniporter in vivo have not been limited to MCU, the channel-forming unit, as the multi-protein complex includes other components critical for its regulation. In particular, the impact of MICU's gatekeeping function on the uniporter in vivo is of great interest. The first two studies investigating MICU1 deletion in mouse models were recently published (Fig. 3.3) [22, 23]. In the first study, MICU1 whole body deletion resulted in complete perinatal lethality shortly after birth [22]. Permeabilized MEFs derived from the animals showed increased calcium uptake at low concentrations of cytoplasmic calcium. No gross anatomical or morphological defects were observed in the knockout animals, despite a trend toward fewer neurons in the nucleus ambigous and nucleus facialis which regulates breathing. The authors suspected this defect might have contributed to the observed perinatal mortality. The authors then deleted MICU1 in the liver through tail-vein injection of hepatocyte-specific Cre. Following a 70% partial hepatectomy, they observed that liver damage was more extensive with MICU1 deletion and regeneration was abrogated. This was hypothesized to occur from mitochondrial calcium overload, with subsequent increased mPTP opening. Thus, in this model, mitochondrial calcium overload resulting from MICU1 deletion was linked to sustained inflammation and injury.

Shortly thereafter, another report described an independently generated mouse model of MICU1 deletion [23]. In this strain, which was a slightly different background than the other MICU1 mouse model, homozygous deletion of MICU1 resulted in high but not total postnatal lethality. Analysis of over 1000 live births suggested that roughly one in seven MICU1$^{-/-}$ mice survived beyond the first postnatal week [23]. These rare surviving MICU1$^{-/-}$ mice initially exhibited a severe phenotype reminiscent of human patients with loss-of-function MICU1 mutations that included abnormal mitochondrial morphology, decreased muscle ATP levels, elevated serum lactate, and neurological features such as ataxia [38]. Confirming the results of the previous study, mitochondria isolated from these MICU1$^{-/-}$ mice exhibited increased calcium uptake at low extramitochondrial calcium concentrations, and in addition had elevated resting mitochondrial matrix calcium levels. Hence, the predominant in vivo function of MICU1 appears to be gatekeeping of the uniporter to prevent mitochondrial calcium overload. Notably, as these MICU1$^{-/-}$ mice aged, their symptoms improved, coincident with reduced EMRE gene expression. As EMRE is reported to be essential for mitochondrial calcium uptake, a reduction in EMRE would potentially limit entry of calcium into mitochondria, helping to relieve calcium overload. Remarkably, deleting one allele of EMRE in MICU1$^{-/-}$ mice rescued mitochondrial calcium handling and perinatal mortality [23]. These results suggest that the relative expression of uniporter components can modulate mitochondrial calcium uptake with profound physiological consequences. Interestingly, it would appear that MICU1 deletion, characterized by calcium overload, is perfectly compatible with embryogenesis but rapidly fatal immediately after birth. In contrast, MCU deletion, which presumably represents a condition of insufficient calcium levels, is embryonically lethal in a pure background, while mice born in a genetically mixed background have no difficulties after birth, no significant basal phenotype and only a modest defect following various stress conditions (Fig. 3.3).

The Role of the Uniporter in Human Disease

There is relatively little direct information regarding the role of the uniporter in human disease. This most likely reflects the relative recent discovery of the molecular basis for mitochondrial calcium entry. One area in which there is already some, albeit conflicting, data is the role of MCU in cancer. Given the potential role of mitochondrial calcium in mediating cell death, it would appear likely that alterations in uniporter activity could alter tumor cell resistance to apoptotic or necrotic cell death. One early report demonstrated that expression of MCU was negatively regulated by miR-25, a microRNA that is often upregulated in human tumors (see Chap. 23) [39]. Indeed, analyzing human colon samples, and the corresponding surrounding normal mucosa, these authors were able to identify that tumors had high levels of miR-25, and corresponding low levels of MCU protein expression [39]. In heterologous tumor cell lines, expression of miR-25 resulted in decreased MCU expression and reduced sensitivity to apoptotic stimuli. This pathway may also be operative in non-malignant diseases. A recent report has also demonstrated reduction of MCU in pulmonary arterial hypertension (PAH), a devastating and ultimately fatal disease characterized by excessive smooth muscle proliferation in the pulmonary vasculature. This report described a reduction in MCU expression in smooth muscle cells (SMCs) derived from patients with PAH when these cells were compared to SMCs from healthy controls [40]. Again, this reduction appeared to be mediated by increased expression of miR-25, as well as another microRNA, miR-138. Remarkably, administering nebulized anti-miRs directed against miR-25 and miR-138 restored MCU expression and reversed PAH in an animal model of this condition [40]. While the previous reports would indicate that reduced MCU expression appears to contribute to cancer cell growth, or the abnormal proliferation of SMCs, there is also evidence for the opposite phenotype. For instance, analysis of human breast cancer samples revealed that there was a positive correlation between MCU expression and breast cancer tumor size and lymph node infiltration [41]. Concurrently, this same clinical data set revealed a negative correlation between expression of the dominant negative MCUb and tumor size. Thus, in breast cancer, it would appear that MCU expression might fuel disease progression. The authors propose that this may result from increased MCU expression leading to augmented mtROS production and the subsequent redox-dependent stabilization of HIF-1α [41]. Again, in a clinical data of some 500 human breast cancer patients, a significant correlation was found between MCU expression and the expression of both HIF-1α and multiple, potentially oncogenic, HIF targets.

The strongest connection between the uniporter and human disease comes not from alterations in MCU expression, but rather from analysis of inherited mutations in MICU1. A cohort of pediatric patients was shown to have loss-of-function mutations in MICU1 leading to the absence of detectable MICU1 protein expression in these individuals [38]. The manifestation of MICU1 loss was characterized by proximal muscle weakness, with evidence of ongoing muscle damage (e.g. elevated serum creatine kinase levels). This was confirmed by muscle biopsies which demonstrated increased central nuclei with clusters of regenerating fibers [38]. Over half of the subjects exhibited extrapyramidal signs that included tremors, chorea, and dystonic posturing. In some patients, there was evidence of ataxia, microcephaly, and ocular dysfunction (i.e. ptosis, opthalmaplegia, and optic atrophy). While the constellation of myopathic and neurological symptoms suggests a potential mitochondrial basis, it is of interest to note that none of the patients appeared to have other common abnormalities such as deafness, heart failure, or diabetes, often associated with primary mitochondrial diseases. Sequencing of affected individuals revealed various MICU1 mutations, often affecting splice donor sites and leading to nonsense decay of the mutant mRNA. Fibroblasts derived from these individuals revealed evidence for increased mitochondrial calcium levels, with evidence of mitochondrial fragmentation [38]. A second, more recent report described two cousins, each of whom contained a large deletion in exon 1 of MICU1. Their clinical course was slightly different, although it did include muscle weakness with high serum creatine kinase levels [42]. For these two cousins, however, most of their symptomatology was episodic. In one case, minimal exercise would precipitate becoming pale and sweaty, followed by lethargy to the point of being noncommunicative and unarousable. The other cousin had similar exercise-induced lethargy and confusion that would last for hours, if not days. Again, fibroblasts from these affected patients showed no MICU1 expression [42]. These cells exhibited normal mitochondrial function but altered mitochondrial calcium uptake. Interestingly, the episodic nature of these symptoms is reminiscent of other channelopathies that often underlie other paroxysmal conditions such as migraines, epilepsy and sudden cardiac death [43].

Conclusion

Since its molecular discovery a mere 5 years ago, remarkable progress has been made in discerning the in vivo role of the mitochondrial calcium uniporter. This analysis has ranged from analyzing uniporter function in a variety of model organisms, to more standard mouse model systems. While important phenotypes have been discerned, perhaps the most surprising result is that viable mice can be produced that lack any functional uniporter [44, 45]. Perhaps, in retrospect, the absence of a uniporter in yeast might have provided some guidance that mitochondria can function without this element of acute calcium regulation. Additional models and analysis are clearly required to understand the other components (e.g. EMRE, MICU2, etc.), as well as how all these components interact. The observation that the perinatal lethality of MICU1 can be rescued by deleting one allele of EMRE [23] suggests that the creation of additional mouse models will allow for important and potentially insightful combinatorial models. Additional information is also needed to explain what appears to be a clear distinction between acute and chronic deletion of uniporter components. There seems to be some functional remodeling and rewiring that occurs following chronic MCU deletion, for example, that allows cells to physiologically adapt. In some ways, this may represent a form of retrograde signaling, namely a pathway by which a mitochondrial perturbation can ultimately result in nuclear transcriptional changes [46]. All these and other related questions suggest that the next 5 years may provide further insight into the in vivo role of the uniporter in addition to the exciting progress already made.

References

1. Glancy B, Balaban RS. Role of mitochondrial Ca2+ in the regulation of cellular energetics. Biochemistry. 2012;51:2959–73.
2. Rasola A, Bernardi P. Mitochondrial permeability transition in Ca(2+)-dependent apoptosis and necrosis. Cell Calcium. 2011;50:222–33.
3. Baughman JM, Perocchi F, Girgis HS, Plovanich M, Belcher-Timme CA, Sancak Y, Bao XR, Strittmatter L, Goldberger O, Bogorad RL, Koteliansky V, Mootha VK. Integrative genomics identifies MCU as an essential component of the mitochondrial calcium uniporter. Nature. 2011;476:341–5.

4. De Stefani D, Raffaello A, Teardo E, Szabo I, Rizzuto R. A forty-kilodalton protein of the inner membrane is the mitochondrial calcium uniporter. Nature. 2011;476:336–40.
5. De Stefani D, Rizzuto R, Pozzan T. Enjoy the trip: calcium in mitochondria back and forth. Annu Rev Biochem. 2016;85:161–92.
6. Perocchi F, Gohil VM, Girgis HS, Bao XR, McCombs JE, Palmer AE, Mootha VK. MICU1 encodes a mitochondrial EF hand protein required for Ca(2+) uptake. Nature. 2010;467:291–6.
7. Csordas G, Golenar T, Seifert EL, Kamer KJ, Sancak Y, Perocchi F, Moffat C, Weaver D, de la Fuente PS, Bogorad R, Koteliansky V, Adijanto J, Mootha VK, Hajnoczky G. MICU1 controls both the threshold and cooperative activation of the mitochondrial Ca(2)(+) uniporter. Cell Metab. 2013;17:976–87.
8. Kamer KJ, Mootha VK. MICU1 and MICU2 play nonredundant roles in the regulation of the mitochondrial calcium uniporter. EMBO Rep. 2014;15:299–307.
9. Mallilankaraman K, Doonan P, Cardenas C, Chandramoorthy HC, Muller M, Miller R, Hoffman NE, Gandhirajan RK, Molgo J, Birnbaum MJ, Rothberg BS, Mak DO, Foskett JK, Madesh M. MICU1 is an essential gatekeeper for MCU-mediated mitochondrial Ca(2+) uptake that regulates cell survival. Cell. 2012;151:630–44.
10. Wang L, Yang X, Li S, Wang Z, Liu Y, Feng J, Zhu Y, Shen Y. Structural and mechanistic insights into MICU1 regulation of mitochondrial calcium uptake. EMBO J. 2014;33:594–604.
11. Sancak Y, Markhard AL, Kitami T, Kovacs-Bogdan E, Kamer KJ, Udeshi ND, Carr SA, Chaudhuri D, Clapham DE, Li AA, Calvo SE, Goldberger O, Mootha VK. EMRE is an essential component of the mitochondrial calcium uniporter complex. Science. 2013;342:1379–82.
12. Fieni F, Lee SB, Jan YN, Kirichok Y. Activity of the mitochondrial calcium uniporter varies greatly between tissues. Nat Commun. 2012;3:1317.
13. Bick AG, Calvo SE, Mootha VK. Evolutionary diversity of the mitochondrial calcium uniporter. Science. 2012;336:886.
14. Carafoli E, Lehninger AL. A survey of the interaction of calcium ions with mitochondria from different tissues and species. Biochem J. 1971;122:681–90.
15. Kovacs-Bogdan E, Sancak Y, Kamer KJ, Plovanich M, Jambhekar A, Huber RJ, Myre MA, Blower MD, Mootha VK. Reconstitution of the mitochondrial calcium uniporter in yeast. Proc Natl Acad Sci U S A. 2014;111:8985–90.
16. Huang G, Vercesi AE, Docampo R. Essential regulation of cell bioenergetics in Trypanosoma brucei by the mitochondrial calcium uniporter. Nat Commun. 2013;4:2865.
17. van Hellemond JJ, Opperdoes FR, Tielens AG. The extraordinary mitochondrion and unusual citric acid cycle in Trypanosoma brucei. Biochem Soc Trans. 2005;33:967–71.
18. Pan X, Liu J, Nguyen T, Liu C, Sun J, Teng Y, Fergusson MM, Rovira II, Allen M, Springer DA, Aponte AM, Gucek M, Balaban RS, Murphy E, Finkel T. The physiological role of mitochondrial calcium revealed by mice lacking the mitochondrial calcium uniporter. Nat Cell Biol. 2013;15:1464–72.
19. Stael S, Wurzinger B, Mair A, Mehlmer N, Vothknecht UC, Teige M. Plant organellar calcium signalling: an emerging field. J Exp Bot. 2012;63:1525–42.
20. Kreimer G, Melkonian M, Holtum JA, Latzko E. Characterization of calcium fluxes across the envelope of intact spinach chloroplasts. Planta. 1985;166:515–23.
21. Wagner S, Behera S, De Bortoli S, Logan DC, Fuchs P, Carraretto L, Teardo E, Cendron L, Nietzel T, Fussl M, Doccula FG, Navazio L, Fricker MD, Van Aken O, Finkemeier I, Meyer AJ, Szabo I, Costa A, Schwarzlander M. The EF-hand Ca2+ binding protein MICU choreographs mitochondrial Ca2+ dynamics in Arabidopsis. Plant Cell. 2015;27:3190–212.
22. Antony AN, Paillard M, Moffat C, Juskeviciute E, Correnti J, Bolon B, Rubin E, Csordas G, Seifert EL, Hoek JB, Hajnoczky G. MICU1 regulation of mitochondrial Ca(2+) uptake dictates survival and tissue regeneration. Nat Commun. 2016;7:10955.
23. Liu JC, Liu J, Holmstrom KM, Menazza S, Parks RJ, Fergusson MM, Yu ZX, Springer DA, Halsey C, Liu C, Murphy E, Finkel T. MICU1 serves as a molecular gatekeeper to prevent in vivo mitochondrial calcium overload. Cell Rep. 2016;16:1561–73.

24. Xu S, Chisholm AD. *C. elegans* epidermal wounding induces a mitochondrial ROS burst that promotes wound repair. Dev Cell. 2014;31:48–60.
25. Xu S, Chisholm AD. A Galphaq-Ca(2)(+) signaling pathway promotes actin-mediated epidermal wound closure in *C. elegans*. Curr Biol. 2011;21:1960–7.
26. Hirose K, Hockenbery DM, Rubel EW. Reactive oxygen species in chick hair cells after gentamicin exposure in vitro. Hear Res. 1997;104:1–14.
27. Clerici WJ, Hensley K, DiMartino DL, Butterfield DA. Direct detection of ototoxicant-induced reactive oxygen species generation in cochlear explants. Hear Res. 1996;98:116–24.
28. Esterberg R, Linbo T, Pickett SB, Wu P, Ou HC, Rubel EW, Raible DW. Mitochondrial calcium uptake underlies ROS generation during aminoglycoside-induced hair cell death. J Clin Invest. 2016;126:3556–66.
29. Walkinshaw E, Gai Y, Farkas C, Richter D, Nicholas E, Keleman K, Davis RL. Identification of genes that promote or inhibit olfactory memory formation in Drosophila. Genetics. 2015;199:1173–82.
30. Drago I, Davis RL. Inhibiting the mitochondrial calcium uniporter during development impairs memory in adult Drosophila. Cell Rep. 2016;16:2763–76.
31. Holmstrom KM, Pan X, Liu JC, Menazza S, Liu J, Nguyen TT, Pan H, Parks RJ, Anderson S, Noguchi A, Springer D, Murphy E, Finkel T. Assessment of cardiac function in mice lacking the mitochondrial calcium uniporter. J Mol Cell Cardiol. 2015;85:178–82.
32. Williams GS, Boyman L, Lederer WJ. Mitochondrial calcium and the regulation of metabolism in the heart. J Mol Cell Cardiol. 2015;78:35–45.
33. Wu Y, Rasmussen TP, Koval OM, Joiner ML, Hall DD, Chen B, Luczak ED, Wang Q, Rokita AG, Wehrens XH, Song LS, Anderson ME. The mitochondrial uniporter controls fight or flight heart rate increases. Nat Commun. 2015;6:6081.
34. Oxenoid K, Dong Y, Cao C, Cui T, Sancak Y, Markhard AL, Grabarek Z, Kong L, Liu Z, Ouyang B, Cong Y, Mootha VK, Chou JJ. Architecture of the mitochondrial calcium uniporter. Nature. 2016;533:269–73.
35. Rasmussen TP, Wu Y, Joiner ML, Koval OM, Wilson NR, Luczak ED, Wang Q, Chen B, Gao Z, Zhu Z, Wagner BA, Soto J, McCormick ML, Kutschke W, Weiss RM, Yu L, Boudreau RL, Abel ED, Zhan F, Spitz DR, Buettner GR, Song LS, Zingman LV, Anderson ME. Inhibition of MCU forces extramitochondrial adaptations governing physiological and pathological stress responses in heart. Proc Natl Acad Sci U S A. 2015;112:9129–34.
36. Kwong JQ, Lu X, Correll RN, Schwanekamp JA, Vagnozzi RJ, Sargent MA, York AJ, Zhang J, Bers DM, Molkentin JD. The mitochondrial calcium uniporter selectively matches metabolic output to acute contractile stress in the heart. Cell Rep. 2015;12:15–22.
37. Luongo TS, Lambert JP, Yuan A, Zhang X, Gross P, Song J, Shanmughapriya S, Gao E, Jain M, Houser SR, Koch WJ, Cheung JY, Madesh M, Elrod JW. The mitochondrial calcium uniporter matches energetic supply with cardiac workload during stress and modulates permeability transition. Cell Rep. 2015;12:23–34.
38. Logan CV, Szabadkai G, Sharpe JA, Parry DA, Torelli S, Childs AM, Kriek M, Phadke R, Johnson CA, Roberts NY, Bonthron DT, Pysden KA, Whyte T, Munteanu I, Foley AR, Wheway G, Szymanska K, Natarajan S, Abdelhamed ZA, Morgan JE, Roper H, Santen GW, Niks EH, van der Pol WL, Lindhout D, Raffaello A, De Stefani D, den Dunnen JT, Sun Y, Ginjaar I, Sewry CA, Hurles M, Rizzuto R, Duchen MR, Muntoni F, Sheridan E. Loss-of-function mutations in MICU1 cause a brain and muscle disorder linked to primary alterations in mitochondrial calcium signaling. Nat Genet 2014;46:188–193.
39. Marchi S, Lupini L, Patergnani S, Rimessi A, Missiroli S, Bonora M, Bononi A, Corra F, Giorgi C, De Marchi E, Poletti F, Gafa R, Lanza G, Negrini M, Rizzuto R, Pinton P. Downregulation of the mitochondrial calcium uniporter by cancer-related miR-25. Curr Biol. 2013;23:58–63.
40. Hong Z, Chen KH, Dasgupta A, Potus F, Dunham-Snary K, Bonnet S, Tian L, Fu J, Breuils-Bonnet S, Provencher S, Wu D, Mewburn J, Ormiston ML, Archer SL. miR-138 and miR-25

downregulate MCU, causing pulmonary arterial hypertension's cancer phenotype. Am J Respir Crit Care Med. 2016;195:515–529.

41. Tosatto A, Sommaggio R, Kummerow C, Bentham RB, Blacker TS, Berecz T, Duchen MR, Rosato A, Bogeski I, Szabadkai G, Rizzuto R, Mammucari C. The mitochondrial calcium uniporter regulates breast cancer progression via HIF-1alpha. EMBO Mol Med. 2016;8:569–85.
42. Lewis-Smith D, Kamer KJ, Griffin H, Childs AM, Pysden K, Titov D, Duff J, Pyle A, Taylor RW, Yu-Wai-Man P, Ramesh V, Horvath R, Mootha VK, Chinnery PF. Homozygous deletion in MICU1 presenting with fatigue and lethargy in childhood. Neurol Genet. 2016;2:e59.
43. Spillane J, Kullmann DM, Hanna MG. Genetic neurological channelopathies: molecular genetics and clinical phenotypes. J Neurol Neurosurg Psychiatry. 2016;87:37–48.
44. Herzig S, Maundrell K, Martinou JC. Life without the mitochondrial calcium uniporter. Nat Cell Biol. 2013;15:1398–400.
45. Harrington JL, Murphy E. The mitochondrial calcium uniporter: mice can live and die without it. J Mol Cell Cardiol. 2015;78:46–53.
46. Bohovych I, Khalimonchuk O. Sending out an SOS: mitochondria as a signaling hub. Front Cell Dev Biol. 2016;4:109.

Chapter 4
Mitochondrial Bioenergetics and Dysfunction in Failing Heart

Freya L. Sheeran and Salvatore Pepe

In order to perform work, the myocardium must obtain a constant supply of cellular energy in the form of ATP. The heart is one of the highest ATP-consuming organs in the body, turning over more than its weight in ATP per day. Approximately 95% of ATP is produced by the mitochondrion, primarily through oxidative phosphorylation and to a lesser extent, oxidation of glucose and amino acids [1, 2]. Thus mitochondria play a crucial bioenergetic role in healthy heart function. Intertwined with this, mitochondria are dynamic regulators of cellular adaptation and homeostasis, cell survival and cell death [3].

Heart failure due to systolic dysfunction of varying etiology has been generally defined as a syndrome featuring progressive inability of the heart to supply sufficient blood flow to the body, resulting in symptoms such as shortness of breath and exercise intolerance as the body becomes starved of oxygen and essential nutrients. The idea that a shortage of energy supply was at the centre of cardiac dysfunction in heart failure has been proposed for decades, but was brought to the forefront by a series of papers published by Ingwall [4–7] and others [8, 9] in the 1990s, enabling real-time patient measures through the use of nuclear magnetic resonance (NMR) techniques. Through these studies it was demonstrated that levels of ATP and the energy storage molecule creatine phosphate (PCr) were around 30% lower in the tissues of heart failure patients [9, 10], correlating with a decline in cardiac function [9, 10], with PCr:ATP ratios shown to be an strong predictor of patient mortality, similar to NYHA class alone [5].

F.L. Sheeran
Heart Research, Murdoch Children's Research Institute, Melbourne, Australia

Department of Paediatrics, University of Melbourne, Melbourne, Australia

Royal Children's Hospital, Melbourne, Australia

S. Pepe (✉)
Department of Cardiology, Royal Children's Hospital,
50 Flemington Road, VIC 3052 Melbourne, Australia

Heart Research, Murdoch Children's Research Institute, Melbourne, Australia

Department of Paediatrics, University of Melbourne, Melbourne, Australia

Royal Children's Hospital, Melbourne, Australia
e-mail: Salvatore.Pepe@mcri.edu.au

G. Santulli (ed.), *Mitochondrial Dynamics in Cardiovascular Medicine*,
Advances in Experimental Medicine and Biology 982,
DOI 10.1007/978-3-319-55330-6_4

Although mitochondrial dysfunction, as a key contributor to the energy deficit underlying diminishing cardiac output, was identified early from animal studies, direct evidence from humans studies has been more limited and hampered by sampling access, methodological approach and biopsy quality [11]. Indeed studies of isolated mitochondrial preparations from human failing myocardium have reported mixed findings in that high quality, freshly isolated mitochondrial preparations can exhibit tightly coupled respiration under strictly regulated in vitro conditions [11]. Such findings suggest that meticulous isolation methods may select for high quality mitochondria with relatively intact structural and functional respiratory chain

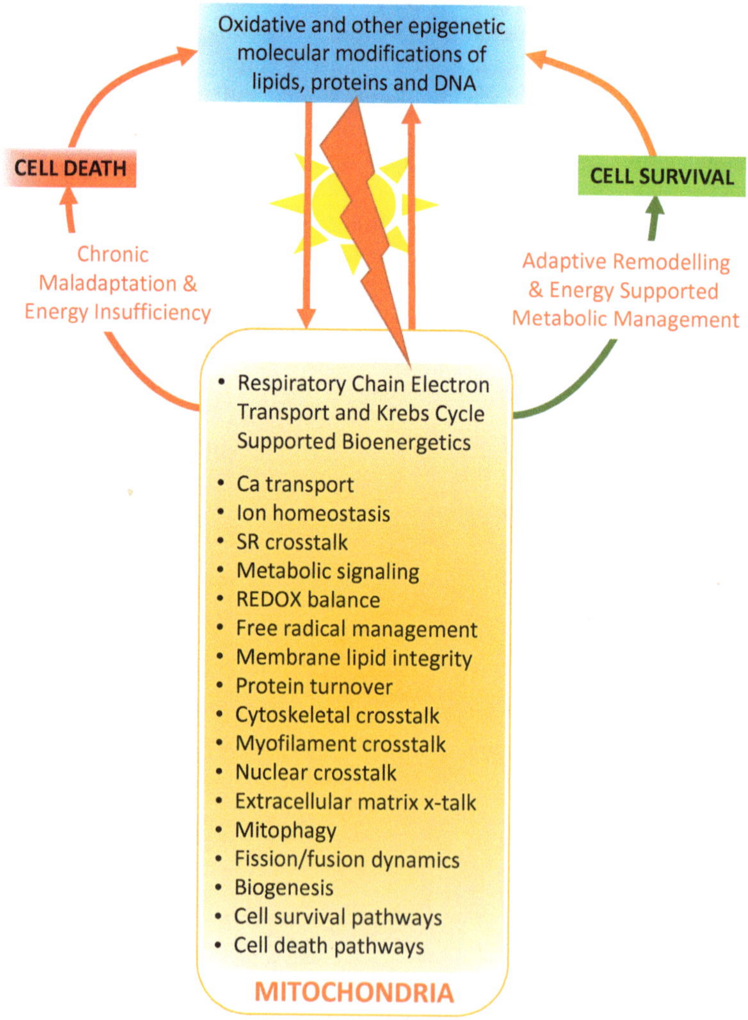

Fig. 4.1 The general scheme highlighting post-translational modifications contributing to mitochondrial and myocardial contractile dysfunction in the progression of systolic heart failure

integrity in vitro. However, this also indicates that in situ study of mitochondria may be more revealing in heart failure because there are numerous other mitochondrial and extra-mitochondrial components beyond respiratory protein complex electron transport regulating bioenergetic function that may be dysfunctional in systolic heart failure.

Progressively worsening energy deficit has profound effects on cellular homeostasis which compounds myocardial dysfunction, further contributing to loss of cardiac function. More than two thirds of energy produced by the cell is utilized for contractile function, with the remainder required for regulation of Ca^{2+} homeostasis, membrane potential, sarcoplasmic reticulum Ca^{2+} handling, as well as other ion channel function, cell maintenance and repair [2]. Restricted energy for contractile function potentially limits capacity for calcium handling, resulting in accumulation of calcium in the sarcoplasmic reticulum, decreased activation of Ca^{2+} -dependent mitochondrial Krebs enzymes as well as altered cell signalling pathways such as SIRT1 which are involved in mitochondrial turnover and biogenesis [12]. Furthermore, increased Ca^{2+} also predisposes the mitochondrion to activation of the permeability transition pore, leading to cell death. All of these above indicators have been reported in heart failure [13, 14]. As seen, disruption of one pathway potentially has a multitude of effects on other signalling, altering the normal regulatory homeostasis of cell function.

In this chapter, we briefly discuss the various aspects contributing to a decline in energy production in the failing heart with particular emphasis on the role of mitochondria. As generally summarized by the scheme in Fig. 4.1, specific focus will include post-translational modification and dysfunction of proteins of the oxidative phosphorylation and Krebs pathways, the role of reactive oxygen species in targeting membrane proteins and lipids, metabolic substrate 'switch', mitochondrial membrane lipid composition as well as the contribution of these factors to cell signalling involving mitochondrial turnover and cell death. While the chapter provides a brief introductory overview, it also highlights that considerable more work is required to augment our understanding of the complex, multiple functional roles of mitochondria, and to target their specific components in the translation of therapeutics to counter heart failure.

Post-Translational Modification of Mitochondrial Proteins Leading to Dysfunction

Mitochondrial ATP synthesis involves a number of distinct pathways within various cellular compartments and is regulated at numerous hierarchical levels [1–3]. Glucose is initially oxidized to form pyruvate by glycolysis within the cytoplasm, resulting in the formation of two ATP molecules. Pyruvate then enters the mitochondrion through the inner membrane mitochondrial pyruvate carrier (MPC), which forms a major regulatory checkpoint for choice of substrate metabolism. Pyruvate is then further broken down through eight enzymatic steps of the Krebs cycle within the mitochondrial matrix, resulting in the reduction of metabolic intermediates NADH and $FADH_2$. Oxidative phosphorylation completes the production of 38 molecules of ATP per mole of glucose through a series of four electron transporting complexes (CI, CII, CIII, CIV), whereby NADH or $FADH_2$ is oxidized by complexes I or II respectively, resulting in the flow of electrons through the chain to convert O_2 to water (complex IV). Protons are extruded at complexes I, III and IV into the intermembrane space, which are subsequently channelled back into the matrix via ATP synthase, driving the synthesis of ATP. This process is tightly regulated, with close coupling of electron transport, membrane potential and ATP synthesis.

Lowered rates of coupled ATP synthesis have been reported in both animal models [15] and human heart failure tissues [16]. As mitochondrial metabolism contributes to around 95% of cardiac ATP supply, predominantly through oxidative phosphorylation, it could reasonably be surmised that disruption of mitochondrial energetics and metabolism lies at the heart of ATP deficit. Decreased activities of individual complexes, namely I, III and IV have been reported in human heart failure tissues [17–21]. Although altered gene expression of electron transport chain protein subunits has not been uniformly reported [18], whether lowered activity was a result of reduced mitochondrial content overall or due to specific protein modification causing dysfunction was unclear. However, we recently reported that the diminished activities of mitochondrial complexes I and IV in end-stage heart failure were due to specific post-translational modifications of key protein subunits, rather than a loss of mitochondrial content per se [17].

Raised Oxidative Stress in Heart Failure Contributing to a Decline in Mitochondrial Function

Under normal cellular conditions, around 2–4% of oxygen is incompletely converted to water, resulting in the formation of superoxide, predominantly at complexes I and III [22–24]. In the healthy cell, endogenous antioxidant enzyme systems match the rate of reactive oxygen species (ROS) production by converting superoxide to hydrogen peroxide and then further processed to form water and oxygen. However, in the scenario of excessive ROS production or a deficit of antioxidant systems, as occurs in advanced age, oxidative damage to cellular proteins, lipids and DNA can accumulate. Under certain conditions, superoxide is able to react with nearby membrane phospholipids, including mitochondrial membranes, resulting in the formation of lipid-soluble, highly reactive aldehyde products such as malondialdehyde (MDA) or 4-hydroxy-2-nonenal (HNE) inducing further damage to lipids, proteins and even mitochondrial DNA. Raised levels of ROS and lipid peroxidation products have been measured in the tissues [17, 25, 26] and pericardial fluid [27] of heart failure patients, and have shown to correlate with contractile dysfunction [27]. In addition, protein carbonyls, formed by the oxidation of sensitive amino acid groups such as the cysteine and lysine residues, have been reported to be elevated in a range of human diseases, but only more recently has a strong correlation between myofilament oxidation and systolic failure been reported in human end-stage heart failure [28]. The most abundant myofilament modifications augmented in end-stage systolic failure patients have been identified to involve carbonylation, disulphide cross-bridge formation and S-nitrosylation of actin and tropomyosin. These structural modifications of the contractile apparatus impede contractile function which is progressively exacerbated as more mitochondria are themselves impacted by oxidative modifications and reactive intermediates that can no longer be adequately cleared in the face of elevated oxidative work stress.

As HNE is formed by superoxide-induced oxidation of membrane polyunsaturated fatty acids (PUFA) such as arachidonic acid, not surprisingly, raised HNE levels have been measured after ischemia-reperfusion [29–31] and in LV tissues and plasma of heart failure patients [32–34]. HNE attacks proteins at histidine, cysteine and lysine residues, resulting in disruption of three-dimensional folding and interfering with enzymatic function [29]. When applied directly to mitochondrial preparations, HNE inhibits state III coupled respiration, resulting in a loss of ATP production [33]. In isolated rat cardiomyocytes, HNE treatment increased ROS levels, resulting in calcium overload leading to hypercontracture [34]. In our study, we demonstrated that specific protein subunits within complexes I, III and IV were targets of post-translational oxidative modification by HNE, carbonylation and protein nitration [17], particularly those associated with redox centres. Selective targeting of specific oxidative phosphorylation protein subunit sites has also been demonstrated in other models of cardiomyopathy, while increased binding by HNE has been measured in several Krebs cycle enzymes, particularly those associated with complex I, following ischemia-reperfusion injury [35–37]. Notably, in fibroblasts isolated from patients with complex I deficiency, increased superoxide levels inversely correlated with CI activity and were also linked to loss of mitochondrial membrane potential [38].

Influence of Krebs Cycle Enzymes on Metabolic Flux

As remarked by Sadek et al. [37], loss of oxidative phosphorylation activity alone is not sufficient to fully account for the deficit in respiratory rates, rather previous studies indicate that NADH levels have a more direct effect on respiration as NADH 'feeds' the respiratory chain [39]. Three of the eight Krebs cycle enzymes; α-ketoglutarate dehydrogenase (KGDH), isocitrate dehydrogenase (ICDH) and malate dehydrogenase (MDH) directly supply NADH to complex I, of which the first two are also regulators of Krebs cycle flux. Due to the close spatial arrangement between complex I and these particular Krebs enzymes, we postulated that in the failing heart, ROS produced from complex I may also affect closely-associated Krebs cycle enzymes. Indeed, as predicted, activities of ICDH and MDH were lower in human failing tissues [17], consistent with impairment seen following ischemia-reperfusion injury [37]. Studies in spontaneously-hypertensive rat strains which progress to failure have demonstrated a loss of ICDH activity in the early hypertrophic phase, which was associated with increased HNE modification. Furthermore, dose-response studies in isolated mitochondria show a direct correlation between HNE binding and loss of ICDH activity, highlighting that ICDH is a direct target and is inactivated by HNE in hypertrophy [40]. Myocardial adaptation following myocardial infarction in rats involves early perturbations of Krebs cycle flux and concomitantly correlates with diminished contractile function, which subsequently progresses to also involve a loss of pyruvate dehydrogenase flux and worsening systolic ejection fractions [41]. Activity of aconitase, a Krebs cycle enzyme that is highly vulnerable to oxidative stress, is also significantly lower in the failing heart [17], highlighting the role of oxidative stress in the pathology of heart failure. Notably, these studies highlight specific loss of activity in both oxidative phosphorylation and Krebs cycle enzymes that is selective and not due to changes in protein abundance [17], but rather post-translational modifications such as those induced by oxidative injury.

In addition to the overall energetic flux, alterations to the NADH/NAD ratio has a profound influence on REDOX status and other downstream pathways, in particular that of the sirtuin pathways. NADH also serves to replenish NADPH stores, which are essential for the reduction of glutathione, one of the most important cellular antioxidant systems. Also crucial to REDOX and mitochondrial NADPH pools is a mitochondrial protein complex that has had limited study in the human heart failure setting, nicotinamide nucleotide transhydrogenase (NADPH-transhydrogenase, Nnt) [42]. One role of Nnt is the maintenance of NADPH-dependent ratio of reduced to oxidised glutathione, and other NADPH-dependent thiols such as thioredoxin, peroxiredoxins and glutaredoxin, in the regulation of cellular oxidative stress. Nnt activity is coupled to the mitochondrial proton motive force during ATP synthesis and is a major source of NADPH. We have previously reported that relative to non-failing myocardium, Nnt activity is markedly diminished due to protein carbonylation, concomitant with diminished glutathione reductase activity, lower NADPH and GSH/GSSG and augmented oxidized glutathione in chronic end-stage heart failure patients [42].

Sirtuin 1 (SIRT1) also has an important role in regulating energy metabolism, as a nuclear metabolic sensor, SIRT1 directly couples the cellular metabolic status (via NAD^+) to the regulation of gene expression via deacetylation of histones, transcription factors and co-factors [12]. SIRT1 can also shuttle between the nucleus and cytosol in response to various environmental stimuli, thus acting as a sensor for the overall energetic status of the cell, primarily through the availability of NAD^+, which is involved in both redox and metabolic reactions. SIRT1 also plays an important role in activating cell survival machinery. SIRT1 expression has shown to be reduced in mouse hearts subjected to ischemia-reperfusion (IR), while preconditioning increased SIRT1 and antioxidant levels and lowered expression of pro-apoptotic factors. In the same study, mice overexpressing SIRT1 showed improved post IR recovery, while SIRT1 induced expression of protective factors and antioxidants such as Mn-SOD, thioredoxin 1, and Bcl-xL, while negatively regulating pro-apoptotic factors such as Bax and cleaved caspase-3 [43]. Intravenous infusion of 10–20 mg/kg NAD^+ into rats has furthermore shown to reduce post IR infarct size by up to 85%, and significantly reduce apoptosis following ischemia-reperfusion by decreasing levels of Bax, cleaved caspase 3 and TUNEL staining, while increasing levels of pro-survival Bcl [44].

Inhibition of oxidative phosphorylation and Krebs cycle enzymes, as may occur following IR injury or heart failure, may prevent the downstream oxidation of NADH by the respiratory chain, resulting in a chemically reduced system and a build-up of NADH. This could be predicted to result in a lowering of the NAD^+/NADH ratio, which in turn would lower expression of SIRT1. Indeed, measures of SIRT1 expression in cardiomyocytes isolated from human failing hearts have shown SIRT1 levels to be reduced to around 55% of that in healthy controls. This was accompanied by increases in pro-apoptotic markers such as Bax, increased numbers of TUNEL-positive cardiomyocytes, oxidative stress (8-OH-dG) and lowered levels of antioxidants such as Mn-SOD, thioredoxin 1 and anti-apoptotic Bcl-xL in advanced failure [14]. Furthermore, extensive studies in complex I-deficient ($NDUFS4^{-/-}$ mice showed a significant decrease in the NAD^+/NADH ratio, despite similar levels of NAD^+, which was accompanied by increased sensitivity to calcium-induced mitochondrial permeability transition and pore opening. Mitochondrial permeability transition sensitivity could be alleviated by treatment with nicotinamide mononucleotide (NMN), an NAD^+ precursor which has been shown to raise NAD^+ levels, thus restoring the NAD^+/NADH ratio [45]. Hence, NAD^+ and energetic status may have a profound effect on cell survival and contribute to the pathogenesis of heart failure. An important implication in the underlying dysfunction that contributes to the augmented NADH and decreased NAD^+/NADH ratio in advanced left ventricular failure relates to the sirtuin deacetylases which require NAD+ for their activity. Subsequent decline in sirtuin deacetylase function may limit the regulation of acetylation such that a multitude of lysine-rich proteins eventually become hyperacetylated in human heart failure, in particular these include regulators of the mitochondrial permeability transition pore, (cyclophilin D, oligomycin sensitive conferral protein site of the adenosine triphosphate synthase) and components of the malate-aspartate shuttle [45, 46]. Thus hyperacetylation of critical mitochondrial proteins further exacerbates loss of REDOX regulation in heart failure.

PDH and Substrate Availability

One of the characteristic effects of cardiac hypertrophy, and to a greater extent, heart failure is the shift towards glycolysis, and away from fatty acid oxidation. The effect of this change is to create a more efficient means of energy supply. However, this comes at the expense of energy output, with glycolysis producing two molecules of ATP per molecule of glucose as opposed to 38 ATPs per molecule during oxidative phosphorylation. Substrate selection within the cell is governed at multiple levels; at the genetic level through expression of metabolic enzymes and receptors, as well as allosteric and substrate feedback regulation. One of the major regulators of this process is the peroxisome proliferator-activated receptors (PPAR), which govern expression of key enzymes and receptors involved in energy metabolism such as glucose receptors GLUT1 and GLUT4, fatty acid oxidation enzymes and pyruvate dehydrogenase kinases (PDKs) [47, 48]. Choice of substrate selection also occurs at 'gateways' to major metabolic pathways, such as pyruvate entry into the mitochondrion and at key crossroads between two pathways.

Pyruvate dehydrogenase (PDH) sits at the crossroad between the glycolytic and fatty acid oxidation pathways, catalysing the conversion of pyruvate from glycolysis to Acetyl CoA, and is therefore a major regulator of substrate selection for ATP synthesis. Pyruvate entry into the mitochondrion is regulated via the inner membrane MPC, whose molecular identity has only recently been discovered [49–51]. The PDH enzyme complex within the mitochondrial matrix is itself regulated by a group of four inhibitory kinases, each of which has varying affinities for various metabolic intermediates such as NADH, ATP and Acetyl CoA. When the intermediates are in a highly reduced state (e.g. high [NADH]/[NAD] or [Acetyl CoA]/CoA], PDKs are activated, inhibiting the PDH complex and thus limiting downstream metabolism through the Krebs cycle [51]. In turn, PDH phosphatases (PDPs) phosphorylate and hence activate PDH. PDP1 is activated in response to Ca^{2+}, producing an increase in PDH activity and therefore energy availability in times of high energy demand such as muscle contraction. PDH can therefore provide rapid metabolic regulation in response to altered energy demands.

In addition to regulatory and allosteric feedback mechanisms, post-translational regulation of proteins through acetylation and other mechanisms has only been recognized more recently and plays a pivotal role in the regulation of mitochondrial energetics. One such protein under this form of regulation is the MPC protein. MPC regulates the entry of pyruvate into the mitochondrion as an inner membrane channel and thus serves as a checkpoint to regulate metabolic flux. Protein binding and deacetylation of MPC1 by sirtuin 3 (SIRT3), which is involved in mitochondrial homeostasis, enhances its activity [52]. MPC1 protein expression is deleted or underexpressed by up to 85% in a number of cancer cell lines, showing a strong correlation with poor prognosis, which is consistent with an increased reliance on glycolytic metabolism in these cells [53]. Recent results from our own lab indicate a decrease in MPC1 and MPC2 expression in the failing heart (unpublished- in preparation). Notably, MPC protein expression is crucial to conferring a survival advantage in post-ischemic infarcted tissues [54]. In a separate study, suppression of MPC1 produced cells which were able to maintain Krebs cycle activity and cell growth via alternative pathways to Acetyl CoA, demonstrating that inhibition of the MPC pathway produced greater metabolic flexibility compared to other downstream pathways [55]. Loss of the MPC1 protein in cardiac-restricted knockout mice has reportedly resulted in mice with normal cardiac function at 8 weeks [56], despite several markers of hypertrophy, but which only develop overt systolic dysfunction by the age of 18 weeks. Thus this study demonstrates that pyruvate uptake is essential for normal cardiac function [56].

Increased expression of PDK2 and PDK4 has been measured in the tissues of heart failure patients, reflecting an increased reliance on glycolytic metabolism, similar to that of fetal phenotype [57]. However, gene expression profiling reveals that it is not a reversion to a fetal profile per se which occurs in heart failure, but rather a downregulation of adult heart isoforms which occurs, allowing the fetal markers to dominate [57]. Conversely, overexpression of PDK4 in a mouse model in itself showed no noticeable effect on cardiac function. However, when introduced in a mouse strain which constitutively expressed calcineurin, which causes hypertrophy, this resulted in increased fatty acid oxidation, cardiac fibrosis and a dramatic rise in mortality [58]. Doxorubicin, a cancer drug known to cause cardiomyopathy, inhibits pyruvate transport, possibly by interacting with cardiolipin molecules surrounding the pyruvate carrier [59], while a decline in pyruvate transport concomitant with lowered cardiolipin levels reportedly occurs in the hypothyroid rat [60]. Thus loss of metabolic flexibility can have profound effects on cardiac function and survival.

Metabolic remodelling involves a progressive alteration in substrate supply, initially by activation and inactivation of pathways involved in fuel selection to provide the most efficient source for that particular environment. At the second level, genetic alterations lead to adaptation and eventually maladaptation of the heart involving metabolic inflexibility, such that that heart is no longer able to modify its metabolism based on it metabolic state and condition [61]. The shift in gene expression to a fetal-like pattern is associated with concomitant limited energy availability due to increased reliance on glucose-based metabolism that is unable to supply sufficient ATP to meet total energy needs.

Cardiolipin and Oxidative Phosphorylation Supercomplex Proteins

Cardiolipin is synthesized de novo within the mitochondrion and is resides almost exclusively within the mitochondrial inner membrane [62, 63]. Cardiolipin (CL) contains four acyl chains attached to a glycerophosphate backbone. Newly synthesized CL contains a mixture of acyl chains, but is enriched post-synthesis to contain four linoleic acid (18:2) side chains by two separate pathways, one of which is the acetyltransferase Tafazzin (TAZ), to form the predominant species tetralinoleoyl-cardiolipin (L_4CL). As they are embedded within the mitochondrial inner membrane, many membrane proteins including complexes I, III, IV, the adenine nucleotide translocase (ANT) and ATP synthesis require a close molecular association with CL for optimal function, in fact, detergent solubilized protein/CL complexes stripped of CL result in loss of enzymatic function [62, 63]. Furthermore, ischemia-reperfusion in isolated rat hearts results in losses of CI, CIII or CIV activity concurrent with decreased CL levels [64–66], with raised mitochondrial ROS following reperfusion contributing to both protein and lipid peroxidation. Notably, in these same studies, enzyme activity could be restored following enrichment of mitochondrial preparations with CL liposomes, indicating CL played a critical role in respiration. Not only is CL required for enzymatic activity, but CL must be in the L_4CL configuration for optimal activity to occur. It is precisely for this reason that hearts from Barth syndrome patients, in which there is a mutation in the TAZ gene, often suffer from severe dysfunctional oxidative phosphorylation leading to cardiomyopathy [67, 68]. Lowered total cardiolipin, in particular the L_4CL species, have been reported in the tissues of heart failure patients [69].

Cardiolipin is not only required for optimal activity of electron transport chain protein complexes, but has more recently been attributed as the structural 'glue' which holds the mitochondrial respiratory chain together. As opposed to earlier stepwise 'linear' arrangement models of electron transfer, these mitochondrial protein complexes appear to be arranged as 'supercomplexes', in particular those of complexes I/III/IV and the ANT/ATP synthase [70, 71]. These supercomplexes provide stability and maximize efficiency of electron transfer, providing optimal spatial orientation between various enzyme components of the respiratory chain [70]. Studies in a canine model of heart failure have demonstrated an increased dissociation of supercomplex formation and a greater proportion of ETC complex enzymes in loose formation, resulting in decreased capacity for electron transfer and ATP synthesis [72]. Together with a decreased enzyme activity by virtue of lowered CL availability, CL loss in heart failure further exacerbates the problem by decreasing supercomplex formation.

Cardiolipin and 'the Pore'

Mitochondria possess an intrinsic mechanism which initiates cell death following lethal cell damage. Cell death may occur by necrotic or apoptotic cellular pathways, of which the latter occurs by a programmed pathway involving numerous cellular messengers and activators. One mitochondrial pathway involves the opening of a large pore spanning the outer and inner membranes, resulting in the release of cytochrome *c* from the intermembrane space and the initiation of cell death by apoptosis. The mitochondrial permeability transition pore is understood to involve the adenine nucleotide translocase (ANT) and voltage-dependent anion channel (VDAC) as well as attached cyclophilin D [73]. Trigger factors that augment sensitivity for permeability transition pore opening include high [Ca^{2+}], ROS, hyperacetylation or ATP depletion, all of which may be present under conditions of heart failure [13, 46, 73]. Recent studies have also attributed a role for cardiolipin in the regulation of mitophagy, the controlled process of removal of defective mitochondria. Studies in neuronal cells have demonstrated cardiolipin externalization to the outer membrane acts as a trigger for mitophagy-induced cell death, with knockout of CL synthesis decreasing the delivery of mitochondria to autophagosomes [74]. Furthermore, knockout of the TAZ gene in mouse embryonic fibroblasts (MEFs), results in defective mitophagy and reduced mitochondrial degradation in these cells, accompanied by impaired oxidative phosphorylation and increased oxidative stress, which may partially explain the raised numbers of dysfunctional mitochondria present in tissues of Barth Syndrome patients [75].

Biogenesis Versus Mitophagy

Whether mitochondrial biogenesis and their breakdown (mitophagy) play a significant role in heart failure has only recently been examined [3]. Gross morphological changes such as tissue fibrosis and hypertrophy have been well described, however whether the surviving tissue contains a reduced proportion of mitochondria has only recently been determined. Karamanlidis [76] and associates have described a loss of mitochondrial DNA in human heart failure, attributed not to an alteration in genetic control (expression of PGC-1α), but rather due to increased oxidative damage to DNA, as measured by a 50% increase in 8-OH-dG. Furthermore, mitochondrial content depends on the etiology of heart failure, with mitochondrial DNA content being increased in patients with dilated cardiomyopathy, yet unchanged in those from ischemic heart disease [77]. Mitochondria also had a greater percentage of mutations and deletions caused by oxidative damage in the dilated cardiomyopathic hearts compared to ischemic heart disease. Despite this, both heart failure subgroups showed lower oxidative phosphorylation capacity. It could be surmised that increasing mitochondrial content serves to increase oxidative phosphorylation capacity, however, in the case of the abovementioned study [77], increasing mitochondrial content actually served to raise ROS levels, further exacerbating the damage. Patients with defective oxidative phosphorylation capacity due to a genetic mutation in one of the electron transport chain proteins frequently display increased numbers of morphologically irregular mitochondria as a compensatory mechanism [77, 78]. However, due to the tight spatial arrangement of the myofibrils, the increased number of mitochondria may push myofibrillar proteins out of alignment, thus providing additional interference with contractile function [78]. Thus heart failure progression can feature a heterogeneously localized increase in the total numbers of defective mitochondria, interspersed with a gradually diminishing pool of normally functioning mitochondria, struggling to sustain cardiac output.

Conclusion

The failing heart is characterized by a complex interplay between defective mitochondrial proteins, oxidative damage and other molecular modifications to cellular proteins, lipids and DNA and altered signalling pathways, resulting in insufficient energy production. This state progressively diminishes the capacity of the cell to carry out normal energy turnover, cellular membrane and extracellular matrix turnover, ion homeostasis, signalling and contractile function. Restoring the energetic status, reducing cellular damage and augmenting lipid and protein turnover are key to targeting treatments for systolic heart failure. Our current understanding of mitochondrial dysfunction in heart failure identifies a plethora of targets that must be tackled concomitantly, many specifically mitochondrial, only some of which have been discussed in this brief review.

Acknowledgements The Authors acknowledge research funding supported from the National Health and Medical Research Council of Australia, the National Heart Foundation of Australia, the Royal Children's Hospital 1000 Foundation, HeartKids Australia and the Victorian Government's Operational Infrastructure Support Program.

References

1. Wisneski JA, Gertz EW, Neese RA, Gruenke LD, Morris DL, Craig JC. Metabolic fate of extracted glucose in normal human myocardium. J Clin Invest. 1985;76:1819–27.
2. Stanley WC, Recchia FA, Lopaschuk GD. Myocardial substrate metabolism in the normal and failing heart. Physiol Rev. 2005;85:1093–129.
3. Verdejo HE, del Campo A, Troncoso R, Gutierrez T, Toro B, Quiroga C, Pedrozo Z, Munoz JP, Garcia L, Castro PF, Lavandero S. Mitochondria, myocardial remodeling, and cardiovascular disease. Curr Hypertens Rep. 2012;14:532–9.
4. Ingwall JS, Weiss RG. Is the failing heart energy starved? On using chemical energy to support cardiac function. Circ Res. 2004;95:135–45.
5. Neubauer S, Horn M, Cramer M, Harre K, Newell JB, Peters W, Pabst T, Ertl G, Hahn D, Ingwall JS, Kochsiek K. Myocardial phosphocreatine-to-ATP ratio is a predictor of mortality in patients with dilated cardiomyopathy. Circulation. 1997;96:2190–6.
6. Liao R, Nascimben L, Friedrich J, Gwathmey JK, Ingwall JS. Decreased energy reserve in an animal model of dilated cardiomyopathy. Relationship to contractile performance. Circ Res. 1996;78:893–902.
7. Nascimben L, Ingwall JS, Pauletto P, Friedrich J, Gwathmey JK, Saks V, Pessina AC, Allen PD. Creatine kinase system in failing and nonfailing human myocardium. Circulation. 1996;94:1894–901.
8. Conway MA, Allis J, Ouwerkerk R, Niioka T, Rajagopalan B, Radda GK. Detection of low phosphocreatine to ATP ratio in failing hypertrophied human myocardium by ^{31}P magnetic resonance spectroscopy. Lancet (London, England). 1991;338:973–6.
9. Beer M, Seyfarth T, Sandstede J, Landschutz W, Lipke C, Kostler H, von Kienlin M, Harre K, Hahn D, Neubauer S. Absolute concentrations of high-energy phosphate metabolites in normal, hypertrophied, and failing human myocardium measured noninvasively with (31) P-SLOOP magnetic resonance spectroscopy. J Am Coll Cardiol. 2002;40:1267–74.
10. Starling RC, Hammer DF, Altschuld RA. Human myocardial ATP content and in vivo contractile function. Mol Cell Biochem. 1998;180:171–7.
11. Cordero-Reyes AM, Gupte AA, Youker KA, Loebe M, Hsueh WA, Torre-Amione G, Taegtmeyer H, Hamilton DJ. Freshly isolated mitochondria from failing human hearts exhibit preserved respiratory function. J Mol Cell Cardiol. 2014;68:98–105.
12. Li X. SIRT1 and energy metabolism. Acta Biochim Biophys Sin. 2013;45:51–60.
13. Santulli G, Nakashima R, Yuan Q, Marks AR. Intracellular calcium release channels: an update. J Physiol. 2017 (in press). doi: 10.1113/JP272781.
14. Lu TM, Tsai JY, Chen YC, Huang CY, Hsu HL, Weng CF, Shih CC, Hsu CP. Downregulation of Sirt1 as aging change in advanced heart failure. J Biomed Sci. 2014;21:57.
15. Sharov VG, Goussev A, Lesch M, Goldstein S, Sabbah HN. Abnormal mitochondrial function in myocardium of dogs with chronic heart failure. J Mol Cell Cardiol. 1998;30:1757–62.

16. Sharov VG, Todor AV, Silverman N, Goldstein S, Sabbah HN. Abnormal mitochondrial respiration in failed human myocardium. J Mol Cell Cardiol. 2000;32:2361–7.
17. Sheeran FL, Pepe S. Posttranslational modifications and dysfunction of mitochondrial enzymes in human heart failure. Am J Physiol Endocrinol Metab. 2016;311:E449–60.
18. Scheubel RJ, Tostlebe M, Simm A, Rohrbach S, Prondzinsky R, Gellerich FN, Silber RE, Holtz J. Dysfunction of mitochondrial respiratory chain complex I in human failing myocardium is not due to disturbed mitochondrial gene expression. J Am Coll Cardiol. 2002;40:2174–81.
19. Jarreta D, Orus J, Barrientos A, Miro O, Roig E, Heras M, Moraes CT, Cardellach F, Casademont J. Mitochondrial function in heart muscle from patients with idiopathic dilated cardiomyopathy. Cardiovasc Res. 2000;45:860–5.
20. Quigley AF, Kapsa RM, Esmore D, Hale G, Byrne E. Mitochondrial respiratory chain activity in idiopathic dilated cardiomyopathy. J Card Fail. 2000;6:47–55.
21. Sheeran FL, Pepe S. Energy deficiency in the failing heart: linking increased reactive oxygen species and disruption of oxidative phosphorylation rate. Biochim Biophys Acta. 1757;2006:543–52.
22. Turrens JF, Alexandre A, Lehninger AL. Ubisemiquinone is the electron donor for superoxide formation by complex III of heart mitochondria. Arch Biochem Biophys. 1985;237:408–14.
23. Turrens JF. Superoxide production by the mitochondrial respiratory chain. Biosci Rep. 1997;17:3–8.
24. Viola HM, Hool LC. Qo site of mitochondrial complex III is the source of increased superoxide after transient exposure to hydrogen peroxide. J Mol Cell Cardiol. 2010;49:875–85.
25. Sam F, Kerstetter DL, Pimental DR, Mulukutla S, Tabaee A, Bristow MR, Colucci WS, Sawyer DB. Increased reactive oxygen species production and functional alterations in antioxidant enzymes in human failing myocardium. J Card Fail. 2005;11:473–80.
26. Nakamura K, Kusano K, Nakamura Y, Kakishita M, Ohta K, Nagase S, Yamamoto M, Miyaji K, Saito H, Morita H, Emori T, Matsubara H, Toyokuni S, Ohe T. Carvedilol decreases elevated oxidative stress in human failing myocardium. Circulation. 2002;105:2867–71.
27. Mallat Z, Philip I, Lebret M, Chatel D, Maclouf J, Tedgui A. Elevated levels of 8-iso-prostaglandin F2alpha in pericardial fluid of patients with heart failure: a potential role for in vivo oxidant stress in ventricular dilatation and progression to heart failure. Circulation. 1998;97:1536–9.
28. Canton M, Menazza S, Sheeran FL, Polverino de Laureto P, Di Lisa F, Pepe S. Oxidation of myofibrillar proteins in human heart failure. J Am Coll Cardiol. 2011;57:300–9.
29. Csala M, Kardon T, Legeza B, Lizak B, Mandl J, Margittai E, Puskas F, Szaraz P, Szelenyi P, Banhegyi G. On the role of 4-hydroxynonenal in health and disease. Biochim Biophys Acta. 2015;1852:826–38.
30. Lakatta EG, Sollott SJ, Pepe S. The old heart: operating on the edge. Novartis Found Symp 2001;235:172–196; discussion 196–201, 217–220.
31. Pepe S. Effect of dietary polyunsaturated fatty acids on age-related changes in cardiac mitochondrial membranes. Exp Gerontol. 2005;40:751–8.
32. Mak S, Lehotay DC, Yazdanpanah M, Azevedo ER, Liu PP, Newton GE. Unsaturated aldehydes including 4-OH-nonenal are elevated in patients with congestive heart failure. J Card Fail. 2000;6:108–14.
33. Humphries KM, Yoo Y, Szweda LI. Inhibition of NADH-linked mitochondrial respiration by 4-hydroxy-2-nonenal. Biochemistry. 1998;37:552–7.
34. Nakamura K, Miura D, Kusano KF, Fujimoto Y, Sumita-Yoshikawa W, Fuke S, Nishii N, Nagase S, Hata Y, Morita H, Matsubara H, Ohe T, Ito H. 4-Hydroxy-2-nonenal induces calcium overload via the generation of reactive oxygen species in isolated rat cardiac myocytes. J Card Fail. 2009;15:709–16.
35. Ichiki T, Tanaka M, Kobayashi M, Sugiyama N, Suzuki H, Nishikimi M, Ohnishi T, Nonaka I, Wada Y, Ozawa T. Disproportionate deficiency of iron-sulfur clusters and subunits of complex I in mitochondrial encephalomyopathy. Pediatr Res. 1989;25:194–201.

36. Taylor SW, Fahy E, Murray J, Capaldi RA, Ghosh SS. Oxidative post-translational modification of tryptophan residues in cardiac mitochondrial proteins. J Biol Chem. 2003;278:19587–90.
37. Sadek HA, Humphries KM, Szweda PA, Szweda LI. Selective inactivation of redox-sensitive mitochondrial enzymes during cardiac reperfusion. Arch Biochem Biophys. 2002;406:222–8.
38. Distelmaier F, Koopman WJ, van den Heuvel LP, Rodenburg RJ, Mayatepek E, Willems PH, Smeitink JA. Mitochondrial complex I deficiency: from organelle dysfunction to clinical disease. Brain. 2009;132:833–42.
39. Moreno-Sanchez R, Hogue BA, Hansford RG. Influence of NAD-linked dehydrogenase activity on flux through oxidative phosphorylation. Biochem J. 1990;268:421–8.
40. Benderdour M, Charron G, DeBlois D, Comte B, Des RC. Cardiac mitochondrial NADP+-isocitrate dehydrogenase is inactivated through 4-hydroxynonenal adduct formation: an event that precedes hypertrophy development. J Biol Chem. 2003;278:45154–9.
41. Dodd MS, Atherton HJ, Carr CA, Stuckey DJ, West JA, Griffin JL, Radda GK, Clarke K, Heather LC, Tyler DJ. Impaired in vivo mitochondrial Krebs cycle activity after myocardial infarction assessed using hyperpolarized magnetic resonance spectroscopy. Circ Cardiovasc Imaging. 2014;7:895–904.
42. Sheeran FL, Rydström J, Shakhparonov MI, Pestov NB, Pepe S. Diminished NADPH transhydrogenase activity and mitochondrial redox regulation in human failing myocardium. Biochim Biophys Acta. 1797;2010:1138–48.
43. Hsu CP, Zhai P, Yamamoto T, Maejima Y, Matsushima S, Hariharan N, Shao D, Takagi H, Oka S, Sadoshima J. Silent information regulator 1 protects the heart from ischemia/reperfusion. Circulation. 2010;122:2170–82.
44. Zhang Y, Wang B, Fu X, Guan S, Han W, Zhang J, Gan Q, Fang W, Ying W, Qu X. Exogenous NAD(+) administration significantly protects against myocardial ischemia/reperfusion injury in rat model. Am J Transl Res. 2016;8:3342–50.
45. Karamanlidis G, Lee CF, Garcia-Menendez L, Kolwicz Jr SC, Suthammarak W, Gong G, Sedensky MM, Morgan PG, Wang W, Tian R. Mitochondrial complex I deficiency increases protein acetylation and accelerates heart failure. Cell Metab. 2013;18:239–50.
46. Lee CF, Chavez JD, Garcia-Menendez L, Choi Y, Roe ND, Chiao YA, Edgar JS, Goo YA, Goodlett DR, Bruce JE, Tian R. Normalization of NAD+ redox balance as a therapy for heart failure. Circulation. 2016;134:883–94.
47. Degenhardt T, Saramaki A, Malinen M, Rieck M, Vaisanen S, Huotari A, Herzig KH, Muller R, Carlberg C. Three members of the human pyruvate dehydrogenase kinase gene family are direct targets of the peroxisome proliferator-activated receptor beta/delta. J Mol Biol. 2007;372:341–55.
48. Barger PM, Kelly DP. PPAR signaling in the control of cardiac energy metabolism. Trends Cardiovasc Med. 2000;10:238–45.
49. Bricker DK, Taylor EB, Schell JC, Orsak T, Boutron A, Chen YC, Cox JE, Cardon CM, Van Vranken JG, Dephoure N, Redin C, Boudina S, Gygi SP, Brivet M, Thummel CS, Rutter J. A mitochondrial pyruvate carrier required for pyruvate uptake in yeast, Drosophila, and humans. Science. 2012;337:96–100.
50. Herzig S, Raemy E, Montessuit S, Veuthey JL, Zamboni N, Westermann B, Kunji ER, Martinou JC. Identification and functional expression of the mitochondrial pyruvate carrier. Science. 2012;337:93–6.
51. Bowker-Kinley MM, Davis WI, Wu P, Harris RA, Popov KM. Evidence for existence of tissue-specific regulation of the mammalian pyruvate dehydrogenase complex. Biochem J. 1998;329(Pt 1):191–6.
52. Liang L, Li Q, Huang L, Li D, Li X. Sirt3 binds to and deacetylates mitochondrial pyruvate carrier 1 to enhance its activity. Biochem Biophys Res Commun. 2015;468:807–12.
53. Schell JC, Olson KA, Jiang L, Hawkins AJ, Van Vranken JG, Xie J, Egnatchik RA, Earl EG, DeBerardinis RJ, Rutter J. A role for the mitochondrial pyruvate carrier as a repressor of the Warburg effect and colon cancer cell growth. Mol Cell. 2014;56:400–13.

54. Fernandez-Caggiano M, Prysyazhna O, Barallobre-Barreiro J, CalvinoSantos R, Aldama Lopez G, Generosa Crespo-Leiro M, Eaton P, Domenech N. Analysis of mitochondrial proteins in the surviving myocardium after ischemia identifies mitochondrial pyruvate carrier expression as possible mediator of tissue viability. Mol Cell Proteomics. 2016;15:246–55.
55. Vacanti NM, Divakaruni AS, Green CR, Parker SJ, Henry RR, Ciaraldi TP, Murphy AN, Metallo CM. Regulation of substrate utilization by the mitochondrial pyruvate carrier. Mol Cell. 2014;56:425–35.
56. Zhang Y, Cox J, Rutter J,E, Dale Abel ED. Mitochondrial pyruvate transport is required for the cardiac adaptation to pressure overload. Circulation. 2016;134(Suppl.1):A17659.
57. Razeghi P, Young ME, Alcorn JL, Moravec CS, Frazier OH, Taegtmeyer H. Metabolic gene expression in fetal and failing human heart. Circulation. 2001;104:2923–31.
58. Zhao G, Jeoung NH, Burgess SC, Rosaaen-Stowe KA, Inagaki T, Latif S, Shelton JM, McAnally J, Bassel-Duby R, Harris RA, Richardson JA, Kliewer SA. Overexpression of pyruvate dehydrogenase kinase 4 in heart perturbs metabolism and exacerbates calcineurin-induced cardiomyopathy. Am J Physiol Heart Circ Physiol. 2008;294:H936–43.
59. Paradies G, Ruggiero FM. The effect of doxorubicin on the transport of pyruvate in rat-heart mitochondria. Biochem Biophys Res Commun. 1988;156:1302–7.
60. Paradies G, Ruggiero FM. Decreased activity of the pyruvate translocator and changes in the lipid composition in heart mitochondria from hypothyroid rats. Arch Biochem Biophys. 1989;269:595–602.
61. Taegtmeyer H. Switching metabolic genes to build a better heart. Circulation. 2002;106:2043–5.
62. Fry M, Green DE. Cardiolipin requirement for electron transfer in complex I and III of the mitochondrial respiratory chain. J Biol Chem. 1981;256:1874–80.
63. Fry M, Green DE. Cardiolipin requirement by cytochrome oxidase and the catalytic role of phospholipid. Biochem Biophys Res Commun. 1980;93:1238–46.
64. Paradies G, Petrosillo G, Pistolese M, Di Venosa N, Federici A, Ruggiero FM. Decrease in mitochondrial complex I activity in ischemic/reperfused rat heart: involvement of reactive oxygen species and cardiolipin. Circ Res. 2004;94:53–9.
65. Petrosillo G, Ruggiero FM, Di Venosa N, Paradies G. Decreased complex III activity in mitochondria isolated from rat heart subjected to ischemia and reperfusion: role of reactive oxygen species and cardiolipin. FASEB J. 2003;17:714–6.
66. Paradies G, Petrosillo G, Pistolese M, Di Venosa N, Serena D, Ruggiero FM. Lipid peroxidation and alterations to oxidative metabolism in mitochondria isolated from rat heart subjected to ischemia and reperfusion. Free Radic Biol Med. 1999;27:42–50.
67. Wang G, McCain ML, Yang L, He A, Pasqualini FS, Agarwal A, Yuan H, Jiang D, Zhang D, Zangi L, Geva J, Roberts AE, Ma Q, Ding J, Chen J, Wang DZ, Li K, Wang J, Wanders RJ, Kulik W, Vaz FM, Laflamme MA, Murry CE, Chien KR, Kelley RI, Church GM, Parker KK, Pu WT. Modeling the mitochondrial cardiomyopathy of Barth syndrome with induced pluripotent stem cell and heart-on-chip technologies. Nat Med. 2014;20:616–23.
68. Raval KK, Kamp TJ. Cardiomyopathy, mitochondria and Barth syndrome: iPSCs reveal a connection. Nat Med. 2014;20:585–6.
69. Sparagna GC, Chicco AJ, Murphy RC, Bristow MR, Johnson CA, Rees ML, Maxey ML, McCune SA, Moore RL. Loss of cardiac tetralinoleoyl cardiolipin in human and experimental heart failure. J Lipid Res. 2007;48:1559–70.
70. Pfeiffer K, Gohil V, Stuart RA, Hunte C, Brandt U, Greenberg ML, Schagger H. Cardiolipin stabilizes respiratory chain supercomplexes. J Biol Chem. 2003;278:52873–80.
71. Schagger H, Pfeiffer K. Supercomplexes in the respiratory chains of yeast and mammalian mitochondria. EMBO J. 2000;19:1777–83.
72. Rosca MG, Vazquez EJ, Kerner J, Parland W, Chandler MP, Stanley W, Sabbah HN, Hoppel CL. Cardiac mitochondria in heart failure: decrease in respirasomes and oxidative phosphorylation. Cardiovasc Res. 2008;80:30–9.
73. Bernardi P. The mitochondrial permeability transition pore: a mystery solved? Front Physiol. 2013;4:95. doi:10.3389/fphys.2013.00095.

74. Chu CT, Ji J, Dagda RK, Jiang JF, Tyurina YY, Kapralov AA, Tyurin VA, Yanamala N, Shrivastava IH, Mohammadyani D, Qiang Wang KZ, Zhu J, Klein-Seetharaman J, Balasubramanian K, Amoscato AA, Borisenko G, Huang Z, Gusdon AM, Cheikhi A, Steer EK, Wang R, Baty C, Watkins S, Bahar I, Bayir H, Kagan VE. Cardiolipin externalization to the outer mitochondrial membrane acts as an elimination signal for mitophagy in neuronal cells. Nat Cell Biol. 2013;15:1197–205.
75. Hsu P, Liu X, Zhang J, Wang HG, Ye JM, Shi Y. Cardiolipin remodeling by TAZ/tafazzin is selectively required for the initiation of mitophagy. Autophagy. 2015;11:643–52.
76. Karamanlidis G, Nascimben L, Couper GS, Shekar PS, del Monte F, Tian R. Defective DNA replication impairs mitochondrial biogenesis in human failing hearts. Circ Res. 2010;106:1541–8.
77. Ahuja P, Wanagat J, Wang Z, Wang Y, Liem DA, Ping P, Antoshechkin IA, Margulies KB, Maclellan WR. Divergent mitochondrial biogenesis responses in human cardiomyopathy. Circulation. 2013;127:1957–67.
78. Sebastiani M, Giordano C, Nediani C, Travaglini C, Borchi E, Zani M, Feccia M, Mancini M, Petrozza V, Cossarizza A, Gallo P, Taylor RW, d'Amati G. Induction of mitochondrial biogenesis is a maladaptive mechanism in mitochondrial cardiomyopathies. J Am Coll Cardiol. 2007;50:1362–9.

Chapter 5
Mitochondrial Mutations in Cardiac Disorders

Sung Ryul Lee and Jin Han

Mitochondria are indispensable organelles for generating adenosine triphosphate (ATP) in eukaryotic cells via the electron transport chain (ETC) and oxidative phosphorylation system (OXPHOS). Additionally, numerous biological functions, including ATP transport, production of heat, metal homeostasis, and stress signaling and defense responses, involve mitochondria [1–4]. The rhythmic contractions of the heart ensure oxygenation of the organs in the body and their functions depend on mitochondrial aerobic energy production by ETC and OXPHOS [5]. Thus, mitochondrial diseases preferentially affect the heart, as impaired cardiac conditions are associated with mitochondrial dysfunction due to defects in either OXPHOS or ETC [6–8]. Numerous efforts to identify the genetic determinants of cardiovascular disease (CVD), including high blood pressure, coronary heart disease, cardiomyopathy, heart failure, and stroke, have been directed primarily to nuclear genomes [9]. For example, nuclear gene mutations that regulate mtDNA maintenance and replication, such as mitochondrial transcription factor A (TFAM), mtDNA polymerase γ (POLG) as a Mutator model, and PEO1 (Twinkle) as a Deletor model [10], have been increasingly recognized for their cardiac involvement. The occurrence and progress of CVD is affected by a combination of genetic, environmental, and behavioral factors that influence an individual's susceptibility to known disease risk factors [11]. In addition, numerous forms of CVD are not completely understood by the anatomic paradigm of medicine and Mendelian genetics [11–13]. Therefore, the

S.R. Lee
Department of Integrated Biomedical Science, Cardiovascular and Metabolic Disease Center, College of Medicine, Inje University, Busan 47392, South Korea
e-mail: lsr1113@inje.ac.kr

J. Han (✉)
National Research Laboratory for Mitochondrial Signaling, Cardiovascular and Metabolic Disease Center, Department of Physiology, College of Medicine, Inje University, Busan 47392, South Korea
e-mail: phyhanj@inje.ac.kr

G. Santulli (ed.), *Mitochondrial Dynamics in Cardiovascular Medicine*, Advances in Experimental Medicine and Biology 982,
DOI 10.1007/978-3-319-55330-6_5

common disease-common variant hypothesis is difficult to apply to most common human age-related diseases such as diabetes, metabolic syndrome, and neuronal disorders [11, 12, 14]. Mitochondrial dysfunction due to mtDNA mutations appears to play a role in the development of cardiac disorders and severe exercise limitation [6, 8, 15–18]. Remarkable advances have been made in understanding the basic mitochondrial genetics and the relationship between inherited mutations and/or sporadic mutations and disease manifestation by mitochondrial diseases, as well as in identifying acquired mtDNA mutations in CVD, cancer, and aging [19]. However, the phenotypic display and genetic background of mitochondrial disorders are extraordinarily diverse [13].

Mitochondrial cardiomyopathy (MCM) is defined as an OXPHOS disease characterized by abnormal number, structure, or functions of cardiac mitochondria [7, 20–22] and may be resulted in cardiac manifestations, such as: (1) structural alterations (hypertrophic or dilated cardiomyopathy, left ventricular myocardial noncompaction), (2) disturbance in heart rate and/or rhythm (ventricular tachycardia, Wolff-Parkinson-White syndrome, heart block, or arrhythmias), and (3) final functional impairment as heart failure [23–27]. Mitochondrial diseases and mtDNA mutations are not restricted to mitochondria. Two pathways have been suggested to explain the etiology of MCM. Primary MCMs are caused by mutations in mtDNA or nuclear DNA [28]. Secondary MCMs are induced by a primary insult affecting mitochondrial function either indirectly or by damaging the mitochondrial genome. Inadequate levels of mtDNA and its mutations can cause defects in the synthesis of key subunits of ETC complexes. In this chapter, we summarize the current understanding of mtDNA mutations, focusing on cardiac disorders, and briefly address the epigenetic modifications of mtDNA for future studies on cardiac disorders. Finally, current strategies for overcoming the effects of mitochondrial mutations and/or dysfunction on cardiac disorders are addressed.

Mitochondrial DNA and Its Genetics

General Aspects of Mitochondria

Mitochondria are spherical, double membrane-bound semiautonomous organelles. Unlike developing non-cardiac cells or cardiomyocytes, mitochondria in adult cardiac myocytes show distinct distributions: (1) interfibrillar, (2) subsarcolemmal, and (3) perinuclear populations [29]. Thus, adult cardiac mitochondria may be relatively firmly fixed and are subjected to very low-amplitude fluctuations or vibrations [30, 31]. In the adult heart, the machinery required to alter mitochondrial shape in terms of mitochondrial fission involves fission proteins, but their physiological functions are extensively under investigations [29]. Mitochondria have their own DNA located in its matrix within the inner membrane (Figs. 5.1 and 5.2 and Table 5.1).

Since most mtDNA-encoded proteins are extremely hydrophobic peptides, they cannot easily enter the mitochondria without the help of mitochondrial translocation machinery [32]. Molecules smaller than 5 kD can pass through the mitochondrial outer membrane through porins. However, larger proteins containing mitochondria signal sequence can only pass through multiprotein translocase complexes such as the transporter inner membrane, transporter outer membrane, and other transporter proteins [33–35]. The TCA cycle, which produces NADH and $FADH_2$, moves electrons into the ETC. All the components of the ETC and ATP synthase complex are located in the inner mitochondrial membrane [36]. Energy production in the mitochondria is coordinated by these steps and thus any defects in this complex process, in terms of quantity and/or functionality, can cause mitochondrial dysfunction [36].

Mitochondria are composed of over 1,000 proteins [37–39], but mtDNA encodes only 13 proteins (Fig. 5.1 and Table 5.1). Nearly all protein components involved in mitochondrial biogenesis, including transcription, replication, and protein synthesis, are encoded by nuclear DNA (Fig. 5.3). Mitochondria also contain several proteases that control protein turnover to match the need for removal of defective peptides such as misfolded proteins, and regulate gene expression in response to various cellular signals [34, 40, 41]. Mitochondrial dysfunction can be caused by the mitochondria, nucleus, or both (Tables 5.1 and 5.2), as the maintenance of mitochondrial function is finely controlled by protein synthesis and degradation. Therefore, mitochondrial dysfunction can result from; (1) single OXPHOS complex deficiencies due to mtDNA or nuclear mutations, (2) multiple OXPHOS deficiencies, (3) coenzyme Q10 deficiency, (4) mitochondrial depletion syndrome, (5) disorders of 3-methylglutaconic aciduria, (6) disorders of mitochondrial fatty acid β-oxidation, and (7) disorders of intramitochondrial organic acid metabolism (due to deficiency of propionyl-CoA carboxylase or methylmalonyl-CoA mutase or its cofactor adenosylcobalamin) with secondary mitochondrial dysfunction [1, 45, 46]. These mitochondrial dysfunctions inhibit mitochondrial O_2 consumption, collapse the membrane potential, and reduce ATP levels by causing an imbalance between energy intake and expenditure [47]. Additionally, mitochondrial dysfunction can impair other numerous aspects of cell metabolism and homeostasis, such as cellular calcium regulation and apoptosis [48]. Dysfunctional mitochondria can be removed from the cell by mitophagic processes, which are controlled by nuclear-encoded proteins [13, 49, 50].

mtDNA

Many aspects of mtDNA differ from those of nuclear DNA [19, 51]. As shown in Fig. 5.1, the human mtDNA is 16,569 base pair (bp) in length and is characterized by high gene density (nuclear DNA and mtDNA; ~1 per 40,000 bp and ~1 per 450 bp, respectively) and differences in codon use [19, 40, 44, 52]. Except for the D-loop containing the control elements for the transcription and replication of mtDNA, mammalian mtDNA is highly compact without introns. mtDNA is packaged into nucleoid structures (Fig. 5.2) composed of TFAM, single-stranded DNA-binding protein (mtSSB), and twinkle, which is a mitochondrial DNA helicase [1, 40, 53, 54]. Typically, one or two copies of mtDNA are packaged with the help of

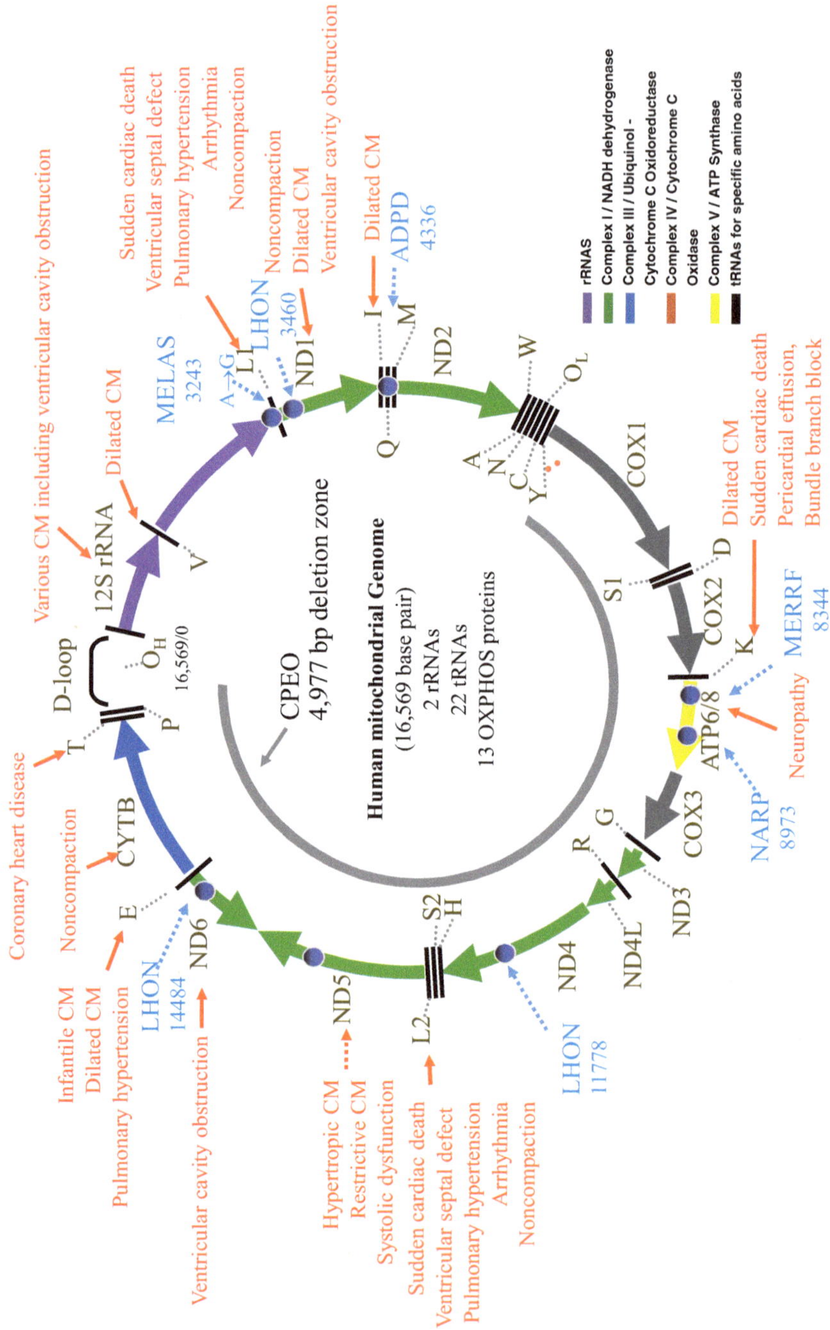
Human mitochondrial Genome
(16,569 base pair)
2 rRNAs
22 tRNAs
13 OXPHOS proteins
CPEO
4,977 bp deletion zone
D-loop
O_H
16,569/0
12S rRNA
MELAS
3243
A→G
LHON
3460
ADPD
4336
MERRF
8344
A→G
NARP
8973
LHON
11778
LHON
14484
ND1
ND2
COX1
COX2
ATP6/8
COX3
ND3
ND4L
ND4
ND5
ND6
CYTB
O_L
T
P
V
L1
I
M
Q
W
A
N
C
Y
S1
D
K
G
R
S2
H
L2
E
Various CM including ventricular cavity obstruction
Dilated CM
Sudden cardiac death
Ventricular septal defect
Pulmonary hypertension
Arrhythmia
Noncompaction
Noncompaction
Dilated CM
Ventricular cavity obstruction
Dilated CM
Dilated CM
Sudden cardiac death
Pericardial effusion,
Bundle branch block
Neuropathy
Hypertropic CM
Restrictive CM
Systolic dysfunction
Sudden cardiac death
Ventricular septal defect
Pulmonary hypertension
Arrhythmia
Noncompaction
Ventricular cavity obstruction
Infantile CM
Dilated CM
Pulmonary hypertension
Noncompaction
Coronary heart disease
rRNAS
Complex I / NADH dehydrogenase
Complex III / Ubiquinol - Cytochrome C Oxidoreductase
Complex IV / Cytochrome C Oxidase
Complex V / ATP Synthase
tRNAs for specific amino acids

Fig. 5.1 Mitochondrial DNA (*mtDNA*) and pathogenic mtDNA mutations associated with cardiac manifestations. Mitochondrial respiratory chain complexes are composed of approximately over 80 polypeptides. Human mtDNA codes for 7 of the 43 subunits of complex I, one of the 11 subunits of complex III (cytochrome b, CYTB), 3 of the 13 subunits of complex IV (COX I, II, III), and 2 of 16 subunits of complex V (ATPases 6 and 8). It also codes for 2 rRNAs and 22 tRNAs for mitochondrial protein synthesis. Mutations in mtDNA can occur in any regions, while mutations in the D-loop region are typically not retained because of its important regulatory roles in transcription, translation, and replication. Some examples of mtDNA mutations associated with known human genetic diseases and cardiac manifestations are marked as *blue* (*broken arrow*) and *red* (*solid arrow*), respectively. *Abbreviations: ADPD* late-onset Alzheimer's disease, *COX* cytochrome C oxidoreductase, *CPEO* chronic progressive external ophthalmoplegia, *LHON* Leber's hereditary optic neuropathy, *MELAS* mitochondrial encephalomyopathy, lactic acidosis, and stroke-like episodes, *MERRF* myoclonic epilepsy with ragged red fibers, *NARP* neuropathy, ataxia, and retinitis pigmentosa, *ND1-6* component of rotenone-sensitive NADH-ubiquinone oxidoreductase (complex I), O_H heavy strand origin of replication, OXPHOS mitochondrial oxidative phosphorylation, *P* promoter of heavy and light strands

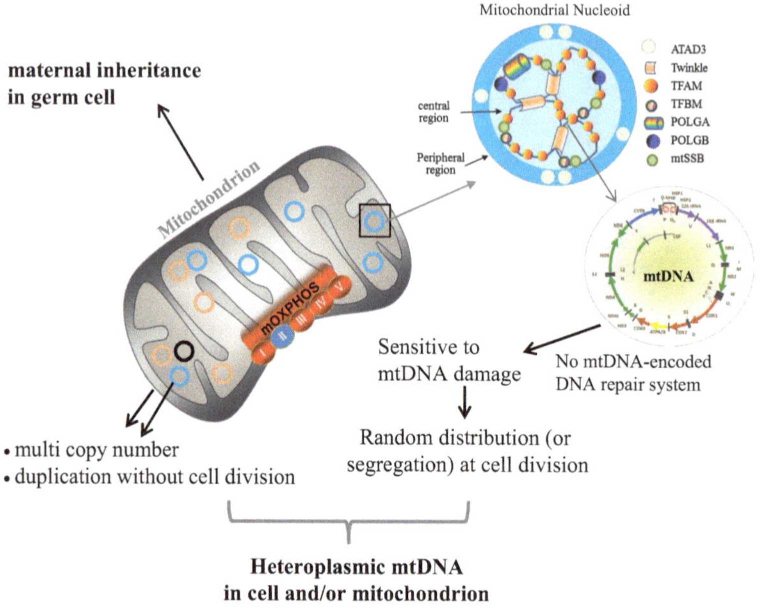

Fig. 5.2 Characteristics of mitochondrial genetics. One cell may contain several thousand mitochondria, and each mitochondrion contains one to ten copies of mtDNA. The replication of mtDNA is less dependent on nuclear DNA division and thus shows continuous replicative behavior. The mtDNA is encapsulated by the mitochondrial nucleoid. In the nucleoid structure, mitochondrial transcription factors, such as TFAM, TFBM, POLGA, POLGB, MTSSB, and Twinkle, are located in the central region along with the mtDNA, whereas proteins have an indirect relationship with mtDNA (e.g. ATAD3) and are arranged peripherally. This nucleoid structure results in a random distribution of mtDNA during cell division. mtDNA encounters more mutation-prone conditions because of its proximity to ROS-producing sites, limited presence of mtDNA repair proteins encoded by nuclear DNA, continuous replication state, and morphological changes during fission and fusion processes. In the mitochondria, there is less selective pressure to propagate wild-type mtDNA, and thus mtDNA mutations easily survive and are randomly distributed. Additionally, over 1,000 mitochondrial proteins and substrates are supplied from the cytosol, and the nucleus can modulate mitochondrial homeostasis, and therefore mitochondrial dysfunctions are not solely affected by mtDNA mutations. These factors result in more complexity in mitochondrial genetics, unlike Mendelian genetics of the nucleus. This characteristic of mitochondrial genetics results in complex phenomena as well as maternal inheritance in germline cells as a heteroplasmy and threshold effect in phenotypic presentations. *Abbreviations: ATAD3* ATPase family AAA Domain-containing protein 3, *mtDNA* mitochondrial DNA, *mtSSB* mitochondrial single-stranded DNA-binding protein, *POLGA* mtDNA-specific DNA polymerase gamma A (catalytic subunit), *POLGB* mtDNA polymerase γ accessory subunit, *TFBM* mitochondrial transcription factor B

interacting factors into a mean size of ~100 nm nucleoid [55, 56]. Within mitochondria, mtDNA is associated with a variety of interacting proteins involved in its packaging, replication, and transcription. Through the action of these proteins, mtDNA is integrated into mitochondrial bioenergetics and exhibit signaling capabilities [57, 58]. If these mtDNA-interacting proteins are affected by insult or stress, mtDNA can be lost.

Table 5.1 Genes in the mitochondrial respiratory chain. Mitochondrial respiratory chain complexes are composed of approximately 80 polypeptides constituting the electron transport chain: 13 are encoded by mtDNA and the remaining are synthesized in the cytosol and translocated into the mitochondria. Mutations in respiratory chain complexes result in single or multiple complex deficiencies originating from either sporadic mutations or inheritance. Italic characters indicate disease-causing mitochondrial mutations

	Mitochondrial respiratory chain				
	Complex I	Complex II	Complex III	Complex IV	Complex V
	NADH ubiquinone oxidoreductase	Succinate ubiquinone oxidoreductase	Ubiquionol-cytochrome c oxidoreductase	Cytochrome c oxidase	F_0F_1 ATPase
mtDNA-encoded	7	0	1	3	2
Nuclear DNA-encoded subunit	~40	4	10	10	~15
Diseases or phenotypes related to defects	DCM HCM Leigh's Syndrome Leukodystrophy *LHON* LHON and dystonia LVNC *MELAS* WPW	DCM HCM Leigh's Syndrome LVNC Paraganglioma Pheochromocytoma	*DCM* Encephalomyopathy GRACILE *HCM* *HiCM* Leigh's Syndrome Septo-optic dysplasia Sporadic myopathy	*ALS-likesyndrome* Cardioencephalomyopathy DCM Encephalomyopathy HCM Hepatopathy HiCM Leigh's Syndrome Leukodystrophy and tubulopathy *Sporadic anemia* *Sporadic Myopathy*	*FBSN* *HCM* *LVNC* *MILS* *NARP*

ALS amyotrophic lateral sclerosis, *FBSN* familial bilateral striatal necrosis, *DCM* dilated cardiomyopathy, *GRACILE* growth retardation, aminoaciduria, lactic acidosis, and early death, *HCM* hypertrophic cardiomyopathy, *HiCM* histiocytoid cardiomyopathy, *LHON* Leber's hereditary optic neuropathy, *LVNC* left ventricular noncompaction, *MELAS* Mitochondrial myopathy, encephalopathy, lactic acidosis and stroke-like episodes syndrome, *MILS* maternally inherited Leigh's syndrome, *NARP* neurogenic weakness with ataxia and retinitis pigmentosa, *WPW* Wolff-Parkinson-White syndrome

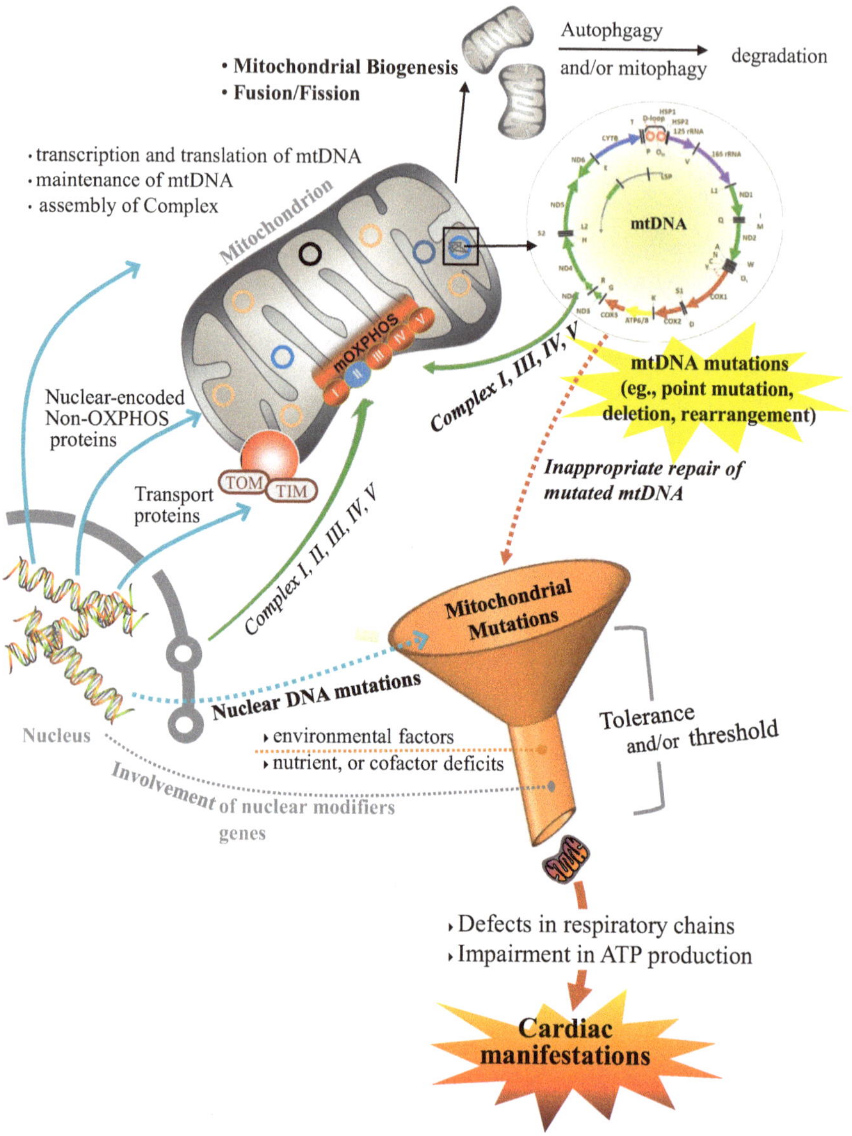

Fig. 5.3 Mitochondrial biogenesis and threshold effects of mitochondrial mutations. Mitochondrial mutations originate from nuclear DNA, mtDNA, or both. Most mitochondrial proteins are encoded by nuclear DNA and imported into mitochondria. In addition to the mitochondrial origin of mtDNA mutations, the occurrence of mtDNA mutations is highly affected by the status of nuclear DNA. Mitophagy can remove damaged mitochondria, but currently the specific removal of mitochondria containing mtDNA mutations is not well-understood. mtDNA mutations can be easily produced in a sporadic manner and these mutations do not always cause mitochondrial dysfunction because of threshold effects, indicating that phenotypic expression of mtDNA mutations must accumulate in mtDNA before presenting as phenotypic or functional anomalies. This threshold level varies among different tissues. Perturbations in mitochondrial metabolism also causes dysfunction in the OXPHOS system, or mtDNA mutations may stimulate increased mitochondrial biogenesis to compensate for mitochondrial dysfunction. The incomplete or indirect exhibition of mtDNA mutations can be influenced by pharmaceutical or environmental exposures and nutrient or cofactor deficits. In addition, nuclear modifier genes, which often have at least two alleles, one that exacerbates disease and another that suppresses the disease, have been investigated to better understand the incomplete penetrance of mitochondrial mutations

Table 5.2 Diagnosis of suspected mitochondrial diseases

Mutation or deficiency	Method	Sample
Mitochondrial DNA mutations	Affymetrix MitoChip 2.0 DGGE dHPLC NGS PCR/RFLP analysis PCR/ASO dot blot analysis Southern blot hybridization analysis	Tissue biopsy
Nuclear DNA mutations	Exome sequencing PCR/RFLP analysis [42–44] Southern blot hybridization analysis	Tissue biopsy
NADH-coenzyme Q reductase (complex I) deficiency	Oximetry (polarographic measurement)	Tissue biopsy, permeabilized fiber, mitochondria
NADH ferricyanide reductase (complex I)	Spectrophotometry	
Succinate cytochrome-c reductase (complex II and/or III) deficiency	Spectrophotometry Oximetry (polarographic)	
Co Q-cytochrome-c reductase (complex III) deficiency	Spectrophotometry Oximetry (polarographic)	
Cytochrome-c oxidase (complex IV)	Spectrophotometry Oximetry (polarographic) Enzyme histochemical assay	Tissue biopsy
>2% subsarcolemmal collection of mitochondria (ragged red fiber) in skeletal muscle	Gomori trichrome staining	Tissue biopsy (frozen section)
Structural abnormalities in mitochondria (number, shape, or size)	Electron microscopy	Tissue biopsy
Other biochemical tests		
Lactate, pyruvate, phosphocreatine measurement	Enzymatic method	Blood or CSF
Amino acid analysis	Tandem mass spectrometry (MS/MS)	Blood or CSF
Organic acid analysis	Column gas chromatography with mass spectrometry	Blood or CSF

ASO allele-specific oligonucleotide, *CSF* cerebrospinal fluid, *DGGE* denaturing gradient gel electrophoresis, *dHPLC* denaturing high-performance liquid chromatography, *NGS* next-generation sequencing, *PCR* polymerase chain reaction, *RFLP* restriction fragment length polymorphism

Biogenesis of Mitochondria: Replication, Translation, and Transcription of mtDNA

Mitochondrial biogenesis can be defined as an increase of mitochondrial mass and number of pre-existing mitochondria [59]. Mitochondrial biogenesis (Fig. 5.3) is coupled with several quality control pathways to maintain mitochondrial integrity via antioxidants, mtDNA repair systems [60], and mitochondrial unfolded protein response systems [49, 61–64]. Albeit nuclear DNA synthesis is highly dependent on the cell cycle, replication of mtDNA is not dependent on cell cycle [40, 65]. The entire mtDNA molecule can be replicated within 90 min (DNA synthesis rate; 180–270 bp/min), based on in vitro minimal mtDNA replisomes consisting of nuclear-encoded POLG consisting of a catalytic subunit with 5→3′ exonuclease activity (PolgA) and processivity subunit (PolgB), mtSSB, and twinkle [40, 44, 66, 67]. Two possible models of mtDNA replication have been suggested: one is a strand-asynchronous mode initiating at the heavy strand origin (O_H) [66], while the other is a leading-lagging strand initiating at the same location [68].

However, the exact mechanism of human mtDNA replication should be further elucidated. Because numerous proteins are involved in mtDNA replication, a loss in mtDNA number may reflect defects in mtDNA replication factor activity [69]. The transcription of mtDNA is initiated from one of the two promoters on the heavy strand (HSP_1 and HSP_2) and from a single promoter on the light strand through dedicated mitochondrial RNA polymerase, TFAM, and 2B [40, 70]. Except for the D-loop region, approximately 93% of mtDNA must be entirely transcribed to produce properly functioning OXPHOS in cells [52]. Mature mRNA is produced by subsequent processing and modifications in the polycistronic transcript [40, 71]. The quantity of each mRNA, rRNA, and tRNA is not equal, possibly because of post-transcriptional modifications affecting transcript processing, maturation, and stability [72, 73]. Conceivably, coincidence of replication and transcription of mtDNA, which may involve head-on collisions between transcription and replication machineries, can be coupled via the action of a mitochondrial transcription elongation factor (TEFM) [40, 72, 74–76]. This factor may serve as a molecular switch that controls the replication of mtDNA or regulates the number or increases the transcription rate [76]. Translation of bicistronic transcripts [42] from mtDNA are produced by RNA-poor mitoribosomes [50]. In addition, partial stop codons in some transcripts that are polyadenylated [73] and deviate from the standard genetic code [19] are present in mitochondria.

Mitochondrial Genetics

Typically, cell contains a single nuclear genome; in contrast, the mtDNA population in cell or mitochondrion is not homogeneous. Unlike Mendelian genetics governing the nuclear genome, the characteristics of mitochondrial genetics are as follows: (1) nearly all maternal inheritance of disease mutation, (2) polyploidy nature of mitochondrial genome, (3) structure of mitochondrial nucleoid, (4) clonal expansion of mtDNA even in the same mitochondrion, and (5) random mitochondrial segregation [77] occur even in the same cell and thus the ratio of mutant to normal mtDNA can be shifted during the segregation (Fig. 5.2). Clonal expansion of mtDNA and random segregation of mitochondria lead to heteroplasmy, which is defined as the presence of at least two populations of mtDNA molecules such as normal (wild-type) and mutated DNA [34]. These characteristics of mitochondrial genetics may explain the age-related and even tissue-related variability in mtDNA-related disorders [78].

Mitochondrial DNA Mutations

The heart must effectively respond to variations in the physiological demand by balancing ATP production and metabolic responsiveness [79]. The first mtDNA mutations were identified several decades ago [80, 81] and significant advances have been made in understanding mtDNA defects involved in the progress of mitochondrial disease [27, 44]. Examples are depicted in Table 5.3. mtDNA variants can be classified according to clinical relevance as recent deleterious mutations, ancient adaptive mtDNA mutations, or somatic mtDNA mutations accumulated over time in tissues [82].

In population genetics, the variability introduced into mtDNA sequences by mutations has been used to identify family members from various points in human history. In disease-based epidemiological studies, the prevalence of mtDNA disease was found to be ~1:5,000 and heteroplasmic mtDNA mutations were found in 1:200 newborns [84, 85]. Compared to nuclear DNA genes, mtDNA genes have a very high sequence evolution rate and mutational load (e.g., nuclear DNA vs. mtDNA; 2.5×10^{-8} vs. 3×10^{-6}, $\sim 2.7 \times 10^{-5}$ per base per generation) because of the continuous replication state and high number of mtDNA [82, 86–88]. Both nuclear DNA- and mtDNA-encoded factors co-evolve in response to mutation rate differences through protein–protein interactions in the OXPHOS system, protein-RNA within the mitochondrial ribosome, and protein-DNA involved in mitochondrial replication and transcription [89].

Possible Causes of mtDNA Mutations

Although DNA is a relatively stable molecule, spontaneous decomposition of the nucleic acid structure of mtDNA and numerous reactive oxygen species (ROS) and resulting products may be involved in DNA modifications and mutagenesis [90–92]. Oxidative stress, which is defined as increased production of ROS and accumulation of their chemically active metabolites over the threshold level, leads to spontaneous deamination of mtDNA, base loss, and other modifications [93–96]. Among these, oxidative modification of mtDNA in either sugar (e.g., 8-oxoguanine) or DNA bases is the most common and prevalent type [92]. Oxidative mtDNA damage has been suggested as a causal factor in mtDNA mutation, but this hypothesis is not consistent with the results of recent studies [97, 98]. For example, in oxoguanine DNA glycosylase (OGG1) null mice, OGG1 deficiency did not cause mitochondrial respiratory dysfunction in the heart and liver, even when 8-oxo-2′-deoxyguanosine was produced in excess [99]. Moreover, mice expressing POLG with low fidelity

Table 5.3 Syndromic mitochondrial mutations and onset of cardiovascular manifestations. Genetic defects in nuclear DNA and mtDNA lead to mitochondrial dysfunction. The clinical spectrum in mtDNA disease is wide within both individuals and families. The clinical phenotypes of mitochondrial disorders are extremely heterogeneous and the onset of clinical manifestations is largely depended on the context of mutations in mtDNA and/or nuclear DNA

Mitochondrial diseases	mtDNA mutations (status)	Phenotype or characteristic	Onest
mtDNA depletion syndromes	Nuclear-encoded DNA mutation	Severe muscle weakness, progressive encephalopathy, or liver failure	Early infancy/childhood
Chronic progressive external ophthalmoplegia (CPEO)	Single or multiple mtDNA deletions (heteroplasmic)	Arrhythmia myopathy and fatigue	Late childhood or adult life
Kearns-Sayre syndrome (KSS)	Single, large-scale mtDNA deletion (heteroplasmic)	Arrhythmia, Third-degree heart block	Early infancy/childhood
Leber's Hereditary optic neuropathy (LHON)	Complex I gene mutations (hetero- or homoplasmic)	Cardiomyopathy accompanied by Wolff-Parkinson-White (WPW) syndrome (an aberrant conduction defect)	Late childhood or adult life
Leigh Syndrome (subacute necrotizing encephalomyelopathy)	Complex I, IV, and V gene mutations (heteroplasmic)	Cardiomyopathy and arrhythmia	Early infancy/childhood
Mitochondrial encephalomyopathy with lactic acidosis and stroke like episodes (MELAS)	mt-tRNA gene mutations (heteroplasmic)	Dilated cardiomyopathy and left ventricular hypertrabeculation, Hypertrophic cardiomyopathy	Late childhood or adult life
Myoclonic epilepsy with ragged red fibers (MERRF)	mt-tRNA gene mutations (heteroplasmic)	Arrhythmia, Dilated cardiomyopathy	Late childhood or adult life
Maternally inherited diabetes and deafness (MIDD)	mt-tRNA gene mutations (heteroplasmic)	Left ventricular hypertrabeculation and arrhythmia	From childhood to late adulthood
Neurogenic muscle weakness, ataxia, and retinitis pigmentosa (NARP)	Complex V mutations (heteroplasmic)	Cardiomyopathy	Late childhood and adult life
Barth Syndrome (lethal infantile cardiomyopathy)	Nuclear-encoded tafazzin gene mutation	Arrhythmia, Left ventricular non-compaction cardiomyopathy	Early infancy/childhood
Exercise intolerance [83]	mt-tRNA gene mutations and/or Complex III defects (heteroplasmic)	Fatigue, muscle weakness	Late childhood and adult life

showed significant accumulation of mtDNA mutations, which shortened the lifespan of mice without remarkable levels of oxidative stress [6, 100, 101].

Several mitochondrion-specific conditions may cause mtDNA to be particularly vulnerable to mutations [92, 102]: (1) the mtDNA is localized close to the inner membrane, in which ROS are produced. During mtDNA replication, the lagging strand of mtDNA is single-stranded for a long time, and thus spontaneous mutations can be easily introduced compared to that in double-stranded DNA. (2) mtDNA in the nucleoid is not extensively condensed and cannot be efficiently protected from deleterious insults. Whether the mitochondrial nucleoid is in proximity to the site of ROS production remains controversial [55]. (3) Covalent modifications of mtDNA in mitochondria cannot be readily detected and repaired by the mtDNA repair system because mtDNA does not encode for proteins involved in mtDNA repair, and mitochondrial POLG has low fidelity [13, 103]. As described above, the continuous replicative nature of mtDNA makes it susceptible to mutations, which could be passed on by clonal expansion and mitochondrial segregation during cell division or renewal [65]. Thus, it has been speculated that the continuous replication of mtDNA is of pathogenic importance not only in patients with mtDNA depletion, but also in patients with different pathogenic mtDNA mutations [104]. Independently, metabolic impairments lead to alterations in mitochondrial bioenergetics, strengthening mtDNA damage, increasing the mutation rate, perturbing mtDNA replication, and mitophagy, and eventually leading to the accumulation of somatic mtDNA mutations [82]. In diabetes, metabolic stress mediated by cytokines, inflammation, and oxidative insults cause a loss in mtDNA number rather than rearrangement and/or point mutation of mtDNA [57]. In addition, an imbalance of nucleotide pools can induce mtDNA depletion, as large portions of nucleotides are imported into the mitochondria and the mitochondrial inner membrane is less permeable to charged molecules [105]. Therefore, an insufficient supply of nucleotides in the mitochondria may lead to misincorporation of bases in replicating mtDNA.

mtDNA Repair System in Mitochondria

Clearly, a reduced repair capacity will result in an increased number of mtDNA mutations and risk of disease progression [106]. Several DNA repair systems function in the nucleus: direct reversal of mutated bases [107], base excision repair [92, 103], mismatch repair, nucleotide excision repair, and DNA double-stranded break repair [103]. However, mitochondria do not contain all of these repair systems. The major repair pathway in mitochondria is the base excision repair system which involves OGG1, POLG, and DNA ligase for recognition and incision, DNA synthesis, and final ligation of the single-stranded nick, respectively [92, 103, 108]. Considering the high rate of mtDNA mutations, oxidative stress may directly or indirectly deteriorate the mtDNA repair system. For example, when cells are exposed to H_2O_2, or possibly other oxidants, the potential of the repair system is suppressed, indirectly increasing the mutation rate [109]. In addition, several polymorphisms in antioxidant enzyme-encoded genes have been identified, which result in a range of effects such as no effect on enzyme activity to complete absence of the gene product. This indicates that polymorphisms in antioxidant defense genes modify the risk of disease development due to mtDNA mutations [110].

mtDNA Rearrangement

The occurrence of mtDNA rearrangement events may be a semi-random process that is relatively unique to each cell. Possible mtDNA rearrangements can be categorized into three types: (1) deletion mainly occurring between the origins of replication, O_H and O_L, which are typically flanked by two 13-bp direct repeats [111], (2) F-type rearrangement (e.g., tandem duplication), and (3) R-type rearrangement that joins two differently oriented mtDNA strands [112–116]. A large-scale rearrangements of mtDNA can result in partial deletions (ΔmtDNA), which vary in size from 1.3 to 8 kb and encompass several genes [117], partial duplications, or rarely triplication; these are typically heteroplasmic [118]. As shown in Fig. 5.2 and Table 5.4, deletions in mtDNA without duplication are more frequently detected in nondividing tissue, such as cardiac muscle cells rather than blood cells, and frequently results in a mosaic pattern [34, 115, 119–121]. mtDNA can be deleted during replication by slipped-strand homology in repetitive sequences or the failure to repair double-stranded breaks [92]. Irradiation [122, 123], chemical agents such as *t*-butyl hydroperoxide [124], and ultraviolet light can also generate double-strand breaks and thus mtDNA rearrangements [113–116]. Loss of *cis*-acting elements required for mtDNA replication, such as the D-loop, may be related to the frequency of mtDNA deletion [34]. Although numerous mtDNA deletions have been detected, the molecular mechanism leading to deletion of mtDNA, particularly large-scale deletions, is poorly understood. Partial mtDNA duplication, which is produced from coexisting ΔmtDNA, is not deleterious. However, at least one family with mitochondrial disease showed duplicated mtDNA with no detectable ΔmtDNA [118].

mtDNA Point Mutations

Point mutations can result in deletion, insertion, or substitution to affect single sites. DNA substitution mutations include transitions and transversions. Transitions are interchanges of bases of similar shape, either between purines or between pyrimidines. Transversions are interchanges of purine for pyrimidine bases. In previous studies, the most frequently detected mutations are transition mutations (http://www.mitomap.org), which is consistent with misincorporation by POLG [125] or deamination of cytidine and adenosine [126] as primary mutagenic events in mtDNA abasic sites [116]. Single- and double-strand breaks as well as 8-oxoguanine DNA adducts are considered to be major mutagenic lesions generated by ROS [116]. The most rapid point mutation observed is a G→A transition at bp 10,068 of the mitochondrial genome. This mutation occurs at a rate of 10^{-6} mutations per genome doubling [127]. For point mutations related to cardiac disorders, mutations in tRNA genes of mtDNA are often detected (Fig. 5.1 and Table 5.4), but are much better tolerated than those that directly affect the functioning of complex I, whose defects are the most frequently observed [128]. In the steady state, all new mutations will be copied, destroyed, and therefore stochastically redistributed as in post-mitotic cells. In addition to the occurrence of functional changes, primary mutations occurring in mtDNA can affect nearly any step of mt-mRNA maturation, including mt-tRNA, mt-rRNA, and their modification processing, which are governed by nuclear-encoded proteins [73].

Table 5.4 Cardiac manifestations with primary mtDNA mutations. Cardiac manifestation are structural, functional, or both. The most frequent cardiac manifestation mediated by mtDNA mutations is cardiomyopathy [7] such as hypertrophic-, dilated-, or restricted cardiomyopathy. The list below does not show the full spectrum of mtDNA mutations, but rather some examples

Cardiac manifestation	mtDNA mutation
Hypertrophic cardiomyopathy	mt-tRNA$^{\text{Leu, Glu, Lys, Val, or Glu}}$, m.4797 C→M, m.8728 T→Y, mt-12s rRNA, NADH:Ubiquinone Oxidoreductase 1(ND1), ND5, ATP Synthase 8, other mutations in nuclear genes
Dilated cardiomyopathy	mt-tRNA$^{\text{Glu, Ile, or Val}}$, mtDNA deletion and/or depletion, ND1
Restrictive cardiomyopathy	mtDNA deletion, ND5
Noncompaction (left ventricular hypertrabeculation)	mt-tRNA^{Leu} , m.8381 A→G, ND1, cytochrome B
Histiocytoid cardiomyopathy	m.8344A→G, cytochrome B
Myocardial fibrosis	mtDNA deletion
Arrhythmia	
Sinus arrhythmias	mt-tRNA^{Leu}
Atrial fibrillation/atrial flutter (AFIB/AFLU)	Mt-tRNA^{Leu}, mtDNA depletion
Wolff–Parkinson–White (WPW)-syndrome	Mt-tRNA$^{\text{Leu or Lys}}$, ND5
Atrio-ventricular-Block	Single mtDNA deletion, Mt-tRNA^{Leu}
Bundle branch block	mt-tRNA$^{\text{Leu or Lys}}$, single mtDNA deletion
QT-prolongation	Single mtDNA deletion
Ventricular tachycardia	Single mtDNA deletion
Heart failure/systolic dysfunction	mt-tRNA^{Leu}, ND5
Pulmonary hypertension	mt-tRNA$^{\text{Leu, or Glu}}$, mtDNA deletion
Dilation of aortic root	mtDNA deletion and/or mtDNA depletion
Pericardial effusion	mt-tRNA^{Lys}
Coronary heart disease	mt-tRNA^{Thr}
Autonomic nerve fibers	mt-tRNA^{Leu}
Patent foramen ovale	mt-tRNA^{Leu}
Ventricular septal defect	mt-tRNA^{Leu} , mtDNA depletion
Hypoplastic left heart syndrome	mtDNA depletion
Ventricular cavity obstruction	mt-12sRNA, ND1, ND4, ND6
Sudden cardiac death	mt-tRNA$^{\text{Leu or Lys}}$, mtDNA depletion

Depletion of mtDNA or Reduction in mtDNA Number

Controlling the amount of mtDNA per cell may be the first line regulation for OXPHOS activity during development [129]. Mutation of nuclear genes involved both in mtDNA replication and maintenance can increase quantitative (e.g., mtDNA depletion) or qualitative (e.g., multiple mtDNA deletions) mtDNA mutations [130]. Interestingly, human diseases caused by excess mtDNA content compared to nuclear DNA are rare [131]. A reduction or loss in mtDNA number (or content) plays a role in a range of cardiac manifestations (Tables 5.3 and 5.4) and aging [132, 133]. Different deletions may accumulate in different tissues of the same individual with a variable range [134], and thus phenotypic presentation varies between individuals (Table 5.4). Although mtDNA deletion is common, it is unclear how mtDNA deletions are tolerated and propagated in somatic cells (Fig. 5.3). One proposed mechanism involves the mitochondrial unfolded protein response, which is regulated by the transcription factor ATFS-1. The activation of ATFS-1 caused by OXPHOS defects transmits or maintains the deleterious mtDNA in an attempt to recover OXPHOS activity by promoting mitochondrial biogenesis and dynamics [135].

Epigenetic Modification in mtDNA

Epigenetics is defined as a stably heritable phenotype resulting from changes in a chromosome through posttranslational modifications, histone variants, and DNA methylation without direct alterations in the DNA sequence [136–138]. For example, disruption of nuclear DNA methylation are widely detected in human diseases [139]. Similarly to in the nucleus, the genetic expression of mtDNA can be altered by epigenetic modification of mtDNA or nucleoid structure-forming components, as mtDNA molecules are organized into nucleoid structures and opening of the closed structure may be involved in regulating mtDNA transcription and replication [40]. In addition, an increasing number of studies has examined mtDNA methylation produced in response to oxidative stress, while other mitochondria epigenetic fields, such as the study of mitochondrial non-coding RNAs, are predominantly limited to basic science studies [136, 137]. However, it is necessary to understand these epigenetic modifications in mtDNA to fine-tune mitochondrial genetics and its disease involvement.

Possible Outcome of mtDNA Mutations and Its Considerations

The distinction between variants with point mutations and variants with large deletions is important for interpreting mitochondrial mutations (Fig. 5.1 and Table 5.2). Inheritance of mitochondrial DNA mutations depends on the mutation type, such as only blockage of replication or possible propagation of a novel mutation [92, 106, 140], and requires months or years for manifestation [34]. In the absence of amino acid substitution, conservative amino acid changes, or mutation in noncoding regions of the mitochondrial genome, with the exception of specific promoters or replication origin region [34, 141], a large number of base changes in mtDNA will not be manifestoed. In addition, tolerable mutations in mtDNA do not impair mitochondrial function (Fig. 5.3) because normal mitochondria remain present, and a tissue-specific threshold effect occurs until the transition to a pathological condition [142, 143]. Therefore, discriminating whether an identified mtDNA variant is a deleterious mutation or polymorphism is complex. For mtDNA mutations involved in cardiac disorders, point mutations in mitochondrial protein-coding genes (Fig. 5.1 and Table 5.4) preferentially affect OXPHOS function, particularly in complexes I and IV [144]. A large number of mutations in non-protein coding genes of mtDNA have been mapped to mt-tRNA loci possessing a high frequency of point mutations (Fig. 5.1), despite accounting for just 5–10% of the mtDNA (Fig. 5.1). Mutations in mt-tRNA may impair overall mitochondrial translation by reducing the availability of functional mt-tRNAs [73, 145] and are functionally recessive. Furthermore, mischarging of mt-tRNAs occurs more frequently as a consequence of certain mutations and leads to a gain or loss of function of the protein involved [73, 146]. Notably, a dominant form of mt-tRNA mutation exists [147] and certain nucleotide changes that are not pathogenic may modulate the effects of other deleterious mtDNA mutations [148]. Interestingly, point mutations in mtDNA-encoded OXPHOS often escape the rules of mitochondrial genetics in that they affect single individuals without maternal inheritance (Fig. 5.3) and single tissues such as the skeletal muscle [149]. Several human disease-causing mutations both in the mtDNA and nuclear DNA were identified as common polymorphisms, suggesting that the pre-occurrence of compensatory mutations either within the same gene or in epistatically interacting genes exist [89]. When mitochondrial metabolism is perturbed because of a dysfunctional OXPHOS system or mtDNA mutations, mitochondrial biogenesis coupled with a peroxisome proliferator-activated receptor γ coactivator 1α may be increased to compensate for mitochondrial dysfunction [150]. In addition, nuclear modifier genes, which often have at least two alleles, one of which exacerbates disease and the other which suppresses the disease, have been investigated to better understand the incomplete penetrance of mitochondrial mutations [151].

Mitochondrial DNA Mutations and Cardiac Disorders

The heart is primarily composed of cardiomyocytes, which contain high numbers of mitochondria (thousands per cell), accounting for approximately 20–40% of the cell volume [36]. When the energy supply becomes limiting due to mitochondrial dysfunction, the infrastructure slowly deteriorates, and more severe consequences may be triggered by catastrophic failures at localized sites [134]. Accordingly, an impaired energy balance from mitochondria can be broadly conceived as both a cause and effect of heart dysfunction [152, 153]. Using advanced technologies, the number of detected mitochondrial diseases has increased (Table 5.2). Recently, it has been increasingly recognized that primary and secondary mitochondrial disorders are closely associated with cardiac defects [7, 8, 21, 141, 154–157]. mtDNA mutations are clearly an important cause of genetic diseases [7, 147]. Various mtDNA mutations have been detected in heart tissue (Tables 5.3 and 5.4); the frequent symptoms found in cardiac manifestations are muscle fatigue and weakness, which do not always receive attention in busy clinics [26]. Patients with specific mtDNA mutations may present with different cardiac phenotypes and similar cardiac involvement can occur in patients with different mtDNA mutations [154]. In addition, the frequency, sex distribution, age at onset, other affected organs, or affected cardiac tissue of mtDNA mutations are diverse, even for similar mtDNA mutations [6, 7, 84, 141]. A more detailed description of all known pathogenic mtDNA mutations is beyond the scope of this chapter and a current compendium can be found on the MitoMAP database [158] or other resources including Support and Advocacy groups, databases and analysis pipelines, research and patient networks. For mtDNA mutations associated with cardiac involvement, there are three main clinical groups: asymptomatic, arrhythmias, and cardiomyopathies. Generally, cardiac disorders involving mtDNA mutations including mtDNA depletion may be in the form of structural defects, functional lesions, or both [6, 7, 129]. The frequency of cardiomyopathy in children with mitochondrial disease is greater than 20–40% [45, 159]. Among cardiac manifestations associated with mtDNA mutations (Table 5.4), hypertrophic cardiomyopathy is a common mitochondrial mutation, occurring in 40% of patients, which is higher than in dilated cardiomyopathy [45, 159, 160]. Mitochondrial cardiomyopathies may also present as left ventricular noncompaction or as dilated, histiocytoid, or restrictive cardiomyopathies (Tables 5.3 and 5.4), or may be associated with endocardial fibroelastosis [45]. Histiocytoid cardiomyopathy, previously regarded to result from abnormal Purkinje fibers within the cardiac conduction system, is now considered to be a primary genetic cardiomyopathy [161] because of its cardiac manifestation of a mitochondrial disorder caused by mtDNA mutation.

Mitochondrial disorder-associated cardiac manifestations can be categorized as hereditary or sporadic disorders [162]. Unlike mitochondrial disorders associated with neurodegeneration or neurological disorders that are largely noticeable [107, 143, 163], the onset the mitochondrial disorders due to mutations in nuclear genes occurs in early childhood, whereas disorders caused by mtDNA mutations are observed in late childhood or adulthood [4, 5]. Mitochondrial diseases occurring in children support the hypothesis that mitochondrial disease can be manifested any tissue and show various signs at different ages [118]. Childhood mitochondrial disease is typically more severe than adult-onset disease and includes progressive neurological, cardiac, and liver dysfunction. The onset of clinical symptoms, phenotypic variability, and variable penetrance of mitochondrial diseases are governed by a number of factors [44]. First, because of the presence of heteroplasmy, a minimum critical proportion of mutated mtDNA (60–90% mutant to wild-type DNA) is necessary before biochemical defects and tissue dysfunction become apparent (Fig. 5.2).

This difference in the threshold level of individuals is related to the low correlation between clinical severity and proportion of mtDNA mutations [151]. Second, eukaryotic cells contain genetically different mitochondria and thus randomly segregate during mitosis to generate some cells containing only mutant or nonmutant organelles. Considering the continuous replicative state of mitochondria and their DNAs, the proportion of mtDNA mutations in cells differs. Third, during aging, mtDNA mutations can accumulate to very high level in nondividing cells such as muscle and neuron cells. This can be achieved by the lower regulation of mtDNA replication through random intracellular drift, resulting in the clonal expansion of single mutant events during human life [164]. Finally, the phenomenon of mtDNA bottleneck occurs in germline cells. In addition, mtDNA aggregates or nucleoids may be involved in rapid changes in the mitochondrial DNA allele frequency between generations [145]. Diseases caused by nuclear genes that affect mtDNA stability are an interesting field of mitochondrial disorders, involving both cellular genomes [1, 85, 165, 166]. Therefore, a primary nuclear gene defect can cause secondary mtDNA loss or deletion formation. The mitochondrial manifestations originated from nuclear gene defect may resemble those caused by mtDNA mutations, but follow a Mendelian inheritance style, such as autosomal dominant progressive external ophthalmoplegia and paralysis of the motor nerves of the eye [166]. Overall, precise prediction or interpretation from mtDNA mutation to malfunction is not an easy task because of mtDNA heteroplasmy, poor genotype-phenotype correlation in many patients, and complex interactions between the nucleus and mtDNA [19, 167–170].

Diagnostic Approaches and Their Considerations

The human heart contains a greater number of mitochondria and exhibits higher ETC activity than the skeletal muscle, but the production of ATP by OXPHOS activity is lower than in muscle (heart; 40% vs. muscle; 80%) [36]. This is because of the much higher each-step activation (ESA) of OXPHOS in the skeletal muscle compared to in the heart [171]. In rodents, the heart shows higher OXPHOS activity than the muscle. In addition, human mitochondria show lower calcium accumulation than that in rodents [36]. These differences between human and rodent hearts must be taken into account. Compensatory mechanisms can maintain cardiac output, even in cases of heart failure. Thus, the time of diagnosis or identification of mtDNA is a clinically critical factor because it is not clear whether alterations in mitochondrial function are a primary or a secondary effect [36].

Based on previously described causes of mitochondrial dysfunction, several diagnostic approaches have been suggested (Table 5.2). The best approach for assessing dysfunction depends on the sample source (e.g., mitochondria, cell, or in vivo), qualitative or quantitative assessment, and routine accessibility [36, 43, 44, 119, 172]. Biochemical analysis for suspected mitochondrial mutations includes: (1) lactate, pyruvate, and phosphocreatine analysis. Clinical testing can be performed to identify a mitochondrial disease by serial measurement of lactate, pyruvate, and alanine after carbohydrate loading with glucose or fructose [43]. Fasting studies can also provide clues to reveal a tendency towards hypoglycemia or secondary fatty acid oxidation defects; (2) amino acid analysis; (3) organic acid analysis; and (4) quantification of coenzyme Q10 and carnitine [43]. More directed biochemical assays can find deficiencies in NADH-coenzyme Q reductase (complex I) and cytochrome-c oxidase (complex IV), whereas the activity of complex II is less affected because it is entirely encoded by nuclear DNA [173, 174]. However, the activities of complexes I and III can be easily assessed using histochemical methods [119]. The presence of cytochrome C oxidase-negative myocytes and increased staining for succinate dehydrogenase may reveal respiratory chain defects and mitochondrial proliferation. Based on the morphological changes induced by possible mitochondrial dysfunction or compensation responses, myofibrillar myopathies are characterized by myofibrillar disorganization beginning at the Z-discs and are genetically caused by mutations in nuclear genes [160, 175, 176]. Cardiomyocytes in affected myocardial tissue show paracrystalline inclusions, enlarged mitochondria, or concentric onion-like cristae [23]. Using transmission electron microscopy and serial block face scanning electron microscopy with 3D reconstruction, more detailed ultrastructural abnormalities of the mitochondria including matrix compartmentalization, nanotunelling, or donut-shaped mitochondria were identified [22]. As depicted in Table 5.2, the molecular diagnosis of suspected mitochondrial disease using cutting-edge sequencing techniques has evolved rapidly over the past two decades [34, 120, 177]. However, these techniques cannot resolve some diagnostic dilemmas found in the genetics of mitochondria because of difficulties in determining the potential pathogenicity of mtDNA variants [82, 178]. Restricting the loss of mtDNA to specific tissues remains challenging, as variable tissue involvement has been reported in a single family [166]. Apart from mutations in the mtDNA sequence, metabolomics, which is based on nuclear magnetic resonance or mass spectrometry, has been applied to understand the comprehensive profiling of metabolites and other small molecules in cells and tissues. This method identifies clinical biomarkers of mtDNA mutation related to established cardiometabolic risk factors [179, 180]. As a less specific test for determining mitochondrial capacity, exercise testing, including bicycle and treadmill ergometry with a combination of biochemical and oxygen consumption tests, can be helpful for identifying possible mitochondrial defects [181]. Experimentally, a transmitochondrial hybrid, which can be generated by fusing a cytoplast containing mutated mtDNA with a cell line lacking mtDNA, can be used to identify defects in OXPHOS and the threshold effect of newly identified mtDNA mutations [142, 182].

Therapeutic Approach

When mitochondrial disease is known or suspected, cardiac examination should be directed towards eliciting signs of cardiac dysfunctions including heart failure (Table 5.4) and signs of hypertension [23]. The susceptibility to cardiac disorders, rate of disease progression, or response to pharmacological therapy depends on the mutations in nuclear DNA, mtDNA, or both [183]. For the management of mtDNA mutation-associated cardiac disorders, treatment strategies are currently based on supportive or imaginary treatment of the symptoms without genetic corrections of the affected mitochondria or cells [6, 26, 46, 64, 183–186]. First, based on the mitochondrial heteroplasmy, mitochondrial biogenesis is considered to be a set of molecular events by which cells replace or increase their mitochondria through enhanced mitochondrial proliferation. At this point, the modulation of peroxisome proliferator-activated receptor γ-coactivator 1α and AMP-dependent kinase are thought to increase mitochondrial biogenesis [6, 60, 187, 188]. Another method may exist for regulating mitochondrial biogenesis through putative mitochondrial-localizing hormone receptors [64, 188–190]. Second, interventional efforts targeting the direct modulation of OXPHOS activity have been extensively investigated in an experimental setting and studies are being conducted to translate these results to human applications [46, 188]. To relieve biochemical defects in the mitochondria, studies have attempted to bypass the pathogenic condition by removing noxious metabolites or suppressing their production [191]. Third, many mitochondrial disorders exhibit mutations in mt-tRNA genes and thus screening of regulatory molecules that can stabilize mt-tRNAs should be considered [25, 187, 192]. Finally, various genetic tools for gene therapy aimed to correct, bypass, switch heteroplasmy, and avoid mtDNA transmission have gained attention in recent years and have shown promising results [18, 25, 188, 192–195]. However, the clinical applications of these methods are limited because safe delivery of the therapeutic agent into mitochondria in an organ-specific manner is required [195, 196]. As a non-pharmacological intervention, exercise can be used as another interventional regime for treating mitochondrial mutations-related cardiac disorders because exercise training improves enzyme activity, mitochondrial biogenesis, and mitochondria quality [19, 197]. Recent findings support that exercise can induce translocation of tumor suppressor protein p53 to the mitochondria and stimulate mtDNA mutation repair independent of POLG and mitochondrial biogenesis [198]. From the perspective of genetic counseling, improving genetic advice through prenatal analysis using amniocentesis and chorionic villus biopsy and the development of new strategies for preventing the transmission of mtDNA into next generation [199, 200], such as oocyte manipulation including spindle transfer, pronuclear transfer, or polar body transfer, is a very important aspect of disease management [201]. Overall, current therapeutic regimes have not been successful. Thus, therapeutic regimens for mitochondrial disorders manifesting in cardiac disorders require more extensive development based on mitochondrial genetics, nuclear modifier genes, and quality control systems [202].

Conclusion

Error-prone conditions in the mitochondria can result in a continuously increasing mutational load of mtDNA. Despite the high variability of phenotypic presentations, studies focused on cardiac manifestation caused by mtDNA mutation are clinically important [7]. At least three factors can perturb mitochondrial bioenergetics and result in cardiac disorders: (1) variation in the mtDNA sequence (e.g., inherited or spontaneous mutations), (2) variation in sequences of over 1,000 nuclear DNA-encoded mitochondrial proteins or in the expression of these genes (e.g., deleterious mutations and epigenetic changes) [203], or (3) susceptibility to varied environmental conditions (e.g., energy resource and demand, mitochondrial toxins). The critical role of mutations in mtDNA should be stressed with respect to human genetic diseases and cardiac disorders. Further studies are required to investigate the genetics of mtDNA, and continued study of this fascinating genome will expand our understanding of cardiac disorders and other mtDNA mutation-associated complications.

Acknowledgements We apologize for the vast number of outstanding publications that could not be cited because of space limitations. This work was supported by the Priority Research Centers Program (2010-0020224) and the Basic Science Research Program (2015R1A2A1A13001900 and 2015R1D1A3A01015596) through the National Research Foundation of Korea (NRF), funded by the Ministry of Education, Science, and Technology.

References

1. Kaniak-Golik A, Skoneczna A. Mitochondria-nucleus network for genome stability. Free Radic Biol Med. 2015;82:73–104.
2. da Cunha FM, Torelli NQ, Kowaltowski AJ. Mitochondrial retrograde signaling: triggers, pathways, and outcomes. Oxidative Med Cell Longev. 2015;2015:482582.
3. McBride HM, Neuspiel M, Wasiak S. Mitochondria: more than just a powerhouse. Curr Biol. 2006;16:R551–60.
4. Picard M, Wallace DC, Burelle Y. The rise of mitochondria in medicine. Mitochondrion. 2016;30:105–16.
5. Chaudhary KR, El-Sikhry H, Seubert JM. Mitochondria and the aging heart. J Geriatr Cardiol JGC. 2011;8:159–67.
6. Schwarz K, Siddiqi N, Singh S, Neil CJ, Dawson DK, Frenneaux MP. The breathing heart – mitochondrial respiratory chain dysfunction in cardiac disease. Int J Cardiol. 2014;171:134–43.
7. Finsterer J, Kothari S. Cardiac manifestations of primary mitochondrial disorders. Int J Cardiol. 2014;177:754–63.
8. Lee S, Kim N, Noh Y, Xu Z, Ko K, Rhee B, Han J. Mitochondrial DNA, mitochondrial dysfunction, and cardiac manifestations. Front Biosci (Landmark Ed). 2016;21:1410–26.
9. Wang Q. Molecular genetics of coronary artery disease. Curr Opin Cardiol. 2005;20:182–8.
10. Tyynismaa H, Suomalainen A. Mouse models of mitochondrial DNA defects and their relevance for human disease. EMBO Rep. 2009;10:137–43.
11. Krzywanski DM, Moellering DR, Fetterman JL, Dunham-Snary KJ, Sammy MJ, Ballinger SW. The mitochondrial paradigm for cardiovascular disease susceptibility and cellular function: a complementary concept to Mendelian genetics. Lab Investig. 2011;91:1122–35.
12. Wang J, Lin F, Guo LL, Xiong XJ, Fan X. Cardiovascular disease, mitochondria, and traditional chinese medicine. Evid Based Complement Alternat Med. 2015;2015:143145.
13. Scheibye-Knudsen M, Fang EF, Croteau DL, Wilson Iii DM, Bohr VA. Protecting the mitochondrial powerhouse. Trends Cell Biol. 2015;25:158–70.
14. Wallace DC. A mitochondrial bioenergetic etiology of disease. J Clin Invest. 2013;123:1405–12.

15. Pikó L, Hougham AJ, Bulpitt KJ. Studies of sequence heterogeneity of mitochondrial DNA from rat and mouse tissues: evidence for an increased frequency of deletions/additions with aging. Mech Ageing Dev. 1988;43:279–93.
16. Stewart JB, Larsson N-G. Keeping mtDNA in shape between generations. PLoS Genet. 2014;10:e1004670.
17. Golob MJ, Tian L, Wang Z, Zimmerman TA, Caneba CA, Hacker TA, Song G, Chesler NC. Mitochondria DNA mutations cause sex-dependent development of hypertension and alterations in cardiovascular function. J Biomech. 2015;48:405–12.
18. Farrar GJ, Chadderton N, Kenna PF, Millington-Ward S. Mitochondrial disorders: aetiologies, models systems, and candidate therapies. Trends Genet. 2013;29:488–97.
19. Taylor RW, Turnbull DM. Mitochondrial dna mutations in human disease. Nat Rev Genet. 2005;6:389–402.
20. El-Hattab AW, Scaglia F. Mitochondrial cardiomyopathies. Front Cardiovasc Med. 2016;3:25.
21. Ritzenthaler T, Luis D, Hullin T, Fayssoil A. Cardiac manifestations of mitochondrial diseases. Presse Med. 2015;44:492–7.
22. Vincent AE, Ng YS, White K, Davey T, Mannella C, Falkous G, Feeney C, Schaefer AM, McFarland R, Gorman GS, Taylor RW, Turnbull DM, Picard M. The spectrum of mitochondrial ultrastructural defects in mitochondrial myopathy. Sci Report. 2016;6:30610.
23. Meyers DE, Basha HI, Koenig MK. Mitochondrial cardiomyopathy: pathophysiology, diagnosis, and management. Tex Heart Inst J. 2013;40:385–94.
24. Magner M, Kolarova H, Honzik T, Svandova I, Zeman J. Clinical manifestation of mitochondrial diseases. Dev Period Med. 2015;19:441–9.
25. Lightowlers RN, Taylor RW, Turnbull DM. Mutations causing mitochondrial disease: what is new and what challenges remain? Science. 2015;349:1494–9.
26. Chinnery PF. Mitochondrial disease in adults: what's old and what's new? EMBO Mol Med. 2015;7:1503–12.
27. Brunel-Guitton C, Levtova A, Sasarman F. Mitochondrial diseases and cardiomyopathies. Can J Cardiol. 2015;31:1360–76.
28. Finsterer J. Cardiogenetics, neurogenetics, and pathogenetics of left ventricular hypertrabeculation/noncompaction. Pediatr Cardiol. 2009;30:659–81.
29. Ong SB, Hausenloy DJ. Mitochondrial morphology and cardiovascular disease. Cardiovasc Res. 2010;88:16–29.
30. Beraud N, Pelloux S, Usson Y, Kuznetsov AV, Ronot X, Tourneur Y, Saks V. Mitochondrial dynamics in heart cells: very low amplitude high frequency fluctuations in adult cardiomyocytes and flow motion in non beating Hl-1 cells. J Bioenerg Biomembr. 2009;41:195–214.
31. Kuznetsov AV, Hermann M, Saks V, Hengster P, Margreiter R. The cell-type specificity of mitochondrial dynamics. Int J Biochem Cell Biol. 2009;41:1928–39.
32. MacKenzie JA, Payne RM. Mitochondrial protein import and human health and disease. Biochim Biophys Acta (BBA) – Mol Basis Dis. 2007;1772:509–23.
33. Suzuki T, Nagao A, Suzuki T. Human mitochondrial tRNAs: biogenesis, function, structural aspects, and diseases. Annu Rev Genet. 2011;45:299–329.
34. Scheffler IE. Mitochondria. 2nd ed. New Jersey: Wiley; 2008.
35. Horvath SE, Rampelt H, Oeljeklaus S, Warscheid B, van der Laan M, Pfanner N. Role of membrane contact sites in protein import into mitochondria. Protein Sci. 2015;24:277–97.
36. Lemieux H, Hoppel CL. Mitochondria in the human heart. J Bioenerg Biomembr. 2009;41:99–106.
37. Meisinger C, Sickmann A, Pfanner N. The mitochondrial proteome: from inventory to function. Cell. 2008;134:22–4.
38. Johnson DT, Harris RA, French S, Blair PV, You J, Bemis KG, Wang M, Balaban RS. Tissue heterogeneity of the mammalian mitochondrial proteome. Am J Phys Cell Phys. 2007;292:C689–97.
39. Mootha VK, Bunkenborg J, Olsen JV, Hjerrild M, Wisniewski JR, Stahl E, Bolouri MS, Ray HN, Sihag S, Kamal M, Patterson N, Lander ES, Mann M. Integrated analysis of protein

composition, tissue diversity, and gene regulation in mouse mitochondria. Cell. 2003;115:629–40.
40. Falkenberg M, Larsson N-G, Gustafsson CM. DNA replication and transcription in mammalian mitochondria. Annu Rev Biochem. 2007;76:679–99.
41. Chen H, Chan DC. Emerging functions of mammalian mitochondrial fusion and fission. Hum Mol Genet. 2005;14(2):R283–9.
42. Naini A, Shanske S. Detection of mutations in mtDNA. Methods Cell Biol. 2007;80:437–63.
43. Haas RH, Parikh S, Falk MJ, Saneto RP, Wolf NI, Darin N, Wong LJ, Cohen BH, Naviaux RK. The in-depth evaluation of suspected mitochondrial disease. Mol Genet Metab. 2008;94:16–37.
44. Tuppen HAL, Blakely EL, Turnbull DM, Taylor RW. Mitochondrial DNA mutations and human disease. Biochim Biophys Acta (BBA) – Bioenerg. 2010;1797:113–28.
45. Holmgren D, Wahlander H, Eriksson BO, Oldfors A, Holme E, Tulinius M. Cardiomyopathy in children with mitochondrial disease; clinical course and cardiological findings. Eur Heart J. 2003;24:280–8.
46. Klopstock T, Metz G, Yu-Wai-Man P, Büchner B, Gallenmüller C, Bailie M, Nwali N, Griffiths PG, von Livonius B, Reznicek L, Rouleau J, Coppard N, Meier T, Chinnery PF. Persistence of the treatment effect of idebenone in Leber's hereditary optic neuropathy. Brain. 2013;136:e230.
47. Pieczenik SR, Neustadt J. Mitochondrial dysfunction and molecular pathways of disease. Exp Mol Pathol. 2007;83:84–92.
48. Brand MD, Nicholls DG. Assessing mitochondrial dysfunction in cells. Biochem J. 2011;435:297–312.
49. Held NM, Houtkooper RH. Mitochondrial quality control pathways as determinants of metabolic health. BioEssays. 2015;37:867–76.
50. Youle RJ, Narendra DP. Mechanisms of mitophagy. Nat Rev Mol Cell Biol. 2011;12:9–14.
51. Schwartz M, Vissing J. Paternal inheritance of mitochondrial DNA. N Engl J Med. 2002;347:576–80.
52. Anderson S, Bankier AT, Barrell BG, de Bruijn MHL, Coulson AR, Drouin J, Eperon IC, Nierlich DP, Roe BA, Sanger F, Schreier PH, Smith AJH, Staden R, Young IG. Sequence and organization of the human mitochondrial genome. Nature. 1981;290:457–65.
53. Kucej M, Butow RA. Evolutionary tinkering with mitochondrial nucleoids. Trends Cell Biol. 2007;17:586–92.
54. Wang Y, Bogenhagen DF. Human mitochondrial DNA nucleoids are linked to protein folding machinery and metabolic enzymes at the mitochondrial inner membrane. J Biol Chem. 2006;281:25791–802.
55. Brown TA, Tkachuk AN, Shtengel G, Kopek BG, Bogenhagen DF, Hess HF, Clayton DA. Superresolution fluorescence imaging of mitochondrial nucleoids reveals their spatial range, limits, and membrane interaction. Mol Cell Biol. 2011;31:4994–5010.
56. Kukat C, Wurm CA, Spåhr H, Falkenberg M, Larsson N-G, Jakobs S. Super-resolution microscopy reveals that mammalian mitochondrial nucleoids have a uniform size and frequently contain a single copy of mtDNA. Proc Natl Acad Sci. 2011;108:13534–9.
57. Gilkerson R. Commentary: mitochondrial DNA damage and loss in diabetes. Diabetes Metab Res Rev. 2016;32:672–4.
58. Shokolenko I, Venediktova N, Bochkareva A, Wilson GL, Alexeyev MF. Oxidative stress induces degradation of mitochondrial DNA. Nucleic Acids Res. 2009;37:2539–48.
59. Jornayvaz FR, Shulman GI. Regulation of mitochondrial biogenesis. Essays Biochem. 2010;47:69–84. doi:10.1042/bse0470069.
60. Valero T. Mitochondrial biogenesis: pharmacological approaches. Curr Pharm Des. 2014;20:5507–9.
61. Arnould T, Michel S, Renard P. Mitochondria retrograde signaling and the UPR(mt): where are we in mammals? Int J Mol Sci. 2015;16:18224–51.

62. Biala AK, Dhingra R, Kirshenbaum LA. Mitochondrial dynamics: orchestrating the journey to advanced age. J Mol Cell Cardiol. 2015;83:37–43.
63. Stotland A, Gottlieb RA. Mitochondrial quality control: easy come, easy go. Biochim Biophys Acta (BBA) – Mol Cell Res. 2015;1853:2802–11.
64. Suliman HB, Piantadosi CA. Mitochondrial quality control as a therapeutic target. Pharmacol Rev. 2016;68:20–48.
65. Bogenhagen D, Clayton DA. Mouse L cell mitochondrial DNA molecules are selected randomly for replication throughout the cell cycle. Cell. 1977;11:719–27.
66. Clayton DA. Replication of animal mitochondrial DNA. Cell. 1982;28:693–705.
67. Korhonen JA, Pham XH, Pellegrini M, Falkenberg M. Reconstitution of a minimal mtDNA replisome in vitro. EMBO J. 2004;23:2423–9.
68. Fish J, Raule N, Attardi G. Discovery of a major D-loop replication origin reveals two modes of human mtDNA synthesis. Science. 2004;306:2098–101.
69. St John JC, Facucho-Oliveira J, Jiang Y, Kelly R, Salah R. Mitochondrial DNA transmission, replication and inheritance: a journey from the gamete through the embryo and into offspring and embryonic stem cells. Hum Reprod Update. 2010;16:488–509.
70. Van Haute L, Pearce SF, Powell CA, D'Souza AR, Nicholls TJ, Minczuk M. Mitochondrial transcript maturation and its disorders. J Inherit Metab Dis. 2015;38:655–80.
71. Fernandez-Silva P, Enriquez JA, Montoya J. Replication and transcription of mammalian mitochondrial DNA. Exp Physiol. 2003;88:41–56.
72. Rackham O, Mercer TR, Filipovska A. The human mitochondrial transcriptome and the RNA-binding proteins that regulate its expression. Wiley Interdiscip Rev RNA. 2012;3:675–95.
73. Yarham JW, Elson JL, Blakely EL, McFarland R, Taylor RW. Mitochondrial tRNA mutations and disease. Wiley Interdiscip Rev RNA. 2010;1:304–24.
74. Bonawitz ND, Clayton DA, Shadel GS. Initiation and beyond: multiple functions of the human mitochondrial transcription machinery. Mol Cell. 2006;24:813–25.
75. Pomerantz RT, O'Donnell M. What happens when replication and transcription complexes collide? Cell Cycle. 2010;9:2537–43.
76. Agaronyan K, Morozov YI, Anikin M, Temiakov D. Mitochondrial biology. Replication-transcription switch in human mitochondria. Science. 2015;347:548–51.
77. Dunbar DR, Moonie PA, Jacobs HT, Holt IJ. Different cellular backgrounds confer a marked advantage to either mutant or wild-type mitochondrial genomes. Proc Natl Acad Sci. 1995;92:6562–6.
78. DiMauro S, Schon EA. Mitochondrial respiratory-chain diseases. N Engl J Med. 2003;348:2656–68.
79. Huss JM, Kelly DP. Mitochondrial energy metabolism in heart failure: a question of balance. J Clin Invest. 2005;115:547–55.
80. Nass S, Nass MM. Intramitochondrial fibers with DNA characteristics. II. Enzymatic and other hydrolytic treatments. J Cell Biol. 1963;19:613–29.
81. Holt IJ, Harding AE, Morgan-Hughes JA. Deletions of muscle mitochondrial DNA in patients with mitochondrial myopathies. Nature. 1988;331:717–9.
82. Wallace DC, Chalkia D. Mitochondrial DNA genetics and the heteroplasmy conundrum in evolution and disease. Cold Spring Harb Perspect Biol. 2013;5:a021220.
83. Andreu AL, Hanna MG, Reichmann H, Bruno C, Penn AS, Tanji K, Pallotti F, Iwata S, Bonilla E, Lach B, Morgan-Hughes J, DiMauro S. Exercise intolerance due to mutations in the cytochrome b gene of mitochondrial DNA. N Engl J Med. 1999;341:1037–44.
84. Elliott HR, Samuels DC, Eden JA, Relton CL, Chinnery PF. Pathogenic mitochondrial DNA mutations are common in the general population. Am J Hum Genet. 2008;83:254–60.
85. Gorman GS, Schaefer AM, Ng Y, Gomez N, Blakely EL, Alston CL, Feeney C, Horvath R, Yu-Wai-Man P, Chinnery PF, Taylor RW, Turnbull DM, McFarland R. Prevalence of nuclear and mitochondrial DNA mutations related to adult mitochondrial disease. Ann Neurol. 2015;77:753–9.

86. Nachman MW, Brown WM, Stoneking M, Aquadro CF. Nonneutral mitochondrial DNA variation in humans and chimpanzees. Genetics. 1996;142:953–63.
87. Brown WM, George M, Wilson AC. Rapid evolution of animal mitochondrial DNA. Proc Natl Acad Sci. 1979;76:1967–71.
88. Itsara LS, Kennedy SR, Fox EJ, Yu S, Hewitt JJ, Sanchez-Contreras M, Cardozo-Pelaez F, Pallanck LJ. Oxidative stress is not a major contributor to somatic mitochondrial DNA mutations. PLoS Genet. 2014;10:e1003974.
89. Levin L, Blumberg A, Barshad G, Mishmar D. Mito-nuclear co-evolution: the positive and negative sides of functional ancient mutations. Front Genet. 2014;5:448.
90. De Bont R, van Larebeke N. Endogenous DNA damage in humans: a review of quantitative data. Mutagenesis. 2004;19:169–85.
91. de Mendoza C, Sanchez-Conde M, Ribera E, Domingo P, Soriano V. Could mitochondrial DNA quantitation be a surrogate marker for drug mitochondrial toxicity? AIDS Rev. 2004;6:169–80.
92. Muftuoglu M, Mori MP, de Souza-Pinto NC. Formation and repair of oxidative damage in the mitochondrial DNA. Mitochondrion. 2014;17:164–81.
93. Davey GP, Peuchen S, Clark JB. Energy thresholds in brain mitochondria. Potential involvement in neurodegeneration. J Biol Chem. 1998;273:12753–7.
94. Leon LJ, Gustafsson AB. Staying young at heart: autophagy and adaptation to cardiac aging. J Mol Cell Cardiol. 2016;95:78–85.
95. Jones DP. Disruption of mitochondrial redox circuitry in oxidative stress. Chem Biol Interact. 2006;163:38–53.
96. Sies H, Jones D. Oxidative stress*. In: Fink G, editor. Encyclopedia of stress. 2nd ed. New York: Academic; 2007. p. 45–8.
97. Payne BAI, Chinnery PF. Mitochondrial dysfunction in aging: much progress but many unresolved questions. Biochim Biophys Acta. 1847;2015:1347–53.
98. Gruber J, Schaffer S, Halliwell B. The mitochondrial free radical theory of ageing – where do we stand? Front Biosci. 2008;13:6554–79.
99. Stuart JA, Bourque BM, de Souza-Pinto NC, Bohr VA. No evidence of mitochondrial respiratory dysfunction in OGG1-null mice deficient in removal of 8-oxodeoxyguanine from mitochondrial DNA. Free Radic Biol Med. 2005;38:737–45.
100. Trifunovic A, Hansson A, Wredenberg A, Rovio AT, Dufour E, Khvorostov I, Spelbrink JN, Wibom R, Jacobs HT, Larsson N-G. Somatic mtDNA mutations cause aging phenotypes without affecting reactive oxygen species production. Proc Natl Acad Sci U S A. 2005;102:17993–8.
101. Kujoth GC, Hiona A, Pugh TD, Someya S, Panzer K, Wohlgemuth SE, Hofer T, Seo AY, Sullivan R, Jobling WA, Morrow JD, Van Remmen H, Sedivy JM, Yamasoba T, Tanokura M, Weindruch R, Leeuwenburgh C, Prolla TA. Mitochondrial DNA mutations, oxidative stress, and apoptosis in mammalian aging. Science. 2005;309:481–4.
102. Richter C, Park JW, Ames BN. Normal oxidative damage to mitochondrial and nuclear DNA is extensive. Proc Natl Acad Sci. 1988;85:6465–7.
103. Szczepanowska K, Trifunovic A. Different faces of mitochondrial DNA mutators. Biochim Biophys Acta. 1847;2015:1362–72.
104. Larsson NG, Oldfors A, Holme E, Clayton DA. Low levels of mitochondrial transcription factor a in mitochondrial DNA depletion. Biochem Biophys Res Commun. 1994;200: 1374–81.
105. Young MJ, Copeland WC. Human mitochondrial DNA replication machinery and disease. Curr Opin Genet Dev. 2016;38:52–62.
106. Cooke MS, Evans MD, Dizdaroglu M, Lunec J. Oxidative DNA damage: mechanisms, mutation, and disease. FASEB J. 2003;17:1195–214.
107. Gredilla R, Bohr VA, Stevnsner T. Mitochondrial DNA repair and association with aging – an update. Exp Gerontol. 2010;45:478–88.
108. Gredilla R. DNA damage and base excision repair in mitochondria and their role in aging. J Aging Res. 2010;2011:257093.

109. Hu JJ, Dubin N, Kurland D, Ma BL, Roush GC. The effects of hydrogen peroxide on DNA repair activities. Mutat Res. 1995;336:193–201.
110. Forsberg L, de Faire U, Morgenstern R. Oxidative stress, human genetic variation, and disease. Arch Biochem Biophys. 2001;389:84–93.
111. Samuels DC, Schon EA, Chinnery PF. Two direct repeats cause most human mtDNA deletions. Trends Genet. 2004;20:393–8.
112. Chen Y, Liu C, Parker WD, Chen H, Beach TG, Liu X, Serrano GE, Lu Y, Huang J, Yang K, Wang C. Mitochondrial DNA rearrangement spectrum in brain tissue of Alzheimer's disease: analysis of 13 cases. PLoS One. 2016;11:e0154582.
113. Szumiel I. Ionizing radiation-induced oxidative stress, epigenetic changes and genomic instability: the pivotal role of mitochondria. Int J Radiat Biol. 2015;91:1–12.
114. Yang N, Galick H, Wallace SS. Attempted base excision repair of ionizing radiation damage in human lymphoblastoid cells produces lethal and mutagenic double strand breaks. DNA Repair. 2004;3:1323–34.
115. Kraytsberg Y, Kudryavtseva E, McKee AC, Geula C, Kowall NW, Khrapko K. Mitochondrial DNA deletions are abundant and cause functional impairment in aged human substantia nigra neurons. Nat Genet. 2006;38:518–20.
116. Bjelland S, Seeberg E. Mutagenicity, toxicity and repair of DNA base damage induced by oxidation. Mutat Res/Fundam Mol Mech Mutagen. 2003;531:37–80.
117. Schon EA, Rizzuto R, Moraes CT, Nakase H, Zeviani M, DiMauro S. A direct repeat is a hotspot for large-scale deletion of human mitochondrial DNA. Science. 1989;244:346–9.
118. Spinazzola A. Mitochondrial DNA mutations and depletion in pediatric medicine. Semin Fetal Neonatal Med. 2011;16:190–6.
119. Taylor RW, Schaefer AM, Barron MJ, McFarland R, Turnbull DM. The diagnosis of mitochondrial muscle disease. Neuromuscul Disord. 2004;14:237–45.
120. Haack TB, Haberberger B, Frisch E-M, Wieland T, Iuso A, Gorza M, Strecker V, Graf E, Mayr JA, Herberg U, Hennermann JB, Klopstock T, Kuhn KA, Ahting U, Sperl W, Wilichowski E, Hoffmann GF, Tesarova M, Hansikova H, Zeman J, Plecko B, Zeviani M, Wittig I, Strom TM, Schuelke M, Freisinger P, Meitinger T, Prokisch H. Molecular diagnosis in mitochondrial complex I deficiency using exome sequencing. J Med Genet. 2012;49:277–83.
121. Zaragoza MV, Brandon MC, Diegoli M, Arbustini E, Wallace DC. Mitochondrial cardiomyopathies: how to identify candidate pathogenic mutations by mitochondrial DNA sequencing, MITOMASTER and phylogeny. Eur J Hum Genet. 2011;19:200–7.
122. Prithivirajsingh S, Story MD, Bergh SA, Geara FB, Kian Ang K, Ismail SM, Stevens CW, Buchholz TA, Brock WA. Accumulation of the common mitochondrial DNA deletion induced by ionizing radiation. FEBS Lett. 2004;571:227–32.
123. Berneburg M, Grether-Beck S, Kürten V, Ruzicka T, Briviba K, Sies H, Krutmann J. Singlet oxygen mediates the UVA-induced generation of the photoaging-associated mitochondrial common deletion. J Biol Chem. 1999;274:15345–9.
124. Dumont P, Burton M, Chen QM, Gonos ES, Frippiat C, Mazarati J-B, Eliaers F, Remacle J, Toussaint O. Induction of replicative senescence biomarkers by sublethal oxidative stresses in normal human fibroblast. Free Radic Biol Med. 2000;28:361–73.
125. Zheng W, Khrapko K, Coller HA, Thilly WG, Copeland WC. Origins of human mitochondrial point mutations as DNA polymerase γ-mediated errors. Mutat Res/Fundam Mol Mech Mutagen. 2006;599:11–20.
126. Kennedy SR, Salk JJ, Schmitt MW, Loeb LA. Ultra-sensitive sequencing reveals an age-related increase in somatic mitochondrial mutations that are inconsistent with oxidative damage. PLoS Genet. 2013;9:e1003794.
127. Marcelino LA, Thilly WG. Mitochondrial mutagenesis in human cells and tissues. Mutat Res. 1999;434:177–203.
128. Bannwarth S, Procaccio V, Lebre AS, Jardel C, Chaussenot A, Hoarau C, Maoulida H, Charrier N, Gai X, Xie HM, Ferre M, Fragaki K, Hardy G, Mousson de Camaret B, Marlin S, Dhaenens CM, Slama A, Rocher C, Paul Bonnefont J, Rotig A, Aoutil N, Gilleron M, Desquiret-Dumas V, Reynier P, Ceresuela J, Jonard L, Devos A, Espil-Taris C, Martinez D,

Gaignard P, Le Quan Sang KH, Amati-Bonneau P, Falk MJ, Florentz C, Chabrol B, Durand-Zaleski I, Paquis-Flucklinger V. Prevalence of rare mitochondrial DNA mutations in mitochondrial disorders. J Med Genet. 2013;50:704–14.
129. Morten KJ, Ashley N, Wijburg F, Hadzic N, Parr J, Jayawant S, Adams S, Bindoff L, Bakker HD, Mieli-Vergani G, Zeviani M, Poulton J. Liver mtDNA content increases during development: a comparison of methods and the importance of age- and tissue-specific controls for the diagnosis of mtDNA depletion. Mitochondrion. 2007;7:386–95.
130. El-Hattab AW, Scaglia F. Mitochondrial DNA depletion syndromes: review and updates of genetic basis, manifestations, and therapeutic options. Neurotherapeutics. 2013;10:186–98.
131. Clay Montier LL, Deng JJ, Bai Y. Number matters: control of mammalian mitochondrial DNA copy number. J Genet Genomics. 2009;36:125–31.
132. Herrera A, Garcia I, Gaytan N, Jones E, Maldonado A, Gilkerson R. Endangered species: mitochondrial DNA loss as a mechanism of human disease. Front Biosci (Schol Ed). 2015;7:109–24.
133. McCormick EM, Kenyon L, Falk MJ. Desmin common mutation is associated with multisystemic disease manifestations and depletion of mitochondria and mitochondrial DNA. Front Genet. 2015;6:199.
134. Scheffler IE. Mitochondrial mutations and disease. Mitochondria: Wiley; 2007. p. 345–416.
135. Lin YF, Schulz AM, Pellegrino MW, Lu Y, Shaham S, Haynes CM. Maintenance and propagation of a deleterious mitochondrial genome by the mitochondrial unfolded protein response. Nature. 2016;533:416–9.
136. Chinnery PF, Elliott HR, Hudson G, Samuels DC, Relton CL. Epigenetics, epidemiology and mitochondrial DNA diseases. Int J Epidemiol. 2012;41:177–87.
137. Lambertini L, Byun HM. Mitochondrial epigenetics and environmental exposure. Curr Environ Health Rep. 2016;3:214.
138. Hernandez-Aguilera A, Fernandez-Arroyo S, Cuyas E, Luciano-Mateo F, Cabre N, Camps J, Lopez-Miranda J, Menendez JA, Joven J. Epigenetics and nutrition-related epidemics of metabolic diseases: current perspectives and challenges. Food Chem Toxicol. 2016;96: 191–204.
139. Shen H, Laird PW. Interplay between the cancer genome and epigenome. Cell. 2013;153: 38–55.
140. Kauppila JHK, Stewart JB. Mitochondrial DNA: radically free of free-radical driven mutations. Biochim Biophys Acta (BBA) – Biomembr. 2015;1847:1354–61.
141. Limongelli G, Masarone D, D'Alessandro R, Elliott PM. Mitochondrial diseases and the heart: an overview of molecular basis, diagnosis, treatment and clinical course. Futur Cardiol. 2012;8:71–88.
142. Wallace DC. Mitochondrial diseases in man and mouse. Science. 1999;283:1482–8.
143. Mancuso C, Scapagini G, Curro D, Giuffrida Stella AM, De Marco C, Butterfield DA, Calabrese V. Mitochondrial dysfunction, free radical generation and cellular stress response in neurodegenerative disorders. Front Biosci. 2007;12:1107–23.
144. DiMauro S. Mitochondrial myopathies. Curr Opin Rheumatol. 2006;18:636–41.
145. Khrapko K. Two ways to make a mtDNA bottleneck. Nat Genet. 2008;40:134–5.
146. Sasarman F, Antonicka H, Shoubridge EA. The A3243G tRNALeu(UUR) MELAS mutation causes amino acid misincorporation and a combined respiratory chain assembly defect partially suppressed by overexpression of EFTu and EFG2. Hum Mol Genet. 2008;17: 3697–707.
147. Sacconi S, Salviati L, Nishigaki Y, Walker WF, Hernandez-Rosa E, Trevisson E, Delplace S, Desnuelle C, Shanske S, Hirano M, Schon EA, Bonilla E, De Vivo DC, DiMauro S, Davidson MM. A functionally dominant mitochondrial DNA mutation. Hum Mol Genet. 2008;17:1814–20.
148. Swalwell H, Blakely EL, Sutton R, Tonska K, Elstner M, He L, Taivassalo T, Burns DK, Turnbull DM, Haller RG, Davidson MM, Taylor RW. A homoplasmic mtDNA variant can influence the phenotype of the pathogenic m.7472Cins MTTS1 mutation: are two mutations better than one? Eur J Hum Genet. 2008;16:1265–74.

149. DiMauro S. Mitochondrial diseases. Biochim Biophys Acta Biomembr. 2004;1658:80–8.
150. Ahuja P, Wanagat J, Wang Z, Wang Y, Liem DA, Ping P, Antoshechkin IA, Margulies KB, MacLellan WR. Divergent mitochondrial biogenesis responses in human cardiomyopathy. Circulation. 2013;127:1957–67.
151. Chen C, Chen Y, Guan MX. A peep into mitochondrial disorder: multifaceted from mitochondrial DNA mutations to nuclear gene modulation. Protein Cell. 2015;6:862–70.
152. Wajner M, Amaral AU. Mitochondrial dysfunction in fatty acid oxidation disorders: insights from human and animal studies. Biosci Rep. 2016;36:e00281.
153. Anan R, Nakagawa M, Miyata M, Higuchi I, Nakao S, Suehara M, Osame M, Tanaka H. Cardiac involvement in mitochondrial diseases. A study on 17 patients with documented mitochondrial DNA defects. Circulation. 1995;91:955–61.
154. Bates MGD, Bourke JP, Giordano C, d'Amati G, Turnbull DM, Taylor RW. Cardiac involvement in mitochondrial DNA disease: clinical spectrum, diagnosis, and management. Eur Heart J. 2012;33:3023–33.
155. Montaigne D, Pentiah AD. Mitochondrial cardiomyopathy and related arrhythmias. Card Electrophysiol Clin. 2015;7:293–301.
156. Pohjoismäki JLO, Boettger T, Liu Z, Goffart S, Szibor M, Braun T. Oxidative stress during mitochondrial biogenesis compromises mtDNA integrity in growing hearts and induces a global DNA repair response. Nucleic Acids Res. 2012;40:6595–607.
157. Dominic EA, Ramezani A, Anker SD, Verma M, Mehta N, Rao M. Mitochondrial cytopathies and cardiovascular disease. Heart. 2014;100:611–8.
158. Mitomap. A human mitochondrial genome database. 2009. http://www.mitomaporg.
159. Scaglia F, Towbin JA, Craigen WJ, Belmont JW, Smith EO, Neish SR, Ware SM, Hunter JV, Fernbach SD, Vladutiu GD, Wong LJ, Vogel H. Clinical spectrum, morbidity, and mortality in 113 pediatric patients with mitochondrial disease. Pediatrics. 2004;114:925–31.
160. Brega A, Narula J, Arbustini E. Functional, structural, and genetic mitochondrial abnormalities in myocardial diseases. J Nucl Cardiol. 2001;8:89–97.
161. Finsterer J. Histiocytoid cardiomyopathy: a mitochondrial disorder. Clin Cardiol. 2008;31: 225–7.
162. Finsterer J, Harbo HF, Baets J, Van Broeckhoven C, Di Donato S, Fontaine B, De Jonghe P, Lossos A, Lynch T, Mariotti C, Schols L, Spinazzola A, Szolnoki Z, Tabrizi SJ, Tallaksen CM, Zeviani M, Burgunder JM, Gasser T. EFNS guidelines on the molecular diagnosis of mitochondrial disorders. Eur J Neurol. 2009;16:1255–64.
163. Schapira AHV. Mitochondrial diseases. Lancet. 2012;379:1825–34.
164. Elson JL, Samuels DC, Turnbull DM, Chinnery PF. Random intracellular drift explains the clonal expansion of mitochondrial DNA mutations with age. Am J Hum Genet. 2001;68:802–6.
165. Copeland WC. Defects of mitochondrial DNA replication. J Child Neurol. 2014;29: 1216–24.
166. Suomalainen A, Kaukonen J. Diseases caused by nuclear genes affecting mtDNA stability. Am J Med Genet. 2001;106:53–61.
167. Trifunovic A, Wredenberg A, Falkenberg M, Spelbrink JN, Rovio AT, Bruder CE, Bohlooly YM, Gidlof S, Oldfors A, Wibom R, Tornell J, Jacobs HT, Larsson NG. Premature ageing in mice expressing defective mitochondrial DNA polymerase. Nature. 2004;429:417–23.
168. Dunbar DR, Moonie PA, Jacobs HT, Holt IJ. Different cellular backgrounds confer a marked advantage to either mutant or wild-type mitochondrial genomes. Proc Natl Acad Sci U S A. 1995;92:6562–6.
169. Wallace DC. Diseases of the mitochondrial DNA. Annu Rev Biochem. 1992;61:1175–212.
170. Fulka Jr J, Fulka H, John JC. Transmission of mitochondrial DNA disorders: possibilities for the elimination of mutated mitochondria. Cloning Stem Cells. 2007;9:47–50.
171. Korzeniewski B. Faster and stronger manifestation of mitochondrial diseases in skeletal muscle than in heart related to cytosolic inorganic phosphate (Pi) accumulation. J Appl Physiol (1985). 2016;121:424–37.
172. Manuscript R. Apoptotic properties of the type 1 interferon induced family of human mitochondrial membrane ISG12 proteins, Manuscript ID:BOC.201600034.R1. 2016-08-05.

173. Arbustini E, Diegoli M, Fasani R, Grasso M, Morbini P, Banchieri N, Bellini O, Dal Bello B, Pilotto A, Magrini G, Campana C, Fortina P, Gavazzi A, Narula J, Vigano M. Mitochondrial DNA mutations and mitochondrial abnormalities in dilated cardiomyopathy. Am J Pathol. 1998;153:1501–10.
174. Yaplito-Lee J, Weintraub R, Jamsen K, Chow CW, Thorburn DR, Boneh A. Cardiac manifestations in oxidative phosphorylation disorders of childhood. J Pediatr. 2007;150:407–11.
175. Jackson S, Schaefer J, Meinhardt M, Reichmann H. Mitochondrial abnormalities in the myofibrillar myopathies. Eur J Neurol. 2015;22:1429–35.
176. Leone O, Veinot JP, Angelini A, Baandrup UT, Basso C, Berry G, Bruneval P, Burke M, Butany J, Calabrese F, d'Amati G, Edwards WD, Fallon JT, Fishbein MC, Gallagher PJ, Halushka MK, McManus B, Pucci A, Rodriguez ER, Saffitz JE, Sheppard MN, Steenbergen C, Stone JR, Tan C, Thiene G, van der Wal AC, Winters GL. 2011 consensus statement on endomyocardial biopsy from the Association for European Cardiovascular Pathology and the Society for Cardiovascular Pathology. Cardiovasc Pathol. 2012;21:245–74.
177. McCormick E, Place E, Falk MJ. Molecular genetic testing for mitochondrial disease: from one generation to the next. Neurotherapeutics. 2013;10:251–61.
178. McFarland R, Elson JL, Taylor RW, Howell N, Turnbull DM. Assigning pathogenicity to mitochondrial tRNA mutations: when "definitely maybe" is not good enough. Trends Genet. 2004;20:591–6.
179. Roberts LD, Gerszten RE. Toward new biomarkers of cardiometabolic diseases. Cell Metab. 2013;18:43–50.
180. Rhee EP, Gerszten RE. Metabolomics and cardiovascular biomarker discovery. Clin Chem. 2012;58:139–47.
181. Tarnopolsky M. Exercise testing as a diagnostic entity in mitochondrial myopathies. Mitochondrion. 2004;4:529–42.
182. Sciacco M, Bonilla E, Schon EA, DiMauro S, Moraes CT. Distribution of wild-type and common deletion forms of mtDNA in normal and respiration-deficient muscle fibers from patients with mitochondrial myopathy. Hum Mol Genet. 1994;3:13–9.
183. Skrzynia C, Berg JS, Willis MS, Jensen BC. Genetics and heart failure: a concise guide for the clinician. Curr Cardiol Rev. 2015;11:10–7.
184. Russell O, Turnbull D. Mitochondrial DNA disease – molecular insights and potential routes to a cure. Exp Cell Res. 2014;325:38–43.
185. Wang W, Karamanlidis G, Tian R. Novel targets for mitochondrial medicine. Sci Transl Med. 2016;8:326rv3.
186. Finsterer J, Bindu PS. Therapeutic strategies for mitochondrial disorders. Pediatr Neurol. 2015;52:302–13.
187. Scarpulla RC, Vega RB, Kelly DP. Transcriptional integration of mitochondrial biogenesis. Trends Endocrinol Metab TEM. 2012;23:459–66.
188. Scarpelli M, Todeschini A, Rinaldi F, Rota S, Padovani A, Filosto M. Strategies for treating mitochondrial disorders: an update. Mol Genet Metab. 2014;113:253–60.
189. Whitaker RM, Corum D, Beeson CC, Schnellmann RG. Mitochondrial biogenesis as a pharmacological target: a new approach to acute and chronic diseases. Annu Rev Pharmacol Toxicol. 2016;56:229.
190. Lee SR, Kim HK, Song IS, Youm J, Dizon LA, Jeong SH, Ko TH, Hye-Jin H, Ko KS, Rhee BD, Kim N, Han J. Glucocorticoids and their receptors: insights into specific roles in mitochondria. Prog Biophys Mol Biol. 2013;112:44–54.
191. De Stefano N, Matthews PM, Ford B, Genge A, Karpati G, Arnold DL. Short-term dichloroacetate treatment improves indices of cerebral metabolism in patients with mitochondrial disorders. Neurology. 1995;45:1193–8.
192. Taivassalo T, Gardner JL, Taylor RW, Schaefer AM, Newman J, Barron MJ, Haller RG, Turnbull DM. Endurance training and detraining in mitochondrial myopathies due to single large-scale mtDNA deletions. Brain. 2006;129:3391–401.
193. Lam BL, Feuer WJ, Schiffman JC, Porciatti V, Vandenbroucke R, Rosa PR, Gregori G, Guy J. Trial end points and natural history in patients with G11778A leber hereditary optic neuropathy: preparation for gene therapy clinical trial. JAMA Ophthalmol. 2014;132:428–36.

194. Gammage P, Van Haute L, Minczuk M. Engineered mtZFNs for manipulation of human mitochondrial DNA heteroplasmy. In: McKenzie M, editor. Mitochondrial DNA. New York: Springer; 2016. p. 145–62.
195. Dovydenko I, Tarassov I, Venyaminova A, Entelis N. Method of carrier-free delivery of therapeutic RNA importable into human mitochondria: lipophilic conjugates with cleavable bonds. Biomaterials. 2016;76:408–17.
196. Smith RA, Hartley RC, Murphy MP. Mitochondria-targeted small molecule therapeutics and probes. Antioxid Redox Signal. 2011;15:3021–38.
197. Clark KM, Bindoff LA, Lightowlers RN, Andrews RM, Griffiths PG, Johnson MA, Brierley EJ, Turnbull DM. Reversal of a mitochondrial DNA defect in human skeletal muscle. Nat Genet. 1997;16(3):222–4.
198. Safdar A, Khrapko K, Flynn JM, Saleem A, De Lisio M, Johnston AP, Kratysberg Y, Samjoo IA, Kitaoka Y, Ogborn DI, Little JP, Raha S, Parise G, Akhtar M, Hettinga BP, Rowe GC, Arany Z, Prolla TA, Tarnopolsky MA. Exercise-induced mitochondrial p53 repairs mtDNA mutations in mutator mice. Skelet Muscle. 2015;6:7.
199. Mitalipov S, Wolf DP. Clinical and ethical implications of mitochondrial gene transfer. Trends Endocrinol Metab. 2014;25:5–7.
200. McCully JD, Levitsky S, Del Nido PJ, Cowan DB. Mitochondrial transplantation for therapeutic use. Clin Transl Med. 2016;5:16.
201. Lee HC, Wei YH. Mitochondria and aging. Adv Exp Med Biol. 2012;942:311–27.
202. Patananan AN, Wu TH, Chiou PY, Teitell MA. Modifying the mitochondrial genome. Cell Metab. 2016;23:785–96.
203. Koopman WJ, Willems PH, Smeitink JA. Monogenic mitochondrial disorders. N Engl J Med. 2012;366:1132–41.

Chapter 6
Mitochondrial Function in Non-ischemic Heart Failure

Anisha A. Gupte and Dale J. Hamilton

Non-ischemic heart failure (HF) is a complex disorder arising from cardiac stressors such as pressure overload that are unrelated to coronary artery disease. The syndrome comprises of several features including metabolic abnormalities, energy deficit and concomitant myocardial contractile dysfunction. HF affects about 6–10% of the population worldwide, and the 5-year prognosis is >50% [1]. Non-ischemic HF is a broad syndrome comprising of various etiologies, differences in chronicity of the condition and many idiopathic etiologies that are never understood. However, the end-stage manifestation of impaired cardiac function is common and similar to ischemic HF. Current treatments of neurohormonal blockage or end-stage unloading have significantly improved quality of life but prognosis still remains poor, worse than that of some cancers. However, some of these current treatments have indicated to us that the failing myocardium although metabolically impaired, is capable of reinvigorating once provided with the appropriate metabolic conditions. It is therefore important to understand the metabolic alterations in the failing myocardium and the function of mitochondria that determine the major aspects of cardiac metabolism.

A.A. Gupte, PhD (✉)
Center for Metabolism and Bioenergetics Research, Houston Methodist Research Institute, Weill Cornell Medical College, Houston, TX, USA
e-mail: aagupte@houstonmethodist.org

D.J. Hamilton, MD, FACP
Center for Metabolism and Bioenergetics Research, Houston Methodist Research Institute, Weill Cornell Medical College, Houston, TX, USA

Houston Methodist, Department of Medicine, Houston, TX, USA
e-mail: djhamilton@houstonmethodist.org

G. Santulli (ed.), *Mitochondrial Dynamics in Cardiovascular Medicine*, Advances in Experimental Medicine and Biology 982,
DOI 10.1007/978-3-319-55330-6_6

Mitochondrial Function in Different Types of Non-ischemic Cardiomyopathy

Non-ischemic heart failure is categorized into four major types, all of which are unrelated to coronary artery disease: (1) Dilated cardiomyopathy (DCM), (2) Hypertrophic Cardiomyopathy, (3) Restrictive cardiomyopathy, (4) Arrhythmogenic Right Ventricular Dysplasia.

1. **Dilated cardiomyopathy (DCM)** DCM, the most common type of non-ischemic CM, is characterized by dilation of the left ventricle (LV) and eventual weakening of the LV wall. It may also be accompanied by fluid retention or congestive HF. DCM can result from heart valve diseases, chronic hypertension, viral infections, diabetes, and peripartum cardiomyopathy but most cases are diagnosed as idiopathic.

Mitochondrial dysfunction has been described in human DCM, stemming from multiple foci, including reduced specific activities of mitochondrial enzymes [2], mtDNA damage [2], mitochondria copy number [3] etc. It has been suggested that suppression of mitochondrial complex activities in DCM may be secondary to metabolic insults associated with DCM [4]. Diabetic cardiomyopathy is classified in this category of non-ischemic HF and has been shown to have mitochondrial dysfunction in both human and animal studies. In diabetic and healthy men, Diamant and colleagues [5] demonstrated diastolic dysfunction which was associated with a reduction in cardiac phosphocreatine/ATP (PCr/ATP) ratios. A similar reduction in cardiac PCr/ATP ratio has also been shown in diabetic patients with normal cardiac function [6], indicating that changes in mitochondrial function may precede impairment of cardiac function. Altered lipid metabolism, increased oxidative stress and altered mitochondrial structure has been reported the diabetic human myocardium, much of which is caused by the diabetic metabolic environment. Using permeabilized atrial cardiac myofibers, Anderson et al. demonstrated impaired mitochondrial respiratory capacity in patients with type 2 diabetes [7]. Even those with HbA1c as low as 6% showed mitochondrial dysfunction in cardiac myofibers, and there was an inverse relationship between HbA1C and mitochondrial respiratory capacity. Mitochondrial dysfunction was associated with greater H2O2 emission, depletion of glutathione, and increased levels of hydroxynonenal-modified proteins and 3-nitrotyrosine-modified proteins [8]. Further, a metabolic shift from fatty acid + glucose to high dependence on fatty acids as the primary fuel has been shown in obese and diabetic hearts [9]. Increased dependence on fatty acids has been postulated to be an underlying cause of increased oxidative stress, incomplete fatty acid oxidation, and thereby inefficient mitochondrial energy production [10].

2. **Hypertrophic cardiomyopathy (HCM)** is characterized by hypertrophy or thickening of the ventricular walls primarily at the septum, left ventricular stiffness, mitral valve insufficiency and disarrayed myocyte structural organization. HCM can be caused by gene mutations, chronic high blood pressure, aging or

unknown causes. Although most mutations associated with HCM are in genes coding for sarcomeric proteins, the PCr/ATP ratios are reduced in HCM both with pre-hypertrophic stage and established LV hypertrophy. This suggests bioenergetics deficiency as a cause of myocardial remodeling [11]. Patients with mutations in mtDNA can have similar cardiac phenotypes as HCM patients with sarcomeric protein mutations [11], further supportive of a mitochondrial role in HCM pathogenesis.

3. **Restrictive cardiomyopathy** and
4. **Arrhythmogenic Right Ventricular Dysplasia** are rare forms of cardiomyopathy characterized by extensive fibrosis. Very little is known about mitochondrial function of these classes of cardiomyopathy, and will not be discussed in this chapter.

Mitochondrial Function in Diastolic Heart Failure

Diastolic heart failure or heart failure with preserved ejection fraction (HFpEF) is characterized by extensive fibrosis, myocardial and myocyte stiffening, abnormal diastolic relaxation and left ventricular (LV) filling [12]. It is commonly seen in post-menopausal women and the elderly. Although it is as prevalent as HF with reduced EF (HFrEF), in contrast to HFrEF, there are no effective treatments for HFpEF. Mitochondrial function is undoubtedly important for myocardial contraction, but diastolic relaxation is also an active, ATP-dependent process that requires efficiently functioning mitochondria. Multiple studies have reiterated the fact that the creatine phosphate/ATP ratio, reflecting the balance of energy consumption and supply in the heart, is reduced in patients with HFpEF [13], hypertensive DD [14], diabetic patients with DD [6] and normotensive patients with DD but well-controlled diabetes [5]. In support, mitochondrial dysfunction has been implicated in DD in animal models [15]. Availability of ATP can influence the rate and extent of myocardial relaxation in the following ways: (1) ATP is required for function of the ion pumps (sarcoplasmic reticular Ca2+ ATPase) that transfer Ca2+ out of the cytosol during diastole; (2) ATP provides energy for the sarcolemmal Na-K-ATP pump required for achieving CM relaxation; (3) ATP interacts with actomyosin to cause its dissociation to actin and myosin, a step critical to relaxation [16]. ATP is indeed required for both contraction and relaxation, but when there is a small depletion in amounts of ATP, the actin-myosin complexes can still form due to the high affinity of the myosin head for ATP, but the dissociation of actin and myosin is reduced since it requires higher concentrations of ATP. Thus, smaller depletions in ATP have a greater impact on diastolic than systolic function, whereas larger depletions in ATP affect contraction and relaxation. Further, increases in cytosolic ADP levels in absence of any other myofilament function regulators are sufficient to slow cross-bridge cycling and cause impaired diastolic relaxation [17].

Differentiating Mitochondrial Function in Ischemic and Non-ischemic Heart Failure

While it is well known that the failing is energy deprived in both ischemic and non-ischemic etiologies of HF, it has been difficult to differentiate whether the mitochondrial bioenergetics in the two etiologies are affected similarly. Given the differences in the underlying pathologies- ischemic oxygen deprivation versus non-ischemic pressure overload, acute ischemic insult versus chronic non-ischemic stress, and more, it would be expected that the mitochondrial function in two etiologies would be widely different. However, distinguishing it has not been easy, primarily due to availability of animal models that mimic the co-morbidities and chronicity of metabolic stress that is seen in human non-ischemic HF, the difficulty in timely and sufficient procurement of human heart tissue for detailed mitochondrial analyses, the complicated data interpretation due to various co-morbidities associated with non-ischemic HF, and finally the end-stage view of human heart tissue samples that may be ultimately similar in mitochondrial function despite following different pathways.

However, some recent studies have shed light on the details of mitochondrial functional aberrations in non-ischemic and ischemic HFs. Park et al. [18] were the first to make a direct comparison of mitochondrial function between ischemic and non-ischemic HF in humans. Using non-failing hearts from rejected donors as controls, Park and colleagues found decreasing mitochondrial OXPHOS from control to non-ischemic to ischemic HF mitochondria when function was expressed per milligram of heart tissue [18]. However, when normalized to the citrate synthase activity, a measure of mitochondrial numbers, the difference between non-ischemic HF and ischemic HF mitochondria was negated. Therefore, more than differences in OXPHOS, the difference in mitochondrial content is more significant between the etiologies. The mitochondria from ischemic HF also had greater proton leak, attenuated OXPHOS efficiency, and greater levels of free radicals. However, the OXPHOS normalized to citrate synthase activity was less than control mitochondria for both non-ischemic HF and ischemic HF mitochondria. While comparing mitochondrial function from well-perfused and chronically ischemic regions of the same heart, Stride et al. [19] found that maximal oxidative phosphorylation capacity was diminished, levels of reactive oxygen species (ROS) were increased and expression of antioxidants were decreased in ischemic compared with non-ischemic myocardial regions, although the degree of mitochondrial coupling was the same. Lai et al. [20] performed myocardial transcriptomic and metabolomic profiling of compensated hypertrophy achieved by transverse aortic constriction (TAC) versus decompensated hypertrophy and heart failure achieved by TAC plus apical myocardial infarction. They found that expression of most mitochondrial genes did not differ between the two conditions, with the exception of decreasing fatty acid transport and oxidation genes in ischemic HF. Greater differences were seen in the metabolomics profiles than transcriptomic profiles of the two conditions, suggesting more post-translational differences that result in different metabolic phenotypes. Similar to our studies in human myocardial mitochondria [21],

Lai et al. [20] also suggest bottlenecks of carbon substrate flux into the Krebs cycle rather than overt mitochondrial failure in HF.

Changes in Myocardial Substrate Utilization in Non-ischemic HF

Most forms of non-ischemic HF are associated with substrate switching to greater glucose utilization [22–24]. These changes are driven by maladaptive changes of the myocardium such as altered enzymes and substrate flux, and not due to limited availability of fatty acid substrates, which may actually be in excess [25]. On the other hand, the diabetic heart may show greater dependence on fatty acids due to cardiac insulin resistance (Fig. 6.1). The shift towards glucose metabolism improves myocardial contractile efficiency by increasing the stoichiometric ratio of ATP production to oxygen consumption in addition to minimizing oxidative losses through mitochondrial respiratory chain uncoupling associated with free fatty acid (FFA) metabolism [26]. This adaptation may have greater relevance in ischemic HF where oxygen supply is limited. Impaired ability of glucose uptake or utilization is

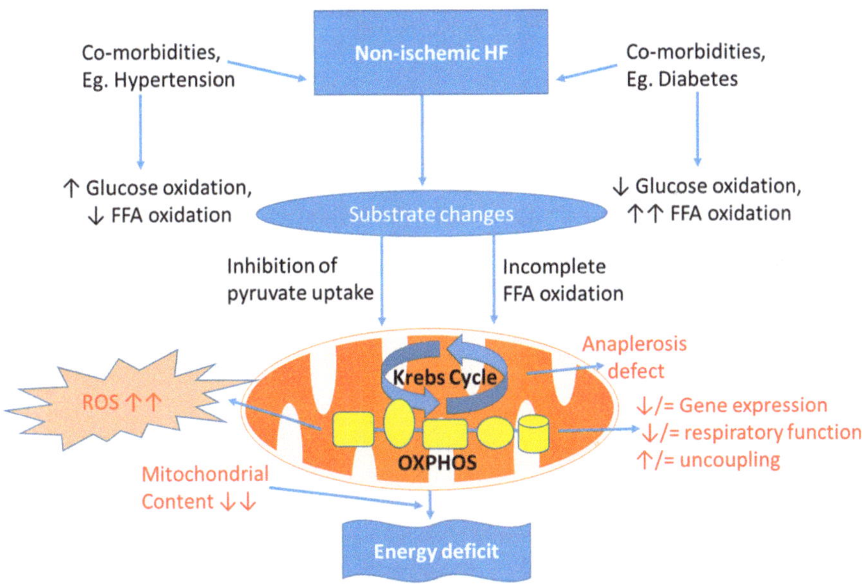

Fig. 6.1 Nodes at which mitochondrial function is impacted in non-ischemic heart failure. Non-ischemic HF and the associated co-morbidities have varied impacts on the substrate preference at the level of cellular uptake and mitochondrial oxidation. While there may or may not be gross mitochondrial OXPHOS limitations in non-ischemic HF, the combination of altered substrates, reduced mitochondrial content and altered mitochondrial flux together result in significant cardiac energy depletion

detrimental to myocardial function as shown in animal studies of GLUT4 deletion [27]. With progression of HF, a compensatory hyperadrenergic state leads to elevation of blood free fatty acid levels and subsequent onset of insulin resistance (see Chap. 25). Increased dependence on fatty acid metabolism increases oxygen consumption, decreased cardiac efficiency and feeds into a vicious cycle of metabolic impairment. Despite increased fatty acid oxidation, the myocardium remains energy depleted and excessive ROS production associated with FFA oxidation may partly underlie mitochondrial dysfunction in HF. Also, FFA oxidation is not complete in the failing myocardium, leading to incomplete FFA oxidation and changes in levels of intermediates of FFA oxidation such as acylcarnitines that further impair mitochondrial function [10, 21]. Finally a decline in utilization of both fatty acids and glucose in invariably observed with further progression of HF, with increased myocardial oxygen consumption but enhanced energy depletion in the myocardium.

Mitochondrial Function Determination in Animal Models of Non-ischemic Heart Failure

Transverse Aortic Constriction (TAC) is the most common animal model of heart failure, in which a fixed aortic constriction causes an abrupt increase in LV afterload, acute hemodynamic instability with reduced EF and early post-operative mortality [28]. This technique allows the study of effects of pressure overload to isolated to the heart, independent of systemic changes. However, the lack of slow progressive pressure overload and absence of systemic metabolic stressors commonly seen in non-ischemic HF, reduces its relevance to humans.

The myocardium consists of two populations of mitochondria, classified based on location: sub-sarcolemmal mitochondria (SSM) and intra-myofibrillar mitochondria (IFM). Previous studies [29, 30] have indicated a role for SSMs in ATP production for non-contractile, supportive cellular functions such as protein synthesis etc., whereas IFMs are the subpopulation of mitochondria that provides ATP for contraction of the myofibrils. To get a complete understanding of the bioenergetics of the failing myocardium, it is therefore important to determine the function of both SSM and IFM. Pressure overload differentially affects respiratory capacity of IFM and SSM in a rat model of TAC [30]. At the basal level, mitochondrial state 3 levels were found to be 50% higher in the IFM than the SSM, whereas ADP:O ratio and ADP-limited state respiration were comparable in both populations. However, TAC significantly impaired respiratory rates in both IFM and SSM, but only to a small extent in the SSM. Thus, after 20 weeks of pressure overload, no differences were seen in the respiratory levels of IFM and SSM. Pressure overload also decreased mitochondrial content as demonstrated by reduced total citrate synthase activity. These studies confirm the effects of pressure overload on cardiac mitochondria and also indicate that the effects are differential among the two subpopulations. Interestingly, electron microscopy showed that the cardiac IFM were smaller in size than SSMs in rats, and FACS analyses showed that IFM had less internal structural complexity than SSM [30].

The role and differential effects of HF on the mitochondrial subpopulations has been controversial, largely due to the isolation techniques and methods of respiration assessment. It is also evident that different co-morbidities associated with the non-ischemic HF may have different effects on the function of the subpopulations of cardiac mitochondria. For instance, type 2 diabetes, calcium stress, volume overload in the heart primarily affect the respiratory capacity of the SSM and not that of the IFM [31–33], whereas type 1 diabetes and cardiomyopathy affect only IFM [34, 35], and ischemia and aging affect both populations with a greater influence of ischemia on SSM [36], and of aging on IFM [37]. One disadvantage of studying isolated IFM or SSM has been due to the differential centrifugation and enzyme treatment process used in the isolation. It is likely that these processes selectively isolate structurally intact mitochondria that are not damaged in HF [38]. To circumvent this problem, many studies use permeabilized fibers or skinned fibers assuming assessment of function of both populations [39, 40]. However, this technique may preferentially assess SSM due to their peripheral location right below the cell membrane whereas the IFM may be readily accessible for substrates due to the greater diffusion distance [41]. There is also a significant increase in production of ROS by mitochondria in non-ischemic HF [42]. Consequently, scavenging mitochondrial ROS by mitochondrial targeted catalase in mice with TAC-induced heart failure, showed reversal of most of the mitochondrial proteome changes in heart failure. The mtCatalase also induced a subset of unique changes that represent processes adaptive to increased workload and metabolic demands of pressure overload which included preservation of fatty acid metabolism, attenuation of apoptosis etc. [43]. However, while implementing pressure overload can induce compensated hypertrophy, it is thought that it may not necessarily result in HF. Thus, Faerber et al. tested a minimally invasive TAC through a ministernotomy without intubation to show induction of hypertrophy, HF symptoms in 57% of the mice undergoing the surgery, accompanied by reduced oxidative capacity and suppression of PGC1a expression [44].

Non-surgical Mouse Model of Non-ischemic Heart Failure to Study Mitochondrial Function

Most of our understanding of the mitochondrial mechanisms underlying HF comes primarily from animal models of acute, ischemic HF induced by coronary artery ligation, intermittent ischemia-reperfusion, pressure overload hypertrophy induced by TAC and genetically engineered models of cardiomyopathy. Recently, our group characterized the transcriptomic and mitochondrial functional profile of mice treated with a combination of the hypertensive agent Angiotensin II and a nitric oxide inhibitor L-N^G-nitroarginine methyl ester (L-NAME) [42]. This model is independent of any surgical trauma and develops heart failure (Ejection Fraction, EF <40%) in 4 weeks with the combination treatment. Either treatment alone did not result in the same severity of heart failure. The mice with L-NAME_+ AngII treatment served as a good mouse model of non-ischemic HF induced by pressure overload to study detailed changes in cardiac mitochondrial function.

The SSM from the myocardium of the non-ischemic HF mice with AngII + L-NAME treatment showed markedly reduced state 3 rates in response to pyruvate-malate (PM) and palmitoyl-carnitine-malate (PCM) substrates [42]. There was an overall reduction in Respiratory Control Ratios (a measurement of mitochondrial efficiency) in response to PM, PCM and glutamate-malate (GM). However, SSM function was unchanged between control and HF mice for the complex II substrate succinate for both SSM and IFM. Although the IFM from HF mice showed reduced state 3 respiratory rates compared with control mice for PM, PCM and GM, the overall RCR was not significantly different, likely to due to compensatory changes in the basal state 2 respiratory rates. Therefore, although the heart is energy-deficient in these HF mice, the greatest suppression of mitochondrial activity is seen in SSM mitochondria which are believed to have a more supportive role rather than the IFM which are thought to have a direct impact on contraction. However, permeabilized cardiac fibers which consist of a combination of IFMs and SSMs showed decreased mitochondrial function in HF mice than control mice [42], confirming an overall compromise of mitochondrial function that contributes to energy depletion in these mice. However, these data also showed that there wasn't a dramatic inhibition of all mitochondrial function, since once the isolated mitochondria were provided with the appropriate buffers and abundant substrates ex vivo, their function was moderately reduced compared with control mice, but not completely blocked.

Thus, the dramatic reduction in contractility of the heart in non-ischemic HF cannot be explained by small changes in mitochondrial function measured in isolated mitochondria or fibers. It is therefore likely that the energy deficit stems from a combination of direct mitochondrial deficiencies and impairment of upstream processes that lead to formation of intermediates in the Krebs cycle and NADH or FADH2 into the electron transport chain. The decrease in overall cardiac bioenergetics in the failing non-ischemic myocardium is also due to reduced mitochondrial content, which has been reported by several groups [18, 19, 45]. This is further accompanied by greater ROS production and overall increase in oxidative stress in the myocardium, which in most part is attributed to the inefficient mitochondrial function [43]. This is supported by the fact that mice with mitochondrial targeted catalase but not peroxisomal catalase are protected from AngII-induced oxidative stress and cardiomyopathy [46]. But oxidative stress is also contributed by impaired anti-oxidant mechanisms, however it is thought that dysfunctional mitochondria are the initial stimulus to initiate the vicious cycle of excessive ROS [47]. It is also likely that in comparison with ischemic HF, the chronicity of non-ischemic HF allows adaptation of the mitochondria to the changing metabolic milieu. In support, many pathways leading to substrates for mitochondria have been shown to be down-regulated in non-ischemic HF. For instance, fatty acid metabolism, including fatty acid oxidation has been shown to reduce in non-ischemic HF independent of changes in OXPHOS genes or function of the mitochondria.

Human Studies in Non-ischemic HF

The understanding of the mechanisms and consequences mitochondrial functional changes in HF would form the basis for innovative personalized metabolic recovery therapies for human HF. Stride and colleagues reported markedly diminished OXPHOS capacity in human HF [45]. Other investigators reported similar findings [7, 48]. However these studies included respiratory assessment of mitochondrial function on atrial or atrial appendage samples rather than ventricular wall specimens. In the study by Sharov et al. [49], investigators procured sub-endocardial and sub-epicardial samples from explanted hearts and examined skinned fibers which were saponinized and studied with a Clark electrode in an oxygraph cell. They demonstrated no difference between control and HF samples in basal respiration rates but showed depressed state 3 ADP stimulated respiration in HF samples. Coupled respiration however was present [49]. Our studies of mitochondrial isolates from left ventricular wall tissue demonstrated persistence of a population of mitochondria that were capable of highly coupled oxidative phosphorylation (OXPHOS) [50, 51]. No major differences were noted in mitochondrial function for end-stage non-ischemic and ischemic HF. However, metabolomics and gene expression studies showed an overall marked down-regulation of the myocardial bioenergetics in human HF [21]. In support, some investigators have documented functional mitochondria in failing myocardial tissue but with reduced OXPHOS capacity due to lower mitochondrial content [3, 18]. Although reduced in number and capable of function these mitochondrial are not meeting energy requirements for the failing heart. The fact remains that the failing myocardium is energy deprived [52] and PCr/ATP ratios progressively decreased as HF severity increases. However, as noted by a letter commentary to a review of the topic, it is not clear as to whether there is a fuel supply or energy transfer problem [53, 54]. What is becoming clear is that there remains a population of mitochondria capable of coupled respiration and therefore there is potential for recovery. Using pre- and post- paired samples, it has been shown that mechanical unloading with left ventricular assist devices results in metabolic recovery [21]. It is thus likely, that the energy deficit is initiated in mechanisms upstream of the mitochondria, perhaps at the levels of substrate oxidation/uptake and the overall number of mitochondria, rather than the biochemical properties of the mitochondria themselves.

Mechanistic understandings of the source of energy depletion could lead to therapeutic metabolic interventions. As example metabolomic studies from our laboratory revealed impaired pyruvate handling and resulting in an anaplerosis problem [21]. In ventricular wall tissue from failing human hearts, pyruvate concentrations were elevated while downstream Krebs cycle intermediates citrate, succinate, fumarate and malate were depleted. This in turn results in reduced delivery of reduced compounds NADH and FADH to complex I and succinate to complex II of the mitochondrial electron transport chain. The mitochondrial isolates from the same tissue were coupled and intact [50]. Mechanical unloading with a left ventricular device reversed the problem [21]. The tissue pyruvate concentration on pre- and post-LVAD paired samples declined and the Krebs intermediate increased. Perhaps then with mechanistic understandings, we can hope to improve substrate processing and energy transfer to fuel the failing heart.

Therapeutic Implications of Understanding Mitochondrial Function in Non-ischemic HF

The rapidly increasing prevalence and limited success of conventional treatments urgently calls for novel treatment strategies. Given the large metabolic changes in non-ischemic HF, metabolism-targeted therapeutic strategies are very attractive. We and other have shown that the failing myocardium although metabolically impacted, remains viable with the possibility of reversal of metabolic function [21, 55]. Contractile failure is largely a manifestation of adaptive responses of viable myocytes to various forms of non-lethal stressors which is reversible either upon the removal of stressors or reversing the mal-adaptive changes [55]. Treatments such as left ventricular assist devices have shown improvement in mitochondrial and metabolic function in the failing myocardium [21]. However, it might be important to apply these metabolism-targeted therapies before attaining end-stage HF which may be more difficult to reverse completely at the level of contractile function. The following important areas of potential metabolism-targeted therapies have the greatest likelihood of success: enhancing mitochondrial biogenesis, decreasing mitochondrial oxidative stress, rescuing iron and calcium homeostasis, reversing changes in substrate preferences [55, 56]. Strategies to enhance mitochondrial function may include activation of the master regulator of mitochondrial biogenesis Peroxisome proliferator-activated receptor gamma coactivator 1-alpha (PGC1α) [57], activation of adenosine monophosphate-activated kinase (AMPK) which drives mitochondrial biogenesis in exercise-induced non-pathological hypertrophy [58], activation of the nitric oxide pathway [59], increasing estrogen receptor signaling [60]. To improve mitochondrial oxidative stress status, direct targeting of ROS inside the mitochondria rather than systemically would have greater benefits, since results from large scale clinical trials of systemic antioxidants such as Vitamin E were disappointing [61, 62]. Mitochondrial ROS scavengers or inhibitors of ROS production such as MitoQ [63], Szeto-Schiller peptides [64] and manganese superoxide dismutase [65] or mitochondrial catalase mimetics [43] have shown potential of cardiac protection in animal models of non-ischemic HF. Selective chelation of mitochondrial free iron can be achieved with compounds such as deferiprone, with some benefits reported in improvement of mitochondrial function in non-cardiac conditions of mitochondrial dysfunction [66]. Another metabolic approach to improve cardiac function in non-ischemic HF will be to target the substrate changes. Enhancing glucose utilization and lowering fatty acid utilization in HF has been shown to be beneficial in HF progression. Inhibitors of fatty acid metabolism such as CPT1 inhibitors Etomoxir and Perhexiline have shown limited but promising effects in humans [67, 68]. Other inhibitors of fatty acid oxidation such as Dichloroacetate, Ranolazine, trimetazidine, and activators of the metabolism regulator AMPK such as AICAR may be considered as potential therapies. Taken together, detailed studies of mitochondrial function in non-ischemic HF myocardium present a strong potential in designing personalized medicine based on the degree of mitochondrial function, mitochondrial content and substrate changes.

References

1. Writing Group M, Lloyd-Jones D, Adams RJ, Brown TM, Carnethon M, Dai S, De Simone G, Ferguson TB, Ford E, Furie K, Gillespie C, Go A, Greenlund K, Haase N, Hailpern S, Ho PM, Howard V, Kissela B, Kittner S, Lackland D, Lisabeth L, Marelli A, McDermott MM, Meigs J, Mozaffarian D, Mussolino M, Nichol G, Roger VL, Rosamond W, Sacco R, Sorlie P, Roger VL, Thom T, Wasserthiel-Smoller S, Wong ND, Wylie-Rosett J, American Heart Association Statistics C, Stroke Statistics S. Heart disease and stroke statistics – 2010 update: a report from the American Heart Association. Circulation. 2010;121:e46–e215.
2. Marin-Garcia J, Goldenthal MJ, Pierpont ME, Ananthakrishnan R. Impaired mitochondrial function in idiopathic dilated cardiomyopathy: biochemical and molecular analysis. J Card Fail. 1995;1:285–91.
3. Ahuja P, Wanagat J, Wang Z, Wang Y, Liem DA, Ping P, Antoshechkin IA, Margulies KB, Maclellan WR. Divergent mitochondrial biogenesis responses in human cardiomyopathy. Circulation. 2013;127:1957–67.
4. Jarreta D, Orus J, Barrientos A, Miro O, Roig E, Heras M, Moraes CT, Cardellach F, Casademont J. Mitochondrial function in heart muscle from patients with idiopathic dilated cardiomyopathy. Cardiovasc Res. 2000;45:860–5.
5. Diamant M, Lamb HJ, Groeneveld Y, Endert EL, Smit JW, Bax JJ, Romijn JA, de Roos A, Radder JK. Diastolic dysfunction is associated with altered myocardial metabolism in asymptomatic normotensive patients with well-controlled type 2 diabetes mellitus. J Am Coll Cardiol. 2003;42:328–35.
6. Scheuermann-Freestone M, Madsen PL, Manners D, Blamire AM, Buckingham RE, Styles P, Radda GK, Neubauer S, Clarke K. Abnormal cardiac and skeletal muscle energy metabolism in patients with type 2 diabetes. Circulation. 2003;107:3040–6.
7. Anderson EJ, Kypson AP, Rodriguez E, Anderson CA, Lehr EJ, Neufer PD. Substrate-specific derangements in mitochondrial metabolism and redox balance in the atrium of the type 2 diabetic human heart. J Am Coll Cardiol. 2009;54:1891–8.
8. Anderson EJ, Lustig ME, Boyle KE, Woodlief TL, Kane DA, Lin CT, Price 3rd JW, Kang L, Rabinovitch PS, Szeto HH, Houmard JA, Cortright RN, Wasserman DH, Neufer PD. Mitochondrial H2O2 emission and cellular redox state link excess fat intake to insulin resistance in both rodents and humans. J Clin Invest. 2009;119:573–81.
9. Lopaschuk GD, Ussher JR, Folmes CD, Jaswal JS, Stanley WC. Myocardial fatty acid metabolism in health and disease. Physiol Rev. 2010;90:207–58.
10. Koves TR, Ussher JR, Noland RC, Slentz D, Mosedale M, Ilkayeva O, Bain J, Stevens R, Dyck JR, Newgard CB, Lopaschuk GD, Muoio DM. Mitochondrial overload and incomplete fatty acid oxidation contribute to skeletal muscle insulin resistance. Cell Metab. 2008;7:45–56.
11. Vakrou S, Abraham MR. Hypertrophic cardiomyopathy: a heart in need of an energy bar? Front Physiol. 2014;5:309.
12. Loffredo FS, Nikolova AP, Pancoast JR, Lee RT. Heart failure with preserved ejection fraction: molecular pathways of the aging myocardium. Circ Res. 2014;115:97–107.
13. Phan TT, Abozguia K, Nallur Shivu G, Mahadevan G, Ahmed I, Williams L, Dwivedi G, Patel K, Steendijk P, Ashrafian H, Henning A, Frenneaux M. Heart failure with preserved ejection fraction is characterized by dynamic impairment of active relaxation and contraction of the left ventricle on exercise and associated with myocardial energy deficiency. J Am Coll Cardiol. 2009;54:402–9.

14. Lamb HJ, Beyerbacht HP, van der Laarse A, Stoel BC, Doornbos J, van der Wall EE, de Roos A. Diastolic dysfunction in hypertensive heart disease is associated with altered myocardial metabolism. Circulation. 1999;99:2261–7.
15. Flarsheim CE, Grupp IL, Matlib MA. Mitochondrial dysfunction accompanies diastolic dysfunction in diabetic rat heart. Am J Phys. 1996;271:H192–202.
16. Morgan JP. Cellular mechanisms of diastolic dysfunction. UpToDate; Wolters Kluwer. 2013.
17. Ingwall JS, Weiss RG. Is the failing heart energy starved? On using chemical energy to support cardiac function. Circ Res. 2004;95:135–45.
18. Park SY, Trinity JD, Gifford JR, Diakos NA, McCreath L, Drakos S, Richardson RS. Mitochondrial function in heart failure: the impact of ischemic and non-ischemic etiology. Int J Cardiol. 2016;220:711–7.
19. Stride N, Larsen S, Hey-Mogensen M, Hansen CN, Prats C, Steinbruchel D, Kober L, Dela F. Impaired mitochondrial function in chronically ischemic human heart. Am J Physiol Heart Circ Physiol. 2013;304:H1407–14.
20. Lai L, Leone TC, Keller MP, Martin OJ, Broman AT, Nigro J, Kapoor K, Koves TR, Stevens R, Ilkayeva OR, Vega RB, Attie AD, Muoio DM, Kelly DP. Energy metabolic reprogramming in the hypertrophied and early stage failing heart: a multisystems approach. Circ Heart Fail. 2014;7:1022–31.
21. Gupte AA, Hamilton DJ, Cordero-Reyes AM, Youker KA, Yin Z, Estep JD, Stevens RD, Wenner B, Ilkayeva O, Loebe M, Peterson LE, Lyon CJ, Wong ST, Newgard CB, Torre-Amione G, Taegtmeyer H, Hsueh WA. Mechanical unloading promotes myocardial energy recovery in human heart failure. Circ Cardiovasc Genet. 2014;7:266–76.
22. Tuunanen H, Engblom E, Naum A, Scheinin M, Nagren K, Airaksinen J, Nuutila P, Iozzo P, Ukkonen H, Knuuti J. Decreased myocardial free fatty acid uptake in patients with idiopathic dilated cardiomyopathy: evidence of relationship with insulin resistance and left ventricular dysfunction. J Card Fail. 2006;12:644–52.
23. Razeghi P, Young ME, Alcorn JL, Moravec CS, Frazier OH, Taegtmeyer H. Metabolic gene expression in fetal and failing human heart. Circulation. 2001;104:2923–31.
24. Taegtmeyer H, Sen S, Vela D. Return to the fetal gene program: a suggested metabolic link to gene expression in the heart. Ann N Y Acad Sci. 2010;1188:191–8.
25. Rajabi M, Kassiotis C, Razeghi P, Taegtmeyer H. Return to the fetal gene program protects the stressed heart: a strong hypothesis. Heart Fail Rev. 2007;12:331–43.
26. Morrow DA, Givertz MM. Modulation of myocardial energetics: emerging evidence for a therapeutic target in cardiovascular disease. Circulation. 2005;112:3218–21.
27. Domenighetti AA, Danes VR, Curl CL, Favaloro JM, Proietto J, Delbridge LM. Targeted GLUT-4 deficiency in the heart induces cardiomyocyte hypertrophy and impaired contractility linked with Ca(2+) and proton flux dysregulation. J Mol Cell Cardiol. 2010;48:663–72.
28. Houser SR, Margulies KB, Murphy AM, Spinale FG, Francis GS, Prabhu SD, Rockman HA, Kass DA, Molkentin JD, Sussman MA, Koch WJ, American Heart Association Council on Basic Cardiovascular Sciences CoCC, Council on Functional G, Translational B. Animal models of heart failure: a scientific statement from the American Heart Association. Circ Res. 2012;111:131–50.
29. Riva A, Tandler B, Loffredo F, Vazquez E, Hoppel C. Structural differences in two biochemically defined populations of cardiac mitochondria. Am J Physiol Heart Circ Physiol. 2005;289:H868–72.
30. Schwarzer M, Schrepper A, Amorim PA, Osterholt M, Doenst T. Pressure overload differentially affects respiratory capacity in interfibrillar and subsarcolemmal mitochondria. Am J Physiol Heart Circ Physiol. 2013;304:H529–37.
31. Dabkowski ER, Baseler WA, Williamson CL, Powell M, Razunguzwa TT, Frisbee JC, Hollander JM. Mitochondrial dysfunction in the type 2 diabetic heart is associated with alterations in spatially distinct mitochondrial proteomes. Am J Physiol Heart Circ Physiol. 2010;299:H529–40.

32. Holmuhamedov EL, Oberlin A, Short K, Terzic A, Jahangir A. Cardiac subsarcolemmal and interfibrillar mitochondria display distinct responsiveness to protection by diazoxide. PLoS One. 2012;7:e44667.
33. Ulasova E, Gladden JD, Chen Y, Zheng J, Pat B, Bradley W, Powell P, Zmijewski JW, Zelickson BR, Ballinger SW, Darley-Usmar V, Dell'italia LJ. Loss of interstitial collagen causes structural and functional alterations of cardiomyocyte subsarcolemmal mitochondria in acute volume overload. J Mol Cell Cardiol. 2011;50:147–56.
34. Dabkowski ER, Williamson CL, Bukowski VC, Chapman RS, Leonard SS, Peer CJ, Callery PS, Hollander JM. Diabetic cardiomyopathy-associated dysfunction in spatially distinct mitochondrial subpopulations. Am J Physiol Heart Circ Physiol. 2009;296:H359–69.
35. Hoppel CL, Tandler B, Fujioka H, Riva A. Dynamic organization of mitochondria in human heart and in myocardial disease. Int J Biochem Cell Biol. 2009;41:1949–56.
36. Kajiyama K, Pauly DF, Hughes H, Yoon SB, Entman ML, McMillin-Wood JB. Protection by verapamil of mitochondrial glutathione equilibrium and phospholipid changes during reperfusion of ischemic canine myocardium. Circ Res. 1987;61:301–10.
37. Fannin SW, Lesnefsky EJ, Slabe TJ, Hassan MO, Hoppel CL. Aging selectively decreases oxidative capacity in rat heart interfibrillar mitochondria. Arch Biochem Biophys. 1999;372:399–407.
38. Chemnitius JM, Manglitz T, Kloeppel M, Doenst T, Schwartz P, Kreuzer H, Zech R. Rapid preparation of subsarcolemmal and interfibrillar mitochondrial subpopulations from cardiac muscle. Int J Biochem. 1993;25:589–96.
39. de Waard MC, van der Velden J, Bito V, Ozdemir S, Biesmans L, Boontje NM, Dekkers DH, Schoonderwoerd K, Schuurbiers HC, de Crom R, Stienen GJ, Sipido KR, Lamers JM, Duncker DJ. Early exercise training normalizes myofilament function and attenuates left ventricular pump dysfunction in mice with a large myocardial infarction. Circ Res. 2007;100:1079–88.
40. Boudina S, Sena S, Theobald H, Sheng X, Wright JJ, Hu XX, Aziz S, Johnson JI, Bugger H, Zaha VG, Abel ED. Mitochondrial energetics in the heart in obesity-related diabetes: direct evidence for increased uncoupled respiration and activation of uncoupling proteins. Diabetes. 2007;56:2457–66.
41. Sperl W, Skladal D, Gnaiger E, Wyss M, Mayr U, Hager J, Gellerich FN. High resolution respirometry of permeabilized skeletal muscle fibers in the diagnosis of neuromuscular disorders. Mol Cell Biochem. 1997;174:71–8.
42. Hamilton DJ, Zhang A, Li S, Cao TN, Smith JA, Vedula I, Cordero-Reyes AM, Youker KA, Torre-Amione G, Gupte AA. Combination of angiotensin II and l-NG-nitroarginine methyl ester exacerbates mitochondrial dysfunction and oxidative stress to cause heart failure. Am J Physiol Heart Circ Physiol. 2016;310:H667–80.
43. Dai DF, Hsieh EJ, Liu Y, Chen T, Beyer RP, Chin MT, MacCoss MJ, Rabinovitch PS. Mitochondrial proteome remodelling in pressure overload-induced heart failure: the role of mitochondrial oxidative stress. Cardiovasc Res. 2012;93:79–88.
44. Faerber G, Barreto-Perreia F, Schoepe M, Gilsbach R, Schrepper A, Schwarzer M, Mohr FW, Hein L, Doenst T. Induction of heart failure by minimally invasive aortic constriction in mice: reduced peroxisome proliferator-activated receptor gamma coactivator levels and mitochondrial dysfunction. J Thorac Cardiovasc Surg. 2011;141:492–500, 500 e1.
45. Stride N, Larsen S, Hey-Mogensen M, Sander K, Lund JT, Gustafsson F, Kober L, Dela F. Decreased mitochondrial oxidative phosphorylation capacity in the human heart with left ventricular systolic dysfunction. Eur J Heart Fail. 2013;15:150–7.
46. Wanagat J, Dai DF, Rabinovitch P. Mitochondrial oxidative stress and mammalian healthspan. Mech Ageing Dev. 2010;131:527–35.
47. White CR, Brock TA, Chang LY, Crapo J, Briscoe P, Ku D, Bradley WA, Gianturco SH, Gore J, Freeman BA, et al. Superoxide and peroxynitrite in atherosclerosis. Proc Natl Acad Sci U S A. 1994;91:1044–8.
48. Lemieux H, Semsroth S, Antretter H, Hofer D, Gnaiger E. Mitochondrial respiratory control and early defects of oxidative phosphorylation in the failing human heart. Int J Biochem Cell Biol. 2011;43:1729–38.

49. Sharov VG, Todor AV, Silverman N, Goldstein S, Sabbah HN. Abnormal mitochondrial respiration in failed human myocardium. J Mol Cell Cardiol. 2000;32:2361–7.
50. Cordero-Reyes AM, Gupte AA, Youker KA, Loebe M, Hsueh WA, Torre-Amione G, Taegtmeyer H, Hamilton DJ. Freshly isolated mitochondria from failing human hearts exhibit preserved respiratory function. J Mol Cell Cardiol. 2014;68:98–105.
51. Gupte AA, Cordero-Reyes AM, Youker KA, Matsunami RK, Engler DA, Li S, Loebe M, Ashrith G, Torre-Amione G, Hamilton DJ. Differential mitochondrial function in remodeled right and nonremodeled left ventricles in pulmonary hypertension. J Card Fail. 2016;22:73–81.
52. Neubauer S, Horn M, Cramer M, Harre K, Newell JB, Peters W, Pabst T, Ertl G, Hahn D, Ingwall JS, Kochsiek K. Myocardial phosphocreatine-to-ATP ratio is a predictor of mortality in patients with dilated cardiomyopathy. Circulation. 1997;96:2190–6.
53. Neubauer S. The failing heart – an engine out of fuel. N Engl J Med. 2007;356:1140–51.
54. Taegtmeyer H. The failing heart. N Engl J Med. 2007;356:2545–6; author reply 2546.
55. Ardehali H, Sabbah HN, Burke MA, Sarma S, Liu PP, Cleland JG, Maggioni A, Fonarow GC, Abel ED, Campia U, Gheorghiade M. Targeting myocardial substrate metabolism in heart failure: potential for new therapies. Eur J Heart Fail. 2012;14:120–9.
56. Santulli G, Xie W, Reiken SR, Marks AR. Mitochondrial calcium overload is a key determinant in heart failure. Proc Natl Acad Sci USA. 2015;112:11389–94.
57. Sihag S, Cresci S, Li AY, Sucharov CC, Lehman JJ. PGC-1alpha and ERRalpha target gene downregulation is a signature of the failing human heart. J Mol Cell Cardiol. 2009;46: 201–12.
58. Maillet M, van Berlo JH, Molkentin JD. Molecular basis of physiological heart growth: fundamental concepts and new players. Nat Rev Mol Cell Biol. 2013;14:38–48.
59. Clementi E, Nisoli E. Nitric oxide and mitochondrial biogenesis: a key to long-term regulation of cellular metabolism. Comp Biochem Physiol A Mol Integr Physiol. 2005;142:102–10.
60. Mattingly KA, Ivanova MM, Riggs KA, Wickramasinghe NS, Barch MJ, Klinge CM. Estradiol stimulates transcription of nuclear respiratory factor-1 and increases mitochondrial biogenesis. Mol Endocrinol. 2008;22:609–22.
61. Steinhubl SR. Why have antioxidants failed in clinical trials? Am J Cardiol. 2008;101: 14D–9D.
62. Salonen JT, Nyyssonen K, Salonen R, Lakka HM, Kaikkonen J, Porkkala-Sarataho E, Voutilainen S, Lakka TA, Rissanen T, Leskinen L, Tuomainen TP, Valkonen VP, Ristonmaa U, Poulsen HE. Antioxidant Supplementation in Atherosclerosis Prevention (ASAP) study: a randomized trial of the effect of vitamins E and C on 3-year progression of carotid atherosclerosis. J Intern Med. 2000;248:377–86.
63. Graham D, Huynh NN, Hamilton CA, Beattie E, Smith RA, Cocheme HM, Murphy MP, Dominiczak AF. Mitochondria-targeted antioxidant MitoQ10 improves endothelial function and attenuates cardiac hypertrophy. Hypertension. 2009;54:322–8.
64. Dai DF, Chen T, Szeto H, Nieves-Cintron M, Kutyavin V, Santana LF, Rabinovitch PS. Mitochondrial targeted antioxidant peptide ameliorates hypertensive cardiomyopathy. J Am Coll Cardiol. 2011;58:73–82.
65. Omar BA, McCord JM. Interstitial equilibration of superoxide dismutase correlates with its protective effect in the isolated rabbit heart. J Mol Cell Cardiol. 1991;23:149–59.
66. Boddaert N, Le Quan Sang KH, Rotig A, Leroy-Willig A, Gallet S, Brunelle F, Sidi D, Thalabard JC, Munnich A, Cabantchik ZI. Selective iron chelation in Friedreich ataxia: biologic and clinical implications. Blood. 2007;110:401–8.
67. Schmidt-Schweda S, Holubarsch C. First clinical trial with etomoxir in patients with chronic congestive heart failure. Clin Sci (Lond). 2000;99:27–35.
68. Lee L, Campbell R, Scheuermann-Freestone M, Taylor R, Gunaruwan P, Williams L, Ashrafian H, Horowitz J, Fraser AG, Clarke K, Frenneaux M. Metabolic modulation with perhexiline in chronic heart failure: a randomized, controlled trial of short-term use of a novel treatment. Circulation. 2005;112:3280–8.

Chapter 7
Mitochondria in Ischemic Heart Disease

L. Maximilian Buja

Ischemic heart disease (IHD) is a leading cause of morbidity and mortality worldwide. IHD generally results from pathological interaction between diseased coronary arteries and the myocardium that requires abundant blood flow. Imbalance in coronary blood flow and myocardial demand leads clinically to the development of an acute coronary syndrome (ACS) [1]. At a fundamental level, IHD involves altered biology of the cardiomyocyte (CMC). Increasing evidence has focused on the mitochondrion as a key organelle in the response of the CMC to ischemic injury [2]. This evidence is reviewed in this chapter in the context of the overall pathobiology of IHD.

Basic Concepts of Ischemic Heart Disease

Thrombotic occlusion of an atherosclerotic coronary artery leads to ischemia to the subtended segment of myocardium due to marked reduction in oxygen and metabolic substrate delivery [1]. The ischemic myocardium is subject to complex metabolic and functional changes that lead to progressive injury to CMC and the microvasculature. Contractile function is lost within seconds. Ischemic CMC can become electrically unstable and become the source of an ectopic focus which can lead to ventricular fibrillation and sudden death. With profound ischemia, CMC become progressively impaired and transition from reversible to irreversible injury by 15–20 min. Irreversible injury then progresses in a wavefront pattern extending out from the ischemic subendocardium to the subepicardium leading to a complete myocardial infarct after about 3 h [3–5].

L. Maximilian Buja, MD (✉)
McGovern Medical School, The University of Texas Health Science Center at Houston (UTHealth), Houston, TX, USA
e-mail: l.maximilian.buja@uth.tmc.edu

G. Santulli (ed.), *Mitochondrial Dynamics in Cardiovascular Medicine*,
Advances in Experimental Medicine and Biology 982,
DOI 10.1007/978-3-319-55330-6_7

Some processes can profoundly alter the response of the myocardium during evolving myocardial infarction [6–11]. If instituted in a timely manner, reperfusion significantly limits the magnitude of irreversible myocardial damage. However, there may be some reperfusion injury in the form of impaired contractile function of spared myocardium (stunning) and lethal reperfusion injury to a population of severely impaired CMC at the time of reperfusion. Another process with a major impact on myocardial ischemia is myocardial conditioning. Myocardial preconditioning refers to the retardation of the development of irreversible myocardial injury when brief periods of coronary occlusion and reperfusion precede a prolonged coronary occlusion. After a latent period, the phenomenon returns 24 h later as a second window of protection (SWOP). Postconditioning also achieves some degree of protection when gradual or intermittent reperfusion rather than immediate full reperfusion is provided after a bout of coronary occlusion. Preconditioning and postconditioning at a distance refer to the salutary effects on the evolution of myocardial ischemic injury produced by episodes of transient ischemia in skeletal muscle.

Myocardium and Its Cardiomyocytes

The myocardium is organized around CMCs and supporting structures [9]. CMCs are large cells (average diameter of 14 μm and length three to four times diameter) which constitute 80% of the volume of the myocardium but 20% of the total number of cells, most of which are non-CMC [9]. The myocardium contains a microvasculature including arterioles, capillaries and venules, all lined by endothelium. The capillaries are tightly aligned with the CMC in a one-to-one ratio. The CMC have one or two (25%) central nuclei containing chromosomes with deoxyribonucleic acid (DNA), abundant myofibrils arranged in sarcomeric contractile units and abundant mitochondria for energy production (Fig. 7.1). The plasma membrane (sarcolemma) has invaginations, the T tubules, at the level of each sarcomere, the T tubules with adjacent elements of smooth endoplasmic reticulum arranged in diads and triads to facilitate calcium flux and excitation-contraction coupling. Adjacent CMC are connected end-to-end by specialized plasma membrane, the intercalated discs, and side-to-side by desmosomes.

Because the CMC are specialized for continuous contraction and relaxation, the CMC have a very high energy requirement. This energy requirement is met by the large number of mitochondria which are organelles specialized to conduct oxidative phosphorylation. The mitochondria of CMC have large numbers of tightly paced cristae formed from invaginations of the inner mitochondrial membrane reflecting their high energy output. Maintenance of the electrical potential difference ($\Delta_{\Psi m}$) across the inner mitochondrial membrane is essential for oxidative phosphorylation and ATP generation [9].

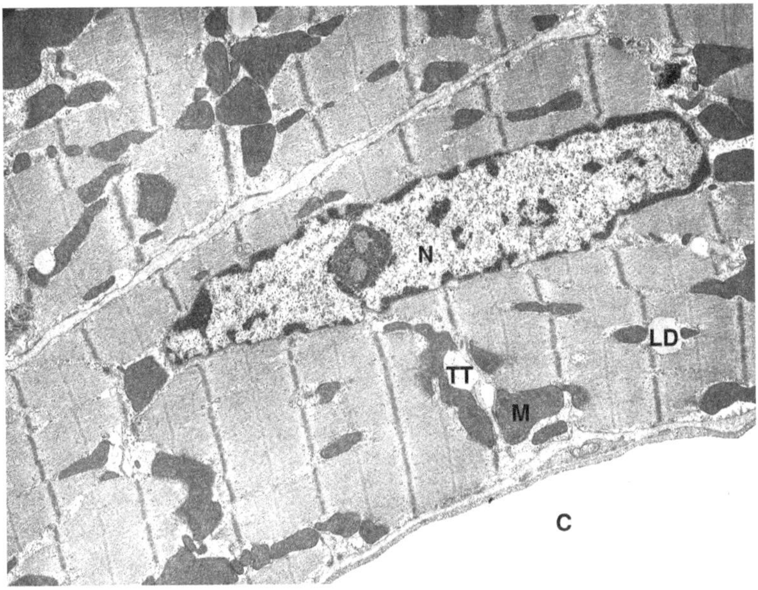

Fig. 7.1 Ultrastructure of the normal mammalian myocardium. The typical cardiomyocyte has a central nucleus (*N*), compact cytoplasm, a sarcolemma (plasma membrane) with invaginations called transverse tubules (*TT*), myofibrils arranged into sarcomeres, a few lipid droplets (*LD*), glycogen granules and numerous mitochondria (*M*) containing numerous densely packed invaginations of the inner membranes, the cristae. In close proximity to the cardiomyocytes are capillaries (*C*). Electron micrograph, ×11,000 (This figure is reproduced from reference [2] with permission)

Modes of CMC Injury and Death

In the last several years, research has shown that CMC have the molecular mechanisms to undergo all of the major modes of cell injury and death. These include oncosis, apoptosis and autophagy [9, 10, 12, 13]. The two of major importance in myocardial ischemic injury and infarction are oncosis and apoptosis [9, 10, 12–15]. Oncosis is manifest by cellular and organellar swelling due to progressive membrane dysfunction and damage due to ischemic or toxic injury coupled with rapid energy depletion reflected by rapid loss of adenosine triphosphate (ATP). Apoptosis is triggered by various physiologic and pathologic stimuli and is characterized by cellular and organellar shrinkage with subsequent fragmentation. Apoptosis is mediated by activation of a cascade of caspase enzymes (cysteine-aspartate proteases) via extrinsic or intrinsic pathways; at least partial preservation of ATP is required. Cell death due to apoptosis is characterized morphologically by cell shrinkage with intact plasma membrane and biochemically by caspase activation and discrete double-stranded DNA breaks. Although oncosis and apoptosis have been termed accidental and programmed cell death, respectively, oncosis can be a relatively uncontrolled process (oncotic necrosis) but often involves signaling pathways that mediate a more regulated process of necroptosis. Many forms of oncosis as well as apoptosis follow programmed patterns mediated by the activation of cell surface receptors, and both involve activation of distinctive gene profiles [14, 15].

Autophagy involves segregation of cellular components, including proteins and mitochondria, in autophagic vacuoles, merger of the autophagic vacuoles with lysosomes to form autophagolysosomes, and subsequent degradation of the constituents in the vacuoles [9]. Depending on whether the autophagic process is regulated or not, autophagy can serve as a mechanism of cell survival following stress or another distinctive form of programmed cell death [9].

Basic Pathobiology of Myocardial Ischemic Injury

Basic information about modes of cell injury and death have been integrated with our knowledge of CMC injury occurring in myocardial infarction. Ischemic cardiomyocyte injury is characterized by progressive membrane damage with a component of cell swelling [3–5]. Ischemic membrane damage has been shown to progress from discrete alterations in specific membrane pumps and ion channels to an intermediate stage of less selective and increasing membrane permeability with more severe ionic disturbances including increased influx of calcium ions to a final stage of membrane rupture [12]. The membrane damage is mediated by activation of phospholipases and proteases and accumulation of toxic metabolites. Mitochondrial swelling and calcium accumulation are prominent components of evolving CMC injury (Figs. 7.2 and 7.3). This pattern of cell injury is typical of the oncotic pattern of cell injury and death [12–15]. With the recognition of apoptosis as another important mode of cell injury and death, the role of apoptosis in ischemic CMC injury also must be considered [12–15].

CMC have been demonstrated to contain the molecular mechanisms to activate apoptotic pathways as well as pathways leading to progressive membrane damage even though CMC undergoing irreversible ischemic injury do not show the classic morphological features of apoptosis [14–21]. This has led to the conclusion that cardiomyocyte ischemic injury and death is a hybrid form of cell injury in which the terminal events are dominated by oncotic ultrastructure [10, 14, 15]. Perturbation of apoptosis with caspase inhibitors and genetic manipulation produces partial but not complete reduction in infarct size [16]. Thus in evolving myocardial infarction, multiple modes of cell death participate including oncotic necrosis, necroptosis, and apoptosis.

Autophagy has been shown to be capable of modulating cardiomyocyte cell injury and acute myocardial ischemic injury. Autophagy has also been found to have a role in cardiomyocyte survival in the setting of hibernating myocardium, a state of chronic ischemic myocardium associated with decreased myocardial function [22–24]. In heart failure, increased loss of CMC occurs by multiple modes of injury, namely, oncotic necrosis, necroptosis, apoptosis and autophagy, and this CMC loss

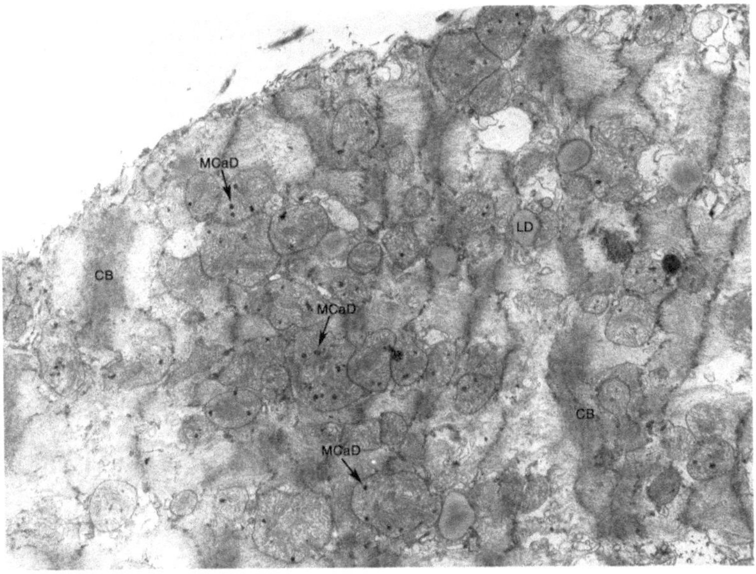

Fig. 7.2 Ultrastructure of ischemic cardiomyocyte with features of contraction band formation and calcium overload as seen in the peripheral zone of an evolving canine myocardial infarct. Note the foci of myofibrils condensed into bands (*CB*), the lipid droplets (*LD*), and granular mitochondrial calcium phosphate deposits (*MCaD*). Electron micrograph, ×6,500 (This figure is reproduced from reference [8] *Dialogues in Cardiovascular Medicine* with the permission of the publisher Les Laboratories Servier, Suresnes, France)

contributes importantly to the progression of heart failure [25–28]. Both in ischemic injury and heart failure, the rate and magnitude of ATP depletion is a key determinant of the pathway of CMC injury and death [14, 15, 28].

Key Role of Mitochondria in CMC Injury

A number of distinct and overlapping subcellular pathways involving cell membrane death receptors, endoplasmic reticulum and mitochondria can be involved in the development of oncotic and apoptotic cell death [14–21]. Nevertheless, recent studies have implicated the mitochondria have a critical role in the pathogenesis of cell injury (Figs. 7.4 and 7.5) [21–28]. The biochemical and ultrastructural changes occurring in cardiac mitochondria during the evolution of myocardial ischemic injury are well documented [3–5]. Mitochondria are dynamic organelles that constantly undergo regulated processes of fusion, fission and substrate trafficking [29, 30]. In stressed cells, mitochondria can be subject to deleterious and beneficial effects due to activation of death channels and salvage pathways, respectively [2, 16–21, 28, 31–33]. The mitochondrial death channels include the mitochondrial permeability transition pore (mPTP) and a putative mitochondrial apoptosis channel

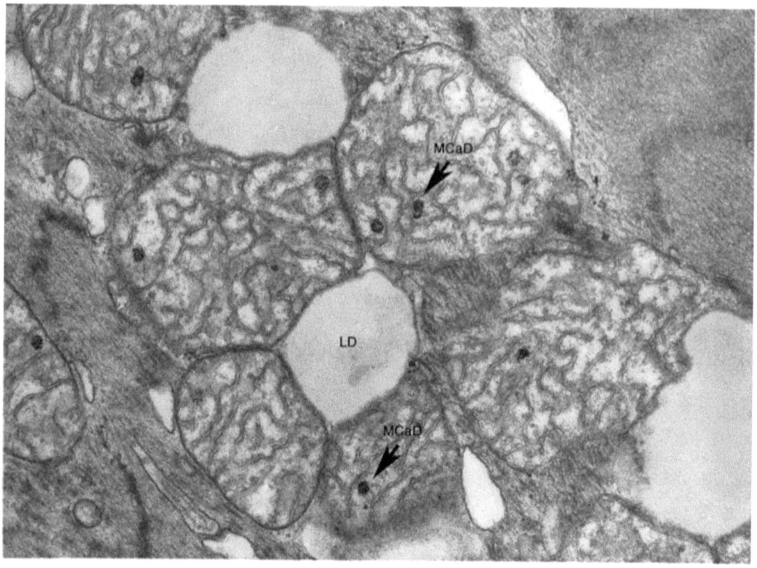

Fig. 7.3 Ultrastructural detail of ischemic cardiomyocyte with lipid deposits (*LD*) and swollen mitochondria containing very electron-dense, annular-granular calcium phosphate deposits (*MCaD*) Electron micrograph, ×26,000. As ATP is reduced, mitochondrial oxidative capacity is decreased, leading to an accumulation of reesterified fatty acids as lipid droplets. As the sarcolemmal function becomes impaired, an increase in calcium influx occurs. With reperfusion, further calcium influx is coupled with an oxidative burst generating toxic oxygen-based radicals. The excess calcium triggers hypercontraction of the myofibrils manifest as contraction bands. The overload of calcium ions and toxic oxygen-based radicals leads to opening of the mitochondrial permeability transition pore (*mPTP*) , loss of membrane potential, swelling and collapse of ATP generation (This figure is reproduced from reference [8] *Dialogues in Cardiovascular Medicine* with the permission of the publisher Les Laboratories Servier, Suresnes, France)

(mAC). The mPTP is a voltage-dependent channel that is regulated by calcium and oxidative stress. The function of the mPTP is influenced by several proteins, especially the voltage-dependent anion channel (VDAC), the adenine nucleotide translocator (ANT) and cyclophilin D (CypD). The VDAC is located in the outer membrane, the ANT in the inner membrane, and CypD on the matrix side of the inner membrane. These proteins span the inner and outer mitochondrial membranes and provide a path for transport between the mitochondrial matrix to the cytoplasm. The opening of the mPTP results in the immediate loss of the electrical potential difference across the inner membrane with resultant cessation of ATP synthesis, influx of solute, and mitochondrial swelling [2, 28, 31–33]. Recent studies have determined a key role for another molecule, the F_0F_1ATPase, in interacting with the other component molecules and leading to the mPTP formation [28]. Modulation of mitochondrial proteins, including a modest increase in expression of uncoupling protein 2 (UCP-2), can cause a rapid decline in mitochondrial membrane potential and ATP level resulting in oncosis [34].

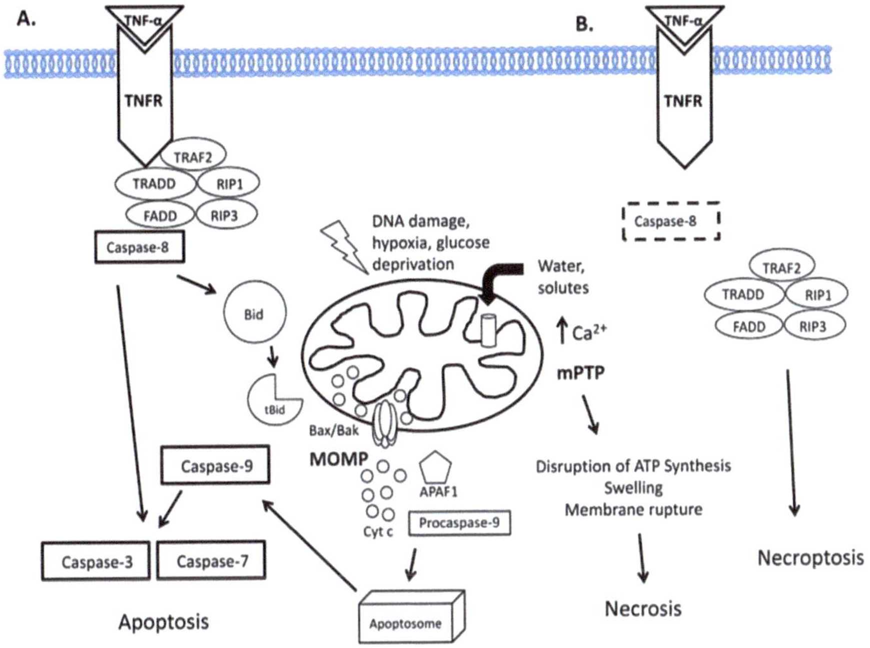

Fig. 7.4 Cell death pathways. (**a**) Apoptosis. In the extrinsic death receptor pathway, binding of an extracellular ligand such as TNF-α to its death receptor TNFR recruits adaptor molecules to form the death-inducing signaling complex (*DISC*). Caspase-8 cleaves and activates effector caspase-3 and -7. In the intrinsic mitochondrial pathway, intracellular stress signals such as DNA damage, hypoxia, or glucose deprivation activate the pro-apoptotic Bax and Bak. During mitochondrial outer membrane permeabilization, Bax and Bak form pores that allow the release of apoptogens such as cytochrome *c* into the cytosol. Cytochrome *c* and APAF1 form the apoptosome which recruits and activates caspase-9 and leads to caspase-3 and -7 activation. Death-receptor mediated caspase-8 activation can also cleave Bid to form truncated Bid (tBid) and activate the intrinsic pathway. (**b**) Necrosis/Necroptosis. Excessive influx of Ca^{2+} into the mitochondria leads to opening of the mitochondrial permeability transition pore (mPTP) and disruption of the proton gradient and ATP synthesis. The influx of water and solutes into mitochondria leads to swelling and membrane rupture. Regulated necrosis, or necroptosis, occurs in the absence of caspase-3 where the RIP1-RIP3 complex facilitates necroptosis (This figure is reproduced from reference [21] with permission)

Apoptosis can be initiated by an extrinsic pathway involving activation of certain membrane receptors with death domains and intrinsic pathways involving the endoplasmic reticulum and the mitochondria. There are levels of interaction between these pathways. A key event in many forms of apoptosis is mitochondrial outer membrane permeabilization (MOMP) produced by activation of the mAC [2, 16–21, 28, 31–33]. Outer mitochondrial membrane integrity, including the putative mAC, is regulated by multiple interactions between proteins of the Bcl-2 faimly, including Bax, Bid, Bcl-2 and Bcl-X_L. Opening of the MOMP leads to release of cytochrome c and other molecules which join with cytoplasmic components including procaspase 9, apoptotic protease activating factor-1 (Apaf-1) and dATP to form

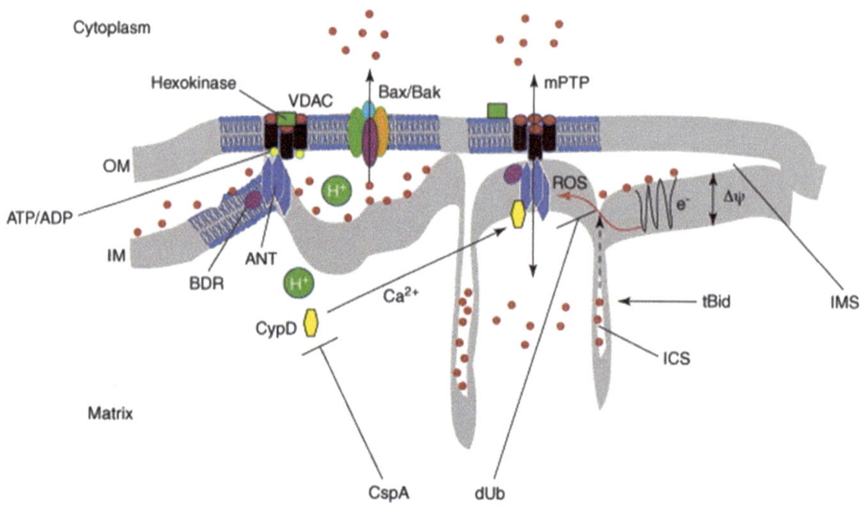

Fig. 7.5 The mitochondrial death channels. The mPTP involved in oncotic necrosis and necroptosis has several component or associated proteins including the voltage-dependent anion channel (*VDAC*), adenine nucleotide translocase (*ANT*), cyclophilin-D (*CypD*), the benzodiazepine receptor (*BDR*) and hexokinase. The specific composition of a mitochondrial apoptosis channel (*mAC*) has not been determined but it is clear that the interactions of the Bcl-2 family of proteins, including Bax, Bak, and tBid with the outer mitochondrial membrane control the mitochondrial outer membrane permeabilization (*MOMP*). The mPTP and mAC (or Bax channel) are both regulated by Bcl-2 proteins and are opened by calcium ionophores and oxidative stress. When the mAC is opened and cytochrome c is released, this triggers apoptosis. When ischemia-reperfusion leads to an increase in matrix calcium levels above a certain threshold, the mPTP opening occurs. CypD and reactive oxygen species (*ROS*) can lower this threshold. Once the mPTP is opened, the mitochondrial membrane potential is rapidly lost ($\Delta_{\psi m}$) (This figure is reproduced from reference [31] with permission)

the apoptosome. The apoptosome then triggers the activation of caspase 3 and other effector caspases in the cytoplasm leading to downstream effects on the nucleus [2, 21, 28].

The mitochondrial pathway rather than the extrinsic pathway appears to be dominant in the apoptosis component of myocardial ischemic injury. Confirmatory evidence has been provided to show that the extent of myocardial ischemia and reperfusion damage was reduced in transgenic mice overexpressing the anti-apoptotic Bcl-2 but not in transgenenic mice overexpressing a truncated form of the surface death receptor [35]. The interrelationships between apoptosis and oncosis has been confirmed by models involving initial activation of apoptosis followed by oncosis and, conversely, oncosis followed by apoptosis [16–19]. At the level of the mitochondria, the oncotic trigger of loss of the mPTP can occur in close temporal relationship to cytochrome c release via the mAC.

In ischemic CMC, when the electron transport system in mitochondria stops functioning due to lack of oxygen, this state leaves various components of the electron transport chain primed to generate oxygen free radicals when oxygen returns.

Indeed with reperfusion a massive diversion of electrons from the electron transport system leads to the generation of oxygen radicals, the oxidative burst. Simultaneously, a large influx of calcium occurs. When the mitochondria are exposed to excess oxygen radicals and calcium, the result is opening of the mPTP and collapse of mitochondrial function [11]. This is a key event in ischemic reperfusion injury. A dual role for autophagy also has been identified. Activation of autophagy during the early phase of ischemia is cardioprotective, whereas delated or late activation of autophagy during reperfusion is detrimental and promotes cell death [28].

Conditioning is a complex phenomenon initiated by activation of various G-protein coupled receptors by various autocoids, including adenosine, bradykinin, and opioids, that are released during the brief periods of ischemia and reperfusion [36–38]. Activation of these receptors is followed by activation of a complex signaling cascade, including multiple kinases, that leads to opening of potassium channels in the mitochondrial membrane and maintenance of the mPTP and the electrical potential of the inner mitochondrial membrane. Preservation of mitochondrial function and ATP production is an absolute requirement for the protective effect of conditioning.

A component of the protective effect of conditioning involves the activation of a reperfusion injury salvage kinase (RISK) pathway in the mitochondria [10, 11, 39, 40]. One component of the pathway involves phosphatidylinositol-3 kinase (PI-3 K) acting on Akt (protein kinase B) and the mammalian target of rapamycin (mTOR). The other component involves mitogen-associated protein kinase (MAPK) and p42/p44 extracellular signal-related kinase (ERK). The two arms of the pathway converge on p70s6 kinase to activate glycogen synthase kinase β which acts to prevent opening of the mPTP. An isoform of the ATP-sensitive potassium channel (K^{+}_{ATP}) also regulates mPTP opening. A component of autophagy, mitophagy, contributes to the protective effect of ischemic preconditioning [28–30].

Therapeutic Considerations

Because of the clinical importance of preserving myocardium, extensive effort has been aimed at developing pharmacological approaches to limiting myocardial ischemic damage. This area of research has been frustrated by initially promising experimental work being followed by negative or equivocal clinical trials in patients. Although myocardial ischemia – reperfusion injury is clearly mediated by calcium overload, toxic oxygen-based molecules, inflammatory mediators and neutrophils, clinical trials of agents aimed against these components of ischemic injury have not been found to be convincingly effective [10, 11]. Deficiencies in the design of the experimental work and complexities of confirming efficacy in clinical trials are important factors for these results [10, 11, 41]. The major advances in salvage of ischemic myocardium have come from the introduction of thrombolytic therapy and percutaneous coronary interventions (PCI) rather than from pharmacological treatments. Nevertheless, increased understanding of the pathobiology of

Table 7.1. Targeting mitochondria in cardiomyocytes and the heart

Target	Inhibitors	Mechanism
Direct action via mPTP	CsA an its analogues	Prevents opening of mPTP pores via mPTP via cyclophilin D (CypD) interaction; side toxicity due to interaction with other cyclophilins
	N-Me-4-Ile-CsA (NIM811)	Prevents mPTP opening via CypD interaction; more specific than CsA
	Sangliferhrin A	Prevents opening of mPTP pores via CypD interaction at different sites from CsA
	BKA and ADP	Induces *m* conformation of ANT
	H^+ and divalent cations ($Mg^{2+}$$\cdot$$Mn^{2+}$$\cdot$$Sr^{2+}$$\cdot$$Ba^{2+}$)	Antagonize Ca^{2+} binding to ANT
Indirect action – mPTP opening	Reactive oxygen species (ROS) scavengers (e.g., propofol, pyruvate, MCI-186)	Prevent accumulation of ROS and oxidation of critical thiol groups on ANT
Pharmacological preconditioning		Prevent oxidative stress, normalize mitochondrial metabolism
	Adenosine	Induces desensitization of mPTP by causing ischemic preconditioning (IPC)
	Nicorandil	Mimics IPC, induces desesitization by opening $mito_K$ATP
	Diazoxide	Similar mechanism
	Bepridil	Similar mechanism
	Isoflurane	Induces IPC-like condition, inhibits complex I and glycogen synthase kinase 3p
Mitochondrial dysfunction	Coenzyme Q	Cofactor in electron transfer; reduces mitochondrial ROS
	L-carnitine	Involved in fatty acid transport into mitochondrial inner membrane
	MITO Q (mitoquinone)	Antioxidant
Other targets	SS-31 (i.e.,MTP-131 and Bendavia)	Binding to inner membrane cardiolipin, preserving oxidativePhosphorylation reducing ROS
	TMD#7538	Inhibits both MOMP and mPTP

Reproduced from Goldenthal [28]

myocardial ischemic injury and the key role of the mitochondria has led to renewed interest and promise in novel therapeutic approaches, used alone or in combination with PCI, for the treatment of patients with acute ischemic heart disease [9–11, 28, 31–33, 42–58].

The prototype for a pharmacological approach targeting the mitochondria is cyclosporine A (CsA) (Table 7.1) [28]. CsA is known to inhibit the formation and opening of the mPTP by binding to CypD. Positive results were obtained in a proof-of-concept small clinical trial of patients with evolving myocardial infarction

[48, 49]. Other pharmacological agents are being designed and tested for their ability to activate the RISK pathway in the mitochondria and to enhance mitochondrial energy metabolism [11, 40, 54, 55]. Pharmacological agents under recent or active investigation include adenosine, atrial naturetic peptide, beta adrenergic blocking drugs, nitrates, phosphodiesterase-5 inhibitors, and supersaturated oxygen therapy, and antioxidant peptides targeted at mitochondria [54–58].

In experimental models, caspase inhibitors and other supressors of apoptosis have been shown to reduce myocardial infarct size [50]. However, approaches to retard apoptosis during evolving myocardial ischemic injury in patients have not yet been demonstrated. Other agents, designated as necrostatins, are being developed for activity against receptors involved in the hybrid necroptosis form of injury [50, 51]. Other approaches directed to the non-myocytic components of the myocardium, including endothelial cells, fibroblasts, and extracellular matrix are being explored [52].

The resurgence of interest in reducing ischemic and reperfusion injury has led to the formation of an NIH-sponsored consortium for preclinical assessment of cardioprotective therapies (CAESAR) [59–61]. Recently, however, several larger multicenter trials have reported lack of protective effect with CsA and remote ischemic pre-conditioning [62–66]. It has been pointed out that the effect of CsA is indirect and mediated by binding of CsA to the matrix protein CyP-D with the net effect being desensitization rather than true block of the pore. This has given rise to strong interest in developing potentially more effective inhibitors of the mPTP (Table 7.1) [28].

Conclusions

The cumulative effects of episodes of ACS and subsequent extent of myocardial infarction have major significance for the prognosis of patients with ischemic heart disease. Acutely, extensive myocardial infarction is associated with acute heart failure, cardiogenic shock and malignant arrhythmias [1]. Chronically, the cumulative effect of ischemic damage leads to chronic heart failure [9]. Chronic heart failure is mediated not only by the loss of myocardium but by progressive pathological remodeling in the viable myocardium [6–10]. Coronary artery stenting and coronary artery bypass grafting, unless introduced early, may not be effective in stabilizing or reversing the process which leads to fixed structural dilatation of the heart and is manifest clinically as ischemic cardiomyopathy [9]. Contradictory or negative results of interventional studies highlight the difficulties of translating basic knowledge of pathobiology into effective clinical therapy and the need for continuing effort in that regard [67, 68]. Nevertheless, continued effort at developing pharmacological and pathophysiological approaches to reducing ischemic damage is worthy of pursuit. Further development of more effective approaches to preserving mitochondrial integrity should continue to be pursued.

References

1. Buja LM, Vander Heide RS. Pathobiology of ischemic heart disease: past, present and future. Cardiovasc Pathol. 2016;25:214–20.
2. Buja LM. The pathobiology of acute coronary syndromes. Clinical implications and central role of the mitochondria. Tex Heart Inst J. 2013;40:221–8.
3. Reimer KA, Ideker RE. Myocardial ischemia and infarction: anatomic and biochemical substrates for ischemic cell death and ventricular arrhythmias. Hum Pathol. 1987;18:462–75.
4. Reimer KA, Jennings RB. Myocardial ischemia, hypoxia, and infarction. In: Fozzard HA, Haber E, Jennings RB, Katz AM, Morgan HE, editors. The heart and cardiovascular system: scientific foundations, vol. 2. 2nd ed. New York: Raven Press; 1992. p. 1875–973.
5. Jennings RB, Reimer KA. The cell biology of acute myocardial ischemia. Annu Rev Med. 1991;42:225–46.
6. Buja LM. Modulation of the myocardial response to ischemia. Lab Investig. 1998;78: 1345–73.
7. Buja LM. Myocardial ischemia and reperfusion injury. Cardiovasc Pathol. 2005;14:170–5.
8. Willerson JT, Buja LM. Myocardial reperfusion: biology, benefits and consequences. Dialogues Cardiovasc Med. 2006;11:267–78.
9. Buja LM, Vela D. Cardiomyocyte death and renewal in the normal and diseased heart. Cardiovasc Pathol. 2008;17:349–74.
10. Buja LM, Weerasinghe P. Unresolved issues in myocardial reperfusion injury. Cardiovasc Pathol. 2010;19:29–35.
11. Yellon DM, Hausenloy DJ. Myocardial reperfusion injury. N Engl J Med. 2007;357: 1121–35.
12. Buja LM, Eigenbrodt ML, Eigenbrodt EH. Apoptosis and necrosis. Basic types and mechanisms of cell death. Arch Pathol Lab Med. 1993;117:1208–14.
13. Majno G, Joris I. Apoptosis, oncosis and necrosis. An overview of cell death. Am J Pathol. 1995;146:3–15.
14. Weerasinghe P, Hallock S, Brown R, Loose DS, Buja LM. A model for cardiomyocyte cell death: insights into mechanisms of oncosis. Exp Mol Pathol. 2013;94:289–300.
15. Weerasinghe P, Buja LM. Oncosis: an important non-apoptotic mode of cell death. Exp Mol Pathol. 2012;93:302–8.
16. Foo RS, Mani K, Kitsis RN. Death begets failure in the heart. J Clin Invest. 2005;115: 565–71.
17. Whelan RS, Kaplinskiy V, Kitsis RN. Cell death in the pathogenesis of heart disease: mechanisms and significance. Annu Rev Physiol. 2010;72:19–44.
18. Kung G, Konstantinidis K, Kitsis RN. Programmed necrosis, not apoptosis, in the heart. Circ Res. 2011;108:1017–36.
19. Konstantinidis K, Whelan RS, Kitsis RN. Mechanisms of cell death in heart disease. Arterioscler Thromb Vasc Biol. 2012;32:1552–65.
20. Gottlieb RA. Cell death pathways in acute I/R injury. J Cardiovasc Pharmacol Ther. 2011;16:233–8.
21. Orogo AM, Gustafsson AB. Cell death in the myocardium: my heart won't go on. IUBMB Life. 2013;65:651–6.
22. Gottlieb RA, Mentzer Jr RM. Autophagy: an affair of the heart. Heart Fail Rev. 2013;18: 575–84.
23. Hickson-Bick DLM, Jones C, Buja LM. Stimulation of mitochondrial biogenesis and autophagy by lipopolysaccharide in the neonatal rat cardiomyocyte protects against programmed cell death. J Mol Cell Cardiol. 2008;44:411–8.
24. Yan L, Sadoshima J, Vatner DE, Vatner SV. Autophagy in ischemic preconditioning and hibernating myocardium. Autophagy. 2009;5:709–12.
25. Marín-García J. Cell death in the pathogenesis and progression of heart failure. Heart Fail Rev. 2016;21:117–21.

26. Moe GW, Marín-García J. Role of cell death in the progression of heart failure. Heart Fail Rev. 2016;21:157–67.
27. Adameova A, Goncalvesova E, Szobi A, Dhalla NS. Necroptotic cell death in heart failure: relevance and proposed mechanisms. Heart Fail Rev. 2016;21:213–21.
28. Goldenthal MJ. Mitochondrial involvement in myocyte death and heart failure. Heart Fail Rev. 2016;21:137–55.
29. Marín-García J, Akhmedov AT, Moe GW. Mitochondria in heart failure: the emerging role of mitochondrial dynamics. Heart Fail Rev. 2013;18:439–56.
30. Mariín-García J, Akhmedov AT. Mitochondrial dynamics and cell death in heart failure. Heart Fail Rev. 2016;21:123–36.
31. Webster KA. Programmed death as a therapeutic target to reduce myocardial infarction. Trends Pharmacol Sci. 2007;9:492–9.
32. Webster KA. Mitochondrial death channels. Am Sci. 2009;97:384–91.
33. Webster KA. Mitochondrial membrane permeabilization and cell death during myocardial infarction: roles of calcium and reactive oxygen species. Futur Cardiol. 2012;8:863–84.
34. Mills EM, Xu D, Fergusson MM, Combs CA, Xu Y, Finkel T. Regulation of cellular oncosis by uncoupling protein 2. J Biochem. 2002;277:27385–92.
35. Kristen AV, Ackermann K, Buss S, Lehmann L, Schnabel PA, Katus HA, Hardt S. Inhibition of apoptosis by the intrinsic but not the extrinsic apoptotic pathway in myocardial ischemia-reperfusion. Cardiovasc Pathol. 2013;22:280–6.
36. Yellon DM, Downey JM. Preconditioning the myocardium: from cellular physiology to clinical cardiology. Physiol Rev. 2003;83:1113–51.
37. Downey JM, Davis AM, Cohen MV. Signaling pathways in ischemic preconditioning. Heart Fail Rev. 2007;12:181–8.
38. Yang X, Cohen MV, Downey JM. Mechanism of cardioprotection by early ischemic preconditioning. Cardiovasc Drugs Ther. 2010;24:225–34.
39. Skyschally A, Gent S, Amanakis G, Schulte C, Kleinbongard P, Heusch G. Across-species transfer of protection by remote ischemic preconditioning with species-specific myocardial signal transduction by reperfusion injury salvage kinase and survival activating factor enhancement pathways. Circ Res. 2015;117:279–88.
40. Smith CCT, Yellon DM. Adipocytokines, cardiovascular pathophysiology and myocardial protection. Pharmacol Therap. 2011;129:206–19.
41. Miura T, Miki T. Limitation of myocardial infarct size in the clinical setting: current status and challenges in translating animal experiments into clinical therapy. Basic Res Cardiol. 2008;103:501–13.
42. Vander Heide R. Clinically useful cardioprotection: ischemic preconditioning then and now. J Cardiovasc Pharmacol Ther. 2011;16:251–4.
43. Vander Heide RS, Steenbergen C. Cardioprotection and myocardial reperfusion: pitfalls to clinical application. Circ Res. 2013;113:464–77.
44. Frohlich GM, Meier P, White SK, Yellon DM, Hausenloy DJ. Myocardial reperfusion injury: looking beyond primary PCI. Eur Heart J. 2013;34:1714–22.
45. Perricone AJ, Vander Heide RS. Novel therapeutic strategies for ischemic heart disease. Pharmacol Res. 2014;89:36–45.
46. Heusch G. Molecular basis of cardioprotection: signal transduction in ischemic pre-, post-, and remote conditioning. Circ Res. 2015;116:674–99.
47. Ibanez B, Heusch G, Ovize M, Van de Werf F. Evolving therapies for myocardial ischemia/reperfusion injury. J Am Coll Cardiol. 2015;65:1454–71.
48. Piot C, Croisille P, Gahide C, et al. Effect of cyclosporine on reperfusion injury in acute myocardial infarction. N Engl J Med. 2008;359:473–81.
49. Morel O, Perret T, Delarche N, Labèque JN, Jouve B, Elbaz M, Piot C, Ovize M. Pharmacologic approaches to reperfusion therapy. Cardiovasc Res. 2012;94:246–52.
50. Oerlemans MI, Koudstall S, Chamuleau SA, de Kleijn DP, Doevendans PA, Sluijter JP. Targeting cell death in the reperfused heart: pharmacological approaches for cardioprotection. Int J Cardiol. 2013;165:410–22.

51. Koudstall S, Oerlemans MI, Van der Spoel TI, Janssen AW, Hoefer IE, Doevendans PA, Sluijter JP, Chamuleau SA. Necrostatin-1 alleviates reperfusion injury following acute myocardial infarction in pigs. Eur J Clin Investig. 2015;45:150–9.
52. Bell RM, Yellon DM. Conditioning the whole heart – not just the cardiomyocyte. J Mol Cell Cardiol. 2012;53:24–32.
53. Hausenloy DJ, Boston-Griffiths E, Yellon DM. Cardioprotection during cardiac surgery. Cardiovasc Res. 2012;94:253–65.
54. Sharma V, Bell RM, Yellon DM. Targeting reperfusion injury in acute myocardial infarction: a review of reperfusion injury pharmacotherapy. Expert Opin Pharmacother. 2012;12:1153–75.
55. Gerczuk PZ, Kloner RA. An update on cardioprotection: a review of the latest adjunctive therapies to limit myocardial infarction size in clinical trials. J Am Coll Cardiol. 2012;59: 969–78.
56. Santulli G, Xie W, Reiken SR, Marks AR. Mitochondrial calcium overload is a key determinant in heart failure. Proc Natl Acad Sci USA. 2015;112:11389–94.
57. Rocha M, Hernandez-Mijares A, Garcia-Malpartida K, Bañuls C, Bellod L, Victor VM. Mitochondria-targeted antioxidant peptides. Curr Pharm Des. 2010;16:3124–31.
58. Szeto HH, Schiller PW. Novel therapies targeting inner mitochondrial membrane – from discoveries to clinical development. Pharm Res. 2011;28:2669–79.
59. Kloner RA, Schwartz Longacre L. State of the science of cardioprotection: challenges and opportunities – proceedings of the 2010 NHLBI workshop on cardioprotection. J Cardiovasc Pharmacol Ther. 2011;16:223–32.
60. Schwartz Longacre L, Kloner RA, Arai AE, et al. New horizons in cardioprotection: Recommendations from the 2010 national heart, lung, and blood institute workshop. Circulation. 2011;124:1172–9.
61. Fernández-Jiménez R, Ibanez B. CAESAR: one step beyond in the construction of a translational bridge for cardioprotection. Circ Res. 2015;116:554–6.
62. Cung TT, Morel O, Cayla G, et al. Cyclosporine before PCI in patients with acute myocardial infarction. N Engl J Med. 2015;373:1021–31.
63. Hausenloy DJ, Kharbanda R, Rahbek Schmidt M, et al. Effect of remote ischaemic conditioning on clinical outcomes in patients presenting with an ST-segment elevation myocardial infarction undergoing primary percutaneous coronary intervention. Eur Heart J. 2015;36: 1846–8.
64. Meybohm P, Bein B, Brosteanu O, et al. A multicenter trial of remote ischemic preconditioning for heart surgery. N Engl J Med. 2015;373:1397–407.
65. Hausenloy DJ, Candilio L, Evans R, et al. Remote ischemic preconditioning and outcomes of cardiac surgery. N Engl J Med. 2015;373:1408–17.
66. Zaugg M, Lucchinetti E. Remote ischemic preconditioning in cardiac surgery – ineffective and risky? N Engl J Med. 2015;373:1470–2.
67. Bulluck H, Yellon DM, Hausenloy DJ. Reducing myocardial infarct size: challenges and future opportunities. Heart. 2016;102:341–8.
68. Hausenloy DJ, Yellon DM. Ischaemic conditioning and reperfusion injury. Nature Rev/ Cardiol. 2016;13:193–209.

Chapter 8
Mitochondrial Bioenergetics During Ischemia and Reperfusion

Alicia E. Consolini, María I. Ragone, Patricia Bonazzola, and Germán A. Colareda

During recent years many works have described the morphological interaction between mitochondria and sarcoplasmic reticulum (SR) with different methodologies, such as co-localization with Ca^{2+} fluorofores and confocal microscopy in cardiomyocytes or detecting electron-dense bridges between the terminal cisternae of the SR and mitochondria by transmission electron microscopy which remain still during the isolation [1–4]. Nevertheless, these models are far from the pathophysiological situation of ischemia/reperfusion (I/R) and they do not provide a good estimation of the functional interaction between mitochondria and SR. Alternatively, the mechano-calorimetrical approach provides a tool to study this interaction in whole hearts exposed to no-flow ischemia and reperfusion, by means of selective pharmacological tools.

A.E. Consolini (✉) • G.A. Colareda
Grupo de Farmacología Experimental y Energética Cardíaca (GFEYEC), Departamento de Ciencias Biológicas, Facultad de Ciencias Exactas, Universidad Nacional de La Plata (UNLP), La Plata, 1900 Buenos Aires, Argentina
e-mail: dinamia@biol.unlp.edu.ar

M.I. Ragone
Grupo de Farmacología Experimental y Energética Cardíaca (GFEYEC), Departamento de Ciencias Biológicas, Facultad de Ciencias Exactas, Universidad Nacional de La Plata (UNLP), Buenos Aires, Argentina

Consejo Nacional de Investigaciones Científicas y Técnicas de la República Argentina (CONICET), Argentina

P. Bonazzola
Instituto de Investigaciones Cardiológicas, Universidad de Buenos Aires (UBA-CONICET), Buenos Aires, Argentina

Consejo Nacional de Investigaciones Científicas y Técnicas de la República Argentina (CONICET), Argentina

G. Santulli (ed.), *Mitochondrial Dynamics in Cardiovascular Medicine*, Advances in Experimental Medicine and Biology 982,
DOI 10.1007/978-3-319-55330-6_8

Calorimetry was first used to estimate cardiac energetic of beating and resting state in papillary muscles from different species [5, 6] and then in entire perfused hearts [7]. The heat fractions associated to the Ca^{2+} activation (binding to troponin-C and ATP hydrolysis for cytosolic removal), to the actomyosin ATP-cycling and to the metabolic recovery of ATP levels were recognized by the different methodologies [6, 8–11]. The flow calorimeter provided a tool to understand Ca^{2+} movements [7, 12–14]. The thermopiles were used with papillary muscles under hyperthyroid state to study cardiac heat fractions under this pathological influence [15] and in senility [16]. More recently, a new work-loop microcalorimeter is being used with trabeculae from diabetic rabbits to study the influence of this pathology on the efficiency-afterload relationship [17, 18]. Nevertheless, only the flow calorimeter for entire perfused hearts is able to measure heat rate under no-flow or low-flow ischemia, as models to study the consequences of ischemia/reperfusion in entire hearts [19–26].

Mitochondria release an important fraction of the total heat flow, both in resting and during steady-state beating, because they provide metabolism and resynthesis of the ATP continuously consumed by ionic pumps and myofilaments. But part of the mitochondrial heat output is related to the maintaining of their own electrochemical gradient ($\Delta\Psi m$) and Ca^{2+} transport. The control of mitochondrial calcium concentration ($[Ca^{2+}]m$) is a key to regulate the activity of some enzymes of the tricarboxylic acid cycle (TCA) and adjust metabolism and ATP synthesis to the cardiac cytosolic demand [16, 27]. The $[Ca^{2+}]m$ mainly depends on the uptake by the uniporter (UCam) driven by the $\Delta\Psi m$ and the extrusion by the mitochondrial Na^+/Ca^{2+} exchanger (mNCX), in a balance which finally determines that $[Ca^{2+}]m$ oscillates in beat-to-beat transients [2, 28, 29]. There are species-dependent differences in the level of free mitochondrial Ca^{2+}, which have been attributed to the cytosolic $[Na^+]$, since this ion is the main driver of mitochondrial Ca^{2+} extrusion via the mNCX [30, 31]. Maack and colleagues demonstrated that inhibition of UCam with Ru360 in guinea-pig myocytes reduced the mitochondrial Ca^{2+}-transients but symmetrically increased the cytosolic Ca^{2+}-transients measured respectively by confocal microscopy with the fluorofores Rhod-2 and Indo-1 in a patch-pipette. In addition, the inhibition of mNCX with CGP-37157 produced the opposite changes [30]. Considering the interaction between mitochondria and SR, it was shown that Ca^{2+} released by the RyR2 at about 100 μM is able to induce the UCam uptake with the consequent increase in $[Ca^{2+}]m$ [32]. Also, physiologically induced mitochondrial reactive oxygen species (ROS) can regulate the SR release by RyR2 [33]. Thus, Ca^{2+} and ROS participate in the functional interaction as parts of a fine and adjusted control of energetic balance between metabolism and ATP consumption.

During ischemia, the total heat rate (Ht) is reduced at the time in which cellular metabolism and oxidative phosphorylation are inhibited. During reperfusion, the reintroduction of oxygen allows the ATP synthesis, so that the ionic pumps can be reactivated. But cytosolic Ca^{2+} is quickly increased, mostly by activation of the Na^+/H^+-exchanger (NHE) and the consequent rise in cytosolic $[Na^+]$ and Ca^{2+} influx by the reverse mode of sarcolemmal NCX (SL-NCX) [34, 35]. These events become in diastolic contracture during the first minutes of reperfusion, and very low contractility (P) with a great increase in total heat release (Ht), thus resulting in a heart with very low total muscle economy (P/Ht). The rise in diastolic cytosolic Ca^{2+} during the ischemia and reperfusion period leads to mitochondrial Ca^{2+} accumulation, particularly when the aerobic metabolism is recovered. The consequent dysfunction depends mostly on the period of ischemia, and goes from the myocardial stunning, when post-ischemic contractility is partially reduced, to the myocardial infarction, when contractility is strongly reduced and cellular necrosis and apoptosis are developed. Also, the ischemic insult on the electron transport chain results in an increased mitochondrial ROS generation. The Ca^{2+} overload and ROS could open the mitochondrial permeability transition pore (mPTP), which further compromises the cellular energetic [36]. In long periods of ischemia mitochondria have a central role in cell death, since they are accumulating Ca^{2+} and they are responsible for the synthesis of ROS, events that could trigger the activation of mPTP [37]. Nevertheless, certain role of the mitochondrial Ca^{2+} transporters to regulate cellular Ca^{2+} homeostasis was proposed to induce cardioprotection, such as the inhibition of the mNCX [38], the UCam blockade or the activation of channels mK_{Ca} [39] and mK_{ATP} [40]. The common mechanism is the prevention of the mitochondrial depolarization and the matrix Ca^{2+} overload, which in turn reduce mPTP activation [39]. Recently, the interaction between SR and mitochondria was characterized by reducing the functional contact points at the complex between IP3R1 receptors and the mitochondrial VDAC1/Grp75 Ca^{2+} channels, which protected the cardiomyocytes by preventing the mitocondrial Ca^{2+} overload and apoptosis [41]. On the other hand, in HL-1 type cardiomyocytes it was demonstrated that the inhibition of the mNCX reduced the SR Ca^{2+} content and slowed the spontaneous SR Ca^{2+} leak, which is a trigger of automaticity and arrhythmogenesis [42]. In conclusion, different evidences suggest two directions for Ca^{2+} regulation between SR and mitochondria, which must be dependent on the degree of dysfunction: that from the SR to mitochondria generally triggers Ca^{2+} overload, while that from mitochondria to SR seems towards regulate Ca^{2+} release and leak.

Energetical Approach to the Functional Interaction Between Mitochondria and Sarcoplasmic Reticulum in Post-ischemic Stunning

In previous works we studied the cardiac stunning induced by a relatively short period of ischemia and reperfusion, and the mechanisms of cardioprotective strategies. Here, a new analysis was done to revise the several evidences about the functional interaction between mitochondria and SR. For this purpose, perfused entire hearts inside a flow-calorimeter were treated with selective mitochondrial inhibitors before ischemia and then reperfused with Krebs containing 10 mM caffeine-36 mM Na^+ (Krebs-caff-low Na) without changing [Ca^{2+}] of initial perfusion (2 mM). The high concentraction of caffeine was frequently used in cardiomyocytes to estimate the content of Ca^{2+} in SR, because it induces a permanent open state in the ryanodine receptors (RyR2) [43]. In rat and guinea-pig perfused hearts, 10 mM caffeine induces a contracture which is potentiated by reducing [Na^+] to 36 mM, the minimal needed to maintain membrane functionality in whole hearts [20]. The condition of low Na^+ minimizes Ca^{2+} removal by the sarcolemmal sodium/calcium exchanger (SL-NCX). Thus, the slow relaxation of the contracture mainly depends on mitochondrial Ca^{2+} uptake and less on the sarcolemmal Ca^{2+}-ATPase, since neither SL-NCX nor the SR are functionally efficient to remove cytosolic Ca^{2+} under 10 mM caffeine-36 mM Na^+. Nevertheless, SERCA is continuously stimulated to uptake the Ca^{2+} lost by the leaky state of RyR2, producing a Ca^{2+} cycling with ATP consumption and heat release. Figure 8.1 shows the ventricular of contracture (ΔLVP) and heat rate (ΔHt) produced over the ischemic levels by reperfusing hearts with Krebs-caff-low Na. In this protocol, the caffeine-induced contracture (CIC) during the reperfusion followed a period of 45 min ischemia at 30 °C. This period was enough to induce stunning in reperfused hearts, since contractility was recovered to 64% ± 10% of initial at the end of R, without infarct [20]. The ratio of area-under-curves during the CIC (AUC-ΔHt/AUC-ΔLVP) allow to estimate the energetical consumption (EC) necessary to maintain the CIC (Table 8.1). When hearts were treated before ischemia with 10 μM clonazepam (Clzp), a selective inhibitor of the mNCX [38], the stunning was slightly worsen because the post-ischemic contractile recovery (PICR) reached 46% ± 8% of initial P. Also, the CIC of these hearts treated with Clzp was reduced and the relaxation accelerated, while Ht was simultaneously increased (Fig. 8.1a, b). In consequence Clzp reduced the AUC- ΔLVP and increased both the AUC-ΔHt and the EC (Table 8.1). These results suggest that in the post-ischemia mitochondria must be regulating the Ca^{2+} content of SR through the extrusion by the mNCX. A Ca^{2+} cycling from the mNCX to SERCA and from RyR2 to the UCam may be happening in the myocytes, and it will be exacerbated under 10 mM caffeine-36 mM Na^+.

The partial blockade of mNCX seems to unbalance this cycling, with the consequent predominance of the mitochondrial Ca^{2+} uptake which drives to reduction of cytosolic Ca^{2+} and SR uptake (Fig. 8.2). This mechanism explains the quicker relaxation of CIC and the increase in the heat rate associated to a higher mitochondrial metabolism necessary to maintain the ΔΨm dissipated by the higher [Ca^{2+}]m. This direction of Ca^{2+} regulation in stunning is opposite to that described in more severe situations of ischemia/reperfusion, explained as a SR Ca^{2+} overload that triggers mitochondrial dysfunction with opening of the mPTP [44]. Nevertheless, we have found several evidences of the regulation from mitochondria to the SR content that will be described here. The experimental differences are mainly due to the duration of ischemia, and express different degrees of mitochondrial response to hypoxia or ischemia, which go from prevention of dysfunction in short periods of oxygen-deficiency to the triggering of mPTP opening and cellular apoptosis in situations of Ca^{2+} overload or long periods of oxygen deprivation. In our results of hearts treated before ischemia with Clzp in control Krebs (C + Clzp), the fall in the SR Ca^{2+} content (suggested by the fall in CIC) explains a tendency to reduce the PICR in beating hearts [20].

The same direction of functional interaction from mitochondria to SR was better characterized in the cardioprotective effects of a model of cold cardioplegic solution, which is used in surgery to stop heart in diastole, consisting in perfusing

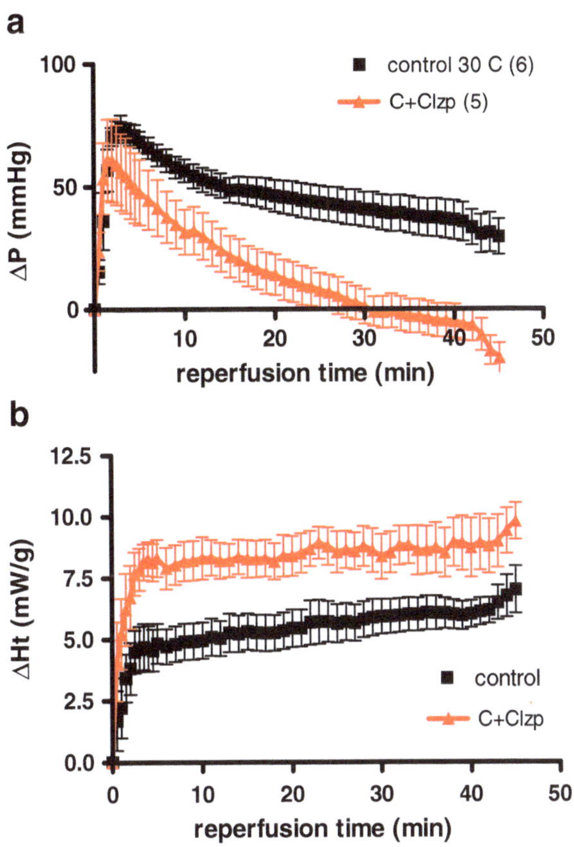

Fig. 8.1 Effects of reperfusing ischemic rat hearts at 30 °C with Krebs-10 mM caffeine-36 mM Na^+, after treatment with 10 μM clonazepam (*Clzp*) or without treatment (*control*). In (**a**) changes in left ventricular pressure (ΔLVP, in mm Hg), in (**b**) changes in the heat rate production (Δ*Ht*, in mW/g), both over the ischemic value

Krebs-25 mM K-0.5 mM Ca^{2+} (CPG) before ischemia, in rat and guinea-pig isolated hearts [20, 24]. In rats, CPG improved the PICR up to 90% ± 10% of initial pressure development (P) in the model of moderate stunning (non-treated C-hearts recovered up to 64% ± 10% of initial P) during the steady beating in reperfusion; CPG also increased total muscle economy (P/Ht) and prevented the diastolic contracture [20]. Moreover, when ischemic hearts previously treated with CPG were reperfused with Krebs-caff-low Na, the peak of the CIC and AUC-ΔLVP were similar to those of non-treated C-hearts (Fig. 8.3a, b), while the simultaneous AUC-ΔHt and the EC were slightly reduced in comparison to the C group (Table 8.1). Comparison of Figs. 8.1a and 8.3a suggests that the peak of SR Ca^{2+} release was slightly increased by CPG, which can explain the increase in PICR. But Ca^{2+} cycling between cytosol and SR (AUC-ΔLVP) and the energetic consumption (EC) for maintaining the CIC during R remain similar in C and CPG hearts. Nevertheless, a Ca^{2+}-cycling dependent of $[Ca^{2+}]o$ has been suggested from the increased resting heat rate in cardioplegic non-ischemic hearts, which was sensitive to verapamil and hypoxia. It was interpreted as a cytosolic Ca^{2+} removal in response to Ca^{2+} entered through L-channels upon depolarization [22, 45].

Table 8.1 Area under the curves of intraventricular pressure (ΔLVP) and heat rate (ΔHt) increments over the value of the end of ischemia, during the reperfusion with Krebs-10 mM caffeine- 36 mM Na^+ on rat hearts with the different treatments described in the text

Treatment	AUC-ΔLVP (mmHg. min)	AUC-Δ*Ht* (mW. min/g)	EC: AUC-Δ*Ht*/AUC-ΔLVP
Moderate stunning at 30 °C			
C (6)	2076.1 ± 263.0	243.9 ± 30.0	0.13 ± 0.03
C + Clzp (5)	831.9 ± 298.2	376.2 ± 33.7	0.80 ± 0.29
CPG non-ischemic (4)	2835.6 ± 624.8	357.9 ± 215.4	0.13 ± 0.07
CPG (13)	2389.1 ± 191.1	206.5 ± 27.8	0.10 ± 0.02
CPG + Clzp (5)	1675.0 ± 393.4	203.2 ± 64.7	0.10 ± 0.02
CPG + KBR (10)	3082.9 ± 427.0	260.1 ± 23.6	0.096 ± 0.01
CPG + KBR + Pyr (6)	2791.9 ± 371.3	265.3 ± 31.5	0.10 ± 0.01
C + Gst females (6)	2419.6 ± 544.0	237.7 ± 36.8	0.21 ± 0.11
C + Gst males (10)	2336.1 ± 255.9	169.6 ± 21.5	0.08 ± 0.01
C + OV + Gst males (9)	2580.0 ± 376.5	479.1 ± 89.1	0.21 ± 0.05
C + Gst + Pyr females (6)	2553.0 ± 481.6	392.5 ± 46.9	0.18 ± 0.03
Moderate stunning at 37 °C			
C (12)	1705.9 ± 236.5	455.9 ± 95.8	0.26 ± 0.02
C + Clzp (5)	568.3 ± 32.7	427.0 ± 7.2	0.76 ± 0.05
C HpT (6)	918.3 ± 17.3	455.2 ± 16.6	0.49 ± 0.01
C+ Clzp HpT (5)	569.9 ± 28.75	353.0 ± 34.7	0.64 ± 0.09
C HypoT	1129.6 ± 108.7	413.1 ± 25.2	0.38 ± 0.04
Severe stunning at 37 °C			
C (11)	2428.9 ± 193.7	296.8 ± 49.0	0.13 ± 0.02
C+ GST 5 mg/kg (ip) males (5)	1865.6 ± 184.4	349.5 ± 30.9	0.19 ± 0.02

Control (C)

How does ischemia modify the SR Ca^{2+} content? Evidences were obtained by comparing the effects of perfusing Krebs-caff-low Na in hearts treated with CPG, with and without an ischemic period [20]. The CIC in non-ischemic CPG hearts had a profile different to that of ischemic hearts, with a low initial peak growing afterwards (Fig. 8.3a, b) in a pattern with higher AUC-ΔLVP, AUC-ΔHt and EC than those from the ischemic CPG hearts (Table 8.1). This pattern suggests that ischemia reduced the SR Ca^{2+} content for reperfusion, which may be caused by a leak of Ca^{2+}. This mechanism explains the diastolic contracture seen during the end of I and still more during the first minutes of R. This interpretation agrees with that of Valverde et al. [46] who demonstrated that in the mice hearts ischemia induced the leak of Ca^{2+} from SR during the first minutes of reperfusion, which becomes in diastolic contracture and reduced contractility (PICR). The fact that in non-ischemic rat hearts the relaxation of the CIC was slower than in the reperfused ischemic hearts indicates that restitution of mitochondrial metabolism after a brief ischemia would stimulate Ca^{2+} uptake, thus reducing cytosolic Ca^{2+}. The increase in mitochondrial

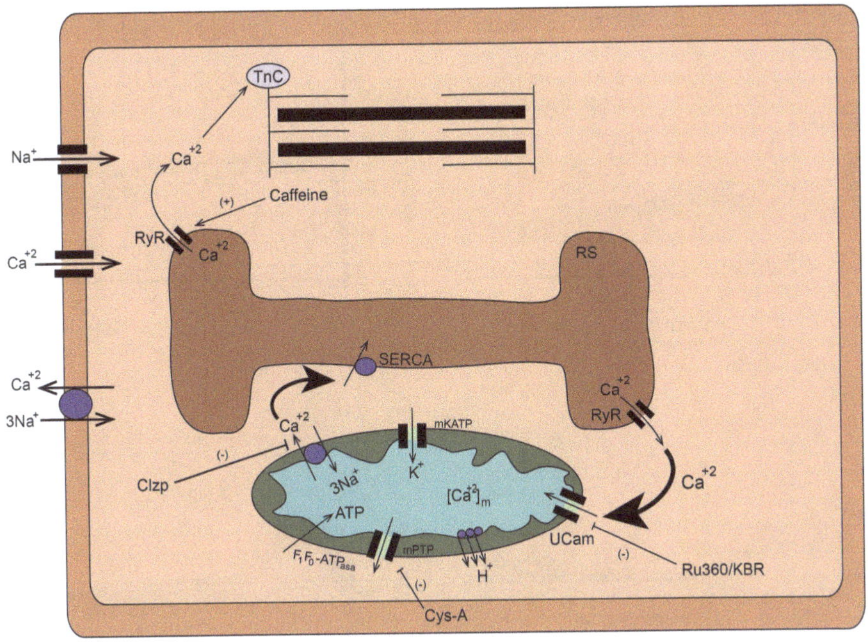

Fig. 8.2 Schematic representation of a cardiac cell with the cellular transporters which could influence in the functional interaction between mitochondria and sarcoplasmic reticulum (SR) and the specific drugs described in this work

Ca^{2+} uptake induces the activity of the metabolic enzymes with the consequent rise in the heat release. This was evidenced from the higher AUC-ΔHt in ischemic than in non-ischemic CPG hearts (Table 8.1).

The inhibition of mNCX with Clzp also reduced the cardioprotection of CPG, since PICR reached up to 50 % ± 6.4% [20]. This effect can be attributed to a reduction in the SR Ca^{2+} content, since Clzp reduced the CIC (Fig. 8.3a) and the AUC-LVP of hearts with CPG, without changing the EC (Table 8.1). Again, results show that PICR depends on the SR Ca^{2+} content regulated by the mitochondrial Ca^{2+} extrusion through the mNCX. Moreover, blockade of UCam with 1 μM Ru-360 during the CPG perfusion also reduced PICR of rat hearts (to 39.9% ± 11.1%) and increased Ht, with reduction of muscle economy (P/Ht) to about 24% of initial economy during R (CPG hearts recovered up to 59% of initial muscle economy) but without an important diastolic contracture [23]. The effects suggest that uptake of mitochondrial Ca^{2+} by UCam (triggered by the cytosolic Ca^{2+} transient) is essential to determine $[Ca^{2+}]m$, metabolism and ATP resynthesis, which in turns maintains contractility. But simultaneously, the UCam blockade may be increasing cytosolic Ca^{2+} whose removal seems to be energetically more expensive. In another work [22] we found that in cardioplegic hearts whose UCam was blocked with 5 μM KB-R7943 (CPG + KBR) the CIC was higher than that obtained in CPG hearts, with similar Ht (Fig. 8.3c, d) and AUC-ΔHt, by which the EC was reduced from 0.12 to 0.05

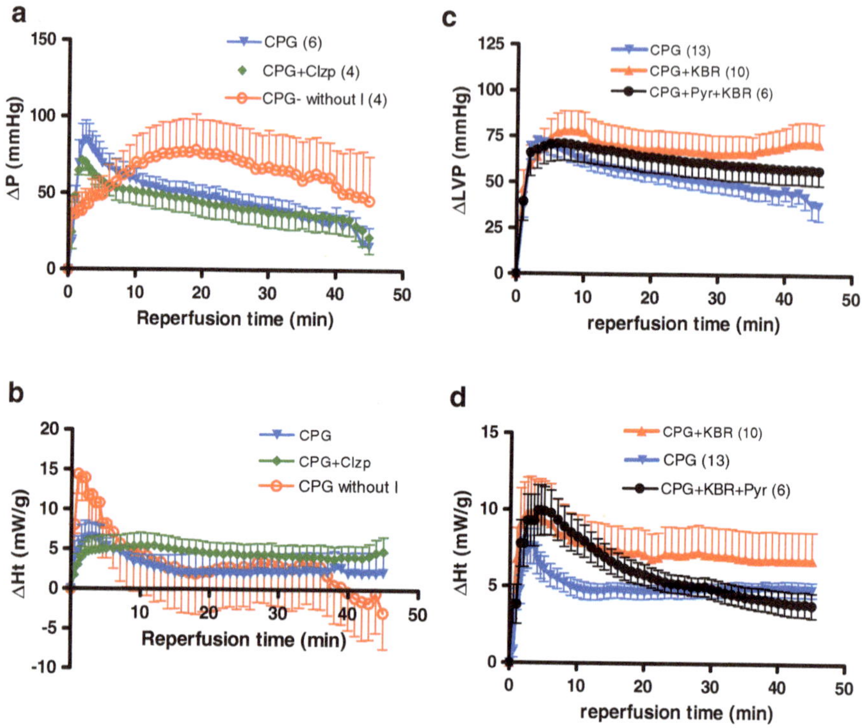

Fig. 8.3 Effects of cardioplegia (*CPG*, Krebs-25 mM K-0.5 mM Ca) and of mNCX blockade with 10 μM clonazepam (Clzp), non-ischemia, UCam blockade with KBR or UCam stimulation with pyruvate (*Pyr*) on the contracture (ΔLVP, in **a** and **c**) and the heat rate (*ΔHt*, in **b** and **d**) produced during reperfusion with Krebs-10 mM caffeine-36 mM Na^+

(Table 8.1). The increase of CIC can be explained by the fact that UCam inhibition leaves more Ca^{2+} available in cytosol for myofilaments during CIC [22]. Although KBR was first known as a blocker of the sarcolemmal NCX, then it was described as inhibitor of the UCam [47] and blocker of L-type Ca channels [48] all of them at the same IC50 of 5 μM, but only the inhibition of UCam explains this effect. Then, both types of results obtained when UCam is blocked under CPG with Ru360 or KBR suggested that the UCam is necessary to determine ATP resynthesis and contractility and contributes to reduce diastolic cytosolic Ca^{2+}. The blockade of UCam with KBR reduced the EC in agreement with the lower metabolic rate and Ca^{2+} removal both of which become in a higher contracture.

When adding the aerobic substrate **pyruvate** (Pyr, at 10 mM) to the perfusion with CPG + KBR, the effect due to UCam blockade was partially reverted, since the CIC and the AUC-ΔLVP were reduced from the value of CPG + KBR towards the value of CPG (Fig. 8.3c). The AUC-ΔHt was not different to that of CPG + KBR but it was higher than that of CPG, while the EC was similar for the three conditions

(Fig. 8.3d and Table 8.1). Results suggest that Pyr reduced the CIC due to a lower $[Ca^{2+}]i$ consequent to the stimulation of mitochondrial Ca^{2+} uptake. Pyr did not induce changes in the energetic consumption of CPG + KBR because the extra $[Ca^{2+}]m$ also increases the mitochondrial metabolism. The effects of pyruvate agree with those attributed to the increase in the cytosolic phosphorylation potential [49, 50]. Thus, a better cardiac energetic state improves the SR calcium handling and release, with the consequent positive inotropic effect [51, 52]. Also, Pyr increased myofilament calcium sensitivity [53]. Nevertheless, in rat hearts perfused at 30 °C, Pyr had induced negative inotropism with prolongation of the relaxation, both of which were reversed by previous treatment with KBR. The negative inotropic effect of Pyr was interpreted by a combination of acidosis and stimulation of UCam [54]. At the metabolic level, Pyr is an aerobic substrate that induces acidosis caused by a symporter for H^+ and Pyr uptake, a process similar to the mitochondrial pyruvate carrier (MPC) [55]. Intracellular acidification produces direct inhibition of the RyR2 channel [52, 56], prolonged relaxation [57] and lenghtening of the decay of Ca^{2+} signals in the transient and in the caffeine-induced rapid cooling contractures in guinea-pig cardiomyocytes, associated to a reduced Ca^{2+} efflux through the SL-NCX [58]. Interestingly, the reversion of the negative inotropism of Pyr by KBR suggested that Pyr must be decreasing the Ca^{2+} available for contraction when it stimulates the mitochondrial Ca^{2+} uptake through the UCam [54]. On mithocondria Pyr acts as a substrate to the electron transport chain and contributes to generate of the proton-motive gradient, which is the driving force for the Ca^{2+} uptake by the UCam. The rise in $[Ca^{2+}]m$ stimulates the mitochondrial metabolism, which is evident from the increase in Ht [54]. The mitochondrial mechanism of Pyr was also evidenced in isolated cardiomyocytes loaded with Rhod-2 (in conditions of 4 °C to favour mitochondrial accumulation). The increase of $[Ca^{2+}]m$ after perfusion Pyr 10 mM was interpreted from the increase of the relative fluorescence signal (F/Fo). When myocytes were perfused with CPG, both signals, Rhod-2 (mitochondrial Ca^{2+}) and Fluo-4 (cytosolic Ca^{2+}), were also increased in agreement to the activation of L-type Ca^{2+} channels during the cardioplegic depolarization and consequent Ca^{2+} influx to cytosol and mitochondria [22, 24, 54]. But the addition of Pyr 10 mM to CPG induced reduction of Fluo-4 signal and further increase of Rhod-2 signal [54], suggesting that Pyr favoured the UCam uptake. In conclusion, in the effects of Pyr, mitochondria influences on Ca^{2+} movements and contractility, with energetic consequences. The cardioprotective effects were strongly improved when 10 mM Pyr was present before, during, and after ischemia in hearts at 37 °C [54], in accordance with the temperature dependences of metabolic reactions, phosphorylation potential, and Ca^{2+} uptake of SR.

In another work we studied the effects of the **phytoestrogen genistein** (Gst) in the models of stunning consequent to I/R. This phytoesterol is one of the soy isoflavones, recommended for states of estrogenic deficiency. It had been described that Gst reduced myocardial infarct in ovariectomized rats [59], and in rabbit hearts with coronary occlusion when it was used as post-conditioner [60]. In ischemic rabbit hearts Gst had also reduced the infarct size but without improving contractility [61].

There were not previous studies of Gst on the stunning models, while in oxygenated papillary muscles or cardiomyocytes there were described negative or positive inotropism depending on the animal species and attributed to blockade of Ca^{2+} and K^{+} channels [62] or increase in sarcorreticular Ca^{2+} uptake [63, 64]. First we evaluated the effects on female and male rat hearts with the model of stunning at 30 °C. Unexpectedly, Gst reduced PICR in males and recovered the pre-ischemic P in females [26]. Looking forward the mechanism, the PICR of males was increased by previous blockade of phosphatases with sodium ortho-vanadate (OV), suggesting that the negative inotropism was due to the typical inhibition of tyrosine kinases by Gst, which affects some mechanisms especially Ca^{2+}-channels. Nevertheless, this effect was not seen in females, because it was counteracted by Gst stimulation of the SR Ca^{2+} content. When it was evaluated by reperfusing hearts with Krebs-10 mM caffeine-36 mM Na^{+}, we found that AUC-ΔLVP of the CIC was similar for females and males, in the abscence and the presence of Gst (Table 8.1). But the heat release was lightly higher with Gst, as well as the AUC-ΔHt and the EC (Table 8.1). Still more they increased in hearts treated with OV and with 10 mM pyruvate (Pyr), the last one due to the stimulation of Ca^{2+} uptake and metabolism [26]. Gst reduced the relaxation rate of CIC in both female and male hearts, which indicates that Gst inhibits the UCam, since it was partially reduced by Pyr. Results suggest that inhibition of UCam by Gst contributes to the diastolic contracture during I/R, among the other mechanisms of Gst [26]. More recently, results still not published showed that Gst was really cardioprotective, more in males than in females rats, in a model of severe stunning at 37 °C when it was administered by injection at 5 mg/Kg the day before (Fig. 8.4a). Nevertheless, when evaluated whether the effect was due to an increased SR Ca^{2+} content, the hearts from rats treated with 5 mg/Kg Gst and reperfused with Krebs-caff-low Na slightly reduced the CIC (Fig. 8.4b), and increased EC (Table 8.1) suggesting that Gst could contribute to a lower diastolic Ca^{2+} leak from SR during I/R, thus reducing the diastolic contracture and dysfunction. But this mechanism seems to be less important by comparison with the improvement of contractility observed during reperfusion. So, other mechanisms in vivo have to be explored to understand Gst cardioprotection.

Energetical Comparison Between Ischemia/Reperfusion and Hypoxia/Reoxygenation

Certainly, there are differences between I/R and hypoxia-reoxygenation (H/Rox). Although both pathological situations share the decrease in oxygen supply as a common event, there are some others such as accumulation of metabolites, profound acidosis and strong calcium overload which are the heritage of the I/R injury [65, 66].

A reduction in oxygen supply to the myocardium (hypoxia) drives to an energetic dysfunction at mitochondrial level so that ATP production becomes insufficient

Fig. 8.4 Effects of 5 mg/Kg of the phytoestrogen genistein on the maximal pressure development (P, as % of initial) in female (FRH) and male (MRH) rat hearts exposed to severe stunning (30 min ischemia at 37 °C) (**a**), and on the contracture (ΔLVP) produced during reperfusion with Krebs-10 mM caffeine-36 mM Na⁺ of the ischemic rat hearts (**b**)

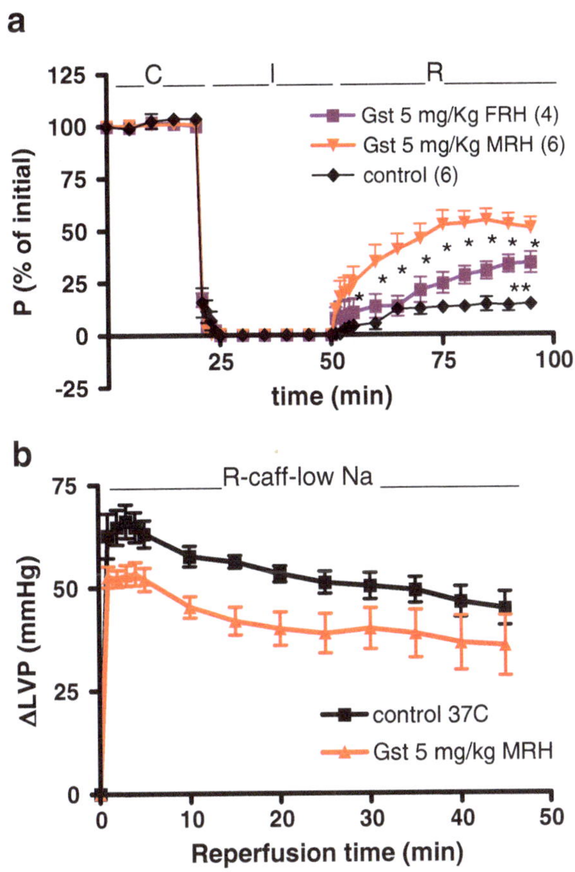

for the imposed demand. The energetic dysfunction affects not only post-ischemic or post-hypoxic myocardial contractile function but also ionic homeostasis, being that of calcium the most relevant. The heart is able to change from aerobic to anaerobic metabolism for obtain energy from the ATP, producing lactic acid and hence cytosolic acidosis. While perfusion is maintained, metabolites and protons produced are continuously eliminated from the cells and consequently the possibility for a strong cytosolic and mitochondrial calcium overload during reoxygenation is decreased respect to post-ischemic reperfusion. Nevertheless, the generation of reactive oxygen species (ROS) during hypoxia and reoxygenation or I/R [67] would contribute to modify the activity of some calcium transporters and proteins that are involved in contractile recovery after a hypoxic or ischemic episode. On this connection, it was described that the activity of the ryanodine receptor (RyR) release channel in the SR is increased by oxidizing conditions [68, 69]. In addition it was reported that although low levels of ROS increases SERCA2a activity, in the model of myocardial severe I/R ROS contributes to inactivate it [70, 71]. Furthermore, the reported direct oxidizing effect of ROS on the protein titin increases the passive stiffness of heart myofibrils [72]. Of course, the irreversible modification of these activities depends on the extension and time of the injury.

The model of H/Rox would allow to study only the oxygen-dependent processes that are included within the I/R event. Since I/R and H/Rox differ from one another, the role of mitochondria and its relationship with the sarcoplasmic reticulum in Ca^{2+} handling is expected to be different accordingly.

At the first sight, the model of H/Rox appears to be less dangerous than that of I/R. In fact, unpublished experiments carried out in our laboratory showed that to get similar post hypoxic contractile recovery than that obtained after 45 min ischemia at 30 °C, it was necessary to spin out the time of exposure to hypoxic perfusion to 60 min. On the other hand, at the end of the hypoxic period contractile activity was still remaining (about 25%, see Fig. 8.5a) while contractility decreased to zero after a few minutes of ischemia. In both situations the total muscle economy at the end of reoxygenation/reperfusion was similarly decreased (see Figs. 8.4 and 8.5b and reference [54]) although in the H/Rox model it seems to be faster. The mechanical and energetic responses to reoxygenation in the presence of Krebs caff-low Na^{+} also differ from that of I/R responses. Thus, the CIC on reoxygenation increased during the first minutes and then decreased toward the last hypoxic value (Fig. 8.5c). The relaxation rate of the CIC was higher than that observed in post-ischemic similar perfusion in identical period (compare Figs. 8.1a and 8.5c). Unlike that observed in post-ischemic CIC, after 10 min of post-hypoxic CIC spontaneous contractile activity appeared and stayed during the rest of reoxygenation period. This mechanical performance was accompanied by a transient increase in total heat production which reached a peak, then decreased (Fig. 8.5d) and later increased again at the time in which spontaneous contractile activity was developed. Once again, Ht behavior was different from that seen in post-ischemic perfusion (compare Figs. 8.1b and 8.5d).

These results suggest the following scenario that is compatible with a lesser mitochondrial calcium overload under H/Rox as compared with I/R [73, 74], yielding more feasible conditions for faster recovery. On reoxygenation with caffeine and low sodium, SR calcium content is massively released activating both, contracture and metabolism at mitochondrial level, which increase heat production. Afterwards, contracture relax quickly accompanied by a partial decrease of Ht suggesting that cytosolic Ca^{2+} is removed not only by mitochondria (without calcium accumulation above a deleterious value for uncoupling) but also partially by sarcolemmal Ca^{2+} -ATPase, according that mitochondrial ATP production is reestablished. Although SERCA may be also removing Ca^{2+} and releasing heat by ATP consumption, the effect of caffeine on RyR must be reducing the net uptake of Ca^{2+} by the SR. The relaxation of CIC suggest that H/Rox affect less the Ca^{2+} homeostasis than I/R, thus permitting the development of spontaneous contractile activity later on reoxygenation even in the presence of caffeine, which was absent in the I/R model.

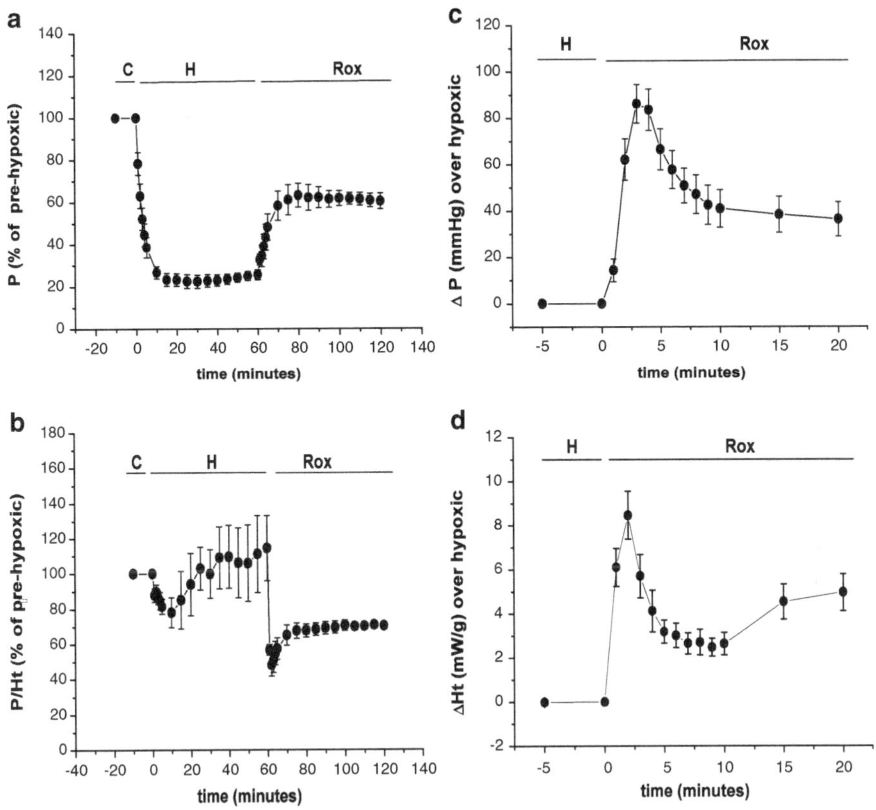

Fig. 8.5 Contractile recovery (**a**) and total muscle economy (P/Ht) (**b**) during hypoxia (*H*) and reoxygenation (*Rox*) of rat hearts. (**c**) Caffeine-low Na induced contracture, and (**d**) total heat production (Ht) over hypoxic value upon reoxygenation with Krebs-10 mM caffeine-36 mM Na^+

Influences of Temperature and Duration of Ischemia on the Mit/SR Functional Interaction

Temperature affects many cellular processes with different rates, more to the primary or secondary active transporters than to channels, binding or passive movements [43]. Obviously, it is possible to obtain two models of moderate stunning with about the same degree of contractile recovery at different temperatures with different periods of ischemia, such us 45 min I at 30 °C and 20 min I at 37 °C in our experiments [20, 25]. Nevertheless, the cellular processes to maintain Ca^{2+} homeostasis are not necessarily functioning at the same degree. Comparing the Mit/RS functional interaction in both models by reperfusing them with Krebs-10 mM caffeine-36 mM Na^+, the model of moderate stunning at 37 °C developed lower peak of LVP and AUC-ΔLVP than the CIC at 30 °C (Fig. 8.6a and Table 8.1). Results suggest that the Ca^{2+} content of SR was reduced at higher and physiological temperature. Thus, temperature increased the rate of cytosolic Ca^{2+} removal by the cellular transporters still active under caffeine – low Na^+ media (mainly SL-Ca-ATPase and UCam, but much less the SL-NCX). Results agree with some reports on rabbit hearts, in which all Ca^{2+} transporters have Q10 of 2–3 [43]. The low temperature mainly reduces the activity of Na,K-ATPase and increases $[Na^+]i$, consequently reducing Ca^{2+} efflux through the SL-NCX becoming in a higher SR Ca^{2+}content. Evidences of this influence were provided by the magnitude of Ca^{2+} transients and caffeine- and cold-induced contractures [43, 75, 76]. More recently, in rat cardiomyocytes exposed to 10 min of simulated ischemia and reperfusion, it was described the same temperature dependence, since hypothermia to 34 and 30 °C gradually increased Ca^{2+} transients and cell shortening either before or after ischemia [77]. Also, the authors showed that at 30 °C the relationship between contractility and $[Ca^{2+}]o$ was steeper than at 37 °C, and they attributed this effect of hypothermia to the increase in Ca^{2+}-sensitivity of myofilaments. All these effects of temperature remained during reperfusion in spite of the reduction in contractility and Ca^{2+} transient characteristics of the stunning.

The influence of temperature on the rate of Ca^{2+} removal was evidenced by the values of Ht associated to the CIC, which at 37 °C were higher than at 30 °C (compare both models of moderate stunning at the start of CIC in Fig. 8.6b). Consequently, the AUC-Ht was higher in the moderate stunning model at 37 than that at 30 °C, as well as the EC (Table 8.1). The comparison suggests that maintaining the CIC was energetically more expensive at 37 °C than at 30 °C. The difference in Ht may be mostly indicating the increase in Ca^{2+} removal rate, since ΔLVP are not very different especially at the end of CIC. Considering the energetic equivalents from the enthalpic changes of each transporter [78, 79], it is possible to calculate the energy for the futile Ca^{2+} cycling through SERCA induced by caffeine, which moves about 85 nmol $Ca^{2+}.g^{-1}.s^{-1}$ in rat hearts [43]. This cycling would release a heat rate of at least 3.5 mW/g, since SERCA moves 2 Ca^{2+}/ATP and the enthalpic change is 80 kJ/mol ATP. Moreover, the sarcolemmal Ca^{2+}-ATPase would remove Ca^{2+} at maximal rate during the CIC, while the SL-NCX would be slow under the low Na^+-media. The heat release of sarcolemmal Ca^{2+}-pumps can be calculated in 0.3 mW/g (from the described maximal flux of 3.7 nmol $Ca^{2+}g^{-1}$ sec^{-1} [43]). Then, the difference in heat released between both models of moderate stunning at 37 °C vs 30 °C (ΔHt of 7–3.5 mW/g in Fig. 8.6b) is higher than the heat rate calculated as the sum of both pumps, energy and may include the Ca^{2+} uptake to mitochondria. The flows of UCam were estimated under physiological and overload conditions in isolated mitochondria [80], respectively in 32 to 350 nmol $Ca^{2+}.g^{-1}.s^{-1}$. Considering the energetic equivalent of 477 $kJ.mol^{-1}$ O_2 and that the electron chain moves 12 H^+ by O_2 and 2 H^+ by Ca^{2+}, [78]) mitochondria could dissipate from 3 up to 31 $mW.g^{-1}$. Thus the increase in heat release during CIC at 37 as compared with 30 °C is related to the rise in Ca^{2+} fluxes by the pumps and the mitochondrial uptake. In conclusion, when temperature increases the CIC is lower and less economical, because of the increased Ca^{2+} cycling through SERCA, sarcolemmal Ca^{2+}-ATPase and the mitochondrial Ca^{2+} uptake.

At 37 °C we also assessed a model of severe stunning (30 min I/45 min R) which produced a PICR of 14.5 % ± 2.4% of initial P, and reduced the pre-ischemic P/Ht

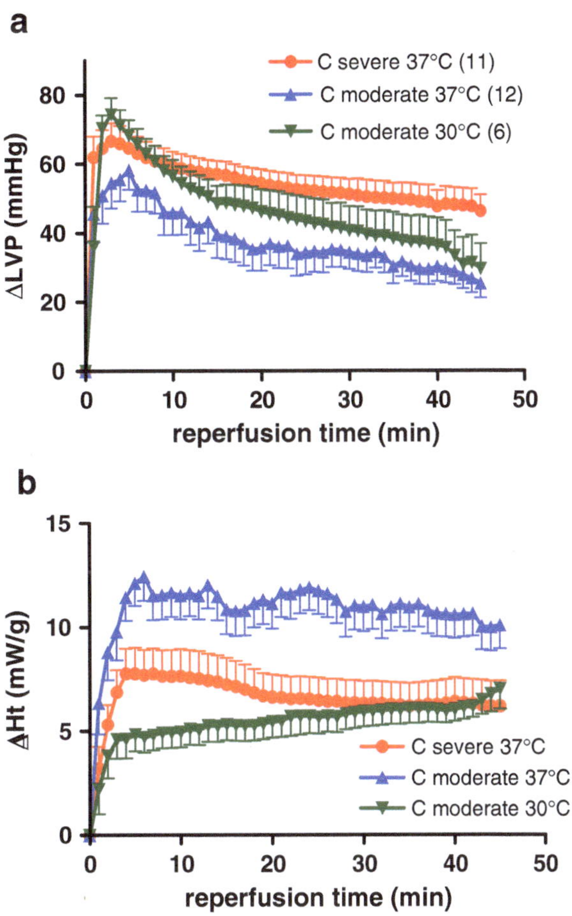

Fig. 8.6 Effects of temperature and degree of stunning on the contracture (Δ*LVP*) (**a**) and the heat rate (Δ*Ht*) (**b**) produced during reperfusion with Krebs-10 mM caffeine-36 mM Na^+ of ischemic rat hearts

to 19.6% ± 5.5% at the end of R. When we studied the role of SR in this model by reperfusing Krebs-caff-low Na, the peak of the CIC was similar but relaxation was slower than the respective values of the CIC in the model of moderate stunning at the same temperature of 37 °C (see Fig. 8.6). The ΔHt associated to CIC was lower than that of the moderate stunning model, and the respective values of AUC and the EC in Table 8.1 show that the CIC was associated to a lower energy consumption. These results suggest that after a longer period of ischemia mitochondria are in a worse energetic situation, which explains certain difficulty to uptake Ca^{2+}, with a slower relaxation and reduced ΔHt. That is, the energetic consumption (EC) for maintaining the CIC was lower in the severe stunning because of mitochondria have reduced electrochemical gradient, with lower Ca^{2+} uptake and metabolism than in the moderate stunning model.

Effects of Thyroid State on the Ischema/Reperfusion Dysfunction

Hyperthyroidism is considered a risk factor for angor, since when it is severe and untreated it induces tachycardia, high cardiac workload, and risk of atrial fibrillation related to the up-regulation of β-adrenergic receptors [81]. Some patients suffer symptoms of cardiac ischemia because of the high oxygen demand related to the increased cardiac output [82]. Nevertheless, there were described cardiac benefits of levothyroxine (T4) or liothyronine (T3) therapies under critical transient conditions of myocardial stunning consequent to regional or global I/R [83–85]. T4 or T3 were recommended to obtain a rapid restoration of energy stores and myocardial function after organ transplantation [85]. In a previous work we showed that hyperthyroidism (*HpT*), induced in rats by daily injection of 20 $\mu g.kg^{-1}$T3 during 15 days, developed higher PICR and improved total muscle economy (P/Ht) with lower diastolic contracture (ΔLVEDP) than the euthyroid (*EuT*) hearts [25]. To understand mitochondrial bioenergetics and the functional interaction between SR and mitochondria in this pathology, *HpT* hearts exposed to a model of moderate stunning (20 min ischemia at 37 °C) were reperfused with Krebs-caff-low Na^+. Fig. 8.7a shows that *EuT* hearts developed a higher initial contracture than *HpT* hearts and higher AUC-ΔLVP (Table 8.1). This result suggested that *HpT* hearts had a lower Ca^{2+} release from SR than *EuT* hearts. Simultaneously, the ΔHt (Fig. 8.7b) and AUC-ΔHt (Table 8.1) were similar for *HpT* and *EuT*, indicating that Ca-cycling between SERCA and mitochondria was similar in *EuT* and *HpT* hearts. Thus, a lower Ca^{2+} leak from SR could be the responsible for the reduced LVEDP and higher PICR of *HpT* hearts regarding the *EuT* ones.

As in previous conditions, the functional interaction between mitochondria and SR was assessed by blocking the mNCX with 10 μM clonazepam (Clzp) before and during ischemia. In beating *HpT* hearts exposed to 20 min ischemia, Clzp reduced contractile recovery (from 108.8% ± 11.6% to 13.0% ± 3.8% of initial P in the abscence and the presence of Clzp, respectively) while almost improved the PICR in the *EuT* hearts (from 77.5% ± 3.2% to 98% ± 14%, respectively in the absence and presence of Clzp) [25]. Nevertheless, during the reperfusion with Krebs-caff-low Na Clzp reduced the CIC in both conditions, but much more in *EuT* than in *HpT* hearts, as well as AUC-ΔLVP (Fig. 8.7a and Table 8.1). These effects were not followed by significant differences in the AUC-ΔHt, whereby the energetic consumption (EC) during CIC was increased by Clzp (Fig. 8.7b and Table 8.1).

These results suggested that mitochondria regulate the SR Ca^{2+} content by functional interaction from the mNCX to SERCA in both conditions, *EuT* and *HpT*, but the improvement of PICR may be higher in *HpT* because they had attenuated the SR Ca^{2+} leak at the start of R. The rise in EC induced by Clzp in *HpT* and *EuT* clearly shows that mitochondria increased their metabolism when Ca^{2+} extrusion was reduced by Clzp. This is in agreement with the increase in the free mitochondrial Ca^{2+} found during perfusion with Clzp in isolated non-ischemic cardiomyocytes from *HpT* rats [25]. Cells were loaded with Rhod-2 (at 4 °C and washed at 37 °C, as

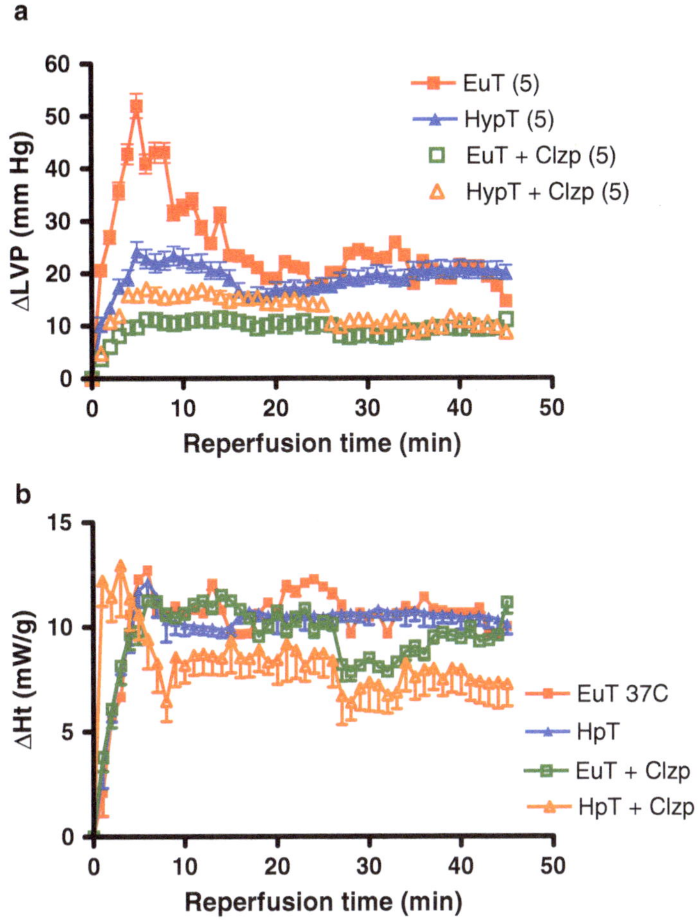

Fig. 8.7 Effects of hyperthyroidism (*HpT*) and blockade of mNCX with 10 μM clonazepam, in comparison with euthyroid rat hearts (*EuT*), on the contracture (Δ*LVP*) (**a**) and the heat rate (Δ*Ht*) (**b**) produced during reperfusion with Krebs-10 mM caffeine-36 mM Na^+ of ischemic rat hearts exposed to moderate stunning (20 min ischemia at 37 °C)

proposed by Trollinger et al. [86]) to measure the fluorometic relative signals (ΔF/Fo) over time under confocal microscopy [25]. Also, either the potentiation of PICR and the rise in the Rhod-2 signal produced by Clzp were partially reversed by addition of ouabain, showing that the high cytosolic Na^+ induced by inhibition of the Na/K-ATPase favored the Ca^{2+} extrusion by mNCX, thus opposing to the inhibition of Clzp [25]. Results show the important role of mitochondrial transporters of Ca^{2+} in the mechano-energetical recovery of the hyperthyroid heart exposed to I/R.

The different responses to Clzp of PICR from *EuT* and *HpT* hearts suggest that there have to be a different threshold of Ca^{2+} for triggering the mitochondrial dysfunction, and this one would be lower in *HpT* than in *EuT* hearts. The possibility that the opening of the mPTP would participate in the dysfunction induced by Clzp in post-ischemic *HpT* hearts was evaluated by perfusing 0.2 μM cyclosporine-A (Cys-A) during all the I/R protocol. The activation of mPTP causes loss of mitochondrial calcium content, swelling and dysfunction [44], and occurs during the first minutes of R [87]. Effectively, Cys-A improved the PICR of *HpT* hearts treated with Clzp up to 66.4% ± 15.4%, without altering the PICR in the *EuT* analogue group [25]. This result confirmed that a same degree of mNCX blockade induced dysfunction by mPTP opening in *HpT* hearts but not in the *EuT* ones.

Moreover, the role of mKATP channels was explored in *HpT* hearts, by blocking them with 100 μM 5-hydroxidecanoate (5-HD), which reduced PICR up to 66.6% ± 8.3% in *HpT* hearts and 46.9% ± 4.0% in the *EuT* ones [25]. Thus, the hyperpolarizing role of mKATP resulted more critical in *HpT* than in *EuT* hearts during the moderate stunning. It is known that at 100 μM, 5-HD selectively blocks the mK_{ATP} channels [88] and induces mitochondrial energy dissipation to maintain the $\Delta\Psi m$ in beating hearts [89]. Accordingly, we have previously shown that in rat hearts in resting state by perfusing CPG, 5-HD increased Ht before and after I [23]. In *HpT* hearts the treatment with 5HD reduced also the muscle economy (P/Ht) but not in *EuT* ventricles, suggesting again that mitochondria from *HpT* are more sensitive to dysfunction when mKATP are blocked that those of *EuT*. However, 5HD induced diastolic contracture in both, *HpT* and *EuT* hearts, suggesting that the mKATP support part of the mitochondrial Ca^{2+} uptake which contributes to maintain a low diastolic $[Ca^{2+}]i$ in both groups.

The UCam was also essential for the coupling between mitochondrial activity and contractile recovery, without differences between *HpT* and *EuT* hearts, since blocking it selectively with 1 μM Ru-360 [90, 91] reduced PICR to about 17% of initial P and P/Ht to 23% and 13% of the initial P/Ht, respectively [25]. Then, results show that blockade of UCam reduced the post-ischemic contractility as far as muscle economy, suggesting that mitochondrial metabolism was reduced as far as the sarcorreticular Ca^{2+} release, which in turn determines contractility. In this way, Ca^{2+} uptake through UCam regulates $[Ca^{2+}]m$ and metabolism, and consequently the extrusion by the mNCX, which finally influenced the SR Ca^{2+} content and release, and the contractile recovery (Fig. 8.2). Moreover, Ru-360 increased the diastolic contracture during reperfusion in both *HpT* and *EuT* hearts, suggesting again that mitochondria contributes to regulate the diastosolic $[Ca^{2+}]i$ in both groups. A similar potentiation of hypercontracture was described in cardiomyocytes exposed to simulated chemical ischemia, in which blockade of UCam prevented the mitochondrial overload and mPTP opening but developed a hypercontracture associated to high cytosolic $[Ca^{2+}]$ that drove to apoptosis [92].

On the other hand, in a model of severe stunning, obtained with 30 min of I and 45 min R, the cardioprotection of hyperthyroidism was missing. Figure 8.8 shows still not published results of *EuT* and *HpT* hearts which similarly reduced their PICR and muscle economy (P/Ht) to about 15%. The muscle economy (P/Ht) has the same pattern, although the *HpT* hearts developed a lower diastolic contracture than the *EuT* ones. In both thyroid states, the dysfunction can be attributed to the opening of mPTP, since perfusion of Cys-A during the whole I/R period improved

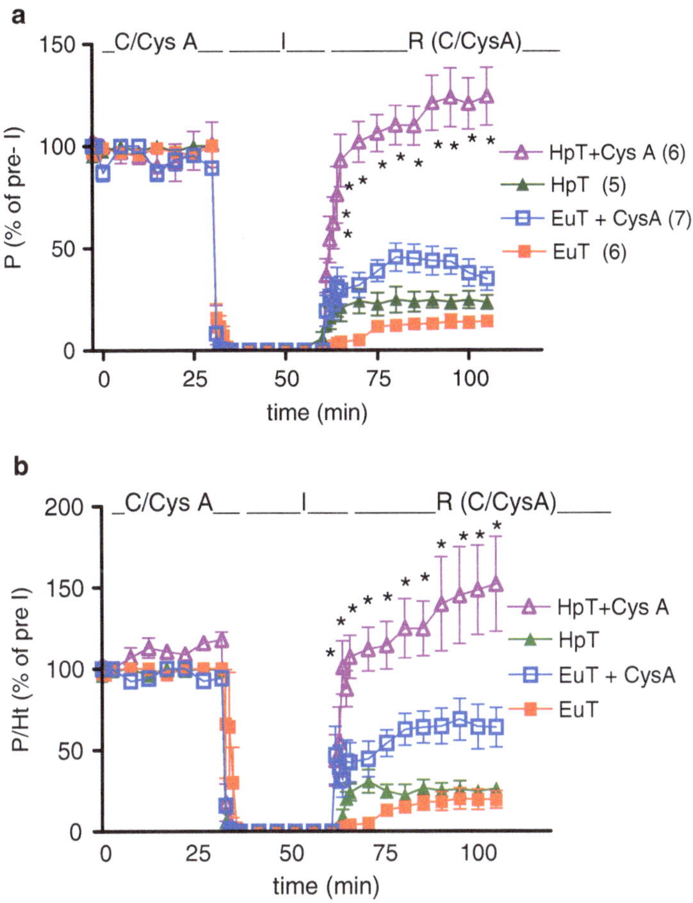

Fig. 8.8 Effects of hyperthyroidism (*HpT*) on the maximal pressure development (*P*, as % of initial) (**a**) and on the total muscle economy (*P/Ht*, as % of initial) of rat hearts exposed to severe stunning (30 min ischemia at 37 °C). Note that the dysfunction was reduced by perfusion of 0.2 μM cyclosporine-A during the I/R in both hyperthyroid (*HpT*) and euthyroid (*EuT*) hearts

PICR and muscle economy, more in *HpT* than in *EuT* hearts (Fig. 8.8). As in previous works, Cys-A did not prevented the diastolic contracture, and this had been previously attributed to non-specific binding of Cys-A [23, 87, 93].

There are different evidences about cardiovascular effects of thyroid hormones, which differ with the type and degree of disease. Low T3 has been considered a beneficial adaptive mechanism activated under stress conditions. Several studies have shown that the low T3 syndrome has adverse prognostic on various cardiac disorders, and T3 is reduced after myocardial infarct. But many of the cardiac alterations observed in subclinical hypothyroidism are reversed once thyroid function has been normalized [94]. The several thyroid effects affect many of the cellular mechanisms for regulating Ca^{2+} and contractility in heart. Because of that, the study of thyroid alteration on different degrees of ischemia gives new understanding to the influence of thyroid hormones on prevention of angor. Surely, a different situation will be found in myocardial infarct. It is known that thyroid hormones induce genomic and non-genomic effects, which regulate cellular targets, some of them positively (SERCA, Na^+,K^+-ATPase, α-myosin, β receptors, K_V channels, Na^+-channels, troponin I, adenilate cyclase, TRα1, PI3K-Akt) and others negatively (phospholamban, SL-NCX) [42, 95–98]. Thyroid hormones are known as regulators of mitochondrial biogenesis and function, such as opening of the protective mKATP channels due to up-regulation of mitochondrial transcription factor A (mtTFA) and peroxisome proliferator activated receptor gamma coactivator 1 alpha [99]. Moreover, it was reported a correlation between the rate of Ca^{2+} uptake to SR and the PLB/SERCA2a ratio through the hypothyroid, euthyroid, and hyperthyroid hearts, thus determining cardiac contractility [100]. In spite of the expected increase in relaxation rate and energy consumption of SERCA, in our basal conditions of hyperthyroid isolated rat hearts beating at 37 °C, the initial Ht (15.4 ± 0.9 mW/g) was not significantly higher than that of *EuT* rats (17.8 ± 0.9 mW/g) as well as the initial muscle economy (P/Ht) [25]. The effect on SERCA is partially due to the well-known up-regulation of the β-adrenergic receptors [81, 101]. But our protocols in isolated rat hearts did not include adrenergic agonists, in a way that β-adrenergic pathway would not be involved. Also, rat hearts have a more important function and store of SR than rabbits and humans [43], by which it seems less probable to find an increase in SERCA activity by the treatment of 15 days with T3.

Finally, in agreement with the clinical evidences about short periods of **hypothyroidism**, our preliminar results show that under a model of severe stunning (30 min I/45 min R) hearts from rats treated with methimazol 0.02% during 15 days (*HypoT*) had cardioprotection respect to the euthyroid ones (Fig. 8.9). The post-ischemic contractility (PICR) and the muscle economy (P/Ht) were increased and diastolic contracture was reduced in *HypoT* in comparison with *EuT* hearts.

Diminished expression of the SERCA associated to hypothyroidism can be responsible for the lower SR load and consequently a reduced Ca^{2+} leak during the start of R, which reduce diastolic contracture. On this sense, it has been recently reported that the transient and abrupt increase in cytosolic Ca^{2+} which occurs at reperfusion in mice hearts was due to changes in the SR release, in a way that the increased Ca^{2+} sparks in ischemia transform into Ca^{2+} waves during reperfusion, which become in arrhythmia [102]. The high economy (P/Ht) suggests that metabolism would be reduced in *HypoT* rat hearts, resulting in a more efficient post-ischemic heart in comparison to euthyroids. This is in part due to the better diastolic Ca^{2+} removal which avoids diastolic contracture during I/R and improves Ca^{2+} release during the beat. Nevertheless, other experiments will be needed to understand why the *HypoT* hearts are so protected against a severe ischemia and no so much in moderate stunning. Probably other mechanisms could modify the Ca^{2+} homeostasis and signal transduction pathways upon such dysfunction. In other aspects, it was described that hearts from hypothyroid rats resulted more resistant to I/R by a reduced activation of p38 MAPK, in an effect similar but not additive to preconditioning [103].

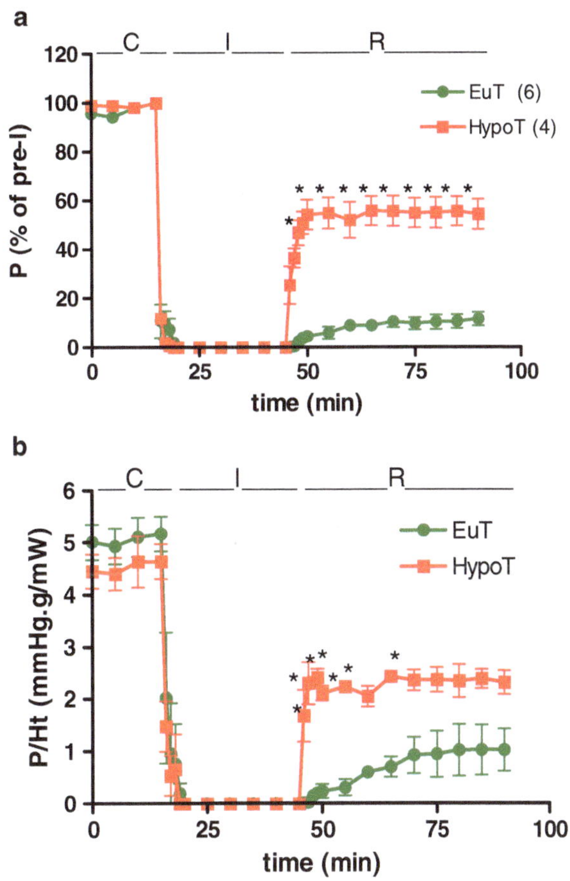

Fig. 8.9 Cardioprotective effects of hypothyroidism (*HypoT*) on the maximal pressure development (*P*, as % of initial) (**a**) and on the total muscle economy (*P/Ht*, as % of initial) of rat hearts exposed to severe stunning (30 min ischemia at 37 °C)

Conclusions

The knowledge of mechanisms and the mitochondrial role during a relatively short period of ischemia/reperfusion is important to find strategies to in order to prevent the dysfunction when a heart suffer a brief episode of coronary obstruction. The energetic approach to study isolated entire hearts provides a tool to evaluate in situ the mitochondrial functions, such as metabolism, resynthesis of ATP and Ca^{2+} fluxes which finally determine the $[Ca^{2+}]$m and metabolism. But calorimetry also permits the evaluation of other exothermic mechanisms that regulate Ca^{2+} homeostasis and contractility, such as sarco-reticular and sarcolemmal transporters and the actomyosin interaction. This review showed several evidences that we experimentally obtained about the functional interaction between mitochondria and SR in models of stunning produced by transient episodes of ischemia and reperfusion. The functional interactions under these conditions differ from that described for the I/R with infarct or H/Rox, and show a regulation of Ca^{2+} from mitochondria towards the SR, affecting the content and release during reperfusion, which determines the contractile recovery. Moreover, results suggested the existence of a critical level of $[Ca^{2+}]m$ to trigger mPTP opening and dysfunction, that is different upon thyroid diseases or cardioprotective interventions.

References

1. Csordás G, Renken C, Várnai P, Walter L, Weaver D, Buttle KF, Balla T, Mannella CA, Hajnóczky G. Structural and functional features and significance of the physical linkage between ER and mitochondria. J Cell Biol. 2006;174:915–21.
2. Griffiths EJ. Mitochondrial calcium transport in the heart: physiological and pathological roles. J Mol Cell Cardiol. 2009;46:789–803.
3. Ruiz-Meana M, Abellán A, Miró-Casas E, Agulló E, García-Dorado D. Role of sarcoplasmic reticulum in mitochondrial permeability transtion and cardiomyocyte death during reperfusion. Am J Physiol Heart Circ Physiol. 2009;297:H1281–9.
4. Guzun R, Kaambre T, Bagur R, Grichine A, Usson Y, Varikmaa M, Anmann T, Tepp K, Timohhina N, Shevchuk I, Chekulayev V, Boucher F, Dos Santos P, Schlattner U, Wallimann T, Kuznetsov AV, Dzeja P, Aliev M, Saks V. Modular organization of cardiac energy metabolism: energy conversion, transfer and feedback regulation. Acta Physiol (Oxford). 2015;213:84–106.
5. Gibbs CL, Loiselle DL. The energy output of tetanized cardiac muscle: species differences. Pflugers Arch. 1978;373:31–8.
6. Mulieri LA, Alpert NR. Activation heat and latency relaxation in relation to calcium movement in skeletal and cardiac muscle. Can J Physiol Pharmacol. 1982;60:529–41.
7. Ponce-Hornos JE, Ricchiuti NV, Langer GA. On-line calorimetry in the arterially perfused rabbit intraventricular septum. Am J Phys. 1982;243:H289–95.
8. Loiselle DS. Cardiac basal and activation metabolism. Basic Res Cardiol. 1987;82:37–50.
9. Gibbs CL, Loiselle DL, Wendt IR. Activation heat in rabbit cardiac muscle. J Physiol Lond. 1988;395:115–30.
10. Holubarsch C, Hasenfuss G, Just H, Alpert NR. Positive inotropism and myocardial energetic: influence of beta receptor agonist stimulation, phosphodiesterase inhibition, and ouabain. Cardiovasc Res. 1994;28:994–1002.
11. Ponce-Hornos JE, Bonazzola P, Marengo FD, Consolini AE, Márquez MT. Tension-dependent and tension-independent energy components of heart contraction. Pflügers Arch Eur J Physiol. 1995;429:841–51.
12. Ponce-Hornos JE, Taquini AC. Calcium effects on contractility and heat production in mammalian myocardium. Am J Phys. 1986;251:H127–32.
13. Ponce-Hornos JE, Marquez MT, Bonazzola P. Influence of extracellular potassium on energetics of resting heart muscle. Am J Phys. 1992;262:H1081–7.
14. Consolini AE, Márquez MT, Ponce-Hornos JE. Energetics of heart muscle contraction under high K+ perfusion: verapamil and Ca^{+2} effects. Am J Phys. 1997;273:H2343–50.
15. Arruda AP, Da-Silva WS, Carvalho DP, De Meis L. Hyperthyroidism increases the uncoupled ATPase activity and heat production by the sarcoplasmic reticulum Ca^{2+}-ATPase. Biochem J. 2003;375:753–60.
16. Kiriazis H, Gibbs CL. Effects of aging on the work output and efficiency of rat papillary muscle. Cardiovasc Res. 2000;48:111–9.
17. Han J-C, Taberner AJ, Kirton RS, Nielsen PM, Smith NP, Loiselle DS. A unique micromechanocalorimeter for simultaneous measurement of heat rate and force production of cardiac trabeculae carneae. J Appl Physiol. 2009;107:946–51.

18. Han JC, Tran K, Nielsen PM, Taberner AJ, Loiselle DS. Streptozotocin-induced diabetes prolongs twitch duration without affecting the energetic of isolated ventricular trabeculae. Cardiovasc Diabetol. 2014;13:79.
19. Consolini AE, Márquez MT, Ponce-Hornos JE. A comparison no-flow and low flow ischemia in the rat heart: an energetic study. Can J Physiol Pharmacol. 2001;79:551–8.
20. Consolini AE, Ragone MI, Conforti P, Volonté MG. Mitochondrial role in ischemia-reperfusion of rat hearts exposed to high-K+ cardioplegia and clonazepam: energetic and contractile consequences. Can J Physiol Pharmacol. 2007;85:483–96.
21. Consolini AE, Bonazzola P. Energetics of Ca^{2+} homeostasis during ischemia-reperfusion on neonatal rat hearts under high-[K^+] cardioplegia. Can J Physiol Pharmacol. 2008;86:866–79.
22. Consolini AE, Ragone MI, Bonazzola P. Mitochondrial and cytosolic calcium in rat hearts under high-$K^{(+)}$ cardioplegia and pyruvate: mechano-energetic performance. Can J Physiol Pharmacol. 2011;89:485–96.
23. Ragone MI, Consolini AE. Role of the mitochondrial Ca^{2+} transporters in the high-[K^+](o) cardioprotection of rat hearts under ischemia and reperfusion: a mechano-energetic study. J Cardiovasc Pharmacol. 2009;54:213–22.
24. Ragone MI, Torres NS, Consolini AE. Energetic study of cardioplegic hearts under ischaemia/reperfusion and [$Ca^{(2+)}$] changes in cardiomyocytes of guinea-pig: mitochondrial role. Acta Physiol. 2013;207:369–85.
25. Ragone MI, Bonazzola P, Colareda GA, Consolini AE. Cardioprotective effect of hyperthyroidism on the stunned rat heart during ischaemia-reperfusion: energetics and role of mitochondria. Exp Physiol. 2015;100(6):680–97.
26. Colareda G, Ragone MI, Consolini AE. Sex differences in the mechano-energetic effects of genistein on stunned rat and guinea pig hearts. Clin Exp Pharmacol Physiol. 2016;43: 102–15.
27. Gunter TE, Buntinas L, Sparagna G, Eliseev R, Gunter K. Mitochondrial calcium transport: mechanisms and functions. Cell Calcium. 2000;28:285–96.
28. Liu T, O'Rourke B. Enhancing mitochondrial Ca^{2+} uptake in myocytes from failing hearts restores energy supply and demand matching. Circ Res. 2008;103:279–88.
29. Dedkova EN, Seidlmayer LK, Blatter LA. Mitochondria-mediated cardioprotection by trimetazidine in rabbit heart failure. J Mol Cell Cardiol. 2013;59:41–54.
30. Maack C, Cortassa S, Aon MA, Ganesan AN, Liu T, O'Rourke B. Elevated cytosolic Na^+ decreases mitochondrial Ca^{2+} uptake during excitation-contraction coupling and impairs energetic adaptation in cardiac myocytes. Circ Res. 2006;99:172–82.
31. Liu T, Brown DA, O'Rourke B. Role of mitochondrial dysfunction in cardiac glycoside toxicity. J Mol Cell Cardiol. 2010;49:728–36.
32. Santulli G, Marks AR. Essential roles of intracellular calcium release channels in muscle, brain, metabolism, and aging. Curr Mol Pharmacol. 2015;8:206–22.
33. Ruiz-Meana M, Fernandez-Sanz C, Garcia-Dorado D. The SR-mitochondria interaction: a new player in cardiac pathophysiology. Cardiovasc Res. 2010;88:30–9.
34. Tani M, Neely JR. Intermittent perfusion of ischemic myocardium. Possible mechanisms of protective effects on mechanical function in isolated rat heart. Circulation. 1990;82:536–48.
35. Schafer C, Ladilov Y, Inserte J, Schafer M, Haffner S, García-Dorado D, Piper HM. Role of the reverse mode of the Na^+/Ca^{2+} exchanger in reoxygenation induced cardiomyocyte injury. Cardiovasc Res. 2001;51:241–50.
36. Murphy E, Steenbergen C. Mechanisms underlying acute protection from cardiac ischemia-reperfusion injury. Physiol Rev. 2008;88:581–609.
37. Perrelli MG, Tullio F, Angotti C, Cerra MC, Angelone T, Tota B, Alloatti G, Penna C, Pagliaro P. Catestatin reduces myocardial ischaemia/reperfusion injury: involvement of PI3K/Akt, PKCs, mitochondrial KATP channels and ROS signalling. Pflugers Arch. 2013;465: 1031–40.

38. Cox DA, Matlib MA. Modulation of intramitochondrial free Ca^{2+} concentration by antagonists of Na/Ca exchange. Trends Pharmacol Sci. 1993;14:408–13.
39. Nishida S, Satoh H. Possible involvement of Ca activated K channels, SK channel, in the quercetin-induced vasodilatation. Korean J Physiol Pharmacol. 2009;13:361–5.
40. O'Rourke B, Blatter LA. Mitochondrial Ca+2 uptake: tortoise or hare? J Mol Cell Cardiol. 2009;46:767–74.
41. Paillard M, Tubbs E, Thiebaut PA, Gomez L, Fauconnier J, Da Silva CC, Teixeira G, Mewton N, Belaidi E, Durand A, Abrial M, Lacampagne A, Rieusset J, Ovize M. Depressing mitochondria-reticulum interactions protects cardiomyocytes from lethal hypoxia-reoxygenation injury. Circulation. 2013;128:1555–65.
42. Takeuchi A, Kim B, Matsuoka S. The mitochondrial Na^+-Ca^{2+} exchanger, NCLX, regulates automaticity of HL-1 cardiomyocytes. Sci Rep. 2013;3:2766.
43. Santulli G, Xie W, Reiken SR, Marks AR. Mitochondrial calcium overload is a key determinant in heart failure. Proc Natl Acad Sci USA. 2015;112:11389–94.
44. Sinha D, D'Silva P. Chaperoning mitochondrial permeability transition: regulation of transition pore complex by a J-protein, DnaJC15. Cell Death Dis. 2014;6(5):e1101.
45. Márquez MT, Consolini AE, Bonazzola P, Ponce-Hornos JE. The energetics of the quiescent heart muscle: high potassium cardioplegic solution and the influence of calcium and hypoxia on the rat heart. Acta Physiol Scand. 1997;160:229–33.
46. Valverde CA, Kornyeyev D, Ferreiro M, Petrosky AD, Mattiazzi A, Escobar AL. Transient Ca^{2+} depletion of the sarcoplasmic reticulum at the onset of reperfusion. Cardiovasc Res. 2010;85:671–80.
47. Santo-Domingo J, Vay L, Hernández-Sanmiguel E, Lobatón CD, Moreno A, Montero M, Alvarez J. The plasma membrane Na^+/Ca^{2+} exchange inhibitor KBR7943 is also a potent inhibitor of the mitochondrial Ca^{2+} uniporter. Br J Pharmacol. 2007;151:647–54.
48. Ouardouz M, Zamponi GW, Barr W, Kiedrowski L, Stys PK. Protection of ischemic rat spinal cord white matter: Dual action of KB-R7943 on Na^+/Ca^{2+} exchange and L-type Ca^{2+} channels. Neuropharmacology. 2005;48:566–75.
49. Bünger R, Mallet RT, Hartman DA. Pyruvate-enhanced phosphorylation potential and inotropism in normoxic and postischemic isolated working heart. Near-complete prevention of reperfusion contractile failure. Eur J Biochem. 1989;180:221–33.
50. Zweier JL, Jacobus WE. Substrate-induced alterations of high energy phosphate metabolism and contractile function in the perfused heart. J Biol Chem. 1987;262:8015–21.
51. Martin BJ, Valdivia HH, Bünger R, Lasley RD, Mentzer Jr RM. Pyruvate augments calcium transients and cell shortening in rat ventricular myocytes. Am J Phys. 1998;274:H8–17.
52. Zima AV, Kockskämper J, Mejia-Alvarez R, Blatter LA. Pyruvate modulates cardiac sarcoplasmic reticulum Ca^{2+} release in rats via mitochondria-dependent and -independent mechanisms. J Physiol. 2003;550:765–83.
53. Torres CA, Varian KD, Canan CH, Davis JP, Janssen PM. The positive inotropic effect of pyruvate involves an increase in myofilament calcium sensitivity. PLoS One. 2013;8:e63608.
54. Bonazzola P, Ragone MI, Consolini AC. Effects of pyruvate on the energetics of rat ventricles stunned by ischemia-reperfusion. Can J Physiol Pharmacol. 2014;92:386–98.
55. Halestrap AP, Price NT. The proton-linked monocarboxylate transporter (MCT) family: structure, function and regulation. Biochem J. 1999;2:281–99.
56. Kockskämper J, Zima AV, Blatter LA. Modulation of sarcoplasmic reticulum Ca^{2+} release by glycolysis in cat atrial myocytes. J Physiol. 2005;564:697–714.
57. Toller W, Wölkart G, Stranz C, Metzler H, Brunner F. Contractile action of levosimendan and epinephrine during acidosis. Eur J Pharmacol. 2005;507:199–209.
58. Terracciano CM, MacLeod KT. Effects of acidosis on Na^+/Ca^{2+} exchange and consequences for relaxation in guinea pig cardiac myocytes. Am J Phys. 1994;267:H477–87.
59. Deodato B, Altavilla D, Squadrito G, Campo GM, Arlotta M, Minutoli L. Cardioprotection by the phytoestrogen genistein in experimental myocardial ischemia-reperfusion injury. Br J Pharmacol. 1999;128:1683–90.

60. Tissier R, Waintraub X, Couvreur N, Gervais M, Bruneval P, Mandet C, Zini R, Enriquez B, Berdeaux A, Ghaleh B. Pharmacological postconditioning with the phytoestrogen genistein. J Mol Cell Cardiol. 2007;42:79–87.
61. Ji E, Wang C, He R. Effects of genistein on intracellular free-calcium concentration in guinea pig ventricular myocytes. Acta Physiol Sin. 2004;56:204–9.
62. Belevych AE, Warrier AS, Harvey RD. Genistein inhibits cardiac L-type Ca^{2+} channel activity by a tyrosine kinase-independent mechanism. Mol Pharmacol. 2002;62:554–65.
63. Li H, Zhang Y, Tian Z, Qiu X, Gu J, Wu J. Genistein stimulates myocardial contractility in guinea pigs by different subcellular mechanisms. Eur J Pharmacol. 2008;597:70–4.
64. Liew R, Stagg MA, Chan J, Collins P, MacLeod KT. Gender determines the acute actions of genistein on intracellular calcium regulation in the guinea pig heart. Cardiovasc Res. 2004;61:66–76.
65. Van Wylen DG. Effect of ischemic preconditioning on interstitial purine metabolite and lactate accumulation during myocardial ischemia. Circulation. 1994;89:2283–9.
66. Clanachan AS. Contribution of protons to post-ischemic Na^{+} and Ca^{2+} overload and left ventricular mechanical dysfunction. J Cardiovasc Electrophysiol. 2006;17:s141–8.
67. Stowe D, Camara KS. Mitochondrial reactive oxygen species production in excitable cells: modulators of mitochondrial and cell function. Antioxid Redox Signal. 2009;11:1373–414.
68. Zima AV, Blatter LA. Redox regulation of cardiac calcium channels and transporters. Cardiovasc Res. 2006;71:310–21.
69. Hidalgo C, Donoso P. Cross-talk between calcium and redox signalling: from molecular mechanisms to health implications. Antioxid Redox Signal. 2008;10:1275–312.
70. Tang WH, Kravtsov GM, Sauert M, Tong XY, Hou XY, Wong TM, Chung SK, Man Chung SS. Polyol pathway impairs the function of SERCA and RyR in ischemic–reperfused rat hearts by increasing oxidative modifications of these proteins. J Mol Cell Cardiol. 2010;49:58–69.
71. Kuster GM, Lancel S, Zhang J, Communal C, Trucillo MP, Lim CC, Pfister O, Weinberg EO, Cohen RA, Liao R, Siwik DA, Colucci WS. Redox mediated reciprocal regulation of SERCA and Na^{+}–Ca^{2+} exchanger contributes to sarcoplasmic reticulum Ca^{2+} depletion in cardiac myocytes. Free Radic Biol Med. 2010;48:1182–7.
72. Grützner A, Garcia-Manyes S, Kötter S, Badilla CL, Fernandez JM, Linke WA. Modulation of titin-based stiffness by disulfide bonding in the cardiac titin N2-B unique sequence. Biophys J. 2009;97:825–34.
73. Crompton M. The role of Ca^{+2} in the function and dysfunction of heart mitochondria. In: Langer GA, editor. Calcium and the heart. New York: Raven Press Ltd.; 1990. p. 167–98.
74. Finkel T, Menazza S, Holmström KM, Parks RJ, Liu J, Sun J, Liu J, Pan X, Murphy E. The ins and outs of mitochondrial calcium. Circ Res. 2015;116:1810–9.
75. Santulli G, Lewis DR, Marks AR. Physiology and pathophysiology of excitation–contraction coupling: the functional role of ryanodine receptor. J Muscle Res Cell Motil. 2017 (in press); doi: 10.1007/s10974-017-9470-z.
76. Santulli G, Nakashima R, Yuan Q, Marks AR. Intracellular calcium release channels: an update. J Physiol. 2017 (in press); doi: 10.1113/JP272781.
77. Ristagno G, Tantillo S, Sun S, Harry Weil M, Tang W. Hypothermia improves ventricular myocyte contractility under conditions of normal perfusion and after an interval of ischemia. Resuscitation. 2010;81:898–903.
78. Curtin N, Woledge R. Energy changes and muscular contraction. Physiol Rev. 1978;58:690–761.
79. Ponce-Hornos JE, Parker JM, Langer GA. Heat production in isolated heart myocytes: differences among species. Am J Phys. 1990;258:H880–6.
80. Arieli Y, Gursahani H, Eaton MM, Hernandez LA, Schaefer S. Gender modulation of Ca($^{2+}$) uptake in cardiac mitochondria. J Mol Cell Cardiol. 2004;37:507–13.
81. Nyirenda MJ, Clark DN, Finlayson AR. Thyroid disease and increased cardiovascular risk. Thyroid. 2005;15:718–24.

82. Klein I, Ojamaa K. Thyroid hormone and the cardiovascular system. N Engl J Med. 2001;44:501–9.
83. Heusch G. The regional myocardial flow-function relationship: a framework for an understanding of acute ischemia, hibernation, stunning and coronary microembolization. 1980. Circ Res. 2013;112:1535–7.
84. Siribaddana S. Cardiac dysfunction in the CABG patient. Curr Opin Pharmacol. 2012;12:166–71.
85. Novitzky D, Cooper DKC. Thyroid hormone and the stunned myocardium. J Endocrinol. 2014;223:R1–8.
86. Trollinger DE, Wayne E, Lemasters J. Selective loading of Rhod 2 into mitochondria shows mitochondrial Ca^{2+} transients during the contractile cycle in adult rabbit cardiac myocytes. Biochem Biophys Res Commun. 1997;236:738–42.
87. Griffiths EJ, Halestrap AP. Protection by Cyclosporin A of ischemia/reperfusion-induced damage in isolated rat hearts. J Mol Cell Cardiol. 1993;25:1461–9.
88. Murata M, Akao M, O'Rourke B, Marban E. Mitochondrial ATP-sensitive potassium channels attenuate matrix Ca^{2+} overload during simulated ischemia and reperfusion. Possible mechanism of cardioprotection. Circ Res. 2001;89:891–8.
89. Garlid KD, Costa ADT, Quinlan CL, Pierre SV, Dos Santos P. Cardioprotective signaling to mitochondria. J Mol Cell Cardiol. 2009;46:858–66.
90. García-Rivas GJ, Guerrero-Hernández A, Guerrero-Serna G, Rodríguez-Zavala JS, Zazueta C. Inhibition of the mitochondrial calcium uniporter by the oxo-bridged dinuclear ruthenium amine complex (Ru 360) prevents from irreversible injury in postischemic rat heart. FEBS J. 2005;272:3477–88.
91. Foskett JK, Philipson B. The mitochondrial Ca^{2+} uniporter complex. J Mol Cell Cardiol. 2015;78:3–8.
92. Ruiz-Meana M, Garcia-Dorado D, Miró-Casas E, Abellán A, Soler-Soler J. Mitochondrial Ca^{2+} uptake during simulated ischemia does not affect permeability transition pore opening upon simulated reperfusion. Cardiovasc Res. 2006;71:715–24.
93. Xie C, Kauffman J, Akar FG. Functional crosstalk between the mitochondrial PTP and KATP channels determine arrhythmic vulnerability to oxidative stress. Front Physiol. 2014;5:264.
94. Nicolini G, Pitto L, Kusmic C, Balzan S, Sabatino L, Iervasi G, Forini F. New insights into mechanisms of cardioprotection mediated by thyroid hormones. J Thyroid Res. 2013;2013:264387. doi: 10.1155/2013/264387. Epub 2013 Mar 10.
95. Cernohorsky J, Kolar F, Pelouch V, Korecky B, Vetter R. Thyroid control of sarcolemmal Na^{+}/Ca^{2+} exchanger and SR Ca^{2+}ATPase in developing rat heart. Am J Phys. 1998;275:H264–73.
96. Kenessey A, Ojamaa K. Thyroid hormone stimulates protein synthesis in the cardiomyocyte by activating the Akt-mTOR and p70S6K pathways. J Biol Chem. 2006;281:20666–72.
97. Klein I, Danzi S. Thyroid disease and the heart. Circulation. 2007;116:1725–35.
98. Arsanjani R, McCarren M, Bahl JJ, Goldman S. Translational potential of thyroid hormone and its analogs. J Mol Cell Cardiol. 2011;51:506–11.
99. Forini F, Lionetti V, Ardehali H, Pucci A, Cecchetti F, Ghanefar M, Nicolini G, Ichikawa Y, Nannipieri M, Recchia FA, Iervasi G. Early long-term LT3 replacement rescues mitochondria and prevents ischemic cardiac remodeling in rats. J Cell Mol Med. 2011;15:514–24.
100. Kiss E, Jakab G, Kranias EG, Edes I. Thyroid hormone induced alterations in phospholamban protein expression: regulatory effects on sarcoplasmic reticulum Ca^{2+} transport and myocardial relaxation. Circ Res. 1994;75:245–51.
101. Tielens ET, Forder JR, Chatham JC, Marrelli SP, Ladenson PW. Acute L-triiodothyronine administration potentiates inotropic responses to beta-adrenergic stimulation in the isolated perfused rat heart. Cardiovasc Res. 1996;32:306–10.
102. Mattiazzi A, Argenziano M, Aguilar-Sanchez Y, Mazzocchi G, Escobar AL. Ca^{2+} Sparks and Ca^{2+} waves are the subcellular events underlying Ca^{2+} overload during ischemia and reperfusion in perfused intact hearts. J Mol Cell Cardiol. 2015;79:69–78.

103. Mourouzis I, Dimopoulos A, Saranteas T, Tsinarakis N, Livadarou E, Spanou D, Kokkinos AD, Xinaris C, Pantos C, Cokkinos DV. Ischemic preconditioning fails to confer additional protection against ischemia-reperfusion injury in the hypothyroid rat heart. Physiol Res. 2009;58:29–38.

Chapter 9
Mechanistic Role of mPTP in Ischemia-Reperfusion Injury

Giampaolo Morciano, Massimo Bonora, Gianluca Campo, Giorgio Aquila, Paola Rizzo, Carlotta Giorgi, Mariusz R. Wieckowski, and Paolo Pinton

Mechanical reperfusion with percutaneous coronary intervention (PCI) represents the gold standard therapy for patients presenting ST-segment elevation myocardial infarction (STEMI) and significantly reduces mortality after acute event. Recent years have seen a significant reduction of 1-year mortality from STEMI, from 15–20% to 5–10% [1, 2]. The "hub and spoke" network for primary PCI and the optimization of techniques, materials, and antithrombotic agents can explain this finding. Unfortunately, improvements in myocardial salvage in some patients have remained small despite successful coronary revascularization.

Reperfusion injury (RI) has significantly hindered the efforts to further optimize STEMI treatment further [3, 4]. RI has been studied for over 30 years and it has been defined as cardiomyocyte damage secondary to myocardial restoration of blood flow [5]. RI is associated with larger infarct size (IS), higher degree of systolic dysfunction, impaired left ventricle ejection fraction (LVEF) and poor prognoses [3]. RI may be responsible for up to 40–50% of the final IS and remains a

G. Morciano • M. Bonora • C. Giorgi • P. Pinton (✉)
Department of Morphology, Surgery and Experimental Medicine, Section of General Pathology, Interdisciplinary Center for the Study of Inflammation (ICSI), University of Ferrara, Ferrara, Italy

Laboratory for Technologies of Advanced Therapies (LTTA), University of Ferrara, Ferrara, Italy
e-mail: paolo.pinton@unife.it

G. Campo
Cardiovascular Institute, Azienda Ospedaliera Universitaria di Ferrara, Ferrara, Italy

G. Aquila • P. Rizzo
Department of Medical Sciences, University of Ferrara, Ferrara, Italy

M.R. Wieckowski
Department of Biochemistry, Nencki Institute of Experimental Biology, Warsaw, Poland

G. Santulli (ed.), *Mitochondrial Dynamics in Cardiovascular Medicine*, Advances in Experimental Medicine and Biology 982,
DOI 10.1007/978-3-319-55330-6_9

complex phenomenon involving different molecular, cellular and clinical factors that culminate in the genesis of the mitochondrial permeability transition pore (mPTP), which is prompted to open during coronary reperfusion [3, 4]. Reperfusion arrhythmias, myocardial stunning, microvascular obstruction and intramyocardial hemorrhage should be considered different aspects of RI, but we will not discuss them here. In this chapter, we provide an update of recent findings concerning RI. In particular, we focus on mPTP structure and on mPTP involvement in RI genesis. We also summarize the principal aspects of RI in humans, focusing specifically on patients with STEMI and on the mechanistic role played by the mPTP in this pathology, as well as the main pitfalls associated with the application of cardioprotective strategies and therapies in daily clinical practice.

Unmasking Reperfusion Injury: Tools for Quantifying Damage

IS is a well-established independent predictor of poor prognosis after STEMI [6]. Several factors determine IS, and up to 50% of the final IS may be caused by RI if no therapeutic interventions are implemented (Fig. 9.1) [3, 6, 7]. There are currently no efficient therapies for preventing myocardial RI, only promising targets, as discussed later in the chapter.

Cardiomyocytes suffer and die from RI due to mPTP opening at the time of reperfusion in hearts already damaged from sustained ischemia [8]. Quantification of cell death by mPTP opening and identification of damage secondary to RI in daily clinical practice remain complex and misleading. The easiest and most used methods for quantifying the myocardial damage are the evaluation of final LVEF or the area under the curve of the release of specific cardiac markers (e.g., CK-MB and troponin). Thus, they are used to quantify IS reductions. Nevertheless, they are nonspecific and cannot be used to differentiate IS due to ischemia from IS due to RI. Otherwise, both electrocardiogram (ST-segment resolution) and coronary artery angiography (myocardial blush grade, MBG) can be used to assess reperfusion outcomes [9, 10]. TIMI (thrombolysis in myocardial infarction) scores at 0–1 flow, ST-segment resolution <30% and/or MBG <2 are markers of poor/absent reperfusion and negative prognoses [9, 10]. The last two indices are significantly more specific than the previous indices but are insufficient for reliably distinguishing among all patients with RI. Majidi and colleagues reported that ventricular arrhythmia bursts (VABs) are associated with larger ISs in patients with similar ST-segment recoveries and MBGs [11]. The authors speculated that VABs reflect myocellular injury in reperfusion settings and that the combination of angiographic and electrocardiogram parameters of epicardial, microvascular and cellular responses may provide a more predictive biosignature of optimal reperfusion [11].

Cardiac magnetic resonance imaging (cMRI) has been used successfully in recent studies to evaluate reperfusion outcomes and to quantify RI, indeed a large and multicenter STEMI population reperfused by primary PCI, cMRI parameters showed an independent and incremental prognostic information in addition to

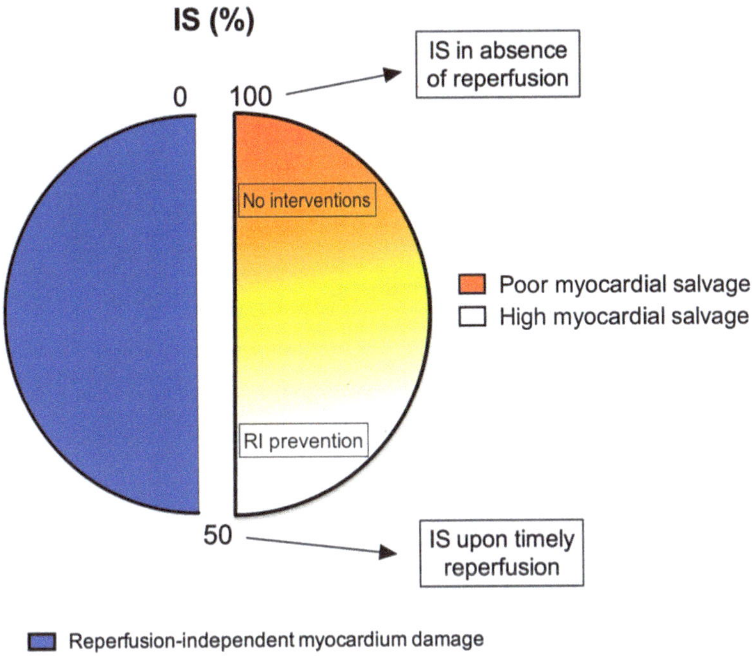

Fig. 9.1 Schematic representation of the final infarct size in percentage (whole graph) in a heart experiencing a myocardial infarction. In the absence of reperfusion, the area at risk represents the entire ischemic area, which has a large final infarct size (100%, whole graph). Successful and timely PCI, which is also responsible for RI, results into 50% of the final IS (*right part*, chromatic scale). Chromatic scale representing the reperfusion-dependent myocardial damage illustrates the possibility of saving the myocardium by preventing and modulating the 50% of the IS caused by clinical intervention-induced RI. The other half of the graph (*blue part*) involves reperfusion-independent mechanisms not reviewed in this chapter. *IS* infarct size, *RI* reperfusion injury

clinical risk scores and LVEF [12]. cMRI is usually performed within the first week after primary PCI [13] and late gadolinium enhancement (LGE) is used to quantify ISs. The latter may be used to calculate the myocardial salvage index (MSI) defined as the difference between the area at risk (AAR) and the normalized LGE divided by the AAR. MSI represents the myocardium that has suffered but survived an ischemic insult and effectively depicts the cardioprotection target [14–16].

Mitochondrial Permeability Transition Pore Contribution in Reperfusion Injury Mechanisms

Oxidative stress, inflammation, calcium (Ca^{2+}) overload, and hypercontracture are significantly implicated in RI genesis (Fig. 9.2). mPTP opening is widely considered a terminal step of RI [17, 18], which begins with an ischemic event. This mPTP

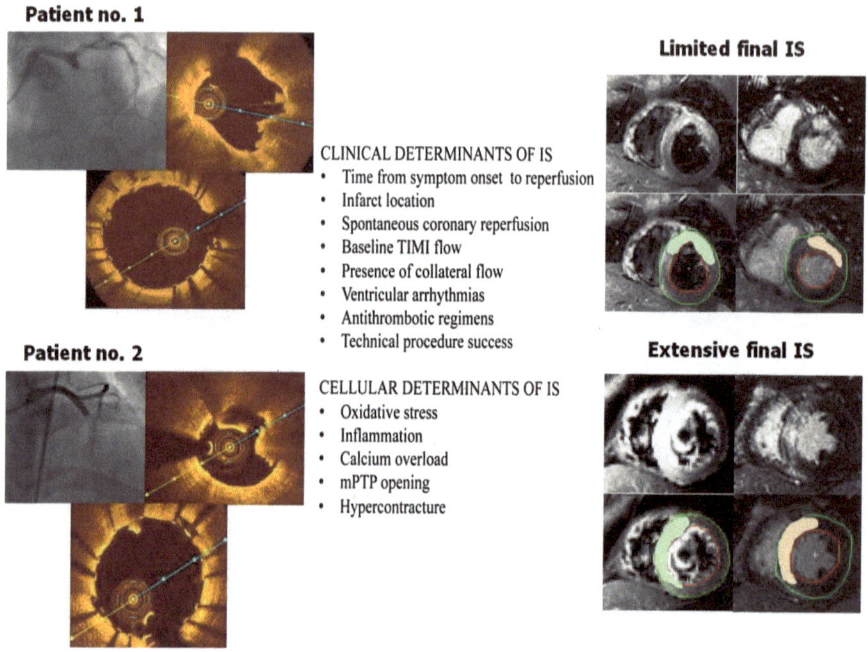

Fig. 9.2 Clinical and cellular determinants of final IS. On the *left*, the angiographic and optical coherence imaging (before and after stent implantation) of two STEMI patients successfully treated with mechanical reperfusion. Cardiac magnetic resonance imaging (on the *right*) reveals the corresponding areas at risk (similar between patients) and the final infarct size (different). *IS* infarct size, *TIMI* thrombolysis in myocardial infarction, *mPTP* mitochondrial permeability transition pore

priming phase was identified by Griffiths and Halestrap in 1995, as these authors proved that mPTP occurs at the time of reperfusion using the mitochondrial "Hot DOG" entrapment technique. They demonstrated that some mitochondria may undergo mPTP opening and closure in ischemic-reperfused hearts [19], as their experimental procedures demonstrated that the extent of 2-deoxy[^{3}H]glucose (DOG) uptake increases until the period of ischemia preceding reperfusion, corresponding to a maximum empirical interval of 30–40 min [19]. During ischemia, which is characterized by the absence of oxygen resulting in progressive ATP depletion, cell metabolism occurs mainly via the anaerobic route, thus lowering the pH. A compensatory mechanism involving the Na^+/H^+ exchanger is activated to counterbalance this condition, leading to a large influx of sodium ions, which reduces the uptake of Ca^{2+} by the endoplasmic reticulum (ER). Thus, cardiomyocytes are subjected to Ca^{2+} overload, reactive oxygen species (ROS) production and long-chain fatty acid accumulation.

These factors increase RI susceptibility but do not cause mPTP opening because a critical value, the pH, remains low. Previous studies have demonstrated that H^+ inhibits Ca^{2+} binding to the mPTP trigger site [20]. Although the specificity of the

Table 9.1 Schematic subdivision of main intracellular mPTP modulators in negative regulators and inducers

Negative regulators	Refs.	Inducers	Refs.
ADP	[20]	Matrix Ca^{2+} overload	[22]
ATP	[20]	P_i	[23]
Mg^{2+}	[20]	O_2^-	[24]
Low pH	[25]	Oxidized thiols	[26]
High Ψ_M	[27]	Peroxided lipids	[28]
Bcl-2[a]	[29]	High pH	[25]
Bcl-XL[a]	[30]	Low Ψ_M	[27]
HXK I-II[a]	[31]	Bax[b]	[32]
		Bak	[33]
		Bad[c]	[34]
		GSK3-β[d]	[35]
		mtCypD	[36]

GSK3-β glycogen synthase kinase 3 beta, *HXK* hexokinase, *mtCypD* mitochondrial cyclophilin D, P_i inorganic phosphate, *Refs* references, Ψ_M mitochondrial membrane potential
[a]Interaction between the protein and VDAC1 promotes cytoprotection
[b]Required interaction with ANT1 in mPTP opening
[c]Displacement of VDAC1 from Bcl-2,
[d]Modification of HXK-VDAC1 interaction

matrix-localized trigger site for Ca^{2+} is absolute [21], there are many intracellular factors that can modulate the mPTP activity (see Table 9.1 for details). The Ca^{2+} concentration required for mPTP opening is highly dependent on the prevailing conditions within the cell, which can change mPTP Ca^{2+} sensitivity.

The cells that survive ischemic insults die from damage generated by coronary reperfusion. Crompton and co-workers clarified the relationship between the mPTP and RI, demonstrating that oxidative stress and Ca^{2+} overload are critical factors in mPTP opening and RI progression. When reperfusion occurs, the respiratory chain is suddenly exposed to oxygen, leading to oxidative stress. Ca^{2+} accumulates due to rapid mitochondrial membrane potential restoration, and the acidic pH is neutralized. All of these processes induce mPTP opening (Fig. 9.3) [37]. This large pore in the inner mitochondrial membrane (IMM) allows the free passage of all molecules <1.5 kDa into the mitochondrial matrix. The IMM becomes freely permeable to protons, effectively uncoupling oxidative phosphorylation and disrupting ATP production. ATPase reversal also occurs, causing the breakdown of cytosolic ATP generated via glycolysis. Energy metabolism is further impaired, resulting in a continuous cycle of increasing Ca^{2+} deregulation and mPTP opening, which leads to osmotic swelling and damage and mitochondrial disruption. The recent discovery of the mitochondrial Ca^{2+} uniporter (MCU) has highlighted the relationship between Ca^{2+} overload and mPTP opening. The MCU and its associated regulators (see Chaps. 2 and 3) are believed to play key roles in the accumulation of large amounts of Ca^{2+} [38]. In an early study, mitochondria from $MCU^{-/-}$ mice exhibited no Ca^{2+}-induced mPTP opening, suggesting that the MCU is required for this pathway. In

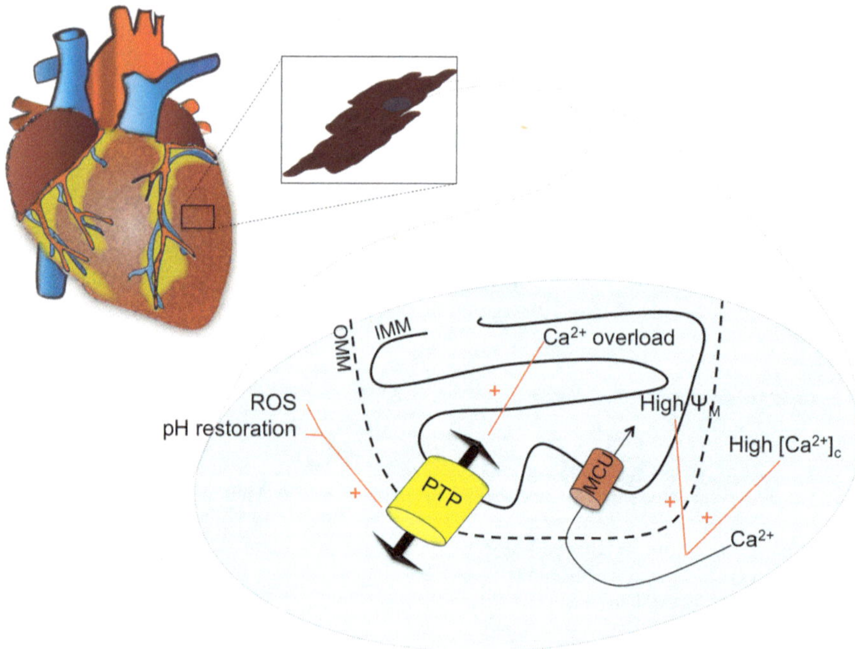

Fig. 9.3 Simplified schematic depicting mPTP involvement in cell death during reperfusion following MI. Restoration of blood flow upon an ischemic event causes RI. Depending on RI severity, cardiomyocytes (*top*) experience oxidative stress, pH increases and matrix Ca^{2+} overload (*bottom*). The diagram depicts all variables that facilitate cell death by mPTP opening. *Red lines* with red plus symbols indicate positive regulation of mPTP opening and Ca^{2+} uptake by the indicated factors. *IMM* inner mitochondrial membrane, *MCU* mitochondrial Ca^{2+} uniporter, *OMM* outer mitochondrial membrane, *PTP* permeability transition pore, *ROS* reactive oxygen species, Ψ_M mitochondrial membrane potential

contrast, recent studies involving $MCU^{-/-}$ mouse hearts subjected to RI revealed that the absence of the MCU does not protect the heart from RI, as these studies noted little difference between hearts of this genotype and wild-type hearts with respect to the incidence of RI [39]. These findings may be explained by the existence of compensatory mechanisms facilitating Ca^{2+}-independent cell death processes involving other molecules that form or regulate the pore when the MCU is absent. However, these aspects require further research.

Mitochondrial permeability transition (MPT) is a critical determinant of RI and is thus responsible for the necrotic and apoptotic cell death processes exhibiting differential contributions to infarct sizes despite their being regulated by many of the same intermediates. Necrosis comprises membrane rupture, cell swelling and cellular debris and intracellular enzyme (lactate dehydrogenase and troponin I) release. Neutrophils migrate to damaged areas and induce an inflammatory response that worsens the pathological condition. In contrast, apoptosis leads to important morphological and structural changes, including blebbing, cell shrinkage and nuclear fragmentation, without inflammatory response activation [40].

Most of the data regarding these processes have been obtained via studies involving cardiac RI animal models (e.g., rabbit, rat and murine) and have indicated that necrosis occurs at a higher rate during the ischemic phase of MI because of the severity of the insults to which cardiomyocytes are subjected. During ischemic events, bursts of cell death occur in the subendocardium and progress toward the epicardium in a transmural manner [41] peaking approximately 24 h after MI [42]. Necrosis can propagate through enhanced gap junction networks by promoting the spread of contraction band necrosis in the at-risk area [43] and forms the core of the infarct—an area exhibiting irreversible damage—following ischemic insults [44]. Apoptosis, together with autophagy and inflammation, is a reversible process and can be manipulated to allow cardiomyocytes to survive during MI. The duration and kinetics of mPTP opening and the percentage of mitochondria that experience the "open state" in a cell are considered determining factors of the pathological states of reperfused tissues. Prolonged pore opening produces significant cell death waves, whereas short and transient opening allows the cell to either recover completely or initiate the apoptotic pathway via cytochrome c release and caspase 9 and caspase 3 activation. This protease mediates the proteolytic cleavage of a wide range of proteins involved in the rearrangement of the cytoskeleton, the plasma membrane and the nucleus. Numerous reports indicate that apoptosis culminates at the time of reperfusion [44–48] and reaches the periphery of the necrotic core, where the damage inflicted by the insult is less severe. However, only few apoptotic cells are present in non-infarcted myocardial areas [49]. Timely reperfusion is required in clinical practice for obvious reasons but also provides the energy necessary for the completion of the apoptotic process (Fig. 9.3). In addition, several mPTP inhibitors, such as Cyclosporine A (CsA) [50], NIM811 [51], Sanghliferin [52] and Debio-025 [53], reportedly protected the heart from RI-induced cell death in experimental models, indicating that mtCypD (Table 9.1) binding is the most promising target in cardioprotection. Furthermore, genetic mtCypD deletion also provides potent protection against heart IRI, as mtCypD knockout mice have exhibited significant decreases in ISs in previous studies. mPTP opening is not fully dependent on mtCypD, as it can occur even in the setting of mtCypD genetic deletion or CsA-mediated mtCypD inhibition due to higher Ca^{2+} loads or oxidative stress [54]. Thus, targeting other mPTP components may be an attractive strategy for improving cardiac recovery following MI (see the paragraph "Cardioprotection against reperfusion injury: current clinical applications and new attempts at targeting mitochondrial functions and the mPTP"). However, there are many confounding factors that make it difficult to assess if targeted therapies reduce cardiomyocyte death in patients suffering from MI, including age, gender, occlusion localization, TIMI flow, comorbidities and door-to-balloon time. All of these factors may influence apoptosis and necrosis levels; thus, it is important to develop and validate new techniques to better define ISs (see the paragraph "Unmasking reperfusion injury: tools for quantifying damage") in studies utilizing strict inclusion/exclusion criteria with respect to patient enrollment [55].

Mitochondrial Permeability Transition Pore: Molecular Structure

Numerous reports from the late 1980s to the early 2000s indicated that the mPTP was most likely a supramolecular structure assembled in the IMM by proteins representing mitochondrial contact sites between the outer and inner mitochondrial membranes. The list of proteins interacting in the mPTP core is long and includes adenosine nucleotide transporter (ANT), voltage-dependent anion-selective channel (VDAC), mtCypD, hexokinase, creatine kinase, Bak, Bax, Bcl-2, Bcl-x_L, benzodiazepine receptor, glycogen synthase kinase 3 beta (GSK3-β), PKCε, PKG, p53 and complement component 1 Q subcomponent-binding protein (C1QBP) [4, 56]. Accumulating data suggest that the mPTP is not the result of the opening of a pre-existing pore but rather is formed by misfolded mitochondrial membranes proteins modified by oxidative damage [57, 58]. Two distinguished papers published in 2004 and 2007 completely revised our understanding of mPTP architecture. First, Kokoszka et al. presented strong data indicating that mitochondria lacking ANT can still exhibit permeability transition phenomena [59]—indicating that these phenomena are not regulated by ANT ligands—and that cells without ANT can undergo apoptosis initiated by TNF-α and Fas [59]. Second, Baines et al. reported that VDACs (VDAC1, VDAC2 and VDAC3) are not critical for mPTP formation [60], as these authors observed that VDAC-deficient cells can undergo Bax- and Bid-dependent cell death, similar to the ANT-deficient model. Taken together, these data indicate that ANT and VDACs are non-essential mPTP structural components but play roles in mPTP regulation. Interestingly, similar experiments performed with transgenic mice lacking the peptidylprolyl isomerase f (*Ppif*) gene confirmed that mtCypD is the mPTP component responsible for its sensitivity to CsA [54].

However, mtCypD is involved in mPTP regulation (significantly increasing the threshold for Ca^{2+}-induced mPTP opening), not mPTP formation. Based on this evidence, Halestrap et al. proposed that phosphate inorganic carrier (PiC) forms the core of the mPTP [61], a hypothesis supported by other groups [23, 62]. However, later studies by Halestrap's group regarding PiC silencing in HeLa cells revealed that decreasing phosphate carrier levels by up to 70% had no effect on Ca^{2+}-induced mPTP opening [63]. In contrast, Kwong et al. reported that complete PiC genetic deletion desensitizes the mPTP. Thus, PiC involvement in mPTP function cannot be excluded [64]. An important breakthrough regarding mPTP structure occurred at the beginning of 2013, as Pinton et al. observed that the c subunit of mitochondrial ATP synthase plays a critical role in mPTP phenomena (Fig. 9.4) [65, 66]. Detailed studies by Jonas et al. supported the idea that a ring composed of c subunits (c-ring) is the best candidate to form the mPTP core (Fig. 9.4) [67]. These authors demonstrated that purified and reconstituted human c subunit rings can form voltage-sensitive channels. However, neither CsA nor high Ca^{2+} has an effect on purified c subunit ring channel properties, indicating that the c subunit ring alone has no components that regulate mPTP opening and closing. Only when the authors used a complete ATP synthase monomer in the presence of mtCypD did classical mPTP

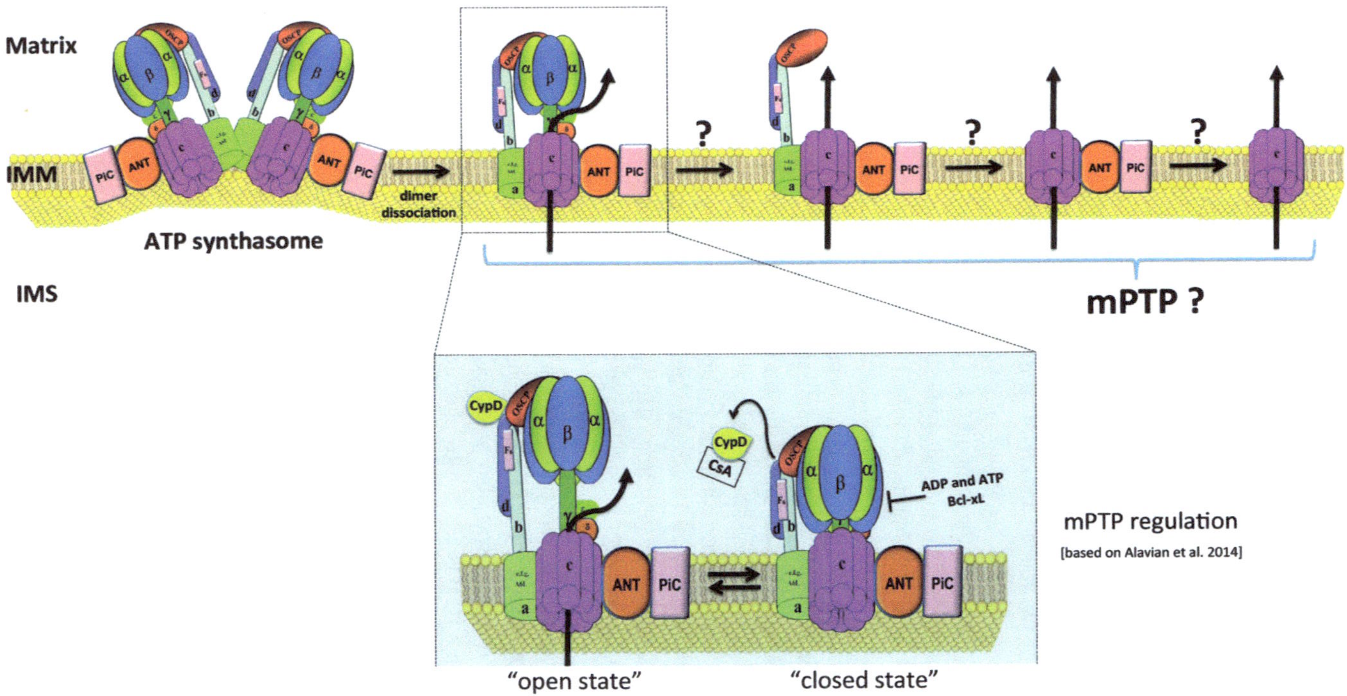

Fig. 9.4 The new-look mitochondrial permeability transition pore. The findings of recent studies indicate that the mPTP is formed via contributions from the mitochondrial ATP synthase c subunit ring under circumstances favorable for mPTP opening. Mitochondrial ATP synthase dimerizes with ANT and PiC to form an ATP synthasome. It is unknown whether the c subunit ring operates as an mPTP alone, in collaboration with other ATP synthase subunits (as a whole monomer or as part of the F_O domain comprising the *a*, *e*, *f*, *g*, and A6L subunits) or with ANT and PiC alone. The lower panel describes ways in which the mPTP may be regulated and thus shift from the open to the closed state (prepared based on [67]). *ATP* adenosine triphosphate, *IMM* inner mitochondrial membrane, *IMS* intermembrane space, *mPTP* mitochondrial permeability transition pore

regulation occur, indicating that Ca^{2+} and CsA (via mtCypD) regulate pore opening/closing by interacting with a peripheral ATP synthase component (probably in combination with the OSCP subunit). Moreover, mutations of highly conserved glycines responsible for optimal c subunit packing within the c-ring cause increases in pore conductance and decreases in sensitivity to CsA, suggesting that impaired c subunit packing may have a large impact on mPTP permeability [67]. Interestingly, adding a purified ATP synthase β-subunit (but not a δ-, ε- or γ-subunit) to the reconstituted c subunit ring decreased pore conductance, indicating that the β-subunit can directly regulate the mPTP. This result explains the well-known inhibitory effects exerted by ADP, as well as the effects exerted by the anti-apoptotic Bcl-xL proteins, on mPTP opening [68]. Contemporaneously, Azarashvili et al. proposed a new mPTP opening/closing regulatory mechanism based on c subunit phosphorylation/dephosphorylation [69]. In their model, PKA-mediated c subunit phosphorylation is responsible for conformational changes in the c subunit that may affect the interactions among the c subunits in the c-ring and may also interfere with the interactions between these c subunits and the stalk subunits of the ATP synthase [69]. These observations suggest that the simplest mPTP model may entail the pore comprising a partially decomposed ATP synthase (e.g., physical decoupling of F_1 from F_O) (see Fig. 9.4 for details). This possibility was confirmed by the existence of free c subunit oligomers that were independent of F_1 components in mitochondria exhibiting Ca^{2+}-induced swelling [67]. However, the presence of free c subunit rings interacting with central and/or lateral stalks has not been confirmed. Therefore, whether physical dissociation of F_1 from F_O is necessary to observe mPTP phenomena, at least in low-conductance mode, in which the inhibitory effects of the β-subunit can be modulated by changing the distance between F_1 and F_O, remains unknown. The latest works of Jonas's and Saris's groups demonstrating that c subunit rings reconstituted in liposomes exhibit channel activity support the original hypothesis that the c subunit is a crucial element responsible for mPTP formation [66–69]. The involvement of ATP synthase in mPTP formation was also proposed by Giorgio et al.; however, their model, in which the mPTP is formed by the dimeric form of ATP synthase [70], seems to fail due to some inconsistencies (e.g., mPTP opening can be detected in Rho0 cells [71]; however, the level of the dimeric form of the ATP synthase in these cells is extremely low [72]). In contrast, Pinton's group developed a model in which the mPTP comprises the monomeric form of ATP synthase, which forms via dimer dissociation (Fig. 9.4). Experimental data tie its activity to a specific c-ring conformation and indicate that variations in the highly conserved glycine zipper domain of the c subunit induced by c subunit-encoding genes have the capacity to influence mPTP activity in various conditions.

The following evidence supports the hypothesis that the mitochondrial ATP synthase is involved in mPTP formation and that the dissociation of the dimeric form causes MPT: Ψ_M and pH influence mPTP opening and also regulate ATP synthesis via ATP synthase, which interacts with ANT and PiC, forming the so-called ATP synthasome. Bcl-2 protein family members physically or functionally interact with ATP synthase (Bcl-x_L inhibits mPTP while enhancing ATP synthesis by ATP synthase), which is sensitive to the oxidation of specific cysteine residues. ATP synthase binds mtCypD (via the OSCP subunit and subunit d), and the c subunit binds Ca^{2+}. The c subunit has pore-forming properties, and its dephosphorylation and overexpression promote mPTP opening, whereas its transient depletion prevents mPTP opening.

Implications for F_O ATP Synthase c Subunit in Reperfusion Injury: Preliminary Findings Require Further Researches

As mentioned previously, the authors of cell culture studies hypothesized that the long-sought molecular pore of the mPTP is the F_O ATP synthase c subunit [65, 67, 69, 73]. However, the direct involvement of the c subunit in cardiovascular pathology has not been investigated to date, as the c subunit was only recently demonstrated to play a significant role in RI by pre-clinical and preliminary clinical studies.

First, unpublished experiments involving a cardiac animal model of RI demonstrated that targeting the c-ring with selective known inhibitors at the time of reperfusion protected the heart from apoptotic cell death via mPTP inhibition as efficient as that facilitated by the known MPT inhibitor CsA, thus reducing ISs. Second, a very interesting study conducted by Campo G. and co-workers [74] attempted to assess c subunit levels after the STEMI onset in a cohort of 158 patients successfully treated with primary PCI. They observed increases in serum c subunit circulation early after STEMI onset and determined that c subunit protein levels were an independent predictor of all surrogate endpoints of myocardial reperfusion. Elevated c subunit levels were significantly related to a worse prognosis characterized by poor ST-segment resolution values, TIMI myocardial perfusion grades and TIMI frame counts, as well as elevated cardiac marker levels [74]. Although additional studies are necessary to investigate this issue, the collected data clearly indicate that patients with higher F_O ATP synthase c subunit values, indicators of hyper-responsive mPTP activity at the moment of reperfusion, are at higher risk for RI.

Cardioprotection Against Reperfusion Injury: Current Clinical Applications and New Attempts at Targeting Mitochondrial Functions and mPTP

Preventing myocardial RI in STEMI patients is not a simple task. MI is a multifactorial disease influenced by multiple confounding factors and comprising many players, such as cardiomyocytes, microvasculature, inflammation and platelets, that all significantly contribute to its pathology. Lack of knowledge regarding the mechanisms underlying RI and its associated targets has not helped clinicians to achieve the above aim. Several cardioprotective strategies against RI have been developed and tested in STEMI patients. These strategies, including preconditioning, postconditioning, hypothermia and hyperoxemia, are effective but should be considered mechanical cardioprotection strategies [75, 76]. Mitochondrial function is considered a crucial mediator of RI and cardiomyocyte death (as widely reviewed in [4, 77]), because of the role played by mPTP opening in these processes; thus, increasing numbers of clinical studies regarding mitochondrial targets and pathways have been conducted in recent years [55, 78–84]. Unfortunately, these studies were unable to find evidence that administrating mitochondria-targeting experimental drugs limits ISs. However, these studies observed that therapies with wide ranges of action (e.g., mechanical strategies) contributed to significant improvements in the clinical outcomes of STEMI patients. Taken together, these results indicate that individual key factors in the RI picture have not yet been targeted with respect to establishing adequate and selective therapies. In addition, the failures of some randomized clinical trials may be due to issues related to clinical study design.

Analysis of randomized trials regarding the use of CsA and TRO40303 (MITOCARE study) in humans may help us to understand the importance of adequately identifying correct cardioprotection targets. TRO40303 binds to the mitochondrial translocator protein 18 kDa (TSPO) at its cholesterol site. Although in the past TSPO has been proposed to play a role in the mPTP; a recent study by Bernardi et al. revealed that TSPO plays no role in its structure or regulation. Endogenous and synthetic ligands of TSPO do not regulate mPTP activity, and OMM regulates mPTP activity through an MTP-independent mechanism [85]. In the MITOCARE study, no differences in final ISs or enzyme release were observed between the TRO40303 and placebo groups [78]. CsA inhibits mPTP opening by binding to mtCypD. Past animal model studies and human clinical trials regarding CsA [84] demonstrated that CsA administration at the time of reperfusion was associated with smaller infarct sizes than placebo administration. Cung and colleagues later performed a larger clinical trial (CIRCUS) to confirm the above findings but failed to replicate the expected cardioprotective effects of CsA on MI severity [79] likely due to differences in inclusion criteria between the two studies. The CIRCUS trial enrolled patients presenting within 12 h after symptom onset, by which time most cardiomyocytes are necrotic, and there is not much living myocardium to salvage. Therefore, the strategy of inhibiting mPTP opening to reduce apoptosis was useless for the majority of patients, who experienced longer coronary occlusion and more extensive necrosis than their counterparts in the previous study (see the next paragraph for further details).

Ongoing basic research studies are focusing more on identifying new mPTP constituents (e.g., the F_1F_O ATP synthase c subunit [74]) to develop new cardioprotection strategies than on CsA and its known side effects. Explaining how the dimeric F_1F_O ATP synthase can be converted to the nonspecific mPTP channel (Fig. 9.4) and elucidating the molecular composition of the mPTP are necessary for correctly designing pharmacological approaches based on mPTP targeting. With the assistance of transgenic animals and RI models, we will soon be able to identify, characterize and selectively inhibit each component and/or modulator of the mPTP complex. Discerning the relationship between mPTP activity or its genetic determinants and RI (as assessed by cMRI) in patients undergoing primary PCI for STEMI will be of great importance for achieving that aim.

Major Pitfalls in Clinical Strategies

To conclude this last part of the chapter, we briefly discuss some key points regarding the current major pitfalls associated with attempts to translate the promising therapeutic strategies discovered in basic science laboratories into the clinical setting.

Infarct Size Assessment

Standardizing IS assessments is desirable for minimizing discrepancies and permitting comparisons between studies. cMRI should be considered the gold-standard for evaluating IS. cMRI parameters (LGE, AAR and, in particular, MSI) are well-established and are related to patient prognoses [12, 14]. Furthermore, cMRI reliably identifies cardioprotection targets (cardiomyocytes exhibiting sub-lethal injury and surviving). Unfortunately, cMRI exhibits several limitations. For example, cMRI is time-consuming and expensive and is not available in every cardiovascular center. Moreover, cMRI is not feasible in patients with MI complications, patients who are overweight, or patients with pacemakers and/or implantable cardioverter defibrillators. Finally, cMRI interpretation requires skill and expertise, and several conditions may confound its results (e.g., prior MI and/or prior coronary revascularization).

Potential Confounding Factors of Cardioprotection Success

The major determinant of successful reperfusion therapy is time. Short symptom onset-to-balloon times are associated with smaller ISs, better myocardial viability, higher LVEFs, less severe heart failure and better long-term prognoses than longer symptom-onset-to-balloon times [86]. The majority of previous studies enrolled patients with a median symptom onset-to-balloon time of 6–12 h. We therefore speculate that the benefits of cardioprotection are inversely related to time. More cardiomyocytes suffer death than sub-lethal injury after several hours (>6 h) of sustained ischemia, indicating that sustained ischemia may significantly minimize the benefits of any cardioprotective strategies, especially those targeting mPTP opening. Similarly, baseline and final TIMI flow also affect the success of reperfusion therapy. Spontaneous coronary reperfusion is not rare, and approximately 30% of patients exhibit anterograde coronary flow to ischemic regions [87]. However, administration of cardioprotective agents in patients with baseline TIMI scores of 2–3 may be ineffectual because RI has already started. Similarly, the success of PCI affects the outcomes of cardioprotection strategies. The location of the culprit lesion may play a significant role in patient outcomes since it is directly related to the myocardium at risk. Larger areas at risk may obtain more benefit from cardioprotective strategies than smaller areas at risk, as the probability and extent of RI are greater in larger infarct areas than in smaller infarct areas. However, few studies have identified specific culprit lesion locations. Finally, patients with cardiovascular histories (prior MI, effort angina, silent ischemia) should be excluded from any studies evaluating cardioprotection success. cMRI interpretations are misleading in these patients and may demonstrate preconditioning-like effects that mask the impact of cardioprotective strategies.

Timing of Cardioprotective Strategy Administration

The ideal time for cardioprotective strategy administration is the time of first medical contact. Early treatment may protect against RI. Unfortunately, early treatment is not always possible in clinical practice. Some mechanical strategies require time and specific instruments that are not available in ambulances or emergency rooms. In addition, the timing of cardioprotective strategies varies significantly between studies. Treatment after the beginning RI (after mPTP opening) may be useless, and the exact moment of mPTP opening, as well as whether it is a reversible process, is unknown. Future research should clarify these points to permit better optimization of study protocols.

Methods

MPT is a fascinating phenomenon and a dangerous intersection in the development of several diseases. It requires a careful and deepen investigation not only in living cells, but also directly in patients. In the ischemic heart diseases, studies involving mPTP in patients are very limited or absent. Thus, if a slice of basic research is studying intensively the molecular composition of mPTP complex with encouraging results [65, 67, 69], simultaneous informative data should be obtained from patients. In this section of the chapter we briefly suggest methods to study mPTP contribution in MI pathology. The investigation of mPTP component(s) behavior in MI and its(their) correlation with standard endpoints of STEMI in patients is a crucial step in understanding how use this(these) potential target(s) in cardioprotection field.

Study Population

Identifying the right cohort of patients is of primary importance. Subjects aged 18–85 years, admitted to the hospital with anterior STEMI which undergo first-time successful PCI (final TIMI flow 3) with an onset to balloon inflation time <4 h and >0.1 mV ST-segment elevation in at least two contiguous precordial leads should be enrolled. Furthermore, the presence of a proximal/mid left anterior descending (LAD) occlusion with TIMI flow 0–1 and no visible evidence of significant coronary collateral flow are essential to study reperfusion damage due to mPTP opening. Major exclusion criteria include a history of prior MI and previous heart failure.

Study Endpoints

The primary endpoint of the studies should be IS and cMRI complete of MVO, LGE and MSI indexes considered the strongest reference points. Second, adverse events in the clinical follow up should be recorded.

mPTP Contribution

Circulating mPTP component(s) and its(their) expression pattern in cells collected from enrolled STEMI patients should be quantified and characterized for mPTP opening kinetics, respectively. Measuring how a putative mPTP component influences the activity of the complex should take into account three different methods: the cobalt-calcein (Co^{2+}–calcein) assay, mitochondrial membrane depolarization, and the swelling technique. The Co^{2+}–calcein assay is the most direct and sensitive experiment for measuring mPTP opening in living cells and can be used in a wide range of cytotype and to study mPTP involvement in many pathological conditions. Since mPTP opening leads to the loss of the proton gradient across membranes, the second suggested assay is the mitochondrial membrane depolarization. The third assay to assess mPTP opening is the measurement of mitochondrial network integrity. Osmotic shock induced by mPTP opening allows for the uptake of solutes into the mitochondrial matrix and concomitant swelling of the inner mitochondrial membrane. This swelling causes mitochondria matrix expansion, which results in rupture of the OMM with loss of mitochondrial network integrity [88].

Statistical Analysis

Biological and clinical data correlation will improve our knowledge on the issue of mPTP contribution in reperfusion injury mechanisms.

Conclusion

Cardioprotection is an area of ongoing active research, especially with the recent discoveries of several components involved in RI, as well as several mechanisms underlying RI occurrence. There remain no effective definitive therapeutic strategies for preventing RI despite numerous failed attempts to devise such strategies; therefore, future studies intended to better characterize mPTP structure, function and regulation and identify agents/strategies that can target its components are mandatory. These studies may improve clinical outcomes in STEMI patients in the future.

Acknowledgements PP is grateful to Camilla degli Scrovegni for continuous support. PP is supported by the Italian Ministry of Education, University and Research (COFIN no. 20129JLHSY_002, FIRB no. RBAP11FXBC_002, and Futuro in Ricerca no. RBFR10EGVP_001), the Italian Cystic Fibrosis Research Foundation (19/2014) and Telethon (GGP15219/B). PP and CG are supported by local funds from the University of Ferrara and the Italian Ministry of Health, as well as by the Italian Association for Cancer Research (AIRC: IG- 18624 and MFAG-13521).

References

1. Campo G, Saia F, Guastaroba P, Marchesini J, Varani E, Manari A, Ottani F, Tondi S, De Palma R, Marzocchi A. Prognostic impact of hospital readmissions after primary percutaneous coronary intervention. Arch Intern Med. 2011;171:1948–9.
2. McManus DD, Piacentine SM, Lessard D, Gore JM, Yarzebski J, Spencer FA, Goldberg RJ. Thirty-year (1975 to 2005) trends in the incidence rates, clinical features, treatment practices, and short-term outcomes of patients <55 years of age hospitalized with an initial acute myocardial infarction. Am J Cardiol. 2011;108:477–82.
3. Frohlich GM, Meier P, White SK, Yellon DM, Hausenloy DJ. Myocardial reperfusion injury: looking beyond primary PCI. Eur Heart J. 2013;34:1714–22.
4. Morciano G, Giorgi C, Bonora M, Punzetti S, Pavasini R, Wieckowski MR, Campo G, Pinton P. Molecular identity of the mitochondrial permeability transition pore and its role in ischemia-reperfusion injury. J Mol Cell Cardiol. 2015;78:142–53.
5. Braunwald E, Kloner RA. Myocardial reperfusion: a double-edged sword? J Clin Invest. 1985;76:1713–9.
6. Takemura G, Nakagawa M, Kanamori H, Minatoguchi S, Fujiwara H. Benefits of reperfusion beyond infarct size limitation. Cardiovasc Res. 2009;83:269–76.
7. Yetgin T, Manintveld OC, Duncker DJ, van der Giessen WJ. Postconditioning against ischaemia-reperfusion injury: ready for wide application in patients? Neth Heart J Mon J Neth Soc Cardiol Neth Heart Found. 2010;18:389–92.
8. Penna C, Perrelli MG, Pagliaro P. Mitochondrial pathways, permeability transition pore, and redox signaling in cardioprotection: therapeutic implications. Antioxid Redox Signal. 2013;18:556–99.
9. Henriques JP, Zijlstra F, vant Hof AW, de Boer MJ, Dambrink JH, Gosselink M, Hoorntje JC, Suryapranata H. Angiographic assessment of reperfusion in acute myocardial infarction by myocardial blush grade. Circulation. 2003;107:2115–9.
10. Kumar S, Sivagangabalan G, Hsieh C, Ryding AD, Narayan A, Chan H, Burgess DC, Ong AT, Sadick N, Kovoor P. Predictive value of ST resolution analysis performed immediately versus at ninety minutes after primary percutaneous coronary intervention. Am J Cardiol. 2010;105:467–74.
11. Majidi M, Kosinski AS, Al-Khatib SM, Smolders L, Cristea E, Lansky AJ, Stone GW, Mehran R, Gibbons RJ, Crijns HJ, Wellens HJ, Gorgels AP, Krucoff MW. Implications of ventricular arrhythmia "bursts" with normal epicardial flow, myocardial blush, and ST-segment recovery in anterior ST-elevation myocardial infarction reperfusion: a biosignature of direct myocellular injury "downstream of downstream". Eur Heart J Acute Cardiovasc Care. 2015;4:51–9.
12. Eitel I, de Waha S, Wohrle J, Fuernau G, Lurz P, Pauschinger M, Desch S, Schuler G, Thiele H. Comprehensive prognosis assessment by CMR imaging after ST-segment elevation myocardial infarction. J Am Coll Cardiol. 2014;64:1217–26.
13. Bonanad C, Monmeneu JV, Lopez-Lereu MP, Hervas A, de Dios E, Gavara J, Nunez J, Minana G, Husser O, Paya A, Racugno P, Garcia-Blas S, Chorro FJ, Bodi V. Prediction of long-term major events soon after a first ST-segment elevation myocardial infarction by cardiovascular magnetic resonance. Eur J Radiol. 2016;85:585–92.

14. Eitel I, Desch S, Fuernau G, Hildebrand L, Gutberlet M, Schuler G, Thiele H. Prognostic significance and determinants of myocardial salvage assessed by cardiovascular magnetic resonance in acute reperfused myocardial infarction. J Am Coll Cardiol. 2010;55:2470–9.
15. Friedrich MG, Abdel-Aty H, Taylor A, Schulz-Menger J, Messroghli D, Dietz R. The salvaged area at risk in reperfused acute myocardial infarction as visualized by cardiovascular magnetic resonance. J Am Coll Cardiol. 2008;51:1581–7.
16. Kim RJ, Wu E, Rafael A, Chen EL, Parker MA, Simonetti O, Klocke FJ, Bonow RO, Judd RM. The use of contrast-enhanced magnetic resonance imaging to identify reversible myocardial dysfunction. N Engl J Med. 2000;343:1445–53.
17. Crompton M, Costi A, Hayat L. Evidence for the presence of a reversible Ca2+-dependent pore activated by oxidative stress in heart mitochondria. Biochem J. 1987;245:915–8.
18. Halestrap AP, Clarke SJ, Javadov SA. Mitochondrial permeability transition pore opening during myocardial reperfusion – a target for cardioprotection. Cardiovasc Res. 2004;61:372–85.
19. Griffiths EJ, Halestrap AP. Mitochondrial non-specific pores remain closed during cardiac ischaemia, but open upon reperfusion. Biochem J. 1995;307(Pt 1):93–8.
20. Haworth RA, Hunter DR. The Ca2+-induced membrane transition in mitochondria. II. Nature of the Ca2+ trigger site. Arch Biochem Biophys. 1979;195:460–7.
21. Bernardi P, Vassanelli S, Veronese P, Colonna R, Szabo I, Zoratti M. Modulation of the mitochondrial permeability transition pore. Effect of protons and divalent cations. J Biol Chem. 1992;267:2934–9.
22. Baumgartner HK, Gerasimenko JV, Thorne C, Ferdek P, Pozzan T, Tepikin AV, Petersen OH, Sutton R, Watson AJ, Gerasimenko OV. Calcium elevation in mitochondria is the main Ca2+ requirement for mitochondrial permeability transition pore (mPTP) opening. J Biol Chem. 2009;284:20796–803.
23. Schroers A, Kramer R, Wohlrab H. The reversible antiport-uniport conversion of the phosphate carrier from yeast mitochondria depends on the presence of a single cysteine. J Biol Chem. 1997;272:10558–64.
24. Takeyama N, Matsuo N, Tanaka T. Oxidative damage to mitochondria is mediated by the Ca(2+)-dependent inner-membrane permeability transition. Biochem J. 1993;294(Pt 3):719–25.
25. Petronilli V, Cola C, Bernardi P. Modulation of the mitochondrial cyclosporin A-sensitive permeability transition pore. II. The minimal requirements for pore induction underscore a key role for transmembrane electrical potential, matrix pH, and matrix Ca2+. J Biol Chem. 1993;268:1011–6.
26. Petronilli V, Costantini P, Scorrano L, Colonna R, Passamonti S, Bernardi P. The voltage sensor of the mitochondrial permeability transition pore is tuned by the oxidation-reduction state of vicinal thiols. Increase of the gating potential by oxidants and its reversal by reducing agents. J Biol Chem. 1994;269:16638–42.
27. Javadov S, Karmazyn M. Mitochondrial permeability transition pore opening as an endpoint to initiate cell death and as a putative target for cardioprotection. Cell Physiol Biochem Int J Exp Cell Physiol Biochem Pharmacol. 2007;20:1–22.
28. Negrelo Newton AP, Cadena SM, Merlin Rocha ME, Skare Carnieri EG, Martinelli de Oliveira MB, et al. Toxicol Lett. 2004;151:407–16.
29. Chen Q, Xu H, Xu A, Ross T, Bowler E, Hu Y, Lesnefsky EJ. Inhibition of Bcl-2 sensitizes mitochondrial permeability transition pore (MPTP) opening in ischemia-damaged mitochondria. PLoS One. 2015;10:e0118834.
30. Arbel N, Ben-Hail D, Shoshan-Barmatz V. Mediation of the antiapoptotic activity of Bcl-xL protein upon interaction with VDAC1 protein. J Biol Chem. 2012;287:23152–61.
31. Smeele KM, Southworth R, Wu R, Xie C, Nederlof R, Warley A, Nelson JK, van Horssen P, van den Wijngaard JP, Heikkinen S, Laakso M, Koeman A, Siebes M, Eerbeek O, Akar FG, Ardehali H, Hollmann MW, Zuurbier CJ. Disruption of hexokinase II-mitochondrial binding blocks ischemic preconditioning and causes rapid cardiac necrosis. Circ Res. 2011;108:1165–9.

32. Marzo I, Brenner C, Zamzami N, Jurgensmeier JM, Susin SA, Vieira HL, Prevost MC, Xie Z, Matsuyama S, Reed JC, Kroemer G. Bax and adenine nucleotide translocator cooperate in the mitochondrial control of apoptosis. Science. 1998;281:2027–31.
33. Narita M, Shimizu S, Ito T, Chittenden T, Lutz RJ, Matsuda H, Tsujimoto Y. Bax interacts with the permeability transition pore to induce permeability transition and cytochrome c release in isolated mitochondria. Proc Natl Acad Sci U S A. 1998;95:14681–6.
34. Roy SS, Madesh M, Davies E, Antonsson B, Danial N, Hajnoczky G. Bad targets the permeability transition pore independent of Bax or Bak to switch between Ca2+-dependent cell survival and death. Mol Cell. 2009;33:377–88.
35. Murphy E, Steenbergen C. Inhibition of GSK-3beta as a target for cardioprotection: the importance of timing, location, duration and degree of inhibition. Expert Opin Ther Targets. 2005;9:447–56.
36. Giorgio V, Bisetto E, Soriano ME, Dabbeni-Sala F, Basso E, Petronilli V, Forte MA, Bernardi P, Lippe G. Cyclophilin D modulates mitochondrial F0F1-ATP synthase by interacting with the lateral stalk of the complex. J Biol Chem. 2009;284:33982–8.
37. Murphy E, Steenbergen C. What makes the mitochondria a killer? Can we condition them to be less destructive? Biochim Biophys Acta. 1813;2011:1302–8.
38. Marchi S, Pinton P. The mitochondrial calcium uniporter complex: molecular components, structure and physiopathological implications. J Physiol. 2014;592:829–39.
39. Pan X, Liu J, Nguyen T, Liu C, Sun J, Teng Y, Fergusson MM, Rovira II, Allen M, Springer DA, Aponte AM, Gucek M, Balaban RS, Murphy E, Finkel T. The physiological role of mitochondrial calcium revealed by mice lacking the mitochondrial calcium uniporter. Nat Cell Biol. 2013;15:1464–72.
40. Giorgi C, Romagnoli A, Pinton P, Rizzuto R. Ca2+ signaling, mitochondria and cell death. Curr Mol Med. 2008;8:119–30.
41. Reimer KA, Jennings RB. The “wavefront phenomenon” of myocardial ischemic cell death. II. Transmural progression of necrosis within the framework of ischemic bed size (myocardium at risk) and collateral flow. Lab Investig J Tech Methods Pathol. 1979;40:633–44.
42. Kajstura J, Cheng W, Reiss K, Clark WA, Sonnenblick EH, Krajewski S, Reed JC, Olivetti G, Anversa P. Apoptotic and necrotic myocyte cell deaths are independent contributing variables of infarct size in rats. Lab Investig J Tech Methods Pathol. 1996;74:86–107.
43. Shintani-Ishida K, Unuma K, Yoshida K. Ischemia enhances translocation of connexin43 and gap junction intercellular communication, thereby propagating contraction band necrosis after reperfusion. Circ J Off J Jpn Circ Soc. 2009;73:1661–8.
44. Fliss H, Gattinger D. Apoptosis in ischemic and reperfused rat myocardium. Circ Res. 1996;79:949–56.
45. Anversa P, Cheng W, Liu Y, Leri A, Redaelli G, Kajstura J. Apoptosis and myocardial infarction. Basic Res Cardiol. 1998;93(Suppl 3):8–12.
46. Dumont EA, Hofstra L, van Heerde WL, van den Eijnde S, Doevendans PA, DeMuinck E, Daemen MA, Smits JF, Frederik P, Wellens HJ, Daemen MJ, Reutelingsperger CP. Cardiomyocyte death induced by myocardial ischemia and reperfusion: measurement with recombinant human annexin-V in a mouse model. Circulation. 2000;102:1564–8.
47. Dumont EA, Reutelingsperger CP, Smits JF, Daemen MJ, Doevendans PA, Wellens HJ, Hofstra L. Real-time imaging of apoptotic cell-membrane changes at the single-cell level in the beating murine heart. Nat Med. 2001;7:1352–5.
48. Gottlieb RA, Burleson KO, Kloner RA, Babior BM, Engler RL. Reperfusion injury induces apoptosis in rabbit cardiomyocytes. J Clin Invest. 1994;94:1621–8.
49. Saraste A, Pulkki K, Kallajoki M, Henriksen K, Parvinen M, Voipio-Pulkki LM. Apoptosis in human acute myocardial infarction. Circulation. 1997;95:320–3.
50. Halestrap AP, Connern CP, Griffiths EJ, Kerr PM. Cyclosporin A binding to mitochondrial cyclophilin inhibits the permeability transition pore and protects hearts from ischaemia/reperfusion injury. Mol Cell Biochem. 1997;174:167–72.

51. Argaud L, Gateau-Roesch O, Raisky O, Loufouat J, Robert D, Ovize M. Postconditioning inhibits mitochondrial permeability transition. Circulation. 2005;111:194–7.
52. Clarke SJ, McStay GP, Halestrap AP. Sanglifehrin A acts as a potent inhibitor of the mitochondrial permeability transition and reperfusion injury of the heart by binding to cyclophilin-D at a different site from cyclosporin A. J Biol Chem. 2002;277:34793–9.
53. Gomez L, Thibault H, Gharib A, Dumont JM, Vuagniaux G, Scalfaro P, Derumeaux G, Ovize M. Inhibition of mitochondrial permeability transition improves functional recovery and reduces mortality following acute myocardial infarction in mice. Am J Physiol Heart Circ Physiol. 2007;293:H1654–61.
54. Basso E, Fante L, Fowlkes J, Petronilli V, Forte MA, Bernardi P. Properties of the permeability transition pore in mitochondria devoid of Cyclophilin D. J Biol Chem. 2005;280:18558–61.
55. Gibson CM, Giugliano RP, Kloner RA, Bode C, Tendera M, Janosi A, Merkely B, Godlewski J, Halaby R, Korjian S, Daaboul Y, Chakrabarti AK, Spielman K, Neal BJ, Weaver WD. EMBRACE STEMI study: a phase 2a trial to evaluate the safety, tolerability, and efficacy of intravenous MTP-131 on reperfusion injury in patients undergoing primary percutaneous coronary intervention. Eur Heart J. 2016;37:1296–303.
56. Bonora M, Wieckowsk MR, Chinopoulos C, Kepp O, Kroemer G, Galluzzi L, Pinton P. Molecular mechanisms of cell death: central implication of ATP synthase in mitochondrial permeability transition. Oncogene. 2015;34:1608.
57. He L, Lemasters JJ. Regulated and unregulated mitochondrial permeability transition pores: a new paradigm of pore structure and function? FEBS Lett. 2002;512:1–7.
58. Kowaltowski AJ, Castilho RF, Vercesi AE. Mitochondrial permeability transition and oxidative stress. FEBS Lett. 2001;495:12–5.
59. Kokoszka JE, Waymire KG, Levy SE, Sligh JE, Cai J, Jones DP, MacGregor GR, Wallace DC. The ADP/ATP translocator is not essential for the mitochondrial permeability transition pore. Nature. 2004;427:461–5.
60. Baines CP, Kaiser RA, Sheiko T, Craigen WJ, Molkentin JD. Voltage-dependent anion channels are dispensable for mitochondrial-dependent cell death. Nat Cell Biol. 2007;9:550–5.
61. Leung AW, Varanyuwatana P, Halestrap AP. The mitochondrial phosphate carrier interacts with cyclophilin D and may play a key role in the permeability transition. J Biol Chem. 2008;283:26312–23.
62. Alcala S, Klee M, Fernandez J, Fleischer A, Pimentel-Muinos FX. A high-throughput screening for mammalian cell death effectors identifies the mitochondrial phosphate carrier as a regulator of cytochrome c release. Oncogene. 2008;27:44–54.
63. Varanyuwatana P, Halestrap AP. The roles of phosphate and the phosphate carrier in the mitochondrial permeability transition pore. Mitochondrion. 2012;12:120–5.
64. Kwong JQ, Davis J, Baines CP, Sargent MA, Karch J, Wang X, Huang T, Molkentin JD. Genetic deletion of the mitochondrial phosphate carrier desensitizes the mitochondrial permeability transition pore and causes cardiomyopathy. Cell Death Differ. 2014;21:1209–17.
65. Bonora M, Bononi A, De Marchi E, Giorgi C, Lebiedzinska M, Marchi S, Patergnani S, Rimessi A, Suski JM, Wojtala A, Wieckowski MR, Kroemer G, Galluzzi L, Pinton P. Role of the c subunit of the FO ATP synthase in mitochondrial permeability transition. Cell Cycle. 2013;12:674–83.
66. De Marchi E, Bonora M, Giorgi C, Pinton P. The mitochondrial permeability transition pore is a dispensable element for mitochondrial calcium efflux. Cell Calcium. 2014;56:1–13.
67. Alavian KN, Beutner G, Lazrove E, Sacchetti S, Park HA, Licznerski P, Li H, Nabili P, Hockensmith K, Graham M, Porter Jr GA, Jonas EA. An uncoupling channel within the c-subunit ring of the F1FO ATP synthase is the mitochondrial permeability transition pore. Proc Natl Acad Sci U S A. 2014;111:10580–5.
68. Alavian KN, Li H, Collis L, Bonanni L, Zeng L, Sacchetti S, Lazrove E, Nabili P, Flaherty B, Graham M, Chen Y, Messerli SM, Mariggio MA, Rahner C, McNay E, Shore GC, Smith PJ, Hardwick JM, Jonas EA. Bcl-xL regulates metabolic efficiency of neurons through interaction with the mitochondrial F1FO ATP synthase. Nat Cell Biol. 2011;13:1224–33.

69. Azarashvili T, Odinokova I, Bakunts A, Ternovsky V, Krestinina O, Tyynela J, Saris NE. Potential role of subunit c of F0F1-ATPase and subunit c of storage body in the mitochondrial permeability transition. Effect of the phosphorylation status of subunit c on pore opening. Cell Calcium. 2014;55:69–77.
70. Giorgio V, von Stockum S, Antoniel M, Fabbro A, Fogolari F, Forte M, Glick GD, Petronilli V, Zoratti M, Szabo I, Lippe G, Bernardi P. Dimers of mitochondrial ATP synthase form the permeability transition pore. Proc Natl Acad Sci U S A. 2013;110:5887–92.
71. Masgras I, Rasola A, Bernardi P. Induction of the permeability transition pore in cells depleted of mitochondrial DNA. Biochim Biophys Acta. 2012;1817:1860–6.
72. Wittig I, Meyer B, Heide H, Steger M, Bleier L, Wumaier Z, Karas M, Schagger H. Assembly and oligomerization of human ATP synthase lacking mitochondrial subunits a and A6L. Biochim Biophys Acta. 2010;1797:1004–11.
73. Bonora M, Bravo-San Pedro JM, Kroemer G, Galluzzi L, Pinton P. Novel insights into the mitochondrial permeability transition. Cell Cycle. 2014;13:2666–70.
74. Campo G, Morciano G, Pavasini R, Bonora M, Sbano L, Biscaglia S, Bovolenta M, Pinotti M, Punzetti S, Rizzo P, Aquila G, Giorgi C, Ferrari R, Pinton P. Fo ATP synthase C subunit serum levels in patients with ST-segment elevation myocardial infarction: preliminary findings. Int J Cardiol. 2016;221:993–7.
75. Hausenloy DJ, Mwamure PK, Venugopal V, Harris J, Barnard M, Grundy E, Ashley E, Vichare S, Di Salvo C, Kolvekar S, Hayward M, Keogh B, MacAllister RJ, Yellon DM. Effect of remote ischaemic preconditioning on myocardial injury in patients undergoing coronary artery bypass graft surgery: a randomised controlled trial. Lancet. 2007;370:575–9.
76. Zhao ZQ, Corvera JS, Halkos ME, Kerendi F, Wang NP, Guyton RA, Vinten-Johansen J. Inhibition of myocardial injury by ischemic postconditioning during reperfusion: comparison with ischemic preconditioning. Am J Physiol Heart Circ Physiol. 2003;285:H579–88.
77. Bulluck H, Yellon DM, Hausenloy DJ. Reducing myocardial infarct size: challenges and future opportunities. Heart. 2016;102:341–8.
78. Atar D, Arheden H, Berdeaux A, Bonnet JL, Carlsson M, Clemmensen P, Cuvier V, Danchin N, Dubois-Rande JL, Engblom H, Erlinge D, Firat H, Halvorsen S, Hansen HS, Hauke W, Heiberg E, Koul S, Larsen AI, Le Corvoisier P, Nordrehaug JE, Paganelli F, Pruss RM, Rousseau H, Schaller S, Sonou G, Tuseth V, Veys J, Vicaut E, Jensen SE. Effect of intravenous TRO40303 as an adjunct to primary percutaneous coronary intervention for acute ST-elevation myocardial infarction: MITOCARE study results. Eur Heart J. 2015;36:112–9.
79. Cung TT, Morel O, Cayla G, Rioufol G, Garcia-Dorado D, Angoulvant D, Bonnefoy-Cudraz E, Guerin P, Elbaz M, Delarche N, Coste P, Vanzetto G, Metge M, Aupetit JF, Jouve B, Motreff P, Tron C, Labeque JN, Steg PG, Cottin Y, Range G, Clerc J, Claeys MJ, Coussement P, Prunier F, Moulin F, Roth O, Belle L, Dubois P, Barragan P, Gilard M, Piot C, Colin P, De Poli F, Morice MC, Ider O, Dubois-Rande JL, Unterseeh T, Le Breton H, Beard T, Blanchard D, Grollier G, Malquarti V, Staat P, Sudre A, Elmer E, Hansson MJ, Bergerot C, Boussaha I, Jossan C, Derumeaux G, Mewton N, Ovize M. Cyclosporine before PCI in patients with acute myocardial infarction. N Engl J Med. 2015;373:1021–31.
80. Kitakaze M, Asakura M, Kim J, Shintani Y, Asanuma H, Hamasaki T, Seguchi O, Myoishi M, Minamino T, Ohara T, Nagai Y, Nanto S, Watanabe K, Fukuzawa S, Hirayama A, Nakamura N, Kimura K, Fujii K, Ishihara M, Saito Y, Tomoike H, Kitamura S, Investigators JW. Human atrial natriuretic peptide and nicorandil as adjuncts to reperfusion treatment for acute myocardial infarction (J-WIND): two randomised trials. Lancet. 2007;370:1483–93.
81. Lincoff AM, Roe M, Aylward P, Galla J, Rynkiewicz A, Guetta V, Zelizko M, Kleiman N, White H, McErlean E, Erlinge D, Laine M, Dos Santos Ferreira JM, Goodman S, Mehta S, Atar D, Suryapranata H, Jensen SE, Forster T, Fernandez-Ortiz A, Schoors D, Radke P, Belli G, Brennan D, Bell G, Krucoff M, Investigators PA. Inhibition of delta-protein kinase C by delcasertib as an adjunct to primary percutaneous coronary intervention for acute anterior ST-segment elevation myocardial infarction: results of the PROTECTION AMI Randomized Controlled Trial. Eur Heart J. 2014;35:2516–23.

82. Lonborg J, Vejlstrup N, Kelbaek H, Botker HE, Kim WY, Mathiasen AB, Jorgensen E, Helqvist S, Saunamaki K, Clemmensen P, Holmvang L, Thuesen L, Krusell LR, Jensen JS, Kober L, Treiman M, Holst JJ, Engstrom T. Exenatide reduces reperfusion injury in patients with ST-segment elevation myocardial infarction. Eur Heart J. 2012;33:1491–9.
83. Ottani F, Latini R, Staszewsky L, La Vecchia L, Locuratolo N, Sicuro M, Masson S, Barlera S, Milani V, Lombardi M, Costalunga A, Mollichelli N, Santarelli A, De Cesare N, Sganzerla P, Boi A, Maggioni AP, Limbruno U, Investigators C. Cyclosporine A in reperfused myocardial infarction: the multicenter, controlled, open-label CYCLE trial. J Am Coll Cardiol. 2016;67:365–74.
84. Piot C, Croisille P, Staat P, Thibault H, Rioufol G, Mewton N, Elbelghiti R, Cung TT, Bonnefoy E, Angoulvant D, Macia C, Raczka F, Sportouch C, Gahide G, Finet G, Andre-Fouet X, Revel D, Kirkorian G, Monassier JP, Derumeaux G, Ovize M. Effect of cyclosporine on reperfusion injury in acute myocardial infarction. N Engl J Med. 2008;359:473–81.
85. MacVicar T, Langer T. OPA1 processing in cell death and disease – the long and short of it. J Cell Sci. 2016;129:2297–306.
86. Christian TF, Schwartz RS, Gibbons RJ. Determinants of infarct size in reperfusion therapy for acute myocardial infarction. Circulation. 1992;86:81–90.
87. Christian TF, Milavetz JJ, Miller TD, Clements IP, Holmes DR, Gibbons RJ. Prevalence of spontaneous reperfusion and associated myocardial salvage in patients with acute myocardial infarction. Am Heart J. 1998;135:421–7.
88. Bonora M, Morganti C, Morciano G, Giorgi C, Wieckowski MR, Pinton P. Comprehensive analysis of mitochondrial permeability transition pore activity in living cells using fluorescence-imaging-based techniques. Nat Protoc. 2016;11:1067–80.

Chapter 10
Functional Role of Mitochondria in Arrhythmogenesis

Jessica Gambardella, Daniela Sorriento, Michele Ciccarelli, Carmine Del Giudice, Antonella Fiordelisi, Luigi Napolitano, Bruno Trimarco, Guido Iaccarino, and Gaetano Santulli

A coordinated electrical propagation throughout the heart is necessary for appropriate cardiac function. When such coordination in cardiac action potential (AP) fails, cardiac arrhythmias occur [1]. Several sub-cellular factors contribute to AP heterogeneity, mainly related to sarcolemmal ion channels; however, a growing attention is focused on the action of mitochondria on cell excitability. Indeed, albeit the mitochondrial role in energy production and apoptotic pathways has been determined, the contribution of this highly dynamic organelle in the regulation of excitation–contraction coupling is less clear.

Mitochondria form a functional network within the cardiomyocyte that represents 20–30% of myocardial volume and produces over 95% of cellular ATP. ATP production is made possible by mitochondrial membrane potential ($\Delta\Psi m$) that generate a proton motive force liberating the energy necessary to phosphorylate ADP to ATP [2]. This mechanism is also one of the main sources of reactive oxygen species (ROS) in the cell. In physiological conditions $\Delta\Psi m$ is highly regulated, so that ATP production sufficiently responds to energy demand and ROS does not exceed cell-detoxifying capacity. In response to pathological stimuli, including ischemia and

J. Gambardella
Columbia University Medical Center, New York, NY, USA

Department of Medicine, Surgery and Dentistry, University of Salerno, Baronissi, Italy

D. Sorriento • C. Del Giudice • A. Fiordelisi • L. Napolitano • B. Trimarco
Department of Advanced Biomedical Sciences, "Federico II" University, Naples, Italy

M. Ciccarelli • G. Iaccarino
Department of Medicine, Surgery and Dentistry, University of Salerno, Baronissi, Italy

G. Santulli (✉)
Dept. of Biomedical Advanced Sciences, Federico II University, Naples, Italy

Columbia University Medical Center, New York Presbyterian Hospital—Manhattan, New York, NY, USA
e-mail: gsantulli001@gmail.com

G. Santulli (ed.), *Mitochondrial Dynamics in Cardiovascular Medicine*, Advances in Experimental Medicine and Biology 982,
DOI 10.1007/978-3-319-55330-6_10

structural injury, alterations of $\Delta\Psi m$ cause reduction of ATP generation, and increase of ROS production that once exceeds the detoxifying capacity, leads to oxidative stress.

The involvement of mitochondria in arrhythmogenesis derives from their ability to produce both ATP and ROS; indeed, on one side, mitochondrial dysfunction can affect electrical function of the heart through reduced ATP production, altering sarcolemmal K^+ fluxes via ATP-sensitive potassium channels. On the other side, excessive mitochondrial ROS production can introduce heterogeneity into cardiac action potential since during oxidative stress the oxidation of mitochondrial ion channels can alter their kinetic opening loading to current dispersion and eventually collapse of $\Delta\Psi m$. The ATP- and ROS- based mechanisms are part of the same cellular response: during mitochondrial dysfunction $\Delta\Psi m$ reduction occurs, resulting in attenuated ATP production and increased ROS level; the latter, in turn, induces further mitochondrial dysfunction, with $\Delta\Psi m$ reduction and consequent reduced ATP generation. This vicious cycle, in which different responses act in synergic manner, causes electrophysiological alterations, thereby conferring a central role to mitochondria in arrhythmogenesis.

Functional Role of Mitochondria in Linking Metabolism and Cell Excitability

Several studies have shown the influence of mitochondrial energetic status on the sarcolemmal action potential and heart excitability. The first experimental evidence derives from simultaneous detection of $\Delta\Psi m$ and AP. $\Delta\Psi m$ oscillations, induced by photo-oxidation, coincide with AP oscillations: in particular, during $\Delta\Psi m$ collapse also AP collapses until a cell non-excitable state, while $\Delta\Psi m$ recovery is mirrored by AP recovery. The impact of mitochondria on cellular excitability is mainly mediated by a class of ion channels abundant in cardiac tissue: energy sensing, ATP sensitive K^+ channels on sarcolemmal membrane (sarcKATP). These channels are heteromultimers inhibited by intracellular ATP and activated by ADP, Pi, Mg^{2+} and extremely sensitive to oxidation. Therefore, sarcKATP channels represent a crucial link between electrical function and metabolism, a dynamic relationship that has been confirmed by independent investigators [3, 4] (Fig. 10.1). During metabolic stress, oscillations of sarcKATP current occur in phase with variation of NADH concentration, with losses of $\Delta\Psi m$, and that these oscillations are responsible of reduced length of AP [5]. Therefore, metabolic stress like nutrient deprivation can impair mitochondrial function, with reduction of $\Delta\Psi m$ and consequent diminished production of ATP, which in physiological conditions inhibits sarKATP channels. The result is the activation of these channels with K^+ dispersion and reduction of cell excitability, allowing mitochondria to indirectly affect the AP [6]. Such effect has a citoprotective role for myocytes during ischemia, when mitochondria cannot provide adequate ATP to support cardiac energy demand; thus, the sarKATP-mediated reduction in excitability and calcium (Ca^{2+}) transients attenuates cell death during metabolic stress [7–10].

Mounting evidence supports the mechanistic role of sarKATP in arrhythmogenesis as a sensor of metabolic condition of the cell. Importantly, a reduction in ventricular

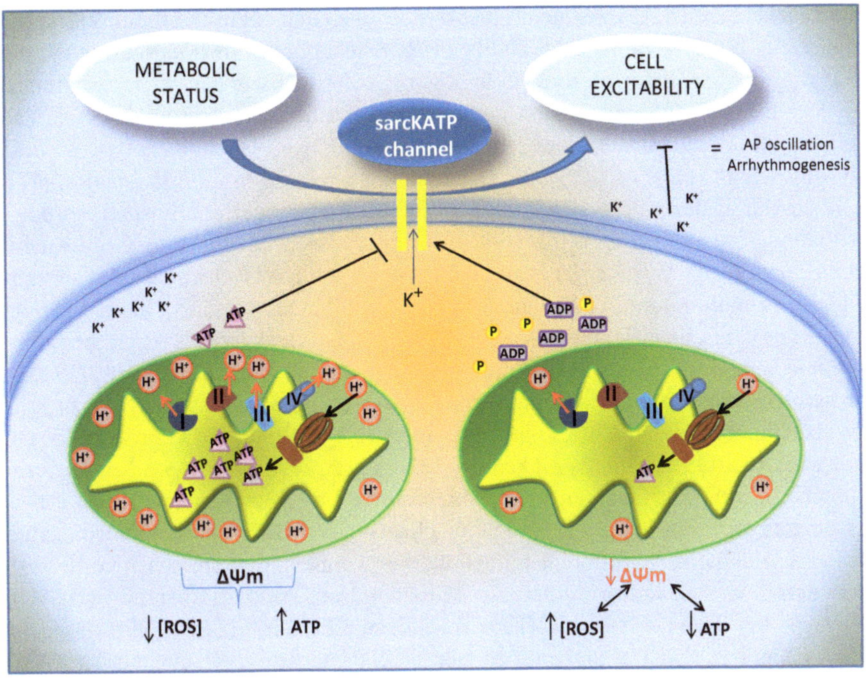

Fig. 10.1 Role of sarcKATP channels in linking cell metabolic status and cell excitability. Mitochondrial dysfunction lead to increase of ROS and reduction of mitochondrial potential with consequent decrease of ATP production. ATP decrease, ATP and Pi activate sarcKATP channel, with K dissipating current and reduction of AP that predispose to arrhythmia

arrhythmia development after blocking sarcKATP channels using HMR1883 has been reported in studies in both animal and human studies [11–14]. However, other studies have shown that the blocking of these channels with glibenclamide is not enough to rescue an arrhythmic phenotype [15]. Because of these conflicting findings further studies are necessary to elucidate the molecular mechanisms that occur before and after activation of these channels, and to clarify the role of several mitochondrial ion channels that could modify their kinetics in response to metabolic stress. The correct dynamic activity of cellular ion channels is synergistically linked to mitochondria, not only for ATP demand but also for mitochondrial ROS generation.

Role of Mitochondrial ROS in the Pathogenesis of Cardiac Arrhythmias

If reduced ATP production as consequence of mitochondrial dysfunction can directly affect the activity of several ion channels, concomitant ROS generation can activate a complex response which in turn amplifies $\Delta\Psi m$ alterations that further inhibit ATP production and increase ROS generation. Sallott and colleagues demonstrated a ROS-dependent oscillation of $\Delta\Psi m$, with collapse of $\Delta\Psi m$ prevented by

ROS scavenger [16, 17]. As mentioned above, given that $\Delta\Psi m$ oscillations promote alterations of myocyte AP, the ability of ROS to affect mitochondrial membrane potential confers them a pro-arrhythmogenic role. In particular such arrhythmogenic effect of ROS depends on complex and multifactorial response of cardiac mitochondria, named ROS-induced ROS release (RIRR), an autocatalytic process by which high levels of ROS induce further ROS release from mitochondria [17]. Mitochondrial-dependent RIRR regulates electro-chemical equilibrium, through oxidation of several proteins that control current fluxes inside the myocyte; the main consequences of oxidative stress are: activation of sarcKATP channels (that are sensible to oxidation, too), with reduction of Na^+ and K^+ currents [18]; altered kinetic of L-Type Ca^{2+} channels; increase of intracellular Ca^{2+} leak from RyR on the sarcoplasmic reticulum [19]. The functional role of mitochondria-dependent RIRR in arrhythmogenesis has been demonstrated also in whole heart, in ex-vivo experiments; stimulation of the heart with H_2O_2 leads to a two-phase ROS production: the first pick of ROS is directly linked to H_2O_2 action, while the second pick corresponds to ROS of endogenous production derived by mitochondrial RIRR and is associated with arrhythmia [20]. The mechanism of RIRR represents a concerted response mediated by different mitochondrial channels, that are sensitive to ROS concentration, including internal Mitochondria Anion Channel (IMAC), Permeability Transition Pore (PTP), Translocator Protein (TSPO), Mitochondrial Ca^{2+} Uniporter (MCU), and mitochondrial ATP- sensitive K^+ channel (mKATP channel). These channels change their activities in response to crescent levels of ROS and then activate RIRR. Despite the specific contribute of each of these channels to RIRR is not fully understood, the important evidence is that mitochondrial ion channels have a central role in the regulation of electrophysiology equilibrium; indeed through them cell excitability is affected by metabolic status of the cell; therefore, they could represent new and innovative therapeutic target for pathological conditions characterized by electrical alterations, typically arrhythmias, mostly in response to metabolic stress.

Mitochondrial Channels and Arrhythmogenesis

The metabolism-excitation axis is finely regulated by several mitochondrial channels including the inner membrane anion channel (IMAC), the mitochondrial permeability transition pore (mPTP) and the translocator protein (TSPO), and by their crosstalk.

IMAC is a channel deputed to anion efflux from mitochondria [21]. Although its chemical structure is not completely determined, it is known to be one of the channels proposed to participate in energy dissipation in mitochondria, determining $\Delta\Psi m$ collapse and consequent AP reduction [22–24]. In vitro, the blockade of IMAC during metabolic stress stabilizes $\Delta\Psi m$ and determines recovery of AP; on the other hand, activation of IMAC accelerates the shortening of AP through RIRR [10]. These results are confirmed in intact mammalian hearts [10, 25, 26]. Indeed, in guinea pig hearts IMAC inhibition reduces the ischemia-induced AP shortening and coincides with decreased tachycardia/fibrillation during reperfusion [10]. This evidence opens the possibility to target IMAC as a primary mitochondrial mediator of ROS amplification, ΔΨm reduction, and AP oscillations, for novel anti-arrhythmic therapeutic strategies.

The electrochemical equilibrium in mitochondria is finely orchestrated by mPTP, a large conductance channel located on the inner mitochondrial membrane that plays a crucial role in the regulation of cell death [27]. Although some studies demonstrate that the inhibition of mPTP is protective for cardiac cells during ischemia, there are controversial results about the role of this channel in determining arrhythmia. In isolated cells, the collapse of ΔΨm induced by laser flash is not inhibited by mPTP blocker, cyclosporine A [4, 16, 28]. A non-protective effect of cyclosporine on arrhythmia was confirmed in rat, pig, and rabbit [10, 25, 29]. However, mPTP is directly involved in RIRR; therefore this channel contributes to the effects of metabolism on the cell excitability, but its inhibition is not enough to prevent arrhythmia [30].

Interestingly, IMAC-mediated RIRR and mPTP-mediated RIRR are both regulated by a third protein, located on the outer mitochondrial membrane, TSPO. Altered expression and activity of this translocator protein have been reported in ischemia/reperfusion injury and myocardial infarction, both of which are considered mayor risk factors for arrhythmias. Growing evidence suggests that TSPO can affect arrhythmic response regulating RIRR, directly through IMAC and indirectly through mPTP. TSPO ligands abolish metabolic and electrophysiological oscillations induced by oxidative stress, and reduce ROS levels in cardiomyocytes [31]; moreover, TSPO inhibition abolishes the second pick of ROS production in response to H_2O_2 that represents the RIRR- derived ROS, suppressing ventricular fibrillation and the frequency of arrhythmogenic triggers [20]. This ability of TSPO to promote RIRR with consequent pro-arrhythmogenic effect, is mediated mainly by its action on IMAC. IMAC is tight regulated by several TSPO -acting -ligands, suggesting a direct interaction between pore- forming subunit of IMAC in the inner-mitochondria-membrane, and the regulatory protein TSPO in out mitochondrial membrane [32–34]. Several studies, demonstrate that TSPO is able to regulate mPTP too, indirectly through interaction with VDAC and ANT [35, 36]. Indeed, TSPO block results in inhibition of ΔΨm depolarization after ROS production, but at the same time also in increase of the cell survival in response to oxidative stress, with reduction of cytochrome C release, caspase-3 activation and DNA fragmentation [37]. This evidence suggests a hierarchal activation pattern of mitochondrial ion channels, with TSPO playing a key regulatory role. In response to moderate levels of ROS, TSPO could mediate activation of IMAC with initial energy dissipation that results in RIRR activation, ΔΨm partial depolarization and consequent AP reduction, ensuring a protective status for the cell characterized by reduced excitability to lower energy demand. When metabolic stress is persistent and ROS levels, derived also from RIRR, mediate extreme oxidative stress, TSPO oxidation leads to the activation of the large conductance mPTP and eventually to irreversible mitochondrial membrane potential depolarization, promoting cell dysfunction and death

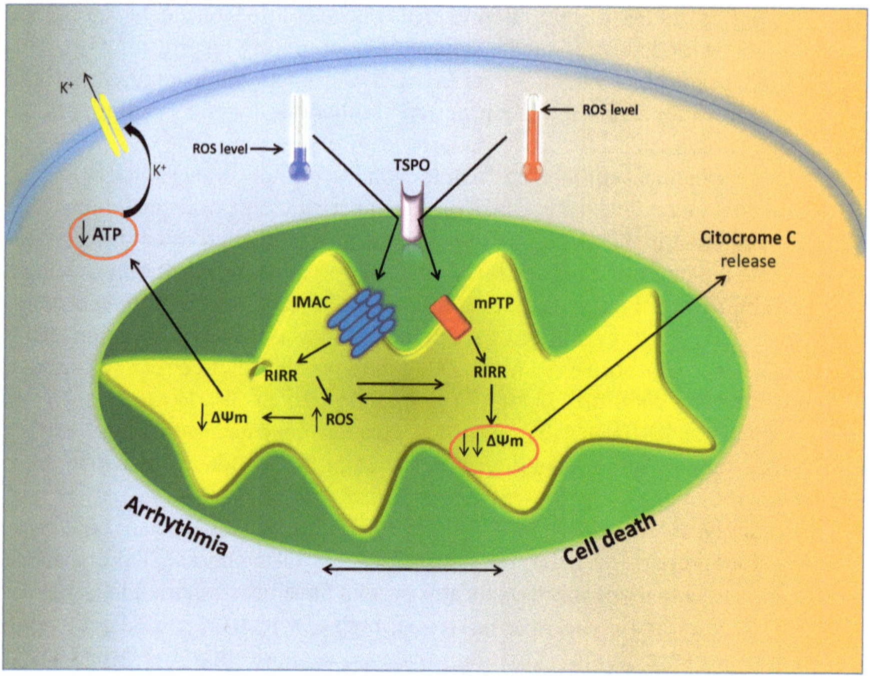

Fig. 10.2 TSPO regulates mitochondrial dissipating energy channel. In presence of low levels of ROS, TSPO is able to activate IMAC with consequent RIRR response that can induce arrhythmia. In presence of high levels of ROS, TSPO becomes able to activate mPTP with terminal collapse of mitochondrial potential and cell death

(Fig. 10.2). This hypothesis confers to mitochondria a central role in determining cell destiny: on one hand this organelle activates cardio-protection at the expense of cell excitability; on the other hand, mitochondrial dysfunction can cause cell death. In this complex response TSPO acts as a sensor of metabolic stress tolerability; not surprisingly, TSPO ligands are promising in preventing ischemia-induced ventricular fibrillation [38].

mitoKATP Channels

A crucial role in arrhythmogenesis is played by a specific class of mitochondrial channels, initially identified in hepatic mitochondria and then found also in the heart: mitochondrial ATP-sensitive potassium channels (mitoKATP) [39]. Little is known about these channels and some studies yielded conflicting results, probably because of the low specificity of the compounds used to target mitoKATP [40–43]. Nevertheless, recent evidence supports a protective role of these channels against arrhythmia, since their opening before metabolic stress (ischemia) induces partial dissipation of $\Delta\Psi m$, reducing the drive force for mitochondrial Ca^{2+} uptake and improving cellular respiration. The ATP sensibility of these channels allow them to behave as sensors of metabolic stress [40–42].

Mechanistic Role of Mitochondrial Ca^{2+} in Arrhythmogenesis

Ca^{2+} homeostasis in mitochondria is mainly ensured by Ca^{2+} influx through MCU (see Chap. 2), and Ca^{2+} efflux through mitochondrial Na^+/Ca^{2+} exchanger [44]. The administration of MCU blockers reduces the incidence of ventricular fibrillation in anesthetized rats [45]. Most likely, the inhibition of MCU opening plays a protective role keeping mitochondrial Ca^{2+} concentration low. Indeed, mitochondrial Ca^{2+} overload can induce mitochondrial dysfunction with increase of open probability of mPTP [45].

Mitochondrial Ca^{2+} overload might trigger opening of the mPTP, causing uncoupling of oxidative phosphorylation, swelling of the mitochondria, and rupture of the mitochondrial outer membrane. It is not easy to define precisely the role of MCU in arrhythmogenesis given the non-specific effects of the compounds used as blockers (ruthenium red and Ru360); however, mitochondrial dysfunction as result of Ca^{2+} overload suggests that these organelles can contribute also indirectly to arrhythmogenesis, through ROS production and $\Delta\Psi m$ collapse in response to high Ca^{2+} influx. In cardiac myocytes, mitochondria and SR are structurally and functionally related, co-localizing in the so-called mitochondrial microdomain; mitochondrial Ca^{2+} influx through MCU is tightly linked to SR Ca^{2+} release [46]. Alterations of RyR2 on SR, and consequent increase of Ca^{2+} leak in the cytosol, are among the causes of cardiac arrhythmias [47]. Mitochondria, while not representing the primary cause, can contribute strongly to the arrhythmic phenotype, being a key component of the following vicious cycle (Fig. 10.3):

RyR alterations → increased Ca^{2+} leak → mitochondrial Ca^{2+} overload → mitochondrial dysfunction → ROS production → RyR2 oxidation → further increase of Ca^{2+} leak.

We have shown the importance of this vicious cycle, indicating that mitochondrial Ca^{2+} overload plays a key role in atrial fibrillation [48], one of the most common arrhythmias [49, 50]. We used murine models characterized by RyR2 mutations causing SR Ca^{2+} leak, and these mice exhibited mitochondrial dysfunction, RyR2 oxidation, high ROS level and atrial fibrillation. The most important evidence is that not only RyR2 pharmacological block, but also inhibition of mitochondrial ROS production, can prevent arrhythmias in these models [48]. Therefore, Ca^{2+} mediated crosstalk between mitochondria and SR represents an interesting field of investigation to identify new and innovative therapeutic strategies for prevention and treatment of arrhythmia.

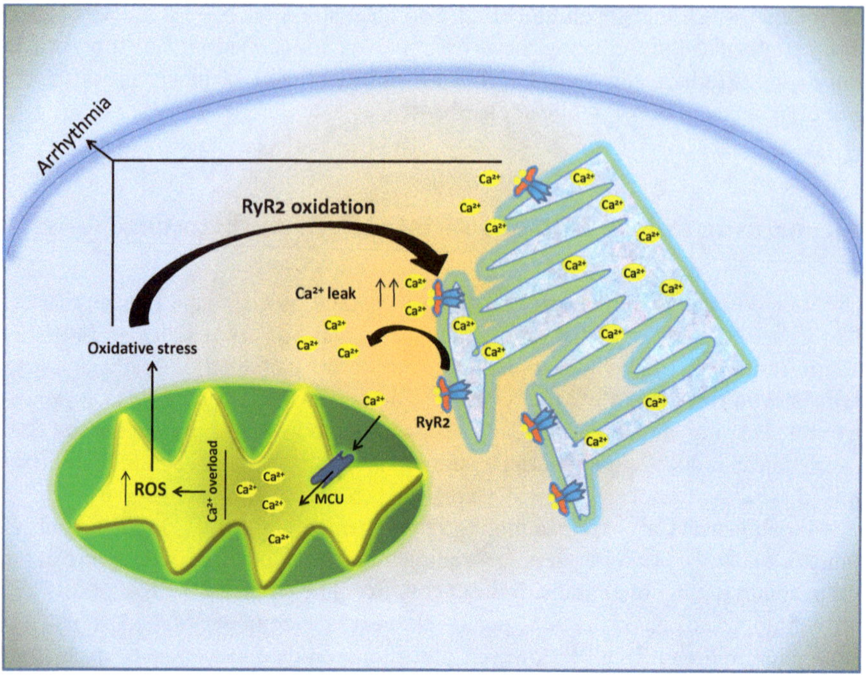

Fig. 10.3 Ca^{2+} dependent pathways and mitochondrial function in arrhythmogenesis. Alterations of Ca^{2+} release from sarcoplasmic reticulum can induce mitochondrial dysfunction as consequence of mitochondrial Ca^{2+} overload. Mitochondrial dysfunction in turn, through means of oxidative stress, can mediate further alterations with increase of Ca^{2+} leak from SR

Mitochondria as Therapeutic Target

As mentioned above, the role of mitochondria in cardiac arrhythmia is the result of a complex network between mitochondrial environment and extramitochondrial environment (cytosol, SR, plasmatic membrane). In particular, these dynamic organelles can contribute to electrical alterations in different circumstances, both when excitability dysregulation is a consequence of metabolic alterations, and when electrical dysfunction is the result of altered Ca^{2+} homeostasis.

Approximately 80% of clinical arrhythmia are consequence of alterations of coronary circulation that induce ischemic events, resulting in mitochondrial metabolic dysfunction [51, 52]. During these events, the mechanisms of RIRR orchestrated by mitochondria are essential, eventually inducing opening of sarcKATP channels with inhomogeneous AP alterations and consequent arrhythmia. After ischemia, restoration of normal blood flow results in additional cardiac damage known as reperfusion injury, with high ROS production that promotes mechanoelectrical dysfunction. In these circumstances, the possibility to modulate mitochondrial response could represent an efficient strategy of intervention to prevent or to treat arrhythmia; for instance, the block of IMAC to inhibit RIRR and prevent AP oscillations, or increasing the anti-oxidant capacity of mitochondria in order to neutralize the first ROS production induced by deprivation of oxygen and metabolite during ischemia, preventing RIRR.

Left ventricular hypertrophy (LVH) is another pathological condition where mitochondrial dysfunction is essential for arrhythmogenesis. During LVH, an increase in $\Delta\Psi m$ occurs, with reduction of apoptosis and bioenergetic alterations since high $\Delta\Psi m$ can interfere with the transport of substrates inside mitochondria with consequent metabolic shift from fatty acid to glucose utilization [53]. The modifications of $\Delta\Psi m$ during LVH, could be attributed to altered expression of mitochondrial uncoupling protein (UCP) that is observed in hypertrophy [54]. Indeed, UCP regulates $\Delta\Psi m$ and ROS production, and mice with low UCP3 develop arrhythmia [55]. Therefore, mitochondrial uncoupling via UCP alterations can induce ventricular fibrillation through AP heterogeneity as consequence of $\Delta\Psi m$ dependent sarcKATP activation [56]. Also in this case, mitochondria can be exploited as therapeutic targets, with possibility to modulate UCP activity to prevent hypertrophy dependent arrhythmia. Moreover, IMAC blockers show to produce beneficial effects in isoproterenol-dependent hypertrophy, most likely via inhibition of RIRR after ROS production, as result of UCP alterations [57].

All these findings indicate that mitochondria could offer several molecular targets for prevention and/or treatment of a broad spectrum of arrhythmic conditions, from ion channels involved in RIRR (IMAC, mPTP), regulatory proteins (TPSO), to channels directly involved in mitochondrial electrical homeostasis and energetic coupling (MCU, UCP). The main problems are represented by difficulties in realizing pharmacological compounds that are specific and efficient ligands of these mitochondrial proteins, since several are able to interact also with other substrates outside mitochondria.

Given the central role of mitochondrial ROS production and RIRR mechanism as primary or secondary cause of cardiac arrhythmias, many investigators are focused on the development of compounds with ROS- scavenger properties, or at least with the ability to increase the anti-oxidant capacity of mitochondria (see Chap. 32). For instance, treatment with superoxide dismutase mimics or mitochondrial-targeted anti-oxidant peptides has been successful in decreasing incidence of arrhythmia [58, 59]. Moreover, administration of N-acetylcysteine to humans after cardiac surgery significantly decreases the probability of arrhythmia development [60].

Further studies are necessary to expand our knowledge on the molecular composition and regulation of mitochondrial targets, in order to develop selective pharmacological compounds able to suppress cardiac arrhythmias.

Acknowledgements Dr. Gaetano Santulli, MD, PhD is supported by the National Institutes of Health (NIH, Grant NIDDK107895).

References

1. Santulli G, Iaccarino G, De Luca N, Trimarco B, Condorelli G. Atrial fibrillation and microRNAs. Front Physiol. 2014;5:15.
2. Mitchell P. Coupling of phosphorylation to electron and hydrogen transfer by a chemi-osmotic type of mechanism. Nature. 1961;191:144–8.
3. Ryu SY, Lee SH, Ho WK. Generation of metabolic oscillations by mitoKATP and ATP synthase during simulated ischemia in ventricular myocytes. J Mol Cell Cardiol. 2005;39: 874–81.
4. Aon MA, Cortassa S, Marban E, O'Rourke B. Synchronized whole cell oscillations in mitochondrial metabolism triggered by a local release of reactive oxygen species in cardiac myocytes. J Biol Chem. 2003;278:44735–44.
5. O'Rourke B, Ramza BM, Marban E. Oscillations of membrane current and excitability driven by metabolic oscillations in heart cells. Science. 1994;265:962–6.
6. Terzic A, Jahangir A, Kurachi Y. Cardiac ATP-sensitive K+ channels: regulation by intracellular nucleotides and K+ channel-opening drugs. Am J Phys. 1995;269:C525–45.
7. Johnson MS, Moore RL, Brown DA. Sex differences in myocardial infarct size are abolished by sarcolemmal KATP channel blockade in rat. Am J Physiol Heart Circ Physiol. 2006;290: H2644–7.
8. Chicco AJ, Johnson MS, Armstrong CJ, Lynch JM, Gardner RT, Fasen GS, Gillenwater CP, Moore RL. Sex-specific and exercise-acquired cardioprotection is abolished by sarcolemmal KATP channel blockade in the rat heart. Am J Physiol Heart Circ Physiol. 2007;292: H2432–7.
9. Brown DA, Chicco AJ, Jew KN, Johnson MS, Lynch JM, Watson PA, Moore RL. Cardioprotection afforded by chronic exercise is mediated by the sarcolemmal, and not the mitochondrial, isoform of the KATP channel in the rat. J Physiol. 2005;569:913–24.
10. Akar FG, Aon MA, Tomaselli GF, O'Rourke B. The mitochondrial origin of postischemic arrhythmias. J Clin Invest. 2005;115:3527–35.
11. Wirth KJ, Rosenstein B, Uhde J, Englert HC, Busch AE, Scholkens BA. ATP-sensitive potassium channel blocker HMR 1883 reduces mortality and ischemia-associated electrocardiographic changes in pigs with coronary occlusion. J Pharmacol Exp Ther. 1999;291:474–81.
12. Vajda S, Baczko I, Lepran I. Selective cardiac plasma-membrane K(ATP) channel inhibition is defibrillatory and improves survival during acute myocardial ischemia and reperfusion. Eur J Pharmacol. 2007;577:115–23.
13. Fischbach PS, White A, Barrett TD, Lucchesi BR. Risk of ventricular proarrhythmia with selective opening of the myocardial sarcolemmal versus mitochondrial ATP-gated potassium channel. J Pharmacol Exp Ther. 2004;309:554–9.
14. Aronson D, Mittleman MA, Burger AJ. Effects of sulfonylurea hypoglycemic agents and adenosine triphosphate dependent potassium channel antagonists on ventricular arrhythmias in patients with decompensated heart failure. Pacing Clin Electrophysiol. 2003;26:1254–61.
15. del Valle HF, Lascano EC, Negroni JA, Crottogini AJ. Glibenclamide effects on reperfusion-induced malignant arrhythmias and left ventricular mechanical recovery from stunning in conscious sheep. Cardiovasc Res. 2001;50:474–85.
16. Zorov DB, Filburn CR, Klotz LO, Zweier JL, Sollott SJ. Reactive oxygen species (ROS)-induced ROS release: a new phenomenon accompanying induction of the mitochondrial permeability transition in cardiac myocytes. J Exp Med. 2000;192:1001–14.
17. Zorov DB, Juhaszova M, Sollott SJ. Mitochondrial ROS-induced ROS release: an update and review. Biochim Biophys Acta. 2006;1757:509–17.
18. Wang J, Wang H, Zhang Y, Gao H, Nattel S, Wang Z. Impairment of HERG K(+) channel function by tumor necrosis factor-alpha: role of reactive oxygen species as a mediator. J Biol Chem. 2004;279:13289–92.
19. Santulli G, Nakashima R, Yuan Q, Marks AR. Intracellular calcium release channels: an update. J Physiol. 2017 (in press). doi: 10.1113/JP272781.

20. Biary N, Xie C, Kauffman J, Akar FG. Biophysical properties and functional consequences of reactive oxygen species (ROS)-induced ROS release in intact myocardium. J Physiol. 2011;589:5167–79.
21. Garlid KD, Beavis AD, Ratkje SK. On the nature of ion leaks in energy-transducing membranes. Biochim Biophys Acta. 1989;976:109–20.
22. Brierley GP. Energy-linked alteration of the permeability of heart mitochondria to chloride and other anions. Biochemistry. 1970;9:697–707.
23. Beavis AD. Properties of the inner membrane anion channel in intact mitochondria. J Bioenerg Biomembr. 1992;24:77–90.
24. Garlid KD, Beavis AD. Evidence for the existence of an inner membrane anion channel in mitochondria. Biochim Biophys Acta. 1986;853:187–204.
25. Brown DA, Aon MA, Akar FG, Liu T, Sorarrain N, O'Rourke B. Effects of 4′-chlorodiazepam on cellular excitation-contraction coupling and ischaemia-reperfusion injury in rabbit heart. Cardiovasc Res. 2008;79:141–9.
26. Brown DA, Aon MA, Frasier CR, Sloan RC, Maloney AH, Anderson EJ, O'Rourke B. Cardiac arrhythmias induced by glutathione oxidation can be inhibited by preventing mitochondrial depolarization. J Mol Cell Cardiol. 2010;48:673–9.
27. McCully JD, Wakiyama H, Hsieh YJ, Jones M, Levitsky S. Differential contribution of necrosis and apoptosis in myocardial ischemia-reperfusion injury. Am J Physiol Heart Circ Physiol. 2004;286:H1923–35.
28. Romashko DN, Marban E, O'Rourke B. Subcellular metabolic transients and mitochondrial redox waves in heart cells. Proc Natl Acad Sci U S A. 1998;95:1618–23.
29. Sato T, Takizawa T, Saito T, Kobayashi S, Hara Y, Nakaya H. Amiodarone inhibits sarcolemmal but not mitochondrial KATP channels in guinea pig ventricular cells. J Pharmacol Exp Ther. 2003;307:955–60.
30. Piot C, Croisille P, Staat P, Thibault H, Rioufol G, Mewton N, Elbelghiti R, Cung TT, Bonnefoy E, Angoulvant D, Macia C, Raczka F, Sportouch C, Gahide G, Finet G, Andre-Fouet X, Revel D, Kirkorian G, Monassier JP, Derumeaux G, Ovize M. Effect of cyclosporine on reperfusion injury in acute myocardial infarction. N Engl J Med. 2008;359:473–81.
31. Aon MA, Cortassa S, O'Rourke B. The fundamental organization of cardiac mitochondria as a network of coupled oscillators. Biophys J. 2006;91:4317–27.
32. Akar FG. Mitochondrial targets for arrhythmia suppression: is there a role for pharmacological intervention? J Interv Card Electrophysiol. 2013;37:249–58.
33. Beavis AD. On the inhibition of the mitochondrial inner membrane anion uniporter by cationic amphiphiles and other drugs. J Biol Chem. 1989;264:1508–15.
34. Kinnally KW, Zorov DB, Antonenko YN, Snyder SH, McEnery MW, Tedeschi H. Mitochondrial benzodiazepine receptor linked to inner membrane ion channels by nanomolar actions of ligands. Proc Natl Acad Sci U S A. 1993;90:1374–8.
35. Sileikyte J, Petronilli V, Zulian A, Dabbeni-Sala F, Tognon G, Nikolov P, Bernardi P, Ricchelli F. Regulation of the inner membrane mitochondrial permeability transition by the outer membrane translocator protein (peripheral benzodiazepine receptor). J Biol Chem. 2011;286:1046–53.
36. McEnery MW, Snowman AM, Trifiletti RR, Snyder SH. Isolation of the mitochondrial benzodiazepine receptor: association with the voltage-dependent anion channel and the adenine nucleotide carrier. Proc Natl Acad Sci U S A. 1992;89:3170–4.
37. Leducq N, Bono F, Sulpice T, Vin V, Janiak P, Fur GL, O'Connor SE, Herbert JM. Role of peripheral benzodiazepine receptors in mitochondrial, cellular, and cardiac damage induced by oxidative stress and ischemia-reperfusion. J Pharmacol Exp Ther. 2003;306:828–37.
38. Motloch LJ, Hu J, Akar FG. The mitochondrial translocator protein and arrhythmogenesis in ischemic heart disease. Oxidative Med Cell Longev. 2015;2015:234104.
39. Paucek P, Mironova G, Mahdi F, Beavis AD, Woldegiorgis G, Garlid KD. Reconstitution and partial purification of the glibenclamide-sensitive, ATP-dependent K+ channel from rat liver and beef heart mitochondria. J Biol Chem. 1992;267:26062–9.

40. Rajesh KG, Sasaguri S, Suzuki R, Xing Y, Maeda H. Ischemic preconditioning prevents reperfusion heart injury in cardiac hypertrophy by activation of mitochondrial KATP channels. Int J Cardiol. 2004;96:41–9.
41. Headrick JP, Willems L, Ashton KJ, Holmgren K, Peart J, Matherne GP. Ischaemic tolerance in aged mouse myocardium: the role of adenosine and effects of A1 adenosine receptor overexpression. J Physiol. 2003;549:823–33.
42. Fryer RM, Hsu AK, Nagase H, Gross GJ. Opioid-induced cardioprotection against myocardial infarction and arrhythmias: mitochondrial versus sarcolemmal ATP-sensitive potassium channels. J Pharmacol Exp Ther. 2000;294:451–7.
43. Schwartz LM, Welch TS, Crago MS. Cardioprotection by multiple preconditioning cycles does not require mitochondrial K(ATP) channels in pigs. Am J Physiol Heart Circ Physiol. 2002;283:H1538–44.
44. Belmonte S, Morad M. 'Pressure-flow'-triggered intracellular Ca2+ transients in rat cardiac myocytes: possible mechanisms and role of mitochondria. J Physiol. 2008;586:1379–97.
45. Garcia-Rivas Gde J, Carvajal K, Correa F, Zazueta C. Ru360, a specific mitochondrial calcium uptake inhibitor, improves cardiac post-ischaemic functional recovery in rats in vivo. Br J Pharmacol. 2006;149:829–37.
46. Santulli G, Pagano G, Sardu C, Xie W, Reiken S, D'Ascia SL, Cannone M, Marziliano N, Trimarco B, Guise TA, Lacampagne A, Marks AR. Calcium release channel RyR2 regulates insulin release and glucose homeostasis. J Clin Invest. 2015;125:4316.
47. Xie W, Santulli G, Guo X, Gao M, Chen BX, Marks AR. Imaging atrial arrhythmic intracellular calcium in intact heart. J Mol Cell Cardiol. 2013;64:120–3.
48. Xie W, Santulli G, Reiken SR, Yuan Q, Osborne BW, Chen BX, Marks AR. Mitochondrial oxidative stress promotes atrial fibrillation. Sci Rep. 2015;5:11427.
49. Santulli G, D'Ascia SL, D'Ascia C. Development of atrial fibrillation in recipients of cardiac resynchronization therapy: the role of atrial reverse remodelling. Can J Cardiol. 2012;28:245 e17; author reply 245 e17–8.
50. D'Ascia SL, D'Ascia C, Marino V, Lombardi A, Santulli R, Chiariello M, Santulli G. Cardiac resynchronisation therapy response predicts occurrence of atrial fibrillation in non-ischaemic dilated cardiomyopathy. Int J Clin Pract. 2011;65:1149–55.
51. Halestrap AP. Calcium, mitochondria and reperfusion injury: a pore way to die. Biochem Soc Trans. 2006;34:232–7.
52. Panth N, Paudel KR, Parajuli K. Reactive oxygen species: a key hallmark of cardiovascular disease. Adv Med. 2016;2016:9152732.
53. Leong HS, Brownsey RW, Kulpa JE, Allard MF. Glycolysis and pyruvate oxidation in cardiac hypertrophy – why so unbalanced? Comp Biochem Physiol A Mol Integr Physiol. 2003;135:499–513.
54. Laskowski KR, Russell 3rd RR. Uncoupling proteins in heart failure. Curr Heart Fail Rep. 2008;5:75–9.
55. Ozcan C, Palmeri M, Horvath TL, Russell KS, Russell 3rd RR. Role of uncoupling protein 3 in ischemia-reperfusion injury, arrhythmias, and preconditioning. Am J Physiol Heart Circ Physiol. 2013;304:H1192–200.
56. Brennan JP, Southworth R, Medina RA, Davidson SM, Duchen MR, Shattock MJ. Mitochondrial uncoupling, with low concentration FCCP, induces ROS-dependent cardioprotection independent of KATP channel activation. Cardiovasc Res. 2006;72:313–21.
57. Jin H, Nass RD, Joudrey PJ, Lyon AR, Chemaly ER, Rapti K, Akar FG. Altered spatiotemporal dynamics of the mitochondrial membrane potential in the hypertrophied heart. Biophys J. 2010;98:2063–71.
58. Konya L, Kekesi V, Juhasz-Nagy S, Feher J. The effect of superoxide dismutase in the myocardium during reperfusion in the dog. Free Radic Biol Med. 1992;13:527–32.
59. Cho J, Won K, Wu D, Soong Y, Liu S, Szeto HH, Hong MK. Potent mitochondria-targeted peptides reduce myocardial infarction in rats. Coron Artery Dis. 2007;18:215–20.
60. Erdogan O. N-acetylcysteine for the prevention of postoperative atrial fibrillation: a prospective, randomized, placebo-controlled pilot study. Eur Heart J 2008;29:1591; author reply 1591.

Chapter 11
Mitochondria and Cardiac Hypertrophy

Heberty di Tarso Fernandes Facundo, Robert Eli Brainard, Francisco Rodrigo de Lemos Caldas, and Aline Maria Brito Lucas

Cardiac tissue adapts to pressure or volume-overload with an enlargement of the myocardium characterized by growth of individual cardiomyocytes (hypertrophy) rather than an increase in cell number (hyperplasia). Long term cardiac hypertrophy, despite being adaptive against hemodynamic overloads, often progresses to heart failure [1]. During hypertrophy a series of intracellular cascades are activated resulting in changes in intracellular Ca^{2+} levels, contractile protein content, reactivation of fetal gene program [2, 3], O-glycosylation of proteins [4] and, oxidative stress [5–10]. The mitochondrion is the power-house of the cell and its dysfunction has been implicated in the pathology of cardiac hypertrophy, primarily, in the transition from cardiac hypertrophy into heart failure. Mitochondria occupy approximately 30% of the cardiomyocytes' mass and functionally correlate with the heart rate, and the oxygen consumption [11]. Additionally, mitochondria play a role in the regulation of intracellular calcium (Ca^{2+}) homeostasis, intracellular signaling, cell death. Furthermore, mitochondria generate >95% of ATP used by the heart [12]. A normal heart is largely fueled by fatty acid oxidation (60–90%) which occurs in the mitochondrial matrix, while the remaining (10–40%) is supplied by carbohydrates [13]. Cardiomyocytes may use other pathways (ketone bodies and amino acids) for ATP generation as well. However, in response to pathological hypertrophy the cardiac metabolism is reprogrammed to rely more on glucose than on fatty acid as a fuel source [12, 14].

H.T.F. Facundo (✉) • A.M.B. Lucas
Faculdade de Medicina, Universidade Federal do Cariri, Barbalha, Brazil
e-mail: heberty.facundo@ufca.edu.br

R.E. Brainard
Valencia College Orlando, Florida, USA

F.R.L. Caldas
Instituto Federal de Educação, Ciência e Tecnologia do Ceará, Juazeiro do Norte, Brazil

G. Santulli (ed.), *Mitochondrial Dynamics in Cardiovascular Medicine*, Advances in Experimental Medicine and Biology 982,
DOI 10.1007/978-3-319-55330-6_11

Even though the hypertrophic phenotype is initially a compensatory mechanism, chronically sustained pressure or volume-overload will deteriorate the heart's pumping capacity. When this occurs, blood flow is not maintained by the heart, which causes a multifactorial clinical syndrome – that is, heart failure, the leading cause of death in the industrialized world [15]. Initially, the heart normalizes the increase in wall tension during systole (in pressure-overload models) and sustains cardiac output by the development of left ventricular concentric hypertrophy. Conversely, volume overload leads to eccentric hypertrophy of the heart. This thickening of the cardiac chambers increases the diffusion distances for oxygen and other sundry substrates, generating a negative impact on the hearts contractile performance [16, 17]. This is partially triggered by anatomical changes including myocardial fibrosis, thickening of the intramyocardial coronary arteries, and perivascular fibrosis [18, 19].

Cardiac mitochondria generate ~95% of the ATP used by the heart, regulate Ca^{2+} homeostasis, participate in signaling, and in some instances trigger cell death [20]. Thus, mitochondria are central organelles responsible for the coordination of energy transduction in cardiomyocytes. Mitochondria, as a consequence of their aerobic metabolism, are the major source of cellular reactive oxygen species (ROS), such as anion superoxide ($O_2^{\cdot -}$) and hydrogen peroxide [21] (H_2O_2). ROS are by-products of the metabolism and are produced by several cellular sources. The electron leak from various sites (mainly complex I and III) of the Electron Transport Chain (ETC) results in reduction of molecular oxygen forming anion superoxide [21–23]. Other sources of intracellular ROS include NADPH oxidases, xanthine oxidase, nitric oxide synthase, lipoxygenases and mieloperoxidases [24]. Their excess generation is counteracted by molecules that eliminate ROS either enzymatically or through the action of antioxidant molecules. The enzymatic system includes superoxide dismutase (converts $O_2^{\cdot -}$ to H_2O_2), catalase and peroxidases (catalyze the breakdown of H_2O_2 to water). The secondary system for ridding the cell of ROS includes vitamins A, E, C and as well as glutathione and thioredoxin, which reduce thiol groups in oxidized proteins [25]. Redox-sensitive signaling pathways play an important role in the pathogenesis of cardiac hypertrophy and heart failure [5–10, 26–28]. For example, H_2O_2 is associated with cardiomyocyte hypertrophy [29], and agonists for G protein-coupled receptors, such as phenylephrine and angiotensin II induce hypertrophy with concomitant increase in ROS formation [7]. Additionally, antagonizing ROS accumulation in heart cells has been found to be beneficial by preventing cardiac hypertrophy [5, 6, 27].

In this chapter we will present and discuss the changes that occur in the mitochondria of the hypertrophic heart, followed by an overview of the potential therapeutic targets and approaches that can possibly reverse these changes and preserve cardiac health. Special attention will be given to adaptations and changes in the mitochondrial electron transport system, substrates metabolism, ROS generation, opening of the mitochondrial permeability transition pore and mitochondrial biogenesis.

Electron Transport Chain in Cardiac Hypertrophy

Mitochondria generate energy by metabolizing dietary carbohydrates, lipids and amino acids. This process is conducted by a series of metabolic pathways that generate energy through oxidative phosphorylation (OXPHOS) in the inner mitochondrial membrane. Mitochondrial energy production is a process in which electrons are passed along four mitochondrial carriers (complexes I, II, III and IV), cytochrome-c and ubiquinone. Once the electrons reach the last carrier, they are utilized by the complex IV for the reduction of oxygen to water. As a result of electron transfer through the carriers, protons are actively pumped into the intermembrane space by the complexes I, III and IV. Protons then flow out of the matrix and into the inter membrane space thereby generating an electrochemical gradient (composed of an electrical potential – $\Delta\psi$ and a protons concentration ratio – ΔpH). This energy is taken by complex V (FO/F1-ATP-synthase) which passively allows protons to flow down their concentration gradient back into the mitochondrial matrix, thus generating energy. This energy is then invested into a bond that joins ADP and P_i which produces ATP. This system will generate the vast majority of cellular ATP that will be either used by the mitochondrion or exported to be used to fuel a multitude of cellular functions [30, 31] (Fig. 11.1).

Maximal mitochondrial oxidative phosphorylation capacity is reduced by cardiac hypertrophy in part due to decreased activity of the respiratory chain complexes [32, 33]. Biochemically, this respiration defect will disturb ATP production by causing a decline in membrane potential and by lowering ATP synthase activity. Additionally, this would lower the substrate oxidation pathways resulting in an accumulation of reducing equivalents. The respiratory chain (complexes I–IV) is the final pathway for the oxidation of reducing equivalents (NADH, $FADH_2$). NADH and $FADH_2$ feed electrons into complex I and II, respectively, which are then transported along the other complexes to eventually reduce O_2 to H_2O. A dysfunctional respiratory chain will lead to an impairment of oxidative phosphorylation resulting not only in decreased ATP production and redox imbalance because of changes in NAD+/NADH, but also leading to increased levels of intracellular Ca^{2+} and increased generation of ROS. Interestingly, up to 2% of the oxygen consumed by the mitochondria may be converted into ROS as byproducts of respiratory chain activity [23]. Mitochondrial complex I contains 45 different subunits and its main role is to transfer electrons from NADH to the electron transport chain. Additionally, this complex has been credited as an important site of electron leakage contributing significantly to ROS formation [34–36]. The transition from compensated LV (left ventricular) hypertrophy to cardiac failure is preceded by complex I and II dysfunction. This is then followed by an increase in ROS and pro-apoptotic markers (Bax/Bcl- 2 ratio) [37]. Other studies have implicated that even at the phase of compensated hypertrophy, the activities of mitochondrial complex I and IV were reduced. This was associated with mitochondrial DNA depletion. These mitochondrial abnormalities were significantly correlated with the degree of cardiac hypertrophy,

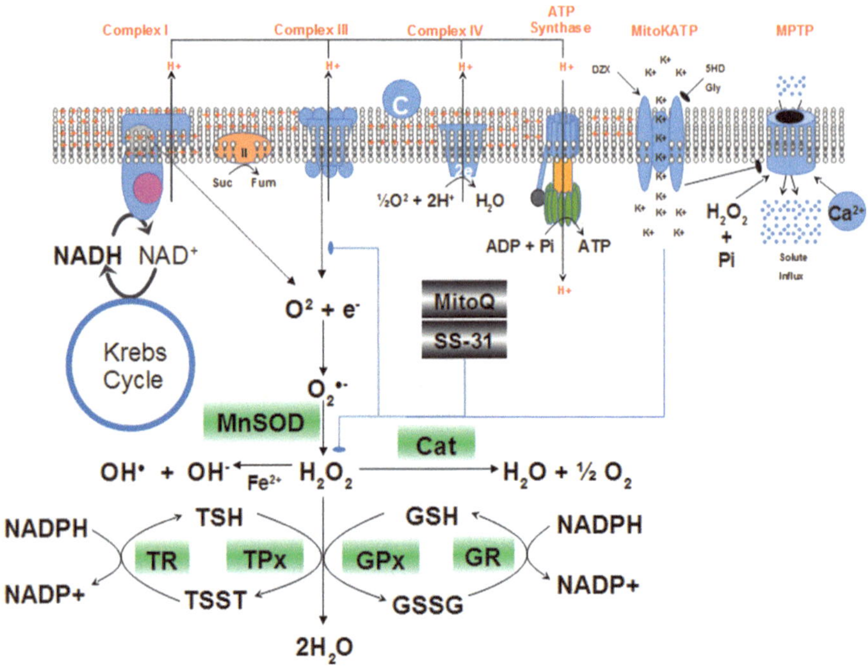

Fig. 11.1 Mitochondrial oxidative phosphorylation and oxidative stress. The enzymes (complex I, II, III, and IV) of the electron transport chain transfer electrons to molecular oxygen (O_2) and pump protons (complex I, III and IV) from the matrix to the intermembrane space. Single electrons may leak to oxygen instead of passing down to complex IV, forming reactive oxygen species (ROS) – anion superoxide ($O_2^{\bullet -}$). Mitochondria possess several enzymatic ROS scavengers: *MnSOD* Manganese Superoxide Dismutase (dismutate $O_2^{\bullet -}$ to H_2O_2), *Cat* Catalase, *Gpx* Glutathione Peroxidase and *Tpx* Thioredoxin Peroxidase that will break H_2O_2 to H_2O. The activities of these two enzymes are coupled to Glutathione Reductase (*GR*) and Thioredoxin Reductase (*TR*). Further, mitochondrial oxidative stress may be interrupted by the Szeto-Schiller peptide (SS-31), mitochondria-targeted ubiquinone (MitoQ10) or by the activity of the mitochondrial ATP-sensitive K^+ channel (mitoKATP). Diazoxide (DZX) opens mitoKATP and 5-hydroxidecanoate (5HD) and Glybenclamide (GLY) close. H_2O_2 together with Ca^{2+}, high phosphate and low adenine nucleotide (not shown on figure) will cause opening of the Mitochondrial Permeability Transition Pore (MPTP)

based on the ratio of heart/body weight [33]. Additionally, one study showed that complex IV activity was decreased by 6 weeks of abdominal aortic banding [32]. Other studies found increased activity of both the nuclear-encoded enzymes of the krebs cycle and the subunits of complex I, III and IV after transverse aortic constriction [38, 39]. However, these studies did not test if the increased complex subunits (by proteomic methods) would impact their individual activity or if they would change the ATP produced by the metabolism of individual substrates.

It is important to highlight that the mitochondrial defect could be at the level of protein import into mitochondria and/or their assembly on complexes in the inner mitochondrial membrane. A direct consequence would be an impaired cardiac oxidative phosphorylation [40]. Mitochondrial proteins (approximately 1,000) are in majority, encoded by nuclear DNA and imported into the mitochondria. More important, mitochondrial function is maintained by the import of proteins into the organelle [41]. In order to be precisely targeted to the mitochondria, proteins must be recognized by receptors through their mitochondrial-targeting signals (Tom70, Tom20 and Tom22) located on the outer mitochondrial membrane. A more detailed description of the mitochondrial entry gate system can be found elsewhere [42]. Interestingly, Tom 70 is down-regulated during pathological cardiac hypertrophy induced by several stimuli (angiotensin-II, isoproterenol and endothelin-1). Conversely, Tom 70 overexpression abolished pathological hypertrophic growth through modulation of ROS production. Strikingly, this protein is also reduced in human pathological cardiac hypertrophy [43]. Therefore, the disturbances in electron transport chain activities may be related to the poor protein import into mitochondria triggered by cardiac hypertrophy. This and other possibilities need to be systematically tested on future studies.

Mitochondrial Metabolism in Hypertrophy

Maintenance of normal cardiac contractility is correlated with the energy (ATP) generation inside the cardiomyocyte. This cell has a low storage of ATP. Mitochondria normally generate energy by burning fatty acyl-coenzyme-A and pyruvate (fatty acid and carbohydrates metabolism, respectively). Additionally, the majority of the ATP used by the heart [12] is produced into mitochondria. The heart is able to efficiently control ATP production by each substrate, which allows for optimized adaptation to changes in oxygen supply, substrate availability, and cardiac workload [44]. This reprogrammed substrate utilization is often associated with altered ATP yield. For instance, the use of fatty acids produces more ATP per mole than glucose oxidation. Additionally, the adult heart predominately utilizes fatty acids as its main substrate for energy production.

Pathological cardiac hypertrophy, once established, leads to a preferential shift in substrate utilization – from fatty acid oxidation to a less efficient glucose oxidation [45, 46]. The first question here is whether this "substrate switch" is a real phenomenon. Firstly, this is measured by the ratio of fatty acid oxidation to glucose oxidation. Therefore, any decrease in fatty acid oxidation would lead to the so-called "substrate switch" even though glucose utilization stays the same. This shift could be considered beneficial because glucose oxidation improves oxygen efficiency [47] but some glucotoxicity may be in effect if glucose uptake increases [48]. Additionally, there is a higher cardiac glucose uptake in animal models of hypertrophy [49, 50]. The increased glucose uptake will feed the well characterized increased glycolysis flux of the hypertrophic heart [49, 51–53].

Some studies have determined that glucose oxidation may be unchanged during cardiac hypertrophy [51, 54] or even reduced [55, 56], while some others investigations have pointed to higher glucose oxidation [57–59]. As evidenced by the various studies, the intricacies of glucose oxidation during cardiac hypertrophy remain to be definitively described. Taking these findings into account, many have proposed an uncoupling of glycolysis and glucose oxidation during hypertrophy [51, 52, 54, 56] as one explanation of these results.

One question that is raised by the unchanged or decreased glucose oxidation is how are mitochondrial NADH levels maintained during decreased glucose and fatty acid oxidation (low acetil-coA formation)? The first possibility is that glycolytic flux will increase cytosolic NADH that, under aerobic conditions, will be delivered to the mitochondria through the malate-aspartate shuttle [60]. However, some studies have failed to show enhanced activity or expression of the key shuttle protein oxoglutarate–malate carrier during hypertrophy [61, 62]. A second possibility is through the introduction of carbons in the Krebs cycle by anaplerotic reactions. In fact, hypertrophied hearts displayed increased anaplerotic flux into the Krebs cycle but this flux bypasses the energy-yielding reactions and lead to a reduced energetic efficiency. Additionally, hypertrophied hearts had an increased anaplerotic flux associated with enhanced expression of malic enzyme (the enzyme that carboxylate pyruvate to malate) [63]. Consistent with this idea, hypertrophied rat hearts had increased malate content [54]. Certainly, both possibilities will have to coordinately operate for an improved mitochondrial number, homeostasis, and substrate use. Therefore, the reprogrammed metabolism of the hypertrophied heart will impact overall ATP synthesis by oxidative phosphorylation [64]. For a schematic view of this see Fig. 11.2.

In the adult heart, fatty acid metabolism is more robust when compared to glucose metabolism. As stated above, cardiac tissue undergoes reprogramming, changing its preference for glucose metabolism with a concomitant decrease in fatty acid oxidation. The decreased fatty acid oxidation observed is due, in part, to the inhibition of carnitine-palmitoyl transferase-1 (CPT-1), which is the rate-limiting enzyme for the entry and oxidation of fatty acids into mitochondria [65]. Further, the mitochondrial uptake and oxidation of long chain fatty acid may be reduced by carnitine deficiency during hypertrophy [66]. This will impair the heart's ability to generate energy since the normal adult heart uses the fatty acid oxidation as a primary energetic source. Furthermore, fatty acid oxidation may be reduced by decreased expression of sarcolemmal fatty acid transporters [67, 68]. Therefore, strategies focusing on enhancing the fatty acid oxidation (or preventing its deficiency) seem reasonable. In the same vein, by sustaining fatty acids as the primary fuel source of the heart, one can prevent the detrimental effect of hypertrophy in cardiac energetics and function [69, 70].

Fatty acid oxidation is transcriptionally regulated by peroxisome proliferator-activated receptor-alpha (PPARα) and its coactivator peroxisome proliferator-activated receptor gamma coactivator 1-alpha (PGC-1α). PPARα activity increasess the activation of the transcriptional program of fatty acid oxidation [71, 72]. Indeed, the activation of PPARα with an agonist during hypertrophy reversed the down regulation of measured PPARα-regulated genes, preventing substrate switching, but contributing to contractile dysfunction [73]. This observed effect could be a direct result of lipotoxicity (increased fatty acid uptake relative to its mitochondrial oxidation) since PPARα activation may induce fatty acid uptake [74]. Others studies using different models have shown mixed results, with some showing improved [75, 76] or relatively modest improvement in cardiac function [77, 78]. It is reasonable to

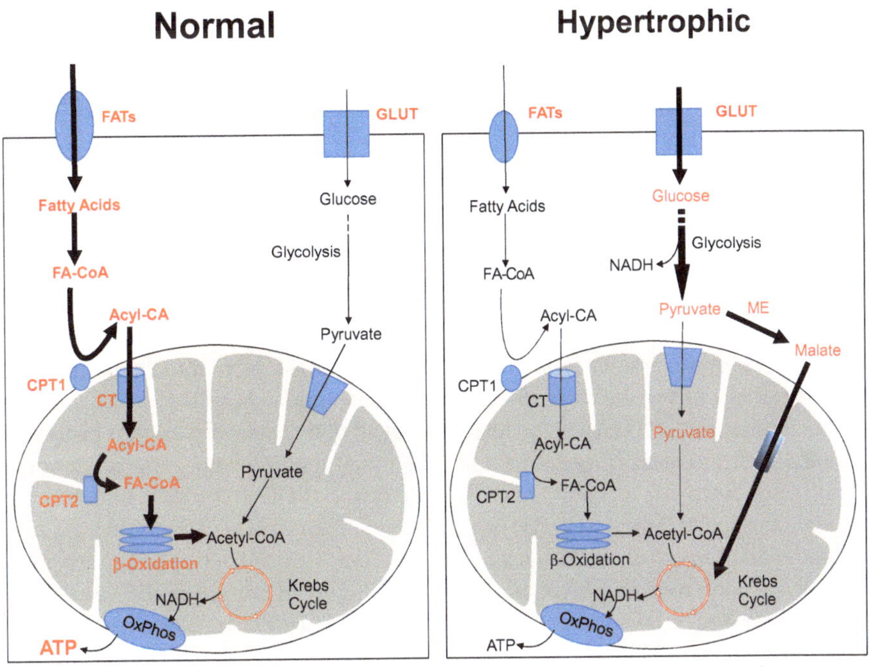

Fig. 11.2 Metabolism in the normal and hypertrophic heart. The metabolic pathways most active in each condition are highlighted in red. Fatty acyl-CoA (FA-CoA – derived from fatty acid) and pyruvate (derived from glucose) are transformed into acetyl coenzyme A (CoA) in the mitochondria. Acetyl-CoA is then fed into the Krebs cycle producing NADH which is then used by the oxidative phosphorylation system (OxPhos) to generate ATP. During hypertrophy, a metabolic shift with up-regulation of glucose utilization and decrease fatty acid metabolism, in part mediated by down-regulation of sarcolemal fatty acid transporters (FATs) and lower carnitine concentrations. This will cause an overall reduction in energy production via OxPhos. The hypertrophied heart has increased anaplerotic flux associated with enhanced expression of malic enzyme, however, this is insufficient to meet the energetic demand of the organ. *CPT* Carnitine palmitoyltransferase 1 and 2, *CT* Carnitine acyltranslocase, *Acyl-CA* Acyl Carnitine, *ME* malic enzyme, *GLUT* Glucose Transporter. *Thick arrows* mean higher flux through that pathway

think that a more direct approach would seek to activate fatty-acid oxidation at the level of the mitochondria.

ATP produced by the mitochondria needs to be transferred from the mitochondria to the cytosol in order to sustain the cardiac excitation-contraction-relaxation cycle. The adenine nucleotide translocator (ANT) is the protein responsible for the transference of ATP from mitochondria to the cytosol. Interestingly, ANT1 knockout mice have enlarged hearts accompanied by mitochondrial proliferation [79]. Additionally, the overexpression of ANT attenuates cardiac dysfunction induced by angiotensin [80]. Therefore, disturbances in ANT activity during hypertrophy would induce a lowering of ATP in the cytosol, thus prompting the "energy-starved heart" condition. This theory links the disturbed mitochondrial metabolism (production of ATP) to the mechanical dysfunction of the heart [81, 82]. In agreement to this theory, ATP (mostly produced by mitochondria) is critical to compensate for the elevated cardiac demand during hypertrophy. Taking into account the previous findings, it appears that one of the primary contributors to cardiac hypertrophy is the defective ATP production brought about by a low rate of fatty acid oxidation coupled with the low yield of ATP produced via glycolysis.

Mitochondrial Biogenesis in Cardiac Hypertrophy

Mitochondrial biogenesis is a complex process triggered by variations in preexisting mitochondria. These mitochondria will grow and divide with changes in size, number and mass. This biogenesis depends primarily on mitochondrial DNA replication as well as synthesis, import, and incorporation of lipids and proteins into organelle structure. It is easy to understand that all of these processes must be tightly coupled and regulated in order to meet the tissue requirements. The peroxisome proliferator-activated receptor gamma co-activator (PGC-1α) discovered first by Spiegelman and collaborators [83] is a master regulator of mitochondrial biogenesis in mammals. This co-activator lacks DNA-binding sites and activates transcription by binding numerous transcription factors. It is highly expressed in the heart, skeletal muscle and brown adipose; tissues with high oxidative capacity [84, 85]. PGC-1α has two structural homologs one called PGC-1β and another called PRC (PGC-1-related coactivator 1). PGC-1β exhibits practically the same properties including the regulation of mitochondrial activity, thermogenesis and cardiac performance [86, 87]. PGC-1α [88, 89] and PGC-1β [90] are transcriptionally repressed during cardiac hypertrophy that is induced by transverse aortic constriction. The repression of PGC-1α has been suggested as being an important phenomenon responsible for driving the shift from fatty-acid oxidation towards glucose oxidation, which contributes to the impaired ATP production which is seen during pressure overload induced hypertrophy and heart failure [66, 89, 91]. PRC expression is not affected by cardiac hypertrophy/heart failure [88].

The absence of PGC-1α causes serious deficiencies on the energetics and function of the heart [92]. ATP production is insufficient due to abnormalities of the expression of mitochondrial enzymes. Moreover, the ablation of PGC-1α had diminished cardiac function after treadmill running [93]. PGC-1α also suppresses the pro-hypertrophic signaling pathway (calcineurin-nuclear factor of activated T cells c4 – NFATc4) which contributes to the protective effect of PGC-1α on cardiomyocytes during hypertrophy [88, 94]. Likewise, PGC-1β deficiency leads to defects on components of the electron transport chain, which contributes to the poor cardiac performance seen after dobutamine stimulation [87]. It is important to understand that the absence of PGC-1α or PGC-1β alone is insufficient to cause any disturbances in contractile function under non-stressed conditions. However, combined deficiency of PGC-1α and PGC-1β revealed an extensive redundancy between these co-activators. These mice died shortly after birth due to decreased mitochondrial number and size combined with reduced cardiac output [95].

Overexpression studies [96–99] have indicated that PGC-1α activates the fatty acid oxidation program and directs oxidative respiration in cardiac tissue. Cardiac-specific (driven by αMHC – α-myosin heavy chain promoter) knockout of PGC-1α, starting at embryonic day 11.5, causes uncontrolled increases in mitochondria number that disturb the sarcomeric structure [97]. These myofilament derangement contribute to cardiac dysfunction [98]. A pitfall on these studies is that they always utilized high-levels of PGC-1α overexpression. However, it is important to point that overexpression of a protein does not necessarily equate increased activity. More recently the effect of moderate overexpression of PGC-1α was tested, but it did not improve cardiac function during chronic pressure overload [100]. One possibility is that PGC-1α is already highly expressed in the heart and any additional attempts to increase its level will not be satisfactory.

Several studies on PGC-1α suppression are based on transcriptional gene expression. It is important to point that gene expression does not necessarily depict its activity. More specifically, the mRNA for MCAD (medium chain acyl coA dehydrogenase – a gene controlled by PGC-1α) was decreased while MCAD protein and activity was not changed during cardiac hypertrophy [101]. Recent studies showed no decrease in PGC-1α protein levels in samples from human failing heart [102]. Besides no defects on PGC-1α protein various abnormalities in mitochondrial biogenesis may be present due to binding and activity impairments. Future strategies must focus targeting the PGC-1α protein as a therapy.

Hypertrophy and Oxidative Stress

Reactive oxygen species are associated with cardiac hypertrophy [6–8, 10, 26–29, 103, 104]. For example, hypertrophy in isolated cardiomyocytes induced by angiotensin II, endothelin-1, alpha-adrenergic agonists, tumor necrosis factor-alpha, ouabain, or cyclic stretch has been shown to elicit increased ROS production [7–9, 27, 105], while utilizing antioxidants inhibited the development of pressure-overload cardiac hypertrophy in mice [5, 6, 27]. Interestingly, the overexpression of mitochondrial-targeted catalase avoids the age-dependent protein oxidation of mitochondrial heart proteins, and also, reduces the age-dependent left ventricular hypertrophy and diastolic dysfunction [106]. Conversely, H_2O_2 (a cell-permeable ROS) causes, at low levels, the hypertrophic phenotype in isolated cardiomyocytes. On the other hand, at higher levels this molecule promotes cell death by necrosis and apoptosis [29]. Also, superoxide dismutase inhibition increases superoxide anion inside cells and activates growth of cardiac cells [107].

The increased evidence for the involvement of oxidative stress in cardiac hypertrophy and heart failure [5–10, 26–28, 106] highlights the need for a better understanding of the mechanisms involved. A better understanding will ultimately help to elucidate new tools to prevent heart failure. Evidence for the role of oxidative stress in cardiac hypertrophy first emerged from a study reporting that angiotensin II induced ROS generation in cardiac myocytes [9]. This hypothesis was confirmed by others showing that the antioxidants could abrogate the observed effect [5, 26, 27, 106]. More specifically, the treatment with antioxidant N-2-mercaptopropionylglycine alleviated cardiac hypertrophy in vivo [5]. This apparent decrease in hypertrophy was associated with less oxidized glutathione. Indeed, intracellular ROS generated during hypertrophy could cause thiol oxidation of cysteines [25]. Moreover, the direct administration of H_2O_2 induced hypertrophy in isolated cardiomyocytes [29]. These ROS are also implicated on the expression of the hypertrophic marker beta-myosin heavy chain via MAPK pathway [108]. Our recent studies pointed toward a simultaneous attenuation of the antioxidant enzymes catalase and SOD during hypertrophy [109]. More importantly, the inhibition of SOD [107] or its cardiac specific knockout [110] is sufficient for the induction of cardiac hypertrophy.

Oxidative stress activates the apoptotic cascade and contributes to maladaptive myocardial remodeling. Additionally, augmented ROS production will stimulate the expression and post-translational activation of matrix metalloproteinases and stimulate cardiac fibroblast proliferation, two events that are central to maladaptive extracellular remodeling [111]. Several hypertrophic signaling molecules are changed by increases in ROS production [112–118]. The ROS cascade culminates in the activation of several hypertrophic signaling molecules including: PKC [112], JNK [119], ERK1/2 [113], AKT [115], MAPKs [117], NF-kB [118]. An interesting molecule is the multifunctional Ca^{2+}/calmodulin-dependent protein kinase II (CaMKII). This protein is central to hypertrophy/heart failure mechanisms [120]. CaMKII may be oxidized in Met residues (281/282) and become activated, altering growth and hypertrophic gene expression [114].

Following CaMKII-mediated phosphorylation, class II HDACs are exported to the cytosol where they can no longer suppress target transcription factors [121, 122]. This lack of suppression will release and activate MEF2/NFAT mediated hypertrophic gene expression [122]. Additionally, HDACs are directly oxidized by ROS that are being exported from the nucleus [123]. Although a direct link between these signaling molecules (specially oxidized CaMKII) and ROS produced by mitochondria during cardiac hypertrophy is still lacking, studies aimed on specifically demonstrating this relationship would be of great value. It is appropriate to affirm that mitochondrial-derived ROS may induce these signaling proteins and regulates the cellular hypertrophic status (Fig. 11.3).

Mitochondrial Oxidative Stress in Cardiac Hypertrophy

ROS are generated inside mitochondria by the leak of electrons from the electron transport chain (mainly complex I and III) to molecular oxygen [21–23] or they can be produced by the activity of NADPH oxidase, [124, 125] among other sources. The leak of electrons to the molecular oxygen will generate the $O_2^{\cdot-}$. The intra-mitochondrial $O_2^{\cdot-}$ it is likely to be very low because of the superoxide dismutase present in the mitochondrial matrix which uses manganese as a cofactor (MnSOD).

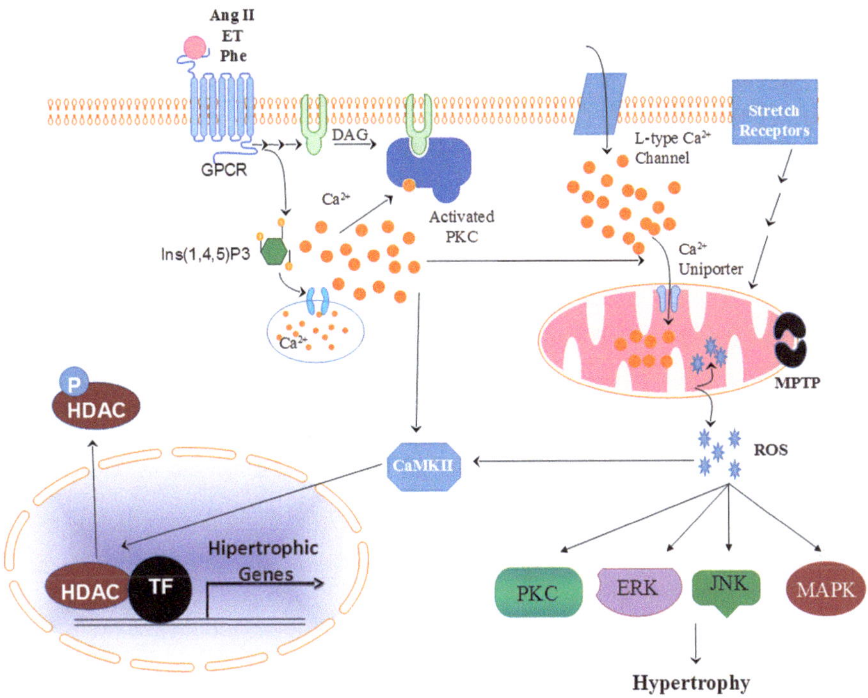

Fig. 11.3 Hypertrophic oxidative signaling. Schematic diagram of key pathways involved in oxidative stress induced by hypertrophy. Pharmacological/neurohumoral signaling or cyclic stretch is sensed by receptors. Receptor activation allows the entrance of Ca^{2+} into the mitochondria and beginning the production of ROS. These ROS will activate direct downstream kinase effectors and together with Ca^{2+} will trigger the opening of the mitochondrial permeability transition pore (MPTP). Additionally, these signals will promote gene transcription (including reactivation of fetal genes) and activation of cellular growth. L-type Ca^{2+} receptors are also activated and contribute to increased intracellular Ca^{2+}. *Ang II* angiotensin II, *Phe* Phenylephrine, *ET* endothelin, *CaMKII* calmodulin-dependent kinase II, *DAG* diacylglycerol, *ERK* extracellular signal regulated kinase, *GPCR* G-protein-coupled receptors, *HDAC* histone deacetylases, *Ins(1,4,5)P3*, inositol-1,4,5-trisphosphate, *JNK* c-Jun N-terminal kinase, *MAPK* mitogen-activated protein kinase, *PDK* phosphoinositide-dependent kinase, *ROS* Reactive Oxygen Species, *TF* Transcription Factor (Part of images from Motifolio drawing toolkits (http://www.motifolio.com/) were utilized in the figure preparation)

This enzyme specifically converts $O_2^{\cdot-}$ to H_2O_2 [126]. This resultant molecule is more stable and may easily diffuse to the inter-membrane space and cytosol, making this a more appropriate compound to be detected in mitochondrial suspensions [127]. On the other hand, not all H_2O_2 produced within the mitochondrial matrix will diffuse out of mitochondria. This molecule can be converted to O_2 and H_20 by catalase [128]. H_2O_2 can be additionally broken down into water by Glutathione peroxidase/glutathione reductase which uses glutathione as a cofactor [129]. Additionally, mitochondria possess thioredoxin peroxidase/redutase that breaks H_2O_2 to H_2O by using thioredoxin as electron donor [130]. If H_2O_2 accumulates its interaction with Fe^{2+} will drive the formation of the highly reactive hydroxyl radical, of which no specific antioxidant system is known (Fig. 11.1).

Mitochondrial damage seems to play a central role in the transition from hypertrophy to heart failure [13]. The activity of the mitochondrial electron transport chain complex has been shown to be impaired in pressure overload-induced hypertrophy leading, consequently, to heart failure [37]. Mitochondrial protection by cardiac-specific deletion of NADPH oxidase isoform 4 (Nox-4), a known source of mitochondrial anion superoxide, leads to resistance against pressure overload-induced hypertrophy [103]. Consistent with these observations, mitochondrial-specific superoxide dismutase knockout mice present high levels of mitochondrial ROS production and die after several weeks of birth from severe dilated cardiomyopathy [116]. One mechanism for the protection of hearts against oxidative damage is through the administration of synthetic antioxidant molecules. For instance, the mitochondrial targeted antioxidant peptide Szeto-Schiller (SS)-31 [131–137] or the mitochondria-targeted ubiquinone, MitoQ10 [138–141]. Thus, the increased ROS during hypertrophy [7–9, 27] and the attenuation of cardiac hypertrophy with antioxidants [5, 26, 27, 106] is a strong point for the implication of mitochondrial oxidative stress on cardiac hypertrophy/heart failure since mitochondria are the main intracellular source of ROS in cardiomyocytes.

NADPH oxidase is another important source of mitochondrial ROS [124, 125]. In fact, treating cells with isoproterenol induces oxidative stress by upregulation by a NOX isoform 4 [125]. Nox-4 is an important source of mitochondrial ROS in myocytes during hypertrophy [103, 124]. NADPH oxidase is the only enzyme designed for the purpose of ROS production. In general, Nox-4 is localized into the mitochondria and is upregulated during Angiotensin II, phenylephrine, or pressure overload-induced hypertrophy [124]. Furthermore, cardiac-specific knockout of Nox4 in mice attenuated cardiac hypertrophy accompanied with less interstitial fibrosis and mitochondrial ROS. The overall cardiac function of these hearts was improved by Nox4 knockout [103].

Cardiac hypertrophy is significantly reduced by increasing mitochondrial antioxidant defenses. For example, mice overexpressing the mitochondrial-targeted catalase had attenuated hypertrophy after pressure-overload triggered by transverse aortic constriction [38] or Angiotensin II treatment [142]. Furthermore, mice lacking one allele of MnSOD or its knockout had enlarged hearts, which supports the concept that mitochondrial ROS is a cause of cardiac hypertrophy [116, 143]. Although these observations prove the point, they are almost ineffective as a therapeutic strategy for humans. A more reliable therapy would be with drugs that will penetrate and target ROS inside mitochondria.

An attractive pharmacological tool for cardiac hypertrophy/heart failure treatment, is the mitochondrial targeted antioxidant peptide Szeto-Schiller (SS)-31 (Fig. 11.1). Peptides are not typically good candidates for the development of new drugs, not only because of their inability to cross cellular membranes, but also because of stability and solubility issues. On the other hand, SS-31 treated mice were resistant against angiotensin II-induced cardiac hypertrophy [133]. Interestingly, SS peptides rapidly accumulate into mitochondria, independent of membrane potential, and avoid ROS production from mitochondrial Complex II and III [137] and lipid peroxidation after ischemia/reperfusion in the cardiac tissue [131]. These peptides possess excellent pharmacokinetic properties, low enzymatic degradation, and rapid accumulation in the heart [134]. Additionally, these peptides increased the ATP content of cardiac tissue subjected to ischemia and reperfusion [131] and restored expression of mitochondrial energy metabolism related genes [135]. Although additional studies must be performed to validate such compounds as therapeutic tools, these compounds can be delivered via intraperitoneal, intravenous, or subcutaneous which allows them to be rapidly distributed to the heart, lung and brain – all highly perfused tissues [136]. Furthermore, the protective action of the peptide SS-31 (MTP-131; Bendavia) is being currently being evaluated in four phase-I and a phase-II clinical trials for ischemic cardiomyopathy [132].

A developed mitochondria-targeted ubiquinone, MitoQ10, is another attractive way to overcome the problem of directly delivering an antioxidant to mitochondria [144]. Indeed, this molecule has been shown to be useful in preventing cardiac hypertrophy [138]. Its antioxidant effect is through its direct oxidation by ROS formed inside mitochondria (Fig. 11.1). The action of mitochondrial complex II recycles the molecule back to its active form. Furthermore, MitoQ10 heavily accumulates within the mitochondrial matrix, driven by membrane potential [140]. MitoQ10 also presented a direct anti-hypertrophic effect in vitro and potentiated the anti-hypertensive effect of losartan [139]. One might question the ability of MitoQ10 to act in the mitochondrial matrix of hypertrophic hearts because this drug is taken up by mitochondria depending on membrane potential integrity. In fact, one possibility is that the delivery of MitoQ might be higher in the normal membrane potential [141]. On the other hand, MitoQ10 had anti-hypertrophic effects and significantly improved the disrupted mitochondrial membrane potential stimulated by cyclic stretch in cardiomyocytes or by volume-overload in rats [105]. Taking these findings into account, a great deal of future research will be needed to prove the significant therapeutic potential of this and other molecules.

The pharmacological opening of a mitochondrial ATP sensitive potassium channel (mitoKATP) attenuates cardiac hypertrophy in vitro [145] and in vivo [109]. Notably, we have shown that the mitochondrial ROS production is decreased by pharmacological mitoKATP opening [146, 147], and although this data is still controversial [148], we believe this is an accurate model displaying the effect of mitoKatp on ROS production in the mitochondria. Therefore, we considered two potential mechanisms for mitoKATP protection against cardiac hypertrophy. First, specific opening of mitoKATP might confer cellular survival and/or protective advantages against the mitochondrial oxidative stress. According to this hypothesis, cells treated with openers of the mitoKATP would induce a favorable redox condition by avoiding the leak of electrons from the electron transport chain to molecular oxygen or by upregulating antioxidant enzymes (Fig. 11.1). These two events ultimately may diminish the hypertrophic phenotype. Secondly, the cardiac protection could occur at the level of the MPTP opening, independent of antioxidant enzyme induction, which preserves mitochondrial homeostasis and oxidative phosphorylation. Interestingly, stimulating mitochondrial potassium flux, through mitoKATP opening, triggers cardioprotection by decreasing mitochondrial [149] and cytosolic [150] Ca^{2+} accumulation. As for other therapeutic drugs, more research must be conducted in order to establish the real impact of the opening of mitoKATP on cardiac hypertrophy and its protective mechanisms.

MPTP Hypertrophy

The transition of hypertrophy to heart failure is accompanied by mitochondrial damage [13]. Perturbed mitochondrial function impairs the ability of this organelle to produce ATP efficiently. It is important to point out that mitochondria produce 90% of all intracellular ATP [151]. Additionally, cardiac Ca^{2+} accumulation may result in increased Ca^{2+}-dependent activation of intracellular hypertrophic factors in cardiac cells [2]. Upregulation of cardiac contractility with isoproterenol induces greater mitochondrial Ca^{2+} load and oxidative stress with consequent mitochondrial permeability transition pore (MPTP) opening [152]. Under normal conditions mitochondria function to balance the cytosolic Ca^{2+} [153]. Ca^{2+} overload during hypertrophy induces cardiac deterioration in part by triggering the opening of the MPTP [154, 155] as a result of increased mitochondrial Ca^{2+} and activation of ROS generation [156]. Mitochondrial Ca^{2+} overload further enhances the generation of ROS by effecting Complex I, complex III, IV of the electrons transport chain [157].

Although the identity of all the proteins of the MPTP is still unknown, the mitochondrial matrix protein Cyclophilin D is the only component of the pore that was genetically verified as present. Cyclophilin D binds to putative pore components on the inner membrane and sensitizes the mitochondria for opening of the MPTP [158, 159]. Interestingly, cardiac hypertrophy leads to MPTP opening with enhanced expression and intramitochondrial translocation of Cyclophilin D [160]. Genetic deletion of cyclophilin D in the mouse protected against massive Ca^{2+} overload and heart failure by overexpression of Beta2a [161]. MPTP opening during cardiac hypertrophy induced by alpha-adrenergic agonists is successfully blocked by Cyclosporin A treatment – a cyclophilin D inhibitor [162]. In addition, mitochondria from volume overload hypertrophied hearts had greater vulnerability to stress-induced opening of the MPTP [155]. Additionally, the net release of Ca^{2+} from the endoplasmic reticulum into the cytosol and its cellular entrance by the L-type Ca^{2+} channel (Fig. 11.3) may lead to respiratory impairment through MPTP opening in mitochondria contributing to cardiac hypertrophy/heart failure.

Some medications used to attenuate the deleterious consequences of cardiac hypertrophy also affect mitochondrial redox and Ca^{2+} levels, which may contribute to their therapeutic efficacy. For example, the blockage of the angiotensin receptor type I (using candesartan) protected mitochondria against swelling [163]. Additionally, beta receptor blocker carvedilol protected heart mitochondria against oxidative stress-induced mitochondrial damage [164], mitochondrial swelling [165] and reverted cardiac hypertrophy [166]. Taken together, these observations suggest that clinically used drugs may improve the overall heart condition by the maintenance of mitochondrial normal function.

Conclusion

Mitochondria are major players in the development of cardiac hypertrophy. Research using cultured cells and animal models has consistently demonstrated that the hypertrophic heart has altered control of mitochondrial function, bioenergetics, and metabolic dysregulations. The induction of hypertrophy (by Angiotensin II (Ang II), activation of alpha (α) and/or beta (β) adrenergic receptors, or pressure/volume overload) has consistently demonstrated mitochondrial dysfunction. The dysfunctional mitochondrial will: (1) generate high levels of ROS with lower antioxidant enzymes. (2) Be more susceptible to Ca^{2+}-induced mitochondrial permeability transition pore (MPTP) and (3) will be energetically impaired producing lower ATP levels. Additionally, hypertrophy impairs mitochondrial biogenesis. Cardiac tissue is highly dependent on mitochondrial homeostasis and number. Some drugs and strategies have been used to target mitochondria (in animal and cellular models) in hopes of avoiding cardiac hypertrophy/heart failure (Fig. 11.4). Future work should be directed on improving their overall cytotoxicity and efficacy. Increased knowledge and improved therapies directed towards mitochondrial function, redox status, biogenesis and energetics will help in understanding and treating the events involved in cardiac hypertrophy.

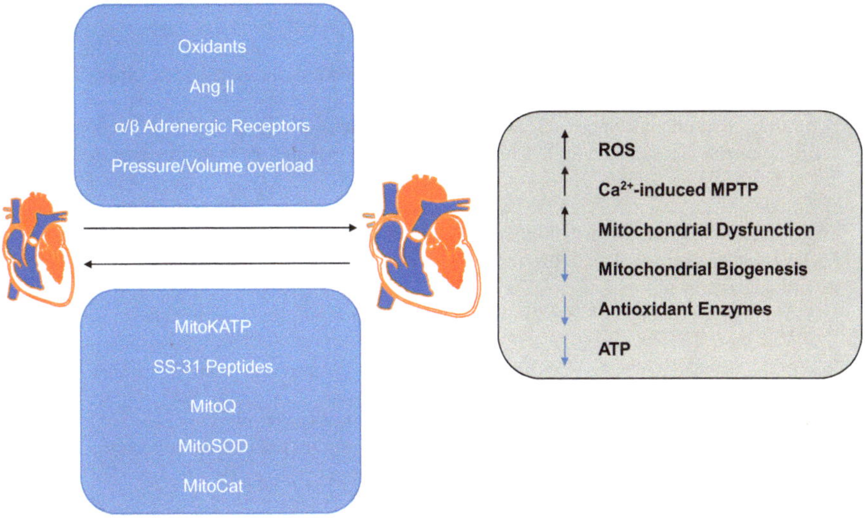

Fig. 11.4 Summary of the mitochondrial dysfunction during cardiac hypertrophy. Oxidants, Angiotensin II (Ang II), activation of alpha (α) and/or beta (β) adrenergic receptors or pressure/volume overload are known inducers of cardiac hypertrophy. The cardiac hypertrophic program includes increases in mitochondrial Reactive Oxygen Species (ROS), Ca^{2+}-induced mitochondrial permeability transition pore (MPTP) and mitochondrial dysfunction. Additionally, there are decreases in mitochondrial biogenesis, antioxidant enzymes and ATP levels. These effects are reversed by mitochondrial ATP-sensitive potassium channel (MitoKATP) opening, antioxidant peptide Szeto-Schiller (SS-31), the mitochondria-targeted ubiquinone (MitoQ) and the overexpression of mitochondrial superoxide dismutase (MitoSOD) or mitochondrial catalase (MitoCat)

References

1. Go AS, Mozaffarian D, Roger VL, Benjamin EJ, Berry JD, Borden WB, et al. Heart disease and stroke statistics – 2013 update: a report from the American Heart Association. Circulation. 2013;127(1):e6–245.
2. Frey N, McKinsey TA, Olson EN. Decoding calcium signals involved in cardiac growth and function. Nat Med. 2000;6(11):1221–7.
3. Molkentin JD, Lu JR, Antos CL, Markham B, Richardson J, Robbins J, et al. A calcineurin-dependent transcriptional pathway for cardiac hypertrophy. Cell. 1998;93(2):215–28.
4. Facundo HT, Brainard RE, Watson LJ, Ngoh GA, Hamid T, Prabhu SD, et al. O-GlcNAc signaling is essential for NFAT-mediated transcriptional reprogramming during cardiomyocyte hypertrophy. Am J Physiol Heart Circ Physiol. 2012;302(10):H2122–30.
5. Date M, Morita T, Yamashita N, Nishida K, Yamaguchi O, Higuchi Y, et al. The antioxidant N-2-mercaptopropionyl glycine attenuates left ventricular hypertrophy in in vivo murine pressure-overload model. J Am Coll Cardiol. 2002;39(5):907–12.
6. Dhalla AK, Singal PK. Antioxidant changes in hypertrophied and failing guinea pig hearts. Am J Phys. 1994;266(4 Pt 2):H1280–5.
7. Hirotani S, Otsu K, Nishida K, Higuchi Y, Morita T, Nakayama H, et al. Involvement of nuclear factor-kappaB and apoptosis signal-regulating kinase 1 in G-protein-coupled receptor agonist-induced cardiomyocyte hypertrophy. Circulation. 2002;105(4):509–15.
8. Pimentel DR, Amin JK, Xiao L, Miller T, Viereck J, Oliver-Krasinski J, et al. Reactive oxygen species mediate amplitude-dependent hypertrophic and apoptotic responses to mechanical stretch in cardiac myocytes. Circ Res. 2001;89(5):453–60.
9. Nakamura K, Fushimi K, Kouchi H, Mihara K, Miyazaki M, Ohe T, et al. Inhibitory effects of antioxidants on neonatal rat cardiac myocyte hypertrophy induced by tumor necrosis factor-alpha and angiotensin II. Circulation. 1998;98(8):794–9.
10. Rajasekaran NS, Connell P, Christians ES, Yan L-J, Taylor RP, Orosz A, et al. Human alpha B-crystallin mutation causes oxido-reductive stress and protein aggregation cardiomyopathy in mice. Cell. 2007;130(3):427–39.
11. Barth E, Stämmler G, Speiser B, Schaper J. Ultrastructural quantitation of mitochondria and myofilaments in cardiac muscle from 10 different animal species including man. J Mol Cell Cardiol. 1992;24(7):669–81.
12. Kolwicz SC, Purohit S, Tian R. Cardiac metabolism and its interactions with contraction, growth, and survival of cardiomyocytes. Circ Res. 2013;113(5):603–16.
13. Osterholt M, Nguyen TD, Schwarzer M, Doenst T. Alterations in mitochondrial function in cardiac hypertrophy and heart failure. Heart Fail Rev. 2013;18(5):645–56.
14. Ashrafian H, Frenneaux MP, Opie LH. Metabolic mechanisms in heart failure. Circulation. 2007;116(4):434–48.
15. Writing Group Members, Mozaffarian D, Benjamin EJ, Go AS, Arnett DK, Blaha MJ, et al. Executive summary: heart disease and stroke statistics – 2016 update: a report from the American Heart Association. Circulation. 2016;133(4):447–54.
16. Anversa P, Ricci R, Olivetti G. Quantitative structural analysis of the myocardium during physiologic growth and induced cardiac hypertrophy: a review. J Am Coll Cardiol. 1986;7(5):1140–9.
17. Friehs I, del Nido PJ. Increased susceptibility of hypertrophied hearts to ischemic injury. Ann Thorac Surg. 2003;75(2):S678–84.
18. Gradman AH, Alfayoumi F. From left ventricular hypertrophy to congestive heart failure: management of hypertensive heart disease. Prog Cardiovasc Dis. 2006;48(5):326–41.
19. De Boer RA, Pinto YM, Van Veldhuisen DJ. The imbalance between oxygen demand and supply as a potential mechanism in the pathophysiology of heart failure: the role of microvascular growth and abnormalities. Microcirculation. 2003;10(2):113–26.
20. Santulli G, Xie W, Reiken SR, Marks AR. Mitochondrial calcium overload is a key determinant in heart failure. Proc Natl Acad Sci U S A. 2015;112:11389–94.

21. Figueira TR, Barros MH, Camargo AA, Castilho RF, Ferreira JCB, Kowaltowski AJ, et al. Mitochondria as a source of reactive oxygen and nitrogen species: from molecular mechanisms to human health. Antioxid Redox Signal. 2013;18(16):2029–74.
22. Turrens JF. Mitochondrial formation of reactive oxygen species. J Physiol. 2003;552(2):335–44.
23. Tahara EB, Navarete FDT, Kowaltowski AJ. Tissue-, substrate-, and site-specific characteristics of mitochondrial reactive oxygen species generation. Free Radic Biol Med. 2009;46(9):1283–97.
24. Madamanchi NR, Runge MS. Redox signaling in cardiovascular health and disease. Free Radic Biol Med. 2013;61C:473–501.
25. Berndt C, Lillig CH, Holmgren A. Thiol-based mechanisms of the thioredoxin and glutaredoxin systems: implications for diseases in the cardiovascular system. Am J Physiol Heart Circ Physiol. 2007;292(3):H1227–36.
26. Dhalla AK, Hill MF, Singal PK. Role of oxidative stress in transition of hypertrophy to heart failure. J Am Coll Cardiol. 1996;28(2):506–14.
27. Xie Z, Kometiani P, Liu J, Li J, Shapiro JI, Askari A. Intracellular reactive oxygen species mediate the linkage of Na+/K± ATPase to hypertrophy and its marker genes in cardiac myocytes. J Biol Chem. 1999;274(27):19323–8.
28. Matsushima S, Kuroda J, Ago T, Zhai P, Park JY, Xie L-H, et al. Increased oxidative stress in the nucleus caused by Nox4 mediates oxidation of HDAC4 and cardiac hypertrophy. Circ Res. 2013;112(4):651–63.
29. Kwon SH, Pimentel DR, Remondino A, Sawyer DB, Colucci WS. H(2)O(2) regulates cardiac myocyte phenotype via concentration-dependent activation of distinct kinase pathways. J Mol Cell Cardiol. 2003;35(6):615–21.
30. Saraste M. Oxidative phosphorylation at the fin de siècle. Science. 1999;283(5407):1488–93.
31. Hatefi Y. The mitochondrial electron transport and oxidative phosphorylation system. Annu Rev Biochem. 1985;54:1015–69.
32. Kindo M, Gerelli S, Bouitbir J, Charles A-L, Zoll J, Hoang Minh T, et al. Pressure overload-induced mild cardiac hypertrophy reduces left ventricular transmural differences in mitochondrial respiratory chain activity and increases oxidative stress. Front Physiol. 2012;3:332.
33. Lin CS, Sun YL, Liu CY. Structural and biochemical evidence of mitochondrial depletion in pigs with hypertrophic cardiomyopathy. Res Vet Sci. 2003;74(3):219–26.
34. Kwong LK, Sohal RS. Substrate and site specificity of hydrogen peroxide generation in mouse mitochondria. Arch Biochem Biophys. 1998;350(1):118–26.
35. St-Pierre J, Buckingham JA, Roebuck SJ, Brand MD. Topology of superoxide production from different sites in the mitochondrial electron transport chain. J Biol Chem. 2002;277(47):44784–90.
36. Grivennikova VG, Vinogradov AD. Generation of superoxide by the mitochondrial complex I. Biochim Biophys Acta. 2006;1757(5–6):553–61.
37. Griffiths ER, Friehs I, Scherr E, Poutias D, McGowan FX, Del Nido PJ. Electron transport chain dysfunction in neonatal pressure-overload hypertrophy precedes cardiomyocyte apoptosis independent of oxidative stress. J Thorac Cardiovasc Surg. 2010;139(6):1609–17.
38. Dai D-F, Hsieh EJ, Liu Y, Chen T, Beyer RP, Chin MT, et al. Mitochondrial proteome remodelling in pressure overload-induced heart failure: the role of mitochondrial oxidative stress. Cardiovasc Res. 2012;93(1):79–88.
39. Bugger H, Schwarzer M, Chen D, Schrepper A, Amorim PA, Schoepe M, et al. Proteomic remodelling of mitochondrial oxidative pathways in pressure overload-induced heart failure. Cardiovasc Res. 2010;85(2):376–84.
40. Rosca MG, Vazquez EJ, Kerner J, Parland W, Chandler MP, Stanley W, et al. Cardiac mitochondria in heart failure: decrease in respirasomes and oxidative phosphorylation. Cardiovasc Res. 2008;80(1):30–9.
41. Schmidt O, Pfanner N, Meisinger C. Mitochondrial protein import: from proteomics to functional mechanisms. Nat Rev Mol Cell Biol. 2010;11(9):655–67.

42. Chacinska A, Koehler CM, Milenkovic D, Lithgow T, Pfanner N. Importing mitochondrial proteins: machineries and mechanisms. PubMed – NCBICell. 2009;138(4):628–44.
43. Li J, Qi M, Li C, Shi D, Zhang D, Xie D, et al. Tom70 serves as a molecular switch to determine pathological cardiac hypertrophy. Cell Res. 2014;24(8):977–93.
44. Taegtmeyer H, Golfman L, Sharma S, Razeghi P, van Arsdall M. Linking gene expression to function: metabolic flexibility in the normal and diseased heart. Ann N Y Acad Sci. 2004;1015:202–13.
45. Bishop SP, Altschuld RA. Increased glycolytic metabolism in cardiac hypertrophy and congestive failure. Am J Phys. 1970;218(1):153–9.
46. Taegtmeyer H, Overturf ML. Effects of moderate hypertension on cardiac function and metabolism in the rabbit. Hypertension. 1988;11(5):416–26.
47. Korvald C, Elvenes OP, Myrmel T. Myocardial substrate metabolism influences left ventricular energetics in vivo. Am J Physiol Heart Circ Physiol. 2000;278(4):H1345–51.
48. Ren J, Gintant GA, Miller RE, Davidoff AJ. High extracellular glucose impairs cardiac E-C coupling in a glycosylation-dependent manner. Am J Phys. 1997;273(6 Pt 2):H2876–83.
49. Nascimben L, Ingwall JS, Lorell BH, Pinz I, Schultz V, Tornheim K, et al. Mechanisms for increased glycolysis in the hypertrophied rat heart. Hypertens (Dallas, Tex 1979). 2004;44(5):662–7.
50. Zhang J, Duncker DJ, Ya X, Zhang Y, Pavek T, Wei H, Merkle H, Uğurbil K, From AHBR. Effect of left ventricular hypertrophy secondary to chronic pressure overload on transmural myocardial 2- deoxyglucose uptake. Circulation. 1995;92(5):1274–83.
51. Allard MF, Wambolt RB, Longnus SL, Grist M, Lydell CP, Parsons HL, et al. Hypertrophied rat hearts are less responsive to the metabolic and functional effects of insulin. Am J Physiol Endocrinol Metab. 2000;279(3):E487–93.
52. El Alaoui-Talibi Z, Guendouz A, Moravec M, Moravec J. Control of oxidative metabolism in volume-overloaded rat hearts: effect of propionyl-L-carnitine. Am J Phys. 1997;272(4 Pt 2):H1615–24. 1997;272:H1615–24
53. Leong HS, Grist M, Parsons H, Wambolt RB, Lopaschuk GD, Brownsey R, et al. Accelerated rates of glycolysis in the hypertrophied heart: are they a methodological artifact? Am J Physiol Endocrinol Metab. 2002;282(5):E1039–45.
54. Sorokina N, O'Donnell JM, McKinney RD, Pound KM, Woldegiorgis G, LaNoue KF, et al. Recruitment of compensatory pathways to sustain oxidative flux with reduced carnitine palmitoyltransferase I activity characterizes inefficiency in energy metabolism in hypertrophied hearts. Circulation. 2007;115(15):2033–41.
55. Doenst T, Pytel G, Schrepper A, Amorim P, Färber G, Shingu Y, et al. Decreased rates of substrate oxidation ex vivo predict the onset of heart failure and contractile dysfunction in rats with pressure overload. Cardiovasc Res. 2010;86(3):461–70.
56. Lydell CP, Chan A, Wambolt RB, Sambandam N, Parsons H, Bondy GP, et al. Pyruvate dehydrogenase and the regulation of glucose oxidation in hypertrophied rat hearts. Cardiovasc Res. 2002;53(4):841–51.
57. Akki A, Smith K, Seymour A-ML. Compensated cardiac hypertrophy is characterised by a decline in palmitate oxidation. Mol Cell Biochem. 2008;311(1–2):215–24.
58. Christe ME, Rodgers RL. Altered glucose and fatty acid oxidation in hearts of the spontaneously hypertensive rat. J Mol Cell Cardiol. 1994;26(10):1371–5.
59. Qanud K, Mamdani M, Pepe M, Khairallah RJ, Gravel J, Lei B, et al. Reverse changes in cardiac substrate oxidation in dogs recovering from heart failure. Am J Physiol Heart Circ Physiol. 2008;295(5):H2098–105.
60. Scholz TD, Koppenhafer SL. Reducing equivalent shuttles in developing porcine myocardium: enhanced capacity in the newborn heart. Pediatr Res. 1995;38(2):221–7.
61. Rupert BE, Segar JL, Schutte BC, Scholz T. Metabolic adaptation of the hypertrophied heart: role of the malate/aspartate and alpha-glycerophosphate shuttles. J Mol Cell Cardiol. 2000;32(12):2287–97. 2000;32(12):2287–97
62. Lewandowski ED, O'donnell JM, Scholz TD, Sorokina N, Buttrick PM. Recruitment of NADH shuttling in pressure-overloaded and hypertrophic rat hearts. Am J Phys Cell Phys. 2007;292(5):C1880–6.

63. Pound KM, Sorokina N, Ballal K, Berkich DA, Fasano M, Lanoue KF, et al. Substrate-enzyme competition attenuates upregulated anaplerotic flux through malic enzyme in hypertrophied rat heart and restores triacylglyceride content: attenuating upregulated anaplerosis in hypertrophy. Circ Res. 2009;104(6):805–12.
64. Allard MF, Schönekess BO, Henning SL, English DR, Lopaschuk GD. Contribution of oxidative metabolism and glycolysis to ATP production in hypertrophied hearts. Am J Phys. 1994;267(2 Pt 2):H742–50.
65. Hue L, Taegtmeyer H. The Randle cycle revisited: a new head for an old hat. Am J Physiol Endocrinol Metab. 2009;297(3):E578–91.
66. Sambandam N, Lopaschuk GD, Brownsey RW, Allard MF. Energy metabolism in the hypertrophied heart. Heart Fail Rev. 2002;7(2):161–73.
67. Schwenk RW, Luiken JJFP, Bonen A, Glatz JFC. Regulation of sarcolemmal glucose and fatty acid transporters in cardiac disease. Cardiovasc Res. 2008;79(2):249–58.
68. van der Vusse GJ, van Bilsen M, Glatz JF. Cardiac fatty acid uptake and transport in health and disease. Cardiovasc Res. 2000;45(2):279–93.
69. Chess DJ, Khairallah RJ, O'Shea KM, Xu W, Stanley WC. A high-fat diet increases adiposity but maintains mitochondrial oxidative enzymes without affecting development of heart failure with pressure overload. Am J Physiol Heart Circ Physiol. 2009;297(5):H1585–93.
70. Okere IC, Young ME, McElfresh TA, Chess DJ, Sharov VG, Sabbah HN, et al. Low carbohydrate/high-fat diet attenuates cardiac hypertrophy, remodeling, and altered gene expression in hypertension. Hypertens (Dallas, Tex 1979). 2006;48(6):1116–23.
71. Hafstad AD, Khalid AM, Hagve M, Lund T, Larsen TS, Severson DL, et al. Cardiac peroxisome proliferator-activated receptor-alpha activation causes increased fatty acid oxidation, reducing efficiency and post-ischaemic functional loss. Cardiovasc Res. 2009;83(3): 519–26.
72. Finck BN, Kelly DP. Peroxisome proliferator-activated receptor gamma coactivator-1 (PGC-1) regulatory cascade in cardiac physiology and disease. Circulation. 2007;115(19):2540–8.
73. Young ME, Laws FA, Goodwin GW, Taegtmeyer H. Reactivation of peroxisome proliferator-activated receptor alpha is associated with contractile dysfunction in hypertrophied rat heart. J Biol Chem. 2001;276(48):44390–5.
74. Yang J, Sambandam N, Han X, Gross RW, Courtois M, Kovacs A, et al. CD36 deficiency rescues lipotoxic cardiomyopathy. Circ Res. 2007;100(8):1208–17.
75. Lebrasseur NK, Duhaney T-AS, De Silva DS, Cui L, Ip PC, Joseph L, et al. Effects of fenofibrate on cardiac remodeling in aldosterone-induced hypertension. Hypertens (Dallas, Tex 1979). 2007;50(3):489–96.
76. Brigadeau F, Gelé P, Wibaux M, Marquié C, Martin-Nizard F, Torpier G, et al. The PPARalpha activator fenofibrate slows down the progression of the left ventricular dysfunction in porcine tachycardia-induced cardiomyopathy. J Cardiovasc Pharmacol. 2007;49(6):408–15.
77. Labinskyy V, Bellomo M, Chandler MP, Young ME, Lionetti V, Qanud K, et al. Chronic activation of peroxisome proliferator-activated receptor-alpha with fenofibrate prevents alterations in cardiac metabolic phenotype without changing the onset of decompensation in pacing-induced heart failure. J Pharmacol Exp Ther. 2007;321(1):165–71.
78. Morgan EE, Rennison JH, Young ME, McElfresh TA, Kung TA, Tserng K-Y, et al. Effects of chronic activation of peroxisome proliferator-activated receptor-alpha or high-fat feeding in a rat infarct model of heart failure. Am J Physiol Heart Circ Physiol. 2006;290(5): H1899–904.
79. Graham BH, Waymire KG, Cottrell B, Trounce IA, MacGregor GR, Wallace DC. A mouse model for mitochondrial myopathy and cardiomyopathy resulting from a deficiency in the heart/muscle isoform of the adenine nucleotide translocator. Nat Genet. 1997;16(3):226–34.
80. Walther T, Tschöpe C, Sterner-Kock A, Westermann D, Heringer-Walther S, Riad A, et al. Accelerated mitochondrial adenosine diphosphate/adenosine triphosphate transport improves hypertension-induced heart disease. Circulation. 2007;115(3):333–44.
81. Ingwall JS, Weiss RG. Is the failing heart energy starved? On using chemical energy to support cardiac function. Circ Res. 2004;95(2):135–45.
82. Neubauer S. The failing heart--an engine out of fuel. N Engl J Med. 2007;356(11):1140–51.

83. Wu Z, Puigserver P, Andersson U, Zhang C, Adelmant G, Mootha V, Troy A, Cinti S, Lowell B, Scarpulla RC, Spiegelman B. Mechanisms controlling mitochondrial biogenesis and respiration through the thermogenic coactivator PGC-1. PubMed – NCBICell. 1999;98(1):115–24.
84. Puigserver P, Wu Z, Park CW, Graves R, Wright M, Spiegelman B. A cold-inducible coactivator of nuclear receptors linked to adaptive thermogenesis. Cell. 1998;92(6):829–39.
85. Esterbauer H, Oberkofler H, Krempler F, Patsch W. Human peroxisome proliferator activated receptor gamma coactivator 1 (PPARGC1) gene: cDNA sequence, genomic organization, chromosomal localization.... PubMed – NCBIGenomics. 1999;62(1):98–102.
86. Lin J, Puigserver P, Donovan J, Tarr P, Spiegelman BM. Peroxisome proliferator-activated receptor gamma coactivator 1beta (PGC-1beta), a novel PGC-1-related transcription coactivator associated with host cell factor. J Biol Chem. 2002;277(3):1645–8.
87. Lelliott CJ, Medina-Gomez G, Petrovic N, Kis A, Feldmann HM, Bjursell M, et al. Ablation of PGC-1beta results in defective mitochondrial activity, thermogenesis, hepatic function, and cardiac performance. PLoS Biol. 2006;4(11):e369.
88. Arany Z, Novikov M, Chin S, Ma Y, Rosenzweig A, Spiegelman BM. Transverse aortic constriction leads to accelerated heart failure in mice lacking PPAR-gamma coactivator 1alpha. Proc Natl Acad Sci U S A. 2006;103(26):10086–91.
89. Garnier A, Fortin D, Deloménie C, Momken I, Veksler V, Ventura-Clapier R. Depressed mitochondrial transcription factors and oxidative capacity in rat failing cardiac and skeletal muscles. J Physiol. 2003;551(Pt 2):491–501.
90. Riehle C, Wende AR, Zaha VG, Pires KM, Wayment B, Olsen C, et al. PGC-1β deficiency accelerates the transition to heart failure in pressure overload hypertrophy. Circ Res. 2011;109(7):783–93.
91. Lehman JJ, Kelly DP. Transcriptional activation of energy metabolic switches in the developing and hypertrophied heart. Clin Exp Pharmacol Physiol. 2002;29(4):339–45.
92. Arany Z, He H, Lin J, Hoyer K, Handschin C, Toka O, Ahmad F, Matsui T, Chin S, Wu PH, Rybkin II, Shelton JM, Manieri M, Cinti S, Schoen FJ, Bassel-Duby R, Rosenzweig A, Ingwall JSSB. Transcriptional coactivator PGC-1 alpha controls the energy state and contractile function of cardiac muscle. PubMed – NCBICell Metab. 2005;4:259–71.
93. Leone TC, Lehman JJ, Finck BN, Schaeffer PJ, Wende AR, Boudina S, et al. PGC-1alpha deficiency causes multi-system energy metabolic derangements: muscle dysfunction, abnormal weight control and hepatic steatosis. PLoS Biol. 2005;3(4):e101.
94. Liu XP, Gao H, Huang XY, Chen YF, Feng XJ, He YH, Li ZM, Liu PQ. Peroxisome proliferator-activated receptor gamma coactivator 1 alpha protects cardiomyocytes from hypertrophy by suppressing calcineurin-nuclear factor of activated T cells c4 signaling pathway. Transl Res. 2015;166(5):459–73.
95. Lai L, Leone TC, Zechner C, Schaeffer PJ, Kelly SM, Flanagan DP, et al. Transcriptional coactivators PGC-1alpha and PGC-lbeta control overlapping programs required for perinatal maturation of the heart. Genes Dev. 2008;22(14):1948–61.
96. Huss JM, Kopp RP, Kelly DP. Peroxisome proliferator-activated receptor coactivator-1alpha (PGC-1alpha) coactivates the cardiac-enriched nuclear receptors estrogen-related receptor-alpha and -gamma. Identification of novel leucine-rich interaction motif within PGC-1alpha. J Biol Chem. 2002;277(43):40265–74.
97. Lehman JJ, Barger PM, Kovacs A, Saffitz JE, Medeiros DM, Kelly DP. Peroxisome proliferator-activated receptor gamma coactivator-1 promotes cardiac mitochondrial biogenesis. J Clin Invest. 2000;106(7):847–56.
98. Russell LK, Mansfield CM, Lehman JJ, Kovacs A, Courtois M, Saffitz JE, et al. Cardiac-specific induction of the transcriptional coactivator peroxisome proliferator-activated receptor gamma coactivator-1alpha promotes mitochondrial biogenesis and reversible cardiomyopathy in a developmental stage-dependent manner. Circ Res. 2004;94(4):525–33.
99. Vega RB, Huss JM, Kelly DP. The coactivator PGC-1 cooperates with peroxisome proliferator-activated receptor alpha in transcriptional control of nuclear genes encoding mitochondrial fatty acid oxidation enzymes. Mol Cell Biol. 2000;20(5):1868–76.

100. Karamanlidis G, Garcia-Menendez L, Kolwicz SC, Lee CF, Tian R. Promoting PGC-1α-driven mitochondrial biogenesis is detrimental in pressure-overloaded mouse hearts. Am J Physiol Heart Circ Physiol. 2014;307(9):H1307–16.
101. Sack MN, Rader TA, Park S, Bastin J, McCune SA, Kelly DP. Fatty acid oxidation enzyme gene expression is downregulated in the failing heart. Circulation. 1996;94(11):2837–42.
102. Hu X, Xu X, Lu Z, Zhang P, Fassett J, Zhang Y, et al. AMP activated protein kinase-α2 regulates expression of estrogen-related receptor-α, a metabolic transcription factor related to heart failure development. Hypertens (Dallas, Tex 1979). 2011;58(4):696–703.
103. Kuroda J, Ago T, Matsushima S, Zhai P, Schneider MD, Sadoshima J. NADPH oxidase 4 (Nox4) is a major source of oxidative stress in the failing heart. Proc Natl Acad Sci U S A. 2010;107(35):15565–70.
104. Yamamoto M, Yang G, Hong C, Liu J, Holle E, Yu X, et al. Inhibition of endogenous thioredoxin in the heart increases oxidative stress and cardiac hypertrophy. J Clin Invest. 2003;112(9):1395–406.
105. Yancey DM, Guichard JL, Ahmed MI, Zhou L, Murphy MP, Johnson MS, et al. Cardiomyocyte mitochondrial oxidative stress and cytoskeletal breakdown in the heart with a primary volume overload. Am J Physiol Heart Circ Physiol. 2015;308(6):H651–63.
106. Dai D-F, Santana LF, Vermulst M, Tomazela DM, Emond MJ, MacCoss MJ, et al. Overexpression of catalase targeted to mitochondria attenuates murine cardiac aging. Circulation. 2009;119(21):2789–97.
107. Siwik DA, Tzortzis JD, Pimental DR, Chang DL, Pagano PJ, Singh K, et al. Inhibition of copper-zinc superoxide dismutase induces cell growth, hypertrophic phenotype, and apoptosis in neonatal rat cardiac myocytes in vitro. Circ Res. 1999;85(2):147–53.
108. Shih NL, Cheng TH, Loh SH, Cheng PY, Wang DL, Chen YS, et al. Reactive oxygen species modulate angiotensin II-induced beta-myosin heavy chain gene expression via Ras/Raf/extracellular signal-regulated kinase pathway in neonatal rat cardiomyocytes. Biochem Biophys Res Commun. 2001;283(1):143–8.
109. Lemos Caldas FR, Rocha Leite IM, Tavarez Filgueiras AB, de Figueiredo Júnior IL, Gomes Marques de Sousa TA, Martins PR, et al. Mitochondrial ATP-sensitive potassium channel opening inhibits isoproterenol-induced cardiac hypertrophy by preventing oxidative damage. J Cardiovasc Pharmacol. 2015;65(4):393–7.
110. Koyama H, Nojiri H, Kawakami S, Sunagawa T, Shirasawa T, Shimizu T. Antioxidants improve the phenotypes of dilated cardiomyopathy and muscle fatigue in mitochondrial superoxide dismutase-deficient mice. Molecules. 2013;18(2):1383–93.
111. Tsutsui H, Kinugawa S, Matsushima S. Oxidative stress and heart failure. Am J Physiol Heart Circ Physiol. 2011;301(6):H2181–90.
112. Adiga IK, Nair RR. Multiple signaling pathways coordinately mediate reactive oxygen species dependent cardiomyocyte hypertrophy. Cell Biochem Funct. 2008;26(3):346–51.
113. Cai J, Yi F-F, Bian Z-Y, Shen D-F, Yang L, Yan L, et al. Crocetin protects against cardiac hypertrophy by blocking MEK-ERK1/2 signalling pathway. J Cell Mol Med. 2009;13(5):909–25.
114. Erickson JR, Joiner ML, Guan X, Kutschke W, Yang J, Oddis CV, Bartlett RK, Lowe JS, O'Donnell SE, Aykin-Burns N, Zimmerman MC, Zimmerman K, Ham AJ, Weiss RM, Spitz DR, Shea MA, Colbran RJ, Mohler PJ, Anderson M. A dynamic pathway for calcium-independent activation of CaMKII by methionine oxidation. Cell. 2008;133(3):462–74.
115. Guo J, Gertsberg Z, Ozgen N, Steinberg SF. p66Shc links alpha1-adrenergic receptors to a reactive oxygen species-dependent AKT-FOXO3A phosphorylation pathway in cardiomyocytes. Circ Res. 2009;104(5):660–9.
116. Li Y, Huang TT, Carlson EJ, Melov S, Ursell PC, Olson JL, et al. Dilated cardiomyopathy and neonatal lethality in mutant mice lacking manganese superoxide dismutase. Nat Genet. 1995;11(4):376–81.
117. Zhang G-X, Kimura S, Nishiyama A, Shokoji T, Rahman M, Yao L, et al. Cardiac oxidative stress in acute and chronic isoproterenol-infused rats. Cardiovasc Res. 2005;65(1):230–8.

118. Zou X-J, Yang L, Yao S-L. Propofol depresses angiotensin II-induced cardiomyocyte hypertrophy in vitro. Exp Biol Med (Maywood). 2008;233(2):200–8.
119. Li H-L, Huang Y, Zhang C-N, Liu G, Wei Y-S, Wang A-B, et al. Epigallocathechin-3 gallate inhibits cardiac hypertrophy through blocking reactive oxidative species-dependent and -independent signal pathways. Free Radic Biol Med. 2006;40(10):1756–75.
120. Anderson ME, Brown JHBD. CaMKII in myocardial hypertrophy and heart failure. J Mol Cell Cardiol. 2011;51(4):468–73.
121. Bossuyt J, Helmstadter K, Wu X, Clements-Jewery H, Haworth RS, Avkiran M, et al. Ca2+/calmodulin-dependent protein kinase IIdelta and protein kinase D overexpression reinforce the histone deacetylase 5 redistribution in heart failure. Circ Res. 2008;102(6):695–702.
122. Backs J, Olson EN. Control of cardiac growth by histone acetylation/deacetylation. Circ Res. 2006;98(1):15–24.
123. Ago T, Liu T, Zhai P, Chen W, Li H, Molkentin JD, et al. A redox-dependent pathway for regulating class II HDACs and cardiac hypertrophy. Cell. 2008;133(6):978–93.
124. Ago T, Kuroda J, Pain J, Fu C, Li H, Sadoshima J. Upregulation of Nox4 by hypertrophic stimuli promotes apoptosis and mitochondrial dysfunction in cardiac myocytes. Circ Res. 2010;106(7):1253–64.
125. Theccanat T, Philip JL, Razzaque AM, Ludmer N, Li J, Xu X, et al. Regulation of cellular oxidative stress and apoptosis by G protein-coupled receptor kinase-2; the role of NADPH oxidase 4. Cell Signal. 2016;28(3):190–203.
126. Fridovich I. Superoxide radical and superoxide dismutases. Annu Rev Biochem. 1995;64:97–112.
127. Kowaltowski AJ, Vercesi A. Mitochondrial damage induced by conditions of oxidative stress. Free Radic Biol Med. 1999;26(3–4):463–71.
128. Radi R, Turrens JF, Chang LY, Bush KM, Crapo JD, Freeman BA. Detection of catalase in rat heart mitochondria. J Biol Chem. 1991;266(32):22028–34.
129. Esworthy RS, Ho YS, Chu F. The Gpx1 gene encodes mitochondrial glutathione peroxidase in the mouse liver. Arch Biochem Biophys. 1997;340(1):59–63.
130. Watabe S, Hiroi T, Yamamoto Y, Fujioka Y, Hasegawa H, Yago N, et al. SP-22 is a thioredoxin-dependent peroxide reductase in mitochondria. Eur J Biochem. 1997;249(1):52–60.
131. Cho J, Won K, Wu D, Soong Y, Liu S, Szeto HH, et al. Potent mitochondria-targeted peptides reduce myocardial infarction in rats. Coron Artery Dis. 2007;18(3):215–20.
132. Chakrabarti AK, Feeney K, Abueg C, Brown DA, Czyz E, Tendera M, Janosi A, Giugliano RP, Kloner RA, Weaver WD, Bode C, Godlewski J, Merkely B, Gibson. Rationale and design of the EMBRACE STEMI study: a phase 2a, randomized, double-blind, placebo-controlled trial to evaluate the safety, tolerability and efficacy of intravenous Bendavia on reperfusion injury in patients treated with standard therapy including primary percutaneous coronary intervention and stenting for ST-segment elevation myocardial infarction. Am Heart J. 2013;165(4):509–14.
133. Dai D-F, Chen T, Szeto H, Nieves-Cintrón M, Kutyavin V, Santana LF, et al. Mitochondrial targeted antioxidant peptide ameliorates hypertensive cardiomyopathy. J Am Coll Cardiol. 2011;58(1):73–82.
134. Szeto HH, Lovelace JL, Fridland G, Soong Y, Fasolo J, Wu D, et al. In vivo pharmacokinetics of selective mu-opioid peptide agonists. J Pharmacol Exp Ther. 2001;298(1):57–61.
135. Shi J, Dai W, Hale SL, Brown DA, Wang M, Han X, et al. Bendavia restores mitochondrial energy metabolism gene expression and suppresses cardiac fibrosis in the border zone of the infarcted heart. Life Sci. 2015;141:170–8.
136. Szeto HH. Mitochondria-targeted cytoprotective peptides for ischemia-reperfusion injury. Antioxid Redox Signal. 2008;10(3):601–19.
137. Zhao K, Zhao G-M, Wu D, Soong Y, Birk AV, Schiller PW, et al. Cell-permeable peptide antioxidants targeted to inner mitochondrial membrane inhibit mitochondrial swelling, oxidative cell death, and reperfusion injury. J Biol Chem. 2004;279(33):34682–90.
138. Graham D, Huynh NN, Hamilton CA, Beattie E, Smith RAJ, Cochemé HM, et al. Mitochondria-targeted antioxidant MitoQ10 improves endothelial function and attenuates cardiac hypertrophy. Hypertension. 2009;54(2):322–8.

139. McLachlan J, Beattie E, Murphy MP, Koh-Tan CHH, Olson E, Beattie W, et al. Combined therapeutic benefit of mitochondria-targeted antioxidant, MitoQ10, and angiotensin receptor blocker, losartan, on cardiovascular function. J Hypertens. 2014;32(3):555–64.
140. Murphy MP, Smith RAJ. Targeting antioxidants to mitochondria by conjugation to lipophilic cations. Annu Rev Pharmacol Toxicol. 2007;47:629–56.
141. Smith RAJ, Murphy MP. Animal and human studies with the mitochondria-targeted antioxidant MitoQ. Ann N Y Acad Sci. 2010;1201:96–103.
142. Dai D-F, Johnson SC, Villarin JJ, Chin MT, Nieves-Cintrón M, Chen T, et al. Mitochondrial oxidative stress mediates angiotensin II-induced cardiac hypertrophy and Galphaq overexpression-induced heart failure. Circ Res. 2011;108(7):837–46.
143. Loch T, Vakhrusheva O, Piotrowska I, Ziolkowski W, Ebelt H, Braun T, et al. Different extent of cardiac malfunction and resistance to oxidative stress in heterozygous and homozygous manganese-dependent superoxide dismutase-mutant mice. Cardiovasc Res. 2009;82(3):448–57.
144. Smith RAJ, Hartley RC, Murphy MP. Mitochondria-targeted small molecule therapeutics and probes. Antioxid Redox Signal. 2011;15(12):3021–38.
145. Xia Y, Rajapurohitam V, Cook MA, Karmazyn M. Inhibition of phenylephrine induced hypertrophy in rat neonatal cardiomyocytes by the mitochondrial KATP channel opener diazoxide. J Mol Cell Cardiol. 2004;37(5):1063–7.
146. Facundo HTF, de Paula JG, Kowaltowski AJ. Mitochondrial ATP-sensitive K+ channels are redox-sensitive pathways that control reactive oxygen species production. Free Radic Biol Med. 2007;42(7):1039–48.
147. Facundo HTF, Carreira RS, de Paula JG, Santos CCX, Ferranti R, Laurindo FRM, et al. Ischemic preconditioning requires increases in reactive oxygen release independent of mitochondrial K+ channel activity. Free Radic Biol Med. 2006;40(3):469–79.
148. Andrukhiv A, Costa AD, West IC, Garlid KD. Opening mitoKATP increases superoxide generation from complex I of the electron transport chain. Am J Physiol Heart Circ Physiol. 2006;291(5):H2067–74.
149. Facundo HTF, de Paula JG, Kowaltowski AJ. Mitochondrial ATP-sensitive K+ channels prevent oxidative stress, permeability transition and cell death. J Bioenerg Biomembr. 2005;37(2):75–82.
150. González G, Zaldívar D, Carrillo E, Hernández A, García M, Sánchez J. Pharmacological preconditioning by diazoxide downregulates cardiac L-type Ca(2+) channels. Br J Pharmacol. 2010;161(5):1172–85.
151. Ventura-Clapier R, Garnier A, Veksler V. Energy metabolism in heart failure. J Physiol. 2004;555(Pt 1):1–13.
152. Izem-Meziane M, Djerdjouri B, Rimbaud S, Caffin F, Fortin D, Garnier A, et al. Catecholamine-induced cardiac mitochondrial dysfunction and mPTP opening: protective effect of curcumin. Am J Physiol Heart Circ Physiol. 2012;302(3):H665–74.
153. Michels G, Khan IF, Endres-Becker J, Rottlaender D, Herzig S, Ruhparwar A, et al. Regulation of the human cardiac mitochondrial Ca2+ uptake by 2 different voltage-gated Ca2+ channels. Circulation. 2009;119(18):2435–43.
154. Javadov S, Karmazyn M. Mitochondrial permeability transition pore opening as an endpoint to initiate cell death and as a putative target for cardioprotection. Cell Physiol Biochem. 2007;20(1–4):1–22.
155. Marcil M, Ascah A, Matas J, Belanger S, Deschepper C, Burelle Y. Compensated volume overload increases the vulnerability of heart mitochondria without affecting their functions in the absence of stress. J Mol Cell Cardiol. 2006;41(6):998–1009.
156. Xie W, Santulli G, Reiken SR, Yuan Q, Osborne BW, Chen BX, Marks AR. Mitochondrial oxidative stress promotes atrial fibrillation. Sci Rep. 2015;5:11427.
157. Xie W, Santulli G, Guo X, Gao M, Chen BX, Marks AR. Imaging atrial arrhythmic intracellular calcium in intact heart. J Mol Cell Cardiol. 2013;64:120–3.
158. Nakagawa T, Shimizu S, Watanabe T, Yamaguchi O, Otsu K, Yamagata H, et al. Cyclophilin D-dependent mitochondrial permeability transition regulates some necrotic but not apoptotic cell death. Nature. 2005;434(7033):652–8.

159. Baines CP, Kaiser RA, Purcell NH, Blair NS, Osinska H, Hambleton MA, et al. Loss of cyclophilin D reveals a critical role for mitochondrial permeability transition in cell death. Nature. 2005;434(7033):658–62.
160. Matas J, Young NT, Bourcier-Lucas C, Ascah A, Marcil M, Deschepper CF, Burelle Y. Increased expression and intramitochondrial translocation of cyclophilin-D associates with increased vulnerability of the permeability transition pore to stress-induced opening during compensated ventricular hypertrophy. J Mol Cell Cardiol. 2009;46(3):420–30.
161. Nakayama H, Chen X, Baines CP, Klevitsky R, Zhang X, Zhang H, et al. Ca2± and mitochondrial-dependent cardiomyocyte necrosis as a primary mediator of heart failure. J Clin Invest. 2007;117(9):2431–44.
162. Javadov S, Baetz D, Rajapurohitam V, Zeidan A, Kirshenbaum LA, Karmazyn M. Antihypertrophic effect of Na+/H+ exchanger isoform 1 inhibition is mediated by reduced mitogen-activated protein kinase activation secondary to improved mitochondrial integrity and decreased generation of mitochondrial-derived reactive oxygen species. J Pharmacol Exp Ther. 2006;317(3):1036–43.
163. Shao J, Iwashita N, Ikeda F, Ogihara T, Uchida T, Shimizu T, Uchino H, Hirose T, Kawamori R, Watada H. Beneficial effects of candesartan, an angiotensin II type 1 receptor blocker, on beta-cell function and morphology in db/db mice. PubMed – NCBIBiochem Biophys Res Commun. 2006;344(4):1224–33.
164. Oliveira PJ, Rolo AP, Palmeira CM, Moreno AJ. Carvedilol reduces mitochondrial damage induced by hypoxanthine/xanthine oxidase: relevance to hypoxia/reoxygenation injury. Cardiovasc Toxicol. 2001;1(3):205–13.
165. Oliveira PJ, Esteves T, Rolo AP, Palmeira CM, Moreno AJM. Carvedilol inhibits the mitochondrial permeability transition by an antioxidant mechanism. Cardiovasc Toxicol. 2004;4(1):11–20.
166. Barone FC, Willette RN, Nelson AH, Ohlstein EH, Brooks DP, Coatney RW. Carvedilol prevents and reverses hypertrophy-induced cardiac dysfunction. Pharmacology. 2007; 80(2–3):166–76.

Chapter 12
Connexin 43 and Mitochondria in Cardiovascular Health and Disease

Kerstin Boengler and Rainer Schulz

Connexin (Cx) proteins are widely expressed in different tissues and are classified according to their molecular weight. Twenty one genes have been identified in the human genome which encode Cx proteins with a molecular weight ranging from 26 to 60 kDa [1]. The heart mainly expresses the Cx isoforms Cx40, Cx43, and Cx45, whereby Cx43 is the most abundant isoform and is found in ventricular cardiomyocytes, but also in the atria, in fibroblasts and in endothelial cells [2]. Cx40 is mainly located in the atria and in coronary endothelial cells, which also express Cx37, and the expression of Cx45 is specific for the conduction system under physiological conditions.

Cx are integral membrane proteins with their amino- and carboxyterminus located in the cytosol. The proteins have four transmembrane domains, two extracellular and one intracellular loop. Six Cx monomers assemble in the Golgi/trans-Golgi network into a so-called connexon or hemichannel which is then – mediated by the cytoskeleton – transported to the plasma membrane. Docking of two hemichannels from neighbouring cells results in the formation of gap junction channels [3, 4]. Gap junctions are assemblies of such closely packed gap junction channels and allow the direct connection between the cytosol of two adjacent cells and the passage of molecules with a molecular weight up to 1,000 Da. Thereby, electrical and metabolic cell coupling is mediated [5]. Cx43 is essential for development since embryos with a homozygous deletion of Cx43 die at birth due to blockage of the right ventricular outflow tract [6].

Due to the short half-life of Cx43 (approximately 1.3 h [7]), the protein has to be regulated at short notice, e.g. via phosphorylation or protein-protein interactions. Cx43 interacts with the scaffold protein zonula occludens 1 (ZO-1) via the PDZ

K. Boengler • R. Schulz (✉)
Institute of Physiology, Justus-Liebig University, Giessen, Germany
e-mail: Kerstin.Boengler@physiologie.med.uni-giessen.de;
Rainer.Schulz@physiologie.med.uni-giessen.de

G. Santulli (ed.), *Mitochondrial Dynamics in Cardiovascular Medicine*,
Advances in Experimental Medicine and Biology 982,
DOI 10.1007/978-3-319-55330-6_12

domain in the carboxyterminus of Cx43 [8]. This protein-protein interaction plays a role in the aggregation of undocked hemichannels into gap junctions, since the inhibition of the Cx43/ZO-1 interaction increases the size of gap junctions [9]. At the intercalated discs, Cx43 interacts with the transmembrane protein 65 (Tmem65), which functionally regulates Cx43 and thereby cellular coupling [10].

Hemichannels not only function as precursors of gap junctions, but also exist as free, non-junctional channels in the plasma membranes of cardiomyocytes [11]. Under physiological conditions, hemichannels are predominantly in a closed state [12], but they open in response to stress conditions, for example changes in calcium (Ca^{2+}) concentrations [11, 13]. Such hemichannel opening leads to the release of adenosine triphosphate (ATP), the uptake of sodium- (Na^{+}) and Ca^{2+} ions, and the efflux of potassium- (K^{+}) ions. The resulting osmotic shifts, energy depletion, and Ca^{2+} overload induce cell swelling and finally cell death [14].

The localization of Cx43 at the plasma membrane is well described and characterized in several models, however, the presence of the protein is not restricted to it, since reports demonstrate the presence of Cx43 within mitochondria [15].

In this chapter, we will review the expression and regulation of Cx43 in the cardiovascular system and focus on the role of the protein – especially mitochondrial Cx43 – in health and disease.

Phosphorylation of Cx43

The extent of cell-cell communication depends on the open-probability of gap junctions, which – upon other factors – is controlled by the phosphorylation status of Cx43. Twenty one phosphorylation sites have been identified within Cx43 [16, 17]. Predominantly the serine residues within the cytoplasmic tail of Cx43 appear to be phosphorylated, although tyrosine phosphorylation at residues 247 and 265 has been reported [18]. The amount of phosphorylated Cx43-residues renders the electrophoretic migration from 41 kDa (unphosphorylated) to about 45 kDa (fully phosphorylated). The phosphorylation of Cx43 regulates e.g. the assembly of Cx43-formed hexamers, trafficking, and turnover of the protein and accordingly electrical and metabolic cell-cell coupling [19, 20]. However, phosphorylation of Cx43 within the cytosolic carboxyterminus is no prerequisite for gap junction formation, since mice with a truncated Cx43 carboxyterminus are principally capable of forming gap junction channels, although with altered electrophysiological properties [21]. Cx43 is the target of multiple kinases e.g. protein kinase A (PKA) [22], protein kinase B (PKB or Akt) [23], protein kinase C (PKC) [24], casein kinase 1 (CK1) [25], Ca^{2+}/calmodulin kinase II (CamKII) [26], mitogen-activated protein kinase (MAPK) [27], $p34^{cdc2}$/cyclin B kinase (p34cdc2) [28], and the pp60src kinase (src) [29]. The phosphorylation of Cx43 by different kinases has consequences for intercellular cell-coupling in the way that Cx43 phosphorylation via src, MAPK, or PKC inhibits, whereas Cx43 phosphorylation via CK1 or Akt enhances cell-cell communication [30] e.g. via regulation of gap junction assembly [31]. The importance of Cx43 phosphorylation for gap junctional and hemichannel function is reviewed e.g. in [16, 32].

The phosphorylation status of Cx43 is not only controlled by protein kinases but represents a dynamic interplay also with protein phosphatases. Compared to the involvement of protein kinases for Cx43 function, less is known about the role of protein phosphatases (for review see [33]. Recent data show that the overexpression of calcineurin results in dephosphorylation and downregulation of Cx43 and this is associated with the development of fibrosis, hypertrophy, and arrhythmias [34]. In addition to phosphorylation, also S-nitrosation of Cx43 is described [35].

Cx43- More Than a Gap Junction Protein

The importance of Cx43 as a gap junctional protein has been analyzed under several physiological and pathophysiological conditions in various species; however, besides the well described localization of Cx43 at the intercalated discs, the protein is also present at other compartments. At the plasma membrane, Cx43 is localized to the periphery of the gap junction plaque – the so-called perinexus – where it is involved e.g. in intermolecular interactions and sodium currents [9, 36].

Furthermore, the carboxyterminus of Cx43 translocates into the nucleus and inhibits the proliferation of HeLa cells [37]. In human endometrial stromal cells, levonorgestrel induces the expression of Cx43 in combination with a nuclear translocation of phosphorylated Cx43 and finally decreases cell proliferation [38].

A recent study demonstrates the presence of Cx43 within exosomes, extracellular vesicles formed and released by various cell types (including cardiac cell lines), which function in intercellular communication at longer distances. Within exosomes, Cx43 forms hexameric channels and is able to mediate interactions between exosomes and target cells [39].

Mitochondria

Whereas the highest amount of Cx43 is found at the plasma membrane, a certain fraction of the protein also resides within mitochondria from endothelial cells [40–42], hepatocytes [43], H9C2 cells [44], and cardiomyocytes [15]. Cx43 is detected in cardiomyocyte mitochondria from mouse, rat, porcine and human origin using antibody-dependent and –independent techniques [15, 45]. Cardiomyocytes contain at least two functionally different mitochondrial subpopulations: the subsarcolemmal mitochondria (SSM), which are located directly beneath the sarcolemma and the interfibrillar mitochondria (IFM), which are present between the myofibrils. IFM have a higher respiratory capacity and a more pronounced Ca^{2+} retention capacity [46, 47]. The analysis of cardiac SSM and IFM for the presence of Cx43 shows that Cx43 is almost exclusively localized in SSM, whereas IFM are devoid of

Cx43 [48, 49]. Cx43 is encoded in the nuclear genome and therefore has to be imported into the mitochondria. This translocation is achieved by an interaction between Cx43 and the heat-shock protein 90 (Hsp90), which delivers proteins to the general mitochondrial import machinery. Here, Cx43 interacts with the translocase of the outer membrane 20 (Tom20), which directs proteins to be imported into the mitochondria to other components of the import machinery, e.g. translocases of the inner mitochondrial membrane. Finally, Cx43 is directed to the inner mitochondrial membrane, with its carboxyterminus oriented towards the intermembrane space [45, 50]. However, one study also demonstrates that Cx43 is present in the outer mitochondrial membrane [51]. Within mitochondria, Cx43 is suggested to interact with the apoptosis-inducing factor, the beta-subunit of the electron-transfer protein [52], and the ATP-sensitive potassium channel subunit Kir6.1 [44]. The amount of mitochondrial Cx43 declines in cardiomyocytes after activation of the N-methyl-D-aspartate receptor 1 via enhanced mitochondrial translocation of the matrix-metalloproteinase 9 [53].

Taking into account that the main function of Cx43 at the plasma membrane is to contribute to cell-cell communication by the formation of gap junctional channels, the question is, whether or not mitochondrial Cx43 also forms channel-like structures. The chemical cross-linking of isolated mitochondria demonstrates Cx43-dependent protein-complexes at a molecular weight corresponding to that of gap junction-enriched membranes which presumably represent Cx43 hexamers. Further evidence for a Cx43-formed channel within mitochondria comes from experiments where the mitochondrial uptake of the Cx43-channel permeable dye Lucifer Yellow is inhibited by the chemically unrelated hemichannel blockers carbenoxolone and heptanol [45]. Carbenoxolone is also effective in reducing Lucifer yellow uptake in rat SSM, but not in Cx43-deficient IFM [35]. The findings that the administration of the Cx43-blockers 18-alpha glycyrrhetinic acid (18αGA), the Cx43-hemichannel-specific peptide Gap19 as well as the genetic ablation of Cx43 reduce the mitochondrial potassium influx further strengthens the hypothesis that Cx43 channels are present in the inner mitochondrial membrane [45, 54].

Since the main function of mitochondria is to generate ATP via the oxidative phosphorylation, the influence of mitochondrial Cx43 on respiration was studied. Both a pharmacological inhibition and a genetic ablation of Cx43 reduce mitochondrial complex 1-mediated oxygen consumption and ATP production, whereas complex 2-mediated respiration is not affected. In contrast, the overexpression of Cx43 within atrial-derived HL-1 cells increases mitochondrial complex 1-, but not complex 2-mediated respiration [55]. Possibly, Cx43 interacts with proteins of complex 1 of the electron transport chain as recently demonstrated [52, 55].

A putative relationship between mitochondrial Cx43 and the mitochondrial permeability transition pore (MPTP) has been suggested. The MPTP represents a large conductance pore – mainly formed by components of the ATP synthase [56] – which is predominantly closed under physiological conditions and opens under various forms of stress (e.g. enhanced amounts of reactive oxygen species (ROS), inorganic phosphate, elevated matrix Ca^{2+}, or a reduced mitochondrial membrane potential) leading to mitochondrial swelling and membrane rupture and ultimately cell death.

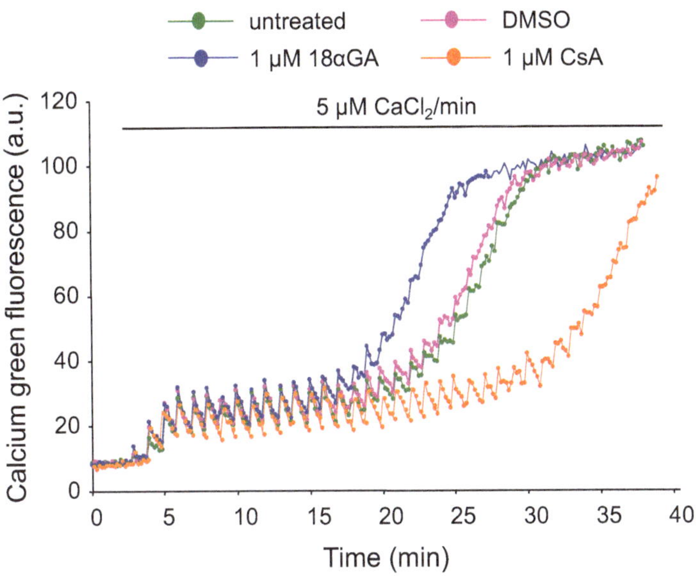

Fig. 12.1 Pharmacological inhibition of Cx43 enhances Ca^{2+}-induced MPTP opening. Original traces showing Ca^{2+} Green 5 N fluorescence (in arbitrary units, a.u.) of 100 μg/ml rat ventricular subsarcolemmal mitochondria (SSM) subjected to increasing $CaCl_2$ concentrations. Five μM $CaCl_2$ was added per minute until MPTP opening (increase in Ca^{2+} Green 5 N fluorescence) occurred. Ca^{2+}-induced MPTP opening was measured in untreated SSM, in the presence of DMSO, the Cx-channel inhibitor 18-alpha glycyrrhetinic acid (18αGA, 1 μM), or the MPTP inhibitor cyclosporin A (CsA, 1 μM)

In SSM isolated from rat hearts, treatment with the Cx43-channel blocking peptide Gap27 reduces the mitochondrial Ca^{2+} retention capacity [57]. Also, treatment of rat SSM with the Cx43 inhibitor 18α-glycyrrhetinic acid (18αGA) induces untimely MPTP opening in response to an extramitochondrial Ca^{2+} stimulus (Fig. 12.1).

Within cardiomyocytes, mitochondria are the predominant source of ROS and a relationship between Cx43 and ROS formation has been established [35, 58], whereby Cx43 inhibition or downregulation decreases the amount of ROS.

The phosphorylation status of mitochondrial Cx43 has not been analyzed in detail yet and therefore, it is not known whether all phosphorylation sites described in gap junctional Cx43 are also phosphorylated in mitochondrial Cx43. However, phosphorylation of mitochondrial Cx43 is found at serine 262 in rat SSM and mitochondria isolated from H9C2 cells [44, 57], whereas Cx43 phosphorylation at serine 368 is detected in mouse, rat and porcine SSM [57, 59–61]. Phosphorylation of Cx43 is not necessarily achieved outside the mitochondria, since PKC is present in mitochondria and the stimulation of such mitochondrial PKC leads to a phosphorylation of Cx43 at serine 262 and serine 368 in rat [57] and mouse SSM (Fig. 12.2).

Activation of mitochondrial PKC renders the organelles more tolerant towards a Ca^{2+} stimulus [57].

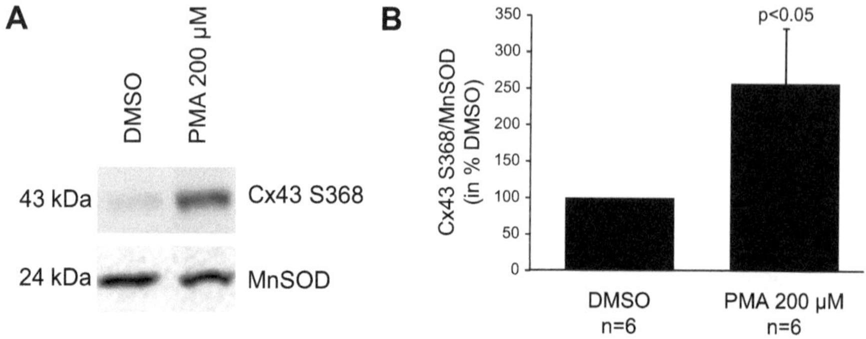

Fig. 12.2 Stimulation of mitochondrial PKC induces Cx43 phosphorylation. (**a**) Subsarcolemmal mitochondria were isolated from the mouse left ventricle and were stimulated for 30 min with 200 μM ATP and 200 μM of the PKC activator PMA or DMSO as solvent. Western blot analysis was performed for Cx43 phosphorylated at serine 368 (Cx43 S368) and the mitochondrial marker protein MnSOD (manganese superoxide dismutase). (**b**) Bar graphs demonstrate the ratio of phosphorylated Cx43 over MnSOD

Cx43 and Risk Factors of Cardiovascular Diseases

Age

The expression of Cx43 is found to decline with age in the hearts of hamsters [62], guinea pigs [63, 64], mouse [65], rats [66, 67] and rabbits [68]. In aged mouse hearts, not only the amount of gap junctional Cx43 but also that of mitochondrial Cx43 is reduced [69]. In addition to such decreased amounts of the protein also enhanced lateralization of the protein is detected in aged rat [70] and rabbit hearts [71] and this correlates with the age-dependent occurrence of arrhythmias in rat hearts [70] and that of atrial fibrillation in patients [72] (see also Chap. 10).

Hypertension and Hypertrophy

The expression of Cx43 in hypertension is mostly investigated in spontaneously hypertensive rats (SHR). Here, Cx43 protein levels either decline [73–75] or increase [76–78] with hypertension. Following abdominal aortic banding, Cx43 is displaced from its usual location at the intercalated discs [79]. In mice, pressure-overload by transverse aortic constriction reduces the expression of Cx43 [80]. The differences between the aforementioned studies may be due to the degree of hypertrophy associated with hypertension, since in porcine myocardium subjected to volume overload Cx43 expression increases in the initial phase, but gradually decreases with the progression of hypertrophy [81]. Also, in the ventricles from pressure-overloaded human hearts with valvular aortic stenosis, Cx43 expression is enhanced in compensated, but reduced in decompensated hypertrophy [82]. The latter is associated with augmented lateralization of the protein. Interventions known to reduce hypertension and/or hypertrophy such as blockage or deletion of angiotensin II receptor I [78, 83], aldosterone antagonists [84], or renin inhibition [75] prevent the changes in Cx43 expression, phosphorylation or lateralization seen with hypertension and hypertrophy.

The expression of mitochondrial Cx43 in hypertension and hypertrophy has not been studied up to now.

Diabetes and Hypercholesterolemia

Diabetes is associated with abnormal conduction and increased risk of arrhythmias in the heart. To elucidate the underlying pathophysiological mechanisms, the expression and phosphorylation of Cx43 is studied, however, the available data give conflicting results. In a pre-diabetes model in rats, Cx43 expression is unaltered and the fraction of Cx43 localized at the intercalated discs increases [85]. Cx43 mRNA and protein levels are either reduced [86–88], unchanged [89, 90], or elevated [91–94] with diabetes. Enhanced lateralization of Cx43 is observed with diabetes in rat hearts with increased expression of Cx43 [91, 93], but also under conditions where the amount of Cx43 is unaltered [89]. The decreased amount of Cx43 in obese-diabetic mice is restored by moderate exercise [88], and the lateralization of Cx43 is normalized by omega-3 polyunsaturated fatty acids [93]. With diabetes, overall [92] and tyrosine [91] Cx43 phosphorylation are found to decrease, but also increased Cx43 phosphorylation is described [86, 93, 95].

Cx43 is detected at the inner membrane of mitochondria from rat retinal endothelial cells [42]. Here, maintenance of the cells under high glucose conditions decreases the level of mitochondrial Cx43.

In rabbits fed a high-cholesterol diet, the expression of Cx43 is increased in left atria [96], but decreased in ventricular tissue, where also the lateralization of Cx43 is enhanced [97]. Such disturbed distribution of Cx43 is also seen in rats fed a high-fat diet, whereby here the expression of Cx43 is unchanged [98]. In this study, the mitochondrial amount of Cx43 decreases in hyperlipidemic animals.

Post-myocardial Infarction

The amount of Cx43 decreases and the localization of Cx43 is disturbed post-myocardial infarction in the border zone and also in the infarcted area of mouse [99, 100], rat [101, 102], and dog [103] hearts. The downregulation of Cx43 is attributed e.g. to the inducible nitric oxide synthase-mediated availability of nitric oxide (NO) [99]. Moreover, the finding that the Cx43 protein level following myocardial infarction is preserved in matrix-metalloproteinase 7 (MMP7)-deficient mice indicates that Cx43 may represent an MMP-7 substrate [100]. The lateralization of Cx43 is limited when mice are given a diet enriched with omega-3 fatty acids prior to myocardial infarction [104]. The transplantation of cardiac stem cells elevates the expression of Cx43 following myocardial infarction in rat hearts [105], and also the administration of a combination of hepatocyte growth factor and insulin-like growth factor improves Cx43 expression and intercellular coupling [102].

The expression of mitochondrial Cx43 post-myocardial infarction has again not been studied so far.

Heart Failure

The level of the Cx43 protein is reduced in an arrhythmogenic model in rabbits [106]. Here, the overexpression of Cx43 in cardiomyocytes from failing hearts to protein levels comparable to that of control cells improves cell coupling and reduces arrhythmias [107]. Additionally, the total amount of Cx43 decreases in failing mouse [108], rat [109, 110], and dog [111] hearts. Moreover, a heterogeneous distribution of Cx43 at the plasma membrane is observed in patients with end-stage heart failure due to ischemic or dilated cardiomyopathy [112, 113]. The analysis of the phosphorylation status of Cx43 shows either dephosphorylation [106, 111] or phosphorylation of the protein at serine 255 [114]. The induction of mitophagy and matrix-metalloproteinase activity is suggested to contribute to the decreased expression of Cx43 in heart failure [115], and the inhibition of the proteasome increases Cx43 amounts in adriamycin-induced heart failure in rats [116].

Data on the role of mitochondrial Cx43 in heart failure are limited. One study addresses the function of mitochondrial Cx43 in a model of doxorubicin cardiotoxicity in H9C2 cells [117]. Here, doxorubicin induces mitochondrial translocation of Cx43. Blocking the Hsp90-mediated import of Cx43 into the mitochondria increases ROS formation and mitochondrial Ca^{2+} content, and enhances depolarization of the mitochondrial membrane compared to doxorubin treatment alone, suggesting that mitochondrial Cx43 is important for the protection against doxorubin-induced cardiotoxicity. In a second study, the induction of dilated cardiomyopathy by furazolidone decreases the amount of Cx43 both in the myocardium as well as in isolated mitochondria [61]. Moreover, mitochondrial Cx43 is dephosphorylated at serine 368. The functional importance of altered mitochondrial Cx43 expression remains unknown.

Cx43 in Ischemia/Reperfusion Injury

Cx43-formed hemichannels open upon metabolic inhibition [118] or ischemia Accordingly, hemichannel inhibition by the specific hemichannel inhibitor Gap19 protects cardiomyocytes from volume overload, increases cell viability, and reduces infarct size following ischemia and reperfusion [14]. The Cx43-hemichannel activity is determined by the phosphorylation status of the protein [119].

The use of less hemichannel-specific inhibitors such as Gap26/Gap27 also reduces myocardial infarction following ischemia/reperfusion [120, 121], and mice in which Cx43 was replaced by Cx32 have smaller infarcts than wildtype mice [122]. The use of heterozygous Cx43-deficient mice in the analysis of myocardial infarction following ischemia/reperfusion demonstrates a reduction of infarct size in one study [123], whereas no effect is seen in other investigations [124, 125]. The further reduction of the Cx43 level in conditional knockout mice results in smaller infarcts after ischemia/reperfusion, whereas non-induced mice (with 50 % Cx43) have similar infarct sizes compared to wildtype mice [126].

Cx43 phosphorylation has been analyzed under conditions of ischemia and reperfusion and seems to be dependent on the cellular ATP content [127]. In particular, ischemia or hypoxia induce dephosphorylation of Cx43 at serine 365 [128] or at serine 325/328/330 [31], but also Cx43 phosphorylation at serine 262, 368 or serine 373 [24, 129, 130]. Serine 365 is suggested to function as a so-called "gatekeeper", since serine 365 phosphorylation prevents serine 368 phosphorylation [128]. The interaction between Cx43 and ZO-1 is regulated by phosphorylation at the Akt target site serine 373, whereby the phosphorylation of Cx43 at serine 373 disturbes the interaction between the two proteins resulting in increased size of gap junctions and enhanced cell-cell communication [129]. The mutation of serine 373 to phosphorylation-insensitive alanine reduces the interaction between Cx43 and 14-3-3- proteins, stabilizes Cx43 at the plasma membrane and thereby avoids Cx43 degradation [130]. During ischemia, a redistribution of Cx43 occurs from the intercalated discs to the lateral sides of the cardiomyocytes [131]. Such lateralized Cx43 is mainly dephosphorylated, whereas phosphorylated Cx43 remains at the gap junctions [2]. Recent data show that ischemia induces Cx43 ubiquitination at the intercalated discs and subsequent autophagic degradation of the protein [132, 133]. Ubiquitination of Cx43 is decreased by the mutation of the phosphorylation site serine 373 to alanine [130]. Additionally, Cx43 is degraded via the proteasome and the lysosome [134].

In H9C2 cells subjected to 12 h hypoxia, the phosphorylation of mitochondrial Cx43 is increased, however, no specific phosphorylation sites are tested [44]. Following a shorter phase of ischemia (30 min ischemia, 5 min reperfusion), an overall dephosphorylation of Cx43 is observed in rat myocardial mitochondria [98]. Also in porcine myocardial mitochondria, ischemia induces a dephosphorylation at the PKC-target site serine 368 [60].

Several pharmacological interventions are known to reduce infarct size after ischemia/reperfusion and some of them impact on the phosphorylation status of Cx43. In canine hearts, administration of 17beta-estradiol decreases myocardial infarction and this effect is associated with an decrease in the overall phosphorylation of Cx43 [135]. Also, inhibition of p38 MAPK induces phosphorylation of Cx43 at serine 368 – eventually via phosphatase inhibition – and reduces infarct size [136]. The sphingosine-1 phosphate (S1P) constituent of high-density lipoptoteins limits cell death following ischemia/reperfusion and this is accompagnied with an increased Cx43 phosphorylation at serine 368. A causal role for this phosphorylation site in cardioprotection by S1P is shown in mice, in which serine 368 of Cx43 is mutated to alanine; here, the infarct size reduction by S1P is lost [137]. Enhanced phosphorylation of Cx43 (total and mitochondrial) at serine 262 and serine 368 is achieved in isolated rat hearts by diazoxide or fibroblast growth factor 2 (FGF-2) [57, 138]. The overexpression of Cx43 in which serine 262 is mutated to alanine in neonatal cardiomyocytes results in augemented cell death following simulated ischemia [138].

Cx43 in Cardioprotection

The phenomena of endogenous conditioning describe the infarct size reduction following sustained ischemia/reperfusion by brief non-lethal periods of ischemia and reperfusion, which are performed either before (pre-), during (per-) or following (post-) the sustained phase of ischemia followed by reperfusion. The protection by such endogenous conditioning is often lost in the presence of cardiovascular risk factors or diseases [139], where total and mitochondrial Cx43 epxression, localization and/or phosphorylation is altered. Ischemic preconditioning (IPC) suppresses chemical coupling of cardiomyocytes via gap junctions during ischemia. Using a protocol of simulated ischemia and reperfusion in vitro, only wildtype cardiomyocytes were protected by IPC, whereas cardiomyocytes isolated from heterozygous Cx43-deficient cells show no increase in cell viability [140]. In heterozygous Cx43-deficient animals, the infarct size reduction by IPC is abolished [124], whereas non-induced conditional Cx43 knockout mice (also having a 50 % reduction of Cx43) demonstrate cardioprotection by IPC [126]. The further reduction of the Cx43 content in these mice, however, abolishes IPC's cardioprotection [126]. In addition, in mice, in which Cx43 is replaced by Cx32, IPC is not effective [122]. When Cx43-formed channels are inhibited by heptanol during the preconditioning cylces of ischemia and reperfusion infarct size reduction by IPC is lost [141].

The analysis of the phosphorylation status of Cx43 reveals that IPC prevents the ischemia-induced dephosphorylation of Cx43 in rat hearts in vitro [142] in vivo [143], or in pig hearts in vivo [60, 144], however, in one study IPC has no impact on Cx43 phosphorylation [145]. Preserved Cx43 phosphorylation is in part due to increased actvity of PKC, since knockout of PKC not only abrogates the protection, but also the maintained Cx43 phosphorylation by IPC [146]. The interaction of Cx43 with protein phosphatases is not affected by IPC [60].

IPC not only affects gap junctional, but also mitochondrial Cx43. The amount of the protein within mitochondria is increased upon IPC in isolated mouse [49] and rat hearts [15, 35], and pig hearts in vivo [15]. In bone marrow stem cells, the overexpression of specifically mitochondrial Cx43 is sufficient to confer cytoprotection [147], pointing to the important role of mitochondrial Cx43 in cell survival.

Pharmacological preconditioning with diazoxide augments mitochondrial Cx43 in a model of hypothermia [148]. When the mitochondrial import of Cx43 is blocked by the use of the Hsp90-inhibitor geldanamycin, the infarct size reduction by diazoxide is lost. Since in this model the gap junctional content of Cx43 remains unaffected, the loss of cardioprotection by diazoxide is associated with the reduced content of mitochondrial Cx43 [50]. In addition to Cx43, geldanamycin also inhibits the mitochondrial import of PKCepsilon, which may phosphorylate Cx43 [149]. The analysis of the mechanims by which mitochondrial Cx43 contributes to diazoxide's cardioprotection demonstrates that the generation of ROS, which, if present in low amounts, function as trigger molecules of preconditioning, is reduced in cardiomyocytes isolated from Cx43-deficient mice [125]. Furthermore, IPC induces S-nitrosation, an important posttranslational modification in cardioprotection [150], predominantly of SSM proteins [49]. One protein with enhanced amounts of S-nitrosation following IPC is mitochondrial Cx43, and the S-nitrosation of mitochondrial Cx43 enhances the permeability of SSM towards potassium ions and increases ROS formation [35]. SSM isolated from wildtype mice show increased respiration when the isolated mitochondria undergo IPC, however, in mice, in which Cx43 is replaced by Cx32, such preserved oxygen consumption is lost [151]. Also, mitochondrial Cx43 may contribute to cardioprotection via its effect on MPTP opening (see above), since the inhibition of MPTP opening reduces myocardial infarction [152].

In remote ischemic preconditioning, where the short periods of ischemia/reperfusion occur in organs other than the heart, sustained Cx43 expression and phosphorylation is oberserved [153].

Ischemic postconditioning induces Cx43 phosphorylation in rat hearts in vivo [154, 155], but also in mitochondria isolated from mice which were subjected to a postconditioning protocol [59]. However, one study demonstrates reduced phosphorylation of mitochondrial Cx43 after postconditioning compared to ischemia/reperfusion alone [156]. Nevertheless, altered Cx43 phosphorylation is not important for the cardioprotection by postconditioning, since heterozygous Cx43-deficient mice still have smaller infarcts than wildtype mice upon a postconditioning stimulus [157].

Conclusion

Cx43, which is mainly known as a gap junction protein, is also present within mitochondria from different cell types. The regulation of Cx43 is achieved – among other factors – by the phosphorylation status of the protein. Several cardiovascular diseases alter the expression and/or localization of gap junctional and mitochondrial Cx43, and dysregulated Cx43 is associated with the loss of cardioprotection. More research is needed to clearly identify such changes in Cx43 expression/regulation in cardiovascular diseases and accordingly to develop strategies to prevent such modifications. The maintenance of the Cx43 expression/phosphorylation under pathophysiological conditions – including that of mitochondrial Cx43 – may represent a therapeutic target to overcome the consequences of cardiovascular diseases.

References

1. Sohl G, Willecke K. Gap junctions and the connexin protein family. Cardiovasc Res. 2004;62:228–32.
2. Severs NJ, Bruce AF, Dupont E, Rothery S. Remodelling of gap junctions and connexin expression in diseased myocardium. Cardiovasc Res. 2008;80:9–19.
3. Saez JC, Berthoud VM, Branes MC, Martinez AD, Beyer EC. Plasma membrane channels formed by connexins: their regulation and functions. Physiol Rev. 2003;83:1359–400.
4. Desplantez T, Dupont E, Severs NJ, Weingart R. Gap junction channels and cardiac impulse propagation. J Membr Biol. 2007;218:13–28.
5. Severs NJ, Dupont E, Coppen SR, Halliday D, Inett E, Baylis D, Rothery S. Remodelling of gap junctions and connexin expression in heart disease. Biochim Biophys Acta. 2004; 1662:138–48.
6. Reaume AG, de Sousa PA, Kulkarni S, Langille BL, Zhu D, Davies TC, Juneja SC, Kidder GM, Rossant J. Cardiac malformation in neonatal mice lacking connexin43. Science. 1995;267:1831–4.
7. Beardslee MA, Laing JG, Beyer EC, Saffitz JE. Rapid turnover of connexin43 in the adult rat heart. Circ Res. 1998;83:629–35.
8. Giepmans BN, Moolenaar WH. The gap junction protein connexin43 interacts with the second PDZ domain of the zona occludens-1 protein. Curr Biol: 1998;8:931–4.
9. Rhett JM, Jourdan J, Gourdie RG. Connexin 43 connexon to gap junction transition is regulated by zonula occludens-1. Mol Biol Cell. 2011;22:1516–28.
10. Sharma P, Abbasi C, Lazic S, Teng AC, Wang D, Dubois N, Ignatchenko V, Wong V, Liu J, Araki T, Tiburcy M, Ackerley C, Zimmermann WH, Hamilton R, Sun Y, Liu PP, Keller G, Stagljar I, Scott IC, Kislinger T, Gramolini AO. Evolutionarily conserved intercalated disc protein Tmem65 regulates cardiac conduction and connexin 43 function. Nat Commun. 2015;6:8391.
11. John S, Cesario D, Weiss JN. Gap junctional hemichannels in the heart. Acta Physiol Scand. 2003;179:23–31.
12. Goodenough DA, Paul DL. Beyond the gap: functions of unpaired connexon channels. Nat Rev Mol Cell Biol. 2003;4:285–94.
13. Saez JC, Leybaert L. Hunting for connexin hemichannels. FEBS Lett. 2014;588:1205–11.
14. Wang N, De Vuyst E, Ponsaerts R, Boengler K, Palacios-Prado N, Wauman J, Lai CP, De Bock M, Decrock E, Bol M, Vinken M, Rogiers V, Tavernier J, Evans WH, Naus CC, Bukauskas FF, Sipido KR, Heusch G, Schulz R, Bultynck G, Leybaert L. Selective inhibition of Cx43 hemichannels by Gap19 and its impact on myocardial ischemia/reperfusion injury. Basic Res Cardiol. 2013;108:309.
15. Boengler K, Dodoni G, Rodriguez-Sinovas A, Cabestrero A, Ruiz-Meana M, Gres P, Konietzka I, Lopez-Iglesias C, Garcia-Dorado D, Di Lisa F, Heusch G, Schulz R. Connexin 43 in cardiomyocyte mitochondria and its increase by ischemic preconditioning. Cardiovasc Res. 2005;67:234–44.
16. Solan JL, Lampe PD. Connexin43 phosphorylation: structural changes and biological effects. Biochem J. 2009;419:261–72.
17. Axelsen LN, Calloe K, Holstein-Rathlou NH, Nielsen MS. Managing the complexity of communication: regulation of gap junctions by post-translational modification. Front Pharmacol. 2013;4:130.
18. Lin R, Warn-Cramer BJ, Kurata WE, Lau AF. V-src phosphorylation of connexin 43 on Tyr247 and Tyr265 disrupts gap junctional communication. J Cell Biol. 2001;154:815–27.
19. van Veen TA, van Rijen HV, Jongsma HJ. Physiology of cardiovascular gap junctions. Adv Cardiol. 2006;42:18–40.
20. Solan JL, Lampe PD. Specific Cx43 phosphorylation events regulate gap junction turnover in vivo. FEBS Lett. 2014;588:1423–9.

21. Maass K, Shibayama J, Chase SE, Willecke K, Delmar M. C-terminal truncation of connexin43 changes number, size, and localization of cardiac gap junction plaques. Circ Res. 2007;101:1283–91.
22. Matsumura K, Mayama T, Lin H, Sakamoto Y, Ogawa K, Imanaga I. Effects of cyclic AMP on the function of the cardiac gap junction during hypoxia. Exp Clin Cardiol. 2006;11:286–93.
23. Dunn CA, Su V, Lau AF, Lampe PD. Activation of Akt, not connexin 43 protein ubiquitination, regulates gap junction stability. J Biol Chem. 2012;287:2600–7.
24. Ek-Vitorin JF, King TJ, Heyman NS, Lampe PD, Burt JM. Selectivity of connexin 43 channels is regulated through protein kinase C-dependent phosphorylation. Circ Res. 2006;98:1498–505.
25. Cooper CD, Lampe PD. Casein kinase 1 regulates connexin-43 gap junction assembly. J Biol Chem. 2002;277:44962–8.
26. Huang RY, Laing JG, Kanter EM, Berthoud VM, Bao M, Rohrs HW, Townsend RR, Yamada KA. Identification of CaMKII phosphorylation sites in Connexin43 by high-resolution mass spectrometry. J Proteome Res. 2011;10:1098–109.
27. Warn-Cramer BJ, Cottrell GT, Burt JM, Lau AF. Regulation of connexin-43 gap junctional intercellular communication by mitogen-activated protein kinase. J Biol Chem. 1998;273:9188–96.
28. Lampe PD, Kurata WE, Warn-Cramer BJ, Lau AF. Formation of a distinct connexin43 phosphoisoform in mitotic cells is dependent upon p34cdc2 kinase. J Cell Sci. 1998;111(Pt 6):833–41.
29. Loo LW, Berestecky JM, Kanemitsu MY, Lau AF. pp60src-mediated phosphorylation of connexin 43, a gap junction protein. J Biol Chem. 1995;270:12751–61.
30. van Veen AA, van Rijen HV, Opthof T. Cardiac gap junction channels: modulation of expression and channel properties. Cardiovasc Res. 2001;51:217–29.
31. Lampe PD, Cooper CD, King TJ, Burt JM. Analysis of Connexin43 phosphorylated at S325, S328 and S330 in normoxic and ischemic heart. J Cell Sci. 2006;119:3435–42.
32. Pogoda K, Kameritsch P, Retamal MA, Vega JL. Regulation of gap junction channels and hemichannels by phosphorylation and redox changes: a revision. BMC Cell Biol. 2016;17(Suppl 1):11.
33. Herve JC, Sarrouilhe D. Protein phosphatase modulation of the intercellular junctional communication: importance in cardiac myocytes. Prog Biophys Mol Biol. 2006;90:225–48.
34. Fontes MS, Raaijmakers AJ, van Doorn T, Kok B, Nieuwenhuis S, van der Nagel R, Vos MA, de Boer TP, van Rijen HV, Bierhuizen MF. Changes in Cx43 and NaV1.5 expression precede the occurrence of substantial fibrosis in calcineurin-induced murine cardiac hypertrophy. PLoS One. 2014;9:e87226.
35. Soetkamp D, Nguyen TT, Menazza S, Hirschhauser C, Hendgen-Cotta UB, Rassaf T, Schluter KD, Boengler K, Murphy E, Schulz R. S-nitrosation of mitochondrial connexin 43 regulates mitochondrial function. Basic Res Cardiol. 2014;109:433.
36. Leo-Macias A, Agullo-Pascual E, Delmar M. The cardiac connexome: non-canonical functions of connexin43 and their role in cardiac arrhythmias. Semin Cell Dev Biol. 2016;50:13–21.
37. Dang X, Doble BW, Kardami E. The carboxy-tail of connexin-43 localizes to the nucleus and inhibits cell growth. Mol Cell Biochem. 2003;242:35–8.
38. Zhao X, Tang X, Ma T, Ding M, Bian L, Chen D, Li Y, Wang L, Zhuang Y, Xie M, Yang D. Levonorgestrel inhibits human endometrial cell proliferation through the upregulation of gap junctional intercellular communication via the nuclear translocation of Ser255 phosphorylated Cx43. Biomed Res Int. 2015;2015:758684.
39. Soares AR, Martins-Marques T, Ribeiro-Rodrigues T, Ferreira JV, Catarino S, Pinho MJ, Zuzarte M, Isabel Anjo S, Manadas B, P G Sluijter J, Pereira P, Girao H. Gap junctional protein Cx43 is involved in the communication between extracellular vesicles and mammalian cells. Sci Rep. 2015;5:13243.

40. Li H, Brodsky S, Kumari S, Valiunas V, Brink P, Kaide J, Nasjletti A, Goligorsky MS. Paradoxical overexpression and translocation of connexin43 in homocysteine-treated endothelial cells. Am J Physiol Heart Circ Physiol. 2002;282:H2124–33.
41. Kiec-Wilk B, Czech U, Janczarska K, Knapp A, Goralska J, Cialowicz U, Malecki MT, Dembinska-Kiec A. Connexin 43 and metabolic effect of fatty acids in stressed endothelial cells. Genes Nutr. 2012;7:257–63.
42. Trudeau K, Muto T, Roy S. Downregulation of mitochondrial connexin 43 by high glucose triggers mitochondrial shape change and cytochrome C release in retinal endothelial cells. Invest Ophthalmol Vis Sci. 2012;53:6675–81.
43. Vinken M, Maes M, Cavill R, Valkenborg D, Ellis JK, Decrock E, Leybaert L, Staes A, Gevaert K, Oliveira AG, Menezes GB, Cogliati B, Dagli ML, Ebbels TM, Witters E, Keun HC, Vanhaecke T, Rogiers V. Proteomic and metabolomic responses to connexin43 silencing in primary hepatocyte cultures. Arch Toxicol. 2013;87:883–94.
44. Waza AA, Andrabi K, Hussain MU. Protein kinase C (PKC) mediated interaction between conexin43 (Cx43) and K(+)(ATP) channel subunit (Kir6.1) in cardiomyocyte mitochondria: implications in cytoprotection against hypoxia induced cell apoptosis. Cell Signal. 2014;26:1909–17.
45. Miro-Casas E, Ruiz-Meana M, Agullo E, Stahlhofen S, Rodriguez-Sinovas A, Cabestrero A, Jorge I, Torre I, Vazquez J, Boengler K, Schulz R, Heusch G, Garcia-Dorado D. Connexin43 in cardiomyocyte mitochondria contributes to mitochondrial potassium uptake. Cardiovasc Res. 2009;83:747–56.
46. Palmer JW, Tandler B, Hoppel CL. Biochemical properties of subsarcolemmal and interfibrillar mitochondria isolated from rat cardiac muscle. J Biol Chem. 1977;252:8731–9.
47. Palmer JW, Tandler B, Hoppel CL. Heterogeneous response of subsarcolemmal heart mitochondria to calcium. Am J Phys. 1986;250:H741–8.
48. Boengler K, Stahlhofen S, van de Sand A, Gres P, Ruiz-Meana M, Garcia-Dorado D, Heusch G, Schulz R. Presence of connexin 43 in subsarcolemmal, but not in interfibrillar cardiomyocyte mitochondria. Basic Res Cardiol. 2009;104:141–7.
49. Sun J, Nguyen T, Aponte AM, Menazza S, Kohr MJ, Roth DM, Patel HH, Murphy E, Steenbergen C. Ischaemic preconditioning preferentially increases protein S-nitrosylation in subsarcolemmal mitochondria. Cardiovasc Res. 2015;106:227–36.
50. Rodriguez-Sinovas A, Boengler K, Cabestrero A, Gres P, Morente M, Ruiz-Meana M, Konietzka I, Miro E, Totzeck A, Heusch G, Schulz R, Garcia-Dorado D. Translocation of connexin 43 to the inner mitochondrial membrane of cardiomyocytes through the heat shock protein 90-dependent TOM pathway and its importance for cardioprotection. Circ Res. 2006;99:93–101.
51. Goubaeva F, Mikami M, Giardina S, Ding B, Abe J, Yang J. Cardiac mitochondrial connexin 43 regulates apoptosis. Biochem Biophys Res Commun. 2007;352:97–103.
52. Denuc A, Nunez E, Calvo E, Loureiro M, Miro-Casas E, Guaras A, Vazquez J, Garcia-Dorado D. New protein-protein interactions of mitochondrial connexin 43 in mouse heart. J Cell Mol Med. 2016;20:794–803.
53. Tyagi N, Vacek JC, Givvimani S, Sen U, Tyagi SC. Cardiac specific deletion of N-methyl-d-aspartate receptor 1 ameliorates mtMMP-9 mediated autophagy/mitophagy in hyperhomocysteinemia. J Recept Signal Transduct Res. 2010;30:78–87.
54. Boengler K, Ungefug E, Heusch G, Leybaert L, Schulz R. Connexin 43 impacts on mitochondrial potassium uptake. Front Pharmacol. 2013;4:73.
55. Boengler K, Ruiz-Meana M, Gent S, Ungefug E, Soetkamp D, Miro-Casas E, Cabestrero A, Fernandez-Sanz C, Semenzato M, Lisa FD, Rohrbach S, Garcia-Dorado D, Heusch G, Schulz R, Mercola M. Mitochondrial connexin 43 impacts on respiratory complex I activity and mitochondrial oxygen consumption. J Cell Mol Med. 2012;16:1649–55.
56. Giorgio V, von Stockum S, Antoniel M, Fabbro A, Fogolari F, Forte M, Glick GD, Petronilli V, Zoratti M, Szabo I, Lippe G, Bernardi P. Dimers of mitochondrial ATP synthase form the permeability transition pore. Proc Natl Acad Sci U S A. 2013;110:5887–92.

57. Srisakuldee W, Makazan Z, Nickel BE, Zhang F, Thliveris JA, Pasumarthi KB, Kardami E. The FGF-2-triggered protection of cardiac subsarcolemmal mitochondria from calcium overload is mitochondrial connexin 43-dependent. Cardiovasc Res. 2014;103:72–80.
58. Matsuyama D, Kawahara K. Oxidative stress-induced formation of a positive-feedback loop for the sustained activation of p38 MAPK leading to the loss of cell division in cardiomyocytes soon after birth. Basic Res Cardiol. 2011;106:815–28.
59. Boengler K, Heusch G, Schulz R. Nuclear-encoded mitochondrial proteins and their role in cardioprotection. Biochim Biophys Acta. 1813;2011:1286–94.
60. Totzeck A, Boengler K, van de Sand A, Konietzka I, Gres P, Garcia-Dorado D, Heusch G, Schulz R. No impact of protein phosphatases on connexin 43 phosphorylation in ischemic preconditioning. Am J Physiol Heart Circ Physiol. 2008;295:H2106–12.
61. Shan H, Wei J, Zhang M, Lin L, Yan R, Zhang R, Zhu YH. Suppression of PKCepsilon-mediated mitochondrial connexin 43 phosphorylation at serine 368 is involved in myocardial mitochondrial dysfunction in a rat model of dilated cardiomyopathy. Mol Med Rep. 2015;11:4720–6.
62. Chen M, Jones DL. Age- and myopathy-dependent changes in connexins of normal and cardiomyopathic Syrian hamster ventricular myocardium. Can J Physiol Pharmacol. 2000;78: 669–78.
63. Jones SA, Lancaster MK, Boyett MR. Ageing-related changes of connexins and conduction within the sinoatrial node. J Physiol. 2004;560:429–37.
64. Jones SA, Lancaster MK. Progressive age-associated activation of JNK associates with conduction disruption in the aged atrium. Mech Ageing Dev. 2015;146–148:72–80.
65. Bonda TA, Szynaka B, Sokolowska M, Dziemidowicz M, Winnicka MM, Chyczewski L, Kaminski KA. Remodeling of the intercalated disc related to aging in the mouse heart. J Cardiol. 2016;68:261–8.
66. Lancaster TS, Jefferson SJ, Korzick DH. Local delivery of a PKCepsilon-activating peptide limits ischemia reperfusion injury in the aged female rat heart. Am J Phys Regul Integr Comp Phys. 2011;301:R1242–9.
67. Watanabe M, Ichinose S, Sunamori M. Age-related changes in gap junctional protein of the rat heart. Exp Clin Cardiol. 2004;9:130–2.
68. Yan J, Kong W, Zhang Q, Beyer EC, Walcott G, Fast VG, Ai X. c-Jun N-terminal kinase activation contributes to reduced connexin43 and development of atrial arrhythmias. Cardiovasc Res. 2013;97:589–97.
69. Boengler K, Konietzka I, Buechert A, Heinen Y, Garcia-Dorado D, Heusch G, Schulz R. Loss of ischemic preconditioning's cardioprotection in aged mouse hearts is associated with reduced gap junctional and mitochondrial levels of connexin 43. Am J Physiol Heart Circ Physiol. 2007;292:H1764–9.
70. Fannin J, Rice KM, Thulluri S, Dornon L, Arvapalli RK, Wehner P, Blough ER. Age-associated alterations of cardiac structure and function in the female F344xBN rat heart. Age (Dordr). 2014;36:9684.
71. Dhein S, Hammerath SB. Aspects of the intercellular communication in aged hearts: effects of the gap junction uncoupler palmitoleic acid. Naunyn Schmiedeberg's Arch Pharmacol. 2001;364:397–408.
72. Kostin S, Klein G, Szalay Z, Hein S, Bauer EP, Schaper J. Structural correlate of atrial fibrillation in human patients. Cardiovasc Res. 2002;54:361–79.
73. Bacova B, Radosinska J, Viczenczova C, Knezl V, Dosenko V, Benova T, Navarova J, Goncalvesova E, van Rooyen J, Weismann P, Slezak J, Tribulova N. Up-regulation of myocardial connexin-43 in spontaneously hypertensive rats fed red palm oil is most likely implicated in its anti-arrhythmic effects. Can J Physiol Pharmacol. 2012;90:1235–45.
74. Benova T, Viczenczova C, Radosinska J, Bacova B, Knezl V, Dosenko V, Weismann P, Zeman M, Navarova J, Tribulova N. Melatonin attenuates hypertension-related proarrhythmic myocardial maladaptation of connexin-43 and propensity of the heart to lethal arrhythmias. Can J Physiol Pharmacol. 2013;91:633–9.

75. Zhang W, Zhao G, Hu X, Wang M, Li H, Ye Y, Du Q, Yao J, Bao Z, Hong W, Fu G, Ge J, Qiu Z. Aliskiren-attenuated myocardium apoptosis via regulation of autophagy and connexin-43 in aged spontaneously hypertensive rats. J Cell Mol Med. 2014;18:1247–56.
76. Chen HJ, Yao L, Chen TG, Yu M, Wang LH, Chen JZ. Atorvastatin prevents connexin43 remodeling in hypertrophied left ventricular myocardium of spontaneously hypertensive rats. Chin Med J. 2007;120:1902–7.
77. Radosinska J, Bacova B, Knezl V, Benova T, Zurmanova J, Soukup T, Arnostova P, Slezak J, Goncalvesova E, Tribulova N. Dietary omega-3 fatty acids attenuate myocardial arrhythmogenic factors and propensity of the heart to lethal arrhythmias in a rodent model of human essential hypertension. J Hypertens. 2013;31:1876–85.
78. Zhao LL, Chen HJ, Chen JZ, Yu M, Ni YL, Zhang WF. Losartan reduced connexin43 expression in left ventricular myocardium of spontaneously hypertensive rats. J Zhejiang Univ Sci B. 2008;9:448–54.
79. Emdad L, Uzzaman M, Takagishi Y, Honjo H, Uchida T, Severs NJ, Kodama I, Murata Y. Gap junction remodeling in hypertrophied left ventricles of aortic-banded rats: prevention by angiotensin II type 1 receptor blockade. J Mol Cell Cardiol. 2001;33:219–31.
80. Alesutan I, Voelkl J, Stockigt F, Mia S, Feger M, Primessnig U, Sopjani M, Munoz C, Borst O, Gawaz M, Pieske B, Metzler B, Heinzel F, Schrickel JW, Lang F. AMP-activated protein kinase alpha1 regulates cardiac gap junction protein connexin 43 and electrical remodeling following pressure overload. Cell Physiol Biochem: Int J Exp Cell Physiol Biochem Pharmacol. 2015;35:406–18.
81. Formigli L, Ibba-Manneschi L, Perna AM, Pacini A, Polidori L, Nediani C, Modesti PA, Nosi D, Tani A, Celli A, Neri-Serneri GG, Quercioli F, Zecchi-Orlandini S. Altered Cx43 expression during myocardial adaptation to acute and chronic volume overloading. Histol Histopathol. 2003;18:359–69.
82. Kostin S, Dammer S, Hein S, Klovekorn WP, Bauer EP, Schaper J. Connexin 43 expression and distribution in compensated and decompensated cardiac hypertrophy in patients with aortic stenosis. Cardiovasc Res. 2004;62:426–36.
83. Yasuno S, Kuwahara K, Kinoshita H, Yamada C, Nakagawa Y, Usami S, Kuwabara Y, Ueshima K, Harada M, Nishikimi T, Nakao K. Angiotensin II type 1a receptor signalling directly contributes to the increased arrhythmogenicity in cardiac hypertrophy. Br J Pharmacol. 2013;170:1384–95.
84. Qu J, Volpicelli FM, Garcia LI, Sandeep N, Zhang J, Marquez-Rosado L, Lampe PD, Fishman GI. Gap junction remodeling and spironolactone-dependent reverse remodeling in the hypertrophied heart. Circ Res. 2009;104:365–71.
85. Axelsen LN, Calloe K, Braunstein TH, Riemann M, Hofgaard JP, Liang B, Jensen CF, Olsen KB, Bartels ED, Baandrup U, Jespersen T, Nielsen LB, Holstein-Rathlou NH, Nielsen MS. Diet-induced pre-diabetes slows cardiac conductance and promotes arrhythmogenesis. Cardiovasc Diabetol. 2015;14:87.
86. Lin H, Ogawa K, Imanaga I, Tribulova N. Remodeling of connexin 43 in the diabetic rat heart. Mol Cell Biochem. 2006;290:69–78.
87. Lin H, Ogawa K, Imanaga I, Tribulova N. Alterations of connexin 43 in the diabetic rat heart. Adv Cardiol. 2006;42:243–54.
88. Veeranki S, Givvimani S, Kundu S, Metreveli N, Pushpakumar S, Tyagi SC. Moderate intensity exercise prevents diabetic cardiomyopathy associated contractile dysfunction through restoration of mitochondrial function and connexin 43 levels in db/db mice. J Mol Cell Cardiol. 2016;92:163–73.
89. Nygren A, Olson ML, Chen KY, Emmett T, Kargacin G, Shimoni Y. Propagation of the cardiac impulse in the diabetic rat heart: reduced conduction reserve. J Physiol. 2007;580:543–60.
90. Stables CL, Musa H, Mitra A, Bhushal S, Deo M, Guerrero-Serna G, Mironov S, Zarzoso M, Vikstrom KL, Cawthorn W, Pandit SV. Reduced Na(+) current density underlies impaired propagation in the diabetic rabbit ventricle. J Mol Cell Cardiol. 2014;69:24–31.

91. Joshi MS, Mihm MJ, Cook AC, Schanbacher BL, Bauer JA. Alterations in connexin 43 during diabetic cardiomyopathy: competition of tyrosine nitration versus phosphorylation. J Diabetes. 2015;7:250–9.
92. Radosinska J, Kurahara LH, Hiraishi K, Viczenczova C, Egan Benova T, Szeiffova Bacova B, Dosenko V, Navarova J, Obsitnik B, Imanaga I, Soukup T, Tribulova N. Modulation of cardiac connexin-43 by omega-3 fatty acid ethyl-ester supplementation demonstrated in spontaneously diabetic rats. Physiol Res/Acad Sci Bohemoslovaca. 2015;64:795–806.
93. Anna Z, Angela S, Barbara B, Jana R, Tamara B, Csilla V, Victor D, Oleksiy M, Narcisa T. Heart-protective effect of n-3 PUFA demonstrated in a rat model of diabetic cardiomyopathy. Mol Cell Biochem. 2014;389:219–27.
94. Howarth FC, Chandler NJ, Kharche S, Tellez JO, Greener ID, Yamanushi TT, Billeter R, Boyett MR, Zhang H, Dobrzynski H. Effects of streptozotocin-induced diabetes on connexin43 mRNA and protein expression in ventricular muscle. Mol Cell Biochem. 2008;319:105–14.
95. Palatinus JA, Gourdie RG. Diabetes increases cryoinjury size with associated effects on Cx43 gap junction function and phosphorylation in the mouse heart. J Diabetes Res. 2016;2016:8789617.
96. Lee KT, Hsieh CC, Tsai WC, Tang PW, Liu IH, Chai CY, Sheu SH, Lai WT. Characteristics of atrial substrates for atrial tachyarrhythmias induced in aged and hypercholesterolemic rabbits. Pacing Clin Electrophysiol: PACE. 2012;35:544–52.
97. Lin LC, Wu CC, Yeh HI, Lu LS, Liu YB, Lin SF, Lee YT. Downregulated myocardial connexin 43 and suppressed contractility in rabbits subjected to a cholesterol-enriched diet. Lab Invest: J Tech Methods Pathol. 2005;85:1224–37.
98. Gorbe A, Varga ZV, Kupai K, Bencsik P, Kocsis GF, Csont T, Boengler K, Schulz R, Ferdinandy P. Cholesterol diet leads to attenuation of ischemic preconditioning-induced cardiac protection: the role of connexin 43. Am J Physiol Heart Circ Physiol. 2011;300: H1907–13.
99. Jackson PE, Feng QP, Jones DL. Nitric oxide depresses connexin 43 after myocardial infarction in mice. Acta Physiol (Oxford). 2008;194:23–33.
100. Lindsey ML, Escobar GP, Mukherjee R, Goshorn DK, Sheats NJ, Bruce JA, Mains IM, Hendrick JK, Hewett KW, Gourdie RG, Matrisian LM, Spinale FG. Matrix metalloproteinase-7 affects connexin-43 levels, electrical conduction, and survival after myocardial infarction. Circulation. 2006;113:2919–28.
101. Matsushita T, Oyamada M, Fujimoto K, Yasuda Y, Masuda S, Wada Y, Oka T, Takamatsu T. Remodeling of cell-cell and cell-extracellular matrix interactions at the border zone of rat myocardial infarcts. Circ Res. 1999;85:1046–55.
102. Savi M, Bocchi L, Rossi S, Frati C, Graiani G, Lagrasta C, Miragoli M, Di Pasquale E, Stirparo GG, Mastrototaro G, Urbanek K, De Angelis A, Macchi E, Stilli D, Quaini F, Musso E. Antiarrhythmic effect of growth factor-supplemented cardiac progenitor cells in chronic infarcted heart. Am J Physiol Heart Circ Physiol. 2016;310:H1622–48.
103. Peters NS, Coromilas J, Severs NJ, Wit AL. Disturbed connexin43 gap junction distribution correlates with the location of reentrant circuits in the epicardial border zone of healing canine infarcts that cause ventricular tachycardia. Circulation. 1997;95:988–96.
104. Baum JR, Dolmatova E, Tan A, Duffy HS. Omega 3 fatty acid inhibition of inflammatory cytokine-mediated Connexin43 regulation in the heart. Front Physiol. 2012;3:272.
105. Hou J, Yan P, Guo T, Xing Y, Zheng S, Zhou C, Huang H, Long H, Zhong T, Wu Q, Wang J, Wang T. Cardiac stem cells transplantation enhances the expression of connexin 43 via the ANG II/AT1R/TGF-beta1 signaling pathway in a rat model of myocardial infarction. Exp Mol Pathol. 2015;99:693–701.
106. Ai X, Pogwizd SM. Connexin 43 downregulation and dephosphorylation in nonischemic heart failure is associated with enhanced colocalized protein phosphatase type 2A. Circ Res. 2005;96:54–63.

107. Ai X, Zhao W, Pogwizd SM. Connexin43 knockdown or overexpression modulates cell coupling in control and failing rabbit left ventricular myocytes. Cardiovasc Res. 2010;85:751–62.
108. Danielson LS, Park DS, Rotllan N, Chamorro-Jorganes A, Guijarro MV, Fernandez-Hernando C, Fishman GI, Phoon CK, Hernando E. Cardiovascular dysregulation of miR-17-92 causes a lethal hypertrophic cardiomyopathy and arrhythmogenesis. FASEB J: Off Publ Fed Am Soc Exp Biol. 2013;27:1460–7.
109. Wang XH, Zhuo XZ, Ni YJ, Gong M, Wang TZ, Lu Q, Ma AQ. Improvement of cardiac function and reversal of gap junction remodeling by Neuregulin-1beta in volume-overloaded rats with heart failure. J Geriatr Cardiol: JGC. 2012;9:172–9.
110. dos Santos DO, Blefari V, Prado FP, Silva CA, Fazan Jr R, Salgado HC, Ramos SG, Prado CM. Reduced expression of adherens and gap junction proteins can have a fundamental role in the development of heart failure following cardiac hypertrophy in rats. Exp Mol Pathol. 2016;100:167–76.
111. Akar FG, Spragg DD, Tunin RS, Kass DA, Tomaselli GF. Mechanisms underlying conduction slowing and arrhythmogenesis in nonischemic dilated cardiomyopathy. Circ Res. 2004;95:717–25.
112. Dupont E, Matsushita T, Kaba RA, Vozzi C, Coppen SR, Khan N, Kaprielian R, Yacoub MH, Severs NJ. Altered connexin expression in human congestive heart failure. J Mol Cell Cardiol. 2001;33:359–71.
113. Kostin S, Rieger M, Dammer S, Hein S, Richter M, Klovekorn WP, Bauer EP, Schaper J. Gap junction remodeling and altered connexin43 expression in the failing human heart. Mol Cell Biochem. 2003;242:135–44.
114. Sato T, Ohkusa T, Honjo H, Suzuki S, Yoshida MA, Ishiguro YS, Nakagawa H, Yamazaki M, Yano M, Kodama I, Matsuzaki M. Altered expression of connexin43 contributes to the arrhythmogenic substrate during the development of heart failure in cardiomyopathic hamster. Am J Physiol Heart Circ Physiol. 2008;294:H1164–73.
115. Givvimani S, Pushpakumar S, Veeranki S, Tyagi SC. Dysregulation of Mfn2 and Drp-1 proteins in heart failure. Can J Physiol Pharmacol. 2014;92:583–91.
116. Chen G, Zhao J, Liu C, Zhang Y, Huo Y, Zhou L. MG132 proteasome inhibitor upregulates the expression of connexin 43 in rats with adriamycin-induced heart failure. Mol Med Rep. 2015;12:7595–602.
117. Pecoraro M, Sorrentino R, Franceschelli S, Del Pizzo M, Pinto A, Popolo A. Doxorubicin-mediated cardiotoxicity: role of mitochondrial connexin 43. Cardiovasc Toxicol. 2015;15:366–76.
118. Kondo RP, Wang SY, John SA, Weiss JN, Goldhaber JI. Metabolic inhibition activates a non-selective current through connexin hemichannels in isolated ventricular myocytes. J Mol Cell Cardiol. 2000;32:1859–72.
119. Salas D, Puebla C, Lampe PD, Lavandero S, Saez JC. Role of Akt and Ca2+ on cell permeabilization via connexin43 hemichannels induced by metabolic inhibition. Biochim Biophys Acta. 2015;1852:1268–77.
120. Hawat G, Benderdour M, Rousseau G, Baroudi G. Connexin 43 mimetic peptide Gap26 confers protection to intact heart against myocardial ischemia injury. Pflugers Arch – Eur J Physiol. 2010;460:583–92.
121. Hawat G, Helie P, Baroudi G. Single intravenous low-dose injections of connexin 43 mimetic peptides protect ischemic heart in vivo against myocardial infarction. J Mol Cell Cardiol. 2012;53:559–66.
122. Rodriguez-Sinovas A, Sanchez JA, Gonzalez-Loyola A, Barba I, Morente M, Aguilar R, Agullo E, Miro-Casas E, Esquerda N, Ruiz-Meana M, Garcia-Dorado D. Effects of substitution of Cx43 by Cx32 on myocardial energy metabolism, tolerance to ischaemia and preconditioning protection. J Physiol. 2010;588:1139–51.
123. Kanno S, Kovacs A, Yamada KA, Saffitz JE. Connexin43 as a determinant of myocardial infarct size following coronary occlusion in mice. J Am Coll Cardiol. 2003;41:681–6.

124. Schwanke U, Konietzka I, Duschin A, Li X, Schulz R, Heusch G. No ischemic preconditioning in heterozygous connexin43-deficient mice. Am J Physiol Heart Circ Physiol. 2002;283:H1740–2.
125. Heinzel FR, Luo Y, Li X, Boengler K, Buechert A, Garcia-Dorado D, Di Lisa F, Schulz R, Heusch G. Impairment of diazoxide-induced formation of reactive oxygen species and loss of cardioprotection in connexin 43 deficient mice. Circ Res. 2005;97:583–6.
126. Sanchez JA, Rodriguez-Sinovas A, Barba I, Miro-Casas E, Fernandez-Sanz C, Ruiz-Meana M, Alburquerque-Bejar JJ, Garcia-Dorado D. Activation of RISK and SAFE pathways is not involved in the effects of Cx43 deficiency on tolerance to ischemia-reperfusion injury and preconditioning protection. Basic Res Cardiol. 2013;108:351.
127. Turner MS, Haywood GA, Andreka P, You L, Martin PE, Evans WH, Webster KA, Bishopric NH. Reversible connexin 43 dephosphorylation during hypoxia and reoxygenation is linked to cellular ATP levels. Circ Res. 2004;95:726–33.
128. Solan JL, Marquez-Rosado L, Sorgen PL, Thornton PJ, Gafken PR, Lampe PD. Phosphorylation at S365 is a gatekeeper event that changes the structure of Cx43 and prevents down-regulation by PKC. J Cell Biol. 2007;179:1301–9.
129. Dunn CA, Lampe PD. Injury-triggered Akt phosphorylation of Cx43: a ZO-1-driven molecular switch that regulates gap junction size. J Cell Sci. 2014;127:455–64.
130. Smyth JW, Zhang SS, Sanchez JM, Lamouille S, Vogan JM, Hesketh GG, Hong T, Tomaselli GF, Shaw RM. A 14-3-3 mode-1 binding motif initiates gap junction internalization during acute cardiac ischemia. Traffic (Copenhagen, Denmark). 2014;15:684–99.
131. Beardslee MA, Lerner DL, Tadros PN, Laing JG, Beyer EC, Yamada KA, Kleber AG, Schuessler RB, Saffitz JE. Dephosphorylation and intracellular redistribution of ventricular connexin43 during electrical uncoupling induced by ischemia. Circ Res. 2000;87:656–62.
132. Martins-Marques T, Catarino S, Marques C, Matafome P, Ribeiro-Rodrigues T, Baptista R, Pereira P, Girao H. Heart ischemia results in connexin43 ubiquitination localized at the intercalated discs. Biochimie. 2015;112:196–201.
133. Martins-Marques T, Catarino S, Zuzarte M, Marques C, Matafome P, Pereira P, Girao H. Ischaemia-induced autophagy leads to degradation of gap junction protein connexin43 in cardiomyocytes. Biochem J. 2015;467:231–45.
134. Laing JG, Tadros PN, Westphale EM, Beyer EC. Degradation of connexin43 gap junctions involves both the proteasome and the lysosome. Exp Cell Res. 1997;236:482–92.
135. Lee TM, Lin MS, Chou TF, Tsai CH, Chang NC. Adjunctive 17beta-estradiol administration reduces infarct size by altered expression of canine myocardial connexin43 protein. Cardiovasc Res. 2004;63:109–17.
136. Surinkaew S, Kumphune S, Chattipakorn S, Chattipakorn N. Inhibition of p38 MAPK during ischemia, but not reperfusion, effectively attenuates fatal arrhythmia in ischemia/reperfusion heart. J Cardiovasc Pharmacol. 2013;61:133–41.
137. Morel S, Christoffersen C, Axelsen LN, Montecucco F, Rochemont V, Frias MA, Mach F, James RW, Naus CC, Chanson M, Lampe PD, Nielsen MS, Nielsen LB, Kwak BR. Sphingosine-1-phosphate reduces ischaemia-reperfusion injury by phosphorylating the gap junction protein Connexin43. Cardiovasc Res. 2016;109:385–96.
138. Srisakuldee W, Jeyaraman MM, Nickel BE, Tanguy S, Jiang ZS, Kardami E. Phosphorylation of connexin-43 at serine 262 promotes a cardiac injury-resistant state. Cardiovasc Res. 2009;83:672–81.
139. Ferdinandy P, Hausenloy DJ, Heusch G, Baxter GF, Schulz R. Interaction of risk factors, comorbidities, and comedications with ischemia/reperfusion injury and cardioprotection by preconditioning, postconditioning, and remote conditioning. Pharmacol Rev. 2014;66:1142–74.
140. Li X, Heinzel FR, Boengler K, Schulz R, Heusch G. Role of connexin 43 in ischemic preconditioning does not involve intercellular communication through gap junctions. J Mol Cell Cardiol. 2004;36:161–3.

141. Li G, Whittaker P, Yao M, Kloner RA, Przyklenk K. The gap junction uncoupler heptanol abrogates infarct size reduction with preconditioning in mouse hearts. Cardiovasc Pathol: Off J Soc Cardiovasc Pathol. 2002;11:158–65.
142. Jain SK, Schuessler RB, Saffitz JE. Mechanisms of delayed electrical uncoupling induced by ischemic preconditioning. Circ Res. 2003;92:1138–44.
143. Hatanaka K, Kawata H, Toyofuku T, Yoshida K. Down-regulation of connexin43 in early myocardial ischemia and protective effect by ischemic preconditioning in rat hearts in vivo. Jpn Heart J. 2004;45:1007–19.
144. Schulz R, Gres P, Skyschally A, Duschin A, Belosjorow S, Konietzka I, Heusch G. Ischemic preconditioning preserves connexin 43 phosphorylation during sustained ischemia in pig hearts in vivo. FASEB J. 2003;17:1355–7.
145. Muhlfeld C, Cetegen C, Freese S, Volkmann R, Hellige G, Vetterlein F. Phosphorylation of extrajunctional Cx43 in ischemic-preconditioned rat hearts. J Surg Res. 2010;162:e1–8.
146. Hund TJ, Lerner DL, Yamada KA, Schuessler RB, Saffitz JE. Protein kinase Cepsilon mediates salutary effects on electrical coupling induced by ischemic preconditioning. Heart Rhythm: Off J Heart Rhythm Soc. 2007;4:1183–93.
147. Lu G, Haider H, Porollo A, Ashraf M. Mitochondria-specific transgenic overexpression of connexin-43 simulates preconditioning-induced cytoprotection of stem cells. Cardiovasc Res. 2010;88:277–86.
148. Yang F, Chen WL, Zheng MZ, Yu GW, Xu HJ, Shen YL, Chen YY. Heat shock protein 90 mediates anti-apoptotic effect of diazoxide by preventing the cleavage of bid in hypothermic preservation rat hearts. J Heart Lung Transplant: Off Pub Int Soc Heart Transplant. 2011;30: 928–34.
149. Budas GR, Churchill EN, Disatnik MH, Sun L, Mochly-Rosen D. Mitochondrial import of PKCepsilon is mediated by HSP90: a role in cardioprotection from ischaemia and reperfusion injury. Cardiovasc Res. 2010;88:83–92.
150. Sun J, Murphy E. Protein S-nitrosylation and cardioprotection. Circ Res. 2010;106:285–96.
151. Ruiz-Meana M, Nunez E, Miro-Casas E, Martinez-Acedo P, Barba I, Rodriguez-Sinovas A, Inserte J, Fernandez-Sanz C, Hernando V, Vazquez J, Garcia-Dorado D. Ischemic preconditioning protects cardiomyocyte mitochondria through mechanisms independent of cytosol. J Mol Cell Cardiol. 2014;68:79–88.
152. Ong SB, Dongworth RK, Cabrera-Fuentes HA, Hausenloy DJ. Role of the MPTP in conditioning the heart – translatability and mechanism. Br J Pharmacol. 2015;172:2074–84.
153. Brandenburger T, Huhn R, Galas A, Pannen BH, Keitel V, Barthel F, Bauer I, Heinen A. Remote ischemic preconditioning preserves connexin 43 phosphorylation in the rat heart in vivo. J Transl Med. 2014;12:228.
154. He H, Li N, Zhao Z, Han F, Wang X, Zeng Y. Ischemic postconditioning improves the expression of cellular membrane connexin 43 and attenuates the reperfusion injury in rat acute myocardial infarction. Biomed Rep. 2015;3:668–74.
155. Wu Y, Gu EW, Zhu Y, Zhang L, Liu XQ, Fang WP. Sufentanil limits the myocardial infarct size by preservation of the phosphorylated connexin 43. Int Immunopharmacol. 2012;13:341–6.
156. Penna C, Perrelli MG, Raimondo S, Tullio F, Merlino A, Moro F, Geuna S, Mancardi D, Pagliaro P. Postconditioning induces an anti-apoptotic effect and preserves mitochondrial integrity in isolated rat hearts. Biochim Biophys Acta. 2009;1787:794–801.
157. Heusch G, Buchert A, Feldhaus S, Schulz R. No loss of cardioprotection by postconditioning in connexin 43-deficient mice. Basic Res Cardiol. 2006;101:354–6.

Chapter 13
Mitochondrial Mechanosensor Microdomains in Cardiovascular Disorders

Michele Miragoli and Aderville Cabassi

The heart is an exquisite electromechanical organ, which synchronizes every contraction with the preceding sinus bioelectrical stimulus. However, the mechanism whereby bioelectrical propagation activates mechanical contraction is not univocal as the parenchymal cardiac cells can generate bioelectricity after mechanical stretch or insult [1]. Cardiomyocytes possess a sort of mechanosensitive related 'cardiac memory', which may adjust their mechanical performance for the subsequent heartbeat (e.g. Frank-Starling phenomenon) [2–4]. It is a common assumption that altered mechanosensitivity initiates electrical instability and arrhythmia in heart failure [5]. While pro-arrhythmic mechanoelectric transduction has been extensively investigated in intact hearts *in-situ*, isolated hearts and in isolated cellular preparations, the initial subcellular mechanisms required for (mechanoelectric) signal transduction and its initiation are a very recent matter of investigation [6–8]. Attention has focused upon different sarcomeric components [9] and, in addition to force generation, several sarcomeric proteins were found to provide mechanosensing and/or signaling functions [10, 11]. Mutations in these sarcomeric or Z-disk complex proteins cause abnormal intracellular Ca^{2+} responses [11]. Cellular remodeling can be both 'structural', thus involving the cell membrane, intracellular organelles, loss of T-tubules [12], the cytoskeleton [13], intercalated disks [14], cellular swelling [15], ryanodine receptors (RyRs) and the sarcoplasmic reticulum (SR) [16] and 'electrical', thus involving ion channel and ion pump redistribution and re-expression [17, 18], adrenergic receptors (de)compartmentation [19],

M. Miragoli (✉)
Department of Medicine and Surgery, University of Parma, Parma 43124, Italy

Humanitas Clinical and Research Center, Rozzano, MI, Italy
e-mail: michele.miragoli@unipr.it

A. Cabassi
Department of Medicine and Surgery, University of Parma, Parma 43124, Italy

G. Santulli (ed.), *Mitochondrial Dynamics in Cardiovascular Medicine*,
Advances in Experimental Medicine and Biology 982,
DOI 10.1007/978-3-319-55330-6_13

altered sequence of electrical activation [20] and complex signalling pathways changes [21].

Mitochondria are individual organelles localized in clusters beneath the sarcolemma (subsarcolemmal mitochondria) and in longitudinal rows within the contractile apparatus (interfibrillar mitochondria) [22]. Cardiac mitochondria are recognized as primary acting subcellular organelles involved in the metabolic changes occurring in ischemic and failing heart, also directly or indirectly participating in the arrhythmogenic mechanisms or altered contractility mainly associated to energy deficiency [22, 23]. Cardiomyocyte in fact houses the highest mitochondrial density (27–35% of cellular volume versus 3–8% in skeletal and smooth muscle cells) of any organ, and the mitochondria exhibit the greatest number of cristae, which enhance oxidative phosphorylation. Mitochondrial dysfunction is of particular importance because of the specific task of these organelles as suppliers of constantly energy demanding cardiomyocytes [24].

Besides, mitochondria change their subcellular location [25, 26] and the interfibrillar mitochondria alignment is altered early following myocardial infarction [27]. Regular alignment of subsarcolemmal mitochondria within the dyad plays a pivotal role in the homeostasis of excitation-contraction coupling [7, 28] together with intracellular calcium (Ca^{2+}) handling [29].

Remodeling includes also the mitochondrial function [30], and metabolism as observed in aging and failing heart but in particular during the development of various cardiovascular diseases when compared to cardiac energetics of a healthy person. In fact, cardiac damages, caused by ischemia, infarction, hypertrophy/dilation due to pressure and volume overloads are characterized by cardiac structural and metabolic remodeling all progressing to the heart failure condition [31, 32].

From the above the following question emerges: does the alteration of microdomain-related mitochondrial alignment play a role in arrhythmogenesis? The answer for this question is thought-provoking as many conventional technologies cannot selectively activate or interrogate mitochondrial mechanosensitivity at the sarcolemmal microdomain level. We started to address this question by employing scanning ion conductance microscopy (SICM) [8, 33] and surface confocal SSICM [34] to resolve cellular topography and subsarcolemmal mitochondria localization in combination with optical recording of intracellular Ca^{2+} transient initiation and propagation.

Mechanical Properties of Cardiac Cell Membrane Interrogated by Scanning Ion Conductance Microscopy

We have applied a hydrojet pressure via the SICM nanopipette with nanometer precision to investigate the subcellular mechanisms underlying mechanically-induced intracellular Ca^{2+} release in heart failure. We scanned the cellular topography of normal and failing rat ventricular cardiomyocytes (mainly from compensatory hypertrophy, derived from the distal zone from the scar) using SICM at different

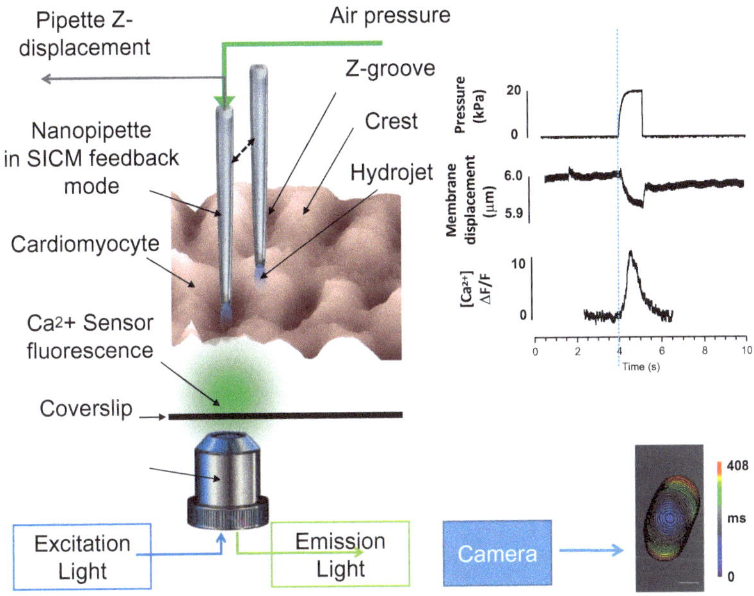

Fig. 13.1 Schematic representation of the experimental approach. After topographical live scan of the cell membrane, the nanopipette was positioned either on the crest or on the grove of a Ca^{2+} -indicator loaded cardiomyocyte cells. Then a hydrojet has been applied at 20 kPa for 2-s and membrane displacement and Ca^{2+} transient initiations and propagations have been recorded (Modified from Miragoli et al. [8], with permission)

time-points following myocardial infarction (MI). Then we positioned the pipette at 200 nm over a precised pre-selected site on the sarcolemma (either a crest or a Z-groove or an area without structure) which was chosen from a previously acquired topographical image (Fig. 13.1).

Then, we applied a localized 20-kPa hydrojet for 2 s. The area indented by the hydrojet is in the range of 0.125 μm^2. We studied rat ventricular cardiomyocytes following MI, progressing towards heart failure. We first obtained a 10×10-µm SICM topographical image of a normal or a failing cardiomyocyte (Fig. 13.1). These scanned topographical images were used to quantify the disruption of surface structural regularity; the images showed that cells progressively change their sarcolemmal regularity. Membrane organization is known to be substantially altered following MI including the disappearance of crests and grooves [12]. Hydrojets delivered on regular Z-grooves in healthy cardiomyocytes caused mechanically-induced intracellular Ca^{2+} release ($MiCa_i$) events, which were spatially confined (Figs. 13.1 and 13.2a -focal). By contrast, $MiCa_i$ propagated throughout the cell in failing cardiomyocytes with irregular Z-grooves (Fig. 13.2a -total). The occurrence of $MiCa_i$ events, transient and amplitude differ compared to AMC where hydrojet had been delivered either on crests or on grooves (Fig. 13.2b, c). Notwithstanding that blocking LTCC does not affect $MiCa_i$ (Fig. 13.2c), such difference cannot be ascribed

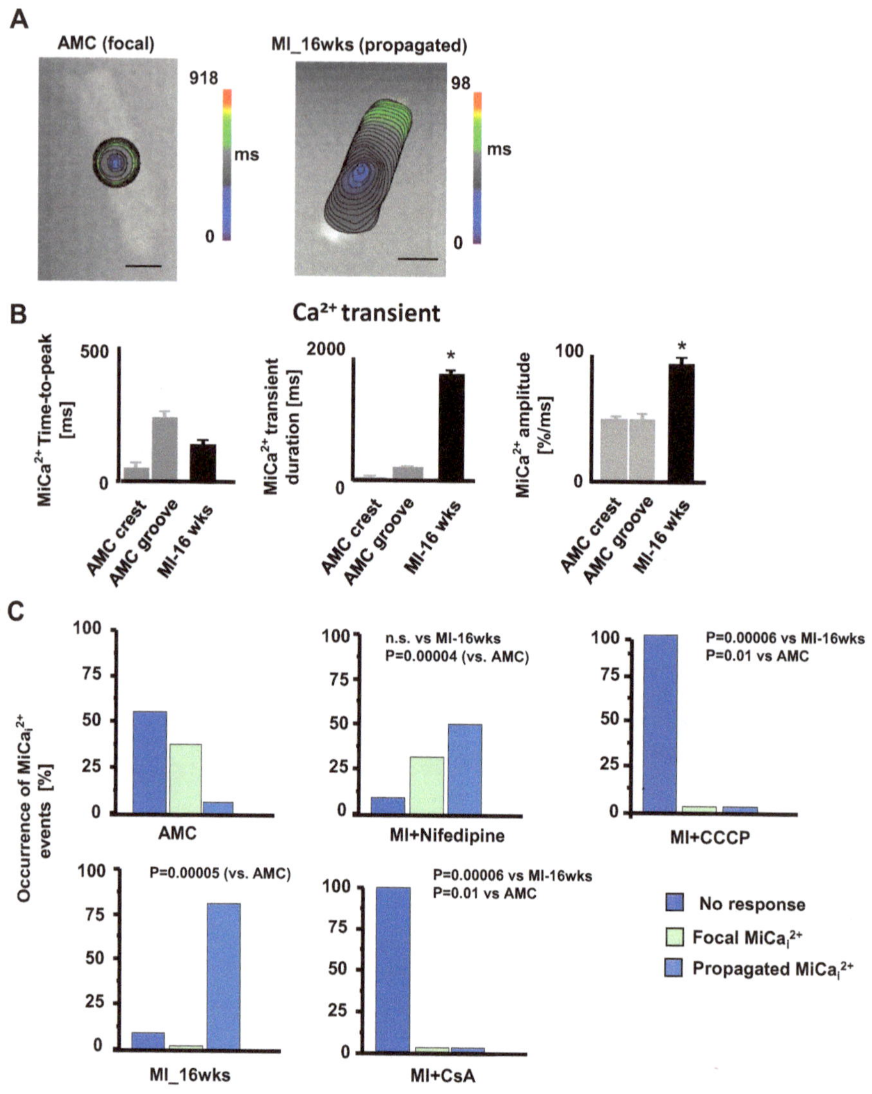

Fig. 13.2 Mechanical induced Ca^{2+} release in failing cardiomyocyte. (**a**) *Left* Hydrojet targeting non-failing structure evoke focal mechanical induced Ca^{2+} release activation (MiCai). (**a**) *Right* Infarcted cardiomyocytes (16-weeks following coronary ligation) response after hydrojet delivering (total MiCai). (**b**) Ca^{2+} transient parameters following crest/groove stimulation (AMC cells) or unstriated regions (MI cells). (**c**) Occurrence of MiCai events (no response, focal and total) in AMC and MI cells (*left*) and in the presence of selective blockers of LTCC (Nifedipine) or mitochondrial function (CCCP and CsA) (Modified from Miragoli et al. [8], with permission)

to LTCC while both (i) uncoupling mitochondrial proton gradient by Carbonyl cyanide m-chlorophenyl hydrazone (CCCP) or (ii) attenuating mitochondria permeability transition by Cyclosporin A (CsA) [35], completely abolished the focal and the propagated conditions.

Mitochondrial Displacement in the Failing Cardiomyocytes

It emerges that (mechanically) Ca^{2+}-induced Ca^{2+} release machinery [36] is still preserved but the existence of another intracellular Ca^{2+} source is supported, and having excluded the SR [8], a possible candidate is the mitochondrion. The involvement of mitochondria in pressure-induced intracellular Ca^{2+} release has been demonstrated previously [37], so we employed both confocal microscopy in combination with SICM (SSCM) and transmission electron microscopy (TEM) to investigate the sub-membrane interaction between dyads and mitochondria in failing cells.

In AMC cardiomyocytes, active TMRM-labeled mitochondria align with crests with a periodic arrangement, which reflects regular arrangement of Z-grooves and T-tubule openings (Fig. 13.3 top), as shown in particular on the TEM panel. Infarcted cells do not display the same regularity but show mitochondrial elongation and fusion [38]. These observations suggest a correlation between mitochondria derangement and the occurrence of propagated $MiCa_i$ and indicate a possible active role of remodeled mitochondria microdomains in $MiCa_i$ initiation.

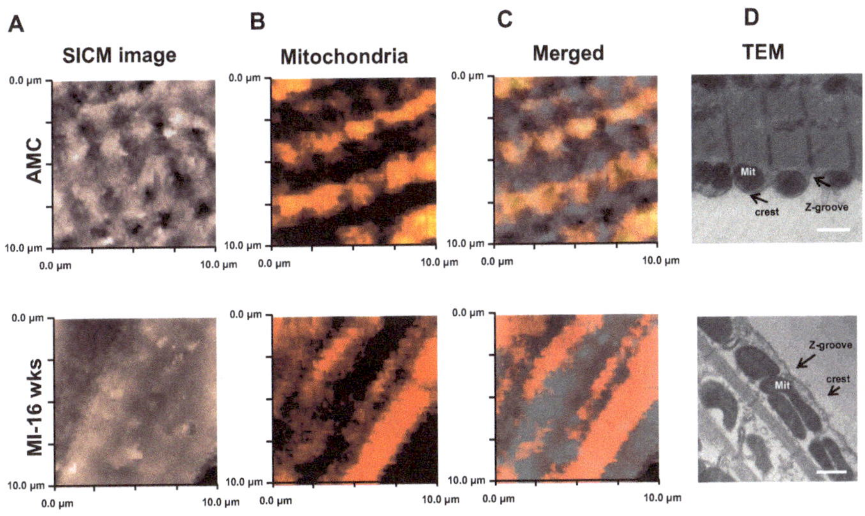

Fig. 13.3 Ultrastructural analysis of mitochondrial displacement in heart failure. (**a**) SICM images of 10 × 10-µm cardiomyocyte region of AMC (*top*) and MI (*bottom*) cells. (**b**) Surface confocal images (obtained by Surface-confocal SICM) of the labelled TMRM mitochondria in (**a**). (**c**) Merged images for SICM topography (**a**) and SICM TMRM-labelled mitochondria (**b**). (**d**) Mitochondrial displacement observed by TEM (Modified from Miragoli et al. [8], with permission)

Microtubules Involvement in Mitochondrial Displacement and MiCai Initiation and Propagation

Recent studies have implicated X-ROS signaling in inducing Ca^{2+} release (independently from mitochondria) by stretching the cell [39].

However, this activation required an intact microtubule network, which is supported by the observation that an increase in the microtubule network compactness (e.g. Duchenne muscular dystrophy) decreases X-ROS signaling [40]. Pharmacologically induced microtubular depolymerization in control cells by colchicine administration produces a functional phenotype which is similar by the one observed in heart failure cells (MI-16 weeks, Fig. 13.4a bottom-b) and increases the likelihood of $MiCa_i$ propagation (Fig. 13.4b), which we see in 69% of the cases. Our data also indicated that colchicine affects the mitochondrial total area (Fig. 13.4c) similar to what we have found in HF cells. We observed a reduction in

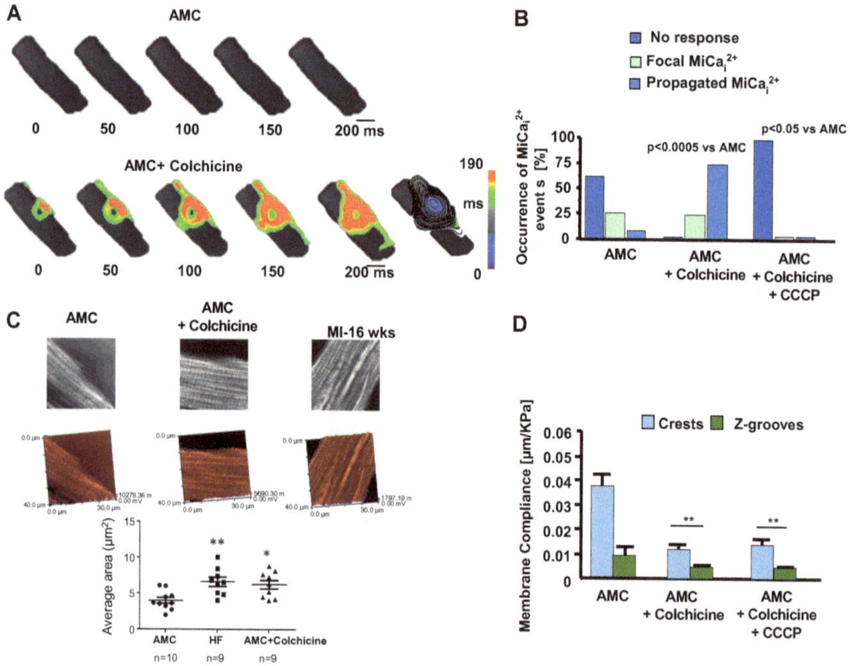

Fig. 13.4 Microtubule network derangement is a prerequisite for MiCai initiation. (**a**) Mechanically induced Ca^{2+} initiation in AMC cell (*top*) and the same cell in the presence of colchicine (*bottom*). (**b**) Occurrence of MiCai events in AMC, AMC+colchicine, AMC+colchcine+CCCP. (**c**) SICM topographical images (*top*) and TMRM-labelled mitochondria position (*middle*) and mitochondria count (*bottom*) in AMC, AMC+colchicine and HF (MI-16 weeks). (**d**) Membrane compliance measured in the three conditions mentioned in (**b**) (Modified from Miragoli et al. [8], with permission)

membrane compliance (measured as Z-displacement during hydrojet delivery, Fig. 13.4d). Moreover, our intracellular findings agree with a large animal study, where colchicine exacerbated chest impact-induced ventricular fibrillation (*commotio cordis*) [1].

Up-regulation of microtubular proteins encountered during cardiac failing [41] is implicated in destabilizing the microtubules network, by affecting T-tubule density and regularity (data not shown here, [8]). Disruption of the microtubule network allowed spreading of the normally constrained $MiCa_i$, mimicking the pattern of $MiCa_i$ propagation observed in failing cardiomyocytes.

Mitochondria and microtubules are in close contact at subsarcolemmal levels since β-tubulin is confined to the perinuclear and interfibrillar spaces and is largely co-localized with the cytoplasmic organelles [42]. In cardiomyocytes the distribution of β-tubulin-2 [43] is restricted to the outer mitochondrial-containing domain which binds to the outer mitochondrial membrane, and this probably also involves microtubular based trans-locators and/or MAPs.

Stretch-Activated Channels: A Secondary Source for Mechanical Induced Ca^{2+} Propagation

Stretch-activated channels from both cardiomyocytes and non-cardiomyocytes cells are well-known to be directly implicated in the mechano-electric transduction [44] and arrhythmias [45]. We sought to investigate their role in the mechanically-induced Ca^{2+} initiation because, in more advanced stages of pathological remodeling following MI. We observed a generation of second 'ectopic' Ca^{2+} waves which arise from remote regions of the cell (Fig. 13.5a, arrows).

In the three conditions presented in Fig. 13.5, we measured the intracellular Ca^{2+} propagation velocity and the real total activation time (TAT). We observed a complex pattern with the $MiCa_i$ 'ripple' starting underneath the pressure site, but after ~1–2 ms an additional remote $MiCa_i$ signal or signals (binary emergence) from the cell periphery follows the initial wave. The latter double or triplet Ca^{2+} wave fronts collide and propagate more rapidly throughout the cell (Fig. 13.5 left and middle panels). One plausible explanation is related to cross bridge cycling [46, 47]. The local sarcomere contraction underneath the pressure site's $MiCa_i$ relaxes while more distant sarcomeres contract, shortening against the lower compliance of the relaxing region, mainly related to stretch-activated channels activations [48].

The shortening of these distant sarcomeres release Ca^{2+} from its cycling troponin [49] into the sarcoplasm. This could manifests as the remote peripheral Ca^{2+} signal. A second conceivable mechanism could be related simply to a stiffer sarcolemma, which leads to transmission of pressure to remote SACs. To further elucidate this mechanism we blocked SACs by streptomycin (not shown) or Gadolinium (Gd^{3+}, selective blockers of SACs). Both interventions abolish this peripheral activation (Fig. 13.5a right panel). To note, TAT is reduced in the presence of multiple activation (Fig. 13.5b, c) because of fast intracellular Ca^{2+} activation (Fig. 13.5b, the cell contraction is completely desynchronized).

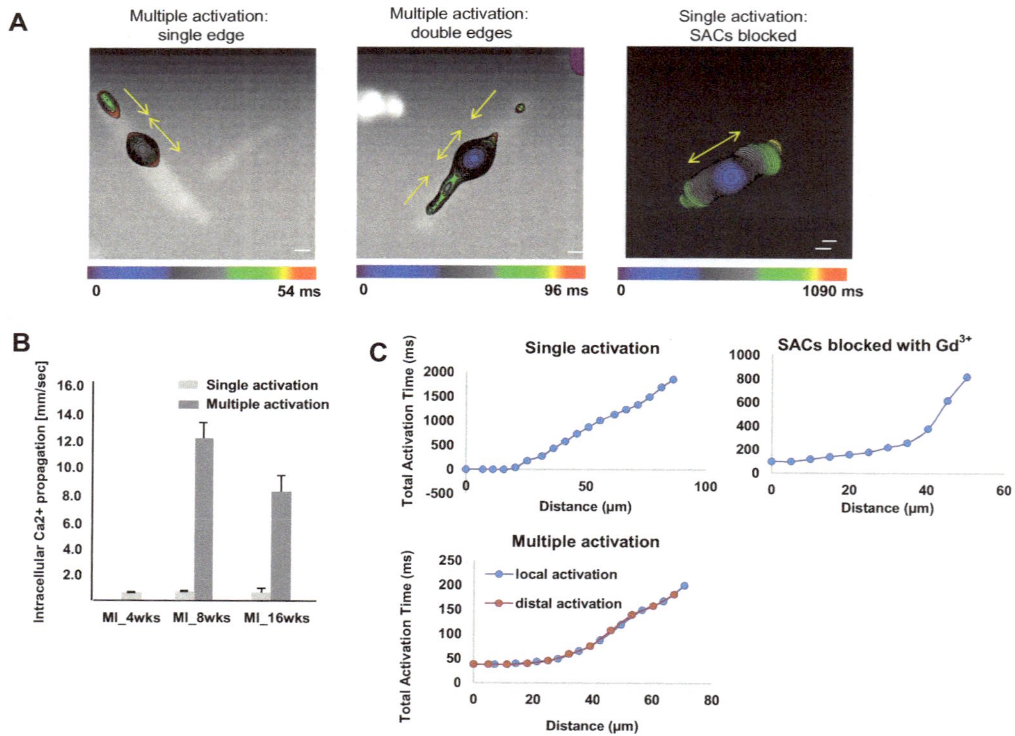

Fig. 13.5 Role of stretch-activated channels (*SACs*) following $MiCa_i$. (**a**) *Left* Secondary remote activation from a single edge (unidirectional *arrow*) following $MiCa_i$ (bidirectional *arrow*). *Middle* Secondary remote activation from the two edges (unidirectional *arrows*) following $MiCa_i$ (bidirectional *arrow*). *Right* Abolishing of multiple activation by selectively blocking of SACs by Gd^{3+}. (**b**) Overall Ca^{2+} intracellular propagation velocity during HF progression. (**c**) Total activation time from normal activation (*top left*), multiple activation (*bottom left*), and in the presence of SACs blocker (*top right*) (Modified from Miragoli et al. [8], with permission)

Mechanosensor Microdomains in Human Failing Cardiomyocytes

We sought to investigate whether MiCa$_i$ phenomena occur in human failing ventricular cardiomyocytes (Fig. 13.6) isolated from patient suffering from dilated cardiomyopathy (DCM). Similar to what has been found in rat cardiomyocytes, SICM imaging revealed a topographical heterogeneity (Fig. 13.6a left) [12]. We observed propagating MiCa$_i$ (Fig. 13.6a right) following pressure application over non-striated areas with a TAT mimicking the rat heart failure model. Mechanically-induced Ca^{2+} initiation occurred in 65% of the cases (Fig. 13.6b); more intriguingly, we measured an upregulation of several microtubules-related proteins. We found that α1C-tubulin (*TUBA1C*), β2A-tubulin (*TUBB2A*), β3-tubulin (*TUBB3),*

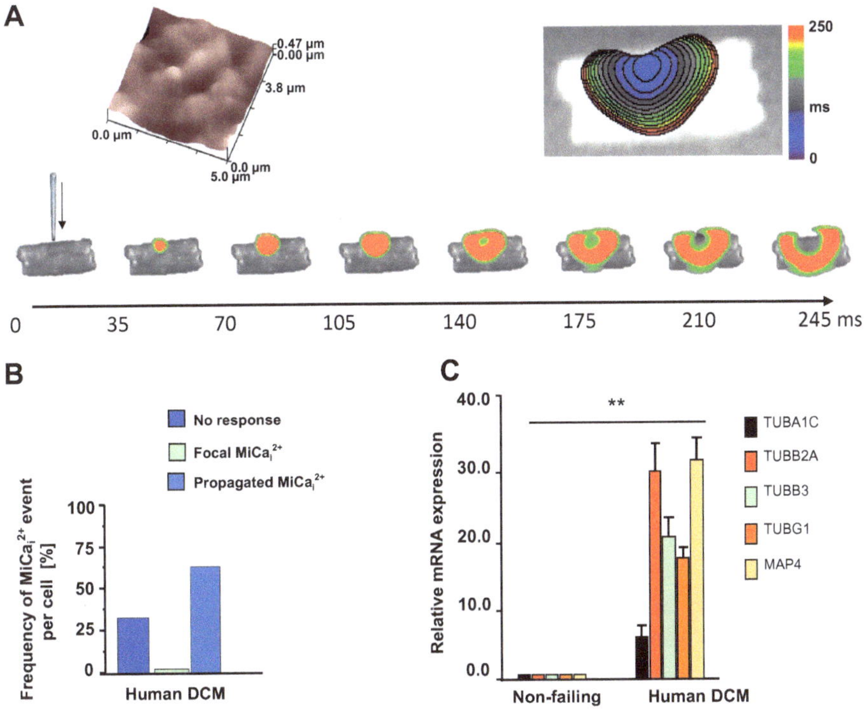

Fig. 13.6 Mechanical induced Calcium initiation in human failing myocytes. (**a**) *Left* Topographical SICM image (5×5 µm) showing partially striated area. *Right* Color-coded isochrones image of Ca^{2+} activation in the same cell. *Bottom* Intracellular MiCa$_i$ initiation and propagation with a TAT of 245 ms. $\Delta F/F_0 = 1\%$. 40× magnification, 0.6 n.a. (**b**) Frequency of MiCa$_i$ events. $n > 12$. (**c**) Relative mRNA expression for microtubules-related proteins. Technical triplicate Normalized to 18 s. mRNA quantities are presented as mean ± SEM (**$p < 0.01$), $n = 7$ (Modified from Miragoli et al. [8], with permission)

γ1-tubulin (*TUBG1*) and microtubule associated protein 4 (*MAP4*) were significantly upregulated as compared to non-failing cardiomyocytes, suggesting a primary role for microtubule disruption, together with mitochondria misalignment, in cellular vulnerability to $MiCa_i$ generation and propagation (Fig. 13.6c).

Ca^{2+} Handling, Mitochondria Metabolism and Function

Among the fundamental factors allowing the adequate integrity of mitochondrial machinery and energy generation fine Ca^{2+} regulation and tuning by cardiomyocyte at the subcellular level is central. Ca^{2+} regulatory mechanisms represent the key elements for a correct functioning of mitochondrial oxidative phosphorylation and energy production and for a controlled reactive oxygen and nitrogen species (ROS and RNS) generation by these organelles both in a healthy as well as in ischemic or failing hearts [50].

The continuous ATP generation from cardiac mitochondria is closely dependent on the oxygen and fuel substrates supplies. ATP generation depends on an efficient oxidative phosphorylation, which generates high amount of ATP. Two-thirds of generated ATP is consumed for cardiac contractions and the rest for the functioning of ion pumps: those linked to intracellular Ca^{2+} handling (e.g. sarcoplasmic reticulum Ca^{2+} -ATPase) of a relevant portion of consumed ATP [51]. In normal heart, fatty acid uptake and utilization is responsible for 70–90% of myocardial ATP produced whereas glucose uptake, utilization and oxidation accounts for remaining 10–30% of cardiac ATP production. Fatty acids generate more ATP than glucose; however, ATP production to oxygen consumption ratio is higher for glucose. Therefore, increased FA oxidation reduces cardiac efficiency (cardiac work/myocardial oxygen consumption). Even in normal hearts, oxidative phosphorylation generates a small amount (1–2%) of electrons leak from electron transfer chain to form reactive oxygen species (ROS). There are several changes observed in mitochondrial function of cardiomyocytes from normal, hypertrophic and failing hearts [24], concerning Ca^{2+} regulation even if a clear and complete picture of the phenomenon is still far. In the early stages of heart failure cardiomyocyte, structural and metabolic remodeling plays compensatory role with a slightly increased fatty acid beta-oxidation. While progressing the heart failure condition, fatty acid beta oxidation and mitochondrial respiratory activity decrease leading to decrease cardiac ATP generation. As a compensation to reduced fatty acid oxidative metabolism, glucose uptake and glycolysis are upregulated, even if absolutely insufficient to compensate the 60% reduction in ATP generation [25].

Noteworthy, mitochondrial Ca^{2+} regulation is fundamental because its signaling is closely implicated in ATP production, NADH generation [52] and activation of metabolic enzymes involved in oxidative phosphorylation [53]. In cardiomyocytes from healthy hearts, the close proximity of endoplasmic reticulum to mitochondria can determine rapid changes of Ca^{2+} after its release from type-2 ryanodine receptor/Ca^{2+} release channel (RyR2) and type 2 inositol 1,4,5-trisphosphate receptor (IP_3R_2), the most important intracellular Ca^{2+} release channels in the heart [16]. RyR2 is essential for cardiac excitation–contraction coupling [16]; such task needs energy in the form of ATP produced primarily by oxidative phosphorylation in mitochondria [54]. Several are the authors that hypothesized that sarcoplasmic Ca^{2+} release via RyR2 and/or IP_3R_2 channels determines mitochondrial Ca^{2+} accumulation. Both mitochondrial excess and deficiency of Ca^{2+} have been reported as determinants of mitochondrial dysfunction [16, 55]. Intracellular Ca^{2+} accumulation in particular can impair mitochondrial function, leading to reduced ATP production and increased release of reactive oxygen species [55].

It has been reported that other Ca^{2+}-dependent pathways may alter oxidative phosphorylation mitochondria, such as Ca^{2+} -dependent protein kinases and calcineurin [56]. Small physiological levels of Ca^{2+} taken up into the mitochondria can regulate mitochondrial metabolism. An increase in contractility needs more increased availability of ATP; increased contractility is linked to higher cytosolic Ca^{2+} transient whose transmission to the mitochondria via Ca^{2+} uptake activates Ca^{2+} -sensitive mitochondrial dehydrogenases and several complexes of electron transport chain [57]. Mitochondrial Na^+-Ca^{2+} exchange has been shown to regulate mitochondrial Ca^{2+} levels linking mitochondrial Ca^{2+} to intracellular sodium. Sodium intracellular rise observed in hypertrophic and heart failure cardiomyocytes is strictly linked to changes of mitochondrial Ca^{2+} concentration that profoundly alter ROS generation, energy production and metabolism [58].

Under pathological conditions of increased Ca^{2+} cytosolic overload as occurs in heart failure, mitochondria after having taking up higher amounts of Ca^{2+}, displayed an opening of mitochondrial permeability transition pore (mPTP) that if maintained can be a trigger for mitochondrial dysfunction, damage and cell death. In our situation blocking mPTP is necessary to avoid MiCai.

Another factor involved in mitochondrial Ca^{2+} handling and energy and ATP generation, concerns the identification of a mitochondrial Ca^{2+} uniporter (MCU) that transport it across the mitochondrial inner membrane (see Chap. 2). Several groups developed knockout mice or mice without a functional MCU to study the role of the channel in modulating metabolism and cell death (see Chap. 3). In mice with MCU knocked out or mutated before birth, the hearts did not show altered metabolism and a decrease in infarct size after ischemia-reperfusion [59]. In adult mice, when uniporter function was loss because of the administration of tamoxifen, a protective effect was observed with smaller infarcts after ischemia and reperfusion [60]. In our situation, blocking mPTP and uniporter abolish $MiCa_i$ initiation and propagation without altering the normal Ca^{2+} -induced-Ca^{2+} release [8]. However, there are still apparent discrepancies in the relationship between changes in MCU expression and metabolic changes and potential protective effects of its deletion in adult mice and further studies are necessary to clarify these phenomena.

Ca^{2+} Microdomains and Mitochondrial ROS and RNS Generation

The relationship between Ca^{2+} regulation and ROS production is strictly interdependent and recent observations opened the understanding of fine regulation at different subcellular levels in cardiomyocytes [25, 61]. It has been shown that ROS play important roles in reshaping local and global Ca^{2+} signal amplitudes and kinetics at different levels of regulation in both physiology and pathology.

First it is well known that cardiomyocytes can generate ROS from several sources, especially mitochondria but other sources such as NAD(P)H oxidase, xanthine oxidase, and uncoupled nitric oxide synthase enzymes are also important [24]. In addition, cardiomyocytes generate not only reactive oxygen species but also reactive nitrogen species (RNS). Enzymatic production of nitric oxide from L-arginine is catalyzed by one of the three nitric oxide (NO)-synthase isoforms (nNOS, iNOS, and eNOS), all expressed in cardiomyocytes [62]. The activity of a major regulator of Ca^{2+} homeostasis SERCA is markedly affected by specific ONOO–mediated tyrosine nitration or cysteine oxidation linking ROS and RNS to cytoplasmic Ca^{2+} level (SERCA) [63]. An increased generation of RNS and ROS has been demonstrated in different settings of myocardial hypertrophy and heart failure in experimental models of hypertension and in heart failure patients [64, 65]. Such a ROS and RNS-mediated functional regulation of Ca^{2+} homeostasis is bidirectional and is related to the type of ROS generated, the amount of ROS and the location of the ROS and RNS subcellular source.

Failing hearts frequently show signs of increased ROS and RNS including lipid peroxidation, protein nitration as well reduction of antioxidants defenses [66].

Ca^{2+} microdomains at their associated dyads can be influenced by mitochondrial superoxide generation. It has been recently demonstrated that an intermittent, quantal generation of superoxide by the mitochondria, named mitochondrial "superoxide flash", active under physiological conditions, affects Ca^{2+} sparks [67]. While superoxide flashes do not normally propagate, there is also an inter-mitochondrial ROS-induced ROS release occurring during pathological conditions that can burst ROS generation. Mitochondrial Ca^{2+} plays a profound role in regulating the aforementioned modes of mitochondrial ROS production both in constitutively or in pathophysiological settings [68].

ROS generation can affect Ca^{2+} -tuning components at the levels of dyads. Under physiological condition, spontaneous Ca^{2+} sparks activity is finely regulated by basal ROS cellular generation [69].

In heart failure the elevated oxidation of RyR increases sarcoplasmic Ca^{2+} leakage during diastole, depleting sarcoplasmic Ca^{2+} content [55]. The depletion of sarcoplasmic reticulum content is also dependent on SERCA oxidation, which tends to inhibit its activity [70]. Diastolic Ca^{2+} increases due to oxidation of RyRs augments the susceptibility to arrhythmogenic afterdepolarizations. Higher generation of ROS and stretch-activated X-ROS leads to oxidize the RyR that generate a greater Ca^{2+} spark response and trigger more Ca^{2+} waves (see Chap. 10).

The bidirectional influence of ROS regulating Ca^{2+} spark activity and in an opposite manner Ca^{2+} modulating ROS generation, indicate a close interaction of Ca^{2+} and ROS and RNS signaling systems. In particular, mitochondrial ROS generation and ROS-mediated damage are strongly involved in the development and progression of HF. Albeit many are the recent discoveries on the relationship between subcellular Ca^{2+} regulation and oxidative stress a complete picture of ROS involvement in heart diseases is still lacking. Future investigations are warranted to unravel how such cellular remodeling affects both Ca^{2+} regulation and ROS generation at different levels from subcellular microdomains ROS generation in terms of X-ROS and superoxide flashes to ROS constitutive and inducible production under physiological and pathological conditions.

Conclusions

The attention paid to the role and function of mitochondria in the context on cardiovascular diseases is well established and studied [71–73]. In this chapter we focus on another aspect, which seems to jeopardize cardiovascular function in a similar manner: their 'belonging' to structural and organized microdomains [74] and their displacement/relocation and fusion [75] when such organization is lost.

The observations described herein have a potential translational element. The mechanically-induced Ca^{2+} changes have electrophysiological consequences that are potentially pro-arrhythmic. While arrhythmia is a multicellular phenomenon, the mechanisms by which mitochondrial relocation and microtubules derangement transmit abnormal $MiCa_i$ can be included in the 'maladaptive electrical and mechanical remodeling', which is known to ultimately predispose the heart to arrhythmias by, for example, inducing Ca^{2+} overload-related triggered activity [76, 77] or Ca^{2+} alternans [78]. As intracellular Ca^{2+} waves contribute to arrhythmogenesis on multicellular scales, the mechanisms that we describe may represent a novel arrhythmogenic substrate for ectopic initiation and propagation. This not only provides mechanistic insights, it also provides new therapeutic targets.

Outlook on Mitochondria Function

Using novel microscopic and super resolution-based technologies will provide new insight into cardiac microdomain levels [79, 80]. Our unique combination of SICM and optical mapping of mechanically-induced impulse propagation in a single cell allow the identification and localization of functional mechano-sensing with nanometer precision in live cells in general, and in particular, cardiomyocytes. Mitochondria are highly dynamic organelles which, depending on the metabolic demands or pathological conditions, frequently change their shape not only in eukaryotic cells but also in yeast or other unicellular organisms [81]. However, their role is not merely related to energetic balance maintenance or signaling. A very recent study has shown that mitochondria translocate from astrocytes to neurons cells in the brain after stroke [82], thus working as 'heterocellular messenger' for the stromal-parenchymal crosstalk. In summary, mitochondrial function, structure and location together with the interaction among cytoskeleton and adjacent cells need to be adequately and jointly addressed for gaining a complete understanding of their role in pathological heart.

References

1. Link MS, Wang PJ, Pandian NG, Bharati S, Udelson JE, Lee MY, Vecchiotti MA, Vander Brink BA, Mirra G, Maron BJ, Estes 3rd NA. An experimental model of sudden death due to low-energy chest-wall impact (commotio cordis). N Engl J Med. 1998;338:1805–11.
2. Ait-Mou Y, Hsu K, Farman GP, Kumar M, Greaser ML, Irving TC, de Tombe PP. Titin strain contributes to the Frank-Starling law of the heart by structural rearrangements of both thin- and thick-filament proteins. Proc Natl Acad Sci U S A. 2016;113:2306–11.
3. Moss RL, Fitzsimons DP. Frank-Starling relationship: long on importance, short on mechanism. Circ Res. 2002;90:11–3.
4. Fassina L, Rozzi G, Rossi S, Scacchi S, Galetti M, Lo Muzio FP, Del Bianco F, Colli Franzone P, Petrilli G, Faggian G, Miragoli M. Cardiac kinematic parameters computed from video of in situ beating heart. Sci Rep. 2017;7:46143 doi:10.1038/srep46143.
5. Kiseleva I, Kamkin A, Wagner KD, Theres H, Ladhoff A, Scholz H, Gunther J, Lab MJ. Mechanoelectric feedback after left ventricular infarction in rats. Cardiovasc Res. 2000;45:370–8.
6. Lammerding J, Kamm RD, Lee RT. Mechanotransduction in cardiac myocytes. Ann N Y Acad Sci. 2004;1015:53–70.
7. Kohlhaas M, Maack C. Calcium release microdomains and mitochondria. Cardiovasc Res. 2013;98:259–68.
8. Miragoli M, Sanchez-Alonso JL, Bhargava A, Wright PT, Sikkel M, Schobesberger S, Diakonov I, Novak P, Castaldi A, Cattaneo P, Lyon AR, Lab MJ, Gorelik J. Microtubule-dependent mitochondria alignment regulates calcium release in response to nanomechanical stimulus in heart myocytes. Cell Rep. 2016;14:140–51.
9. Kim SJ, Iizuka K, Kelly RA, Geng YJ, Bishop SP, Yang G, Kudej A, McConnell BK, Seidman CE, Seidman JG, Vatner SF. An alpha-cardiac myosin heavy chain gene mutation impairs contraction and relaxation function of cardiac myocytes. Am J Physiol. 1999;276:H1780–7.
10. Borg TK, Goldsmith EC, Price R, Carver W, Terracio L, Samarel AM. Specialization at the Z line of cardiac myocytes. Cardiovasc Res. 2000;46:277–85.
11. Knoll R, Hoshijima M, Hoffman HM, Person V, Lorenzen-Schmidt I, Bang ML, Hayashi T, Shiga N, Yasukawa H, Schaper W, McKenna W, Yokoyama M, Schork NJ, Omens JH, McCulloch AD, Kimura A, Gregorio CC, Poller W, Schaper J, Schultheiss HP, Chien KR. The cardiac mechanical stretch sensor machinery involves a Z disc complex that is defective in a subset of human dilated cardiomyopathy. Cell. 2002;111:943–55.
12. Lyon AR, MacLeod KT, Zhang Y, Garcia E, Kanda GK, Lab MJ, Korchev YE, Harding SE, Gorelik J. Loss of T-tubules and other changes to surface topography in ventricular myocytes from failing human and rat heart. Proc Natl Acad Sci U S A. 2009;106:6854–9.
13. Janmey PA, Miller RT. Mechanisms of mechanical signaling in development and disease. J Cell Sci. 2011;124:9–18.
14. Ferreira-Cornwell MC, Luo Y, Narula N, Lenox JM, Lieberman M, Radice GL. Remodeling the intercalated disc leads to cardiomyopathy in mice misexpressing cadherins in the heart. J Cell Sci. 2002;115:1623–34.

15. Yancey DM, Guichard JL, Ahmed MI, Zhou L, Murphy MP, Johnson MS, Benavides GA, Collawn J, Darley-Usmar V, Dell'Italia LJ. Cardiomyocyte mitochondrial oxidative stress and cytoskeletal breakdown in the heart with a primary volume overload. Am J Physiol Heart Circ Physiol. 2015;308:H651–63.
16. Santulli G, Marks AR. Essential roles of intracellular calcium release channels in muscle, brain, metabolism, and aging. Curr Mol Pharmacol. 2015;8:206–22.
17. Yang D, Wang T, Ni Y, Song B, Ning F, Hu P, Luo L, Wang Y, Ma A. Apamin-sensitive K+ current upregulation in volume-overload heart failure is associated with the decreased interaction of CK2 with SK2. J Membr Biol. 2015;248:1181–9.
18. Kuwahara K, Nakao K. New molecular mechanisms for cardiovascular disease: transcriptional pathways and novel therapeutic targets in heart failure. J Pharmacol Sci. 2011;116:337–42.
19. Nikolaev VO, Moshkov A, Lyon AR, Miragoli M, Novak P, Paur H, Lohse MJ, Korchev YE, Harding SE, Gorelik J. Beta2-adrenergic receptor redistribution in heart failure changes cAMP compartmentation. Science. 2010;327:1653–7.
20. Jeyaraj D, Wan X, Ficker E, Stelzer JE, Deschenes I, Liu H, Wilson LD, Decker KF, Said TH, Jain MK, Rudy Y, Rosenbaum DS. Ionic bases for electrical remodeling of the canine cardiac ventricle. Am J Physiol Heart Circ Physiol. 2013;305:H410–9.
21. Rowell J, Koitabashi N, Kass DA, Barth AS. Dynamic gene expression patterns in animal models of early and late heart failure reveal biphasic-bidirectional transcriptional activation of signaling pathways. Physiol Genomics. 2014;46:779–87.
22. Adhihetty PJ, Ljubicic V, Menzies KJ, Hood DA. Differential susceptibility of subsarcolemmal and intermyofibrillar mitochondria to apoptotic stimuli. Am J Physiol Cell Physiol. 2005;289:C994–C1001.
23. Lesnefsky EJ, Tandler B, Ye J, Slabe TJ, Turkaly J, Hoppel CL. Myocardial ischemia decreases oxidative phosphorylation through cytochrome oxidase in subsarcolemmal mitochondria. Am J Physiol. 1997;273:H1544–54.
24. Rosca MG, Hoppel CL. Mitochondrial dysfunction in heart failure. Heart Fail Rev. 2013;18:607–22.
25. Rosca MG, Tandler B, Hoppel CL. Mitochondria in cardiac hypertrophy and heart failure. J Mol Cell Cardiol. 2013;55:31–41.
26. Piquereau J, Caffin F, Novotova M, Lemaire C, Veksler V, Garnier A, Ventura-Clapier R, Joubert F. Mitochondrial dynamics in the adult cardiomyocytes: which roles for a highly specialized cell? Front Physiol. 2013;4:102.
27. Dague E, Genet G, Lachaize V, Guilbeau-Frugier C, Fauconnier J, Mias C, Payre B, Chopinet L, Alsteens D, Kasas S, Severac C, Thireau J, Heymes C, Honton B, Lacampagne A, Pathak A, Senard JM, Gales C. Atomic force and electron microscopic-based study of sarcolemmal surface of living cardiomyocytes unveils unexpected mitochondrial shift in heart failure. J Mol Cell Cardiol. 2014;74:162–72.
28. Ter Keurs HE, Wakayama Y, Sugai Y, Price G, Kagaya Y, Boyden PA, Miura M, Stuyvers BD. Role of sarcomere mechanics and Ca2+ overload in Ca2+ waves and arrhythmias in rat cardiac muscle. Ann N Y Acad Sci. 2006;1080:248–67.
29. Belmonte S, Morad M. Shear fluid-induced Ca2+ release and the role of mitochondria in rat cardiac myocytes. Ann N Y Acad Sci. 2008;1123:58–63.
30. Chaanine AH, Kohlbrenner E, Gamb SI, Guenzel AJ, Klaus KA, Fayyaz AU, Nair KS, Hajjar RJ, Redfield MM. FOXO3a regulates BNIP3 and modulates mitochondrial calcium, dynamics and function in cardiac stress. Am J Physiol Heart Circ Physiol. 2016;311:H1540. ajpheart 00549 2016
31. Doenst T, Pytel G, Schrepper A, Amorim P, Farber G, Shingu Y, Mohr FW, Schwarzer M. Decreased rates of substrate oxidation ex vivo predict the onset of heart failure and contractile dysfunction in rats with pressure overload. Cardiovasc Res. 2010;86:461–70.
32. Sharov VG, Todor AV, Silverman N, Goldstein S, Sabbah HN. Abnormal mitochondrial respiration in failed human myocardium. J Mol Cell Cardiol. 2000;32:2361–7.
33. Miragoli M, Yacoub MH, El-Hamamsy I, Sanchez-Alonso JL, Moshkov A, Mongkoldhumrongkul N, Padala M, Paramagurunathan S, Sarathchandra P, Korchev YE,

Gorelik J, Chester AH. Side-specific mechanical properties of valve endothelial cells. Am J Physiol Heart Circ Physiol. 2014;307:H15–24.

34. Novak P, Shevchuk A, Ruenraroengsak P, Miragoli M, Thorley AJ, Klenerman D, Lab MJ, Tetley TD, Gorelik J, Korchev YE. Imaging single nanoparticle interactions with human lung cells using fast ion conductance microscopy. Nano Lett. 2014;14:1202–7.
35. Sharov VG, Todor A, Khanal S, Imai M, Sabbah HN. Cyclosporine A attenuates mitochondrial permeability transition and improves mitochondrial respiratory function in cardiomyocytes isolated from dogs with heart failure. J Mol Cell Cardiol. 2007;42:150–8.
36. Rusconi F, Ceriotti P, Miragoli M, Carullo P, Salvarani N, Rocchetti M, Di Pasquale E, Rossi S, Tessari M, Caprari S, Cazade M, Kunderfranco P, Chemin J, Bang ML, Polticelli F, Zaza A, Faggian G, Condorelli G, Catalucci D. Peptidomimetic targeting of cavbeta2 overcomes dysregulation of the L-type calcium channel density and recovers cardiac function. Circulation. 2016;134:534–46.
37. Belmonte S, Morad M. 'Pressure-flow'-triggered intracellular Ca2+ transients in rat cardiac myocytes: possible mechanisms and role of mitochondria. J Physiol. 2008;586:1379–97.
38. Knowlton AA, Chen L, Malik ZA. Heart failure and mitochondrial dysfunction: the role of mitochondrial fission/fusion abnormalities and new therapeutic strategies. J Cardiovasc Pharmacol. 2014;63:196–206.
39. Kombairaju P, Kerr JP, Roche JA, Pratt SJ, Lovering RM, Sussan TE, Kim JH, Shi G, Biswal S, Ward CW. Genetic silencing of Nrf2 enhances X-ROS in dysferlin-deficient muscle. Front Physiol. 2014;5:57.
40. Robison P, Caporizzo MA, Ahmadzadeh H, Bogush AI, Chen CY, Margulies KB, Shenoy VB, Prosser BL. Detyrosinated microtubules buckle and bear load in contracting cardiomyocytes. Science. 2016;352:aaf0659.
41. Roos KP, Palmer RE, Miller TW. The role of microtubules in structural remodeling and the progression to heart failure. J Card Fail. 2002;8:S300–10.
42. Saetersdal T, Greve G, Dalen H. Associations between beta-tubulin and mitochondria in adult isolated heart myocytes as shown by immunofluorescence and immunoelectron microscopy. Histochemistry. 1990;95:1–10.
43. Kuznetsov AV, Javadov S, Guzun R, Grimm M, Saks V. Cytoskeleton and regulation of mitochondrial function: the role of beta-tubulin II. Front Physiol. 2013;4:82.
44. Ingber DE. Tensegrity-based mechanosensing from macro to micro. Prog Biophys Mol Biol. 2008;97:163–79.
45. Grand T, Salvarani N, Jousset F, Rohr S. Aggravation of cardiac myofibroblast arrhythmogeneicity by mechanical stress. Cardiovasc Res. 2014;104:489–500.
46. Biesiadecki BJ, Davis JP, Ziolo MT, Janssen PM. Tri-modal regulation of cardiac muscle relaxation; intracellular calcium decline, thin filament deactivation, and cross-bridge cycling kinetics. Biophys Rev. 2014;6:273–89.
47. Stohr A, Friedrich FW, Flenner F, Geertz B, Eder A, Schaaf S, Hirt MN, Uebeler J, Schlossarek S, Carrier L, Hansen A, Eschenhagen T. Contractile abnormalities and altered drug response in engineered heart tissue from Mybpc3-targeted knock-in mice. J Mol Cell Cardiol. 2013;63:189–98.
48. Lorin C, Vogeli I, Niggli E. Dystrophic cardiomyopathy: role of TRPV2 channels in stretch-induced cell damage. Cardiovasc Res. 2015;106:153–62.
49. ter Keurs HE, Zhang YM, Davidoff AW, Boyden PA, Wakayama Y, Miura M. Damage induced arrhythmias: mechanisms and implications. Can J Physiol Pharmacol. 2001;79:73–81.
50. Traaseth N, Elfering S, Solien J, Haynes V, Giulivi C. Role of calcium signaling in the activation of mitochondrial nitric oxide synthase and citric acid cycle. Biochim Biophys Acta. 1658;2004:64–71.
51. Lopaschuk GD, Ussher JR, Folmes CD, Jaswal JS, Stanley WC. Myocardial fatty acid metabolism in health and disease. Physiol Rev. 2010;90:207–58.
52. Glancy B, Willis WT, Chess DJ, Balaban RS. Effect of calcium on the oxidative phosphorylation cascade in skeletal muscle mitochondria. Biochemistry. 2013;52:2793–809.

53. Hoffman NE, Miller BA, Wang J, Elrod JW, Rajan S, Gao E, Song J, Zhang XQ, Hirschler-Laszkiewicz I, Shanmughapriya S, Koch WJ, Feldman AM, Madesh M, Cheung JY. Ca(2)(+) entry via Trpm2 is essential for cardiac myocyte bioenergetics maintenance. Am J Physiol Heart Circ Physiol. 2015;308:H637–50.
54. Nickel A, Loffler J, Maack C. Myocardial energetics in heart failure. Basic Res Cardiol. 2013;108:358.
55. Santulli G, Xie W, Reiken SR, Marks AR. Mitochondrial calcium overload is a key determinant in heart failure. Proc Natl Acad Sci U S A. 2015;112:11389–94.
56. Nowak G. Protein kinase C-alpha and ERK1/2 mediate mitochondrial dysfunction, decreases in active Na+ transport, and cisplatin-induced apoptosis in renal cells. J Biol Chem. 2002;277:43377–88.
57. Denton RM, McCormack JG, Edgell NJ. Role of calcium ions in the regulation of intramitochondrial metabolism. Effects of Na+, Mg2+ and ruthenium red on the Ca2+-stimulated oxidation of oxoglutarate and on pyruvate dehydrogenase activity in intact rat heart mitochondria. Biochem J. 1980;190:107–17.
58. Liu T, O'Rourke B. Regulation of mitochondrial Ca2+ and its effects on energetics and redox balance in normal and failing heart. J Bioenerg Biomembr. 2009;41:127–32.
59. Pan X, Liu J, Nguyen T, Liu C, Sun J, Teng Y, Fergusson MM, Rovira II, Allen M, Springer DA, Aponte AM, Gucek M, Balaban RS, Murphy E, Finkel T. The physiological role of mitochondrial calcium revealed by mice lacking the mitochondrial calcium uniporter. Nat Cell Biol. 2013;15:1464–72.
60. Luongo TS, Lambert JP, Yuan A, Zhang X, Gross P, Song J, Shanmughapriya S, Gao E, Jain M, Houser SR, Koch WJ, Cheung JY, Madesh M, Elrod JW. The mitochondrial calcium uniporter matches energetic supply with cardiac workload during stress and modulates permeability transition. Cell Rep. 2015;12:23–34.
61. Tse G, Yan BP, Chan YW, Tian XY, Huang Y. Reactive oxygen species. Endoplasmic reticulum stress mitochondrial dysfunction: the link card arrhythmogenesis. Front Physiol. 2016;7:313.
62. Seddon M, Shah AM, Casadei B. Cardiomyocytes as effectors of nitric oxide signalling. Cardiovasc Res. 2007;75:315–26.
63. Tocchetti CG, Wang W, Froehlich JP, Huke S, Aon MA, Wilson GM, Di Benedetto G, O'Rourke B, Gao WD, Wink DA, Toscano JP, Zaccolo M, Bers DM, Valdivia HH, Cheng H, Kass DA, Paolocci N. Nitroxyl improves cellular heart function by directly enhancing cardiac sarcoplasmic reticulum Ca2+ cycling. Circ Res. 2007;100:96–104.
64. Cabassi A, Dancelli S, Pattoneri P, Tirabassi G, Quartieri F, Moschini L, Cavazzini S, Maestri R, Lagrasta C, Graiani G, Corradi D, Parenti E, Tedeschi S, Cremaschi E, Coghi P, Vinci S, Fiaccadori E, Borghetti A. Characterization of myocardial hypertrophy in prehypertensive spontaneously hypertensive rats: interaction between adrenergic and nitrosative pathways. J Hypertens. 2007;25:1719–30.
65. Cabassi A, Binno SM, Tedeschi S, Ruzicka V, Dancelli S, Rocco R, Vicini V, Coghi P, Regolisti G, Montanari A, Fiaccadori E, Govoni P, Piepoli M, de Champlain J. Low serum ferroxidase I activity is associated with mortality in heart failure and related to both peroxynitrite-induced cysteine oxidation and tyrosine nitration of ceruloplasmin. Circ Res. 2014;114:1723–32.
66. Cabassi A, Binno SM, Tedeschi S, Graiani G, Galizia C, Bianconcini M, Coghi P, Fellini F, Ruffini L, Govoni P, Piepoli M, Perlini S, Regolisti G, Fiaccadori E. Myeloperoxidase-related chlorination activity is positively associated with circulating ceruloplasmin in chronic heart failure patients: relationship with neurohormonal, inflammatory, and nutritional parameters. Biomed Res Int. 2015;2015:691693.
67. Wang W, Fang H, Groom L, Cheng A, Zhang W, Liu J, Wang X, Li K, Han P, Zheng M, Yin J, Wang W, Mattson MP, Kao JP, Lakatta EG, Sheu SS, Ouyang K, Chen J, Dirksen RT, Cheng H. Superoxide flashes in single mitochondria. Cell. 2008;134:279–90.
68. Zorov DB, Juhaszova M, Sollott SJ. Mitochondrial ROS-induced ROS release: an update and review. Biochim Biophys Acta. 1757;2006:509–17.
69. Zhang H, Gomez AM, Wang X, Yan Y, Zheng M, Cheng H. ROS regulation of microdomain Ca(2+) signalling at the dyads. Cardiovasc Res. 2013;98:248–58.

70. Lancel S, Qin F, Lennon SL, Zhang J, Tong X, Mazzini MJ, Kang YJ, Siwik DA, Cohen RA, Colucci WS. Oxidative posttranslational modifications mediate decreased SERCA activity and myocyte dysfunction in Galphaq-overexpressing mice. Circ Res. 2010;107:228–32.
71. Sharma A, Fonarow GC, Butler J, Ezekowitz JA, Felker GM. Coenzyme Q10 and heart failure: a state-of-the-art review. Circ-Heart Fail. 2016;9:e002639.
72. Jahng JWS, Song E, Sweeney G. Crosstalk between the heart and peripheral organs in heart failure. Exp Mol Med. 2016;48:e217.
73. Wang W, Karamanlidis G, Tian R. Novel targets for mitochondrial medicine. Sci Transl Med. 2016;8:326rv3.
74. Spat A, Fulop L, Koncz P, Szanda G. When is high-Ca2+ microdomain required for mitochondrial Ca2+ uptake? Acta Physiol. 2009;195:139–47.
75. Goldenthal MJ. Mitochondrial involvement in myocyte death and heart failure. Heart Fail Rev. 2016;21:137–55.
76. Wasson S, Reddy HK, Dohrmann ML. Current perspectives of electrical remodeling and its therapeutic implications. J Cardiovasc Pharm Ther. 2004;9:129–44.
77. Adamson PB, Barr RC, Callans DJ, Chen P-S, Lathrop DA, Makielski JC, Nerbonne JM, Nuss HB, Olgin JE, Przywara DA, Rosen MR, Rozanski GJ, Spach MS, Yamada KA. The perplexing complexity of cardiac arrhythmias: beyond electrical remodeling. Heart Rhythm. 2005;2:650.
78. Li Q, Pogwizd SM, Prabhu SD, Zhou LF. Inhibiting Na+/K+ ATPase can impair mitochondrial energetics and induce abnormal Ca2+ cycling and automaticity in Guinea pig cardiomyocytes. PLoS One. 2014;9:e93928.
79. Fu Y, Shaw SA, Naami R, Vuong CL, Basheer WA, Guo XQ, Hong TT. Isoproterenol promotes rapid ryanodine receptor movement to Bridging Integrator 1 (BIN1)-organized dyads. Circulation. 2016;133:388–97.
80. Sanchez-Alonso JL, Bhargava A, O'Hara T, Glukhov AV, Schobesberger S, Bhogal N, Sikkel MB, Mansfield C, Korchev YE, Lyon AR, Punjabi PP, Nikolaev VO, Trayanova NA, Gorelik J. Microdomain-specific modulation of L-type calcium channels leads to triggered ventricular arrhythmia in heart failure. Circ Res. 2016;119:944–55.
81. Zhao J, Lendahl U, Nister M. Regulation of mitochondrial dynamics: convergences and divergences between yeast and vertebrates. Cell Mol Life Sci. 2013;70:951–76.
82. Hayakawa K, Esposito E, Wang XH, Terasaki Y, Liu Y, Xing CH, Ji XM, Lo EH. Transfer of mitochondria from astrocytes to neurons after stroke. Nature. 2016;535:551–5.

Chapter 14
Mechanistic Role of Thioredoxin 2 in Heart Failure

Chaofei Chen, Haixuan Chen*, Huanjiao Jenny Zhou*, Weidong Ji, and Wang Min

Heart failure is one of the primary causes of cardiovascular morbidity and mortality, which is predicted to increase dramatically worldwide in the decades ahead [1]. Many diseases including inherited cardiomyopathy, dilated cardiomyopathy, cardiac hypertrophy, myocardial infarction and diabetic cardiomyopathy can all ultimately lead to heart failure. Studies conducted with human and animal models have suggested that heart failure is closely related with increased levels of ROS [2–5]. ROS induced apoptosis of cardiomyocytes has been causally related to the pathogenesis of heart failure [6, 7]. Mitochondria are a major source of ROS, which produce 85–90 % of cellular ROS [8]. Meanwhile, several antioxidants systems are localized in mitochondria and maintain the intracellular redox balance [9]. Of these, the mitochondrial thioredoxin system, consisting of Trx2, peroxiredoxin 3 (Prx3), and thioredoxin reductase 2 (TrxR2) is the major scavenger of ROS and participates in redox-controlled signaling [10, 11]. A better understanding of the role of the

*Haixuan Chen and Huanjiao Jenny Zhou are contributed equally.

C. Chen • H. Chen • W. Ji
The First Affiliated Hospital, Center for Translational Medicine, Sun Yat-sen University, Guangzhou, China

H.J. Zhou
Interdepartmental Program in Vascular Biology and Therapeutics, Department of Pathology, Yale University School of Medicine, 10 Amistad St, 401B, New Haven, CT 06520, USA

W. Min (✉)
The First Affiliated Hospital, Center for Translational Medicine, Sun Yat-sen University, Guangzhou, China

Interdepartmental Program in Vascular Biology and Therapeutics, Department of Pathology, Yale University School of Medicine, 10 Amistad St, 401B, New Haven, CT 06520, USA
e-mail: wang.min@yale.edu

G. Santulli (ed.), *Mitochondrial Dynamics in Cardiovascular Medicine*, Advances in Experimental Medicine and Biology 982,
DOI 10.1007/978-3-319-55330-6_14

mitochondrial thioredoxin system in heart failure is required for the development of new therapeutic strategies. Here, we provide an overview of current advances in the understanding of mechanistic role of thioredoxin 2 in heart failure.

ROS Generation and Elimination Mechanisms

Heart failure is a physiological state that occurs when the heart is unable to pump enough blood to meet the body's needs. Mitochondria are the primary place to produce adenosine triphosphate (ATP) that supports the pumping activity of the heart. The insufficient energy supplying in cardiac cells leads to heart failure [12–14]. There are variety supportive productions of energy in heart, including glucose, fatty acids, lactate, ketones, and amino acids. Of these metabolic progresses, Nicotinamide adenine dinucleotide hydrogen (NADH), reduced flavin adenine dinucleotide (FADH2) and acetyl-coenzyme (Acetyl-CoA) are produced. Then acetyl-CoA enters citric acid cycle to generate enough amount of ATP to meet the heart needs. Meanwhile, more NADH and FADH2 are byproducts in this progress. However, when the heart has insufficient energy to bump blood, the resultant NADH and FADH2 enter into electron transport chain, which produces ATP by consuming oxygen through oxidative phosphorylation in the heart mitochondria. During this progress, 0.4–4 % of the consumed oxygen is released in the mitochondria as ROS. Thus, the formation of "electron leakage" phenomenon in electron transport chain leads to ROS generation [15–17]. Besides, there are also other ROS-producing systems including numerous NAD(P)H oxidases, xanthine oxidase, the uncoupling of NO synthase [18–21]. The univalent reduction of molecular oxygen will transform into O_2^- by leaking electrons from the mitochondrial electron transport chain. O_2^- turns into H_2O_2 through mitochondria-specific manganese-dependent superoxide dismutase (MnSOD). H_2O_2 could be converted to oxidant $\cdot OH$ via the Fenton reaction without suitable regulation. So as other ROS members, like O_2^- and $\cdot OH$ [22].

Although physiological levels of ROS are benefit for phagocytes to deal with the bacteria and other metabolism, excessive ROS disrupts mitochondrial structure and function, leading to severe derangement of cells or apoptosis in heart [23]. Overexpression of catalase in mitochondria, but not in other cellular compartments, prolongs the life span of mice [18, 19, 24, 25]. Thus, how to deal with the excessive ROS is essential for the heart and other tissues.

There are several cellular mechanisms to regulate the counterbalance formation of ROS, including enzymatic and nonenzymatic pathways [10]. Both catalase and glutathione peroxidase are the members of well-clarified enzymatic pathways and they can catalysis H_2O_2 into H_2O dependently or cooperatively. The superoxide dismutases (SODs) have the gift to change O_2^- into H_2O_2 [26–29]. Additionally, thioredoxin and thioredoxin reductase coordinately form an additional enzymatic antioxidant and redox regulatory system [10]. The thioredoxin system can catalyze the regeneration of other antioxidant molecules to defense against ROS, such as ubiquinone (Q10), lipoic acid, and ascorbic acid. Calculative data show that the heart-specific deletion of thioredoxin or thioredoxin reductase in mice results in myocyte apoptosis and heart developmental abnormalities [23, 30].

Nonenzymatic pathways include intracellular antioxidants such as the vitamins E, C, and β-carotene (a precursor to vitamin A), ubiquinone, lipoic acid, and urate.

Trx2 is an essential gene regulating mitochondria-dependent apoptosis, which also play an antioxidative role in scavenging ROS. Since thioredoxin system is essential for maintaining mitochondrial integrity and reducing ROS generation, we will mainly introduce the role Trs2 in heart failure.

The Thioredoxin (Trx) System

The thioredoxin (Trx) system, consisting of Trx, Trx reductase (TrxR), and peroxidase (Prx), plays an important role in regulating the cellular redox state [31–33]. Trxs are small redox proteins, which have proximate cysteines in their conserved CXXC motif. Reduced Trx reduces the disulfide bonds in target proteins through cysteine thiol-disulfide exchanges. In turn, oxidized Trx is reduced by TrxR and NADPH to facilitate its redox activity. Trx-dependent Prx plays a direct role in scavenging ROS (H_2O_2). There are at least three distinct variants of Trxs in human: thioredoxin-1 (Trx1), mitochondrial Trx2, and Sp-trx [11, 34]. In mammalian cells, Cytosolic Trx1 system is composed of Trx1 reductase (TrxR1), Trx1, and Trx1-dependent peroxidase. The mitochondrial-specific Trx2 system consists of Trx2 reductase (TrxR2), Trx2, and peroxiredoxin-3 (Prx3), which is highly expressed in tissues with great metabolic demand, such as the heart, brain, and liver [31–33]. Besides their character in regulating the cellular redox state, both cytosolic Trx1 and mitochondrial Trx2 can form a complex with a redox-sensitive serine/threonine kinase ASK1, which is activated in response to oxidative stress [35]. When the cellular ROS increase, redox-active cysteine residues in Trxs catalytic center become oxidized, and Trxs are dissociated from the Trx-ASK1 complex, which lead to the autoactivation of ASK1 and induct mitochondrial apoptotic cell death pathway [36, 37].

Roles of Thioredoxin System in Heart Failure

Genetic deletion of Trx1, TrxR1, Trx2, or TrxR2 leads to embryonic lethality likely attributable to the increased cellular oxidative stress [30, 38–40]. Mice with Trx1 homozygous mutants die shortly in embryonic period because of the defects of proliferation ability of cell mass cells [41]. The deletion of TrxR1 in vivo causes early embryonic morbidity and homozygous mutation embryos live a short lifespan and fail to survival [40]. Besides, embryonic fibroblasts with TrxR1 knockout can't undergo further proliferation in vitro. A global deletion Trx1 or TrxR1 display similar growth retardation and reduced cell proliferation. Mice with systemic knockout of TrxR2 show severely anemic, increased apoptosis in the liver, and thinner of the ventricular heart wall, resulting in embryonic lethality at embryonic day 13 (E13) [30]. Trx2 homozygous mutant embryos cannot survive to birth with massive apoptosis and anencephaly [38]. Cells treating with buthionine-sulfoximine in order to delete Trx2 leads to an increased intracellular ROS accumulation, as well as the decreased number of total and mitochondrial glutathione (GSH) level [39]. Heart-specific deletions of TrxR1 or TrxR2 show different phenotypes. Mice with TrxR1 heart-specific deletion are not affected and develop normally and appear healthy [40]. It seems that TrxR1 is not that important for cardiomyocytes formation. On the contrary, the deletion of cardiac cell-specific TrxR2 gene in mice causes fatal dilated cardiomyopathy and died quickly after birth [30]. Of potential clinical relevance, a recent clinical genetic study demonstrates that TrxR2 loss-of-function mutations cause dilated cardiomyopathy [42]. According to these results, it seems that both Trx2/TrxR2 are essential for maintaining cardiac normal function. Since Trx2 has capacities to directly scavenge ROS and catalyze mitochondrial thiol-disulfide exchanges, and this capacity is more obvious than TrxR2, we believe that the expression and activity of Trx2 play a crucial role in maintaining the normal cardiac function. But the mechanistic role of Trx2 in heart development and pathogenesis is not clear. Therefore, we recently have reported the role of Trx2 in human hearts with cardiomyopathy and create a mouse line with cardiac-specific deletion of Trx2 to investigate the intrinsic role in the heart.

Myocardium form patients with severe idiopathic DCM shows significant reduced Trx2 expression compared with that form organ donors with preserved cardiac function [23]. However, TrxR2 and Prx3, the other two components of Trx2 system, are increased in idiopathic DCM. Prx3 is an indicator of oxidative stress, whose function is dependent on the dimerization of Trx2. TrxR2 expression is increased, which possibly results from a compensatory expression for reduced Trx2. Trx2 play an inhibitory effect on ASK1 activity and apoptosis [37]. Consistent with this concept, phosphorylated ASK1 and active caspase-3 are observably increased in idiopathic DCM patients. All these results suggest that Trx2 is important in maintaining myocardial function through inhibiting oxidative stress and ASK1 activity.

Because of embryonic lethality in Trx2 global knockout mice, the function of Trx2 in cardiac function in the post-developmental environment has not been clarified before [30, 38–40]. Therefore, in the cardiac-specific Trx2 knockout (Trx2-cKO) mice model, we also identify that Trx2 is critical to preserving cardiac function through maintaining mitochondrial integrity and function, suppressing ROS production, and preventing ASK1-dependent apoptosis [23]. The Trx2-cKO mice develop a spontaneous DCM, characterized by increased heart size, wall thinning, interstitial fibrosis, and a progressive decline in left ventricular contractile function, resulting in heart failure and death by 4 months of age. At the cellular level, cardiomyocyte shows age-dependent increases of cell size in Trx2-cKO mice hearts. The molecular analysis shows conspicuous expression changes of genes that participate in human heart failure in the Trx-cKO mice. Taken together, these results suggest that Trx2 is essential for maintaining normal cardiac structure and function during early cardiac growth.

Roles of Trx2 in Cardiomyocyte and Heart Failure

Morphological and biochemical analyses of Trx2-cKO hearts indicate that Trx2 deletion results in a progressive loss of mitochondrial integrity and function, leading to mitochondrial swelling, ultrastructural derangement, and reduced respiration and ATP generation. Meanwhile, ROS generation and cellular apoptosis are age-dependent increased in Trx2-cKO mitochondria compared with WT control, revealing that Trx2 deletion directly promotes mitochondrial ROS generation and cellular apoptosis. Both increased ROS and cellular apoptosis have been forcefully associated with the pathological process of hypertrophy and heart failure [2–5, 43, 44]. Besides, Trx2-cKO strongly induces increased ASK1 activity without significant increases in JNK phosphorylation, which is consistent with our previous result that Trx2-ASK1 regulates mitochondrial apoptosis in a JNK-independent way [37]. Consistently, Trx2 deletion cardiomyocytes show a similar cellular phenotype of mitochondrial dysfunction with increased cellular ROS, ASK1 activation, and apoptosis, demonstrating that the cardioprotective function of Trx2 observed in the heart is intrinsic to the cardiomyocyte. Selective ASK1 inhibition reduces ROS generation and apoptosis in cardiomyocytes and retards the progression of DCM and heart failure in Trx2-cKO mice. These results suggest that Trx2 is important for suppressing ROS-induced ASK1 activation and keeping off maladaptive left ventricular (LV) remodeling as a key antioxidant enzyme in the heart.

Both the cytosolic Trx1/TrxR1 and the mitochondrial Trx2/TrxR2 systems play a protective role in the pathogenesis of cardiac dysfunction [7, 45]. Trx1 shows cardioprotective under multiple pathological conditions, which is well established and reviewed previously [46]. Trx1 protects cardiomyocytes mainly through its antioxidative activities and interaction with several critical signaling molecules [47]. Exogenous Trx1 has cardioprotective effects in myocardial ischemia/reperfusion through its antiapoptotic activities [48]. Trx1 gene therapy in infarcted myocardium of diabetic rats reduces ROS generation and cardiomyocytes apoptosis, resulting in enhanced angiogenic signaling and reduced ventricular remodeling [49]. In this model, Trx1 gene therapy significantly increases the expression of anti-apoptotic p38MAPKβ and decreased the expression of proapoptotic p38MAPKα and phosphorylated JNK (p-JNK). Trx1 transgenic mouse hearts overexpressing Trx1 have a protective role against to adriamycin-induced cardiotoxicity by reducing oxidative stress and to ischemia/reperfusion through redox signaling [50, 51]. On the contrary, transgenic mice with cardiac-specific overexpression of a dominant negative mutant (C32S/C35S) of Trx1 exhibit cardiac hypertrophy not DCM or heart failure, in which the activity of endogenous Trx1 is diminished [52]. Trx1 attenuates cardiac hypertrophy through modulating the redox modification and nucleocytoplasmic shuttling of class II histone deacetylases, master regulators of cardiac hypertrophy [53]. However, mitochondrial Trx2 has different functions from the Trx1 in heart. In our study, the cardiac-specific deletion of Trx2 causes severe spontaneous DCM, resulting in heart failure. In agreement with our study, mice with heart-specific ablation of TrxR2 also develop to fatal DCM and morphological abnormalities of cardiomyocytes. [30] Auranofin (an antirheumatic gold compound) is known as an efficient inhibitor of TrxR2. Administration of auranofin leads to poor cardiac recovery following ischemia/reperfusion and increased cardiac apoptosis [54]. Besides, there is a selenocysteine in the active site of TrR2, whose activity relies on the presence of this amino acid. Removing selenium from the diet of rats downregulates the activity of TrxR2 and also results into poor cardiac recovery following ischemia/reperfusion [54]. Analogously, cardiac-specific deletion of TrxR2 in the adult mice results into an excessive vulnerability to ischemia/reperfusion injury [55]. A recent clinical study has indicated that mutations in the mitochondrial TrxR2 gene cause DCM in patients [42]. Two mutations in the highly conserved helices result in acid substitutions and reduced flavin-adenine dinucleotide (FAD) binding of TrxR2. These mutations likely do harm to the function of cardiomyocyte through a dominant negative mechanism. In addition, TrxR2 plays a critical role during the progression of age-related heart failure [46, 56]. These results show that mitochondrial TrxR2, like Trx2, is indispensable to endogenous cardioprotection. As we reported recently, TrxR2 inhibitor auranofin reduces Trx2 protein level but not its mRNA, suggesting that TrxR2 is important for the stability of Trx2 [57]. In Trx2-cKO mice, the expression of mitochondrial antioxidant proteins TrxR2 and superoxide dismutase 2 (SOD2) are increased by compensatory mechanism, while the cytosolic Trx1 and glutathione peroxidase 1 are not specifically altered.

Cardiomyocyte Apoptosis and Heart Diseases

Cardiomyocyte apoptosis is causally related to the pathogenesis of DCM and heart failure [58, 59]. Several studies demonstrate that reduced apoptosis protects the heart from failing [6, 60]. Cardiac-restricted inducible expression of caspase-8 fusion protein lead to very low levels of cardiomyocyte apoptosis, which is sufficient to cause a lethal DCM within 8–24 weeks [59]. In the Trx2-cKO mice, it takes similar time to cause DCM and lethality [23]. Overexpression of Trx2 can inhibit apoptosis induced by t-butylhydroperoxide and etoposide [61]. However, since by cell number 75 % cells in the heart are non-myocytes, the importance of non-myocyte apoptosis in heart may be overlooked and the apoptosis in non-myocytes is greater than in myocytes in heart failure [62, 63]. The antiapoptotic role of Trx2 in other non-myocyte cells in the heart has not been elucidated.

Mitochondrial Dysfunction and Heart Failure

Since the heart is an energy-demanding organ, destruction of cardiac energy metabolism and mitochondrial function have been inexorably linked to cardiac dysfunction [64]. It is increasingly recognized that mitochondrial dysfunction contributes to damaged myocardial energetics and increased oxidative stress in heart failure. Trx2 is a small mitochondrial antioxidant, which protects heart against excessive mitochondrial ROS and keeps the integrity and function of Mitochondria [23]. In the Trx2-cKO mice, the disrupted structure and reduced function most likely directly result in DCM. Trx2-cKO hearts have a progressive loss of mitochondrial integrity and function, as measured by mitochondrial DNA copy number and mass, ultrastructure, mitochondrial permeability, ATP production and oxygen consumption rate. The well-aligned rows of mitochondria are in close proximity to the longitudinally oriented cardiac myofibrils, offering a quick and constant supply of ATP for contractile function [65]. On the contrary, mitochondrial arrays are disorganized and swollen mitochondria are aggregated with mild lysis of the cristae in Trx2-cKO hearts. Besides, the cytochrome c oxidases (COX1/4) expression has a dramatic reduction, which is an important terminal enzyme complex of in the mitochondrial electron transport chain. All these alterations result in the disruption of the mitochondrial membrane potential, significantly reduced ATP production and ROS production. Overexpression of Trx2 enhances mitochondrial membrane potential [66]. Exome sequencing has found a homozygous stop mutation of Trx2 in a 16-year-old adolescent with early-onset neurodegeneration and severe cerebellar atrophy [67]. Fibroblasts from the patient show absence of Trx2 protein, increased ROS, disrupted oxidative stress and oxidative phosphorylation dysfunction. Interestingly, reconstitution the expression of Trx2 restores all the parameters, demonstrating that Trx2 mutations play a causal role in the disease development. However, a Trx2 gene mutation involved in human DCM or heart failure has not been found. Exome sequencing should be done to find the endogenous Trx2 mutations in patients with heart diseases.

Signaling Pathways of Trx2 in Heart Failure

Many extracellular factors have capacities of inducing heart failure, and many of the various downstream signaling pathways that mediate the heart failure growth response to these factors can be activated directly or indirectly by ROS. These include apoptosis-signaling kinase 1 (ASK1), protein kinase C (PKC); the MAPKs p38, c-Jun N-terminal kinase (JNK), and ERK1/2; phosphatidylinositol 3-kinase (PI3K); Protein kinase B (PKB); several tyrosine kinases (e.g., src and FAK); nuclear factor-κB (NF-κB) and calcineurin [68, 69]. For instance, ASK1 protein, a serine/threonine kinase upstream of p38/JNK, plays an important role in mediating apoptosis and hypertrophy in the heart. The reduced form of Trx2 physically interacts with ASK1 and suppresses its kinase activity. Mice with Trx2 heart specific knockout presents ASK1 kinase dramatically up-regulation in the early stage of DCM disease progression. Besides, the morphological and biochemical changes including mitochondrial integrity damage, cellular apoptosis and cardiac dysfunction will emerge in the Trx2-cKO heart with ASK1 overexpression. Additionally, ASK1 is also detected in normal heart mitochondria. A surprising result shows that GS-444217, a selective, small molecule inhibitor of ASK1, preserves Trx2-KO

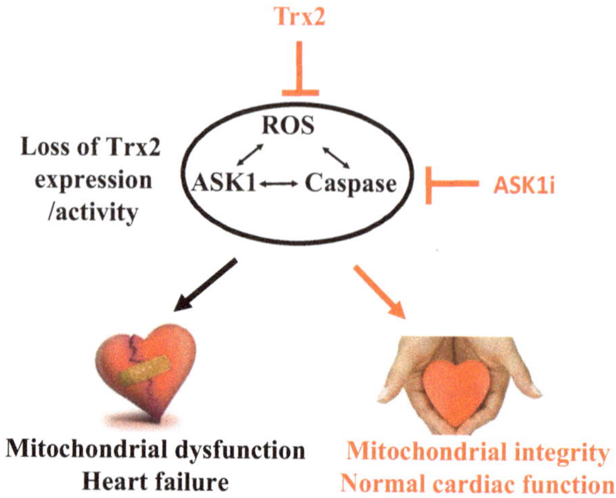

Fig. 14.1 Trx2-cKO Model and therapeutics for the treatment of heart failure. Loss of Trx2 expression or activity in heart induces ASK1 activation, ROS production and caspase-dependent cardiomyocyte apoptosis, leading to mitochondrial dysfunction and heart failure. ASK1 inhibitor could blunt activities of ASK1, ROS and caspases, and rescue heart failure in Trx2-deficient mice

mice from mitochondrial dysfunction, ROS accumulation, and apoptosis in cardiomyocytes. Besides, inhibition of ASK1 maintains mitochondrial integrity and function by blocking ROS production, cellular apoptosis which is induced by Trx2 deletion. It seems that ASK1 activation, mitochondrial dysfunction and ROS production are 3 interdependent issues that may form a vicious positive-feedback cycle, which promotes cardiomyocyte cell apoptosis (Fig. 14.1). Two recent reports have proved that a small molecule inhibitor of ASK1 could also reduce myocardial infarct size in rat and mouse ischemia/reperfusion models [70, 71]. ASK1 inhibition is a potential therapeutic strategy for the treatment of DCM and heart failure.

Conclusion

The heart is a prominent oxygen-consuming organ and the redox states critically affect its function. In this chapter, we have reviewed the main mechanistic role of Trx2 in heart failure through suppressing ROS generation, keeping mitochondria integrity and function, and inhibiting ASK1-dependent cardiomyocytes apoptosis. Although numerous studies have clarified the mechanisms of Trx2 in heart failure, there still remain issues to be solved. Future work in the field needs to better address several key aspects that include (a) the mechanism(s) responsible for Trx2 downregulation in human DCM and heart failure, (b) the role of Trx2 in other nonmyocyte cells in heart, (c) identification and research if genetic mutations or polymorphisms of the Trx2 gene are participated in DCM in humans. A better understanding of the role of the mitochondrial thioredoxin system in the physiological and pathological processes of heart may allow the development of more specific therapeutic strategies for heart diseases.

References

1. Mozaffarian D, Benjamin EJ, Go AS, Arnett DK, Blaha MJ, Cushman M, Das SR, de Ferranti S, Despres JP, Fullerton HJ, Howard VJ, Huffman MD, Isasi CR, Jimenez MC, Judd SE, Kissela BM, Lichtman JH, Lisabeth LD, Liu S, Mackey RH, Magid DJ, McGuire DK, Mohler 3rd ER, Moy CS, Muntner P, Mussolino ME, Nasir K, Neumar RW, Nichol G, Palaniappan L, Pandey DK, Reeves MJ, Rodriguez CJ, Rosamond W, Sorlie PD, Stein J, Towfighi A, Turan TN, Virani SS, Woo D, Yeh RW, Turner MB. Heart disease and stroke statistics-2016 update: a report from the American Heart Association. Circulation. 2016;133:e38–60.
2. Mann DL, Bristow MR. Mechanisms and models in heart failure: the biomechanical model and beyond. Circulation. 2005;111:2837–49.
3. Murdoch CE, Zhang M, Cave AC, Shah AM. NADPH oxidase-dependent redox signalling in cardiac hypertrophy, remodelling and failure. Cardiovasc Res. 2006;71:208–15.
4. Santulli G, Xie W, Reiken SR, Marks AR. Mitochondrial calcium overload is a key determinant in heart failure. Proc Natl Acad Sci USA. 2015;112:11389–94.
5. Giordano FJ. Oxygen, oxidative stress, hypoxia, and heart failure. J Clin Invest. 2005;115: 500–8.
6. Crow MT, Mani K, Nam YJ, Kitsis RN. The mitochondrial death pathway and cardiac myocyte apoptosis. Circ Res. 2004;95:957–70.
7. Ago T, Sadoshima J. Thioredoxin and ventricular remodeling. J Mol Cell Cardiol. 2006;41: 762–73.
8. Aon MA, Stanley BA, Sivakumaran V, Kembro JM, O'Rourke B, Paolocci N, Cortassa S. Glutathione/thioredoxin systems modulate mitochondrial H2O2 emission: an experimental-computational study. J Gen Physiol. 2012;139:479–91.
9. Handy DE, Loscalzo J. Redox regulation of mitochondrial function. Antioxid Redox Signal. 2012;16:1323–67.
10. Nordberg J, Arner ES. Reactive oxygen species, antioxidants, and the mammalian thioredoxin system. Free Radic Biol Med. 2001;31:1287–312.
11. Miranda-Vizuete A, Damdimopoulos AE, Spyrou G. The mitochondrial thioredoxin system. Antioxid Redox Signal. 2000;2:801–10.
12. Sanbe A, Tanonaka K, Kobayasi R, Takeo S. Effects of long-term therapy with ACE inhibitors, captopril, enalapril and trandolapril, on myocardial energy metabolism in rats with heart failure following myocardial infarction. J Mol Cell Cardiol. 1995;27:2209–22.
13. Neubauer S, Horn M, Cramer M, Harre K, Newell JB, Peters W, Pabst T, Ertl G, Hahn D, Ingwall JS, Kochsiek K. Myocardial phosphocreatine-to-ATP ratio is a predictor of mortality in patients with dilated cardiomyopathy. Circulation. 1997;96:2190–6.
14. Beer M, Seyfarth T, Sandstede J, Landschutz W, Lipke C, Kostler H, von Kienlin M, Harre K, Hahn D, Neubauer S. Absolute concentrations of high-energy phosphate metabolites in normal, hypertrophied, and failing human myocardium measured noninvasively with (31) P-SLOOP magnetic resonance spectroscopy. J Am Coll Cardiol. 2002;40:1267–74.
15. Davies KJ. Oxidative stress: the paradox of aerobic life. Biochem Soc Symp. 1995;61:1–31.
16. Miwa S, Brand MD. Mitochondrial matrix reactive oxygen species production is very sensitive to mild uncoupling. Biochem Soc Trans. 2003;31:1300–1.
17. Genova ML, Pich MM, Biondi A, Bernacchia A, Falasca A, Bovina C, Formiggini G, Parenti Castelli G, Lenaz G. Mitochondrial production of oxygen radical species and the role of coenzyme Q as an antioxidant. Exp Biol Med (Maywood). 2003;228:506–13.
18. Ballinger SW, Patterson C, Knight-Lozano CA, Burow DL, Conklin CA, Hu Z, Reuf J, Horaist C, Lebovitz R, Hunter GC, McIntyre K, Runge MS. Mitochondrial integrity and function in atherogenesis. Circulation. 2002;106:544–9.
19. Nishikawa T, Edelstein D, Du XL, Yamagishi S, Matsumura T, Kaneda Y, Yorek MA, Beebe D, Oates PJ, Hammes HP, Giardino I, Brownlee M. Normalizing mitochondrial superoxide production blocks three pathways of hyperglycaemic damage. Nature. 2000;404:787–90.

20. Wolin MS. Interactions of oxidants with vascular signaling systems. Arterioscler Thromb Vasc Biol. 2000;20:1430–42.
21. Cai H. NAD(P)H oxidase-dependent self-propagation of hydrogen peroxide and vascular disease. Circ Res. 2005;96:818–22.
22. Balaban RS, Nemoto S, Finkel T. Mitochondria, oxidants, and aging. Cell. 2005;120:483–95.
23. Huang Q, Zhou HJ, Zhang H, Huang Y, Hinojosa-Kirschenbaum F, Fan P, Yao L, Belardinelli L, Tellides G, Giordano FJ, Budas GR, Min W. Thioredoxin-2 inhibits mitochondrial reactive oxygen species generation and apoptosis stress kinase-1 activity to maintain cardiac function. Circulation. 2015;131:1082–97.
24. Schriner SE, Linford NJ, Martin GM, Treuting P, Ogburn CE, Emond M, Coskun PE, Ladiges W, Wolf N, Van Remmen H, Wallace DC, Rabinovitch PS. Extension of murine life span by overexpression of catalase targeted to mitochondria. Science. 2005;308:1909–11.
25. Ramachandran A, Levonen AL, Brookes PS, Ceaser E, Shiva S, Barone MC, Darley-Usmar V. Mitochondria, nitric oxide, and cardiovascular dysfunction. Free Radic Biol Med. 2002;33: 1465–74.
26. Kirkman HN, Gaetani GF. Catalase: a tetrameric enzyme with four tightly bound molecules of NADPH. Proc Natl Acad Sci USA. 1984;81:4343–7.
27. Kirkman HN, Rolfo M, Ferraris AM, Gaetani GF. Mechanisms of protection of catalase by NADPH. Kinetics and stoichiometry. J Biol Chem. 1999;274:13908–14.
28. Ursini F, Maiorino M, Brigelius-Flohe R, Aumann KD, Roveri A, Schomburg D, Flohe L. Diversity of glutathione peroxidases. Methods Enzymol. 1995;252:38–53.
29. de Haan JB, Bladier C, Lotfi-Miri M, Taylor J, Hutchinson P, Crack PJ, Hertzog P, Kola I. Fibroblasts derived from Gpx1 knockout mice display senescent-like features and are susceptible to H2O2-mediated cell death. Free Radic Biol Med. 2004;36:53–64.
30. Conrad M, Jakupoglu C, Moreno SG, Lippl S, Banjac A, Schneider M, Beck H, Hatzopoulos AK, Just U, Sinowatz F, Schmahl W, Chien KR, Wurst W, Bornkamm GW, Brielmeier M. Essential role for mitochondrial thioredoxin reductase in hematopoiesis, heart development, and heart function. Mol Cell Biol. 2004;24:9414–23.
31. Holmgren A. Antioxidant function of thioredoxin and glutaredoxin systems. Antioxid Redox Signal. 2000;2:811–20.
32. Santos CX, Anilkumar N, Zhang M, Brewer AC, Shah AM. Redox signaling in cardiac myocytes. Free Radic Biol Med. 2011;50:777–93.
33. Lee S, Kim SM, Lee RT. Thioredoxin and thioredoxin target proteins: from molecular mechanisms to functional significance. Antioxid Redox Signal. 2013;18:1165–207.
34. Powis G, Mustacich D, Coon A. The role of the redox protein thioredoxin in cell growth and cancer. Free Radic Biol Med. 2000;29:312–22.
35. Pober JS, Min W, Bradley JR. Mechanisms of endothelial dysfunction, injury, and death. Annu Rev Pathol. 2009;4:71–95.
36. Hatai T, Matsuzawa A, Inoshita S, Mochida Y, Kuroda T, Sakamaki K, Kuida K, Yonehara S, Ichijo H, Takeda K. Execution of apoptosis signal-regulating kinase 1 (ASK1)-induced apoptosis by the mitochondria-dependent caspase activation. J Biol Chem. 2000;275:26576–81.
37. Zhang R, Al-Lamki R, Bai L, Streb JW, Miano JM, Bradley J, Min W. Thioredoxin-2 inhibits mitochondria-located ASK1-mediated apoptosis in a JNK-independent manner. Circ Res. 2004;94:1483–91.
38. Nonn L, Williams RR, Erickson RP, Powis G. The absence of mitochondrial thioredoxin 2 causes massive apoptosis, exencephaly, and early embryonic lethality in homozygous mice. Mol Cell Biol. 2003;23:916–22.
39. Tanaka T, Hosoi F, Yamaguchi-Iwai Y, Nakamura H, Masutani H, Ueda S, Nishiyama A, Takeda S, Wada H, Spyrou G, Yodoi J. Thioredoxin-2 (TRX-2) is an essential gene regulating mitochondria-dependent apoptosis. EMBO J. 2002;21:1695–703.
40. Jakupoglu C, Przemeck GK, Schneider M, Moreno SG, Mayr N, Hatzopoulos AK, de Angelis MH, Wurst W, Bornkamm GW, Brielmeier M, Conrad M. Cytoplasmic thioredoxin reductase is essential for embryogenesis but dispensable for cardiac development. Mol Cell Biol. 2005;25:1980–8.

41. Matsui M, Oshima M, Oshima H, Takaku K, Maruyama T, Yodoi J, Taketo MM. Early embryonic lethality caused by targeted disruption of the mouse thioredoxin gene. Dev Biol. 1996;178:179–85.
42. Sibbing D, Pfeufer A, Perisic T, Mannes AM, Fritz-Wolf K, Unwin S, Sinner MF, Gieger C, Gloeckner CJ, Wichmann HE, Kremmer E, Schafer Z, Walch A, Hinterseer M, Nabauer M, Kaab S, Kastrati A, Schomig A, Meitinger T, Bornkamm GW, Conrad M, von Beckerath N. Mutations in the mitochondrial thioredoxin reductase gene TXNRD2 cause dilated cardiomyopathy. Eur Heart J. 2011;32:1121–33.
43. Byrne JA, Grieve DJ, Cave AC, Shah AM. Oxidative stress and heart failure. Arch Mal Coeur Vaiss. 2003;96:214–21.
44. Grieve DJ, Byrne JA, Cave AC, Shah AM. Role of oxidative stress in cardiac remodelling after myocardial infarction. Heart Lung Circ. 2004;13:132–8.
45. Ago T, Sadoshima J. Thioredoxinl as a negative regulator of cardiac hypertrophy. Antioxid Redox Signal. 2007;9:679–87.
46. Yoshioka J. Thioredoxin reductase 2 (Txnrd2) regulates mitochondrial integrity in the progression of age-related heart failure. J Am Heart Assoc. 2015;4:e002278.
47. Yamawaki H, Haendeler J, Berk BC. Thioredoxin: a key regulator of cardiovascular homeostasis. Circ Res. 2003;93:1029–33.
48. Tao L, Gao E, Bryan NS, Qu Y, Liu HR, Hu A, Christopher TA, Lopez BL, Yodoi J, Koch WJ, Feelisch M, Ma XL. Cardioprotective effects of thioredoxin in myocardial ischemia and reperfusion: role of S-nitrosation [corrected]. Proc Natl Acad Sci U S A. 2004;101:11471–6.
49. Samuel SM, Thirunavukkarasu M, Penumathsa SV, Koneru S, Zhan L, Maulik G, Sudhakaran PR, Maulik N. Thioredoxin-1 gene therapy enhances angiogenic signaling and reduces ventricular remodeling in infarcted myocardium of diabetic rats. Circulation. 2010;121:1244–55.
50. Shioji K, Kishimoto C, Nakamura H, Masutani H, Yuan Z, Oka S, Yodoi J. Overexpression of thioredoxin-1 in transgenic mice attenuates adriamycin-induced cardiotoxicity. Circulation. 2002;106:1403–9.
51. Turoczi T. Thioredoxin redox signaling in the ischemic heart: an insight with transgenic mice overexpressing Trx1. J Mol Cell Cardiol. 2003;35:695–704.
52. Yamamoto M, Yang G, Hong C, Liu J, Holle E, Yu X, Wagner T, Vatner SF, Sadoshima J. Inhibition of endogenous thioredoxin in the heart increases oxidative stress and cardiac hypertrophy. J Clin Invest. 2003;112:1395–406.
53. Ago T, Liu T, Zhai P, Chen W, Li H, Molkentin JD, Vatner SF, Sadoshima J. A redox-dependent pathway for regulating class II HDACs and cardiac hypertrophy. Cell. 2008;133:978–93.
54. Venardos K, Harrison G, Headrick J, Perkins A. Auranofin increases apoptosis and ischaemia-reperfusion injury in the rat isolated heart. Clin Exp Pharmacol Physiol. 2004;31:289–94.
55. Horstkotte J, Perisic T, Schneider M, Lange P, Schroeder M, Kiermayer C, Hinkel R, Ziegler T, Mandal PK, David R, Schulz S, Schmitt S, Widder J, Sinowatz F, Becker BF, Bauersachs J, Naebauer M, Franz WM, Jeremias I, Brielmeier M, Zischka H, Conrad M, Kupatt C. Mitochondrial thioredoxin reductase is essential for early postischemic myocardial protection. Circulation. 2011;124:2892–902.
56. Kiermayer C, Northrup E, Schrewe A, Walch A, de Angelis MH, Schoensiegel F, Zischka H, Prehn C, Adamski J, Bekeredjian R, Ivandic B, Kupatt C, Brielmeier M. Heart-specific knock-out of the mitochondrial thioredoxin reductase (Txnrd2) induces metabolic and contractile dysfunction in the aging myocardium. J Am Heart Assoc. 2015;4:e002153.
57. Chen X, Zhou HJ, Huang Q, Lu L, Min W. Novel action and mechanism of auranofin in inhibition of vascular endothelial growth factor receptor-3-dependent lymphangiogenesis. Anticancer Agents Med Chem. 2014;14:946–54.
58. Narula J, Haider N, Virmani R, DiSalvo TG, Kolodgie FD, Hajjar RJ, Schmidt U, Semigran MJ, Dec GW, Khaw BA. Apoptosis in myocytes in end-stage heart failure. N Engl J Med. 1996;335:1182–9.
59. Wencker D, Chandra M, Nguyen K, Miao W, Garantziotis S, Factor SM, Shirani J, Armstrong RC, Kitsis RN. A mechanistic role for cardiac myocyte apoptosis in heart failure. J Clin Investig. 2003;111:1497–504.

60. Chandrashekhar Y, Sen S, Anway R, Shuros A, Anand I. Long-term caspase inhibition ameliorates apoptosis, reduces myocardial troponin-I cleavage, protects left ventricular function, and attenuates remodeling in rats with myocardial infarction. J Am Coll Cardiol. 2004;43:295–301.
61. Chen Y, Cai J, Murphy TJ, Jones DP. Overexpressed human mitochondrial thioredoxin confers resistance to oxidant-induced apoptosis in human osteosarcoma cells. J Biol Chem. 2002;277: 33242–8.
62. Park M, Shen YT, Gaussin V, Heyndrickx GR, Bartunek J, Resuello RR, Natividad FF, Kitsis RN, Vatner DE, Vatner SF. Apoptosis predominates in nonmyocytes in heart failure. Am J Physiol Heart Circ Physiol. 2009;297:H785–91.
63. Jose Corbalan J, Vatner DE, Vatner SF. Myocardial apoptosis in heart disease: does the emperor have clothes? Basic Res Cardiol. 2016;111:31.
64. Goldenthal MJ. Mitochondrial involvement in myocyte death and heart failure. Heart Fail Rev. 2016;21:137–55.
65. Hom J, Sheu SS. Morphological dynamics of mitochondria – a special emphasis on cardiac muscle cells. J Mol Cell Cardiol. 2009;46:811–20.
66. Damdimopoulos AE, Miranda-Vizuete A, Pelto-Huikko M, Gustafsson JA, Spyrou G. Human mitochondrial thioredoxin. Involvement in mitochondrial membrane potential and cell death. J Biol Chem. 2002;277:33249–57.
67. Holzerova E, Danhauser K, Haack TB, Kremer LS, Melcher M, Ingold I, Kobayashi S, Terrile C, Wolf P, Schaper J, Mayatepek E, Baertling F, Friedmann Angeli JP, Conrad M, Strom TM, Meitinger T, Prokisch H, Distelmaier F. Human thioredoxin 2 deficiency impairs mitochondrial redox homeostasis and causes early-onset neurodegeneration. Brain. 2016;139:346–54.
68. Sabri A, Hughie HH, Lucchesi PA. Regulation of hypertrophic and apoptotic signaling pathways by reactive oxygen species in cardiac myocytes. Antioxid Redox Signal. 2003;5: 731–40.
69. Pajares M, Jimenez-Moreno N, Dias IH, Debelec B, Vucetic M, Fladmark KE, Basaga H, Ribaric S, Milisav I, Cuadrado A. Redox control of protein degradation. Redox Biol. 2015;6: 409–20.
70. Toldo S, Breckenridge DG, Mezzaroma E, Van Tassell BW, Shryock J, Kannan H, Phan D, Budas G, Farkas D, Lesnefsky E, Voelkel N, Abbate A. Inhibition of apoptosis signal-regulating kinase 1 reduces myocardial ischemia-reperfusion injury in the mouse. J Am Heart Assoc. 2012;1:e002360.
71. Gerczuk PZ, Breckenridge DG, Liles JT, Budas GR, Shryock JC, Belardinelli L, Kloner RA, Dai W. An apoptosis signal-regulating kinase 1 inhibitor reduces cardiomyocyte apoptosis and infarct size in a rat ischemia-reperfusion model. J Cardiovasc Pharmacol. 2012;60:276–82.

Chapter 15
Mitochondria in Structural and Functional Cardiac Remodeling

Natalia Torrealba, Pablo Aranguiz, Camila Alonso, Beverly A. Rothermel, and Sergio Lavandero

The heart is the organ responsible for pumping blood to the whole body, allowing thus the distribution of oxygen and nutrients to peripheral tissues and the elimination of toxic products through excretory organs. Cardiac tissue is composed of multiple cell types, including cardiomyocytes, fibroblasts, endothelial cells, vascular smooth muscle cells, macrophages and mast cells. Cardiomyocytes are the specialized cells responsible for myocardial contraction. Although they correspond to only 30–40% of the total number of cells in the heart, cardiomyocytes represent almost 75% of the heart volume [1]. Contractile function is manifest from the fetal stage forward, however, mitotic capacity of cardiomyocytes diminishes in the later stages of embryogenesis and stops almost completely after birth [2]. Therefore, the ability of the heart to respond to physiological or pathological conditions involves cellular remodeling processes rather than cellular proliferation. Cardiac remodeling is classically defined as changes in the structure of cardiac tissue, evidenced by changes in

N. Torrealba • P. Aranguiz • C. Alonso
Advanced Center for Chronic Diseases (ACCDiS), Faculty of Chemical & Pharmaceutical Sciences & Faculty of Medicine, University of Chile, Olivos 1007, Santiago 8380492, Chile

B.A. Rothermel (✉)
Cardiology Division, Department of Internal Medicine, UT Southwestern Medical Center, 6000 Harry Hines Blvd, Dallas, TX 75390, USA

Department of Molecular Biology, University of Texas, Southwestern Medical Center, Dallas, TX, USA
e-mail: Beverly.Rothermel@UTSouthwestern.edu

S. Lavandero (✉)
Advanced Center for Chronic Diseases (ACCDiS), Faculty of Chemical & Pharmaceutical Sciences & Faculty of Medicine, University of Chile, Olivos 1007, Santiago 8380492, Chile

Department of Internal Medicine, University of Texas, Southwestern Medical Center, Dallas, TX, USA
e-mail: slavander@uchile.cl

G. Santulli (ed.), *Mitochondrial Dynamics in Cardiovascular Medicine*, Advances in Experimental Medicine and Biology 982,
DOI 10.1007/978-3-319-55330-6_15

ventricle mass, volume or shape, fibrotic content, vascularization, cellular hypertrophy, and cell death. For instance, when the heart faces sustained work overload, compensatory mechanisms are activated in order to increase the capacity to pump blood. Initial mechanisms involve an increase in the frequency of contraction. When this is not sufficient, cardiac remodeling pathways are triggered. Cardiac hypertrophy can be caused by an increase in the size of individual cardiomyocytes [3] as well as by an increase in fibroblast proliferation (hyperplasia) [4]. It can be classified as adaptive or pathological, depending on the mechanisms activated. In adaptive hypertrophy muscular, vascular, and interstitial compartments maintain normal proportionality and collagen content, thus preserving tissue structure and function. This type of hypertrophy occurs in response to exercise training or during pregnancy, and changes to the heart can revert when the workload normalizes [5]. In contrast, during pathological hypertrophy, intercompartmental proportionality is ultimately lost, leading to structural heterogeneity of the myocardium. Importantly, relative collagen content of the heart increases, leading to myocardial fibrosis and detrimental structural remodeling [6].

Cardiac remodeling also occurs after a myocardial infarction (MI), where a large number of cardiac cells undergo cell death by necrosis, necroptosis, and apoptosis initiated by nutrient and oxygen deprivation. In an attempt to maintain cardiac output surviving cardiomyocytes increase in size, mass and volume. The left ventricle generally undergoes the most pronounced remodeling following MI, because it is the primary chamber responsible for pumping blood to the rest of the body [7]. Similar changes occur in the spared myocardium following damage from ischemia reperfusion (I/R).

Other stimuli leading to cardiac remodeling, and especially to hypertrophy, are elevated blood pressure and volume overload, both regulated primarily through the sympathetic and renin-angiotensin-aldosterone (RAA) systems. The sympathetic system is mediated by the action of norepinephrine and epinephrine on adrenergic receptors, regulating the rate and strength of cardiac contraction, as well as blood pressure in blood vessels [8]. Adrenergic receptors are G protein-coupled receptors that are desensitized upon sustained exposure to agonists, leading to attenuation of receptor responsiveness. Depressed receptor function is strongly associated with ventricular dysfunction, hypertrophy and heart failure [8].

Similarly, the function of the RAA system is to regulate blood pressure. When cardiac function becomes less efficient, secretion of renin by the kidney is increased, leading to an increase in renin- dependent conversion of angiotensinogen into angiotensin I (Ang I). In turn, angiotensin-converting enzyme (ACE) converts Ang I into angiotensin II (Ang II), which binds to angiotensin receptor (AT1) mediating changes in cells of both the vascular system and the heart, primarily involving vasoconstriction of vessels that increases blood pressure. Activation of the RAA is closely associated with the severity of heart failure, dilated cardiomyopathy and left ventricular (LV) hypertrophy [9].

Cardiac fibroblasts are responsible for secreting components of the extracellular matrix (ECM). During cardiac remodeling, fibroblasts can differentiate to myofibroblasts, that express contractile proteins including alpha-smooth muscle actin (α-SMA), and exhibit increased migratory, proliferative and secretory properties [4]. Cardiac myofibroblasts are activated by proinflammatory cytokines (e.g. tumor necrosis factor alpha (TNFα), interleukin-1 (IL-1), interleukin-6 (IL-6), transforming growth factor beta (TGF-β)), vasoactive peptides (e.g. Ang II, endothelin-1, natriuretic peptides) and hormones (e.g. noradrenaline). This leads to increased cell proliferation and migration, enhanced secretion of the previously mentioned cytokines and vasoactive peptides, and secretion of additional growth factors such as VEGF. This can further increase fibrosis, and remodeling of the heart in a feed-forward mechanism. Myofibroblasts are also be activated directly by mechanical stretch and in response to I/R [4].

In addition to changes in organ structure, cardiac remodeling includes changes in structure or function on the subcellular level. Although many organelles in cardiomyocytes undergo remodeling, this chapter will focus on those changes directly related to mitochondria. In particular, we will discuss the ability of mitochondria to undergo dynamic changes that affect both their outer membrane structure (fusion/fission) and their inner membrane structure (cristae remodeling), and how these processes contribute to the development or progression of cardiac pathologies. In addition we will address the cardiovascular consequences of changes to mitochondrial function, including loss of mitochondrial respiration, increase in ROS production, diminished ATP synthesis and the release of pro-death factors into the cytosol.

In the human heart, ATP production is carried out primarily by mitochondria, and is estimated to be as much as 6 kg ATP per day [10]. Although, during the fetal stage, lactate and glucose are the major sources for energy production, the heart undergoes a metabolic remodeling at birth, switching substrate oxidation from glucose to fatty acids. Thus, β-oxidation of free fatty acids and oxidative phosphorylation become the primary mechanisms for ATP production, producing approximately 70% of the ATP consumed by the heart [11].

In addition to ATP production, mitochondria participate in processes involving cell proliferation, differentiation, cellular immunity, calcium regulation, iron homeostasis, lipid metabolism, cellular aging, cell death, production of reactive oxygen species (ROS) and ROS scavenging [12].

In neonatal cardiomyocytes, mitochondria are distributed throughout the cytoplasm and around the nucleus. In contrast, in the adult heart they are packaged and aligned [13], located in parallel, longitudinal rows interleaved with the contractile machinery (interfibrillary mitochondria, IFM) or in monolayers immediately underneath sarcolemma (subsarcolemmal mitochondria, SSM) [14]. These differentially distributed mitochondria, also differ in the morphology of their cristae, being mostly tubular in the IFM and mainly lamelliform in SSM [15]. The contrasting morphologies of the two populations imply metabolic differences, as IFM display a higher rate of oxidative phosphorylation and enzymatic activity than SSM [16]. SSM also exhibit a lower capacity for calcium uptake than IFM [17], potentially leading to increased susceptibility to cytochrome c release and initiation of death signals in this population. Thus these two functionally distinct pools of mitochondria may play different roles in the pathologies leading to cardiac remodeling.

Mitochondria have their own genome (mtDNA), which in mammals is a densely packed double- stranded DNA molecule of 16.6 kb, containing 37 genes encoding 11 messenger RNAs (mRNAs; translated to 13 proteins), 2 ribosomal RNAs (rRNAs) and 22 transfer RNAs (tRNAs) [18]. The proteins encoded in the mitochondrial genome are core constituents of the mitochondrial respiratory complexes I, III, IV and V that are embedded in the inner mitochondrial membrane (IMM) [19]. Similarly, the tRNAs and rRNAs encoded in mtDNA are essential for mitochondrial ribosome assembly the translation of mitochondrial mRNAs [20].

Although mitochondria are the primary organelles responsible for energy production, the nuclear genome is the primary site for encoding the bulk of mitochondrial proteins. Therefore, the two organelles need to communicate closely. This allows nuclear control of mitochondrial ATP production [21] and conversely, changes in mtDNA or mitochondrial function are capable of modifying nuclear gene expression.

Mitochondria are highly dynamic organelles, continuously undergoing repeated cycles of fusion and fission in response to environmental changes and to the metabolic status of the cell (Fig. 15.1). These processes are strictly controlled by the activity of a group of guanosine triphosphatases (GTPases) related to the dynamin family [13]. Mitochondrial fission creates smaller and more circular mitochondria, through the activity of Dynamin-1-like protein (DRP1) and its adapter proteins mitochondrial fission 1 protein (FIS1) and mitochondrial fission factor (MFF) [22]. This process is required for redistribution of mitochondria in mitosis, release of cytochrome c during cell death by apoptosis and for selective mitochondrial degradation (mitophagy) [22]. On the other hand, mitochondrial fusion generates fewer, larger and more elongated mitochondria, in a process regulated by mitofusin (MFN) 1 and 2, both situated in the outer mitochondrial membrane (OMM), and by optical atrophy protein 1 (OPA1), located in the IMM. This process allows the exchange of material (matrix components, damaged mtDNA) and promotes balance in bioenergetics properties (e.g. mitochondrial membrane potential) [11].

Increased oxidative stress, altered calcium load or cellular pH are each capable of impacting mitochondrial dynamics, disrupting mitochondrial membrane potential, and inducing opening of the mitochondrial permeability transition pore (MPTP). This latter process has been widely associated with remodeling of the mitochondrial cristae, especially during I/R [23].

Existing therapies to prevent the progression of cardiac remodeling primarily act at the cardiac or vascular level. These include drugs for reducing hypertension, such as inhibitors of ACE (enalapril), angiotensin receptor blockers (losartan) and beta-blockers (atenolol); and drugs that reduce cholesterol levels like statins and fibrates [4]. However, no drugs currently in use are targeted directly at the causes of subcellular remodeling of mitochondria, warranting further research aimed at understanding the possible relation between the prevention of functional or metabolic remodeling and the prevention or amelioration of cardiac diseases.

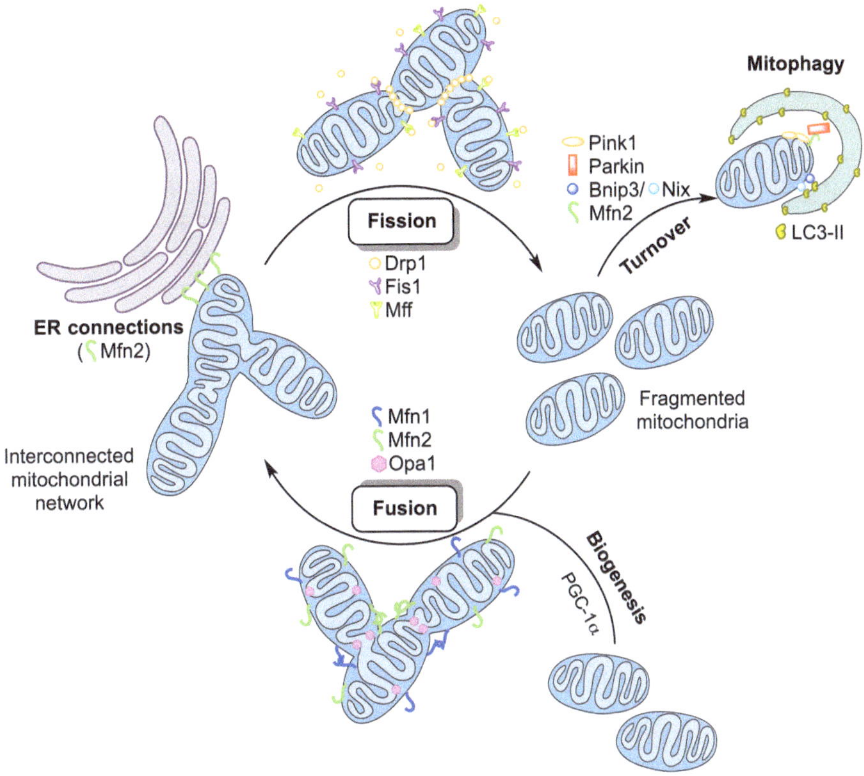

Fig. 15.1 Mitochondrial life cycle. Mitochondrial dynamics enable mitochondria to respond to external stimuli or metabolic signals in a way that maintains proper function. Mitochondrial fusion is regulated by mitofusin (*Mfn*) 1 and 2 and optical atrophy protein 1 (*Opa1*), at the outer and inner mitochondrial membranes, respectively. This process results in a more interconnected mitochondrial network that enhances communication with the endoplasmic reticulum (*ER*) through mitochondria-associated endoplasmic reticulum membranes (*MAM*) and a complex mechanism involving the presence of Mfn2 amongst other proteins, thus enhancing calcium movement from ER to mitochondria and regulating its function. Conversely, Drp1 and its adapter proteins Fis1 and Mff control mitochondrial fission. This process is required for selective mitochondrial turnover (mitophagy), which allows isolation of damaged mitochondria to ensure proper mitochondrial quality control. Here, the mitochondrial kinase Pink1 phosphorylates Mfn2 which allows Parkin E3 ligase bind phosphorylated Mfn2 to catalyze the ubiquitination of several proteins on the OMM that promotes interaction with LC3 at nascent phagophores, initiating mitochondrial autophagy. Nix and Bnip3 are mitophagy regulators that sense myocardial damage and facilitate the elimination of damaged cardiac mitochondria. Peroxisome proliferator activated receptor gamma coactivator 1α (PGC-1α) is a critical factor that coordinates mitochondrial biogenesis and the expression of metabolic genes and therefore critical to the synthesis aspect of mitochondrial dynamics

Mitochondrial Dynamics

The term mitochondrial dynamics comprises at least three processes. One is the capability of mitochondria to undergo fission and fusion processes, changing its network between punctate and fragmented, and elongated and interconnected mitochondria, respectively [6]. Secondly remodeling of the mitochondrial matrix, which influences metabolite diffusion, and thus, mitochondrial metabolism [24]. And finally, mitochondrial motility within the cell, which is mediated by the interplay of mitochondrial fission and fusion proteins with kinesin 1 and 3 motors and the adaptors Milton and Miro (Miro/Milton/kinesin complex), thus allowing interaction of mitochondria with cytoskeletal components to facilitate their movement along microtubules [25].

In cardiac cells, the most prominent events related to mitochondrial dynamics are fission and fusion and remodeling of the cristae. These dynamic processes occur in response to external stimuli and metabolic signals. Changes in mitochondrial morphology have been implicated in cell division, embryonic development, apoptosis, autophagy, metabolism, development, and differentiation [26]. In the heart, disruption of these mechanisms has been identified in multiple cardiac diseases, including cardiac hypertrophy, heart failure, dilated cardiomyopathy, and ischemic heart disease. In addition, modification of these processes has been shown to protect the heart against injury in specific settings [27].

Mitofusins 1 and 2

Mitofusins are present in all tissues, but in the heart the levels of mRNA encoding these proteins are particularly high, with MFN2 in greater abundance than MFN1 [28]. Both are members of the family of transmembrane GTPase and reside in the OMM. They mediate OMM fusion through GTP hydrolysis. MFN1 has a higher GTPase activity than MFN2, making it more efficient in the fusion process [29]. The C- terminal end of the proteins is oriented towards the cytosol and contains both a coiled-coil domain 2 (also called hepta-repeat domain or HR2) and a transmembrane domain. The N-terminal end contains the GTP binding domain and another coiled-coil domain (HR1) [30]. To facilitate fusion of the OMM, the HR2 domains of MFN1 or MFN2 form dimeric complexes with MFN1 or MFN2 on adjacent mitochondrion, allowing tethering of the two mitochondria.

OPA1

OPA1 is expressed throughout the body, but is present in greater quantity in the retina, brain, testis, heart and skeletal muscle [31]. This protein is a large GTPase located in the IMM, whose function is to promote mitochondrial fusion of the IMM and is also involved in remodeling of mitochondrial cristae [32]. The protein is comprised of a N-terminal domain containing a mitochondrial import sequence (MIS, which is removed by a mitochondrial matrix protease when the protein is imported to the mitochondria), a transmembrane domain (TM), a region of alternative splicing, a coiled-coil region, a GTPase domain, a middle domain, and a GTPase effector domain (GED or assembly domain) at the C-terminus [26]. In addition to the GTPase domain, OPA1 requires the presence of MFN1 to induce mitochondrial fusion [33]. OPA1 undergoes post-translational modifications by alternative splicing at exons 4, 4b, and 5b, generating eight human isoforms of OPA1. Four of these isoforms (3, 5, 6, and 8) can be cleaved by YME1L, an intermembrane AAA (ATPase associated with diverse cellular activities) protease, generating shorter forms of OPA1 (S-OPA1), which are soluble and reside in the mitochondrial intermembrane space. The other isoforms of OPA1 (1, 2, 4, and 7) do not normally undergo cleavage and constitute the long forms of OPA1 (L-OPA1), which contain the TM domain, and are anchored into the IMM. Both, long and short isoforms are required for mitochondrial fusion [26]. Under conditions of mitochondrial membrane depolarization, ATP deficiency and apoptosis induction, the L-OPA1 isoforms undergo inducible cleavage by OMA1 generating short forms of OPA1. This prevents the pro-fusion activity of OPA1, resulting in mitochondrial fragmentation. Through this process, depolarized fragmented mitochondria can be removed by mitophagy [34]. In addition, short and long forms of OPA1 form oligomers, keeping cristae junctions tightly closed, thereby opposing cytosolic release of cytochrome c that can initiate apoptosis [35].

DRP1

DRP1 is abundant in brain, heart and muscle [36]. It is located in the cytosol and translocates to the OMM to induce fission or division of mitochondria [26]. DRP1 contains a GTPase domain, a central domain and a GTPase effector domain or assembly domain [36], which regulates the GTPase activity of the protein and directs it to the mitochondria [37]. To initiate fission, DRP1 translocates to the OMM where it binds to the receptor, FIS1 or MFF, undergoes oligomerization, and then begins constriction of the mitochondrial scission site using GTP to power the process [37]. DRP1 is regulated by a diversity of post- translational modifications such as phosphorylation and nitrosylation [38]. Particularly important in cardiac remodeling, dephosphorylation of DRP1 by calcineurin initiates fission, whereas phosphorylation of the same site by PKA opposes it.

FIS1

FIS1 is uniformly distributed in the OMM and is comprised of an N-terminal domain, required for protein–protein interactions, FIS1 oligomerization, and mitochondrial fission; a transmembrane domain and a C-terminal domain exposed to the intermembrane space, that are essential for mitochondrial localization and the induction of apoptosis [39]. It has been shown that, although FIS1 is required for mitochondrial fission, it is not a limiting factor in the process and mitochondrial recruitment of DRP1 can be regulated by other factors [45].

MFF

Mitochondrial fission factor (MFF) mediates both mitochondrial and peroxisome fission and is highly expressed in heart, kidney, liver, brain, muscle, and stomach [40]. It has a C-terminal transmembrane domain essential for its localization to mitochondria, anchoring to the OMM, and the ability to multimerize, exhibiting similar properties to FIS1 [40]. MFF (and specifically its N-terminal cytosolic region) is required for DRP1 recruitment to mitochondria and together MFF and DRP1 are capable of mediating mitochondrial fission independent of FIS1 [41]. This indicates that FIS1 and MFF are independent of one another and may activate fission in response to different stressors. Interestingly, both proteins also mediate fission of peroxisomes [40].

Mitochondrial Elongation Factors

The mitochondrial elongation factor proteins, MIEF1 and MEIF2 are also known as mitochondrial dynamics proteins MiD49 and MiD51 respectively [49]. The levels of MIEF1 mRNA are high in adult human heart, skeletal muscle, pancreas and kidney [42]. The activity of these proteins has not been fully characterized, however, studies indicate that although overexpression of MiD49/MiD51 recruits DRP1 to mitochondria, it actually reduces DRP1's GTP-binding activity, thereby promoting mitochondrial fusion and increasing mitochondrial elongation. Furthermore, knockdown of MiD49/MiD51 increases fragmentation [42] and suggests that these proteins may prevent DRP1's interaction with FIS1, and thus, its fission-related function. In contrast, studies from other investigators [43] indicate that, rather than blocking fission, MiDs might recruit DRP1 to mitochondria and maintain it in an inactive state until a cellular signal triggers fission. Palmer et al. also demonstrated that siRNA knockdown of both MiD49 and MiD51 reduced translocation of DRP1 to mitochondria and resulted in mitochondrial fusion and elongation [44]. Yet, overexpression of MiD49/51 blocked fission by sequestering DRP1 specifically at mitochondria, and promoted MFN1 and MFN2-dependent mitochondrial fusion [45]. Thus, the exact function and effect of these proteins remains to be elucidated.

Pathological Mitochondrial Remodeling

As will be expanded upon later in the chapter, pathological remodeling of mitochondria can be particularly onerous in the heart as it can lead to depletion of ATP and increases in intracellular calcium and ROS, thereby triggering mechanisms such as mitophagy or apoptosis, outcomes often observed in the failing heart or following cardiac I/R. In this sense, proteins involved in mitochondrial fission and fusion are particularly important, as they can be critical for maintaining or restoring mitochondrial integrity and function. Cellular imbalances such as oxidative stress, damage to mtDNA or loss of mitochondrial membrane potential, among others, affect the integrity of mitochondria, leading to energy insufficiency, and thus, an inability to support cell function that may eventually lead to cell death (Fig. 15.2).

Remodeling of mitochondrial cristae is a necessary prerequisite for apoptosis. Loss of cristae structure allows redistribution of cytochrome c in the intermembrane space and its release into the cytoplasm through pores in the OMM generated by Bax and Bak, initiating activation of caspase cascades leading to cell death [46]. In the IMM, OPA1 oligomerizes and stabilizes the morphology of the cristae, preventing remodeling of such structures and thus opposing the release of cytochrome c and subsequent apoptosis [35]. For cytochrome c release to occur, L-OPA1 isoforms are processed to S-OPA1, allowing the disassembly of oligomers [47]. Therefore, OPA1 acts as an anti-apoptotic protein, preventing remodeling, whereas the cleaved forms are pro-apoptotic.

The combined actions of DRP1 and MFF are also important in this process. Otera and colleagues demonstrated that knockdown of DRP1 and MFF compromised mitochondrial fission and apoptosis induced by exogenous stimuli. In turn, MFF overexpression induced extensive mitochondrial fragmentation concomitant with an increased sensitivity to apoptosis, suggesting that mitochondrial fission facilitates apoptotic cristae remodeling [41]. hFIS1 likewise displays dual functionality, participating in mitochondrial fission as a component of normal mitochondrial dynamics, as well as being capable of trigger caspase-dependent cell death causing the release of cytochrome c from mitochondria, in the setting of calcium release from the endoplasmic reticulum (ER) and opening the MPTP [44].

MFN2 is another key player that carries out multiple functions. In addition to its role in mitochondrial fusion, it also modulates mitochondrial apoptotic pathways. Recent studies have shown that MFN2 is also present on ER where it assists in the tethering of ER to mitochondria. This contact between ER and mitochondria, enables calcium signaling, inositol trisphosphate (IP3) signaling, and calcium-dependent induction of apoptotic signaling [48]. MFN2 protein levels and apoptotic cell death increased in rat cardiomyocytes exposed to hydrogen peroxide. Whereas, siRNA depletion of MFN2 suppressed apoptosis [49]. Activation of apoptosis by oxidative stress is likewise reduced in the hearts of mice with a genetic disruption of MFN2 [50]. Therefore, MFN2 appears to play an active role in cardiomyocyte death mediated by oxidative stress [49]. In contrast, in other contexts, MFN2 may exert an anti-apoptotic effect. When cardiomyocytes are treated with ceramide, DRP1/FIS1-dependent mitochondrial fragmentation and apoptosis occurs [51]. When MFN2 is depleted, ceramide's effects are exacerbated (mitochondrial fragmentation, DRP1 co-localization with FIS1, mitochondrial membrane depolarization, cytochrome c release, and cell death), suggesting that MFN2 may exert an anti-apoptotic effect against ceramide [51]. Therefore, the role of MFN2 in the apoptotic process appears to be mediated by the nature of the apoptotic stimulus.

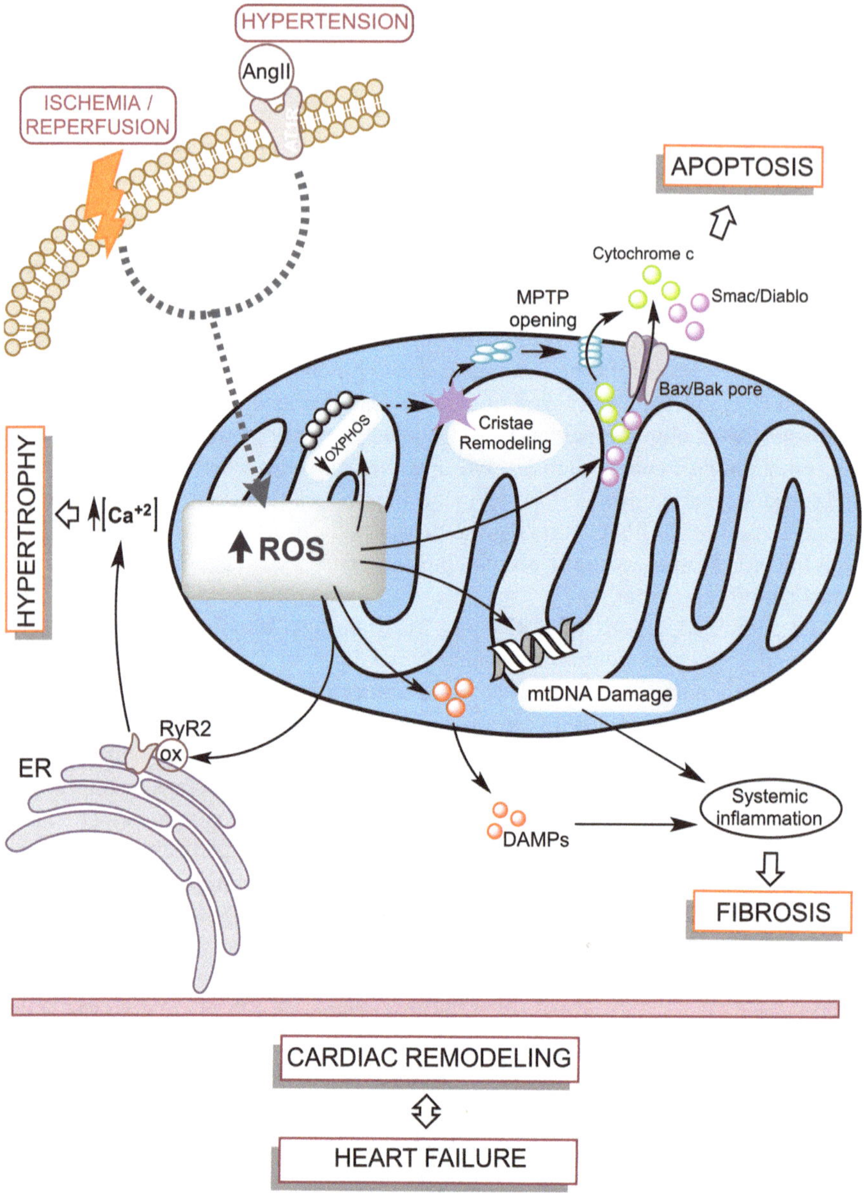

Fig. 15.2 Mitochondrial ROS in cardiac remodeling. Cardiac remodeling is induced by several mitochondrial ROS-related mechanisms. Cardiac pathological conditions like hypertension and ischemia/reperfusion (I/R) both induce an increase in mitochondrial ROS production. This leads to cristae remodeling, which in turn triggers the opening of the mitochondrial permeability transition pore (*MPTP*) and the assembly of the Bax/Bak pore. Both pores allow the release of pro-apoptotic proteins (like cytochrome c and Smac/Diablo), inducing the activation of the intrinsic cell death pathways in cardiac cells. Mitochondrial ROS can also cause the release of intra-mitochondrial damage-associated molecular patterns (*DAMPs*) and damage mitochondrial DNA, leading to systemic inflammation and triggering cardiac fibrosis. ROS generated by mitochondria can increase cytosolic calcium levels through oxidative damage of the ryanodine receptor (*RyR2*), leading to the activation of pro- hypertrophic pathways. Together, the cumulative effects of mitochondrial ROS triggers cardiac remodeling that ultimately leads to chronic heart failure

MARCH5, an E3 ubiquitin ligase associated with the OMM can act as a negative regulator of mitochondrial fission by mediating proteasomal degradation of MiD49 [52]. When cells were subjected to pro-apoptotic stimuli or to stimuli that affects mitochondrial activity, protein levels of MiD49 were decreased in a MARCH5-dependent manner. This data suggests that MARCH5, by increasing MiD49 degradation, inhibits fission and may protect cell from apoptosis induced by stress.

Proteins involved in mitochondrial dynamics are also affected during I/R, a prominent initiator of myocardial remodeling. In this context, Pei and colleagues [53] reported a reduction in MFN2 after I/R. The authors also observed a significant decrease in ATP production and an increase in ROS generation in MFN2 deficient mice subjected to I/R compared to the response of wild type mice. Subsequent suppression of cardiac function and the extent of myocardial fibrosis were likewise greater in the MFN2-deficient mice. They also showed that melatonin can increase MFN2 expression in ischemic myocardium by activating Notch1, thereby preserving mitochondrial integrity, increasing ATP production, reducing ROS production and preventing cardiomyocyte apoptosis after I/R, leading to an improved cardiac function and reduced fibrosis in myocardium, and attenuating cardiac remodeling.

Changes to mitochondrial form and function are a prominent feature of cardiac remodeling is observed in heart failure. Chen et al. [54] documented decreases in OPA1 protein levels in failing hearts, that occurred in conjunction with an increase in smaller and fragmented mitochondria. Mitochondrial fragmentation and reduction in OPA1 protein levels likewise occurs in vitro in H9c2 myoblasts subjected to simulated ischemia [34]. In addition, reduction of OPA1 levels with shRNA increased mitochondrial fragmentation, cytochrome c release, and apoptosis in response to simulated ischemia [54]. All these findings suggest that OPA1 helps limit fission of mitochondria during ischemia, and reduction of OPA1 levels following ischemia may contribute to heart failure progression.

DRP1 plays a central role promoting mitochondrial fragmentation and death during simulated ischemia, which can be prevented by expressing a dominant negative mutant of DRP1 [55, 56]. Treatment with a pharmacological inhibitor of DRP1, mdivi-1, increases the proportion of elongated mitochondria in adult cardiomyocytes, protecting them against simulated I/R in vitro and decreasing the myocardial infarct size in vivo [41]. This suggests that inhibition of DRP1 may be a good therapeutic target to mitigate the damage caused by I/R injury, thus preventing the subsequent pathological remodeling of the heart. Consistent with this, the prohyperthrophic-activation of calcineurin has been shown to recruit DRP1 to mitochondria, thereby stimulating mitochondrial fission, and the depletion of MFN2 to stimulate cardiac hypertrophy [57]. Either calcineruin inhibition or expression of a dominant-negative of Drp1 (K38A) was sufficient to prevent norepinephrine-induced hypertrophy, demonstrating that fission is a requisite step for calcineurin-mediated hypertrophic remodeling [57].

Disatnik and colleagues [58] demonstrated that treatment with p110 (a peptide that selectively inhibits the FIS1/DRP1 interaction, thereby preventing fission) restores mitochondrial functions in different models of cardiac I/R injury, including isolated primary cardiomyocytes, ex vivo heart, and in vivo myocardial infarction models. The authors showed that inhibition of fission during reperfusion preserves cardiac tissue integrity and cardiac function. This highlights the importance of mitochondrial dynamics in the progression of heart disease and supports the notion that intervening in such processes can help stop its progression.

Mitochondrial ROS, Cell Death in Cardiac Remodeling

Mitochondria are broadly recognized as one of the main sources of oxidative stress and ROS production [59]. This is particularly true in cardiomyocytes, in which several studies have shown that approximately 2% of the oxygen reduced during the normal functioning of the electron transport chain is lost as superoxide anion (O2·-) due to leak of the electron transport chain, mainly at the level of complexes I and III [60]. Thus, it is not surprising that mitochondria are involved in many oxidative stress-related cardiac pathologies, including cardiac hypertrophy [61], fibrosis [62], diabetic [63] or dilated [64] cardiomyopathies, I/R damage [53], and heart failure [65]. ROS can impact many signaling pathways. It is therefore capable of triggering a diversity of signaling cascades relevant to cardiomyocyte hypertrophy [66]. Moreover, ROS generation plays an important role in both Ang II and α-adrenergic-receptor induced hypertrophy [67] as well as in the pro-hypertrophic activation of NFκB [68]. ROS can likewise stimulate proliferation of cardiac fibroblast [69] as well as enhancing expression of matrix metalloproteinases (MMPs) [70] and their activation [71]. Based on the pro-hypertrophic action of ROS, antioxidant compounds such as omega-3 fatty acids have been proposed as protective against pressure overload-induced myocardial hypertrophy [72].

I/R damage to the heart is broadly linked to mitochondrial ROS production [53]. During ischemic events, mitochondrial respiration is lowered to near-zero, and energy production through oxidative phosphorylation is almost stopped. Later, when reperfusion is achieved and oxygen flow through the electron transport chain restored abruptly, the initial burst in ROS production is thought to overwhelm cellular antioxidant mechanisms [73]. This increase in mitochondrial generation ROS is thought to underlie most of the cellular damage associated with I/R. The antioxidant compounds picroside II [74], lycopene [75], all-trans retinoic acid [76] and vitamin C [77] have been shown to diminish ROS production during I/R in cardiomyocytes, thereby reducing mitochondrial damage and cellular death. These studies support the contention that increased ROS production activate cellular death pathways to effect cardiac damage during I/R.

Over the course of time, each of the above-discussed cardiac pathologies can lead to heart failure [78]. Thus, heart failure is considered and end-phase condition, triggered by accumulative damage in both the structure and function of the heart [79]. Given the cumulative, progressive nature of the disease, it is perhaps not surprising that ROS is capable of damaging a range of macromolecules relevant to cardiac excitation-contraction coupling and mitochondrial energy production rather than targeting a single substrate [65]. Additional support to idea that ROS plays an fundamental role in the development of heart failure comes from the evidence of changes in uncoupling proteins (UCPs) [65]. For instance, mRNA and protein levels of UCP2 and UCP3 are reported to decrease in the heart during the early stages of heart failure in both animal and human models [80, 81]. UCPs are endogenous proteins capable of dissipating the mitochondrial membrane potential and increasing efficient electron flow through the electron transport chain by transporting of protons across the IMM [82]. UCPs increase oxygen consumption but diminish ROS production, by reducing electron leak responsible for O2·– production [83]. Thus, reduced UCP expression may represent an attempt to preserve oxygen and ATP in the early stages of heart failure that could come at the cost of increasing the capacity for generating ROS. In the later stages of heart failure, however, UCP2 and UCP3 protein levels are reported to increase [84, 85] following the activation of PPARα by free fatty acids in plasma. It is unclear whether this late stage increase in UCP ultimately plays a protective or detrimental role in chronic heart failure [65].

Mitochondrial induction of apoptosis has been broadly associated with cardiac remodeling. Indeed, the frequency of apoptotic cardiomyocytes has been shown to be 10–100-fold higher after myocardial infarction than those observed in control hearts [86]. Mitochondria contain several intrinsic apoptosis- inducing factors, such as cytochrome c, the SMAC/Diablo protein and some of the calpain proteases [87]. Upon internal damage signals, such as DNA fragmentation, accumulation of intracellular toxic compounds, or uncontrolled oxidative stress, mitochondrial inner membrane is permeabilized, thereby releasing these pro-apoptotic factors into the cytosol and triggering the activation of the intrinsic cell death pathway [88]. Oxidative stress in particular, when not controlled by the antioxidant systems, is able to rapidly destabilize the structure of the IMM [89]. Therefore, it is likely that many of the cardiac pathologies linked to increased ROS production involve mitochondrial activation of cell death signals.

One of the main avenues by which mitochondrial pro-apoptotic molecules are released into the cytosol is via a pore formed by the pro-apoptotic Bcl-2 family proteins [90]. Upon apoptosis-inducing stimuli, Bcl-2 proteins (such as Bak and Bax) oligomerize on the IMM, opening a pore that allows the release of pro- apoptotic factors. Another important channel in mitochondria-activated apoptosis is the MPTP [90]. Such channel is thought to be formed by several mitochondrial membrane proteins and is activated upon IMM damage, thereby contributing to cell death [90]. The importance of MPTP opening in cardiac damage is supported by several articles reporting on the potential of MPTP inhibition to diminish apoptotic death, offering cardioprotection in diverse cardiac pathological states [91–93].

Mitophagy in Cardiac Remodeling

Autophagy is a catabolic recycling pathway triggered by various intra or extracellular stimuli that is conserved from yeast to mammals. During autophagy diverse cytosolic constituents are enveloped by double-membrane vesicles named autophagosomes, which later fuse with lysosomes in order to degrade their cargo. Dysregulation in autophagy is associated with a diverse range of pathologies including cardiovascular diseases [94]. Mitochondrial autophagy or mitophagy, which literally means "eating mitochondria," is the term applied to the cellular mechanism for identifying and selectively eliminating dysfunctional mitochondria as part of the overall mitochondrial quality control process. Increasing lines of evidence indicate that autophagy is intimately involved in the survival and death of cardiomyocytes under stress, however it is likely also involved in maintaining normal cardiac function and metabolism (Fig. 15.3). The beneficial effects of autophagy are partly mediated by the elimination of damaged mitochondria and the consequent prevention of mitochondrial dysfunction, oxidative stress increase, and cell death [95].

The most well studied mechanism of mitophagy in cardiomyocytes is the one mediated by the cytosolic E3 ubiquitin ligase Parkin [96] and the mitochondrial membrane kinase PTEN-induced putative kinase-1 (Pink1) [97]. Pink1-Parkin-mediated autophagy participates in mitochondrial quality control and in the maintenance of cardiac function in the heart at baseline. Cardiomyocytes in which Bnip3 is overexpressed are able to eliminate damaged mitochondria through autophagy via a Parkin-dependent mechanism, indicating that Parkin-mediated mitochondrial autophagy exists in cardiomyocytes [98]. The lack of Parkin-mediated mitophagy in heart-specific MFN2 KO mice is compensated by activation of non-selective autophagy in cardiomyocytes [95], thereby allowing cardiomyocytes to maintain mitochondrial quality.

PINK1−/− mice develop LV dysfunction and pathological cardiac hypertrophy [99], which is mediated by increased oxidative stress and dysfunctional mitochondria in cardiomyocytes. In the same line, PINK1 protein levels are significantly decreased in heart failure [99]. Consistently, mouse hearts lacking MFN2 develop cardiac dysfunction with age [100], suggesting that mitophagy regulated by the PINK1-MFN2 pathway plays an important role in maintaining cardiac function at baseline.

Translocation of Parkin to mitochondria and ubiquitination of mitochondrial proteins occurs during the acute phase of I/R, suggesting that Parkin-mediated mitophagy is stimulated [101]. Parkin-deficient mice exhibit a decrease in mitochondrial autophagy, accumulation of dysfunctional mitochondria and a reduced survival rate after I/R [101]. Taken together, these results suggest that mitophagy is stimulated in the heart during the chronic phase of cardiac remodeling and that it would be protective for the heart.

In cardiomyocytes, mitochondrial dynamics play an important role in mitophagy quality control through the process of DRP1-mediated asymmetric fission [102]. Accordingly, mitochondria in the early stages of senescence or those that have sustained moderate damage, will segregate dysfunctional components into one of the two daughter organelles generated by a fission event. The damaged daughter mitochondrion will be promptly identified as such and removed via Pink1-Parkin mediated mitophagy, whereas the healthy daughter will re-join the cellular mitochondrial pool, likely by fusing with other similarly fit mitochondria [103].

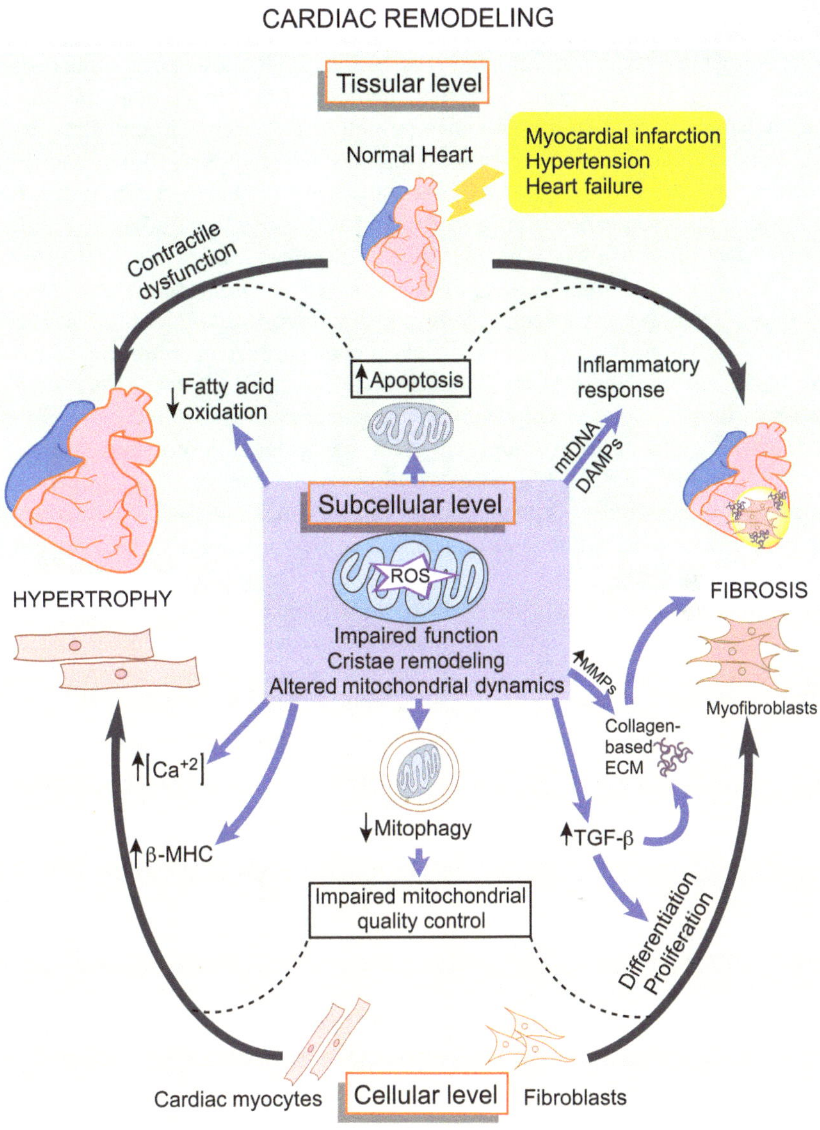

Fig. 15.3 Cardiac remodeling. Cardiac remodeling may occur at tissue, cellular or subcellular levels. At the tissue level, pathological conditions such as myocardial infarction, hypertension or heart failure lead to contractile myocardial dysfunction, cardiac hypertrophy and cardiac fibrosis. At the subcellular level, impaired mitochondrial function, cristae remodeling, altered mitochondrial dynamics and decreased mitophagy, together with increased ROS production, trigger a variety of downstream effects, including increased β-MHC expression, elevated cytosolic calcium, and decreased fatty acid oxidation. Mitochondrial function also impacts collagen-based extracellular matrix homeostasis by increasing the activity and synthesis of matrix metalloproteinases (MMPs) and the levels of TGF-β. Increased TGF-β production leads to differentiation and proliferation of cardiac fibroblasts into myofibroblasts, thus increasing cardiac fibrosis. Similarly, increased mitochondrial ROS production damages mtDNA and leads to the release of DAMPs, triggering an inflammatory response that further enhances cardiac fibrotic remodeling. Finally, damage to mitochondrial function can leads to the activation of cell death pathways, causing further contractile dysfunction, hypertrophy and cardiac fibrosis

The protective effect of mitophagy in the heart generally appears most prominently during stress. During ischemia, autophagy is triggered as an adaptive mechanism providing nutrients and eliminating damaged mitochondria, which could otherwise release damaging ROS and initiate apoptosis [104]. In fact, pharmacological inhibition of autophagy increases cardiomyocyte death by I/R, indicating that autophagy functions as a pro-survival mechanism in such conditions [105]. After reperfusion, triggered cardiac autophagy can be either adaptive or detrimental [94], and the impact of these processes on cardiac remodeling remains to be fully understood. Moreover, in response to myocardial infarction, activation of autophagy in the border zone is increased at an early stage, returning to normal during the chronic phase [106]. The autophagic vacuoles in the border zone are over-sized and contain organelles, including degraded mitochondria, in contrast with normalized autophagosomes which are upregulated in the remote zone during the chronic phase [106]. Pharmacological enhancing of autophagy with rapamycin 2 weeks after coronary ligation, ameliorates cardiac dysfunction and maladaptive remodeling, whereas inhibition of this process with bafilomycin A1 worsens them [106].

In mice, myocardial ischemia induced by permanent coronary artery ligation upregulates p53 and TP53-induced glycolysis and apoptosis regulator (TIGAR). Genetic deletion of p53 or TIGAR in mice stimulates mitophagy and inhibits accumulation of damaged mitochondria and apoptosis, an effect abolished by chloroquine. The authors suggest that the beneficial effects of the p53 or TIGAR downregulation may be mediated by mitophagy [107].

During myocardial reperfusion after a period of myocardial ischemia, mitochondria generate ROS, initiating a feedforward mechanism of oxidative stress, mitochondrial injury and cell death [108]. Selective elimination of damaged mitochondria by mitophagy is predicted to protect cardiomyocytes during reperfusion.

Despite the discussed reports, further investigations are needed in order to introduce modification of autophagy and mitophagy as useful therapeutic targets in the treatment of cardiovascular diseases, and to better understand where and when autophagy is activated or inhibited and how it affects the function of mitochondria in response to a wide variety of cardiovascular stress conditions that leads to heart remodeling.

Mitochondria in Cardiac Pathology

As previously discussed in this chapter, the main causes of cardiac remodeling that lead to heart failure are hypertension and I/R. At a cellular level, the processes that contribute to cardiac remodeling include cardiomyocyte hypertrophy, cell death and increased interstitial fibrosis. In this section, we will discuss the role of mitochondria in this processes and its potential as a therapeutic target.

Mitochondria and Cardiac Hypertrophy

It is generally accepted that mitochondrial dysfunction develops in the failing heart, and in many of the studies the model used was one of pathological cardiac hypertrophy. Cardiac hypertrophy can be divided into pathological or physiological hypertrophy. Thus, if after an initial phase of compensation, the growth response leads to contractile dysfunction, ventricular dilation, and heart failure, hypertrophy is considered pathological [45]. Pathological cardiac hypertrophy is a chronic complex disease that occurs in response to hemodynamic overload and is accompanied by oxidative stress and mitochondrial dysfunction [109]. In most cases of pathological hypertrophy, mitochondrial changes in structure and function have been described. In contrast, physiological cardiac hypertrophy is associated with enhanced mitochondrial function [45].

Myocardial oxidative stress has been implicated in the transition from compensated cardiac hypertrophy to heart failure, and evidence exists to support a role for mitochondrial and non-mitochondrial sources of ROS [45]. Oxidative stress can further impair mitochondrial function by leading to oxidative modifications of mitochondrial proteins, mutations of mtDNA and activation of the MPTP [77].

Pathological cardiac hypertrophy is associated with activation of many signaling pathways, exemplified by the calcineurin/nuclear factor of activated T-cell pathway, histone deacetylases, phosphatidyl inositol 3 kinase (PI3K)/Akt/Forkhead box protein O1 (FoxO1)/ mammalian target of rapamycin (mTOR) signaling networks, the ERK signaling pathway, G-protein-coupled receptor signaling pathways, among others [45, 110]. Change in calcium homeostasis is a critical factor in the deterioration of cardiac hypertrophy. Calcium affects the integrity of cardiomyocyte function in two ways: one is through calcium/calcineurin signaling pathway that directly activates/inactivates calcium/calcineurin dependent kinase [111, 112]. Another is through calcium channels that control the calcium flow in and out of organelles like mitochondria and ER [110]. Under normal conditions, mitochondria play a key role in cytosolic calcium clearance [113].

In vivo studies using pigs with hypertrophic cardiomyopathy revealed swollen cardiac mitochondria with disrupted cristae and substantial mtDNA depletion. Complex I and complex IV activity were also reduced in this model of hypertrophy [54].

Modulation of mitochondrial energy supply may provide means for therapy against cardiac hypertrophy. In cardiomyocytes, mitochondria produce ATP mainly through fatty acid oxidation (FAO) and oxidative phosphorylation. Many groups have described reduced myocardial FAO in pressure overload cardiac hypertrophy and heart failure. However, during the early stages of hypertrophy, which precede heart failure, changes in, or potential mechanisms for reduced FAO are less clear [45].

Currently, there are no effective methods to treat cardiac hypertrophy. It is reasonable to conclude that compensated cardiac hypertrophy is associated with relatively preserved mitochondrial function and that the development of mitochondrial dysfunction occurs in parallel with the development of heart failure. It remains to be definitively proved if adapting mitochondria to the hypertrophic state and maintaining mitochondrial function can be a promising therapeutic method in preventing the deterioration of cardiac hypertrophy and remodeling.

Mitochondrial Metabolism in Cardiac Fibrosis

Cardiac fibrosis is characterized by accumulation of ECM in the myocardium and it is present in most cardiac pathological conditions [114]. Because the adult mammalian myocardium has negligible regenerative capacity, the most extensive fibrotic remodeling of the ventricle is found in diseases associated with acute cardiomyocyte death. Following acute I/R, the sudden loss of a large number of cardiomyocytes triggers an inflammatory reaction, ultimately leading to replacement of dead myocardium with a collagen- based scar [115]. Similarly, pressure overload induced by hypertension, results in extensive cardiac fibrosis that is initially associated with increased stiffness and diastolic dysfunction. A persistent pressure load may lead to ventricular dilation and further diastolic and systolic heart failure [114]. In addition, hypertrophic cardiomyopathy has been associated with the development of significant cardiac fibrosis [116].

Several cell types are implicated in fibrotic remodeling of the heart, either directly by producing matrix proteins (fibroblasts) or indirectly by secreting fibrogenic mediators (macrophages, mast cells, lymphocytes, cardiomyocytes, and vascular cells). In all conditions associated with cardiac fibrosis, fibroblast transdifferentiation into secretory and contractile cells, termed myofibroblasts, is the key cellular event that drives the fibrotic response [117]. Regardless of the aetiology of cardiac injury, myofibroblasts are prominently involved in both reparative and fibrotic processes. Indeed, myofibroblast accumulation in the cardiac interstitium has been reported, not only in myocardial infarction [118] but also in the pressure and volume overloaded myocardium [119].

Oxidative stress has also been implicated in the pathogenesis of cardiac fibrosis, both through direct actions and through its involvement in cytokine and growth factor signaling. ROS are able to directly regulate the quantity and quality of interstitial ECM by modulating both matrix protein expression and metabolism. Inappropriate production of ROS in mitochondria, overwhelming the antioxidant defense systems, and the resulting oxidative damage to mtDNA and mitochondrial proteins have long been recognized as playing a causative role in the development and progression of cardiac remodeling and dysfunction [120]. Recent data support an essential role for the mitochondrial antioxidant enzyme thioredoxin 2 in preserving cardiac function by suppressing mitochondrial ROS production [64]. Mice with cardiac specific deletion of thioredoxin 2 develop progressive dilated cardiomyopathy with impairment of contractile function, heart chamber dilation, and marked interstitial fibrosis [121].

TGF-β1 is required for fibroblast differentiation into a pro-fibrotic myofibroblast phenotype, and it has been established that NADPH oxidase 4 (NOX)-dependent

$O_2\bullet-$ production by TGF-β1 is required for this differentiation [122, 123]. Moreover, ROS generated from mitochondrial electron transport chain complex III are required for TGF-β1–mediated transcription of NOX4 [124] and the initial activation of NOX4 amplifies and sustains TGF-β1–induced oxidative stress. Therefore, prevention of early mitochondrial $O_2\bullet-$ formation and ensuing NOX4 up-regulation might be a key mechanism involved in the anti-fibrotic effect of PETN [125]. In conclusion, long-term PETN treatment targeting superoxide generation and NO• bioavailability most likely prevented the changes of mitochondrial scavenging pathways and progressive fibrotic remodeling, leading to improved cardiac functional performance in congestive heart failure.

Alterations in the balance of MMPs and tissue inhibitors of metalloproteinases (TIMPs) are also involved in myocardial matrix remodeling. MMPs are expressed at very low levels in normal myocardium but markedly increased expression and activity of MMPs have been shown in human and animal hearts during the remodeling process after I/R [126, 127]. Increased oxidative stress activates MMPs and decreases fibrillar collagen synthesis in cardiac fibroblasts [62]. On the other hand, the TGF-β activating effects of ROS may enhance ECM deposition in the cardiac interstitium [128]. In adult rat cardiac fibroblasts, Ang II-stimulated collagen production is mediated through ROS generation [129]. When the mitochondrial bioenergetics and ROS production is pharmacologically restored, up-regulation of MMP9 is completely prevented, while maintaining TIMP1 gene expression, suggesting a reduction of MMP activity and less pronounced ventricular remodeling in this condition [130].

Injured mitochondria possess inflammatory properties, including damage-associated molecular patterns (DAMPs), which might produce cardiomyocyte injury [58]. Cellular injury releases mitochondrial DAMPs into the circulation eliciting a sepsis-like neutrophil mediated organ injury [131]. Furthermore, mitochondrial DAMPs activate the inflammasome, a group of intracellular multiprotein complexes that activates caspase-1 and proinflammatory cytokines [132]. Alternatively, mtDNA that escapes from autophagy leads to inflammatory responses in cardiomyocytes, inducing myocarditis and dilated cardiomyopathy [43]. Hypertension-induced inflammatory mediators may also compromise mitochondrial integrity and function. Progressive increase in interleukin (IL)-1 levels in diabetic rats is associated with decreased activity of the cardiac mitochondrial aldehyde dehydrogenase-2 [76]. Similarly, TNF-α treatment in vitro magnifies morphological changes in mitochondria and decreases membrane potential and ATP production in adipocytes [133], implicating mitochondria as targets of systemic inflammation that can lead to immunologic and fibrotic response. Taken together, these observations implicate mitochondria in the genesis of chronic inflammation in failing hearts.

Mitochondrial dysfunction also alters intracellular calcium homeostasis, which in smooth muscle cells regulates peripheral vascular resistance, influencing the risk of hypertension. Mitochondrial calcium additionally promotes biogenesis by upregulating PGC-1α expression [134]. In addition, increased myocardial expression of mitophagy markers is associated with changes in calcium cycling proteins, contributing to LV interstitial fibrosis and diastolic dysfunction [135]. Hence, mitochondrial abnormalities might also provoke myocardial fibrotic remodeling and dysfunction through alterations in intracellular calcium signaling.

Mitochondria During Hypertension

Myocardial mitochondria are vulnerable to the effects of hypertension, which most commonly impairs mitochondrial structure, bioenergetics, or homeostasis. Although antihypertensive drugs have the capacity to attenuate mitochondrial injury secondary to hypertension, drugs that specifically target the mitochondria may prove more efficacious in ameliorating hypertensive mitochondrial dysfunction or end-organ damage. Structural mitochondrial alterations secondary to hypertension may include decreased mass and density, swelling, as well as cristae remodeling, fragmentation, or loss. Decreased mitochondrial mass and density have ramifications for mitochondrial function, as fewer and smaller-sized mitochondria compromise their oxidative capacity and energy production. Hypertension has been prominently associated with damage and loss of cardiolipin, a phospholipid uniquely found in the IMM and necessary for proper cristae formation [136]. Cardiolipin regulates mitochondrial dynamics and prevents the formation and opening of the MPTP, and release of cytochrome c from into the cytosol where it triggers apoptosis [64].

Several studies provide evidence of hypertension-induced mitochondrial structural and functional abnormalities, including alterations of biogenesis and dynamics, RAAS-induced mitochondrial damage (ATP synthase, and creatine kinase activity are reduced in hypertensive heart mitochondria, while cytochrome c release and caspase-3 expression are upregulated, implying stimulated apoptosis), ROS overproduction, apoptosis, and mtDNA mutations. For example, despite similar volume, cardiac mitochondria from hypertensive rats are disorganized and show reduced number of cristae [74]. Cardiac mitochondria are normally in continuous interaction with the sarcoplasmic reticulum and sarcomeric structures forming 'intracellular energetic units' to allow optimal energy transfer. Thus, alterations in mitochondrial arrangement might impair muscle power and contractile function [137]. Mitochondrial structural changes are also accompanied by reduced mitochondrial respiration, triggered by decreased complex-I activity. The notion that mitochondrial damage compromises cardiomyocyte contractility suggests a possible role of mitochondria in the development of hypertension induced cardiac dysfunction. Conversely, ECM expansion and fibrosis secondary to hypertension may directly contribute to structural mitochondrial abnormalities and contractile dysfunction [75]. Therefore, despite the link between damaged mitochondria and hypertensive heart disease, a cause effect relationship remains to be established.

In hypertensive rats, mRNA levels of the fusion proteins MFN1 and MFN2, and OPA1 were decreased, suggesting a shift towards increased mitochondrial fragmentation. Furthermore, hypertensive rats exhibited reduced enzyme activities and expression of ETC subunits, indicating that mitochondrial function and remodeling are compromised in hypertension-induced cardiac hypertrophy. However, whether mitochondrial alterations are the primary cause or downstream consequence of hypertensive myocardial injury remains obscure.

Mitochondria-targeted antioxidants are novel cell-permeable compounds that may attenuate hypertensive injury by protecting mitochondria from oxidative damage. MitoQ is a ubiquinone derivative that prevents lipid peroxidation and mitochondrial damage in several experimental models. MitoQ also blunts hypertension, improves endothelial function, and reduces cardiac hypertrophy in spontaneously hypertensive rats [69]. Furthermore, its combination with low-dose losartan provides additive therapeutic benefit, by attenuating hypertension and reducing LV hypertrophy in spontaneously hypertensive rats, and by a direct anti-hypertrophic effect on rat cardiomyocytes in vitro [71]. Hence, MitoQ provides both hemodynamic and mitochondria-specific beneficial effects.

Mitochondria During Cardiac I/R Injury

I/R injury occurs when the blood supply to a tissue is blocked for minutes to hours (ischemia) and then restored (reperfusion). This important pathological mechanism underlies a range of disorders, including heart attack and stroke, where the prevention of blood flow to the tissue, for example by a blood clot, is followed by reperfusion when the blockage is removed mechanically or pharmacologically [138].

The current consensus is that a period of ischemia primes the tissue for subsequent damage upon reperfusion. Ischemic cells will eventually die if blood flow is not restored, but it is during reperfusion itself that most I/R damage is initiated. Thus, paradoxically, the essential therapeutic intervention to treat ischemia drives tissue pathophysiology [139]. The first damaging event upon reperfusion is a burst of ROS production from mitochondria [136]. Mitochondrial ROS not only drive acute damage but also initiate the pathology that develops over the minutes, days, and weeks following reperfusion [140]. The initial burst of ROS production upon reperfusion directly causes oxidative damage to mitochondria, thereby disrupting ATP production [141]. The mitochondria are recognized as a key player in cardiomyocyte death after myocardial infarction and cardiomyopathies. The events occurred during cardiac I/R significantly alter mitochondrial structure and function: reduce membrane potential, impair oxidative phosphorylation and decrease high energy phosphate synthesis, dysregulate calcium homeostasis, increase ROS production, matrix swelling and membrane permeability, and release of cytochrome c and other apoptotic factors leading to cell death. These events are initiated during ischemia and extend throughout reperfusion to severely compromise cardiac post-ischemic functional recovery and cell viability [142, 143].

During reperfusion, the reoxygenation of ischemic tissue results in mitochondrial calcium overload and renormalization of intracellular and matrix pH which are accompanied by the prodigious generation of ROS that synergistically induce the opening of the MPTP [144–146]. Together, this mitochondrial disruptions are the principal mechanism of apoptotic/necrotic cardiomyocyte death accounting for the majority of I/R injury [136, 139, 147, 148].

ROS-mediated mitochondrial damage also releases DAMPs, such as mtDNA, which can initiate the sterile inflammatory response [149, 150]. This activation of the innate immune system causes inflammation that contributes to I/R injury and can continue for days after the initial damage [149]. Therefore, during the weeks and months following the initial I/R injury, fibrotic scar tissue will form replacing dead cells along with extensive tissue remodeling [115]. The end result is often an organ that is structurally altered and functions poorly, frequently leading to persistent long-term pathology such as chronic heart disease [117].

Improvement in the clinical management of ischemic heart disease remains elusive despite the discovery of many molecular and cellular mechanisms that may be valuable targets against I/R injury. The importance of mitochondrial bioenergetics and function in contributing not only to cardiac injury but also to reducing cardiac injury is now well recognized. There are several cardioprotective strategies or treatments against I/R injury directed to mitochondria. For instance, cyclosporine administered at the time of reperfusion seems to have a sustained beneficial effect on infarct size reduction, which might improve the post-infarction remodeling [151]. Pharmacological modulation with the mitochondria-targeting peptide bendavia leads to sustainment of $\Delta\Psi m$ in cardiomyocytes during I/R, suggesting that bendavia might act as a direct MPTP blocker [152]. Moreover, mitochondria can be delivered to the ischemic heart through the vasculature and transplantation of these organelles affords an important functional benefit [153].

Conclusions

- Cardiac remodeling not only occurs at the organ or tissue levels, but can be also evidenced at the level of organelles like mitochondria.
- Mitochondrial dynamics remodeling has shown to be a cause or consequence of the development of several cardiac pathologies.
- Increased ROS production derived from mitochondrial metabolism has been associated with the development of cardiac pathologies.
- Mitophagy is stimulated in the heart during the chronic phase of cardiac remodeling and would be protective for the heart.
- The mitochondria is a therapeutic target in cardiac remodeling that leads to heart failure, hypertension and I/R.

Acknowledgements This work was supported by grants from the Comision Nacional de Investigacion Cientıfica y Tecnologica (CONICYT), Chile (FONDAP 15130011 to S.L., Postdoctoral FONDECYT 3160549 to P.A.). N.T. is recipient of a PhD fellowship from CONICYT, Chile. National Institutes of Health U54 HD087351 and HL098051 to B.A.R.

References

1. Vliegen HW, van der Laarse A, Cornelisse CJ, Eulderink F. Myocardial changes in pressure overload-induced left ventricular hypertrophy. A study on tissue composition, polyploidization and multinucleation. Eur Heart J. 1991;12(4):488–94.
2. Nadal-Ginard B, Anversa P, Kajstura J, Leri A. Cardiac stem cells and myocardial regeneration. Novartis Found Symp. 2005;265:142–54; discussion 55–7, 204–11.
3. Ford LE. Heart size. Circ Res. 1976;39(3):297–303.
4. Porter KE, Turner NA. Cardiac fibroblasts: at the heart of myocardial remodeling. Pharmacol Ther. 2009;123(2):255–78.
5. McMullen JR, Izumo S. Role of the insulin-like growth factor 1 (IGF1)/phosphoinositide-3-kinase (PI3K) pathway mediating physiological cardiac hypertrophy. Novartis Found Symp. 2006;274:90–111; discussion -7, 52–5, 272–6.
6. Weber KT, Brilla CG. Pathological hypertrophy and cardiac interstitium. Fibrosis and renin-angiotensin-aldosterone system. Circulation. 1991;83(6):1849–65.
7. Swynghedauw B. Molecular mechanisms of myocardial remodeling. Physiol Rev. 1999;79(1): 215–62.
8. Barki-Harrington L, Perrino C, Rockman HA. Network integration of the adrenergic system in cardiac hypertrophy. Cardiovasc Res. 2004;63(3):391–402.
9. Unger T, Li J. The role of the renin-angiotensin-aldosterone system in heart failure. J Renin-Angiotensin-Aldosterone Syst. 2004;5(Suppl 1):S7–10.
10. Archer SL. Mitochondrial dynamics – mitochondrial fission and fusion in human diseases. N Engl J Med. 2013;369(23):2236–51.
11. Neely JR, Morgan HE. Relationship between carbohydrate and lipid metabolism and the energy balance of heart muscle. Annu Rev Physiol. 1974;36:413–59.
12. Hom J, Sheu SS. Morphological dynamics of mitochondria – a special emphasis on cardiac muscle cells. J Mol Cell Cardiol. 2009;46(6):811–20.
13. Kuzmicic J, Del Campo A, Lopez-Crisosto C, Morales PE, Pennanen C, Bravo-Sagua R, et al. Mitochondrial dynamics: a potential new therapeutic target for heart failure. Rev Esp Cardiol. 2011;64(10):916–23.
14. Hoppel CL, Tandler B, Fujioka H, Riva A. Dynamic organization of mitochondria in human heart and in myocardial disease. Int J Biochem Cell Biol. 2009;41(10):1949–56.
15. Riva A, Tandler B, Loffredo F, Vazquez E, Hoppel C. Structural differences in two biochemically defined populations of cardiac mitochondria. Am J Physiol Heart Circ Physiol. 2005;289(2):H868–72.
16. Palmer JW, Tandler B, Hoppel CL. Biochemical properties of subsarcolemmal and interfibrillar mitochondria isolated from rat cardiac muscle. J Biol Chem. 1977;252(23):8731–9.
17. Palmer JW, Tandler B, Hoppel CL. Heterogeneous response of subsarcolemmal heart mitochondria to calcium. Am J Phys. 1986;250(5 Pt 2):H741–8.
18. Gustafsson CM, Falkenberg M, Larsson NG. Maintenance and Expression of Mammalian Mitochondrial DNA. Annu Rev Biochem. 2016;85:133–60.
19. Friedman JR, Nunnari J. Mitochondrial form and function. Nature. 2014;505(7483): 335–43.
20. Fung S, Nishimura T, Sasarman F, Shoubridge EA. The conserved interaction of C7orf30 with MRPL14 promotes biogenesis of the mitochondrial large ribosomal subunit and mitochondrial translation. Mol Biol Cell. 2013;24(3):184–93.
21. Wallace DC. Genetics: mitochondrial DNA in evolution and disease. Nature. 2016;535(7613): 498–500.
22. Vasquez-Trincado C, Garcia-Carvajal I, Pennanen C, Parra V, Hill JA, Rothermel BA, et al. Mitochondrial dynamics, mitophagy and cardiovascular disease. J Physiol. 2016;594(3): 509–25.
23. Chelli B, Falleni A, Salvetti F, Gremigni V, Lucacchini A, Martini C. Peripheral-type benzodiazepine receptor ligands: mitochondrial permeability transition induction in rat cardiac tissue. Biochem Pharmacol. 2001;61(6):695–705.

24. Scalettar BA, Abney JR, Hackenbrock CR. Dynamics, structure, and function are coupled in the mitochondrial matrix. Proc Natl Acad Sci U S A. 1991;88(18):8057–61.
25. Liu X, Hajnoczky G. Ca2+-dependent regulation of mitochondrial dynamics by the Miro-Milton complex. Int J Biochem Cell Biol. 2009;41(10):1972–6.
26. Liesa M, Palacin M, Zorzano A. Mitochondrial dynamics in mammalian health and disease. Physiol Rev. 2009;89(3):799–845.
27. Ong SB, Hausenloy DJ. Mitochondrial morphology and cardiovascular disease. Cardiovasc Res. 2010;88(1):16–29.
28. Santel A, Frank S, Gaume B, Herrler M, Youle RJ, Fuller MT. Mitofusin-1 protein is a generally expressed mediator of mitochondrial fusion in mammalian cells. J Cell Sci. 2003;116(Pt 13):2763–74.
29. Ishihara N, Eura Y, Mihara K. Mitofusin 1 and 2 play distinct roles in mitochondrial fusion reactions via GTPase activity. J Cell Sci. 2004;117(Pt 26):6535–46.
30. Santel A, Fuller MT. Control of mitochondrial morphology by a human mitofusin. J Cell Sci. 2001;114(Pt 5):867–74.
31. Alexander C, Votruba M, Pesch UE, Thiselton DL, Mayer S, Moore A, et al. OPA1, encoding a dynamin-related GTPase, is mutated in autosomal dominant optic atrophy linked to chromosome 3q28. Nat Genet. 2000;26(2):211–5.
32. Olichon A, Emorine LJ, Descoins E, Pelloquin L, Brichese L, Gas N, et al. The human dynamin- related protein OPA1 is anchored to the mitochondrial inner membrane facing the intermembrane space. FEBS Lett. 2002;523(1–3):171–6.
33. Cipolat S, de Brito OM, Dal Zilio B, Scorrano L. OPA1 requires mitofusin 1 to promote mitochondrial fusion. P Natl Acad Sci USA. 2004;101(45):15927–32.
34. Ehses S, Raschke I, Mancuso G, Bernacchia A, Geimer S, Tondera D, et al. Regulation of OPA1 processing and mitochondrial fusion by m-AAA protease isoenzymes and OMA1. J Cell Biol. 2009;187(7):1023–36.
35. Frezza C, Cipolat S, de Brito OM, Micaroni M, Beznoussenko GV, Rudka T, et al. OPA1 controls apoptotic cristae remodeling independently from mitochondrial fusion. Cell. 2006;126(1):177–89.
36. Smirnova E, Shurland DL, Ryazantsev SN, van der Bliek AM. A human dynamin-related protein controls the distribution of mitochondria. J Cell Biol. 1998;143(2):351–8.
37. Smirnova E, Griparic L, Shurland DL, van der Bliek AM. Dynamin-related protein Drp1 is required for mitochondrial division in mammalian cells. Mol Biol Cell. 2001;12(8): 2245–56.
38. Santel A, Frank S. Shaping mitochondria: the complex posttranslational regulation of the mitochondrial fission protein DRP1. IUBMB Life. 2008;60(7):448–55.
39. Jofuku A, Ishihara N, Mihara K. Analysis of functional domains of rat mitochondrial Fis1, the mitochondrial fission-stimulating protein. Biochem Bioph Res Co. 2005;333(2):650–9.
40. Gandre-Babbe S, van der Bliek AM. The novel tail-anchored membrane protein Mff controls mitochondrial and peroxisomal fission in mammalian cells. Mol Biol Cell. 2008;19(6): 2402–12.
41. Otera H, Wang CX, Cleland MM, Setoguchi K, Yokota S, Youle RJ, et al. Mff is an essential factor for mitochondrial recruitment of Drp1 during mitochondrial fission in mammalian cells. J Cell Biol. 2010;191(6):1141–58.
42. Zhao J, Liu T, Jin S, Wang X, Qu M, Uhlen P, et al. Human MIEF1 recruits Drp1 to mitochondrial outer membranes and promotes mitochondrial fusion rather than fission. EMBO J. 2011;30(14):2762–78.
43. Loson OC, Song Z, Chen H, Chan DC. Fis1, Mff, MiD49, and MiD51 mediate Drp1 recruitment in mitochondrial fission. Mol Biol Cell. 2013;24(5):659–67.
44. Palmer CS, Osellame LD, Laine D, Koutsopoulos OS, Frazier AE, Ryan MT. MiD49 and MiD51, new components of the mitochondrial fission machinery. EMBO Rep. 2011;12(6): 565–73.
45. Palmer CS, Elgass KD, Parton RG, Osellame LD, Stojanovski D, Ryan MT. Adaptor proteins MiD49 and MiD51 can act independently of Mff and Fis1 in Drp1 recruitment and are specific for mitochondrial fission. J Biol Chem. 2013;288(38):27584–93.

46. Tait SW, Green DR. Mitochondria and cell death: outer membrane permeabilization and beyond. Nat Rev Mol Cell Biol. 2010;11(9):621–32.
47. Jiang X, Jiang H, Shen Z, Wang X. Activation of mitochondrial protease OMA1 by Bax and Bak promotes cytochrome c release during apoptosis. Proc Natl Acad Sci U S A. 2014; 111(41):14782–7.
48. de Brito OM, Scorrano L. Mitofusin 2 tethers endoplasmic reticulum to mitochondria. Nature. 2008;456(7222):605–U47.
49. Shen T, Zheng M, Cao CM, Chen CL, Tang J, Zhang WR, et al. Mitofusin-2 is a major determinant of oxidative stress-mediated heart muscle cell apoptosis. J Biol Chem. 2007;282(32):23354–61.
50. Papanicolaou KN, Khairallah RJ, Ngoh GA, Chikando A, Luptak I, O'Shea KM, et al. Mitofusin-2 maintains mitochondrial structure and contributes to stress-induced permeability transition in cardiac myocytes. Mol Cell Biol. 2011;31(6):1309–28.
51. Parra V, Eisner V, Chiong M, Criollo A, Moraga F, Garcia A, et al. Changes in mitochondrial dynamics during ceramide-induced cardiomyocyte early apoptosis. Cardiovasc Res. 2008;77(2):387–97.
52. Xu S, Cherok E, Das S, Li S, Roelofs BA, Ge SX, et al. Mitochondrial E3 ubiquitin ligase MARCH5 controls mitochondrial fission and cell sensitivity to stress-induced apoptosis through regulation of MiD49 protein. Mol Biol Cell. 2016;27(2):349–59.
53. Pei H, Yang Y, Zhao H, Li X, Yang D, Li D, et al. The role of mitochondrial functional proteins in ROS production in ischemic heart diseases. Oxidative Med Cell Longev. 2016;2016:5470457.
54. Chen L, Gong Q, Stice JP, Knowlton AA. Mitochondrial OPA1, apoptosis, and heart failure. Cardiovasc Res. 2009;84(1):91–9.
55. Ong SB, Subrayan S, Lim SY, Yellon DM, Davidson SM, Hausenloy DJ. Inhibiting mitochondrial fission protects the heart against ischemia/reperfusion injury. Circulation. 2010; 121(18):2012–22.
56. Zepeda R, Kuzmicic J, Parra V, Troncoso R, Pennanen C, Riquelme JA, et al. Drp1 loss-of-function reduces cardiomyocyte oxygen dependence protecting the heart from ischemia-reperfusion injury. J Cardiovasc Pharmacol. 2014;63(6):477–87.
57. Pennanen C, Parra V, Lopez-Crisosto C, Morales PE, Del Campo A, Gutierrez T, et al. Mitochondrial fission is required for cardiomyocyte hypertrophy mediated by a Ca2+-calcineurin signaling pathway. J Cell Sci. 2014;127(Pt 12):2659–71.
58. Disatnik MH, Ferreira JC, Campos JC, Gomes KS, Dourado PM, Qi X, et al. Acute inhibition of excessive mitochondrial fission after myocardial infarction prevents long-term cardiac dysfunction. J Am Heart Assoc. 2013;2(5):e000461.
59. Andreyev AY, Kushnareva YE, Starkov AA. Mitochondrial metabolism of reactive oxygen species. Biochem Biokhim. 2005;70(2):200–14.
60. Cadenas E, Boveris A, Ragan CI, Stoppani AO. Production of superoxide radicals and hydrogen peroxide by NADH-ubiquinone reductase and ubiquinol-cytochrome c reductase from beef-heart mitochondria. Arch Biochem Biophys. 1977;180(2):248–57.
61. Sabri A, Hughie HH, Lucchesi PA. Regulation of hypertrophic and apoptotic signaling pathways by reactive oxygen species in cardiac myocytes. Antioxid Redox Signal. 2003;5(6): 731–40.
62. Siwik DA, Pagano PJ, Colucci WS. Oxidative stress regulates collagen synthesis and matrix metalloproteinase activity in cardiac fibroblasts. Am J Physiol Cell Physiol. 2001;280(1): C53–60.
63. Ni R, Zheng D, Xiong S, Hill DJ, Sun T, Gardiner RB, et al. Mitochondrial calpain-1 disrupts ATP synthase and induces superoxide generation in type 1 diabetic hearts: a novel mechanism contributing to diabetic cardiomyopathy. Diabetes. 2016;65(1):255–68.
64. Huang Q, Zhou HJ, Zhang H, Huang Y, Hinojosa-Kirschenbaum F, Fan P, et al. Thioredoxin-2 inhibits mitochondrial reactive oxygen species generation and apoptosis stress kinase-1 activity to maintain cardiac function. Circulation. 2015;131(12):1082–97.

65. Akhmedov AT, Rybin V, Marin-Garcia J. Mitochondrial oxidative metabolism and uncoupling proteins in the failing heart. Heart Fail Rev. 2015;20(2):227–49.
66. Aikawa R, Nagai T, Tanaka M, Zou Y, Ishihara T, Takano H, et al. Reactive oxygen species in mechanical stress-induced cardiac hypertrophy. Biochem Biophys Res Commun. 2001; 289(4):901–7.
67. Tanaka K, Honda M, Takabatake T. Redox regulation of MAPK pathways and cardiac hypertrophy in adult rat cardiac myocyte. J Am Coll Cardiol. 2001;37(2):676–85.
68. Hirotani S, Otsu K, Nishida K, Higuchi Y, Morita T, Nakayama H, et al. Involvement of nuclear factor-kappaB and apoptosis signal-regulating kinase 1 in G-protein-coupled receptor agonist- induced cardiomyocyte hypertrophy. Circulation. 2002;105(4):509–15.
69. Cheng TH, Cheng PY, Shih NL, Chen IB, Wang DL, Chen JJ. Involvement of reactive oxygen species in angiotensin II-induced endothelin-1 gene expression in rat cardiac fibroblasts. J Am Coll Cardiol. 2003;42(10):1845–54.
70. Sorescu D, Griendling KK. Reactive oxygen species, mitochondria, and NAD(P)H oxidases in the development and progression of heart failure. Congest Heart Fail. 2002;8(3):132–40.
71. Spinale FG. Bioactive peptide signaling within the myocardial interstitium and the matrix metalloproteinases. Circ Res. 2002;91(12):1082–4.
72. Ghule AE, Kandhare AD, Jadhav SS, Zanwar AA, Bodhankar SL. Omega-3-fatty acid adds to the protective effect of flax lignan concentrate in pressure overload-induced myocardial hypertrophy in rats via modulation of oxidative stress and apoptosis. Int Immunopharmacol. 2015;28(1):751–63.
73. Zhou T, Chuang CC, Zuo L. Molecular characterization of reactive oxygen species in myocardial ischemia-reperfusion injury. Biomed Res Int. 2015;2015:864946.
74. Li JZ, Yu SY, Mo D, Tang XN, Shao QR. Picroside II inhibits hypoxia/reoxygenation-induced cardiomyocyte apoptosis by ameliorating mitochondrial function through a mecha-nism involving a decrease in reactive oxygen species production. Int J Mol Med. 2015;35(2): 446–52.
75. Xu J, Hu H, Chen B, Yue R, Zhou Z, Liu Y, et al. Lycopene protects against hypoxia/reoxygenation injury by alleviating ER stress induced apoptosis in neonatal mouse cardiomyocytes. PLoS One. 2015;10(8):e0136443.
76. Zhu Z, Zhu J, Zhao X, Yang K, Lu L, Zhang F, et al. All-trans retinoic acid ameliorates myocardial ischemia/reperfusion injury by reducing cardiomyocyte apoptosis. PLoS One. 2015;10(7):e0133414.
77. Hao J, Li WW, Du H, Zhao ZF, Liu F, Lu JC, et al. Role of vitamin C in cardioprotection of ischemia/reperfusion injury by activation of mitochondrial KATP channel. Chem Pharm Bull. 2016;64(6):548–57.
78. Tsutsui H, Kinugawa S, Matsushima S. Oxidative stress and heart failure. Am J Physiol Heart Circ Physiol. 2011;301(6):H2181–90.
79. Anilkumar N, Sirker A, Shah AM. Redox sensitive signaling pathways in cardiac remodeling, hypertrophy and failure. Front Biosci. 2009;14:3168–87.
80. Young ME, Patil S, Ying J, Depre C, Ahuja HS, Shipley GL, et al. Uncoupling protein 3 transcription is regulated by peroxisome proliferator-activated receptor (alpha) in the adult rodent heart. FASEB J: Off Publ Fed Am Soc Exp Biol. 2001;15(3):833–45.
81. Bugger H, Guzman C, Zechner C, Palmeri M, Russell KS, Russell 3rd RR. Uncoupling protein downregulation in doxorubicin-induced heart failure improves mitochondrial coupling but increases reactive oxygen species generation. Cancer Chemother Pharmacol. 2011;67(6): 1381–8.
82. Sastre-Serra J, Valle A, Company MM, Garau I, Oliver J, Roca P. Estrogen down-regulates uncoupling proteins and increases oxidative stress in breast cancer. Free Radic Biol Med. 2010;48(4):506–12.
83. Modriansky M, Gabrielova E. Uncouple my heart: the benefits of inefficiency. J Bioenerg Biomembr. 2009;41(2):133–6.
84. Noma T, Nishiyama A, Mizushige K, Murakami K, Tsuji T, Kohno M, et al. Possible role of uncoupling protein in regulation of myocardial energy metabolism in aortic regurgitation model rats. FASEB J: Off Publ Fed Am Soc Exp Biol. 2001;15(7):1206–8.

85. Murray AJ, Cole MA, Lygate CA, Carr CA, Stuckey DJ, Little SE, et al. Increased mitochondrial uncoupling proteins, respiratory uncoupling and decreased efficiency in the chronically infarcted rat heart. J Mol Cell Cardiol. 2008;44(4):694–700.
86. Guerra S, Leri A, Wang X, Finato N, Di Loreto C, Beltrami CA, et al. Myocyte death in the failing human heart is gender dependent. Circ Res. 1999;85(9):856–66.
87. Kar P, Samanta K, Shaikh S, Chowdhury A, Chakraborti T, Chakraborti S. Mitochondrial calpain system: an overview. Arch Biochem Biophys. 2010;495(1):1–7.
88. Marin-Garcia J. Mitochondrial DNA repair: a novel therapeutic target for heart failure. Heart Fail Rev. 2016;21(5):475–87.
89. Jacob SF, Wurstle ML, Delgado ME, Rehm M. An analysis of the truncated bid- and ROS-dependent spatial propagation of mitochondrial permeabilization waves during apoptosis. J Biol Chem. 2016;291(9):4603–13.
90. Kinnally KW, Antonsson B. A tale of two mitochondrial channels, MAC and PTP, in apoptosis. Apoptosis: Int J Programmed Cell Death. 2007;12(5):857–68.
91. Sun Y, Zhang Y, Yan M, Wu Y, Zheng X. B-type natriuretic peptide-induced cardioprotection against reperfusion is associated with attenuation of mitochondrial permeability transition. Biol Pharm Bull. 2009;32(9):1545–51.
92. Ikeda G, Matoba T, Nakano Y, Nagaoka K, Ishikita A, Nakano K, et al. Nanoparticle-mediated targeting of cyclosporine a enhances cardioprotection against ischemia-reperfusion injury through inhibition of mitochondrial permeability transition pore opening. Sci Rep. 2016;6:20467.
93. Thompson J, Hu Y, Lesnefsky EJ, Chen Q. Activation of mitochondrial calpain and increased cardiac injury: beyond AIF release. Am J Physiol Heart Circ Physiol. 2016;310(3): H376–84.
94. Gatica D, Chiong M, Lavandero S, Klionsky DJ. Molecular mechanisms of autophagy in the cardiovascular system. Circ Res. 2015;116(3):456–67.
95. Saito T, Sadoshima J. Molecular mechanisms of mitochondrial autophagy/mitophagy in the heart. Circ Res. 2015;116(8):1477–90.
96. Kitada T, Asakawa S, Hattori N, Matsumine H, Yamamura Y, Minoshima S, et al. Mutations in the parkin gene cause autosomal recessive juvenile parkinsonism. Nature. 1998;392(6676):605–8.
97. Valente EM, Abou-Sleiman PM, Caputo V, Muqit MM, Harvey K, Gispert S, et al. Hereditary early- onset Parkinson's disease caused by mutations in PINK1. Science. 2004;304(5674): 1158–60.
98. Lee Y, Lee HY, Hanna RA, Gustafsson AB. Mitochondrial autophagy by Bnip3 involves Drp1- mediated mitochondrial fission and recruitment of Parkin in cardiac myocytes. Am J Physiol Heart Circ Physiol. 2011;301(5):H1924–31.
99. Billia F, Hauck L, Konecny F, Rao V, Shen J, Mak TW. PTEN-inducible kinase 1 (PINK1)/ Park6 is indispensable for normal heart function. Proc Natl Acad Sci U S A. 2011;108(23):9572–7.
100. Min Jin SM, Youle RJ. PINK1- and Parkin-mediated mitophagy at a glance. J Cell Sci. 2012;125:795–9.
101. Kubli DA, Zhang X, Lee Y, Hanna RA, Quinsay MN, Nguyen CK, et al. Parkin protein deficiency exacerbates cardiac injury and reduces survival following myocardial infarction. J Biol Chem. 2013;288(2):915–26.
102. Santulli G, Lewis DR, Marks AR. Physiology and pathophysiology of excitation–contraction coupling: the functional role of ryanodine receptor. J Muscle Res Cell Motil. 2017 (in press); doi: 10.1007/s10974-017-9470-z
103. Shao CH, Rozanski GJ, Nagai R, Stockdale FE, Patel KP, Wang M, et al. Carbonylation of myosin heavy chains in rat heart during diabetes. Biochem Pharmacol. 2010;80(2):205–17.
104. Zhang H, Bosch-Marce M, Shimoda LA, Tan YS, Baek JH, Wesley JB, et al. Mitochondrial autophagy is an HIF-1-dependent adaptive metabolic response to hypoxia. J Biol Chem. 2008;283(16):10892–903.

105. Kurian GA, Berenshtein E, Saada A, Chevion M. Rat cardiac mitochondrial sub-populations show distinct features of oxidative phosphorylation during ischemia, reperfusion and ischemic preconditioning. Cell Physiol Biochem. 2012;30(1):83–94.
106. Kanamori H, Takemura G, Goto K, Maruyama R, Tsujimoto A, Ogino A, et al. The role of autophagy emerging in postinfarction cardiac remodelling. Cardiovasc Res. 2011;91(2):330–9.
107. Hoshino A, Matoba S, Iwai-Kanai E, Nakamura H, Kimata M, Nakaoka M, et al. p53-TIGAR axis attenuates mitophagy to exacerbate cardiac damage after ischemia. J Mol Cell Cardiol. 2012;52(1):175–84.
108. Lesnefsky EJ, Moghaddas S, Tandler B, Kerner J, Hoppel CL. Mitochondrial dysfunction in cardiac disease: ischemia – reperfusion, aging, and heart failure. J Mol Cell Cardiol. 2001;33(6):1065–89.
109. Frey N, Katus HA, Olson EN, Hill JA. Hypertrophy of the heart: a new therapeutic target? Circulation. 2004;109(13):1580–9.
110. Zhou LY, Liu JP, Wang K, Gao J, Ding SL, Jiao JQ, et al. Mitochondrial function in cardiac hypertrophy. Int J Cardiol. 2013;167(4):1118–25.
111. Wilkins BJ, Molkentin JD. Calcium-calcineurin signaling in the regulation of cardiac hypertrophy. Biochem Biophys Res Commun. 2004;322(4):1178–91.
112. Maier LS. Ca(2+)/calmodulin-dependent protein kinase II (CaMKII) in the heart. Adv Exp Med Biol. 2012;740:685–702.
113. Michels G, Khan IF, Endres-Becker J, Rottlaender D, Herzig S, Ruhparwar A, et al. Regulation of the human cardiac mitochondrial Ca2+ uptake by 2 different voltage-gated Ca2+ channels. Circulation. 2009;119(18):2435–43.
114. Berk BC, Fujiwara K, Lehoux S. ECM remodeling in hypertensive heart disease. J Clin Invest. 2007;117(3):568–75.
115. Frangogiannis NG. Regulation of the inflammatory response in cardiac repair. Circ Res. 2012;110(1):159–73.
116. Ashrafian H, McKenna WJ, Watkins H. Disease pathways and novel therapeutic targets in hypertrophic cardiomyopathy. Circ Res. 2011;109(1):86–96.
117. Kong P, Christia P, Frangogiannis NG. The pathogenesis of cardiac fibrosis. Cell Mol Life Sci. 2014;71(4):549–74.
118. Willems IE, Havenith MG, De Mey JG, Daemen MJ. The alpha-smooth muscle actin-positive cells in healing human myocardial scars. Am J Pathol. 1994;145(4):868–75.
119. Leslie KO, Taatjes DJ, Schwarz J, von Turkovich M, Low RB. Cardiac myofibroblasts express alpha smooth muscle actin during right ventricular pressure overload in the rabbit. Am J Pathol. 1991;139(1):207–16.
120. Xie M, Burchfield JS, Hill JA. Pathological ventricular remodeling: therapies: part 2 of 2. Circulation. 2013;128(9):1021–30.
121. Widder JD, Fraccarollo D, Galuppo P, Hansen JM, Jones DP, Ertl G, et al. Attenuation of angiotensin II-induced vascular dysfunction and hypertension by overexpression of Thioredoxin 2. Hypertension. 2009;54(2):338–44.
122. Cucoranu I, Clempus R, Dikalova A, Phelan PJ, Ariyan S, Dikalov S, et al. NAD(P)H oxidase 4 mediates transforming growth factor-beta1-induced differentiation of cardiac fibroblasts into myofibroblasts. Circ Res. 2005;97(9):900–7.
123. Hecker L, Vittal R, Jones T, Jagirdar R, Luckhardt TR, Horowitz JC, et al. NADPH oxidase-4 mediates myofibroblast activation and fibrogenic responses to lung injury. Nat Med. 2009;15(9):1077–81.
124. Jain M, Rivera S, Monclus EA, Synenki L, Zirk A, Eisenbart J, et al. Mitochondrial reactive oxygen species regulate transforming growth factor-beta signaling. J Biol Chem. 2013;288(2):770–7.
125. Fraccarollo D, Galuppo P, Neuser J, Bauersachs J, Widder JD. Pentaerythritol tetranitrate targeting myocardial reactive oxygen species production improves left ventricular remodeling and function in rats with ischemic heart failure. Hypertension. 2015;66(5):978–87.
126. Cleutjens JP, Kandala JC, Guarda E, Guntaka RV, Weber KT. Regulation of collagen degradation in the rat myocardium after infarction. J Mol Cell Cardiol. 1995;27(6):1281–92.

127. Tyagi SC, Kumar SG, Haas SJ, Reddy HK, Voelker DJ, Hayden MR, et al. Post-transcriptional regulation of extracellular matrix metalloproteinase in human heart end-stage failure secondary to ischemic cardiomyopathy. J Mol Cell Cardiol. 1996;28(7):1415–28.
128. Barcellos-Hoff MH, Dix TA. Redox-mediated activation of latent transforming growth factor-beta 1. Mol Endocrinol. 1996;10(9):1077–83.
129. Lijnen P, Papparella I, Petrov V, Semplicini A, Fagard R. Angiotensin II-stimulated collagen production in cardiac fibroblasts is mediated by reactive oxygen species. J Hypertens. 2006;24(4):757–66.
130. Shi J, Dai W, Hale SL, Brown DA, Wang M, Han X, et al. Bendavia restores mitochondrial energy metabolism gene expression and suppresses cardiac fibrosis in the border zone of the infarcted heart. Life Sci. 2015;141:170–8.
131. Pei H, Du J, Song X, He L, Zhang Y, Li X, et al. Melatonin prevents adverse myocardial infarction remodeling via Notch1/Mfn2 pathway. Free Radic Biol Med. 2016;97:408–17.
132. Schrader M, Costello JL, Godinho LF, Azadi AS, Islinger M. Proliferation and fission of peroxisomes – an update. Biochim Biophys Acta. 2016;1863(5):971–83.
133. Ansari SB, Kurian GA. Hydrogen sulfide modulates sub-cellular susceptibility to oxidative stress induced by myocardial ischemic reperfusion injury. Chem Biol Interact. 2016;252:28–35.
134. Merkwirth C, Dargazanli S, Tatsuta T, Geimer S, Lower B, Wunderlich FT, et al. Prohibitins control cell proliferation and apoptosis by regulating OPA1-dependent cristae morphogenesis in mitochondria. Genes Dev. 2008;22(4):476–88.
135. Norton M, Ng AC, Baird S, Dumoulin A, Shutt T, Mah N, et al. ROMO1 is an essential redox-dependent regulator of mitochondrial dynamics. Sci Signal. 2014;7(310):ra10.
136. Loor G, Kondapalli J, Iwase H, Chandel NS, Waypa GB, Guzy RD, et al. Mitochondrial oxidant stress triggers cell death in simulated ischemia-reperfusion. Biochim Biophys Acta. 2011;1813(7):1382–94.
137. Zhang P, Lu Y, Yu D, Zhang D, Hu W. TRAP1 provides protection against myocardial ischemia-reperfusion injury by ameliorating mitochondrial dysfunction. Cell Physiol Biochem: Int J Exp Cell Physiol Biochem Pharmacol. 2015;36(5):2072–82.
138. Yellon DM, Hausenloy DJ. Myocardial reperfusion injury. N Engl J Med. 2007;357(11):1121–35.
139. Murphy E, Steenbergen C. Mechanisms underlying acute protection from cardiac ischemia-reperfusion injury. Physiol Rev. 2008;88(2):581–609.
140. Eltzschig HK, Eckle T. Ischemia and reperfusion – from mechanism to translation. Nat Med. 2011;17(11):1391–401.
141. Murphy E, Steenbergen C. Ion transport and energetics during cell death and protection. Physiology (Bethesda). 2008;23:115–23.
142. Santulli G, Nakashima R, Yuan Q, Marks AR. Intracellular calcium release channels: an update. J Physiol. 2017 (in press). doi: 10.1113/JP272781.
143. Stowe DF, Camara AK. Mitochondrial reactive oxygen species production in excitable cells: modulators of mitochondrial and cell function. Antioxid Redox Signal. 2009;11(6):1373–414.
144. Bernardi P, Rasola A, Forte M, Lippe G. The mitochondrial permeability transition pore: channel formation by F-ATP synthase, integration in signal transduction, and role in pathophysiology. Physiol Rev. 2015;95(4):1111–55.
145. Burwell LS, Nadtochiy SM, Tompkins AJ, Young S, Brookes PS. Direct evidence for S-nitrosation of mitochondrial complex I. Biochem J. 2006;394(Pt 3):627–34.
146. Halestrap AP. A pore way to die: the role of mitochondria in reperfusion injury and cardioprotection. Biochem Soc Trans. 2010;38(4):841–60.
147. Ye G, Metreveli NS, Ren J, Epstein PN. Metallothionein prevents diabetes-induced deficits in cardiomyocytes by inhibiting reactive oxygen species production. Diabetes. 2003;52(3):777–83.
148. Baseler WA, Dabkowski ER, Williamson CL, Croston TL, Thapa D, Powell MJ, et al. Proteomic alterations of distinct mitochondrial subpopulations in the type 1 diabetic heart: contribution of protein import dysfunction. Am J Physiol Regul Integr Comp Physiol. 2011;300(2):R186–200.

149. Arslan F, de Kleijn DP, Pasterkamp G. Innate immune signaling in cardiac ischemia. Nat Rev Cardiol. 2011;8(5):292–300.
150. Chen GY, Nunez G. Sterile inflammation: sensing and reacting to damage. Nat Rev Immunol. 2010;10(12):826–37.
151. Mewton N, Croisille P, Gahide G, Rioufol G, Bonnefoy E, Sanchez I, et al. Effect of cyclosporine on left ventricular remodeling after reperfused myocardial infarction. J Am Coll Cardiol. 2010;55(12):1200–5.
152. Kloner RA, Hale SL, Dai W, Gorman RC, Shuto T, Koomalsingh KJ, et al. Reduction of ischemia/reperfusion injury with bendavia, a mitochondria-targeting cytoprotective Peptide. J Am Heart Assoc. 2012;1(3):e001644.
153. Shen X, Zheng S, Metreveli NS, Epstein PN. Protection of cardiac mitochondria by overexpression of MnSOD reduces diabetic cardiomyopathy. Diabetes. 2006;55(3):798–805.

Chapter 16
Functional Role of Nox4 in Autophagy

Maurizio Forte, Silvia Palmerio, Derek Yee, Giacomo Frati, and Sebastiano Sciarretta

NADPH oxidase (Nox) are transmembrane proteins dedicated to the production of reactive oxygen species (ROS), such as superoxide anion (O_2^-) and hydrogen peroxide (H_2O_2) [1, 2]. Importantly, Nox are the only enzymes that generate ROS as their primary function, and not as a byproduct like other enzymes [1, 3]. To date, seven isoforms belonging to NADPH oxidase family have been identified and characterized: Nox1, Nox2 (also known as gp91^{phox}), Nox3, Nox4, Nox5, Dual oxidase (Duox) 1 and Duox2 [1, 2, 4]. A large body of evidence suggests the involvement of Nox in a plethora of cellular processes, such as proliferation, migration, apoptosis, as well as in inflammatory responses and in cellular senescence [1, 2, 5]. Several stimuli have been shown to activate Nox, such as cytokines, growth factors, high glucose, angiotensin II (Ang II), as well as different types of stress [6–10].

It is well known that high levels of ROS cause oxidative stress, which in turn contribute to the oxidation of proteins, nucleic acids and lipids, culminating in cellular damage and death [11–15]. For example, in the cardiovascular system,

M. Forte, PhD
Department of AngioCardioNeurology, IRCCS Neuromed, 86077 Pozzilli, IS, Italy

S. Palmerio, MS
Department of Medical and Surgical Sciences and Biotechnologies, Sapienza University of Rome, Corso della Repubblica 79, 04100 Latina, LT, Italy

D. Yee, MD
Department of Internal Medicine, Washington University School of Medicine, St. Louis, MO, USA

G. Frati, MD • S. Sciarretta, MD, PhD (✉)
Department of AngioCardioNeurology, IRCCS Neuromed, 86077 Pozzilli, IS, Italy

Department of Medical and Surgical Sciences and Biotechnologies, Sapienza University of Rome, Corso della Repubblica 79, 04100 Latina, LT, Italy
e-mail: sebastiano.sciarretta@uniroma1.it

G. Santulli (ed.), *Mitochondrial Dynamics in Cardiovascular Medicine*, Advances in Experimental Medicine and Biology 982,
DOI 10.1007/978-3-319-55330-6_16

ROS promote endothelial dysfunction, vascular inflammation, proliferation and migration of vascular smooth muscle cells (VSMCs). These factors ultimately lead to vascular remodeling and dysfunction, a common feature in several cardiovascular diseases (CVDs) [16–18]. These processes are in part mediated by the reduction of nitric oxide (NO). O_2^- chemically reacts and oxidizes NO, producing peroxynitrite ($ONOO^-$). Peroxynitrite not only enhances pre-existing oxidative stress, but it is also responsible for endothelial nitric oxide synthase (eNOS) uncoupling through the reduction of tetrahydro-L-biopterin (BH4), an essential cofactor for eNOS catalysis [18, 19]. When eNOS becomes uncoupled, it produces mainly $ONOO^-$ and O_2^- rather than NO, creating a vicious cycle that exacerbates oxidative stress [18, 20, 21].

In contrast, when levels of ROS are low or compartmentalized (a condition termed "redox state"), they act as signaling molecules in different pathways, thereby mediating physiological responses through the activation or inhibition of a cascade of downstream molecules involved in several cellular functions, such as proliferation, differentiation and cytoskeleton rearrangement [22–24]. In this regard, ROS act as second messengers downstream of specific ligands and may enhance the activity of transcription factors, or may be involved in post-transcriptional modifications of proteins [22]. Moreover, insufficient or absent generation of ROS may have detrimental consequences since phagocytic ROS are involved in host defense. This is illustrated in patients suffering from X-linked chronic granulomatous disease (CGD); these patients have an increased susceptibility to infections due to their inability to eliminate invading pathogens [25–27].

Autophagy is one of the functions modulated by ROS during cellular stress [28–30]. Autophagy is an evolutionarily-conserved, intracellular degradation process found in all mammalian cells, necessary for the elimination of damaged proteins or long-lived organelles [31, 32]. Autophagy represents a survival mechanism since it reduces the accumulation of misfolded proteins or damaged organelles during cellular stress, and recycles the indispensable substrates necessary for fundamental cellular processes such as ATP and protein synthesis [33, 34]. Studies suggest that ROS activate autophagy whereas reactive nitrogen species (RNS) may inhibit autophagy and autophagic flux [28–30, 35, 36]. The mechanisms through which ROS activate autophagy are not completely understood and remain to be elucidated. For instance, ROS can induce autophagy by directly oxidizing components of the autophagic machinery. However, cells also activate autophagy after increased generation of ROS as a compensatory and survival mechanism to reduce cellular damage and death that excessive oxidative stress may otherwise promote [28]. Recent evidence suggests that Nox may serve as a gatekeeper in regulating the activation of autophagy based on cell type and cellular conditions [37–40]. Furthermore, studies show that ROS are responsible for the induction of autophagy through Nox.

In the next several paragraphs, we will discuss the importance of Nox in the modulation of autophagy, with particular emphasis on the Nox4 isoform in the cardiovascular system. Studies providing evidence of the involvement of Nox4 in other cells will also be highlighted. A description of NADPH oxidase and autophagy biology will be also be briefly described.

NADPH Biology

The first NADPH oxidase was described in phagocytic cells and to date seven different isoforms of NADPH have been identified: Nox1, Nox2 (or gp91^{phox}), Nox3, Nox4, Nox5, Duox 1 and Duox2. All Nox isoforms transfer electrons to molecular oxygen (O_2) to generate O_2^-, using NADPH as an electron donor. Subsequently, O_2^- can be rapidly dismutated into hydrogen peroxide (H_2O_2) [1, 41]. Structurally, Nox isoforms are characterized by at least six transmembrane domains containing a NADPH binding domain in their cytosolic c- terminus. This binding domain represents the catalytic core of Nox. In order to exert its function, Nox isoforms are associated with another membrane bound subunit, called p22^{phox} for Nox1-4 and DuoxA1 and DuoxA2, for Duox1 and Duox2, respectively [4, 42]. p22^{phox} is necessary for the stabilization and the activation of Nox enzymes. Nox5 and Duox present an aminoterminal calcium-binding domain on the cytosolic side of the membrane, whereas Duox alone additionally presents a peroxidase domain on the opposite side of the membrane. Cytosolic subunit partners are also required for Nox activation: p47^{phox}, p67^{phox} and p40^{phox} for Nox2, and NOXA1 for Nox1 and Nox3. In response to various stimuli, these subunits associate with the catalytic subunit then translocate onto the membrane, thereby activating Nox [43, 44]. In contrast, Nox4 appears to be mainly transcriptionally-regulated, although it contains a polymerase δ-interacting protein 2 (POLDIP2) domain that acts as a positive regulator of Nox4 mediating its association with p22^{phox}, as reported in VSMCs [45].

Nox2 activation has been widely studied. Specifically, p47^{phox} (called the "organizer subunit") is initially activated by phosphorylation by protein kinase C (PKC). Once activated, it translocates onto the membrane and physically interacts with p22^{phox} along with p67^{phox} and p40^{phox}. For its final activation, Nox2 requires the involvement of the small cytosolic GTPase Ras-related C3 botulinum toxin substrate (Rac). Different isoforms of Rac have been described, such as Rac2, expressed predominantly in phagocytic cells, and Rac1, found in non-phagocytic cells [46]. Rac has also been shown to be associated with Nox1 and Nox3 [47]. Interestingly, Nox1-3 and Nox5 release O_2^- while Nox4 generates both O_2^- and H_2O_2 [48]. Duoxs generate mainly H_2O_2 [49].

Among the Nox isoforms, Nox2 and Nox4 share 35 % of sequence similarity and also share a similar hydrophobicity profile [1, 2, 50]. However, while Nox2 is mainly expressed on the plasma membrane, Nox4 displays different subcellular localizations [1, 45, 51].

Both Nox2 and Nox4 are widely distributed in the cardiovascular system, such as in endothelial cells, cadiomyocytes, fibroblasts and VSMCs. Nox2 and Nox4 have been demonstrated to play a prominent role in vascular physiology during different stress conditions [6, 52–56]. In this regard, recent evidence suggests the involvement of Nox2 in hypertension, atherosclerosis, as well as cardiac hypertrophy, fibrosis and remodeling. Conversely, the systemic deletion of Nox2, or the inhibition of Nox2 by means of selective inhibitors has been reported to improve vascular physiology, for example, reducing blood pressure and improving endothelial and cardiac function [57–61].

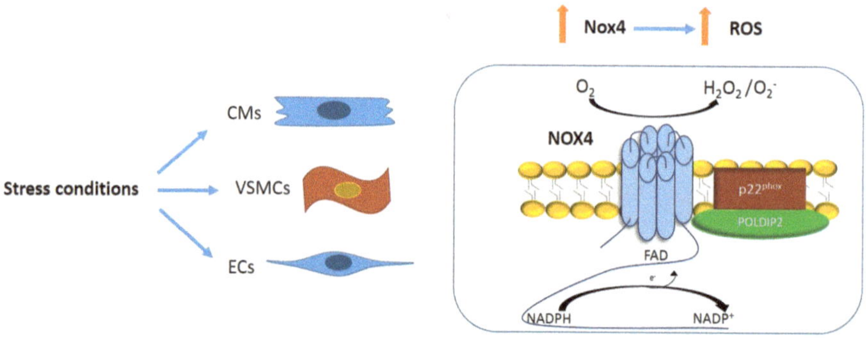

Fig. 16.1 Schematic representation of Nox4 and its associated membrane bound subunits (p22[phox] and POLDIP2). Nox4 displays different subcellular localizations and it is constitutively expressed in cardiomyocytes (*CMs*), vascular smooth muscle cells (*VSMCs*) and endothelial cells (*ECs*). Different stressors are able to enhance Nox4 expression and activity, leading to increased production of ROS

Nox4

Nox4 (Fig. 16.1), first identified as a renal-specific Nox, has also been found to be abundantly expressed in the cardiovascular system [2]. In particular, Nox4 represents the most expressed NADPH oxidase in endothelial and VSMCs [1, 62]. Depending on the cell type and the cell content, Nox4 is variably expressed in different organelles such as mitochondria, the nucleus, and the endoplasmic reticulum (ER). The distinct localizations of Nox4 are due to differences in the signal sequences of its primary structure [63–65]. Alternative splicing, either in the coding sequence of Nox4 or in its mRNA untranslated regions (UTRs), may contribute to the differences in intracellular localization of Nox4. Additionally, post-translational modifications in the forming protein (for example myristoylation and glycosylation) may also change localization [66–68]. In general, Nox4 activity is thought be transcriptionally-regulated: however, recent evidence suggests that POLDIP2 may be also considered a specific regulator of Nox4 [45].

Several stimuli have been demonstrated to activate both Nox2 and Nox4 in the cardiovascular system: G-protein coupled receptor agonists, such as Ang-II and thrombin, growth factors (vascular endothelial growth factor – VEGF; platelet-derived growth factor, PDGF), cytokines (tumor necrosis factor – TNFα), different cellular conditions such as high levels of glucose, glycated products and oxidized low density lipoproteins (LDL), or hypoxia-regeneration and nutrient starvation [7, 69]. TGF-β and E2F have been reported to enhance the expression of Nox4 in cadiomyocytes and in vascular cells under unstressed conditions [70–72], whereas NF-kB enhances expression of Nox4 in vascular cells under conditions of hypoxia [73, 74]. In contrast, thrombin, PDGF and peroxisome proliferator-activated receptor-γ (PPAR-γ) ligands were found to reduce Nox4 expression in VSMCs and in endothelial cells [53, 75, 76].

Similar to Nox2, Nox4 also is activated during stress conditions in the vascular system, and it is believed to be the major source of ROS in cadiomyocytes [77]. Nox4 displays dichotomous effects in cardiac and vascular physiology, being either detrimental or protective.

It has been shown that Nox4 inhibition prevents cardiac hypertrophy under pressure overload [77]. While Nox4 overexpression contributes to diabetes and vascular abnormalities [78–80]. Systemic Nox4 gene deletion increases brain injury in response to both transient and permanent cerebral ischemia [81]. Moreover, aged mice with cardiac specific overexpression of Nox4 showed enhanced cardiac hypertrophy and increased production of O_2^- [63]. Moreover, Nox4 was found to promote neointimal hyperplasia in VSMCs [82, 83]. On the other hand, systemic knockout mice for Nox4 were shown to develop cardiac hypertrophy and dilatation during chronic pressure overload whereas mice with overexpression of Nox4 in cardiomyocytes were protected [70]. Previous work showed that Nox4 deletion increases myocardial ischemic injury and induces cardiac dysfunction in response to starvation [40, 84]. Similarly, overexpression of a dominant-negative form of Nox4 exacerbates ischemic injury in ex vivo experiments [85]. A combined deletion of Nox2 and Nox4 exacerbates reperfusion injury, although a single deletion of these genes was found to be beneficial [86]. In addition, downregulation of Nox4 in the aortas of Ang-II hypertensive mice models was found to increase inflammation and endothelial dysfunction compared to wild type [87]. Similarly, pulmonary arterial hypertension and endothelial dysfunction were reported in knockout Nox4 mice. Nox4 has also been shown to prevent atherosclerosis progression in Nox4 and ApoE double knockout animals [88, 89].

The downstream pathways that mediate the protective functions exerted by Nox4 have been investigated. For example, in cardiomyocytes and in VSMCs, Nox4 activates the nuclear factor (erythroid-derived 2)-like 2 (Nrf2) signaling pathways, regulating the anti-oxidant responses to stress [87, 90]. In this regard, in endothelial cells under ischemic and inflammatory stressors, loss of Nox4 has been reported to impair heme oxygenase-1 (HO-1) expression, a protein strictly associated with inflammation and apoptosis [87]. Moreover, Nox4 was demonstrated to activate eNOS in endothelial cells [91] and to promote vascular cell survival through the activation of the RAS/extracellular signal-regulated kinases (ERK) pathway in response to accumulation of misfolded proteins or oxidized lipoproteins [39]. In addition, Nox4 was seen to promote cardiomyocyte survival in response to stress by inhibiting PP1, thereby promoting the activation of eIF2α signaling [84]. Nox4 induces vascular cell proliferation and migration in response to hypoxia through the stabilization of hypoxia inducible factor (HIF)-α levels [73, 92, 93]. Nox4 was found to promote angiogenesis during pressure overload and autophagy in the heart [40, 70]. Previous work also showed that the beneficial role of Nox4 in promoting angiogenesis in the heart under chronic stress involves increased expression of HIF-α and vascular endothelial growth factor (VEGF) [70].

This dual role of Nox4 is likely to be dependent on the stress condition, on the level and duration of its activation and on the subcellular compartment in which it produces ROS. The pathophysiological function of Nox4 during cardiac stress may also be determined by the cell-type in which Nox4 is activated. For example, global deletion versus cellular-specific deletion of Nox4 leads to different outcomes in response to pressure overload [70, 77]. This is probably dependent on the fact that Nox4-dependent angiogenesis is critical in response to mechanical stress, but it is not affected in the cardiomyocyte-specific knockout model.

Autophagy Biology

Autophagy (from the Greek, “auto” self, “phagein” to eat) is a cellular process in which cytoplasmic elements, such as proteins and organelles, are delivered to and digested in lysosomes/vacuoles (Fig. 16.2). The resulting molecular components – amino acids, lipids, nucleic acids, and carbohydrates – are recycled and reintroduced for cellular metabolism [31, 32]. In this way, autophagy ensures the homeostasis of cells, especially during stress and nutrient starvation. It is an

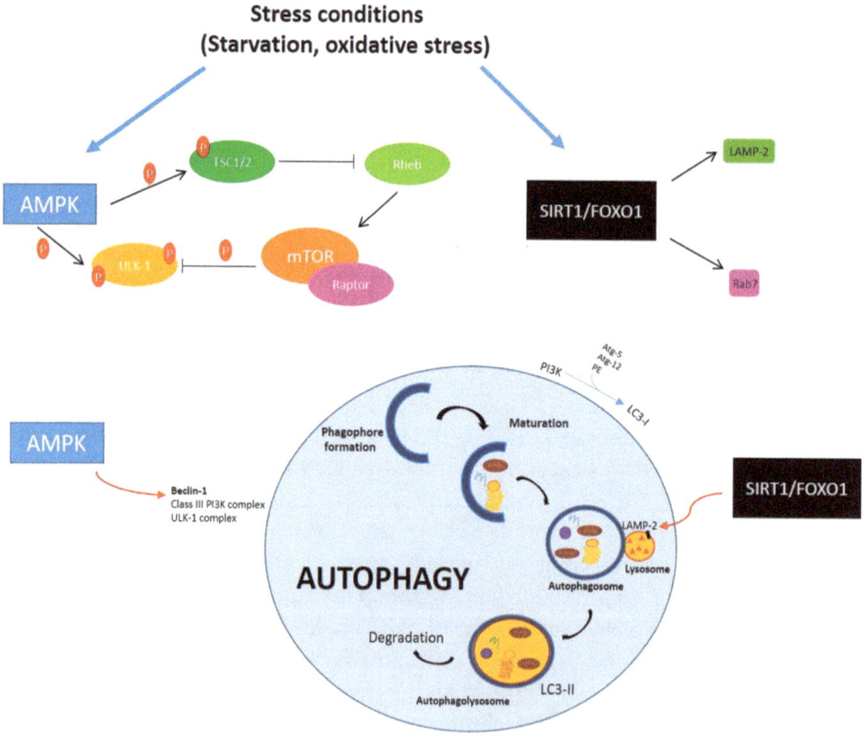

Fig. 16.2 Schematic representation of autophagy. AMPK/Rheb/mammalian target of rapamycin (*mTOR*) and SIRT1/FOXO1 are major signaling regulators of autophagy (*upper*). Several molecules are involved in the autophagic flux, from phagophore to autophagolysosome formation (see text for details)

evolutionarily conserved mechanism, first discovered in yeast and now found to occur in all eukaryotic cells [33].

Three primary forms of autophagy have been characterized: microautophagy, chaperone- mediated autophagy and macroautophagy. Chaperone- mediated autophagy involves the degradation of cytoplasmic proteins with a specific amino acid motif (KFERQ). That amino acid motif is recognized by chaperones which facilitate delivery of the protein into lysosomes through an interaction with the lysosomal membrane protein LAMP2A [94]. In microautophagy the cargoes are internalized directly through the lysosomal membrane [95]. Finally, in macroautophagy cellular components are internalized by double-membrane vesicles called autophagosomes that then deliver them to lysosomes. Functionally, both microautophagy and macroautophagy (hereafter the terms macroautophagy and autophagy will be used synonymously) can be non-selective or selective; non-selective autophagy is a process in which cargoes are not specific and it is required for the turnover of bulk cytoplasm [96]. In contrast, selective autophagy is finely regulated by several mechanisms and refers to the delivery and the digestion of specific cargoes such as mitochondria (mitophagy) and ribosomes (ribophagy). In the past decade, the complex signaling network that regulated the different stages of autophagy has been extensively studied and characterized. However, new molecules involved in autophagy are continuously identified.

The different stages of autophagy can be divided and summarized as it follows: (a) phagophore formation and sequestration of damaged cargoes, (b) autophagosome maturation and (c) autolysosome formation. In the initial step of autophagy, an isolation double-membrane, derived from ER or mitochondria, is formed. Next, it elongates (paghophore), surrounds, and invaginates the cargo, ultimately closing to generate an autophagosome. The recently formed autophagosome then transports the cargo and fuses with a lysosome, generating the autolysosome [97, 98]. Autophagy is a mechanism that is highly dynamic and finely regulated, under a state of constant intracellular flux termed "autophagic flux" [99, 100].

AMPK/Rheb/mammalian target of rapamycin (mTOR), together with SIRT1/FOXO1, represent important pathways involved in the promotion of stress-induced autophagy [33, 101]. Specifically, AMPK induces autophagy through the direct phosphorylation of Unc-51 like kinase (ULK) 1. Additionally, the inhibition of mTOR through the phosphorylation of tuberous sclerosis factor (TSC)-2 or Raptor and Rheb inhibition also induces autophagy. The inhibition of autophagy by mTOR occurs via the phosphorylation of Ulk-1, Ulk-2, Atg13 and focal adhesion kinase family interacting protein of 200 kD (FIP200). Ulk phosphorylates and activates Beclin-1 which is part of a complex containing the autophagy-specific class III phosphatidylinositol 3-kinase (PI3K) which is needed for phagophore elongation. PI3K recruits two additional ubiquitin-like complexes, Atg5-Atg12, as well as phosphatidylethanolamine (PE)-LC3-I for the formation of the autophagosome. PE-LC3-I forms LC3-II, which represents the only autophagy protein associated with the complete autophagosome. LC3-II is then cleaved from PE by Atg4 and returns to the cytoplasm. LAMP-2 and the small GTPase Rab7 regulate autophagolysosome formation, and they are activated by the SIRT1/FOXO1 pathway and by the transcription factor EB [33, 34, 101, 102].

Several findings suggest a crucial role of autophagy in the cardiovascular system. For example, mice with cardiac specific deletion of Atg5 were shown to develop contractile dysfunction and left ventricular dilatation, together with an increased number of dysfunctional mitochondria [103]. Moderate activation of autophagy has also been shown to be protective during pressure overload and chronic myocardial infarction [103, 104]. Interestingly, in aged mice, cardiac autophagy was shown to be reduced, which corresponded to an increased propensity to develop cardiac abnormalities [33, 34]. Moreover, the inhibition of autophagy through cardiac specific downregulation of AMPK or through Rheb overexpression has been found to significantly increase ischemic injury [31, 102, 105]. Mitochondrial autophagy has been reported to protect the heart from ischemia-reperfusion injury in a pathway mediated by the GTPase dynamin-related protein 1 (Drp1) [106]. In contrast, several studies also report that exaggerated activation of autophagy may be the cause of myocardial injury and cardiac remodeling [31, 107, 108].

Overall, the findings may suggest that autophagy at physiologic levels may be considered an adaptive and stress-protective mechanism for cardiovascular pathophysiology; conversely, high levels of autophagy may be the cause of several pathological conditions in response to specific types of stress, likely by inducing autosis [109].

ROS and Autophagy

The role of ROS and radical nitrogen species (RNS) as key signaling molecules is now well recognized. Owing to their ability to oxidize target molecules, ROS and RNS can modulate molecular pathways, for example, by switching on and off proteins by means of post-translational modifications [22–24].

Compelling evidence showed that oxidative stress modulates autophagy under cellular stress conditions [28–30].

The mechanisms through which ROS activate autophagy are still under investigation. Some data suggest that in most cases autophagy is activated as a compensatory and death-limiting mechanism under elevated level of ROS. Other findings suggest instead that ROS activate the SIRT1/FOXO1 pathway [110] or induce autophagy through the upregulation of Beclin1 and Atg5 [111, 112]. Other signaling pathways are involved in the activation of autophagy by ROS; for example, in cardiomyocytes during ischemia-reperfusion, ROS enhance BCL2/adenovirus E1B 19 kDa protein-interacting protein 3 (Bnip-3) expression that mediates Drp-1 translocation to the mitochondria, upregulating autophagy [113, 114]. Moreover, in the heart it has been demonstrated that ROS induce autophagy via NF-kB signaling pathways [115], as well as through the translocation of Nrf2 in the nucleus [116]. Importantly, activation of autophagy in the heart via Nrf2 prevents cardiac dysfunction and necrosis under pressure overload and proteotoxic stress [117]. Nrf2 was also reported to enhance the expression of NDP52, an adaptor protein of autophagic machinery. In addition, it has also been demonstrated that ROS are able to stimulate autophagy by the direct oxidation of the proteins involved in autophagy. In this regard, in ovarian cells H_2O_2 was reported to inactivate Atg4 by oxidation of cysteine residue 81 at its C-terminal portion [118]. The inactivation of Atg4, as mentioned above, prevents the dissociation of LC3-II from the autophagosome membrane.

NADPH Oxidase and Autophagy

Several studies reported that Nox derived-ROS are able to regulate autophagy under different cellular conditions, both in the cardiovascular system, and in other tissues.

For example, in human macrophages and dendritic cells, Nox2 was found to stimulate autophagy, leading to pathogen recognition and phagocytosis regulation. In particular, Nox2 derived ROS were able to increase levels of Atg8/LC3-II bound to phagosomes in a pathway mediated by the Toll-like receptor (TLR). Interestingly, no other proteins of autophagic machinery were found to be associated with phagosomes. Moreover, the Atg8/LC3 coated phagosomes display prolonged antigen presentation by MHC class II molecules, retaining the antigen for a longer period of time [119]. The same findings were also reported in murine cells [38].

Cancer cells also modulate autophagy via Nox activation. In HeLa cells under starvation, autophagy was found to be enhanced concomitantly with Nox activity and ROS generation, which in turn contribute to the activation of the JAK2-STAT3 pathway. JAK-STAT is a well characterized cytokine signaling pathway in which JAK tyrosine kinase phosphorylates STAT, which in turn forms dimers and moves to the cell nucleus leading to transcription activation. Precisely, in HeLa cells under starvation, STAT3 was found to be activated by tyrosine phosphorylation and to localize in cell nuclei, leading to an increased LC3-II formation. Once activated by Nox derived-ROS, STAT3 enhances the expression of interleukin (IL)-6. IL-6 expression contributes to cell survival and inhibition of cell death [120]. However, in this study, authors did not identify the specific Nox isoform responsible for autophagy activation, since in the experiments nonspecific inhibitors of Nox were used. Additionally, the deletion of $p22^{phox}$, the membrane bound subunit, is common in more than one Nox isoform.

Nox2 activation was also found to inhibit autophagy. Abel's group found that high fat diet-induced Nox2 activation inhibits autophagic flux in cardiomyoytes [121]. In skeletal muscles of a mouse model of Duchenne muscular dystrophy (DMD), Nox2 activation was found to inhibit autophagy, contributing to the pathogenesis of DMD. In particular, Nox2-derived ROS activate Src kinase. Src kinase in turn enhances mTOR activity via the PI3K/Akt pathway leading to inhibition of autophagosome formation and impairment of lysosome biogenesis. Additionally, inhibition of Nox2 upregulates autophagy and improves the pathophysiological phenotype of DMD in skeletal muscles with corresponding decrease in serum creatine kinase levels [122]. Authors of this study suggest that the impairment of autophagy via Nox2/Src is based on the particular location of the caveolae within the skeletal muscle sarcolemma [123, 124]. Inhibition of autophagy via Nox2 has been also reported in neuronal cells treated with rotenone, a toxic environmental agent that causes a sporadic form of Parkinson's disease due to its oxidative properties [125]. Interestingly, when cells were treated with rotenone for a short time, autophagic flux was inhibited through the Nox2/Src/PI3K/Akt pathway, independently of the mTOR pathway, thereby leading to a decreased Beclin1-Phosphatidylinositol 3-kinase catalytic subunit type 3 (VPS34) complex formation. Furthermore, exposure time with rotenone was found to decrease lysosomal function and to promote protein accumulation, ultimately promoting cell death [126]. The effect of rotenone with regard to autophagy modulation via Nox2 was also demonstrated in lung cancer cells. Acute exposure to rotenone impairs the autophagic flux via an mTOR-dependent manner whereas chronic exposure increases autophagy [127]. Recently, in a mouse model of epilepsy induced by pentylenetetrazole (PTZ), Nox-derived ROS were found to modulate autophagy in the hippocampal regions. It was further shown that inhibition of autophagy increased the deleterious effects of PTZ [128].

In endothelial cells, Nox2 has also been shown to induce autophagy, exerting both protective and maladaptive functions. For example, in pulmonary artery endothelial cells with persistent pulmonary hypertension, autophagy stimulation via Nox2 derived- ROS impairs angiogenesis and enhances apoptosis. Inhibition of autophagy has been found to lead to a decreased association between Nox2 and $p47^{phox}$ resulting in the improvement of angiogenesis (see Chap. 21). These findings suggest a crosstalk between Nox2 and autophagy in hypertensive pulmonary endothelial cells [129]. In contrast, autophagy induced by Nox2 was found to improve endothelial function in transgenic mice with endothelial overexpression of Nox2. Specifically, the increase in autophagy via Nox2-derived ROS enhances coronary vaso-relaxation and promotes endothelial cell survival in a pathway mediated by CaMKKβ kinase-AMPK. AMPK enhances autophagy through inhibition of mTOR while also increasing eNOS activity and NO bioavailability [37].

Nox4 and Autophagy

As shown for Nox2, Nox4 has also been demonstrated to exert a functional role in autophagy in different cell types. All these studies suggested that Nox4 induces autophagy. Nox4-induced autophagy was reported to enhance cell resistance to chemotherapy. Specifically, epidermal growth factor receptor (EGFR) is upregulated in many types of cancers [130] and the therapeutic strategies for cancer are often to target EGFR with specific EGFR inhibitors. However, cancer cells frequently develop resistance to cytotoxic effects exerted by EGFR inhibitors. For example, in head and neck squamous cancer cells, erlotinib, an EGFR inhibitor, was shown to promote cell survival through activation of autophagy via Nox4-derived H_2O_2. Interestingly, erlotinib was also reported to increase the expression of Nox4 by enhancing its promoter activity and mRNA stability. This suggests that autophagy is often used by cancer cells as a survival mechanism under stress conditions, causing some therapeutic strategies to fail to promote cell death [131]. Conversely, in breast cancer cells (MCF-7), psoralidin, a natural polyphenol coumarin exerting anti-cancer effects, was found to arrest cell proliferation and to induce DNA damage through the activation of Nox4-dependent ROS production. Interestingly, Nox4 was also found to induce autophagy as a compensatory mechanism [132].

Nox4 also promotes autophagy in the cardiovascular system. In human umbilical vein endothelial cells (HUVEC) the signaling pathway through which Nox4 activates autophagy has been reported. In response to misfolded protein accumulation, Nox4-derived H_2O_2 in the ER activates local RAS. RAS phosphorylates ERK promoting autophagy in an mTOR-independent pathway, which in turn contributes to increased endothelial cell survival [133]. 7-ketocholesterol (7-KC), an oxidized lipoprotein, was found to enhance autophagy in aortic smooth muscle cells, promoting cell survival. Specifically, H_2O_2 produced by Nox4 increased the autophagic flux via inhibition of Atg4. Interestingly, in the same study, the inhibition of autophagy was shown to increase ER stress and to enhance cell death [39]. In a rat model of sepsis, Nox4 was found to activate autophagy in response to high doses of lipopolysaccharides (LPS). The natural anti- inflammatory molecule cinnamaldehyde (CA) was found to attenuate Nox4-derived ROS, autophagy and alleviate LPS-induced cardiac dysfunction [134].

Nox4 also appears to elicit a critical pro-autophagic function in response to cardiomyocyte energy deprivation (Fig. 16.3). We previously demonstrated that protein levels of Nox4 are upregulated in the ER in glucose-deprived cardiomyocytes and this effect was paralleled by an increased production of ROS in this organelle. Nox4 appeared to be required for glucose deprivation-induced autophagy through the protein kinase RNA-activated-like ER kinase (PERK)/eukaryotic initiation factor 2α (eIF-2α)/activating transcription factor 4 (ATF4) pathway, a pathway modulating autophagy also in cancer cells [135]. Importantly, Nox4-induced autophagy promoted cardiomyocyte survival, since Nox4 knockdown increased cardiomyocyte cell death in response to glucose deprivation and this effect was rescued by autophagy reactivation. Nox4 also activated autophagy in vivo leading to cardioprotection in response to fasting and prolonged ischemia in the mouse heart [40]. This latter result was recently corroborated by a work from Shah's group, which demonstrated that Nox4 induces cardioprotective effects, also in response to I/R injury, through the activation of the eIF-2α/ATF4 signaling pathway [84].

A recent paper from Sadoshima's group also provided new insights into the mechanisms regulating Nox4 activity and autophagy in response to energy deprivation. FYN, a Src family tyrosine kinase, was shown to co-localize with Nox4 in the mitochondria, nucleus and ER and to phosphorylate it on the tyrosine 566 residue in the C-terminal NADPH binding domain. Phosphorylation on tyrosine 566 impaired the interaction between Nox4 and its membrane domain $p22^{phox}$, leading to enzyme inactivation. FYN appeared to regulate both maladaptive and physiological functions of Nox4. It was found to limit oxidative stress and cardiac remodeling in response to pressure overload by inhibiting Nox4. FYN knockout mice displayed increased

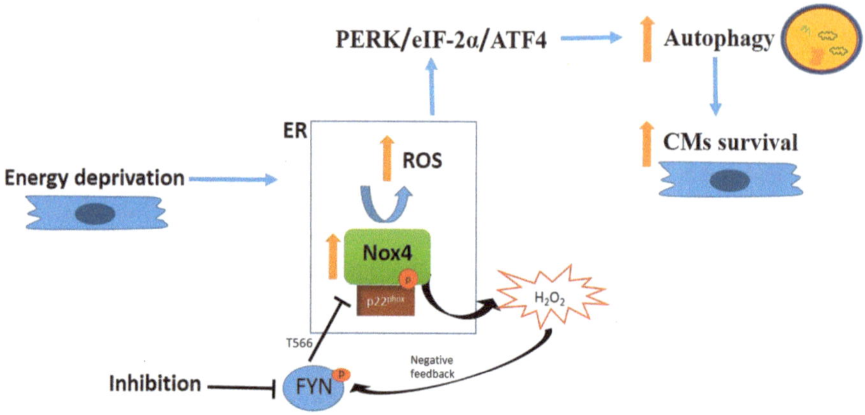

Fig. 16.3 Functional role of Nox4 in autophagy in cardiomyocytes (*CMs*) under energy deprivation. Nox4 is activated in the endoplasmic reticulum (*ER*) of CMs, where it enhances the production of ROS that leads to autophagy activation through the PERK/eIF2α/ATF4 pathway. Autophagy promotes CM survival under energy deprivation. Tyrosine kinase FYN inhibits Nox4 through the phosphorylation of Tyrosine 566 that impairs Nox4 association with its subunit partner $p22^{phox}$. Furthermore, Nox4 modulates FYN activity with a negative feedback mechanism. Inhibition of FYN contributes to CM survival under energy deprivation through autophagy activation (see text for details)

apoptosis in the heart during cardiac remodeling, which was attenuated by the concomitant deletion of Nox4. On the other hand, inhibition of FYN was shown to be beneficial in cardiomyocytes in response to glucose deprivation through the activation of Nox4. Precisely, in cardiomyocytes under glucose deprivation, overexpression of FYN led to decreased levels of autophagy and decreased cell viability whereas downregulation of FYN resulted in autophagy upregulation and cytoprotective effects in a Nox4-dependent manner. These results indicate that FYN activation may attenuate autophagy in cardiomyocytes under glucose deprivation, thereby inhibiting the adaptive response that Nox4 promotes in order to preserve cardiomyocyte survival during energy deprivation [136].

Conclusions

The evidence described in this chapter underlines the functional role of NADPH oxidase in the promotion of autophagy. We discussed experimental data in which Nox-derived ROS play a pivotal role as intracellular messengers, activating or inhibiting several downstream targets, such as autophagy.

Interestingly, the effects of Nox2 and Nox4 in autophagy regulation do not appear to be redundant. In fact, in specific cell types and stress conditions Nox2 was shown to inhibit autophagy, whereas Nox4 was consistently shown to activate the autophagic process. Several reasons may underline this differential role of Nox2 and Nox4 in autophagy regulation. Nox2 produces O_2^- whereas Nox4 mainly produces H_2O_2. It is possible that in specific conditions and cell types O_2^- inhibits autophagy, whereas H_2O_2 mainly exerts pro-autophagic effects. The amount of ROS produced by the two Nox isoforms may also determine their ability to activate or inhibit autophagy. Finally, the different subcellular localization and signaling pathways that are modulated by the two isoforms could be another reason. Future studies are warranted to elucidate this issue. It will also be interesting to investigate in the future the role of Nox2 and Nox4 in the regulation of mitophagy and mitochondrial dynamics in response to stress, particularly in the cardiovascular system.

Autophagy and Nox4 also have a dual role in the regulation of cardiomyocyte survival, promoting either beneficial or deleterious effects. It is possible that the dichotomous effect of autophagy in the regulation of survival or death might in turn determine when Nox4 activation is beneficial and when it is detrimental, since Nox4 appears to activate autophagy in response to stress. Of course, also other mechanisms may explain the dual role of Nox4 in the regulation of cell survival and death. The recent discovery of FYN as a negative regulator of Nox4 in mediating autophagy may open new scenarios on Nox4 biology and ROS-mediated cell signaling and regulation of cell survival.

References

1. Bedard K, Krause K-H. The NOX family of ROS-generating NADPH oxidases: physiology and pathophysiology. Physiol Rev. 2007;87(1):245–313.
2. Lassègue B, San Martín A, Griendling KK. Biochemistry, physiology, and pathophysiology of NADPH oxidases in the cardiovascular system. Circ Res. 2012;110:1364–90.
3. Lambeth JD. NOX enzymes and the biology of reactive oxygen. Nat Rev Immunol. 2004; 4(3):181–9.
4. Leto TL, Morand S, Hurt D, Ueyama T. Targeting and regulation of reactive oxygen species generation by Nox family NADPH oxidases. Antioxid Redox Signal. 2009;11(10):2607–19.
5. Maejima Y, Kuroda J, Matsushima S, Ago T, Sadoshima J. Regulation of myocardial growth and death by NADPH oxidase. J Mol Cell Cardiol. 2011;50:408–16.
6. Lassègue B, Griendling KK. NADPH oxidases: functions and pathologies in the vasculature. Arterioscler Thromb Vasc Biol. 2010;30:653–61.
7. Li JM, Shah AM. Mechanism of endothelial cell NADPH oxidase activation by angiotensin II: role of the p47phox subunit. J Biol Chem. 2003;278(14):12094–100.
8. Frey RS, Rahman A, Kefer JC, Minshall RD, Malik AB. PKCzeta regulates TNF-alpha-induced activation of NADPH oxidase in endothelial cells. Circ Res. 2002;90(9):1012–9.
9. Hwang J, Ing MH, Salazar A, Lassègue B, Griendling K, Navab M, et al. Pulsatile versus oscillatory shear stress regulates NADPH oxidase subunit expression: implication for native LDL oxidation. Circ Res. 2003;93(12):1225–32.
10. Xiao L, Pimentel D, Wang J, Singh K, Colucci W, Sawyer D. Role of reactive oxygen species and NAD(P)H oxidase in alpha(1)-adrenoceptor signaling in adult rat cardiac myocytes. Am J Physiol Cell Physiol. 2002;282:926–34.
11. Thannickal VJ, Fanburg BL. Reactive oxygen species in cell signaling. Am J Physiol Lung Cell Mol Physiol. 2000;279(6):L1005–28.

12. Davies KJ. Oxidative stress: the paradox of aerobic life. Biochem Soc Symp. 1995;61:1–31.
13. Hensley K, Robinson KA, Gabbita SP, Salsman S, Floyd RA. Reactive oxygen species, cell signaling, and cell injury. Free Radic Biol Med. 2000;28(10):1456–62.
14. Hemnani T, Parihar MS. Reactive oxygen species and oxidative DNA damage. Indian J Physiol Pharmacol. 1998;42(4):440–52.
15. Lockwood TD. Redox control of protein degradation. Antioxid Redox Signal. 2000;2(4): 851–78.
16. Madamanchi NR, Vendrov A, Runge MS. Oxidative stress and vascular disease. Arterioscler Thromb Vasc Biol. 2005;25(1):29–38.
17. Cai H, Harrison DG. Endothelial dysfunction in cardiovascular diseases: the role of oxidant stress. Circ Res. 2000;87(10):840–4.
18. Förstermann U. Nitric oxide and oxidative stress in vascular disease. Pflugers Arch Eur J Physiol. 2010;459(6):923–39.
19. Szabó C, Ischiropoulos H, Radi R. Peroxynitrite: biochemistry, pathophysiology and development of therapeutics. Nat Rev Drug Discov. 2007;6(8):662–80.
20. Li H, Förstermann U. Uncoupling of endothelial NO synthase in atherosclerosis and vascular disease. Curr Opin Pharmacol. 2013;13:161–7.
21. Fleming I, Busse R. Signal transduction of eNOS activation. Cardiovasc Res. 1999;43: 532–41.
22. Finkel T. Signal transduction by reactive oxygen species. J Cell Biol. 2011;194:7–15.
23. Dröge W. Free radicals in the physiological control of cell function. Physiol Rev. 2002;82(1): 47–95.
24. Wingler K, Hermans JJR, Schiffers P, Moens AL, Paul M, Schmidt HHHW. NOX1, 2, 4, 5: counting out oxidative stress. Br J Pharmacol. 2011;164:866–83.
25. Segal BH, Leto TL, Gallin JI, Malech HL, Holland SM. Genetic, biochemical, and clinical features of chronic granulomatous disease. Medicine (Baltimore). 2000;79(3):170–200.
26. Dinauer MC, Orkin SH. Chronic granulomatous disease. Annu Rev Med. 1992;43:117–24.
27. Thrasher A, Keep N, Wientjes F, Segal A. Chronic granulomatomatous disease. Biochim Biophys Acta. 1994;1227:1–24.
28. Navarro-Yepes J, Burns M, Anandhan A, Khalimonchuk O, Del Razo LM, Quintanilla-Vega B, et al. Oxidative stress, redox signaling and autophagy: cell death vs survival. Antioxid Redox Signal. 2014;21(1):66–85.
29. Wen X, Wu J, Wang F, Liu B, Huang C, Wei Y. Deconvoluting the role of reactive oxygen species and autophagy in human diseases. Free Radic Biol Med. 2013;65:402–10.
30. Dodson M, Darley-Usmar V, Zhang J. Cellular metabolic and autophagic pathways: traffic control by redox signaling. Free Radic Biol Med. 2013;63:207–21.
31. Matsui Y, Takagi H, Qu X, Abdellatif M, Sakoda H, Asano T, et al. Distinct roles of autophagy in the heart during ischemia and reperfusion: roles of AMP-activated protein kinase and beclin 1 in mediating autophagy. Circ Res. 2007;100(6):914–22.
32. Klionsky DJ, Baehrecke EH, Brumell JH, Chu CT, Codogno P, Cuervo AM, et al. A comprehensive glossary of autophagy-related molecules and processes. (2 nd edition)Autophagy. 2011;7:1273–94.
33. Levine B, Kroemer G. Autophagy in the pathogenesis of disease. Cell. 2008;132:27–42.
34. Sciarretta S, Hariharan N, Monden Y, Zablocki D, Sadoshima J. Is autophagy in response to ischemia and reperfusion protective or detrimental for the heart? Pediatr Cardiol. 2011;32:275–81.
35. Sarkar S, Korolchuk VI, Renna M, Imarisio S, Fleming A, Williams A, et al. Complex inhibitory effects of nitric oxide on autophagy. Mol Cell. 2011;43(1):19–32.
36. Chung KKK, Thomas B, Li X, Pletnikova O, Troncoso JC, Marsh L, et al. S-nitrosylation of parkin regulates ubiquitination and compromises parkin's protective function. Science. 2004;304(5675):1328–31.

37. Shafique E, Choy WC, Liu Y, Feng J, Arthur Lyra BC, Arafah M, et al. Oxidative stress improves coronary endothelial function through activation of the pro-survival kinase AMPK. Aging (Albany NY). 2013;5(7):515–30.
38. Huang J, Canadien V, Lam GY, Steinberg BE, Dinauer MC, Magalhaes MAO, et al. Activation of antibacterial autophagy by NADPH oxidases. Proc Natl Acad Sci U S A. 2009;106(15): 6226–31.
39. He C, Zhu H, Zhang W, Okon I, Wang Q, Li H, et al. 7-ketocholesterol induces autophagy in vascular smooth muscle cells through Nox4 and Atg4B. Am J Pathol. 2013;183(2):626–37.
40. Sciarretta S, Zhai P, Shao D, Zablocki D, Nagarajan N, Terada L, et al. Activation of NADPH oxidase 4 in the endoplasmic reticulum promotes cardiomyocyte autophagy and survival during energy stress through the protein kinase RNA-activated-like endoplasmic reticulum kinase/eukaryotic initiation factor 2alpha/activating transcr. Circ Res. 2013;113:1253–64.
41. Sumimoto H. Structure, regulation and evolution of Nox-family NADPH oxidases that produce reactive oxygen species. FEBS J. 2008;275:3249–77.
42. Opitz N, Drummond G, Selemidis S, Meurer S, Schmidt H. The "A"s and "O"s of NADPH oxidase regulation: a commentary on "Subcellular localization and function of alternatively spliced Noxo1 isoforms". Free Radic Biol Med. 2007;42:175–9.
43. Groemping Y, Lapouge K, Smerdon SJ, Rittinger K. Molecular basis of phosphorylation-induced activation of the NADPH oxidase. Cell. 2003;113(3):343–55.
44. Han CH, Freeman JLR, Lee T, Motalebi SA, Lambeth JD. Regulation of the neutrophil respiratory burst oxidase: identification of an activation domain in p67(phox). J Biol Chem. 1998;273(27):16663–8.
45. Lyle AN, Deshpande NN, Taniyama Y, Seidel-Rogol B, Pounkova L, Du P, et al. Poldip2, a novel regulator of Nox4 and cytoskeletal integrity in vascular smooth muscle cells. Circ Res. 2009;105(3):249–59.
46. Carrizzo A, Forte M, Lembo M, Formisano L, Puca AA, Vecchione C. Rac-1 as a new therapeutic target in cerebro- and cardio-vascular diseases. Curr Drug Targets. 2014;15(13):1231–46.
47. Ueyama T, Geiszt M, Leto TL. Involvement of Rac1 in activation of multicomponent Nox1- and Nox3-based NADPH oxidases. Mol Cell Biol. 2006;26(6):2160–74.
48. Takac I, Schröder K, Zhang L, Lardy B, Anilkumar N, Lambeth J, et al. The E-loop is involved in hydrogen peroxide formation by the NADPH oxidase Nox4. J Biol Chem. 2011;286:13303–13.
49. Hoste C, Dumont JE, Miot F, De Deken X. The type of DUOX-dependent ROS production is dictated by defined sequences in DUOXA. Exp Cell Res. 2012;318(18):2353–64.
50. Montezano AC, Touyz RM. Reactive oxygen species, vascular Noxs, and hypertension: focus on translational and clinical research. Antioxid Redox Signal. 2014;20(1):164–82.
51. Martyn KD, Frederick LM, Von Loehneysen K, Dinauer MC, Knaus UG. Functional analysis of Nox4 reveals unique characteristics compared to other NADPH oxidases. Cell Signal. 2006;18(1):69–82.
52. Dworakowski R, Alom-Ruiz SP, Shah AM. NADPH oxidase-derived reactive oxygen species in the regulation of endothelial phenotype. Pharmacol Rep. 2008;60:21–8.
53. Lassègue B, Sorescu D, Szöcs K, Yin Q, Akers M, Zhang Y, et al. Novel gp91phox homologues in vascular smooth muscle cells : nox1 mediates angiotensin II-induced superoxide formation and redox-sensitive signaling pathways. Circ Res. 2001;88(9):888–94.
54. Pagano PJ, Clark JK, Cifuentes-Pagano ME, Clark SM, Callis GM, Quinn MT. Localization of a constitutively active, phagocyte-like NADPH oxidase in rabbit aortic adventitia: enhancement by angiotensin II. Proc Natl Acad Sci U S A. 1997;94(26):14483–8.
55. Haurani MJ, Pagano PJ. Adventitial fibroblast reactive oxygen species as autacrine and paracrine mediators of remodeling: bellwether for vascular disease? Cardiovasc Res. 2007;75:679–89.
56. Touyz RM, Chen X, Tabet F, Yao G, He G, Quinn MT, et al. Expression of a functionally active gp91phox-containing neutrophil-type NAD(P)H oxidase in smooth muscle cells from human resistance arteries: regulation by angiotensin II. Circ Res. 2002;90(11):1205–13.

57. Mollnau H, Wendt M, Szöcs K, Lassègue B, Schulz E, Oelze M, et al. Effects of angiotensin II infusion on the expression and function of NAD(P)H oxidase and components of nitric oxide/cGMP signaling. Circ Res. 2002;90(4):E58–65.
58. Wang HD, Xu S, Johns DG, Du Y, Quinn MT, Cayatte AJ, et al. Role of NADPH oxidase in the vascular hypertrophic and oxidative stress response to angiotensin II in mice. Circ Res. 2001;88(9):947–53.
59. Wind S, Knut B, Armitage ME, Ashraf T, Arun H, Kumar DJ, et al. Oxidative stress and endothelial dysfunction in aortas of aged spontaneously hypertensive rats by NOX1/2 is reversed by NADPH oxidase inhibition. Hypertension. 2010;56:490–7.
60. Judkins CP, Diep H, Broughton BRS, Mast AE, Hooker EU, Miller AA, et al. Direct evidence of a role for Nox2 in superoxide production, reduced nitric oxide bioavailability, and early atherosclerotic plaque formation in ApoE-/- mice. Am J Physiol Heart Circ Physiol. 2010;298(1):H24–32.
61. Bendall JK, Cave AC, Heymes C, Gall N, Shah AM. Pivotal role of a gp91phox-containing NADPH oxidase in angiotensin II-induced cardiac hypertrophy in mice. Circulation. 2002;105(3):293–6.
62. Ago T, Kitazono T, Ooboshi H, Iyama T, Han YH, Takada J, et al. Nox4 as the major catalytic component of an endothelial NAD(P)H oxidase. Circulation. 2004;109(2):227–33.
63. Ago T, Kuroda J, Pain J, Fu C, Li H, Sadoshima J. Upregulation of Nox4 by hypertrophic stimuli promotes apoptosis and mitochondrial dysfunction in cardiac myocytes. Circ Res. 2010;106(7):1253–64.
64. Matsushima S, Kuroda J, Ago T, Zhai P, Park JY, Xie LH, et al. Increased oxidative stress in the nucleus caused by Nox4 mediates oxidation of HDAC4 and cardiac hypertrophy. Circ Res. 2013;112(4):651–63.
65. Chen K, Kirber MT, Xiao H, Yang Y, Keaney JF. Regulation of ROS signal transduction by NADPH oxidase 4 localization. J Cell Biol. 2008;181(7):1129–39.
66. Anilkumar N, Jose GS, Sawyer I, Santos CXC, Sand C, Brewer AC, et al. A 28-kDa splice variant of NADPH oxidase-4 is nuclear-localized and involved in redox signaling in vascular cells. Arterioscler Thromb Vasc Biol. 2013;33(4):e104–12.
67. Andreassi C, Riccio A. To localize or not to localize: mRNA fate is in 3′UTR ends. Trends Cell Biol. 2009;19:465–74.
68. Colombo S, Longhi R, Alcaro S, Ortuso F, Sprocati T, Flora A, et al. N-myristoylation determines dual targeting of mammalian NADH-cytochrome b(5) reductase to ER and mitochondrial outer membranes by a mechanism of kinetic partitioning. J Cell Biol. 2005;168(5):735–45.
69. Li J-M, Mullen AM, Yun S, Wientjes F, Brouns GY, Thrasher AJ, et al. Essential role of the NADPH oxidase subunit p47(phox) in endothelial cell superoxide production in response to phorbol ester and tumor necrosis factor-alpha. Circ Res. 2002;90(2):143–50.
70. Zhang M, Brewer AC, Schröder K, Santos CXC, Grieve DJ, Wang M, et al. NADPH oxidase-4 mediates protection against chronic load-induced stress in mouse hearts by enhancing angiogenesis. Proc Natl Acad Sci U S A. 2010;107(42):18121–6.
71. Jiang F, Liu GS, Dusting GJ, Chan EC. NADPH oxidase-dependent redox signaling in TGF-beta-mediated fibrotic responses. Redox Biol. 2014;2:267–72.
72. Ismail S, Sturrock A, Wu P, Cahill B, Norman K, Huecksteadt T, et al. NOX4 mediates hypoxia-induced proliferation of human pulmonary artery smooth muscle cells: the role of autocrine production of transforming growth factor-{beta}1 and insulin-like growth factor binding protein-3. Am J Physiol Lung Cell Mol Physiol. 2009;296(3):L489–99.
73. Lu X, Murphy TC, Nanes MS, Hart CM. PPAR{gamma} regulates hypoxia-induced Nox4 expression in human pulmonary artery smooth muscle cells through NF-{kappa}B. Am J Physiol Lung Cell Mol Physiol. 2010;299(4):L559–66.
74. Williams CR, Lu X, Sutliff RL, Hart CM. Rosiglitazone attenuates NF-κB-mediated Nox4 upregulation in hyperglycemia-activated endothelial cells. Am J Physiol Cell Physiol. 2012;303(2):C213–23.

75. Ellmark SHM, Dusting GJ, Ng Tang Fui M, Guzzo-Pernell N, Drummond GR. The contribution of Nox4 to NADPH oxidase activity in mouse vascular smooth muscle. Cardiovasc Res. 2005;65(2):495–504.
76. Hwang J, Kleinhenz DJ, Lassègue B, Griendling KK, Dikalov S, Hart CM. Peroxisome proliferator-activated receptor-gamma ligands regulate endothelial membrane superoxide production. Am J Physiol Cell Physiol. 2005;288(4):C899–905.
77. Kuroda J, Ago T, Matsushima S, Zhai P, Schneider MD, Sadoshima J. NADPH oxidase 4 (Nox4) is a major source of oxidative stress in the failing heart. Proc Natl Acad Sci U S A. 2010;107(35):1–6.
78. Maalouf RM, Eid AA, Gorin YC, Block K, Escobar GP, Bailey S, et al. Nox4-derived reactive oxygen species mediate cardiomyocyte injury in early type 1 diabetes. Am J Physiol Cell Physiol. 2012;302(3):C597–604.
79. Paravicini TM, Touyz RM. NADPH oxidases, reactive oxygen species, and hypertension: clinical implications and therapeutic possibilities. Diabetes Care. 2008;31(Suppl 2):S170–80.
80. Duncan ER, Crossey PA, Walker S, Anilkumar N, Poston L, Douglas G, et al. Effect of endothelium-specific insulin resistance on endothelial function in vivo. Diabetes. 2008;57(12):3307–14.
81. Kleinschnitz C, Grund H, Wingler K, Armitage ME, Jones E, Mittal M, et al. Post-stroke inhibition of induced NADPH Oxidase type 4 prevents oxidative stress and neurodegeneration. PLoS Biol. 2010;8(9):e1000479.
82. Tong X, Hou X, Jourd'Heuil D, Weisbrod RM, Cohen RA. Upregulation of Nox4 by TGFbeta 1 oxidizes SERCA and inhibits NO in arterial smooth muscle of the prediabetic zucker rat. Circ Res. 2010;107(8):975–83.
83. Tong X, Khandelwal AR, Qin Z, Wu X, Chen L, Ago T, et al. Role of smooth muscle Nox4-based NADPH oxidase in neointimal hyperplasia. J Mol Cell Cardiol. 2015;89:185–94.
84. Santos CX, Hafstad AD, Beretta M, Zhang M, Molenaar C, Kopec J, et al. Targeted redox inhibition of protein phosphatase 1 by Nox4 regulates eIF2α-mediated stress signaling. EMBO J. 2016;35(3):319–34.
85. Yu Q, Lee C, Wang W, Karamanlidis G, Kuroda J, Matsushima S, et al. Elimination of NADPH oxidase activity promotes reductive stress and sensitizes the heart to ischemic injury. J Am Heart Assoc. 2014;3(1):e0005.
86. Matsushima S, Kuroda J, Ago T, Zhai P, Ikeda Y, Oka S, et al. Broad suppression of NADPH oxidase activity exacerbates ischemia/reperfusion injury through inadvertent downregulation of hypoxia-inducible factor-1alpha and upregulation of peroxisome proliferator-activated receptor-alpha. Circ Res. 2013;112(8):1135–49.
87. Schröder K, Zhang M, Benkhoff S, Mieth A, Pliquett R, Kosowski J, et al. Nox4 Is a protective reactive oxygen species generating vascular NADPH oxidase. Circ Res. 2012;110(9):1217–25.
88. Schurman C, Rezende F, Kruse C, Yasar Y, Lowe O, Fork C, et al. The NADPH oxidase Nox4 has anti-atherosclerotic functions. Eur Heart J. 2015;36(48):3447–56.
89. Craige SM, Kant S, Reif M, Chen K, Pei Y, Angoff R, et al. Endothelial NADPH oxidase 4 protects ApoE−/− mice from atherosclerotic lesions. Free Radic Biol Med. 2015;89:1–7.
90. Brewer AC, Murray TVA, Arno M, Zhang M, Anilkumar NP, Mann GE, et al. Nox4 regulates Nrf2 and glutathione redox in cardiomyocytes in vivo. Free Radic Biol Med. 2011;51(1):205–15.
91. Craige SM, Chen K, Pei Y, Li C, Huang X, Chen C, et al. NADPH Oxidase 4 promotes endothelial angiogenesis through endothelial nitric oxide synthase activation. Circulation. 2011; 124(6):731–40.
92. Diebold I, Flugel D, Becht S, Belaiba RS, Bonello S, Hess J, et al. The hypoxia-inducible factor-2alpha is stabilized by oxidative stress involving NOX4. Antioxid Redox Signal. 2010;13(4):425–36.

93. Block K, Gorin Y, Hoover P, Williams P, Chelmicki T, Clark RA, et al. NAD(P)H oxidases regulate HIF-2alpha protein expression. J Biol Chem. 2007;282(11):8019–26.
94. Kiffin R, Christian C, Knecht E, Cuervo AM. Activation of chaperone-mediated autophagy during oxidative stress. Mol Biol Cell. 2004;15(11):4829–40.
95. Mijaljica D, Prescott M, Devenish RJ. Microautophagy in mammalian cells: Revisiting a 40-year-old conundrum. Autophagy. 2011;7:673–82.
96. Shintani T, Klionsky DJ. Autophagy in health and disease: a double-edged sword. Science. 2004;306(5698):990–5.
97. Klionsky D, Cregg J, Dunn WJ. A unified nomenclature for yeast autophagy-related genes. Dev Cell. 2003;5:539–45.
98. Feng Y, He D, Yao Z, Klionsky DJ. The machinery of macroautophagy. Cell Res. 2014;24(1):24–41.
99. Galluzzi L, Pietrocola F, Levine B, Kroemer G. Metabolic control of autophagy. Cell. 2014;159:1263–76.
100. Green DR, Galluzzi L, Kroemer G. Metabolic control of cell death. Science. 2014;345(6203):1250256.
101. Ravikumar B, Futter M, Jahreiss L, Korolchuk V, Lichtenberg M, Luo S, et al. Mammalian macroautophagy at a glance. J Cell Sci. 2009;122:1707–11.
102. Sciarretta S, Zhai P, Shao D, Maejima Y, Robbins J, Volpe M, et al. Rheb is a critical regulator of autophagy during myocardial ischemia: pathophysiological implications in obesity and metabolic syndrome. Circulation. 2012;125(9):1134–46.
103. Nakai A, Yamaguchi O, Takeda T, Higuchi Y, Hikoso S, Taniike M, et al. The role of autophagy in cardiomyocytes in the basal state and in response to hemodynamic stress. Nat Med. 2007;13:619–24.
104. Maejima Y, Kyoi S, Zhai P, Liu T, Li H, Ivessa A, et al. Mst1 inhibits autophagy by promoting the interaction between Beclin1 and Bcl-2. Nat Med. 2013;19:1478–88.
105. Sciarretta S, Volpe M, Sadoshima J. NOX4 regulates autophagy during energy deprivation. Autophagy. 2014;10(4):699–701.
106. Ikeda Y, Shirakabe A, Brady C, Zablocki D, Ohishi M, Sadoshima J. Molecular mechanisms mediating mitochondrial dynamics and mitophagy and their functional roles in the cardiovascular system. J Mol Cell Cardiol. 2015;78:116–22.
107. Zhu H, Tannous P, Johnstone JL, Kong Y, Shelton JM, Richardson JA, et al. Cardiac autophagy is a maladaptive response to hemodynamic stress. J Clin Invest. 2007;117(7):1782–93.
108. Zhai P, Sciarretta S, Galeotti J, Volpe M, Sadoshima J. Differential roles of GSK-3{beta} during myocardial ischemia and ischemia/reperfusion. Circ Res. 2011;109:502–11.
109. Liu Y, Shoji-Kawata S, Sumpter RM, Wei Y, Ginet V, Zhang L, et al. Autosis is a Na+,K+-ATPase-regulated form of cell death triggered by autophagy-inducing peptides, starvation, and hypoxia-ischemia. Proc Natl Acad Sci U S A. 2013;110(51):20364–71.
110. Hariharan N, Maejima Y, Nakae J, Paik J, Depinho RA, Sadoshima J. Deacetylation of FoxO by Sirt1 plays an essential role in mediating starvation-induced autophagy in cardiac myocytes. Circ Res. 2010;107(12):1470–82.
111. Zhang H, Bosch-Marce M, Shimoda LA, Yee ST, Jin HB, Wesley JB, et al. Mitochondrial autophagy is an HIF-1-dependent adaptive metabolic response to hypoxia. J Biol Chem. 2008;283(16):10892–903.
112. Bellot G, Garcia-Medina R, Gounon P, Chiche J, Roux D, Pouysségur J, et al. Hypoxia-induced autophagy is mediated through hypoxia-inducible factor induction of BNIP3 and BNIP3L via their BH3 domains. Mol Cell Biol. 2009;29(10):2570–81.
113. Hamacher-Brady A, Brady NR, Logue SE, Sayen MR, Jinno M, Kirshenbaum LA, et al. Response to myocardial ischemia/reperfusion injury involves Bnip3 and autophagy. Cell Death Differ. 2007;14(1):146–57.

114. Lee Y, Lee H-Y, Hanna RA, Gustafsson AB. Mitochondrial autophagy by Bnip3 involves Drp1-mediated mitochondrial fission and recruitment of Parkin in cardiac myocytes. AJP Heart Circ Physiol. 2011;301(5):H1924–31.
115. Zeng M, Wei X, Wu Z, Li W, Li B, Zhen Y, et al. NF-kappaB-mediated induction of autophagy in cardiac ischemia/reperfusion injury. Biochem Biophys Res Commun. 2013;436(2):180–5.
116. Criollo A, Senovilla L, Authier H, Maiuri MC, Morselli E, Vitale I, et al. IKK connects autophagy to major stress pathways. Autophagy. 2010;6(1):189–91.
117. Wang W, Li S, Wang H, Li B, Shao L, Lai Y, et al. Nrf2 enhances myocardial clearance of toxic ubiquitinated proteins. J Mol Cell Cardiol. 2014;72:305–15.
118. Scherz-Shouval R, Shvets E, Fass E, Shorer H, Gil L, Elazar Z. Reactive oxygen species are essential for autophagy and specifically regulate the activity of Atg4. EMBO J. 2007;26(7):1749–60.
119. Romao S, Gasser N, Becker AC, Guhl B, Bajagic M, Vanoaica D, et al. Autophagy proteins stabilize pathogen-containing phagosomes for prolonged MHC II antigen processing. J Cell Biol. 2013;203(5):757–66.
120. Yoon S, Woo SU, Kang JH, Kim K, Kwon MH, Park S, et al. STAT3 transcriptional factor activated by reactive oxygen species induces IL6 in starvation-induced autophagy of cancer cells. Autophagy. 2010;6(8):1125–38.
121. Jaishy B, Zhang Q, Chung HS, Riehle C, Soto J, Jenkins S, et al. Lipid-induced NOX2 activation inhibits autophagic flux by impairing lysosomal enzyme activity. J Lipid Res. 2014;56:319–53.
122. Pal R, Palmieri M, Loehr JA, Li S, Abo-Zahrah R, Monroe TO, et al. Src-dependent impairment of autophagy by oxidative stress in a mouse model of Duchenne muscular dystrophy. Nat Commun. 2014;5:4425.
123. Whitehead NP, Yeung EW, Froehner SC, Allen DG. Skeletal muscle NADPH oxidase is increased and triggers stretch-induced damage in the mdx mouse. PLoS ONE. 2010;5(12):e15354.
124. Gervásio OL, Whitehead NP, Yeung EW, Phillips WD, Allen DG. TRPC1 binds to caveolin-3 and is regulated by Src kinase – role in Duchenne muscular dystrophy. J Cell Sci. 2008;121(Pt 13):2246–55.
125. Uehara T, Nakamura T, Yao D, Shi Z-Q, Gu Z, Ma Y, et al. S-nitrosylated protein-disulphide isomerase links protein misfolding to neurodegeneration. Nature. 2006;441(7092):513–7.
126. Pal R, Bajaj L, Sharma J, Palmieri M, Di Ronza A, Lotfi P, et al. NADPH oxidase promotes Parkinsonian phenotypes by impairing autophagic flux in an mTORC1-independent fashion in a cellular model of Parkinson's disease. Sci Rep. 2016;6:22866.
127. Hu W, Tian H, Yue W, Li L, Li S, Gao C, et al. Rotenone induces apoptosis in human lung cancer cells by regulating autophagic flux. IUBMB Life. 2016;68(5):388–93.
128. Zhu X, Shen K, Bai Y, Zhang A, Xia Z, Chao J, et al. NADPH oxidase activation is required for pentylenetetrazole kindling-induced hippocampal autophagy. Free Radic Biol Med. 2016;94:230–42.
129. Teng R-J, Du J, Welak S, Guan T, Eis A, Shi Y, et al. Cross talk between NADPH oxidase and autophagy in pulmonary artery endothelial cells with intrauterine persistent pulmonary hypertension. Am J Physiol Lung Cell Mol Physiol. 2012;302(7):L651–63.
130. Grandis JR, Tweardy DJ. Elevated levels of transforming growth factor {alpha} and epidermal growth factor receptor messenger RNA are early markers of carcinogenesis in head and neck cancer. Cancer Res. 1993;53(15):3579–84.
131. Sobhakumari A, Schickling BM, Love-Homan L, Raeburn A, Fletcher EVM, Case AJ, et al. NOX4 mediates cytoprotective autophagy induced by the EGFR inhibitor erlotinib in head and neck cancer cells. Toxicol Appl Pharmacol. 2013;272(3):736–45.
132. Ren G, Luo W, Sun W, Niu Y, Ma DL, Leung CH, et al. Psoralidin induced reactive oxygen species (ROS)-dependent DNA damage and protective autophagy mediated by NOX4 in breast cancer cells. Phytomedicine. 2016;23(9):939–47.

133. Wu R-F, Ma Z, Liu Z, Terada LS. Nox4-derived H2O2 mediates endoplasmic reticulum signaling through local Ras activation. Mol Cell Biol. 2010;30(14):3553–68.
134. Zhao H, Zhang M, Zhou F, Cao W, Bi L, Xie Y, et al. Cinnamaldehyde ameliorates LPS-induced cardiac dysfunction via TLR4-NOX4 pathway: the regulation of autophagy and ROS production. J Mol Cell Cardiol. 2016;101:11–24. Epub ahead of print
135. Kouroku Y, Fujita E, Tanida I, Ueno T, Isoai A, Kumagai H, et al. ER stress (PERK/eIF2alpha phosphorylation) mediates the polyglutamine-induced LC3 conversion, an essential step for autophagy formation. Cell Death Differ. 2007;14(2):230–9.
136. Matsushima S, Kuroda J, Zhai P, Liu T, Ikeda S, Nagarajan N, et al. Tyrosine kinase FYN negatively regulates NOX4 in cardiac remodeling. J Clin Invest. 2016;126:3403–16.

Chapter 17
Mitochondrial Ubiquitin Ligase in Cardiovascular Disorders

Tao Yu, Yinfeng Zhang, and Pei-feng Li

The mitochondrion is a double-membrane organelle found in almost all eukaryotic organisms that retain its own small genome, and it is normally in a size between 0.75 and 3 μm in diameter. The word mitochondrion comes from the Greek *μίτος*, *mitos*, i.e. "thread", and *χουδρίου*, *chondrion*, i.e. "granule" or "grain-like". It is considered as the center of cellular energy production, which is commonly called "powerhouse" and supply adenosine triphosphate (ATP) to most of the cells [1]. Mitochondrion is a remarkably dynamic organelle, which undergoes constant migration and morphological changes during physiological as well as pathological processes. Particularly, mitochondrial fission and fusion machinery plays an essential role in the dynamics, division, distribution and morphology of the organelle, while these activities are independent of the cell cycle activity in most of the time [2–5]. Moreover, mitochondria are critically involved in regulating apoptosis, calcium homeostasis, lipid metabolism, aging, and the production of reactive oxygen species, which contribute to series of physiological abnormalities, especially heart diseases, including cardiac hypertrophy, heart failure, dilated cardiomyopathy and ischemia [6–9].

Ubiquitination is a post-translational modification in which a small conserved peptide, ubiquitin (Ub), is appended to target proteins in the cell through a series of complex enzymatic reactions [8]. Ub is a small protein with 76 amino acids, which prefers to join to an amino group on a substrate, either at the free amino terminus or on an internal lysine residue, and that results in protein modification [10]. Further, this tagging activity can influence the target protein's biological half-life, cellular localization and the recognition with its binding partners [11]. Numerous studies demonstrated that this ubiquitin proteasome system (UPS) is an essential regulator

T. Yu • Y. Zhang • P.-f. Li (✉)
Institute of Translational Medicine, Qingdao University,
38 Dengzhou, Qingdao, People's Republic of China
e-mail: peifli@qdu.edu.cn

G. Santulli (ed.), *Mitochondrial Dynamics in Cardiovascular Medicine*,
Advances in Experimental Medicine and Biology 982,
DOI 10.1007/978-3-319-55330-6_17

of protein quality control in all cells, including cardiomyocytes. UPS participates in protein trafficking, protein degradation, cellular signal transduction, cell cycle progression, cell transformation and immune recognition [12].

Recently researches focus on the functions of ubiquitin ligase in mitochondrial system and its significant role and mechanism in regulating mitochondrial dynamics attracts more and more attention. The mitochondrial ubiquitin ligase (MITOL) is also known as MARCH5/RNF153, which is localized in the mitochondrial outer membrane reported to play a functional role in mitochondria and apoptosis. It can potentially modulate mitochondrial fission as well as mitochondrial morphology [13–15]. Mitochondrial ubiquitin ligase contains four transmembrane spanning segments for the binding of mitochondrial fission proteins, namely, human mitochondrial Fission 1 (hFis1), dynamin-related protein 1 (Drp1) and mitofusin 2 [16–18]. MITOL possesses a Plant Homeo-Domain (PHD) motif responsible for E3 ubiquitin ligase activity, which is rapidly degraded by auto-ubiquitination activity in a PHD-dependent manner. MITOL has been shown to control mitochondrial dynamics by regulating mitochondrial fission factors such as mitochondrial fission factor Drp1, microtubule-associated protein 1B and mitofusin2 [13, 16, 18]. However, the regulatory mechanism of MITOL in the mitochondrial dynamics remains unclear. In addition, there are other kinds of ubiquitin ligases, such as MULAN (also known as GIDE) and RNF185 also found in mitochondria.

Critical Role of Mitochondrial Ubiquitin Ligases in the Pathophysiology of Cardiovascular Diseases

As mitochondrial ubiquitin ligases maintain protein quality control and regulating many critical cellular processes, they play an important role in maintaining the health and diseases. A increasing number of evidence proved that mitochondrial ubiquitin ligases play fundamental role in the development of cardiovascular diseases.

Regulation of Mitochondrial Dynamics by MITOL/MARCH5 via Controlling the Balance Between Mitochondrial Fission and Fusion Proteins

Apoptosis plays a vital role in the cardiac and vascular structures during both the early morphogenesis and later developmental stages [19]. However, excessive apoptosis could trigger many cardiovascular diseases such as myocardial infarction, cardiomyopathy, cardiac hypertrophy and anthracycline-induced cardiotoxicity [20]. Cardiomyocytes are enriched with mitochondria, which constantly undergo fusion and fission to maintain organelle health. However, abnormal fusion or fission participates in the regulation of cellular apoptosis [21, 22]. The mitochondrial ubiquitin ligase (MITOL), a E3 Ub ligase, can ubiquitinate a variety of mitochondria-associated proteins [13, 21]. The well-known substrates of MITOL are two mitochondrial fission factors (Fis1 and Drp1), which are essential components of the mitochondrial membrane fission machinery. Drp1 accumulation induces reactive oxygen species (ROS) production and mitochondrial dysfunction [13, 16]. The denatured and aggregated proteins accumulate in mitochondria and disrupt normal mitochondrial function, and that could be degraded by MITOL [23]. For example, overexpression of MITOL ubiquitinate mutant superoxide dismutase1 (mSOD1) and promotes its degradation. This can suppress both accumulation of mSOD1 and mSOD1-induced ROS generation in mitochondria, which indicates that MITOL protects against mitochondrial dysfunction caused by the accumulation of mSOD1 [24]. Wang and colleagues recently found that MITOL plays a protective role against apoptosis in cardiomyocytes by weakening mitochondrial fission through suppression of H2O2 induced accumulation of Drp1 in mitochondria [15]. However, there is a contrary report that MITOL is required for Drp1 dependent mitochondrial division, and the inhibition of MITOL caused mitochondrial fusion. A study by Nagashima and colleagues explained that MITOL-dependent Drp1 regulation could vary depends on the circumstances, such as cell cycle phase or nutritional status [16].

Mitochondrial fusion is mainly regulated by membrane bounded GTPases, mitofusin 1 (MFN1) and mitofusin 2 (MFN2), which mainly control outer membrane fusion [18, 24]. MFN2 is also localized in the endoplasmic reticulum (ER). However, MITOL only interacts with and ubiquitinates mitochondrial MFN2, but not ER associated MFN2 [25]. In addition, researchers found that the C-terminal domain of MITOL is specifically interact with the HR1 domain of MFN2, and MITOL regulates ER tethering to mitochondria by stimulating MFN2 via the ubiquitination of the GTPase domain K192. Thus, knockdown of MITOL can suppress MFN2 complex formation and triggers MFN2 mislocalization and ER membrane dysfunction [25]. Additionally, HDAC6, a cytosolic ubiquitin binding deacetylase, can irritate adaptive mitochondrial fusion under glucose starvation condition [26]. Recently, Kim and colleagues discovered that MITOL is responsible for hypoxia-induced MFN2 degradation in HDAC6 deficient cells, and genetic abolition of HDAC6 in amyotrophic lateral sclerosis model mice suppressed MFN2 degradation even MARCH5 induction is high, which illustrates that HDAC6 is a vital regulator of MFN2 degradation by MITOL [27, 28]. However, the relationship of this pathway and hypoxia induced cardiovascular diseases hasn't been evaluated, thus further studies need to be conducted.

Under mitochondrial stress conditions, the level of MFN1 is significantly increased and its depletion can induce cell death. Interestingly, overexpression of MFN1 also sensitizes cells to apoptotic death, which indicates that a fine tuning of MFN1 levels is necessary for cell survival [29]. Moreover, acetylated MFN1 is elevated under starvation induced mitochondrial stress conditions, thus the balancing of MFN1 levels is maintained by MITOL-mediated control on acetylation of MFN1, which is crucial for cell survival under mitochondria stress conditions [29].

MULAN Has an Important Role in Cell Growth, Cell Death, and Mitophagy

MULAN (for Mitochondrial Ubiquitin Ligase Activator of NF-κB) is another RING-finger E3 Ub ligase anchoring to mitochondrial outer membrane, which can also influence mitochondrial dynamics [27]. NF-κB plays a pivotal role in the regulation of the cell proliferation and survival in cardiovascular systems in addition to its typical role in promoting inflammation in response to pathogens [30]. Tacchi and colleagues cloned and analyzed the expression of MULAN in Atlantic salmon (*Salmo salar*) to confirm its role in the activation of NF-κB pathway [31]. The capability of MULAN to stimulate NF-κB depends on RING domain of MULAN and E3 ubiquitin ligase binding TRAF2. During stress-induced mitochondrial hyperfusion (SIMH), MULAN forms a complex with TRAF2 and modulates its ubiquitylation, signifying that TRAF2 may serve as an ubiquitylated transmitter of NF-κB signaling in this pathway [32, 33]. The endoplasmic reticulum (ER) is the major site of cellular homeostasis regulation, in which the accumulation of misfolded protein could cause stress [31]. Fujita and colleagues discovered that MULAN could activate NF-κB to protect cells from ER stress-induced apoptosis [34]. Furthermore, the serine/threonine kinase Akt (protein kinase B) is known to be involved in the cellular response to various stimuli. The activation of Akt controls multiple biological responses in cardiovascular systems, including protein synthesis, metabolism, cell proliferation and growth, cell-cycle progression, and the suppression of apoptosis [35, 36]. MULAN can degrade Akt and thus inhibits Akt dependent cell proliferation and viability [34].

Interestingly, MULAN is involved in numerous biological pathways including cell growth, cell death, and more recently in mitophagy. In addition, the interaction of Gamma-aminobutyric acid receptor-associated protein (GABARAP) with MULAN–Ube2E3 confirms that MULAN is a vital regulator of mitophagy and it provides a credible mechanism for its function [37]. The nuclear encoded mitochondrial serine protease Omi/HtrA2 is a dual function protein, whose function depends entirely on its subcellular localization. During apoptosis, Omi/HtrA2 is released into the cytoplasm to participate in cell death [38]. Under oxidative stress condition, Omi/HtrA2 degrades MULAN, but inactivation of Omi/HtrA2 protease leads to the deregulation of MULAN and increases mitophagy [39] .

Other Important Mitochondrial Ubiquitin E3 Ligase

Currently, there is no direct evidence for a relationship between other mitochondrial ubiquitin ligase, such as GIDE and RNF185, and cardiovascular disorders. Nevertheless, several studies found the effect of these ligases on cell growth, apoptosis and autophagy in other systems or cell types.

GIDE is a mitochondrial E3 ubiquitin ligase, which was identified by Zhang and colleagues in 2008 [40]. GIDE contains a conserved C-terminal RING finger domain, which is required for E3 ligase activity [41]. JNK is a stress responsive mitogen-activated protein kinase (MAPK), which acts as a pro-apoptotic factor under some circumstances in several cell types as well as in the cardiovascular system [42]. GIDE can activate JNK, and overexpression of GIDE induces apoptosis via a pathway involving activation of caspases. On the contrary, GIDE accelerates cell growth when its expression is low by unknown mechanisms.

RNF185 is another novel mitochondrial ubiquitin E3 ligase, which selectively regulates mitochondrial autophagy in cultured cells [40]. RNF185 promotes LC3II to accumulate in human cell lines as well as the formation of autophagolysosomes. In addition, the Bcl-2 family protein BNIP1 is a substrate of RNF185. The polyubiquitinated BNIP1 recruits autophagy receptor p62, which attaches both ubiquitin and LC3 to link ubiquitination and autophagy [40]. In the ER, misfolded or improperly assembled proteins are exported to the cytoplasm and degraded by the ubiquitin-proteasome pathway through a process called ER-associated degradation (ERAD). RNF18 can manipulate the stability of Cystic fibrosis transmembrane conductance regulator (CFTR), a substrate of ERAD, in a RING and proteasome-dependent manner [43].

Conclusion

The protein quality control is fundamental for the maintenance of normal cellular function. When this equilibrium is collapsed, a variety of diseases, including cardiovascular disorders, will emerge. Accumulating evidence indicates that mitochondrial ubiquitin ligases are prominent in the maintenance of protein quality control. In particular, many studies have illustrated the critical role of mitochondrial ubiquitin ligases in the heart. Despite the regulatory functions of mitochondrial ubiquitin ligases such as MULAN, GIDE, and RNF185 are reported in different cell systems, there is yet no direct indication proving their regulatory role in cardiovascular diseases. The available evidence indicates the crucial role of MITOL in cardiomyocytes and heart disease. Thus, attention to the research on understanding of the relationship between mitochondrial ubiquitin ligase and cardiovascular diseases could offer a potential therapeutic target for cardiac disorders in the future.

References

1. Henze K, Martin W. Evolutionary biology: essence of mitochondria. Nature. 2003; 426:127–8.
2. Yaffe MP. Dynamic mitochondria. Nat Cell Biol. 1999;1:E149–50.
3. Griparic L, van der Bliek AM. The many shapes of mitochondrial membranes. Traffic. 2001;2:235–44.

4. Jensen RE, Hobbs AE, Cerveny KL, Sesaki H. Yeast mitochondrial dynamics: fusion, division, segregation, and shape. Microsc Res Tech. 2000;51:573–83.
5. Shaw JM, Nunnari J. Mitochondrial dynamics and division in budding yeast. Trends Cell Biol. 2002;12:178–84.
6. Xie W, Santulli G, Reiken SR, Yuan Q, Osborne BW, Chen BX, Marks AR. Mitochondrial oxidative stress promotes atrial fibrillation. Sci Rep. 2015;5:11427.
7. Santulli G, Xie W, Reiken SR, Marks AR. Mitochondrial calcium overload is a key determinant in heart failure. Proc Natl Acad Sci U S A. 2015;112:11389–94.
8. Hershko A, Ciechanover A. The ubiquitin system for protein degradation. Annu Rev Biochem. 1992;61:761–807.
9. Zungu M, Schisler JC, Essop MF, McCudden C, Patterson C, Willis MS. Regulation of AMPK by the ubiquitin proteasome system. Am J Pathol. 2011;178:4–11.
10. Pickart CM, Eddins MJ. Ubiquitin: structures, functions, mechanisms. Biochim Biophys Acta. 1695;2004:55–72.
11. Livnat-Levanon N, Glickman MH. Ubiquitin-proteasome system and mitochondria – reciprocity. Biochim Biophys Acta. 2011;1809:80–7.
12. Hershko A, Ciechanover A. The ubiquitin system. Annu Rev Biochem. 1998;67:425–79.
13. Yonashiro R, Ishido S, Kyo S, Fukuda T, Goto E, Matsuki Y, Ohmura-Hoshino M, Sada K, Hotta H, Yamamura H, Inatome R, Yanagi S. A novel mitochondrial ubiquitin ligase plays a critical role in mitochondrial dynamics. EMBO J. 2006;25:3618–26.
14. Nagashima S, Yanagi S. Role of MITOL in mitochondrial dynamics and diseases. Seikagaku J Jpn Biochem Soc. 2014;86:63–7.
15. Wang J, Aung LH, Prabhakar BS, Li P. The mitochondrial ubiquitin ligase plays an anti-apoptotic role in cardiomyocytes by regulating mitochondrial fission. J Cell Mol Med. 2016;20:2278.
16. Nagashima S, Tokuyama T, Yonashiro R, Inatome R, Yanagi S. Roles of mitochondrial ubiquitin ligase MITOL/MARCH5 in mitochondrial dynamics and diseases. J Biochem. 2014;155:273–9.
17. Nakamura N, Kimura Y, Tokuda M, Honda S, Hirose S. MARCH-V is a novel mitofusin 2- and Drp1-binding protein able to change mitochondrial morphology. EMBO Rep. 2006;7:1019–22.
18. Karbowski M, Neutzner A, Youle RJ. The mitochondrial E3 ubiquitin ligase MARCH5 is required for Drp1 dependent mitochondrial division. J Cell Biol. 2007;178:71–84.
19. James TN. Normal and abnormal consequences of apoptosis in the human heart. From postnatal morphogenesis to paroxysmal arrhythmias. Circulation. 1994;90:556–73.
20. Liu W, Wang X, Mei Z, Gong J, Gao X, Zhao Y, Ma J, Xie F, Qian L. Chronic stress promotes the progression of pressure overload-induced cardiac dysfunction through inducing more apoptosis and fibrosis. Physiol Res. 2015;64:325–34.
21. Suen DF, Norris KL, Youle RJ. Mitochondrial dynamics and apoptosis. Genes Dev. 2008;22:1577–90.
22. Fusco A, Santulli G, Sorriento D, Cipolletta E, Garbi C, Dorn GW 2nd, Trimarco B, Feliciello A, Iaccarino G. Mitochondrial localization unveils a novel role for GRK2 in organelle biogenesis. Cell Signal. 2012;24:468–75.
23. Hashimoto M, Rockenstein E, Crews L, Masliah E. Role of protein aggregation in mitochondrial dysfunction and neurodegeneration in Alzheimer's and Parkinson's diseases. Neruomol Med. 2003;4:21–36.
24. Yonashiro R, Sugiura A, Miyachi M, Fukuda T, Matsushita N, Inatome R, Ogata Y, Suzuki T, Dohmae N, Yanagi S. Mitochondrial ubiquitin ligase MITOL ubiquitinates mutant SOD1 and attenuates mutant SOD1-induced reactive oxygen species generation. Mol Biol Cell. 2009;20:4524–30.
25. Koshiba T, Detmer SA, Kaiser JT, Chen H, McCaffery JM, Chan DC. Structural basis of mitochondrial tethering by mitofusin complexes. Science. 2004;305:858–62.
26. Santel A, Fuller MT. Control of mitochondrial morphology by a human mitofusin. J Cell Sci. 2001;114:867–74.

27. Kim HJ, Nagano Y, Choi SJ, Park SY, Kim H, Yao TP, Lee JY. HDAC6 maintains mitochondrial connectivity under hypoxic stress by suppressing MARCH5/MITOL dependent MFN2 degradation. Biochem Biophys Res Commun. 2015;464:1235–40.
28. Sugiura A, Nagashima S, Tokuyama T, Amo T, Matsuki Y, Ishido S, Kudo Y, McBride HM, Fukuda T, Matsushita N, Inatome R, Yanagi S. MITOL regulates endoplasmic reticulum-mitochondria contacts via Mitofusin 2. Mol Cell. 2013;51:20–34.
29. Lee JY, Kapur M, Li M, Choi MC, Choi S, Kim HJ, Kim I, Lee E, Taylor JP, Yao TP. MFN1 deacetylation activates adaptive mitochondrial fusion and protects metabolically challenged mitochondria. J Cell Sci. 2014;127:4954–63.
30. Park YY, Nguyen OT, Kang H, Cho H. MARCH5-mediated quality control on acetylated Mfn1 facilitates mitochondrial homeostasis and cell survival. Cell Death Dis. 2014;5:e1172.
31. Tacchi L, Casadei E, Bickerdike R, Secombes CJ, Martin SA. Cloning and expression analysis of the mitochondrial ubiquitin ligase activator of NF-kappaB (MULAN) in Atlantic salmon (*Salmo salar*). Mol Immunol. 2011;49:558–65.
32. Li W, Bengtson MH, Ulbrich A, Matsuda A, Reddy VA, Orth A, Chanda SK, Batalov S, Joazeiro CA. Genome-wide and functional annotation of human E3 ubiquitin ligases identifies MULAN, a mitochondrial E3 that regulates the organelle's dynamics and signaling. PLoS One. 2008;3:e1487.
33. Karin M, Cao Y, Greten FR, Li ZW. NF-kappaB in cancer: from innocent bystander to major culprit. Nat Rev Cancer. 2002;2:301–10.
34. Fujita H, Aratani S, Fujii R, Yamano Y, Yagishita N, Araya N, Izumi T, Azakami K, Hasegawa D, Nishioka K, Nakajima T. Mitochondrial ubiquitin ligase activator of NF-kappaB regulates NF-kappaB signaling in cells subjected to ER stress. Int J Mol Med. 2016;37:1611–8.
35. Zemirli N, Pourcelot M, Ambroise G, Hatchi E, Vazquez A, Arnoult D. Mitochondrial hyperfusion promotes NF-kappaB activation via the mitochondrial E3 ligase MULAN. FEBS J. 2014;281:3095–112.
36. Sitia R, Braakman I. Quality control in the endoplasmic reticulum protein factory. Nature. 2003;426:891–4.
37. Franke TF. PI3K/Akt: getting it right matters. Oncogene. 2008;27:6473–88.
38. Bae S, Kim SY, Jung JH, Yoon Y, Cha HJ, Lee H, Kim K, Kim J, An IS, Kim J, Um HD, Park IC, Lee SJ, Nam SY, Jin YW, Lee JH, An S. Akt is negatively regulated by the MULAN E3 ligase. Cell Res. 2012;22:873–85.
39. Ambivero CT, Cilenti L, Main S, Zervos AS. Mulan E3 ubiquitin ligase interacts with multiple E2 conjugating enzymes and participates in mitophagy by recruiting GABARAP. Cell Signal. 2014;26:2921–9.
40. Zhang B, Huang J, Li HL, Liu T, Wang YY, Waterman P, Mao AP, Xu LG, Zhai Z, Liu D, Marrack P, Shu HB. GIDE is a mitochondrial E3 ubiquitin ligase that induces apoptosis and slows growth. Cell Res. 2008;18:900–10.
41. Martins LM, Morrison A, Klupsch K, Fedele V, Moisoi N, Teismann P, Abuin A, Grau E, Geppert M, Livi GP, Creasy CL, Martin A, Hargreaves I, Heales SJ, Okada H, Brandner S, Schulz JB, Mak T, Downward J. Neuroprotective role of the Reaper-related serine protease HtrA2/Omi revealed by targeted deletion in mice. Mol Cell Biol. 2004;24:9848–62.
42. Cilenti L, Ambivero CT, Ward N, Alnemri ES, Germain D, Zervos AS. Inactivation of Omi/HtrA2 protease leads to the deregulation of mitochondrial Mulan E3 ubiquitin ligase and increased mitophagy. Biochim Biophys Acta. 1843;2014:1295–307.
43. Deng Y, Ren X, Yang L, Lin Y, Wu X. A JNK-dependent pathway is required for TNFalpha-induced apoptosis. Cell. 2003;115:61–70.

Chapter 18
Nitrite-Nitric Oxide Signaling and Cardioprotection

Matthias Totzeck, Ulrike B. Hendgen-Cotta, and Tienush Rassaf

An acute myocardial infarction remains the leading single cause of death worldwide. Although reperfusion strategies have significantly improved overall mortality rates, reduction of the I/R-associated cardiomyocyte damage in patients remains an unmet goal [1]. Mitochondria as central energy suppliers in cardiac cells are key players in I/R-driven lethal injury. Protection of mitochondrial integrity and function is therefore a pivotal target in cardioprotection. Therapies that modify NO levels protect the heart from lethal myocardial I/R injury but both the resolution of the downstream signaling as well as the translation into clinical practice have not been achieved so far [2, 3]. This pertains both to the exact nature of how the response to the NO donors are transferred to the heart and which exact signaling cascades are activated within the cardiomyocyte and in cardiac mitochondria. Transfer of NO to the cardiomyocyte can be achieved by direct application of so-called NO-donors – compounds that continuously release the free radical NO. Recent experimental evidence also points to a potentially mitochondria-selective NO donor (mito-*S*NO) [4]. NO metabolizes to nitrite and nitrate. These chemically more stable forms distribute throughout the circulation and tissues but may be recycled to NO under hypoxic and ischemic conditions – an activation mechanism, which is canalized by metalloproteins, e.g. myoglobin in the heart [5–12]. Exogenous application of nitrate or nitrite can significantly enhance the bodily provision of these compounds and has been shown to contribute to cardioprotection in experimental studies. Furthermore, endogenous NO production can be enhanced by mechanical maneuvers, particularly rIPC – short alternating phases of extremity ischemia by cuff occlusion followed by reperfusion [3, 5]. Here, we briefly recapitulate the events that characterize myocardial I/R injury. This is followed by an outline of the NO-related signaling in

M. Totzeck • U.B. Hendgen-Cotta • T. Rassaf (✉)
Department of Cardiology and Vascular Medicine, West German Heart and Vascular Center, University Hospital Essen, Essen, Germany
e-mail: tienush.rassaf@uk-essen.de

G. Santulli (ed.), *Mitochondrial Dynamics in Cardiovascular Medicine*, Advances in Experimental Medicine and Biology 982,
DOI 10.1007/978-3-319-55330-6_18

I/R with particular emphasis on mitochondrial functions. Finally, we discuss how NO signaling can be effectively enhanced and how this could translate into cardioprotection.

Myocardial I/R Injury: Underlying Mechanisms

The events leading to myocardial I/R injury are complex and still under intense debate. The three major modifiers of I/R injury are ROS, a deranged Ca^{2+} signaling and the destruction of mitochondrial integrity and function.

Reperfusion therapy by percutaneous coronary intervention or pharmacological thrombolysis rapidly increases intracellular oxygen levels. A subsequent insufficient electron transport over mitochondrial membranes can result in the formation of ROS. This may overcome the availability of endogenous antioxidant mechanisms. The relative imbalance between the generation and the decomposition of ROS is one of the key initiators of I/R injury with consecutive organelle membrane and protein damage and the induction of apoptosis and necrosis [13]. Paralleled by increased ROS concentrations, the levels of bioactive NO appear to decrease during the early reperfusion phase. NO fulfills numerous functions in the heart including scavenging of radicals, modulating immune response, improvement of blood flow and left ventricular function [14–16]. As a consequence, these functions are impaired.

Evidence from experimental studies also implicates that a rapid burst-like increase in Ca^{2+} concentration contributes to the lethal injury of the reperfused myocardium [17–20]. Elevated Ca^{2+} levels can trigger arrhythmias, hypercontracture [21] and the activation of deleterious signaling cascades (see Chap. 10). In addition, hypercontracture and necrosis is not limited to one cell but also disseminated by directed contracture from cell to cell and by cell-to-cell progression through intercellular gap junctions [22].

A third central mediator/target of I/R injury are the mitochondria and specifically the mitochondrial membranes [23–25]. Mitochondria are not only the main source of energy production but can also signal cell death and cardiomyocyte disintegration [26]. However, modification of mitochondrial function is regarded a promising therapeutic option given to modulate functional recovery, reduce infarct size, improve cardiac function and prevent heart failure progression after ischemic injuries in experimental approaches [27]. Results from our studies implicate that administration of exogenous nitrite during myocardial I/R injury yields bioactive NO leading in particular to an *S*-nitrosation modification of mitochondrial complex I. This in turn regulates mitochondrial respiration and limits formation of mitochondria- derived ROS. *S*-nitrosation of complex I is furthermore associated with an adaption of myocardial functions to a reduced oxygen supply — a mechanism which is known as hibernation. Finally, nitrite reduction to NO contributes to a decrease in myocardial necrosis and apoptosis [8, 12, 28, 29].

In addition to the regulation of mitochondrial function by hypoxic nitrite signaling, mitochondrial integrity is maintained by the intraorganellar proteolytic system and the dynamic nature of the mitochondrial population in the cell. Membrane fusion and fission allow mitochondrial content adaption for cell integrity, and damaged mitochondria are selectively removed by a process, termed mitophagy, which protects against cell death [27, 30]. When mitochondria are damaged thus losing membrane potential, the kinase PTEN-induced putative kinase protein 1 (PINK1) accumulates and recruits the E3 ubiquitin ligase parkin, which then ubiquitylates mitochondrial proteins and causes mitochondria to become enclosed by membranes. BNIP3 as well as NIX, both related to the BH3-only family, also contribute to mitophagy in response to hypoxia. The current knowledge of the cellular mechanism behind mitochondrial quality control promotes interest in associated interventions in physiology and diseases such as aging, cancer, degenerative disorders and coronary heart disease. Notably, initial experimental evidence suggests that mitophagy is involved in the protection from ischaemic preconditioning (IPC) applied locally to a myocardial region. This mechanism leads to cardioprotection of the reperfused areas [27, 31]. It is speculated that the inhibition of the formation of mitochondrial membrane pores, e.g. the so-called mitochondrial permeability transition pore (mPTP), are involved in protection from ischemic preconditioning applied to the heart before index ischemia as well as in the regulation of mitophagy to reduce the numbers of dysfunctional organelles [27, 30, 31] (see Chap. 9). Less is known about the relative contribution to cardioprotection when the protective stimulus is applied to a heart-distant organ – rIPC.

Cardiomyocytes die primarily by apoptosis or necrosis with mitochondria at the center of regulation [32]. Apoptosis causes cell shrinkage, cellular and compartment fragmentation, and phagocytosis. The general characteristics of necrosis include cellular swelling and rupture, marked depletion of energy carriers, and an inflammatory response. Newer studies argue that both apoptosis and necrosis are regulated by a complex signaling machinery with overlapping processes. The critical step for apoptosis is the permeabilization of the MOM pore (MOMP). This should occur by oligomerization of BAX and other pro-apoptotic BH3 proteins. Subsequently, a release of apoptogenic factors, e.g. of apoptosis-inducing factor (AIF) which is also cleaved and activated by calpains, and endonuclease G, cause DNA damage [33, 34]. By contrast, the key characteristic of necrotic cell death is the permeabilization of the mitochondrial inner membrane. This may be further substantiated by a destruction of mitochondrial complex I. However, also BAX substantially contributes to necrotic cell death and *Bax* deficient mice show a remarkable reduction in necrosis. Interestingly, BAX-signaling is not only involved as critical step in apoptosis and necrosis, but also furthermore involved in mitochondrial structural dynamism [35].

Mitochondrial morphology and structural dynamism is regulated by a fine equilibrium between fission and fusion, repeated cycles of which re-distribute mitochondrial constituents by separation and fusion of mitochondria [36]. Fission is mediated by dynamin-related protein 1 (DRP1), a GTPase that transits from cytosol to mitochondria. Fusion is controlled by three dynamin-related GTPases: mitofusin

(Mfn1) and Mfn2 in the MOM and optic atrophy (OPA1) in the mitochondrial inner membrane. Interestingly, *Bax* knockout mice (formerly examined to have reduced apoptosis and necrosis) also show smaller, more fragmented mitochondria, which are less susceptible to cell death induction in I/R. Comparable characteristics were found in *Mfn2* knockout mice and cell lines. BAX reconstitution has been shown to cause an increase in mitochondrial fusion with mitochondria much more sensible to I/R related cellular stress. Although BAX is principally involved in both mitochondria-driven cell death and structural dynamism, the relationship between mitochondrial dynamics and cell death is poorly understood [37–42].

NO in Acute I/R Injury

As demonstrated in numerous experimental studies, myocardial I/R injury can be modulated. However, although numerous animal model-based studies showed much-reduced I/R injury, the respective translational clinical trials failed to demonstrate the same benefit in humans. Several of these experimental studies point to a protective role of the NO pool during myocardial I/R (Fig. 18.1) [28–31]. Increase of the circulating and the tissue (cardiac) NO pool exerts protective tissue effects in a mouse model of I/R injury in vivo [28, 30, 32]. The protective effects can derive from a hemeprotein-dependent reduction of nitrite to NO, as demonstrated in mice without myoglobin [30].

NO derived from nitrite reduction via cardiac myoglobin reversibly modulates mitochondrial electron transport, thus decreasing reperfusion-derived oxidative stress and inhibiting cellular apoptosis leading to a smaller final infarct size. In details, under hypoxic conditions the non-enzymatically formed NO binds to the aa_3 side of cytochrome oxidase (complex IV in the mitochondria) and thus inhibits respiration. As a result of this, a reduction in the energy status occurs, which results in an attenuation of the myocardial pump function and consequently, in reduced oxygen consumption. If the myocardium is again adequately supplied with oxygen, the non-enzymatic NO formation via myoglobin ceases and concurrently the inhibition of energy production as well as the restriction (hibernation) of the myocardial pump function also ends. Myoglobin thus appears to assume the role of an oxygen sensor in the myocardium via NO adjusts the myocardial energetics to the diminished oxygen supply [28].

Posttranslational modifications of relevant cardiomyocyte proteins are the second major aspect of hypoxic NO signaling. *S*-nitrosation is the addition of an NO moiety to a protein sulfhydryls. This can effectively change proteins conformation and function [8]. We and others have demonstrated that within the cardiomyocyte, NO reacts with numerous yet incompletely identified proteins [43]. Macrophage-migration inhibitory factor (MIF) plays a very important role in myocardial I/R injury. *S*-nitrosation of MIF was able to enhance the ROS decomposition properties of this molecule resulting in less necrosis and apoptosis in the reperfused myocardium [8]. *S*-nitrosation may further affect mitochondrial elements. It has recently

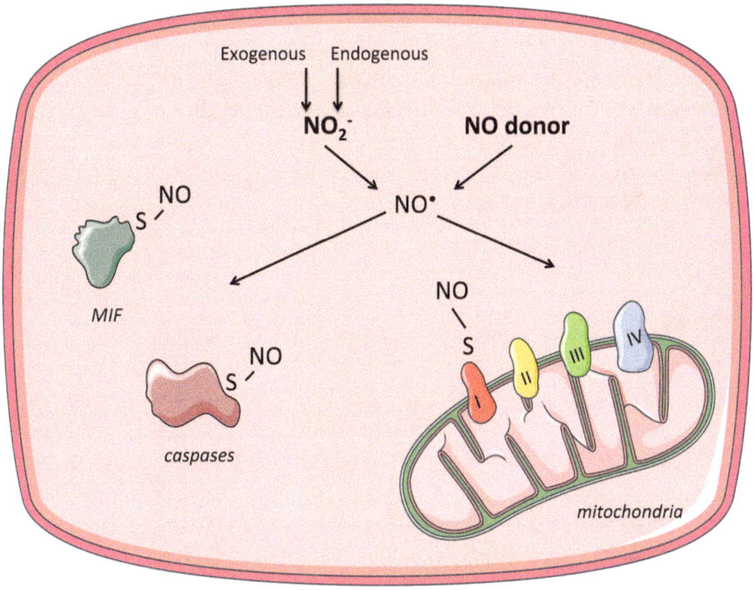

Fig. 18.1 Hypoxic NO signaling in myocardial I/R. Hypoxic NO signaling can be enhanced by e.g. increasing the levels of cardiomyocyte nitrite either by direct intravenous injection of by intake of dietary nitrate (exogenous) or be increase of endogenous NO/nitrite generation through mechanical endothelium stimulation (e.g. by rIPC). Alternatively, NO donors directly release NO which then causes an S-nitrosation modification of cytoplasmatic proteins involved in apoptosis (e.g. caspases) or related to ROS decomposition (e.g. MIF). NO can furthermore modifiy mitochondrial respiratory chain elements by S-nitrosation of complex I

been forwarded that elements of the fusion and fission machinery were significantly altered by NO-dependent modification [44]. While this may relate to a preservation of mitochondrial structure, *S*-nitrosation of mitochondrial complex I, has been significantly associated with a decrease in mitochondria-derived ROS through a decrease in misguided respiration [12, 45]. Taken together, multiple effects of NO have been characterized during I/R. The relative contribution of each of these signaling aspects remains obscure. However, the majority was in support of the notion that NO is cardioprotective in the scope of myocardial I/R.

Delivery of NO

Novel therapeutic strategies are required to alleviate the burden of I/R injury. As demonstrated in numerous infarction studies in animals and a small number of clinical proof-of-concept investigations, I/R injury can be effectively reduced by mechanical and pharmacological interventions [1].

Several pathways have been characterized that may affect the levels of NO in the reperfused myocardium. This ranges from chemical compounds that release NO on a continuous basis (NO donors) to inorganic nitrate and nitrite and mechanical maneuvers in particularly rIPC. While the next section provides and overview of these three entities, the reader is kindly referred to previous reviews discussing the use and benefits of organic nitrates which are characterized by a completely different mechanism of NO release [2].

NO Donors

A wide range of NO-releasing molecules has been forwarded with some interesting pharmacological agents among them. The reader is kindly referred to recent relevant reviews of this topic [2]. However, some very interesting aspects must be mentioned. Recent studies have evaluated compounds that combine standard medications in ischemic heart disease with NO-releasing agents. Among these were NO releasing pravastatin and nitro-aspirin. For both substances preclinical data implicate a therapeutical benefit for the reduction of myocardial I/R injury. Clinical data have to be awaited for.

While these donors cause a more or less continuous release of NO regardless of the localization (circulation, cytoplasma, mitochondria), the mitochondria-selective S-nitrosating agent mito-SNO potentially avoids these caveats by selective activation when crossing mitochondrial borders only [4]. This caused a significant S-nitrosation of mitochondrial complex I, which was reversible and protected the heart in the first few minutes of reperfusion. It remains to be evaluated whether this pertains to the clinical setting.

The Nitrate-Nitrite-NO Pathway

Nitrate and nitrite are now regarded as source for NO under hypoxic, ischemic, or injury conditions [11, 46]. Nitrite is reduced significantly to bioactive NO with decreased oxygen levels and pH by a variety of mechanisms, including reaction with deoxygenated myoglobin, in the heart as well as other heme proteins in solid organs and blood cells [28, 47–49]. Nitrite levels result from the reduction of dietary inorganic nitrate intake, which is reduced to nitrite by oral cavitiy bacteria [50, 51]. A dose of 1.2 nmoles of nitrite was sufficient in reducting myocardial and liver injury in I/R also these cause a moderate increase of 10% of the circulating nitrite levels [52, 53]. Furthermore, mice with decreased endgenous circulating nitrite concentrations are more susceptible to IR injury, an effect that is attenuated by exogenous nitrate [54]. These data suggest that a diet rich in nitrate and nitrite may have profound cytoprotective effects. In addition, eldery participants given dietary nitrate supplementation showed a marked incresase in endothelial function, which is a hallmark of arteriosclerosis, which, in turn, precedes myocardial infaction [55].

rIPC

Mechanical strategies that reduce I/R injury have been named IPC and postcondioning (PostC) [56, 57] They are applied as brief non-deleterious cycles of I/R before (IPC) or after (PostC) the main ischemic event. It is widely accepted that these mechanical interventions condition the heart and activate endogenous, protective modalities [58, 59]. Interestingly, the conditioning stimulus can also be applied to an organ distant (remote) to the heart such as the limb, kidney or intestine (rIPC) [60] and rPostC [61] respectively. The protective assets of IPC and rIPC rely on a distinct cellular signaling cascade, whose individual members can be subdivided into triggers, effectors, mediators and end-effectors [21, 62–66]. Endogenous triggering molecules are e.g. opioids, bradykinin, prostaglandins and adenosine and effects are mostly exerted by G-protein coupled receptors (GPCR) [67–70]. The rIPC downstream signaling cascade is structurally complex [65, 71, 72]. Activation of GPCR leads to an activation of phosphoinositide 3-kinase (PI3K) and further downstream targets [73–78]. Mitochondria are generally regarded to be the end-effector of the IPC signaling cascade [79–81]. Disruption may lead to cell swelling and disintegration of mitochondrial membranes.

rIPC is one of the most effective techniques in rendering the myocardium capable of protecting itself against I/R injury. rIPC is induced via short non-deleterious phases of I/R prior to an index ischemic event. While the cardioprotective effects of rIPC are generally acknowledged, the underlying mechanism and specifically a role for NO remain under intense debate. The signaling initiated by the rIPC stimulus involves a trigger, and a distinct cardiac signal transduction mechanism finally protecting the cardiomyocyte from I/R injury. As triggering pathways both humoral/blood borne factors [82–84] and nerval transmission [85] have been proposed. Given that the cardioprotective effects can be transferred when transfusing blood from preconditioned donor animals to unconditioned recipients in both in vivo and ex vivo preparations [82, 84, 86], a contribution of a blood borne factor appears rather presumable. Using a mouse model of warm liver I/R, it was recently demonstrated that ablation of endothelial NO Snythase (eNOS) abrogates the protective effects seen with rIPC on microscopic liver damage [87], eNOS generates NO which can modulate cardiovascular functions either locally or at a distance when transported as nitrite or nitroso species [88]. Changes in shear stress, e.g. due to an increase in blood flow, are the strongest physiological stimulus of eNOS activity, which is mirrored in higher circulating NO metabolites. In a recent investigation in humans, we determined that a 5 min phase of forearm occlusion massively increased postischemic blood flow in the proximal conduit arteries. This was paralleled by an increase in nitrite in the plasma of these individuals [89]. This rIPC maneuver causes an increase of circulating endogenous nitrite both in an experimental and clinical setting, enhances nitrite levels in the myocardium and thus protects from myocardial I/R injury [5]. The relative importance of this pathway remains to be elucidated.

Conclusion

Mitochondria remain a central target of cardioprotection strategies. This relates to the preservation of mitochondrial structure and function. Experimental evidence points to a benficial role NO in mediating protection during I/R. Therefore an increase of NO in the reperfused myocardium is a desired approach to protect the heart. The strategies to achiev this range from NO donors, inorganic nitrate and nitrite to an endogenouse enhancement of NO/nitrite production, e.g. by rIPC: the relative importance and a potential clinical application must be evaluated in future clinical trials.

References

1. Yellon DM, Hausenloy DJ. Myocardial reperfusion injury. N Engl J Med. 2007;357: 1121–35.
2. Andreadou I, Iliodromitis EK, Rassaf T, Schulz R, Papapetropoulos A, Ferdinandy P. The role of gasotransmitters NO, H2S and CO in myocardial ischaemia/reperfusion injury and cardioprotection by preconditioning, postconditioning and remote conditioning. Br J Pharmacol. 2015;172:1587–606.
3. Heusch G, Botker HE, Przyklenk K, Redington A, Yellon D. Remote ischemic conditioning. J Am Coll Cardiol. 2015;65:177–95.
4. Chouchani ET, Methner C, Nadtochiy SM, Logan A, Pell VR, Ding S, James AM, Cocheme HM, Reinhold J, Lilley KS, Partridge L, Fearnley IM, Robinson AJ, Hartley RC, Smith RA, Krieg T, Brookes PS, Murphy MP. Cardioprotection by S-nitrosation of a cysteine switch on mitochondrial complex I. Nat Med. 2013;19:753–9.
5. Rassaf T, Totzeck M, Hendgen-Cotta UB, Shiva S, Heusch G, Kelm M. Circulating nitrite contributes to cardioprotection by remote ischemic preconditioning. Circ Res. 2014;114:1601–10.
6. Totzeck M, Hendgen-Cotta UB, Kelm M, Rassaf T. Crosstalk between nitrite, myoglobin and reactive oxygen species to regulate vasodilation under hypoxia. PloS one. 2014;9:e105951.
7. Totzeck M, Hendgen-Cotta UB, Luedike P, Berenbrink M, Klare JP, Steinhoff HJ, Semmler D, Shiva S, Williams D, Kipar A, Gladwin MT, Schrader J, Kelm M, Cossins AR, Rassaf T. Nitrite regulates hypoxic vasodilation via myoglobin-dependent nitric oxide generation. Circulation. 2012;126:325–34.
8. Luedike P, Hendgen-Cotta UB, Sobierajski J, Totzeck M, Reeh M, Dewor M, Lue H, Krisp C, Wolters D, Kelm M, Bernhagen J, Rassaf T. Cardioprotection through s-nitros(yl)ation of macrophage migration inhibitory factor. Circulation. 2012;125:1880–9.
9. Totzeck M, Hendgen-Cotta UB, Rammos C, Frommke LM, Knackstedt C, Predel HG, Kelm M, Rassaf T. Higher endogenous nitrite levels are associated with superior exercise capacity in highly trained athletes. Nitric Oxide Biol Chem Off J Nitric Oxide Soc. 2012;27:75–81.
10. Totzeck M, Hendgen-Cotta UB, Rammos C, Petrescu AM, Meyer C, Balzer J, Kelm M, Rassaf T. Assessment of the functional diversity of human myoglobin. Nitric Oxide Biol Chem Off J Nitric Oxide Soc. 2012;26:211–6.
11. Cosby K, Partovi KS, Crawford JH, Patel RP, Reiter CD, Martyr S, Yang BK, Waclawiw MA, Zalos G, Xu X, Huang KT, Shields H, Kim-Shapiro DB, Schechter AN, Cannon 3rd RO, Gladwin MT. Nitrite reduction to nitric oxide by deoxyhemoglobin vasodilates the human circulation. Nat Med. 2003;9:1498–505.
12. Hendgen-Cotta UB, Merx MW, Shiva S, Schmitz J, Becher S, Klare JP, Steinhoff HJ, Goedecke A, Schrader J, Gladwin MT, Kelm M, Rassaf T. Nitrite reductase activity of myoglobin regu-

lates respiration and cellular viability in myocardial ischemia-reperfusion injury. Proc Natl Acad Sci U S A. 2008;105:10256–61.
13. Zweier JL. Measurement of superoxide-derived free radicals in the reperfused heart. Evidence for a free radical mechanism of reperfusion injury. J Biol Chem. 1988;263:1353–7.
14. Penna C, Mancardi D, Rastaldo R, Pagliaro P. Cardioprotection: a radical view free radicals in pre and postconditioning. Biochim Biophys Acta. 1787;2009:781–93.
15. Zweier JL, Talukder MA. The role of oxidants and free radicals in reperfusion injury. Cardiovasc Res. 2006;70:181–90.
16. Rassaf T, Poll LW, Brouzos P, Lauer T, Totzeck M, Kleinbongard P, Gharini P, Andersen K, Schulz R, Heusch G, Mîdder U, Kelm M. Positive effects of nitric oxide on left ventricular function in humans. Eur Heart J. 2006;27:1699–705.
17. Piper HM, Garcia-Dorado D. Prime causes of rapid cardiomyocyte death during reperfusion. Ann Thorac Surg. 1999;68:1913–9.
18. Steenbergen C, Murphy E, Levy L, London RE. Elevation in cytosolic free calcium concentration early in myocardial ischemia in perfused rat heart. Circ Res. 1987;60:700–7.
19. Steenbergen C, Murphy E, Watts JA, London RE. Correlation between cytosolic free calcium, contracture, ATP, and irreversible ischemic injury in perfused rat heart. Circ Res. 1990;66:135–46.
20. Steenbergen C, Perlman ME, London RE, Murphy E. Mechanism of preconditioning. Ionic Alterations Circ Res. 1993;72:112–25.
21. Xie W, Santulli G, Guo X, Gao M, Chen BX, Marks AR. Imaging atrial arrhythmic intracellular calcium in intact heart. J Mol Cell Cardiol. 2013;64:120–3.
22. Garcia-Dorado D, Inserte J, Ruiz-Meana M, Gonzalez MA, Solares J, Julia M, Barrabes JA, Soler-Soler J. Gap junction uncoupler heptanol prevents cell-to-cell progression of hypercontracture and limits necrosis during myocardial reperfusion. Circulation. 1997;96:3579–86.
23. Hausenloy DJ, Maddock HL, Baxter GF, Yellon DM. Inhibiting mitochondrial permeability transition pore opening: a new paradigm for myocardial preconditioning? Cardiovasc Res. 2002;55:534–43.
24. Baines CP. The mitochondrial permeability transition pore and ischemia-reperfusion injury. Basic Res Cardiol. 2009;104:181–8.
25. Heusch G, Boengler K, Schulz R. Inhibition of mitochondrial permeability transition pore opening: the Holy Grail of cardioprotection. Basic Res Cardiol. 2010;105:151–4.
26. Heusch G, Boengler K, Schulz R. Cardioprotection: nitric oxide, protein kinases, and mitochondria. Circulation. 2008;118:1915–9.
27. Hoshino A, Matoba S, Iwai-Kanai E, Nakamura H, Kimata M, Nakaoka M, Katamura M, Okawa Y, Ariyoshi M, Mita Y, Ikeda K, Ueyama T, Okigaki M, Matsubara H. p53-TIGAR axis attenuates mitophagy to exacerbate cardiac damage after ischemia. J Mol Cell Cardiol. 2012;52:175–84.
28. Rassaf T, Flogel U, Drexhage C, Hendgen-Cotta U, Kelm M, Schrader J. Nitrite reductase function of deoxymyoglobin: oxygen sensor and regulator of cardiac energetics and function. Circ Res. 2007;100:1749–54.
29. Totzeck M, Kelm M, Rassaf T. Endothelial damage after resuscitation: reactive oxygen species as possible therapeutic targets? Crit Care Med. 2011;39:1837–9.
30. Ding WX, Yin XM. Mitophagy: mechanisms, pathophysiological roles, and analysis. Biol Chem. 2012;393:547–64.
31. Huang C, Andres AM, Ratliff EP, Hernandez G, Lee P, Gottlieb RA. Preconditioning involves selective mitophagy mediated by Parkin and p62/SQSTM1. PLoS One. 2011;6:e20975.
32. Foo RS, Mani K, Kitsis RN. Death begets failure in the heart. J Clin Invest. 2005;115:565–71.
33. Cao G, Xing J, Xiao X, Liou AK, Gao Y, Yin XM, Clark RS, Graham SH, Chen J. Critical role of calpain I in mitochondrial release of apoptosis-inducing factor in ischemic neuronal injury. J Neurosci Off J Soc Neurosci. 2007;27:9278–93.
34. Ozaki T, Yamashita T, Ishiguro S. Mitochondrial m-calpain plays a role in the release of truncated apoptosis-inducing factor from the mitochondria. Biochim Biophys Acta. 1793;2009:1848–59.

35. Kane A, Peddibhotla S, Maloney P, Mehta A, Hood B, Suyama E, Nguyen K, Vasile S, Leavitt L, Cheltsov A, Salaiwal S, Stonich D, Mangravita-Novo A, Vicchiarelli M, Smith LH, Diwan J, Chung TDY, Pinkerton AB, Hershberger P, Malany S, Kitsis RN. Cardioprotective inhibitors of reperfusion injury. Probe Reports from the NIH Molecular Libraries Program. Bethesda; 2010.
36. Qi Y, Yan L, Yu C, Guo X, Zhou X, Hu X, Huang X, Rao Z, Lou Z, Hu J. Structures of human mitofusin 1 provide insight into mitochondrial tethering. J Cell Biol. 2016;215:621–9.
37. Wang Y, Chen Y, Yan Y, Li X, Chen G, He N, Shen S, Chen G, Zhang C, Liao W, Liao Y, Bin J. Loss of CEACAM1, a tumor-associated factor, attenuates post-infarction cardiac remodeling by inhibiting apoptosis. Sci Rep. 2016;6:21972.
38. Stornaiuolo M, La Regina G, Passacantilli S, Grassia G, Coluccia A, La Pietra V, Giustiniano M, Cassese H, Di Maro S, Brancaccio D, Taliani S, Ialenti A, Silvestri R, Martini C, Novellino E, Marinelli L. Structure-based lead optimization and biological evaluation of BAX direct activators as novel potential anticancer agents. J Med Chem. 2015;58:2135–48.
39. Ma SB, Nguyen TN, Tan I, Ninnis R, Iyer S, Stroud DA, Menard M, Kluck RM, Ryan MT, Dewson G. Bax targets mitochondria by distinct mechanisms before or during apoptotic cell death: a requirement for VDAC2 or Bak for efficient Bax apoptotic function. Cell Death Differ. 2014;21:1925–35.
40. Ishikita A, Matoba T, Ikeda G, Koga J, Mao Y, Nakano K, Takeuchi O, Sadoshima J, Egashira K. Nanoparticle-mediated delivery of mitochondrial division inhibitor 1 to the myocardium protects the heart from ischemia-reperfusion injury through inhibition of mitochondria outer membrane permeabilization: a new therapeutic modality for acute myocardial infarction. J Am Heart Assoc. 2016;5:e003872.
41. Silva Ramos E, Larsson NG, Mourier A. Bioenergetic roles of mitochondrial fusion. Biochim Biophys Acta. 1857;2016:1277–83.
42. Zhang Y, Liu X, Bai J, Tian X, Zhao X, Liu W, Duan X, Shang W, Fan HY, Tong C. Mitoguardin regulates mitochondrial fusion through MitoPLD and is required for neuronal homeostasis. Mol Cell. 2016;61:111–24.
43. Anand P, Hess DT, Stamler JS. Identifying single S-nitrosothiol sites with cardioprotection. Circ Res. 2013;113:849–51.
44. Nakamura T, Lipton SA. S-nitrosylation of critical protein thiols mediates protein misfolding and mitochondrial dysfunction in neurodegenerative diseases. Antioxid Redox Signal. 2011;14:1479–92.
45. Shiva S, Huang Z, Grubina R, Sun J, Ringwood LA, MacArthur PH, Xu X, Murphy E, Darley-Usmar VM, Gladwin MT. Deoxymyoglobin is a nitrite reductase that generates nitric oxide and regulates mitochondrial respiration. Circ Res. 2007;100:654–61.
46. Duranski MR, Greer JJ, Dejam A, Jaganmohan S, Hogg N, Langston W, Patel RP, Yet SF, Wang X, Kevil CG, Gladwin MT, Lefer DJ. Cytoprotective effects of nitrite during in vivo ischemia-reperfusion of the heart and liver. J Clin Invest. 2005;115:1232–40.
47. Huang Z, Shiva S, Kim-Shapiro DB, Patel RP, Ringwood LA, Irby CE, Huang KT, Ho C, Hogg N, Schechter AN, Gladwin MT. Enzymatic function of hemoglobin as a nitrite reductase that produces NO under allosteric control. J Clin Invest. 2005;115:2099–107.
48. Basu S, Azarova NA, Font MD, King SB, Hogg N, Gladwin MT, Shiva S, Kim-Shapiro DB. Nitrite reductase activity of cytochrome c. J Biol Chem. 2008;283:32590–7.
49. Webb A, Bond R, McLean P, Uppal R, Benjamin N, Ahluwalia A. Reduction of nitrite to nitric oxide during ischemia protects against myocardial ischemia-reperfusion damage. Proc Natl Acad Sci U S A. 2004;101:13683–8.
50. Petersson J, Phillipson M, Jansson EA, Patzak A, Lundberg JO, Holm L. Dietary nitrate increases gastric mucosal blood flow and mucosal defense. Am J Physiol Gastrointest Liver Physiol. 2007;292:G718–24.
51. Weitzberg E, Lundberg JO. Nonenzymatic nitric oxide production in humans. Nitric Oxide Biol Chem Off J Nitric Oxide Soc. 1998;2:1–7.

52. Carlstrom M, Larsen FJ, Nystrom T, Hezel M, Borniquel S, Weitzberg E, Lundberg JO. Dietary inorganic nitrate reverses features of metabolic syndrome in endothelial nitric oxide synthase-deficient mice. Proc Natl Acad Sci U S A. 2010;107:17716–20.
53. Rassaf T, Lauer T, Heiss C, Balzer J, Mangold S, Leyendecker T, Rottler J, Drexhage C, Meyer C, Kelm M. Nitric oxide synthase-derived plasma nitrite predicts exercise capacity. Br J Sports Med. 2007;41:669–73. discussion 673
54. Bryan NS, Calvert JW, Elrod JW, Gundewar S, Ji SY, Lefer DJ. Dietary nitrite supplementation protects against myocardial ischemia-reperfusion injury. Proc Natl Acad Sci U S A. 2007;104:19144–9.
55. Rammos C, Hendgen-Cotta UB, Sobierajski J, Bernard A, Kelm M, Rassaf T. Dietary nitrate reverses vascular dysfunction in older adults with moderately increased cardiovascular risk. J Am Coll Cardiol. 2014;63:1584–5.
56. Murry CE, Jennings RB, Reimer KA. Preconditioning with ischemia: a delay of lethal cell injury in ischemic myocardium. Circulation. 1986;74:1124–36.
57. Zhao ZQ, Corvera JS, Halkos ME, Kerendi F, Wang NP, Guyton RA, Vinten-Johansen J. Inhibition of myocardial injury by ischemic postconditioning during reperfusion: comparison with ischemic preconditioning. Am J Physiol Heart Circ Physiol. 2003;285:H579–88.
58. Hausenloy DJ, Yellon DM. Remote ischaemic preconditioning: underlying mechanisms and clinical application. Cardiovasc Res. 2008;79:377–86.
59. Hausenloy DJ, Yellon DM. Preconditioning and postconditioning: new strategies for cardioprotection. Diabetes Obes Metab. 2008;10:451–9.
60. Przyklenk K, Bauer B, Ovize M, Kloner RA, Whittaker P. Regional ischemic 'preconditioning' protects remote virgin myocardium from subsequent sustained coronary occlusion. Circulation. 1993;87:893–9.
61. Kerendi F, Kin H, Halkos ME, Jiang R, Zatta AJ, Zhao ZQ, Guyton RA, Vinten-Johansen J. Remote postconditioning. Brief renal ischemia and reperfusion applied before coronary artery reperfusion reduces myocardial infarct size via endogenous activation of adenosine receptors. Basic Res Cardiol. 2005;100:404–12.
62. Skyschally A, Schulz R, Heusch G. Pathophysiology of myocardial infarction: protection by ischemic pre- and postconditioning. Herz. 2008;33:88–100.
63. Ladilov Y, Haffner S, Balser-Schafer C, Maxeiner H, Piper HM. Cardioprotective effects of KB-R7943: a novel inhibitor of the reverse mode of Na+/Ca2+ exchanger. Am J Phys. 1999;276:H1868–76.
64. Hausenloy DJ, Wynne AM, Yellon DM. Ischemic preconditioning targets the reperfusion phase. Basic Res Cardiol. 2007;102:445–52.
65. Hausenloy DJ, Yellon DM. Preconditioning and postconditioning: underlying mechanisms and clinical application. Atherosclerosis. 2009;204:334–41.
66. Schulz R, Cohen MV, Behrends M, Downey JM, Heusch G. Signal transduction of ischemic preconditioning. Cardiovasc Res. 2001;52:181–98.
67. Hartman JC, Wall TM, Hullinger TG, Shebuski RJ. Reduction of myocardial infarct size in rabbits by ramiprilat: reversal by the bradykinin antagonist HOE 140. J Cardiovasc Pharmacol. 1993;21:996–1003.
68. Schulz R, Post H, Vahlhaus C, Heusch G. Ischemic preconditioning in pigs: a graded phenomenon: its relation to adenosine and bradykinin. Circulation. 1998;98:1022–9.
69. Schulz R, Gres P, Heusch G. Role of endogenous opioids in ischemic preconditioning but not in short-term hibernation in pigs. Am J Physiol Heart Circ Physiol. 2001;280:H2175–81.
70. Gres P, Schulz R, Jansen J, Umschlag C, Heusch G. Involvement of endogenous prostaglandins in ischemic preconditioning in pigs. Cardiovasc Res. 2002;55:626–32.
71. Murphy E, Steenbergen C. Mechanisms underlying acute protection from cardiac ischemia-reperfusion injury. Physiol Rev. 2008;88:581–609.
72. Hausenloy DJ, Tsang A, Mocanu MM, Yellon DM. Ischemic preconditioning protects by activating prosurvival kinases at reperfusion. Am J Physiol Heart Circ Physiol. 2005;288:H971–6.

73. Tong H, Chen W, Steenbergen C, Murphy E. Ischemic preconditioning activates phosphatidylinositol-3-kinase upstream of protein kinase C. Circ Res. 2000;87:309–15.
74. Tong H, Imahashi K, Steenbergen C, Murphy E. Phosphorylation of glycogen synthase kinase-3beta during preconditioning through a phosphatidylinositol-3-kinase – dependent pathway is cardioprotective. Circ Res. 2002;90:377–9.
75. Tong H, Rockman HA, Koch WJ, Steenbergen C, Murphy E. G protein-coupled receptor internalization signaling is required for cardioprotection in ischemic preconditioning. Circ Res. 2004;94:1133–41.
76. Ma XL, Kumar S, Gao F, Louden CS, Lopez BL, Christopher TA, Wang C, Lee JC, Feuerstein GZ, Yue TL. Inhibition of p38 mitogen-activated protein kinase decreases cardiomyocyte apoptosis and improves cardiac function after myocardial ischemia and reperfusion. Circulation. 1999;99:1685–91.
77. Mackay K, Mochly-Rosen D. An inhibitor of p38 mitogen-activated protein kinase protects neonatal cardiac myocytes from ischemia. J Biol Chem. 1999;274:6272–9.
78. Fryer RM, Patel HH, Hsu AK, Gross GJ. Stress-activated protein kinase phosphorylation during cardioprotection in the ischemic myocardium. Am J Physiol Heart Circ Physiol. 2001;281:H1184–92.
79. Hausenloy DJ, Ong SB, Yellon DM. The mitochondrial permeability transition pore as a target for preconditioning and postconditioning. Basic Res Cardiol. 2009;104:189–202.
80. Hausenloy DJ, Yellon DM. The mitochondrial permeability transition pore: its fundamental role in mediating cell death during ischaemia and reperfusion. J Mol Cell Cardiol. 2003;35:339–41.
81. Hausenloy DJ, Yellon DM, Mani-Babu S, Duchen MR. Preconditioning protects by inhibiting the mitochondrial permeability transition. Am J Physiol Heart Circ Physiol. 2004;287:H841–9.
82. Shimizu M, Tropak M, Diaz RJ, Suto F, Surendra H, Kuzmin E, Li J, Gross G, Wilson GJ, Callahan J, Redington AN. Transient limb ischaemia remotely preconditions through a humoral mechanism acting directly on the myocardium: evidence suggesting cross-species protection. Clin Sci. 2009;117:191–200.
83. Dickson EW, Lorbar M, Porcaro WA, Fenton RA, Reinhardt CP, Gysembergh A, Przyklenk K. Rabbit heart can be "preconditioned" via transfer of coronary effluent. Am J Phys. 1999;277:H2451–7.
84. Dickson EW, Reinhardt CP, Renzi FP, Becker RC, Porcaro WA, Heard SO. Ischemic preconditioning may be transferable via whole blood transfusion: preliminary evidence. J Thromb Thrombolysis. 1999;8:123–9.
85. Lim SY, Yellon DM, Hausenloy DJ. The neural and humoral pathways in remote limb ischemic preconditioning. Basic Res Cardiol. 2010;105:651–5.
86. Serejo FC, Rodrigues Jr LF, da Silva Tavares KC, de Carvalho AC, Nascimento JH. Cardioprotective properties of humoral factors released from rat hearts subject to ischemic preconditioning. J Cardiovasc Pharmacol. 2007;49:214–20.
87. Abu-Amara M, Yang SY, Quaglia A, Rowley P, Fuller B, Seifalian A, Davidson B. Role of endothelial nitric oxide synthase in remote ischemic preconditioning of the mouse liver. Liver Transplant Off Publ Am Assoc Study Liver Dis Int Liver Transplant Soc. 2011;17:610–9.
88. Rassaf T, Preik M, Kleinbongard P, Lauer T, Hei C, Strauer BE, Feelisch M, Kelm M. Evidence for in vivo transport of bioactive nitric oxide in human plasma. J Clin Investig. 2002;109:1241–8.
89. Rassaf T, Heiss C, Hendgen-Cotta U, Balzer J, Matern S, Kleinbongard P, Lee A, Lauer T, Kelm M. Plasma nitrite reserve and endothelial function in the human forearm circulation. Free Radic Biol Med. 2006;41:295–301.

Chapter 19
Peripheral Blood Mitochondrial DNA and Myocardial Function

Tatiana Kuznetsova and Judita Knez

A major burden of modern society is the progressive increase in age-associated disorders due to the continuously increasing life expectancy and high prevalence of cardiovascular risk factors such as hypertension, diabetes and obesity. Symptomatic heart failure (HF) is one of the most common age-related diseases, currently affecting an estimated 15 million Europeans and 10,000 people are newly diagnosed every day [1]. In the US, HF contributes to an economic burden close to $40 billion per year and affects 1–2% of the US population [2]. Thus, with the increasing burden of symptomatic HF there is a need for a better early diagnostic and prevention strategies, and for a better treatment regimen for HF patients.

While traditionally associated with the concept of pump failure or reduced left ventricular ejection fraction, it has become widely recognized that HF can occur even when ejection fraction is preserved, constituting the syndrome of HF with preserved ejection fraction or so called diastolic HF [3] which is characterized by impaired left ventricular relaxation and increased left ventricular stiffness. It is important to note that HF is a progressive disorder that begins with cardiovascular risk factors, proceeds to asymptomatic maladaptive left ventricular remodeling and dysfunction, and then evolves into clinically overt HF and disability. Recent population studies revealed a high prevalence (up to 25%) of asymptomatic left ventricular systolic and/or diastolic dysfunction using echocardiographic imaging [4, 5]. Understanding to what extent genetic along with environmental factors influence

T. Kuznetsova, MD, PhD (✉)
Research Unit Hypertension and Cardiovascular Epidemiology, KU Leuven Department of Cardiovascular Sciences, University of Leuven,
Campus Sint Rafaël, Kapucijnenvoer 35, Box 7001, 3000 Leuven, Belgium
e-mail: tatiana.kouznetsova@med.kuleuven.be

J. Knez
Research Unit Hypertension and Cardiovascular Epidemiology, KU Leuven Department of Cardiovascular Sciences, University of Leuven,
Campus Sint Rafaël, Kapucijnenvoer 35, Box 7001, 3000 Leuven, Belgium

Hypertension Division, Department of Internal Medicine, University Clinical Centre Ljubljana, Ljubljana, Slovenia

G. Santulli (ed.), *Mitochondrial Dynamics in Cardiovascular Medicine*,
Advances in Experimental Medicine and Biology 982,
DOI 10.1007/978-3-319-55330-6_19

myocardial function is an important issue in view of the relation of left ventricular systolic and diastolic dysfunction with outcome [5].

The progression of left ventricular dysfunction to HF is associated with changes in cardiac energy metabolism [6]. Mitochondria play a central role in a variety of cell functions, including oxidative energy production, storage of calcium ions, and programmed cell death [7]. Cardiomyocyte function depends on mitochondrial oxidative respiration for energy production almost entirely [8]. Each mitochondrion contains two to ten copies of its own mitochondrial genome (Fig. 19.1) [9]. The number of mitochondrial DNA (mtDNA) copies in cells correlates with the size and number of mitochondria, which change not only under different energy demands but also due to increased oxidative stress and in different pathological conditions [9]. The major difference which makes mtDNA more vulnerable to damage as compare to nuclear DNA is that mtDNA lacks protective histones, chromatin structure, introns and its DNA repair mechanisms work less efficiently.

Accumulation of mtDNA damages can perturb its replication mechanisms and result in decreased overall cellular mitochondrial-to-nuclear DNA ratio (mtDNA content) [10]. The process of depletion of mtDNA content might be caused by both mtDNA production disturbances as well as its degradation. Experimental studies showed that mitochondria carrying mutant mtDNA could be removed by mitophagy [11]. On the other hand, increased reactive oxygen species (ROS) impair mitochondrial function, damage mtDNA replication mechanisms, and result in decreased mtDNA production [12]. Notably, excessive oxidative stress is recognized as an important contributor to the pathophysiology of heart failure [13]. The precise molecular mechanisms and the contribution of each of the processes to changes in mtDNA content remain to be unveiled. Depletion of mtDNA and mitochondrial malfunction has been implicated in pathogenesis of different diseases such as cancer, neurodegeneration, diabetes, and HF [9].

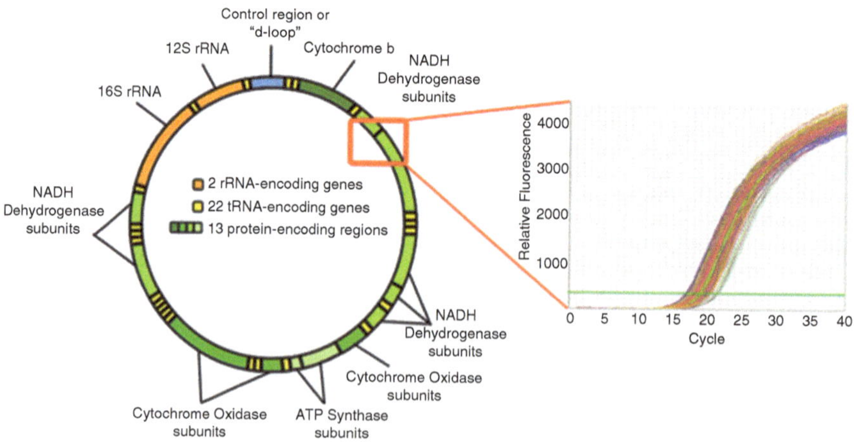

Fig. 19.1 Real time qPCR amplification of stable mitochondrial sequences. The graph shows real time qPCR amplification curves for a single 386 well plate. qPCR, quantitative polymerase chain reaction

Measurements of mtDNA Content in Peripheral Blood Cells

Mitochondrial function in various diseases and conditions could be estimated by assessment of the mtDNA content in specific tissues or in easily accessible peripheral blood cells. The mtDNA content could be measured by determining the ratio of several mitochondrial sequences (for instance, MT-ND1 and MTF3212/R3319) to one or two single-copy housekeeping nuclear genes (for instance, RPLP0) using real time PCR [14, 15]. Both mtDNA primers should not be located around the D-loop since this region is thought to be hypervariable and more prone to deletions or mutations than other mitochondrial genes (Fig. 19.1). The relative mtDNA content is calculated by normalizing cycle threshold (Ct) values of the mitochondrial DNA sequences relative to the nuclear genes (Fig. 19.1).

Recent epidemiological studies including ours described the distribution and determinants of the peripheral blood mtDNA content measured in a general population [16, 17]. Overall, the distribution of blood mtDNA content measured in general population follows a normal distribution curve with range from 0.39 to 3.06 (Fig. 19.2).

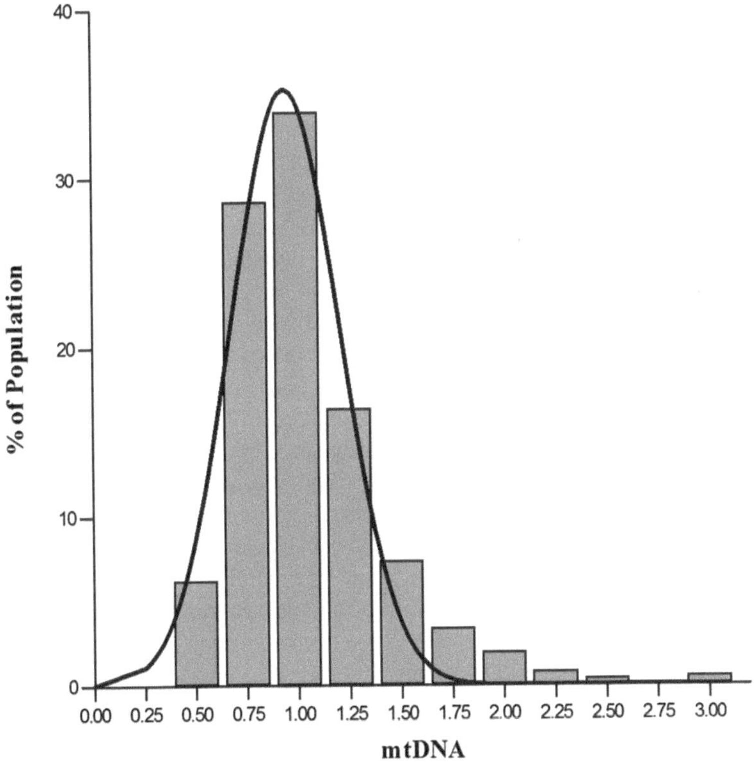

Fig. 19.2 Distribution of the mitochondrial DNA (mtDNA) content in 710 participants recruited from the general population (Reproduced from Knez J. et al. [34] with permission)

Previous studies reported a curvilinear relationship between the relative mtDNA content and age with a significant decline in blood mtDNA content in older subjects [16–19]. Importantly, higher mtDNA content in the elderly subjects was associated with better cognitive function, physical performance [18], and longevity [16, 19]. Women had higher mtDNA content as compared to men [16, 17, 20]. The mtDNA content once it measured in the whole pool of peripheral blood cells was significantly increased with platelet count, whereas it was inversely associated with white blood cell count [17]. Thus, measurement of the mtDNA content in peripheral blood cells could serve as a reliable marker to assess mitochondrial function. Studies involving measurements of the mtDNA content in peripheral blood should take into account these covariates.

Heart Maladaptation in Rodents with mtDNA Depletion

There is strong evidence that mitochondrial DNA depletion may play a key role in the progression to cardiac remodeling and dysfunction. Indeed, previous experimental studies demonstrated that a decrease in myocardial mtDNA content could trigger development of dilated cardiomyopathy in mice [21–27]. Genetic manipulation of key proteins involved in the maintenance, replication and transcription of mtDNA such as mitochondrial transcription factor A (Tfam) or DNA polymerase γ (Pol γ) can trigger the development of HF. Indeed, in the tissue-specific Tfam knockout mice, the decline of respiratory chain function following depletion of mtDNA resulted in mitochondrial cardiomyopathy, characterized by left ventricular dilation and increased cardiac mass [22, 23]. On the other hand, overexpression of Tfam in mice after myocardial infarction ameliorated post-infarct remodeling including reduced left ventricular cavity size, myocytes hypertrophy, interstitial fibrosis, and apoptosis of non-ischemic cardiomyocytes [28]. These beneficial effects of Tfam overexpression were associated with an attenuation of mitochondrial oxidative stress, mtDNA decline, and dysfunction [28].

Transgenic cardiac targeting of human Pol γ harboring the pathogenic mutation was also resulted in depletion of mtDNA, oxidative stress, dilated cardiomyopathy with left ventricular dysfunction, bradycardia, and premature death [26]. In addition, Sahin et al. [27] demonstrated that impaired mitochondrial biogenesis including decline in mtDNA content in the telomere-deficient mice causes maladaptive changes in heart such as an increase of left ventricular diameter, thinning of left ventricular walls, and a decrease of fractional shortening.

Overexpression of mitochondrial antioxidant Prx-3 protein in transgenic mice with myocardial infarction caused alleviation of mitochondrial oxidative stress and, therefore, preservation of mtDNA copy number as compared to wild-type mice [29]. In these transgenic mice left ventricular cavity dilatation and dysfunction were also attenuated compared to wild-type mice, with no significant differences in infarct and aortic pressure between groups [29]. Thus, experimental studies provided solid evidences that mtDNA depletion and mitochondrial dysfunction lead left ventricular cavity size dilatation and dysfunction in rodents.

mtDNA Content in Patients with Symptomatic HF

Recent clinical studies in patients with advanced stages of HF suggested an impaired mitochondrial biogenesis in biopsies obtained from human heart [30–32]. Karamanlidis et al. showed that mtDNA content in cardiomyocytes was decreased by 40% in 23 human end-stages failing hearts as compared to 19 control hearts [30]. Moreover, progressive depletion of mtDNA was observed in cardiomyocytes derived from the right ventricle of patients with congenital heart disease during the transition from hypertrophy to HF [31]. Of notice, in this study decrease in mtDNA content preceded any changes in mitochondrial enzyme activity or protein levels in cardiomyocytes [31]. On the other hand, Ahuja et al. observed divergent mtDNA content in tissues obtained from hearts of 16 patients with dilated and ischemic cardiomyopathy [32]. In patients with dilated cardiomyopathy, the authors observed an increase amount of defective mtDNA. In contrast, end-stage ischemic hearts show a depletion of mtDNA [32].

Huang et al. showed that mtDNA content measured in peripheral blood cells is significantly lower in patients with advanced HF in particular of ischemic origin compared with age-matched controls [33]. In this study, HF patients with lower blood mtDNA had a higher risk for cardiovascular death (adjusted hazard ratio [HR] for 1 standard deviation decrease in mtDNA content = 1.25; $p = 0.023$) and for HF hospitalization (HR = 1.32; $p < 0.001$). Thus, increasing evidences suggest that imbalances in mitochondrial function and associated oxidative stress play an important role in the pathophysiology of congestive HF.

Association of Subclinical Left Ventricular Remodeling and Dysfunction with Peripheral mtDNA Content

Hence, mtDNA content might be an important biomarker in heart disease prediction and to date studies addressing the relationship of mtDNA content with parameters reflecting cardiac structure and function analyzed continuously are scarce. Previously, we investigated in a general population in cross-sectional and longitudinal studies whether echocardiographic indexes of left ventricular structure and function are associated with mtDNA content measured in peripheral blood cells [34]. *First*, in the cross-sectional analysis, we demonstrated that participants with higher blood mtDNA content had smaller left ventricular volumes. *Second*, we observed that baseline mtDNA content was also a significant predictor of longitudinal changes of left ventricular diastolic cavity volume and size. As shown in Fig. 19.3, left ventricular diastolic volume, which overall decreased during the follow-up period, was less likely to decrease in participants with low mtDNA content.

In the Framingham heart study, the presence of certain risk factors in midlife, including hypertension and diabetes mellitus, was also associated with a lesser decrease in left ventricular dimensions with older age [35]. Furthermore, individuals with symptomatic HF are more likely to have greater left ventricular end-diastolic

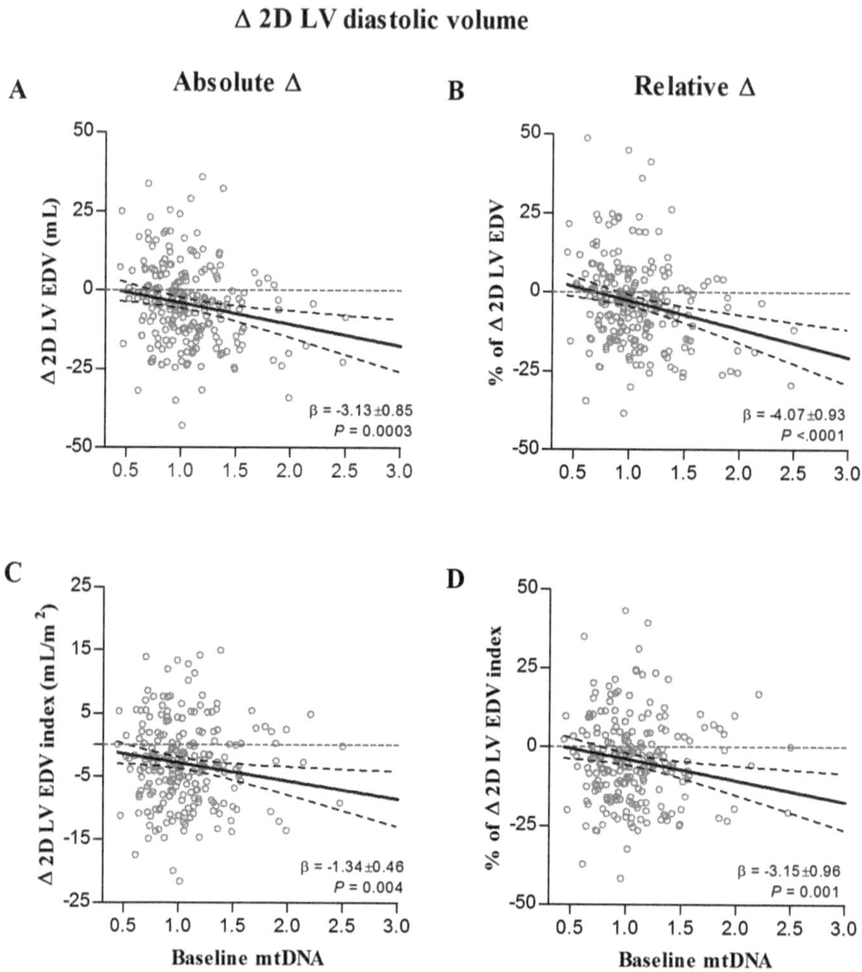

Fig. 19.3 Scatterplots of absolute (panels **a** and **c**) and relative (panels **b** and **d**) changes in 2D left ventricular diastolic volume by baseline mtDNA content in multivariable adjusted analyses. Within-subject changes were calculated by subtracting baseline from follow-up values. Percent changes were calculated by dividing the within-subject changes by the baseline echocardiographic indexes. Model for left ventricular diastolic volume was adjusted for sex, age, body weight and height, systolic blood pressure, use of different classes of antihypertensive drugs, heart rate and changes in heart rate, while accounting for family clusters and follow-up duration. For left ventricular diastolic volume index, model did not include anthropometric covariables. β (±SE) represents the effect size associated with 1-SD increase in baseline mtDNA (+0.37). *mtDNA* mitochondrial DNA content, 2D, two-dimensional; *LV* left ventricle; *EDV* end-diastolic volume (Reproduced from Knez J et al. [34] with permission)

volume than subjects of similar age range but without HF [36]. Therefore, mtDNA content along with other risk factors might contribute to alterations in the typical course of left ventricle remodeling with age [35] and may, in turn, contribute to the risk for overt HF in older age. *Third*, in our study, higher baseline blood mtDNA content predicted less increase in left ventricular mass during the follow-up period (Fig. 19.4). Based on these findings, we might speculate that having a higher level of blood mtDNA content might delay adverse left ventricular remodeling over time

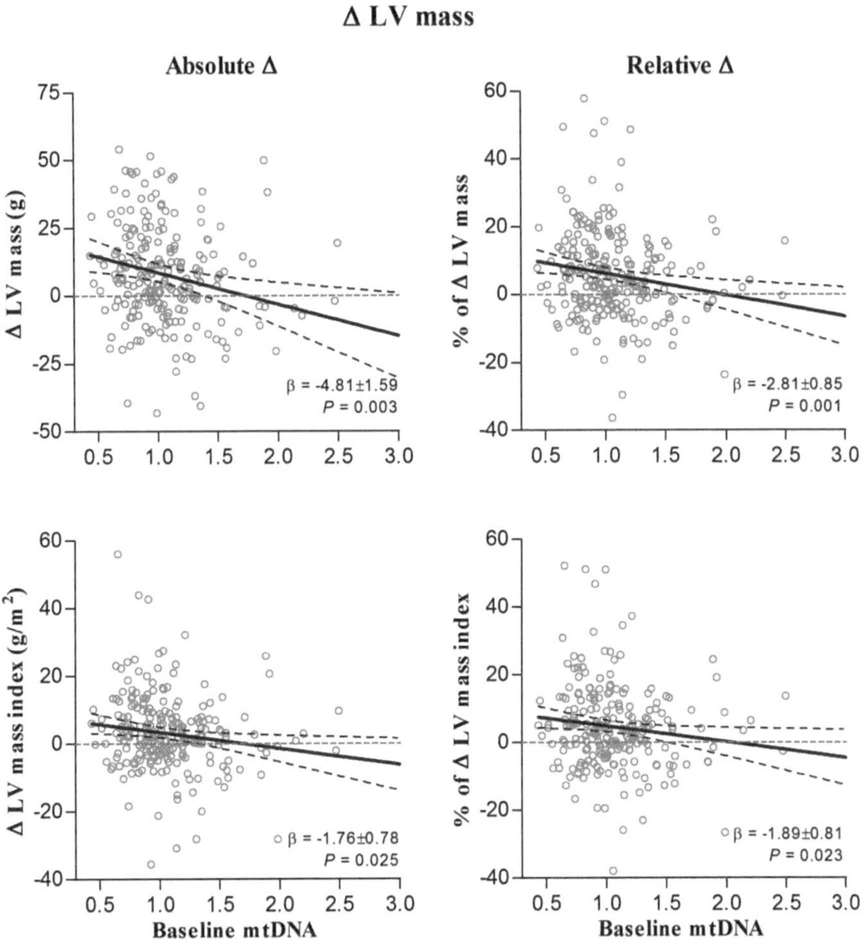

Fig. 19.4 Scatterplots of absolute and relative changes in left ventricular mass index by baseline mtDNA content in multivariable adjusted analyses. Models for left ventricular mass were adjusted for sex, age, systolic blood pressure, use of different classes of antihypertensive drugs, heart rate and changes in heart rate, while accounting for family clusters and follow-up duration. β (±SE) represents the effect size associated with 1-SD increase in baseline mtDNA (+0.37). *mtDNA* mitochondrial DNA content, *2D* two-dimensional, *LV* left ventricle (Reproduced from Knez J et al. [34] with permission)

and, therefore, future cardiovascular events. In line with these findings, several previous population studies found that higher blood mtDNA content in elderly was associated with better outcome and healthy aging [16, 18, 19].

We also found that higher mtDNA content was associated with better left ventricular systolic and diastolic function as reflected by an increase in longitudinal mitral annulus movement during systole (s′ peak) and a decrease in ratio of transmitral flow velocity to early diastolic mitral annular velocity (E/e′ ratio) [34]. Low longitudinal systolic s′ velocity [37] has independent prognostic value in patients with HF [38] and in asymptomatic subjects with cardiovascular risk factors [39]. Combining transmitral flow velocity with early diastolic mitral annular velocity (E/e′ ratio) is frequently used for assessing the LV diastolic filling pressure [40]. High E/e′ independently predicted cardiac mortality in HF patients [41]. In our study, higher blood mtDNA content was associated with better left ventricular systolic function characterized by higher s' peak and better left ventricular diastolic function characterized by lower E/e′.

These findings could have a biological basis related to the role of mitochondrial biogenesis in determining cardiac function and size. Whether these associations can aid in risk stratification for early stages of cardiac remodeling and dysfunction and whether mechanisms governing mtDNA content can be a target for pharmacological interventions require further investigation.

Correlation Between mtDNA Measured in Tissues and Peripheral Blood Cells

As we previously described the mtDNA content could be measured in tissues as well as in easily accessible peripheral blood buffy coat, which contains leukocytes and platelets. Some previous experimental and clinical studies compared mtDNA content in blood cells and in tissue samples. Sahin et al. reported a similar parallel decrease in mtDNA content in myocardial and haematopoietic cells in rapidly ageing mice [27]. In patients with coronary artery disease, Chen et al. found the significant positive association between mtDNA measured in peripheral blood cells and in tissue obtained from atherosclerotic plaque [42]. Moreover, the authors reported that lower blood mtDNA content was associated with higher cardiovascular risk [42]. To our knowledge, so far, only one study reported significant correlation (Pearson $r = 0.72$; $p = 0.019$) between mtDNA contents measured in both peripheral blood cells and cardiomyocytes of 10 end-stages HF patients who received heart transplantation [33]. Overall, these findings suggested that mtDNA measured in peripheral blood cells might in some degree reflect the overall oxidative status and, therefore, might be served as a marker useful for prognostication in patients with HF. However, the assumption that mtDNA content measured in peripheral blood cells reflects in some degree mtDNA content of cardiomyocytes should be further confirmed in clinical studies.

Future Directions

Future experimental and clinical studies are necessary to further elucidate the role of mtDNA in the progression of left ventricular remodeling and dysfunction to symptomatic HF. Because disturbances in mitochondrial biogenesis could lead to the development of HF, mitochondria represent targets for therapeutic interventions [43–45]. Current treatment strategies for HF employ strategies to decreasing workload on heart by lowering volume load, blood pressure and heart rate. An alternative approach to the treatment of HF might be related to targeting mitochondrial plasticity which lead to improving ATP supply and therefore, restoring energy to the failing heart. A recent experimental study in rats showed that administration of a novel mitoprotective agent after ischemic injury can prevent adverse cardiac remodeling and decrease left ventricular dilation [46]. Few clinical trials are currently on-going to evaluate safety, tolerability and efficacy of this agent to treat HF patients. Therefore, pharmacologic substances targeting mechanisms that maintain sufficient level and quality of mtDNA might stabilize mitochondrial function and, therefore, prevent progression of maladaptive left ventricular remodeling and dysfunction.

Conclusion

There is increasing evidence from experimental and clinical studies that mtDNA depletion play a key role in the progression to cardiomyopathy. Recent epidemiological study also demonstrated that higher mtDNA content measured in peripheral blood cells was associated with smaller left ventricular volume and size, better systolic and diastolic function and more favorable trend in longitudinal changes in left heart structure. These findings could have a biological basis related to the role of mitochondrial biogenesis in determining myocardial function. Whether these associations can aid in risk stratification of patients with early and advanced stages of HF and whether mechanisms governing mtDNA content can be a target for pharmacological interventions require further investigation.

Acknowledgments The Fonds voor Wetenschappelijk Onderzoek Vlaanderen, Ministry of the Flemish Community, Brussels, Belgium, supported the Research Unit Hypertension and Cardiovascular Epidemiology (Leuven, Belgium) (grant G.0880.13).

References

1. Remme WJ, McMurray JJ, Rauch B, Zannad F, Keukelaar K, Cohen-Solal A, et al. Public awareness of heart failure in Europe: first results from SHAPE. Eur Heart J. 2005;26(22):2413–21.
2. Lam CS, Donal E, Kraigher-Krainer E, Vasan RS. Epidemiology and clinical course of heart failure with preserved ejection fraction. Eur J Heart Fail. 2011;13(1):18–28.

3. Owan TE, Redfield MM. Epidemiology of diastolic heart failure. Prog Cardiovasc Dis. 2005;47(5):320–32.
4. Kuznetsova T, Herbots L, López B, Jin Y, Richart T, Thijs L, et al. Prevalence of left ventricular diastolic dysfunction in a general population. Circ Heart Fail. 2009;2(2):105–12.
5. Kuznetsova T, Cauwenberghs N, Knez J, Yang WY, Herbots L, D'hooge J, Haddad F, Thijs L, Voigt JU, Staessen J. Additive prognostic value of left ventricular systolic dysfunction in a population-based cohort. Circ Cardiovasc Imaging. 2016;9(7):pii: e004661.
6. Neubauer S. The failing heart – an engine out of fuel. N Engl J Med. 2007;356:1140–51.
7. Xie W, Santulli G, Reiken SR, Yuan Q, Osborne BW, Chen BX, Marks AR. Mitochondrial oxidative stress promotes atrial fibrillation. Sci Rep. 2015;5:11427.
8. Santulli G, Xie W, Reiken SR, Marks AR. Mitochondrial calcium overload is a key determinant in heart failure. Proc Natl Acad Sci USA. 2015;112:11389–94.
9. Clay Montier L, Deng J, Bai Y. Number matters: control of mammalian mitochondrial DNA copy number. J Genet Genomics. 2009;36:125–31.
10. Tewari S, Santos J, Kowluru R. Damaged mitochondrial DNA replication system and the development of diabetic retinopathy. Antioxid Redox Signal. 2012;17(3):492–504.
11. Gomes LC, Scorrano L. High levels of Fis1, a pro-fission mitochondrial protein, trigger autophagy. Biochim Biophys Acta. 2008;1777:860–6.
12. Tewari S, Santos JM, Kowluru RA. Damaged mitochondrial DNA replication system and the development of diabetic retinopathy. Antioxid Redox Signal. 2012;17:492–504.
13. Hafstad AD, Nabeebaccus AA, Shah AM. Novel aspects of ROS signalling in heart failure. Basic Res Cardiol. 2013;108:359.
14. Janssen BG, Munters E, Pieters N, Smeets K, Cox B, Cuypers A, Fierens F, Penders J, Vangronsveld J, Gyselaers W, Nawrot TS. Placental mitochondrial DNA content and particulate air pollution during in utero life. Environ Health Perspect. 2012;120:1346–52.
15. Pieters N, Koppen G, Smeets K, Napierska D, Plusquin M, De Prins S, Van De Weghe H, Nelen V, Cox B, Cuypers A, Hoet P, Schoeters G, Nawrot TS. Decreased mitochondrial DNA content in association with exposure to polycyclic aromatic hydrocarbons in house dust during wintertime: from a population enquiry to cell culture. PLoS One. 2013;8:e63208.
16. Ashar FN, Moes A, Moore AZ, Grove ML, Chaves PH, Coresh J, Newman AB, Matteini AM, Bandeen-Roche K, Boerwinkle E, Walston JD, Arking DE. Association of mitochondrial DNA levels with frailty and all-cause mortality. J Mol Med. 2015;93:177–86.
17. Knez J, Winckelmans E, Plusquin M, Thijs L, Cauwenberghs N, Gu Y, Staessen JA, Nawrot TS, Kuznetsova T. Correlates of peripheral blood mitochondrial DNA content in a general population. Am J Epidemiol. 2016;183:138–46.
18. Mengel-From J, Thinggaard M, Dalgård C, Kyvik KO, Christensen K, Christiansen L. Mitochondrial DNA copy number in peripheral blood cells declines with age and is associated with general health among elderly. Hum Genet. 2014;133:1149–59.
19. He YH, Lu X, Wu H, Cai WW, Yang LQ, Xu LY, Sun HP, Kong QP. Mitochondrial DNA content contributes to healthy aging in Chinese: a study from nonagenarians and centenarians. Neurobiol Aging. 2014;35:1779.
20. Curran JE, Johnson MP, Dyer TD, Göring HH, Kent JW, Charlesworth JC, Borg AJ, Jowett JB, Cole SA, MacCluer JW, Kissebah AH, Moses EK, Blangero J. Genetic determinants of mitochondrial content. Hum Mol Genet. 2007;16:1504–14.
21. Ide T, Tsutsui H, Hayashidani S, Kang D, Suematsu N, Nakamura K, Utsumi H, Hamasaki N, Takeshita A. Mitochondrial DNA damage and dysfunction associated with oxidative stress in failing hearts after myocardial infarction. Circ Res. 2001;88:529–35.
22. Hansson A, Hance N, Dufour E, Rantanen A, Hultenby K, Clayton D, Wibom R, Larsson N. A switch in metabolism precedes increased mitochondrial biogenesis in respiratory chain-deficient mouse hearts. Proc Natl Acad Sci U S A. 2004;101:3136–41.
23. Wang J, Wilhelmsson H, Graff C, Li H, Oldfors A, Rustin P, Brüning J, Kahn C, Clayton D, Barsh G, Thorén P, Larsson N. Dilated cardiomyopathy and atrioventricular conduction blocks induced by heart-specific inactivation of mitochondrial DNA gene expression. Nat Genet. 1999;21:133–7.

24. Trifunovic A, Wredenberg A, Falkenberg M, Spelbrink J, Rovio A, Bruder C, Bohlooly-Y M, Gidlöf S, Oldfors A, Wibom R, Törnell J, Jacobs H, Larsson N. Premature ageing in mice expressing defective mitochondrial DNA polymerase. Nature. 2004;429:417–23.
25. Zhang D, Mott J, Farrar P, Ryerse J, Chang S, Stevens M, Denniger G, Zassenhaus H. Mitochondrial DNA mutations activate the mitochondrial apoptotic pathway and cause dilated cardiomyopathy. Cardiovasc Res. 2003;57:147–57.
26. Lewis W, Day BJ, Kohler JJ, Hosseini SH, Chan SS, Green EC, Haase CP, Keebaugh ES, Long R, Ludaway T, Russ R, Steltzer J, Tioleco N, Santoianni R, Copeland WC. Decreased mtDNA, oxidative stress, cardiomyopathy, and death from transgenic cardiac targeted human mutant polymerase gamma. Lab Investig. 2007;87:326–35.
27. Sahin E, Colla S, Liesa M, Moslehi J, Muller F, Guo M, Cooper M, Kotton D, Fabian A, Walkey C, Maser R, Tonon G, Foerster F, Xiong R, Wang A, Shukla S, Jaskelioff M, Martin E, Heffernan T, Protopopov A, Ivanova E, Mahoney J, Kost-Alimova M, Perry S, Bronson R, Liao R, Mulligan R, Shirihai O, Chin L, DePinho R. Telomere dysfunction induces metabolic and mitochondrial compromise. Nature. 2011;470:359–65.
28. Ikeuchi M, Matsusaka H, Kang D, Matsushima S, Ide T, Kubota T, Fujiwara T, Hamasaki N, Takeshita A, Sunagawa K, Tsutsui H. Overexpression of mitochondrial transcription factor a ameliorates mitochondrial deficiencies and cardiac failure after myocardial infarction. Circulation. 2005;112:683–90.
29. Matsushima S, Ide T, Yamato M, Matsusaka H, Hattori F, Ikeuchi M, Kubota T, Sunagawa K, Hasegawa Y, Kurihara T, Oikawa S, Kinugawa S, Tsutsui H. Overexpression of mitochondrial peroxiredoxin-3 prevents left ventricular remodeling and failure after myocardial infarction in mice. Circulation. 2006;113:1779–86.
30. Karamanlidis G, Nascimben L, Couper G, Shekar P, del Monte F, Tian R. Defective DNA replication impairs mitochondrial biogenesis in human failing hearts. Circ Res. 2010;106:1541–8.
31. Karamanlidis G, Bautista-Hernandez V, Fynn-Thompson F, Del Nido P, Tian R. Impaired mitochondrial biogenesis precedes heart failure in right ventricular hypertrophy in congenital heart disease. Circ Heart Fail. 2011;4:707–13.
32. Ahuja P, Wanagat J, Wang Z, Wang Y, Liem DA, Ping P, Antoshechkin IA, Margulies KB, Maclellan WR. Divergent mitochondrial biogenesis responses in human cardiomyopathy. Circulation. 2013;127:1957–67.
33. Huang J, Tan L, Shen R, Zhang L, Zuo H, Wang DW. Decreased peripheral mitochondrial DNA copy number is associated with the risk of heart failure and long-term outcomes. Medicine (Baltimore). 2016;95:e3323.
34. Knez J, Cauwenberghs N, Thijs L, Winckelmans E, Brguljan-Hitij J, Yang WY, Staessen JA, Nawrot TS, Kuznetsova T. Association of left ventricular structure and function with peripheral blood mitochondrial DNA content in a general population. Int J Cardiol. 2016;214:180–8.
35. Cheng S, Xanthakis V, Sullivan L, Lieb W, Massaro J, Aragam J, Benjamin E, Vasan R. Correlates of echocardiographic indices of cardiac remodeling over the adult life course: longitudinal observations from the Framingham Heart Study. Circulation. 2010;122:570–8.
36. Maurer M, Burkhoff D, Fried L, Gottdiener J, King D, Kitzman D. Ventricular structure and function in hypertensive participants with heart failure and a normal ejection fraction: the Cardiovascular Health Study. J Am Coll Cardiol. 2007;49:972–81.
37. Seo J, Kim D, Kim W, Song J, Kang D, Song J. Peak systolic velocity of mitral annular longitudinal movement measured by pulsed tissue Doppler imaging as an index of global left ventricular contractility. Am J Physiol Heart Circ Physiol. 2010;298:H1608–15.
38. Nikitin N, Loh P, Silva R, Ghosh J, Khaleva O, Goode K, Rigby A, Alamgir F, Clark A, Cleland J. Prognostic value of systolic mitral annular velocity measured with Doppler tissue imaging in patients with chronic heart failure caused by left ventricular systolic dysfunction. Heart. 2006;92:775–9.
39. Di Salvo G, Di Bello V, Salustri A, Antonini-Canterin F, La Carrubba S, Materazzo C, Badano L, Caso P, Pezzano A, Calabrò R, Carerj S. The prognostic value of early left ventricular

longitudinal systolic dysfunction in asymptomatic subjects with cardiovascular risk factors. Clin Cardiol. 2011;34:500–6.
40. Ommen S, Nishimura R, Appleton C, Miller F, Oh J, Redfield M, Tajik A. Clinical utility of Doppler echocardiography and tissue Doppler imaging in the estimation of left ventricular filling pressures: a comparative simultaneous Doppler-catheterization study. Circulation. 2000;102:1788–94.
41. Dokainish H, Zoghbi W, Lakkis N, Ambriz E, Patel R, Quinones M, Nagueh S. Incremental predictive power of B-type natriuretic peptide and tissue Doppler echocardiography in the prognosis of patients with congestive heart failure. J Am Coll Cardiol. 2005;45:1223–6.
42. Chen S, Xie X, Wang Y, Gao Y, Xie X, Yang J, Ye J. Association between leukocyte mitochondrial DNA content and risk of coronary heart disease: a case-control study. Atherosclerosis. 2014;237:220–6.
43. Wang W, Karamanlidis G, Tian R. Novel targets for mitochondrial medicine. Sci Transl Med. 2016.; 178:326rv3.
44. Szeto H, Birk A. Serendipity and the discovery of novel compounds that restore mitochondrial plasticity. Clin Pharmacol Ther. 2014;96:672–83.
45. Wang W, Karamanlidis G, Tian R. Novel targets for mitochondrial medicine. Sci Transl Med. 2016;8(326):326.
46. Dai W, Shi J, Gupta R, Sabbah H, Hale S, Kloner R. Bendavia, a mitochondria-targeting peptide, improves postinfarction cardiac function, prevents adverse left ventricular remodeling, and restores mitochondria-related gene expression in rats. J Cardiovasc Pharmacol. 2014;64:543–53.

Chapter 20
Mitochondrial Proton Leak Plays a Critical Role in Pathogenesis of Cardiovascular Diseases

Jiali Cheng, Gayani Nanayakkara, Ying Shao, Ramon Cueto, Luqiao Wang, William Y. Yang, Ye Tian, Hong Wang, and Xiaofeng Yang

Historically considered as merely cellular "powerhouse" that manufactures ATP and other metabolites, the mitochondrion is increasingly being recognized as a "sentinel" organelle capable of both detecting cellular insults and orchestrating inflammatory responses [1]. Mitochondria are complex organelles, containing their own DNA and composed of a double membrane: outer mitochondrial membrane (OMM) and inner mitochondrial membrane (IMM). This double membrane gives rise to two compartments; (1) the intermembrane space (IMS) located between OMM and IMM, and (2) the matrix (M) located in the internal space created by the IMM itself.

During oxidative phosphorylation, the electrons removed from biological fuels such as glucose and fatty acids goes through a series of electron carrier system

---

Gayani Nanayakkara, Ying Shao, Ramon Cueto, Luqiao Wang, contributed to this work equally.

J. Cheng
Department of Cardiovascular Medicine, The First Affiliate Hospital of Harbin Medical University, Harbin, Heilongjiang Province, China

Centers for Metabolic Disease Research, Cardiovascular Research, & Thrombosis Research, Department of Pharmacology, Lewis Katz School of Medicine at Temple University, 3500 North Broad Street, MERB-1059, Philadelphia, PA 19140, USA

G. Nanayakkara • Y. Shao • R. Cueto • L. Wang • H. Wang • X. Yang, MD, PhD, FAHA (✉)
Centers for Metabolic Disease Research, Cardiovascular Research, & Thrombosis Research, Department of Pharmacology, Lewis Katz School of Medicine at Temple University, 3500 North Broad Street, MERB-1059, Philadelphia, PA 19140, USA
e-mail: xfyang@temple.edu

W.Y. Yang • Y. Tian
Department of Cardiovascular Medicine, The First Affiliate Hospital of Harbin Medical University, Harbin, Heilongjiang Province, China

G. Santulli (ed.), *Mitochondrial Dynamics in Cardiovascular Medicine*, Advances in Experimental Medicine and Biology 982,
DOI 10.1007/978-3-319-55330-6_20

located in IMM, which is widely known as mitochondrial electron transport chain (ETC). The passage of electrons through ETC generate energy; and ultimately these electrons reduce molecular oxygen in to water [2]. According to the chemiosmotic theory introduced by Peter Mitchell in 1961, the energy produced by electron transfer through ETC is used to establish a proton gradient across the IMM that drives mitochondrial ATP production [3]. Simply, chemiosmotic theory explains that the mechanism which propels oxidative phosphorylation is the proton gradient formed across the IMM, further coupling respiratory oxygen to ADP phosphorylation/ATP generation.

The proton gradient is established by pumping protons against their electrochemical gradient across the IMM. This process simultaneously induce a proton gradient (chemical) across the membranes which is known as a protonmotive force (ΔP) and electrical gradient known as mitochondrial membrane potential ($\Delta\Psi_m$) [4].

Mitochondrial Proton Leak Is the Principal, but Not the Only, Mechanism That Incompletely Couples Substrate Oxygen to ATP Generation

The mitochondrial ETC is comprised of a series of redox carriers named: Complex I(NADH: ubiquinone oxidoreductase), complex II (succinate dehydrogenase), complex III (ubiquinol-cytochrome c reductase) and complex IV (cytochrome c oxidase). The exergonic process of electron transport through complex I–IV create ΔP, which will drive the protons back to matrix from IMS via ATP synthase (also known as complex V), during which the ATP are generated with the aid of adenine nucleotide and phosphate carriers. However, protons can migrate to the matrix independent of ATP synthase, a process known as "proton leak". Proton leak can also be defined as the dissipation of ΔP in the presence of ATP synthase inhibitor Oligomycin in both isolated mitochondria and intact cells [5]. As proton leak depicts the protons that migrate in to the matrix without producing ATP, it makes the coupling of substrate oxygen and ATP generation incomplete. Proton leak is the principal, but not the only, mechanism that incompletely couples substrate oxygen to ATP generation (Fig. 20.1). Even though the contribution is insignificant, a phenomenon called "electron slip" is also attributed to incomplete coupling of ATP generation and substrate oxygen as well. Electron slip refers to the process where electrons are transported via ETC without pumping protons to IMS. Therefore, electron slip results in disproportionate increase in oxygen consumption at high ΔP [6, 7].

The exact mechanism of how proton leak takes place is not fully known. The physiological regulation of proton leak is categorized in to two; (1) basal/constitutive proton leak, and (2) regulated/inducible proton leak.

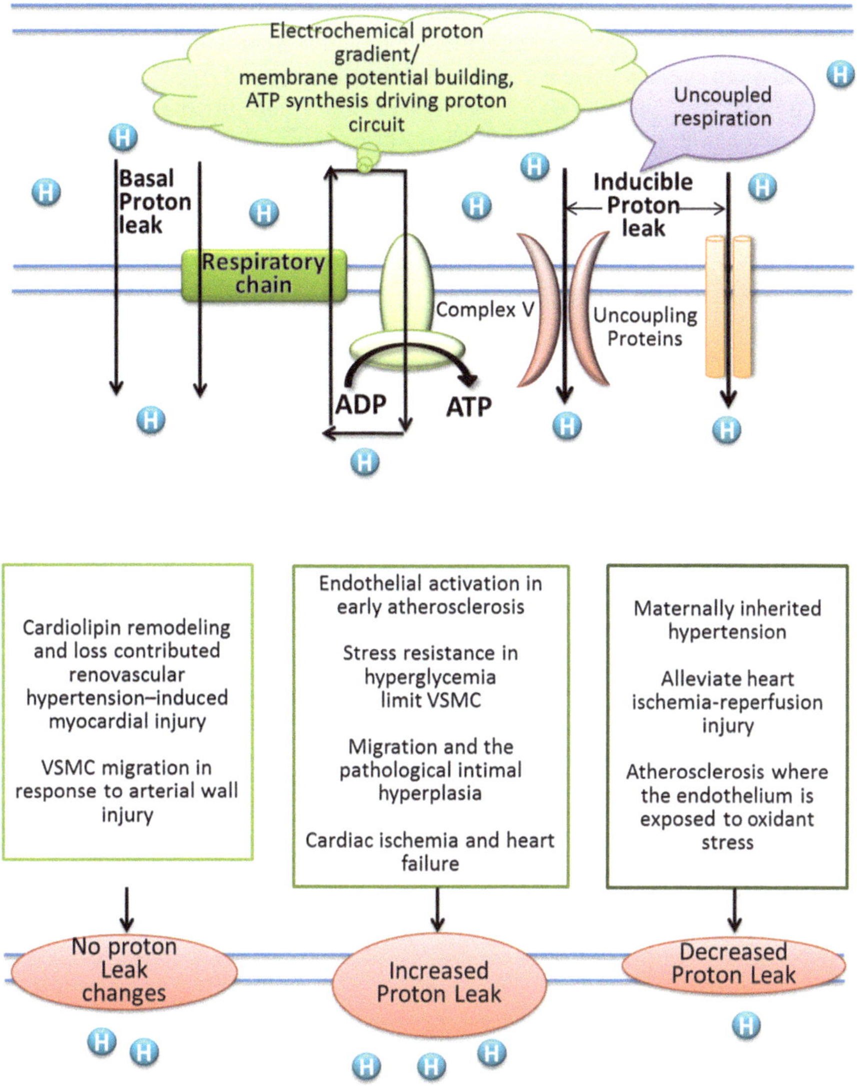

Fig. 20.1 Schematic representation of proton-leak and of the proposed role of UCPs in pathogenesis of cardiovascular disease

Basal Proton Conductance Accounts for Only 5% of Proton Leak

Basal proton conductance is generally unregulated, and largely depends on the fatty acyl composition of the inner membrane phospholipids [8, 9]. However, the proton conductance through the lipid bilayer of the inner membrane accounts for only 5%

of the proton leak. Therefore, this suggests that majority of basal conductance is regulated via mitochondrial inner membrane proteins. Approximately, two-third of the basal proton conductance is correlated to the abundance of ANT (adenine nucleotide translocase), which is a type of mitochondrial anion carrier protein found in the mitochondrial inner membrane [10].

Inducible Proton Leak Is Catalyzed by Specific Mitochondrial Inner Membrane Proteins Such as Uncoupling Proteins (UCPs)

Inducible proton leak is catalyzed by specific mitochondrial inner membrane proteins such as uncoupling proteins (UCPs) [11]. UCPs are a subfamily of the mitochondrial solute carrier family proteins that mediate transport of various metabolites across the IMM. There are five types of UCPs found in mammals, named as UCP1-5. Out of the five UCPs, UCP1 was the first to be identified in brown adipose tissue and had been extensively studied. It had been implicated in converting mitochondrial energy potential to heat, thus mainly regulate adaptive thermogenesis. Even though UCP2 and UCP3 reduce the mitochondrial coupling efficiency, it is evident that their role is largely focused on regulating reactive oxygen species (ROS) levels rather that regulating thermogenesis [12, 13]. UCPs-mediated proton conductance is regulated at molecular, transcriptional, translational and proteolytic levels [14].

Electron Leak Is the Major Causative Factor for Production of Mitochondrial Superoxide

Electron leak refers to exit of electrons from the ETC before they get reduced to water at Complex IV, and is the major causative factor for production of mitochondrial superoxide. Generally, superoxide generation is attributed to Complexes I and III, however, other mitochondrial proteins such as flavin adenine dinucleotide (FAD)-linked pyruvate and α-ketoglutarate dehydrogenase complexes are also reported to generate mitochondrial superoxide [15]. Superoxide is the primary reactive oxygen species (ROS) generated by ETC. Increased level of ROS can be detrimental to overall viability of the cells, therefore, superoxide once generated get rapidly dismutated by superoxide dismutase (SOD) to hydrogen peroxide as a protective mechanism [16]. However, at low concentrations, ROS are involved as signaling molecules in various biological pathways. Under non-pathological conditions, mitochondria are the most significant ROS producers in the cells. Other than mitochondrial ROS, cell membrane bound NADPH (Nicotinamide Adenine Dinucleotide Phosphate) oxidase enzyme complex also contribute to total cellular ROS production during certain pathological conditions [17].

ROS Generation and Proton Leak

There are contradictory reports that link proton leak to ROS generation. Some reports provide evidence that the dissipation of ΔP [18–20] or presence of ADP decreases the mtROS production, while few claims that uncoupling enhances the ROS production [21, 22]. These contradictory evidence might be due to the differences in biological and experimental settings. However, the view that uncoupling decreases mitochondrial ROS production prevails. Therefore, uncoupling may play a protective role by mitigating ROS production in cells. This cyto-protective role of uncoupling was specifically observed in heart under conditions of oxidative stress such as ischemia reperfusion (IR) injury [23], aging [24] and diabetes [25]. Most prominently, UCP1 and UCP2, and also chemical uncouplers offers protection against IR injury. Similarly, it was observed that uncoupling can exert a profound protective effect against toxicity produced in the presence of hyperglycemia in endothelial cells. In addition, protective role of uncoupling against atherosclerosis was reported.

On the other hand, increased level of ROS seems to cause an increase proton leak [18]. It was observed that peroxynitrite, which is a potent inducer of lipid oxidation, increased the proton leak in isolated brain mitochondria [26]. Additionally, superoxide can also enhance the electron leak to a similar extent as peroxynitrite, and was shown to activate UCPs [27]. The UCPs and ANTs are attributed to enhanced proton conductance in the presence of ROS. Therefore, the presence of a protective feedback loop had been suggested, where increased ROS generation activates mechanisms that promote proton leak; and the enhanced proton leak in turn reduce the ROS production limiting further damage to mitochondrial function [18, 27].

"Uncoupling" in Pathogenesis of Cardiovascular Disease

The Magnitude of the Proton Leak and the Mechanism Involved in Mediating the Proton Leak Determine Whether There Is a Protective Effect in IR Injury

IR injury, atherosclerosis and diabetes-mediated cardiovascular complications are known to induce oxidative stress by elevating ROS generation including peroxynitrite, superoxide and hydrogen peroxide that adversely affect the viability of cardiomyocytes. Increased levels of ROS produced during various cardiac pathologies were shown to mediate breakdown of plasma membrane of cardiomyocytes, increased the potential for arrhythmogenesis during a post-ischemic events and also led to apoptosis of myocytes [28, 29] (see Chap. 10). Recent evidence suggests that UCPs exert a protective role against oxidative stress by suppressing ROS.

Diseased heart has a strong association with increased oxidative stress and an impairment of reserve respiratory capacity. During physiological conditions, mitochondria does not use its full bioenergetic capacity unless there is an increase in energy demand during situations where cellular repair and detoxifying ROS is required to maintain cellular function. Mitochondrial reserve respiratory capacity refers to the difference between maximum respiratory capacity and basal respiratory capacity [30]. Therefore, depletion of reserve respiratory capacity indicate the inability of the cells to meet the increased energy demand in response to cellular stress that subsequently affect adversely to the cell viability. Intact rat neonatal ventricular myocytes exhibit exhaustion of the reserve capacity with an increased proton leak when exposed to pathologically relevant concentrations of HNE (4-hydroxynonenal), which is a reactive lipid species accumulated in the heart during ischemia and heart failure [31].

Higher rate of proton leak is associated in mitochondria isolated from hearts subjected to IR [32, 33]. Furthermore, few research groups have shown that using uncoupling agents such as FCCP [carbonyl cyanide p-(tri-fluromethoxy) phenylhydrzone] or DNP (2,4-Dinitrophenol) protects mitochondrial and cardiomyocyte function from IR injury [34, 35]. However, this observation was reversed in mitochondria subjected to ischemic preconditioning (IPC), which is a cardioprotective strategy consisting of brief cycles of IR (3–5 min) to protect the myocardium from the damaging effects of subsequent longer episodes of IR. Surprisingly, mitochondria subjected to protective IPC had a smaller increment in proton conductance after IR compared to non-preconditioned mitochondria with higher proton conductance [32, 33]. This is contradictory to the observation that increased proton conductance exert cardioprotective against IR injury. This observation was explained by differential mechanisms that are responsible for promoting mild proton leak and intensive proton leak. It was observed that generally mild-moderate proton leak during IR injury is regulated by UCPs [33], whereas a larger proton leak is mediated by ANT, which was shown to be a part of mitochondrial permeability transition pore (MPTP) [36]. Opening of MPTP under pathological conditions rapidly dissipate $\Delta\Psi_m$ leading mitochondrial swelling, and mediate cellular apoptosis or necrosis. The larger proton leak mediated by opening of MPTP is detrimental to mitochondria and adversely impact to the cell viability [36]. Therefore, these observations suggest that magnitude of the proton leak and the mechanism involved in mediating the proton leak determine whether there is a protective effect against IR injury.

Uncoupling by UCP2 Preserves Vascular Function in Diet-Induced Obesity Mice as Well as In Diabetes

Oxidative stress plays a central role in development of diabetes-mediated macrovascular and microvascular complications. In bovine aortic endothelial cells, overexpressed UCP1 blocked hyperglycemia-induced ROS production, indicating that uncoupling may play a role in reducing the progression of diabetes induced vascular complications [37]. Further, endothelial UCP2 also function as a sensor and negatively regulate mitochondrial ROS production in response to hyperglycemia, further validating importance of uncoupling in attenuating diabetes induced vascular diseases [38].

Interestingly, it was observed that UCP2 expression was gradually reduced with increased lipid deposition in atherosclerotic lesion in aorta in hyperlipidemic apolipoprotein E deficient ($ApoE^{-/-}$) mice. Furthermore, it was observed that pro-inflammatory tumor necrosis factor-α (TNF-α) can downregulate expression of UCP2 in vasculature and accelerate vascular damage. Therefore, this suggests that lowering the TNF-α expression within the aorta during exposure to vascular stress factors such as hyperinsulinemia may prevent vascular damage [39]. UCP2 preserves endothelial function through increasing endothelial nitric oxide synthase (eNOS) phosphorylation secondary to the inhibition of ROS production in the endothelium of obese diabetic mice. Silencing UCP_2 impairs endothelium-dependent relaxation in aorta and mesenteric arteries from diet-induced obesity mice as well as aortic rings exposed to high glucose [40].

UCP2 in lung endothelial cells was recently reported to play a significant role in progression of pulmonary hypertension. UCP2 ablation in mice causes high right ventricular pressure and right ventricular hypertrophy when subjected to intermittent hypoxia induced pulmonary hypertension. Loss of UCP2 in lung endothelial cells increases mitophagy leading to progression of pulmonary hypertension. Mitophagy refers to selective mitochondrial autophagy, a process initiated by changes in the $\Delta\Psi_m$. Therefore, loss of UCP2 was implicated in deficiency of mitochondria and increased apoptosis in lung endothelial cells that contribute to exacerbation of pulmonary hypertension [41].

Also, in maternally inherited hypertension, it was shown that a specific mutation in mitochondrial $tRNA^{Ala}$ (A5665G) significantly reduced the proton leak. The changes in the mitochondrial proton conductance and bioenergetics led to increased ROS and subsequently elevated blood pressure [42].

However, *Eirin and colleagues* demonstrated that the use of organic peroxide tert-Butyl hydrogen peroxide (tBHP) in isolated rat cardiomyocytes can suppress the basal respiration, maximal respiration and ATP production without altering the proton conductance in mitochondria. Also, the same research group demonstrated that use of MTP (mitochondrial peptide), which protects cardiolipin from peroxidation, ameliorated this effects while keeping the proton leak still intact. Cardiolipin is an inner membrane mitochondrial protein, which tethers cytochrome C to the inner membrane, thus facilitating the electron transport from complex III to complex IV. This protein is highly vulnerable to oxidative damage and subsequent loss, therefore leading to mitochondrial loss and contributing to development of cardiovascular disorders. The authors also demonstrate that restoration of cardiolipin improves cardiac function of pigs subjected to renovascular hypertension [43]. Therefore, all these observations indicate that the response of altering proton conductance depends on the etiology of hypertension.

Proton Leak Regulates ATP Synthesis-Uncoupled Mitochondrial ROS Generation, Which Determines Pathological Activation of Endothelial Cells for Recruitment of Inflammatory Cells

Most interestingly, alterations in mitochondrial morphology were reported to alter the proton conductance as well. Mitochondria are dynamic organelles which undergo constant shape changes by fission and fusion. As the name implies, fission results in smaller mitochondria while fusion results in larger mitochondria. In addition, variations in mitochondrial morphology have been found to contribute to vascular smooth muscle cell migration in response to arterial wall injury. Silencing Drp1 (mitochondrial fission protein dynamin-like protein 1), which is the protein that regulates mitochondrial fission, limits VSMC migration through a markedly increase in proton leak, which suppresses not only ATP synthesis but also ROS production. Similarly, respiration coupling efficiency reduction was also validated in vivo with mitochondrial fission suppression, proposing an effective strategy to inhibit intimal hyperplasia in restenosis and the progression of atherosclerotic lesions [44].

Furthermore, UCP2 was shown to mitigate the atherosclerotic burden as well. *Blanc and colleagues* demonstrated that UCP2 is a protective mediator against early stage of atherosclerosis. The same group had shown that depletion of UCP2 in blood cells significantly increased the lesion size in thoracic aorta with more macrophage infiltration in hyperlipidemic mice. This finding was attributed to increased ROS production and inflammatory responsiveness in UCP2 deficient macrophages [45]. Similarly, another study demonstrated that UCP2 prevents progression of atherosclerosis by exerting antioxidant effect presumably by suppressing NF-κB activation in human aortic endothelial cells (HAECs). Moreover, UCP2 can attenuate apoptotic death of HAECs mediated by pro-atherogenic LPC (lysophosphatidylcholine) [46].

We recently demonstrated that LPC promotes HAEC activation, which is one of the primary steps in disease progression of atherosclerosis [47]. We demonstrated the LPC specifically induces mitochondrial ROS rather than cytosolic ROS generation. Interestingly, we observed that LPC-mediated ROS generation was accompanied by increased proton leak without affecting the overall ATP production. Albeit the proton conductance was increased in LPC treated HAECs, we could not detect any change in the net proton gradient (ΔP) in mitochondria. The fact that ΔP and ATP production was not significantly compromised in LPC-mediated endothelial cell activation indicates that proton conductance through complex V was not impaired. Also, intact ΔP in the presence of an increased proton leak implies that pro-atherogenic stimuli such as LPC may promote ETC activity. Furthermore, we demonstrated that LPC-mediated mitochondrial ROS was responsible for increasing the expression of adhesion molecules such as ICAM-1 in endothelial cells, which initiates atherosclerosis process in vasculature [48]. We have further validated this observation by using mitoTempo, which is a specific mitochondrial ROS inhibitor. Our data implies that inhibition of mitochondrial ROS generation can significantly reduce the atherosclerotic burden in aortas of hyperlipidemic mice. Therefore, the major purpose for LPC-mediated increase in proton leak may be to modulate the mitochondrial ROS generation for signal transduction that ultimately lead to endothelial cell activation, without inducing mitochondrial damage or endothelial cell death [47, 49].

Conclusions

Mitochondria account for one third of the volume of cardiomyocytes and regulate ATP production that is essential to maintain cardiomyocyte health and survival [50]. Therefore, changes in mitochondrial biogenesis, morphology and function may contribute significantly to the development of cardiovascular diseases. Increased ROS can enhance proton leak, and in turn increased proton leak reduces the ROS generation, indicating the existence of a protective feedback loop that helps to ameliorate the detrimental effects caused by ROS on biological systems. The magnitude of proton current and the mechanism of proton conductance determine the level of ROS production, extent of mitochondrial damage and cell viability during cardiac pathologies. Moreover, cellular stresses such as pro-atherogenic stimuli can enhance proton leak without affecting the overall energy efficiency in endothelial cells. This finding indicates that proton leak acts as an indirect mediator of signal transduction by fine tuning ROS production, which acts as a signaling molecule to enhance endothelial activation during early atherogenesis, and contribute to the progression of the disease. Therefore, regulation of proton leak can be a potential therapeutic target for the treatment of many cardiovascular disorders.

References

1. Li X, Fang P, Mai J, Choi ET, Wang H, Yang XF. Targeting mitochondrial reactive oxygen species as novel therapy for inflammatory diseases and cancers. J Hematol Oncol. 2013;6:19.
2. Rich PR, Marechal A. The mitochondrial respiratory chain. Essays Biochem. 2010;47:1–23.
3. Mitchell P. Coupling of phosphorylation to electron and hydrogen transfer by a chemi-osmotic type of mechanism. Nature. 1961;191:144–8.
4. Noji H, Yasuda R, Yoshida M, Kinosita Jr K. Direct observation of the rotation of F1-ATPase. Nature. 1997;386:299–302.
5. Chance B, Williams GR. Respiratory enzymes in oxidative phosphorylation. III. The steady state. J Biol Chem. 1955;217:409–27.
6. Kadenbach B. Intrinsic and extrinsic uncoupling of oxidative phosphorylation. Biochim Biophys Acta. 2003;1604:77–94.
7. Murphy MP. Slip and leak in mitochondrial oxidative phosphorylation. Biochim Biophys Acta. 1989;977:123–41.
8. Brookes PS, Buckingham JA, Tenreiro AM, Hulbert AJ, Brand MD. The proton permeability of the inner membrane of liver mitochondria from ectothermic and endothermic vertebrates

and from obese rats: correlations with standard metabolic rate and phospholipid fatty acid composition. Comp Biochem Physiol B Biochem Mol Biol. 1998;119:325–34.
9. Fontaine EM, Moussa M, Devin A, Garcia J, Ghisolfi J, Rigoulet M, Leverve XM. Effect of polyunsaturated fatty acids deficiency on oxidative phosphorylation in rat liver mitochondria. Biochim Biophys Acta. 1996;1276:181–7.
10. Brand MD, Pakay JL, Ocloo A, Kokoszka J, Wallace DC, Brookes PS, Cornwall EJ. The basal proton conductance of mitochondria depends on adenine nucleotide translocase content. Biochem J. 2005;392:353–62.
11. Parker N, Vidal-Puig A, Brand MD. Stimulation of mitochondrial proton conductance by hydroxynonenal requires a high membrane potential. Biosci Rep. 2008;28:83–8.
12. Arsenijevic D, Onuma H, Pecqueur C, Raimbault S, Manning BS, Miroux B, Couplan E, Alves-Guerra MC, Goubern M, Surwit R, Bouillaud F, Richard D, Collins S, Ricquier D. Disruption of the uncoupling protein-2 gene in mice reveals a role in immunity and reactive oxygen species production. Nat Genet. 2000;26:435–9.
13. Vidal-Puig AJ, Grujic D, Zhang CY, Hagen T, Boss O, Ido Y, Szczepanik A, Wade J, Mootha V, Cortright R, Muoio DM, Lowell BB. Energy metabolism in uncoupling protein 3 gene knockout mice. J Biol Chem. 2000;275:16258–66.
14. Azzu V, Brand MD. The on-off switches of the mitochondrial uncoupling proteins. Trends Biochem Sci. 2010;35:298–307.
15. Starkov AA, Fiskum G, Chinopoulos C, Lorenzo BJ, Browne SE, Patel MS, Beal MF. Mitochondrial alpha-ketoglutarate dehydrogenase complex generates reactive oxygen species. J Neurosci. 2004;24:7779–88.
16. Murphy MP. How mitochondria produce reactive oxygen species. Biochem J. 2009;417:1–13.
17. Santulli G, Xie W, Reiken SR, Marks AR. Mitochondrial calcium overload is a key determinant in heart failure. Proc Natl Acad Sci U S A. 2015;112:11389–94.
18. Brookes PS. Mitochondrial H(+) leak and ROS generation: an odd couple. Free Radic Biol Med. 2005;38:12–23.
19. Turrens JF, Boveris A. Generation of superoxide anion by the NADH dehydrogenase of bovine heart mitochondria. Biochem J. 1980;191:421–7.
20. Herrero A, Barja G. ADP-regulation of mitochondrial free radical production is different with complex I- or complex II-linked substrates: implications for the exercise paradox and brain hypermetabolism. J Bioenerg Biomembr. 1997;29:241–9.
21. Boveris A, Chance B. The mitochondrial generation of hydrogen peroxide. General properties and effect of hyperbaric oxygen. Biochem J. 1973;134:707–16.
22. Starkov AA, Fiskum G. Regulation of brain mitochondrial H2O2 production by membrane potential and NAD(P)H redox state. J Neurochem. 2003;86:1101–7.
23. Ganote CE, Armstrong SC. Effects of CCCP-induced mitochondrial uncoupling and cyclosporin A on cell volume, cell injury and preconditioning protection of isolated rabbit cardiomyocytes. J Mol Cell Cardiol. 2003;35:749–59.
24. Speakman JR, Talbot DA, Selman C, Snart S, McLaren JS, Redman P, Krol E, Jackson DM, Johnson MS, Brand MD. Uncoupled and surviving: individual mice with high metabolism have greater mitochondrial uncoupling and live longer. Aging Cell. 2004;3:87–95.
25. Green K, Brand MD, Murphy MP. Prevention of mitochondrial oxidative damage as a therapeutic strategy in diabetes. Diabetes. 2004;53(Suppl 1):S110–8.
26. Brookes PS, Land JM, Clark JB, Heales SJ. Peroxynitrite and brain mitochondria: evidence for increased proton leak. J Neurochem. 1998;70:2195–202.
27. Echtay KS, Roussel D, St-Pierre J, Jekabsons MB, Cadenas S, Stuart JA, Harper JA, Roebuck SJ, Morrison A, Pickering S, Clapham JC, Brand MD. Superoxide activates mitochondrial uncoupling proteins. Nature. 2002;415:96–9.
28. Zhu X, Zuo L, Cardounel AJ, Zweier JL, He G. Characterization of in vivo tissue redox status, oxygenation, and formation of reactive oxygen species in postischemic myocardium. Antioxid Redox Signal. 2007;9:447–55.

29. Murphy E, Steenbergen C. Mechanisms underlying acute protection from cardiac ischemia-reperfusion injury. Physiol Rev. 2008;88:581–609.
30. Pfleger J, He M, Abdellatif M. Mitochondrial complex II is a source of the reserve respiratory capacity that is regulated by metabolic sensors and promotes cell survival. Cell Death Dis. 2015;6:e1835.
31. Hill BG, Dranka BP, Zou L, Chatham JC, Darley-Usmar VM. Importance of the bioenergetic reserve capacity in response to cardiomyocyte stress induced by 4-hydroxynonenal. Biochem J. 2009;424:99–107.
32. Quarrie R, Cramer BM, Lee DS, Steinbaugh GE, Erdahl W, Pfeiffer DR, Zweier JL, Crestanello JA. Ischemic preconditioning decreases mitochondrial proton leak and reactive oxygen species production in the postischemic heart. J Surg Res. 2011;165:5–14.
33. Nadtochiy SM, Tompkins AJ, Brookes PS. Different mechanisms of mitochondrial proton leak in ischaemia/reperfusion injury and preconditioning: implications for pathology and cardioprotection. Biochem J. 2006;395:611–8.
34. Brennan JP, Southworth R, Medina RA, Davidson SM, Duchen MR, Shattock MJ. Mitochondrial uncoupling, with low concentration FCCP, induces ROS-dependent cardioprotection independent of KATP channel activation. Cardiovasc Res. 2006;72:313–21.
35. Minners J, van den Bos EJ, Yellon DM, Schwalb H, Opie LH, Sack MN. Dinitrophenol, cyclosporin A, and trimetazidine modulate preconditioning in the isolated rat heart: support for a mitochondrial role in cardioprotection. Cardiovasc Res. 2000;47:68–73.
36. Crompton M. The mitochondrial permeability transition pore and its role in cell death. Biochem J. 1999;341(Pt 2):233–49.
37. Nishikawa T, Edelstein D, Du XL, Yamagishi S, Matsumura T, Kaneda Y, Yorek MA, Beebe D, Oates PJ, Hammes HP, Giardino I, Brownlee M. Normalizing mitochondrial superoxide production blocks three pathways of hyperglycaemic damage. Nature. 2000;404:787–90.
38. Koziel A, Sobieraj I, Jarmuszkiewicz W. Increased activity of mitochondrial uncoupling protein 2 improves stress resistance in cultured endothelial cells exposed in vitro to high glucose levels. Am J Physiol Heart Circ Physiol. 2015;309:H147–56.
39. Gomez-Hernandez A, Perdomo L, de las Heras N, Beneit N, Escribano O, Otero YF, Guillen C, Diaz-Castroverde S, Gozalbo-Lopez B, Cachofeiro V, Lahera V, Benito M. Antagonistic effect of TNF-alpha and insulin on uncoupling protein 2 (UCP-2) expression and vascular damage. Cardiovasc Diabetol. 2014;13:108.
40. Tian XY, Wong WT, Xu A, Lu Y, Zhang Y, Wang L, Cheang WS, Wang Y, Yao X, Huang Y. Uncoupling protein-2 protects endothelial function in diet-induced obese mice. Circ Res. 2012;110:1211–6.
41. Haslip M, Dostanic I, Huang Y, Zhang Y, Russell KS, Jurczak MJ, Mannam P, Giordano F, Erzurum SC, Lee PJ. Endothelial uncoupling protein 2 regulates mitophagy and pulmonary hypertension during intermittent hypoxia. Arterioscler Thromb Vasc Biol. 2015;35:1166–78.
42. Jiang P, Wang M, Xue L, Xiao Y, Yu J, Wang H, Yao J, Liu H, Peng Y, Li H, Chen Y, Guan MX. A hypertension-associated tRNAAla mutation alters tRNA metabolism and mitochondrial function. Mol Cell Biol. 2016;36:1920–30.
43. Eirin A, Ebrahimi B, Kwon SH, Fiala JA, Williams BJ, Woollard JR, He Q, Gupta RC, Sabbah HN, Prakash YS, Textor SC, Lerman A, Lerman LO. Restoration of mitochondrial cardiolipin attenuates cardiac damage in swine renovascular hypertension. J Am Heart Assoc. 2016;5:e003118.
44. Wang L, Yu T, Lee H, O'Brien DK, Sesaki H, Yoon Y. Decreasing mitochondrial fission diminishes vascular smooth muscle cell migration and ameliorates intimal hyperplasia. Cardiovasc Res. 2015;106:272–83.
45. Blanc J, Alves-Guerra MC, Esposito B, Rousset S, Gourdy P, Ricquier D, Tedgui A, Miroux B, Mallat Z. Protective role of uncoupling protein 2 in atherosclerosis. Circulation. 2003;107:388–90.
46. Lee KU, Lee IK, Han J, Song DK, Kim YM, Song HS, Kim HS, Lee WJ, Koh EH, Song KH, Han SM, Kim MS, Park IS, Park JY. Effects of recombinant adenovirus-mediated uncoupling protein 2 overexpression on endothelial function and apoptosis. Circ Res. 2005;96:1200–7.

47. Zhang Y, Liu X, Bai J, Tian X, Zhao X, Liu W, Duan X, Shang W, Fan HY, Tong C. Mitoguardin regulates mitochondrial fusion through mitoPLD and is required for neuronal homeostasis. Mol Cell. 2016;61:111–24.
48. Gambardella J, Sardu C, Sacra C, Del Giudice C, Santulli G. Quit smoking to outsmart atherogenesis: molecular mechanisms underlying clinical evidence. Atherosclerosis. 2017;257: 242–5.
49. Li X, Fang P, Yang WY, Chan K, Lavallee M, Xu K, Gao T, Wang H, Yang X. Mitochondrial ROS, uncoupled from ATP synthesis, determine endothelial activation for both physiological recruitment of patrolling cells and pathological recruitment of inflammatory cells. Can J Physiol Pharmacol. 2017;95(3):247–52.
50. Goffart S, von Kleist-Retzow JC, Wiesner RJ. Regulation of mitochondrial proliferation in the heart: power-plant failure contributes to cardiac failure in hypertrophy. Cardiovasc Res. 2004;64:198–207.

Chapter 21
Mitochondria and Angiogenesis

Raluca Marcu, Ying Zheng, and Brian J. Hawkins

During prenatal development, de novo formation of the primitive vascular network to support tissue growth occurs through vasculogenesis by differentiation from mesoderm-derived angioblasts [1]. Angiogenesis ensures expansion and fine remodeling of this vascular network through sprouting of new blood vessels from preexisting structures, anastomosis and pruning [2]. In the postnatal organism, angiogenesis is limited primarily to the physiologic repair of tissue wounding, for the proper functioning of female reproductive organs, and as an adaptive response to exercise training [3, 4]. Nevertheless, abnormal angiogenic vascular growth is frequently associated with the pathogenesis of cancer, inflammation, atherosclerosis, retinopathy, endometriosis and arthritis, while impaired angiogenesis is a common deleterious consequence of ischemic injury, hypertension, and preeclampsia [5]. Pharmacological interventions that target angiogenesis are therefore of great clinical interest for disease treatment where obstructing angiogenesis or promoting revascularization are essential for recovering tissue functions.

Angiogenesis engages a series of highly coordinated events that resolve with the formation of functional, perfusable blood vessels [6]. In the presence of pro-angiogenic and guidance cues (VEGF, netrins, semaphorins) that outperform anti-angiogenic signals, a quiescent endothelial cell known as tip cell is selected and activated based on regulatory VEGF-Dll4-Notch signaling [7]. Following vascular basement membrane degradation, the tip cell guides the migration and proliferation of neighbouring endothelial stalk cells that elongate the vascular sprout under Nrarp/Notch and Wnt signaling control [8]. Subsequent lumen formation,

R. Marcu (✉)
San Raffaele Scientific Institute, Milan, Italy
e-mail: marcu.raluca@hsr.it

Y. Zheng
University of Washington, Seattle, WA, USA

B.J. Hawkins
University of Washington, Seattle, WA, USA

BioLife Solutions Inc., Bothell, WA, USA

G. Santulli (ed.), *Mitochondrial Dynamics in Cardiovascular Medicine*,
Advances in Experimental Medicine and Biology 982,
DOI 10.1007/978-3-319-55330-6_21

recruitment of pericytes, deposition of extracellular matrix and reestablishment of endotheial quiescence contribute to blood vessel maturation [9, 10].

Due to relatively low energetic requirements, baseline performance of quiescent endothelial cells is supported mainly through oxygen-independent glycolysis. It is estimated that endothelial cells acquire 85 % of their total intracellular ATP supply from glycolysis, and that the endothelial glycolytic flux is more than 200-fold higher than the mitochondrial oxidation of glucose, fatty acids or glutamine [11]. In agreement with a predominatly glycolityc metabolism, mitochondrial oxygen utilization in endothelial cells is marginal compared to neurons, hepatocytes, and cardiomyocytes [11]. Accordingly, the mitochondrial content of endothelial cells is low with respect to metabolically active cells: in rat capillary endothelial cells from the hypothalamus, kidney, skin, lung, heart and skeletal muscle, the mitochondrial volumes represent merely 2–5 % of the endothelial cytoplasmic volumes [12] compared to 32 % volume density of mitochondria in rat cardiomiocytes [13]. During angiogenesis, motile, invasive and proliferative behaviour requires increased energetic support, which endothelial cells also accomplish by upregulating glycolysis. Indeed, impairment of glycolysis due to pharmacological or genetic inhibition of phosphofructokinase-2/fructose-2,6-bisphosphatase-3 (PFKB3) compromises vessel sprouting and hyper-branching and reduces pathological angiogenesis in models of age-related macular degeneration, retinopaty of prematurity, and inflammation [11, 14].

However, despite low oxygen metabolism, endothelial mitochondria maintain a considerable bioenergetic reserve capacity that represents an important resource to lessen oxidative stress [15]. Moreover, in conditions of reduced glycolytic rates, endothelial cells retain the ability to switch to oxidative metabolism utilizing glucose, fatty acids, and amino acids [16]. The requirement for functionally intact mitochondria during angiogenesis is clearly demonstrated by impaired neovascularization in aged POLGA mutator mice, a model for mitochondrial dysfunction and aging due to defective mitochondrial DNA polymerase [17]. In POLGA mice, growth factor induced angiogenesis is severely impaired with no adverse effects on established vascular networks, consistent with a role for mitochondrial function specifically during neovessel formation. Growth factors also stimulate production of nitric oxide (NO), which is a potent angiogenic stimulus [18]. While NO can influence multiple angiogenic signaling pathways, a major target is the terminal enzyme of the mitochondrial respiratory chain, mitochondrial cytochrome c oxidase [19]. Both under basal and stimulated conditions, and depending on the calcium influx, NO adjusts the capacity of this enzyme to utilize oxygen and minimizes respiration [20]. Indeed, it is possible that a major role of endothelial NO production may be to inhibit mitochondrial respiration and prompt the increase in glycolysis required to fuel angiogenic vessel growth.

While not preferentially used for bioenergetic purposes, endothelial mitochondria alternatively serve as signaling organelles that are both a source and target of reactive oxygen species (ROS). For example, under NO control, mitochondria contribute to mitochondrial ROS-dependent AMPK activation at low oxygen concentrations, which may play a protective role in endothelial cells [21]. However, the signaling function of mitochondria is not limited to NO targeting and ROS production, but also involves modulation of cytosolic calcium signals, the release of mitochondrial metabolic intermediates and the intrinsic activation of apoptosis [22], all of which can impact the proliferative properties of endothelial cells prerequisite for angiogenesis.

Mitochondrial Functions and Mitochondrial Proteins Involved in Angiogenesis

Despite a repressed mitochondrial energetic function in endothelial cells, other mitochondrial functions are highly involved in maintaining vascular homeostasis and regulating angiogenesis. Among them, mitochondrial ROS production, regulation of intracellular calcium, metabolic regulation and apoptosis are essential in defining the proper balance necessary to ensure the endothelial switch towards a proliferative phenotype required for angiogenesis (Fig. 21.1).

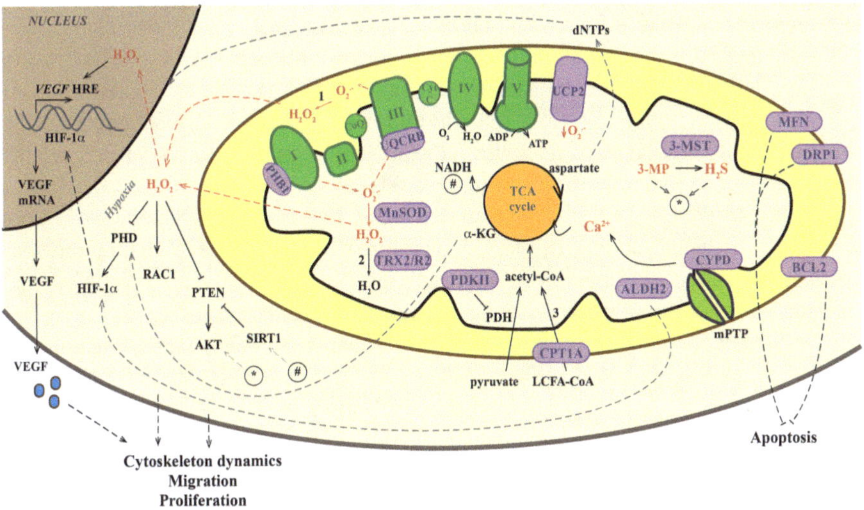

Fig. 21.1 Mitochondrial proteins and angiogenesis. Several mitochondrial proteins (*purple*) and mitochondrial signaling agents (*red*) control angiogenesis by modulating endothelial migration, proliferation and apoptosis. *UQCRB* Ubiquinol-cytochrome c reductase binding protein, *PHB1* Prohibitin 1, *MnSOD* Manganese superoxide dismutase, *TRX2* Thioredoxin 2, *TRXR2* Thioredoxin reductase 2, *UCP2* Uncoupling protein 2, *CPT1A* Carnitine palmitoyltransferase 1A, *PDKII* Pyruvate dehydrogenase kinase II, *PDH* Pyruvate dehydrogenase, *ALDH2* Aldehyde dehydrogenase 2, *3-MP* 3-Mercaptopyruvate, *3-MST* 3-mercaptopyruvate sulfurtransferase, *H_2S* Hydrogen sulfide, *CYPD* Cyclophilin D, *mPTP* Mitochondrial permeability transition pore, *MFN* Mitofusins, *DRP1* Dynamin-related protein 1, *BCL2* B cell lymphoma 2, *1* Copper-zinc superoxide dismutase, *2* Glutaredoxin, *3* β-oxidation; # mitochondrial matrix Ca^{2+} accumulation stimulates TCA cycle activity and NADH production, concomitant with decreased mitochondrial and cytosolic NAD^+/NADH ratio, * 3-MP and H_2S promote AKT activation

Mitochondrial ROS Production

Mitochondrial ROS are products of incomplete oxygen metabolism and despite a longstanding causative involvement in cellular oxidative damage and toxicity, ROS are emerging as physiological regulators of cellular functions [23]. In blood vessels, ROS control shear stress-induced vasodilation, non-vasomotor response to mechanical stimuli and the hypoxia-reoxygenation response, and excessive ROS ensuing mitochondrial dysfunction are responsible for inflammatory vascular reactions leading to cardiovascular disease [24, 25]. The dual character of ROS as both physiologic and pathologic mediators of cellular outcome is determined by the intensity of the pro-oxidant challenge, and as such the balance between mitochondrial ROS production and detoxification is tightly regulated.

Mitochondrial ROS arise from the leak of electrons along the mitochondrial electron transport chain, mainly at complex I (NADH-ubiquinone oxidoreductase) and complex III (ubiquinol-cytochrome c oxidoreductase), followed by partial reduction of molecular oxygen. The proximal reactive species produced is the superoxide anion ($O_2^{\bullet}-$), short-lived and nondiffusible, which is released on both sides of the mitochondrial inner membrane [26]. $O_2^{\bullet}-$ is quickly dismutated into hydrogen peroxide (H_2O_2) by mitochondrial superoxide dismutases (manganese-SOD, MnSOD, in the mitochondrial matrix and copper-zinc-SOD, CuZnSOD, in the intermembrane space), for subsequent removal by mitochondrial peroxidases (glutaredoxin-2, peroxiredoxin-3 and thioredoxin-2). H_2O_2 can also generate the short-lived but highly reactive hydroxyl radical, responsible for most deleterious reactions with lipids, proteins and nucleic acids [27].

Among oxidants, H_2O_2 is the most stable and longest-lived species that is both membrane permeable and diffusible, and therefore suitable to act as signaling messenger. At physiologic concentrations, H_2O_2 can impact signal transduction pathways and activate redox-sensitive transcription factors that drive cellular responses via the reversible posttranslational modification of cysteine residues [23]. With respect to angiogenesis, mitochondrial-derived H_2O_2 proved essential for endothelial migration and proliferation during angiogenesis and wound healing. Specifically, pro-angiogenic stimulation with VEGF drove an upregulation of mitochondrial metabolism and mitochondrial H_2O_2 release in endothelial cells, which targeted the Rho-related small GTPase Rac1, a regulator of cytoskeleton dynamics, and triggered downstream activation of cell survival, migratory and proliferative signals via PAK, Akt, p38MAP and ERK [28]. Conversely, mitochondrial ROS scavenging abrogated VEGF-mediated Rac1 activation and endothelial cell migration, and attenuated carotid artery reendothelialization upon injury in vivo. Mitochondrial ROS are also key in regulating VEGF expression and secretion to initiate vascular sprouting by directly stabilizing nuclear VEGF transcription through oxidative modifications of the VEGF hypoxia response element (VEGF HRE) [29], and indirectly, by ensuring HIF-1α stabilization upon hypoxia to promote VEGF transcription [30].

Due to the difficulty in selectively quantifying minute quantities of discrete oxidant species within the mitochondrial compartment, the involvement of mitochondrial ROS signaling in angiogenesis is often defined by genetic manipulation of mitochondrial antioxidant defenses or by employing mitochondria-targeted antioxidants. Below we provide an overview of the mitochondrial proteins with proven angiogenic functions that are associated with both mitochondrial ROS detoxification (MnSOD, TRX2, TRXR2, and UCP2) and ROS production (UQCRB and PHB1).

Manganese Superoxide Dismutase (MnSOD) The existence of a mitochondrial H_2O_2-directed, redox sensitive, angiogenic pathway is strongly supported by genetic manipulation of MnSOD, the enzyme responsible for superoxide to H_2O_2 dismutation in the mitochondrial matrix. MnSOD is required for normal biological function of tissues, and global MnSOD deletion is lethal in mice because of dilated cardiomyopathy [31]. The endothelium expresses high levels of MnSOD and MnSOD expression is further enhanced upon VEGF via an NADPH oxidase-dependent mechanism [32], and upon inflammatory cytokine challenge (tumor necrosis factor, interleukin-1, lipopolysaccharide) via protein kinase C [33], to protect against oxidative damage. Indeed, targeted scavenging of mitochondrial superoxide by MnSOD overexpression improved vascular function and was protective against hypertension [34], prevented oxidation of low density lipoprotein (LDL) involved in atherosclerosis [35], and ameliorated the development of diabetic retinopathy by counteracting high glucose-induced oxidative mitochondrial DNA damage [36]. Oppositely, decreased MnSOD expression amplified oxidative stress and endothelial dysfunction in the persistent pulmonary hypertension observed in newborns [37].

With respect to vessel growth, enhanced MnSOD expression stimulates angiogenic functions. Explicitly, in lung microvessel cells, overexpression of MnSOD boosted mitochondrial H_2O_2 production, which led to PTEN oxidation and inactivation in favor of PI3K and Akt signaling [38]. Phenotypically, MnSOD increased sprouting angiogenesis and blood vessel development in the chorioallantoic membrane based upon enhanced mitochondrial ROS-dependent VEGF secretion. The pro-angiogenic effect of increased MnSOD expression was counteracted by catalase coexpression that removed excess H_2O_2 [38]. Similarly, the MnSOD mimetic MnTBAP (manganese (III) tetrakis (4-benzoic acid) porphirin) showed a strong pro-angiogenic effect substantiated by increased endothelial migration, tube formation and capillary sprouting [39]. Restored MnSOD expression also accelerated wound healing in diabetic mice in which endothelial progenitor cells were transplanted at the wound site [40]. Impaired wound healing due to defective angiogenesis is a severe complication of diabetes, and a compromised number and functionality of circulating endothelial progenitor cells contribute to defective neovascularization. Endothelial progenitor cells isolated from type 2 diabetic (*db/db*) mice displayed elevated mitochondrial superoxide levels and reduced expression and activity of MnSOD; adenoviral MnSOD gene transfer in these cells hindered mitochondrial superoxide, improved in vitro tube formation and cell adhesion, and enhanced capillary formation and wound closure in vivo [40]. In a similar manner, enhanced MnSOD expression via pharmacologic AMPK activation rescued impaired angiogenic functions of endothelial progenitor cells from streptozotocin-induced type 1 diabetic mice [41]. Conversely, siRNA-mediated knockdown of MnSOD in normal endothelial progenitor cells reduced their angiogenic activity in diabetic wound healing assays [40]. Together, these experiments support a protective role against endothelial oxidative damage and a pro-angiogenic function for MnSOD, which might have important implications for therapeutic angiogenesis when restoring defective vascularization is required to recover organ functions.

Thioredoxin 2 (Trx2) Similar to MnSOD, the mitochondrial antioxidant enzyme Trx2 removes H_2O_2 and, as a result, exerts vaso-protective and pro-angiogenic properties. Cells deficient in Trx2 accumulate endogenous ROS and are highly sensitive to exogenous oxygen radicals, as evidenced by the embryonic lethality of Trx2-deficient mice early in development [42]. In contrast, genetic targeting of Trx2 using the endothelial-specific VE-cadherin promoter (Trx2 TG mice) produced mice that were both viable and fertile [43]. Trx2 TG mice phenotypically exhibited elevated antioxidant levels and reduced ROS (both H_2O_2 and $O_2^{\bullet}-$) and oxidative stress, as well as increased NO levels that enhanced aortic vasodilation. Pathologically, endothelial Trx2 expression also prevented the atherosclerotic plaque development in apolipoprotin E-deficient mice [43]. The pro-angiogenic role of Trx2 was substantiated during post-ischemic recovery in a femoral artery ligation model in Trx2 TG mice: the enhanced perfusion of the ischemic tissue was due to pericyte recruitment that increased capillary formation and vessel maturation [44]. Trx2 counteracted the deleterious effects of ischemia by reducing superoxide stress and blocking ASK1-JNK/p38 signaling and apoptosis to promote cell survival, while also increasing NO bioavailability to promote endothelial migration and proliferation. Noteworthy, while H_2O_2 itself is pro-angiogenic [28, 38], H_2O_2 and $O_2^{\bullet}-$ removal can lead to a similar angiogenic phenotype by increasing the concentration of pro-angiogenic NO [18, 43, 45]. Thus, stimulated expression of the antioxidant defense can facilitate normalization of injured tissue by promoting either ROS-dependent or NO-dependent endothelial proliferation and neovascularization.

Thioredoxin reductase 2 (TrxR2) The positive angiogenic regulatory function of the mitochondrial antioxidant defense on angiogenesis is reinforced by genetic inactivation of the mitochondrial TrxR2. TrxR2 is a selenoenzyme that plays an important role in the redox homeostasis by serving as an electron donor for the peroxiredoxin/thioredoxin system during H_2O_2 removal. Global knockout of TrxR2 in mice is embryonically lethal and cardiac specific TrxR2 inactivation produces offspring that perish shortly after birth due to dilated cardiomyopathy [46], which demonstrates the importance of TrxR2 for H_2O_2 detoxification during development. In endothelial cells, TrxR2 deletion increased H_2O_2 levels, inhibited endothelial nitric oxide synthase activity and suppressed VEGF-induced NO production [47]. With regard to angiogenesis, the loss of TrxR2 and the resultant abnormal redox balance impaired tumor growth and tumor-related angiogenesis by hindering Hif-1α signaling pathway. Immortalized fibroblasts lacking TrxR2 exhibited increased total cellular ROS levels that led to reduced tumor vascularization and smaller tumor mass when implanted subcutaneously into mice [48]. While the effect of TrxR2 on vascularization may at first appear counter to the ability of ROS stabilize Hif-1α [30], TrxR2-deficient tumor cells fail to activate Hif-1α signaling, and increased ROS in turn activate JNK by inhibiting JNK phosphatases that leads to

PHD2 accumulation and Hif-1α degradation [48]. Failure to stabilize Hif-1α decreased VEGF levels which resulted in a delayed angiogenic switch, reduced tumor vascularization and reduced tumor growth. Therefore, inhibition of mitochondrial ROS scavenging leads to deleterious ROS accumulation that interferes with pro-angiogenic Hif-1α and VEGF signaling and limits neovascularization.

Uncoupling Protein 2 (UCP2) Although UCP2 is not a traditional antioxidant defense protein, recent evidence supports a role for UCP2 in depressing mitochondrial ROS production [49]. Uncoupling proteins are located in the inner mitochondrial membrane and translocate protons from the intermembrane space into the mitochondrial matrix, uncoupling ATP synthesis from proton flux through the mitochondrial membrane. The physiologic functions of UCP2 include the regulation of ROS production, prevention of inflammation, and inhibition of cell death, and thus UCP2 plays central roles in cardiovascular and neurodegenerative diseases. Indeed, increasing UCP2 expression and activity protects neurons from damage following traumatic brain injury and during Parkinson's disease, as well as in atherosclerosis and myocardial ischemia [50]. However, it is the regulation of mitochondrial ROS generation in particular that has received recent attention as mild uncoupling caused by UCP2 can lower mitochondrial ROS production and afford protection against oxidative stress damage [49]. From this perspective, impaired UCP2 expression has been causally linked to defective angiogenesis while UCP2 reconstitution rescued the phenotype [51, 52]. Indeed, upregulation of UCP2 in proliferating and angiogenic endothelial cells reduced membrane potential and limited mitochondrial superoxide production that would otherwise have triggered cellular senescence [51]. In vitro, UCP2-null endothelial cells exhibited impaired proliferation and migration, as well as reduced capillary sprouting. The underlying mechanism for defective angiogenesis involved excess superoxide-dependent mitochondrial fragmentation and p53-dependent cellular senescence that limited proliferation. In agreement, UCP2 knockout mice exhibited lower blood flow and recovery during hind limb ischemia-induced angiogenesis, which could be restored by adenoviral reintroduction of UCP2 [51]. Subsequently, targeting UCP2 may constitute an important means for limiting tumor angiogenesis, and promoting mitochondrial uncoupling could enhance the angiogenic response to ischemia.

The pro-angiogenic properties of UCP2 were also noted in response to ischemia-induced AMPK activation, resulting in enhanced UCP2 expression, which decreased mitochondrial ROS production and increased NO availability [52]. AMPK signaling is essential to support neovascularization in response to hypoxic stress based on pro-angiogenic Akt activation and NO release [53]. As mentioned previously, NO promotes new vessel formation by stimulating VEGF to support endothelial migration and proliferation [18]. In excess, mitochondrial $O_2^{\cdot-}$ can rapidly interact with NO to form the peroxynitrite radical ($ONOO^{\cdot-}$), which effectively reduces NO bioavailability and resultant angiogenic signaling [45]. In agreement with AMPK involvement in angiogenesis, endothelial cells isolated from AMPKα1−/− and AMPKα2−/− mice exhibited reduced UCP2 expression and increased oxidative stress that resulted in impaired vascular tube formation, similar to the angiogenic phenotype of UCP2-null cells [52]. AMPK-dependent upregulation of UCP2 expression and angiogenesis were confirmed in ischemic thigh adductor muscles from control mice, but were absent in AMPKα1 −/− and AMPKα2 −/− mice [52]. Together, this evidence demonstrates a positive role for UCP2 in angiogenesis similar to bona fide antioxidant enzymes (MnSOD, Trx2, TrxR2), in that UCP2 protects the vasculature by limiting deleterious ROS accumulation and stimulates vascular growth by increasing NO availability.

Ubiquinol-cytochrome c reductase binding protein (UQCRB) In addition to proteins involved in ROS detoxification, proteins that exert regulatory effects on mitochondrial ROS generation are similarly implicated in angiogenesis. UQCRB is a 13.4 kDa subunit of the mitochondrial respiratory chain complex III (ubiquinol-cytochrome c oxidoreductase) with a role in ensuring complex III assembly and maintenance. A deletion in the nuclear gene encoding UQCRB resulted in reduced respiratory function and decreased complex III enzyme activity and manifested as hypoglycemia and lactic acidosis in patients [54]. In vitro and in vivo data specify a central role for UQCRB in endothelial angiogenesis: UQCRB modulates hypoxia-induced mitochondrial ROS and HIF-1α-mediated signaling in cancer cells, UQCRB controls mitochondrial ROS-mediated VEGFR2 signaling in endothelial cells, and UQCRB is a target of the anti-angiogenic compound terpestacin [55].

In endothelial cells, UQCRB knockdown inhibited VEGF-stimulated mitochondrial superoxide production, which downregulated the activation of VEGFR2 and the downstream mediators ERK and Akt. The net result was a decrease in VEGF-induced endothelial invasiveness [56]. Similarly, pharmacological inhibition of UQCRB with the fungal metabolite terpestacin suppressed mitochondrial ROS generation and VEGF-induced angiogenesis, and potentiated the anti-angiogenic activity of the anti-VEGF monoclonal antibody bevacizumab to inhibit glioblastoma xenograft tumor growth in mice [56]. In vivo UQCRB knockdown in zebrafish also lead to the suppression of VEGF expression and inhibition of angiogeneic sprouting, without affecting vasculogenesis [57], supporting the role of UQCRB in vessel outgrowth versus neovascularization.

Mutations in the UQCRB protein also enhanced mitochondrial ROS generation and increased migration, proliferation and angiogenic activity in HEK293 cells that were attributed to HIF-1α stabilization and increased VEGF secretion. Moreover, treatment of endothelial cells with conditioned media from UQCRB mutant HEK293 cells enhanced endothelial invasiveness [58]. Overexpression of UQCRB also presented with pro-angiogenic activity via the fusion of UQCRB with a protein transduction domain (PTD) resulting in the generation of a potent cell-permeable in vitro and in vivo pro-angiogenic agent (PTD-UQCRB) [59]. PTD-UQCRB localized to the mitochondrial inner membrane and enhanced angiogenesis by stimulating mitochondrial ROS production, increasing HIF-1α stability and VEGF protein levels. PTD-UQCRB significantly increased angiogenesis within Matrigel plugs even in the absence of VEGF and topical application enhanced wound healing by promoting increased microvessel density in the regenerating wounds.

Taken as a whole, the evidence suggests UQCRB expression positively influences angiogenesis by stabilizing HIF-1α and increasing VEGF secretion through an oxidant-dependent mechanism. However, despite the known interaction with complex III, a major site for mitochondrial ROS generation [60], the precise means by which UQCRB influences oxidant generation is not fully understood and warrants further study.

Prohibitin 1 (PHB1) Prohibitins (PHB1 and PHB2) are evolutionarily conserved proteins that localize mainly to the mitochondria, where they form ring-shaped complexes on the inner membrane [61]. They have been functionally linked to a wide range of processes that include transcriptional control, senescence, apoptosis and development. In mitochondria, prohibitins have been associated with mitochondrial biogenesis, mitochondrial dynamics, maintenance of mitochondrial membrane and cristae structure, and the assembly and stabilization of mitochondrial respiratory complexes [62]. While not a constitutive component of the electron transport chain, prohibitin 1 (PHB1) reportedly associates with the mitochondrial Complex I and contributes to angiogenesis by amending ROS levels.

In endothelial cells PHB1 is involved in cellular proliferation, migration and angiogenesis by modulating ROS production at mitochondrial complex I. Specifically, knockdown of PHB1 resulted in mitochondrial depolarization and complex I inhibition that triggered ROS generation and endothelial senescence [63]. Mechanistically, these phenotypic consequences were attributed to a sustained activation of Akt and downstream Rac1, leading to cytoskeletal rearrangements that decreased endothelial migration and the ability to form vascular tubes. A similar effect was observed in an in vivo Matrigel angiogenesis assay, where silencing of PHB1 blocked the formation of functional blood vessels [63]. Taken together, it appears that PHB1 prevents ROS-induced senescence and maintains the angiogenic capacity of endothelial cells. The mechanism by which PHB1 controls ROS production at complex I is unclear. However, PHB1 was reported to associate with complex I subunits, suggesting a possible role for PHB1 in complex I assembly or degradation [64]. Due to its chaperone-like function in stabilization of newly synthesized mitochondrial translation products and respiratory-chain complexes assembly [65] lack of PHB1 might destabilize or misassemble complex I and result in inefficient electron flow and ROS generation. In view of the above evidence, it is clear that the ROS-dependent angiogenic outcome strictly depends on the intensity of the pro-oxidant challenge, with moderate ROS signaling supporting vascular growth via activation of pro-angiogenic pathways, and excessive ROS restricting vascularization via cellular senescence or apoptosis that limit endothelial proliferation. Accordingly, a fine-tuned, complex machinery ensures ROS generation and removal, and therapeutic interventions aimed at regulating the oxidant balance (reviewed in Sect. 2.1 of this chapter) are challenging to achieve.

Mitochondrial Metabolism

As research has expanded beyond the concept of mitochondria as largely ROS generators, recent work has brought endothelial metabolism to the forefront as a viable alternative to growth factor signaling anti-angiogenic strategies [11]. As endothelial cells show a predilection towards glycolysis for endothelial growth, migration and proliferation, pharmacological interventions aimed at limiting the glycolytic pathway to influence neovascularization have been the focus of intense study [14, 66]. In contrast to glycolysis, the contribution of endothelial mitochondrial energetics to angiogenesis has been largely disregarded. This lack of study is despite the fact that the mitochondrial oxidation of glucose and other substrates is a major function of mitochondria and a significantly more efficient means of energy production compared to glycolysis. Mitochondria in endothelial cells are also the site for numerous non-energetic functions. Aside from a role in energy production, mitochondrial metabolism is the primary source for the biological building blocks of proliferating cells, regulates the metabolic flow of substrates, detoxifies many harmful byproducts of cellular processes and generates non-ROS signaling molecules. Each of these processes can impact endothelial functions and may play important primary or secondary roles in angiogenic vessel growth.

Carnitine palmitoyltransferase 1A (CPT1A) To date, fatty acid metabolism appears to be the only mitochondrial energetic pathway to support angiogenic proliferation by providing carbons for de novo nucleotide synthesis required for DNA replication [67]. Mitochondrial fatty acid oxidation (β-oxidation) requires the import of long chain fatty acids (LCFA) in the form of long-chain fatty acyl-CoAs (LCFA-CoA) into the mitochondrial matrix via the carnitine palmitoyltransferase system (CPT) [68]. CPT1A is the most abundant endothelial isoform of CPT1, the mitochondrial outer membrane enzymatic constituent of the CPT transport system, and represents the rate-limiting step in mitochondrial β-oxidation [69]. Genetic and pharmacologic inactivation of CPT1A in the vascular endothelium associated with compromised in vitro and in vivo angiogenesis due to impaired endothelial proliferation [70]. CPT1A silencing decreased vessel sprouting in cultured endothelial cells, and endothelial-specific CPT1A conditional knockout mice presented with reduced vascular branching and radial expansion of the retinal vascular plexus. Similar vascular abnormalities were noted upon pharmacological blockade of CPT1 with etomoxir, which reduced pathological angiogenesis in a model of retinopathy of prematurity. CPT1A-mediated fatty acid transport contributes to endothelial proliferation by providing TCA cycle intermediates (aspartate) required for deoxyribonucleotide synthesis during DNA replication independent of ATP generation [70]. This mechanism of deoxyribonucleotide synthesis appears to be primarily restricted to endothelial cells and fibroblasts, and to a lesser extent select lung and breast cancer cell lines. As such, fatty acid metabolism may constitute a particularly promising target to manipulate angiogenesis in the clinic.

Pyruvate dehydrogenase kinase (PDKII) Unlike fatty acid oxidation, mitochondrial glucose metabolism exerts an anti-angiogenic influence on vessel growth by preventing HIF-1α stabilization. Dichloroacetate (DCA) is an inhibitor of mitochondrial pyruvate dehydrogenase kinase (PDK) that activates pyruvate dehydrogenase (PDH) and increases mitochondrial glucose oxidation. Indeed, DCA-mediated activation of mitochondrial oxidative metabolism in glycolytic cancer cells can induce mitochondrial-directed apoptosis and limit tumor growth [71]. Inhibition of PDK by DCA in cancer cells negatively regulated angiogenesis via the inhibition of HIF-1α and downstream expression of the pro-angiogenic genes VEGF and stromal-derived factor 1 (SDF1) [72]. The effect was demonstrated both in vitro, by treating endothelial cells with tumor cell conditioned media, and in vivo, in rat xenotransplant models of non-small cell lung cancer and breast cancer. Mechanistically, DCA inhibited endothelial tube formation and tumor perfusion both by α-ketoglutarate-dependent activation of prolyl-hydroxylases, which stimulated the degradation HIF-1α, and by increasing mitochondrial H_2O_2 production to inhibit HIF-1α transcriptional activity through p53 [72].

Aldehyde dehydrogenase 2 (ALDH2) Independent of energy metabolism, mitochondrial aldehyde detoxification metabolism is another promising target for angiogenic control due to protective effects against cellular oxidative damage. Aldehyde dehydrogenase 2 (ALDH2) is a mitochondrial matrix enzyme that detoxifies numerous intermediate metabolites containing aldehyde functional groups, including acetaldehyde produced during alcohol metabolism [73]. As such, ALDH2 exerts a protective effect on cells by oxidizing both endogenous aldehydic products arising from lipid peroxidation due to oxidative stress, such as 4-hydroxy-2-nonenal (4-HNE) and malondialdehide (MDA), as well as environmental aldehides, such as acrolein [74]. In animal models, the activation of ALDH2 reduces oxidative stress-induced cardiac following ischemic injury [75, 76].

In endothelial cells, exposure to Alzheimer-associated amyloid Aβ protein evoked a decline in ALDH2 enzymatic activity along with marked mitochondrial dysfunction that included membrane potential loss, cytochrome c release, increased superoxide generation, and the accumulation of 4-HNE adducts [77]. Aβ treatment further impaired cell proliferation and the ability to form pseudocapillary networks on Matrigel, compromised barrier function due to disorganization of VE cadherin adherens junctions and tight junction protein ZO-1, and reduced endogenous FGF-2 expression. Pharmacological ALDH2 activation with Alda-1 reversed mitochondrial dysfunction and restored the endothelial angiogenic phenotype [77]. In contrast, deletion of ALDH2 promoted the accumulation of aldehydic adducts and decreased perfusion recovery in a murine ischemic limb model [78]. The suspended tissue repair and regeneration was due to compromised revascularization and perfusion which the investigators attributed to a HIF-1α/VEGF-dependent mechanism. ALDH2 was found to regulate HIF-1α and the accumulation of its downstream angiogenic protein VEGF. ALDH2 deficiency led to unfavorable vascular sequelae including worsened blood supply after ligation of femoral artery, fewer numbers of capillary and small arteries, overt muscular atrophy, as well as reduced endothelial cell proliferation, migration and tube formation in vitro [78]. Should these findings translate to humans, ALDH2 activation may constitute a particularly promising avenue for the treatment of age-related vasculopathies, including those associated with dementia.

3-mercaptopyruvate sulfurtransferase/H_2S pathway Stimulation of the mitochondrial hydrogen sulfide (H_2S) synthesis pathway is also supportive of endothelial angiogenic proliferation. H_2S is an endogenous gas which can induce vasodilation and angiogenesis via the VEGF pathway [79]. H_2S is mainly generated in the mitochondria from 3-mercaptopyruvate (3MP) by 3-mercaptopyruvate sulfurtransferase (3-MST), and the 3-MP/3-MST/H_2S pathway acts as a physiological stimulator of angiogenesis [80]. Specifically, 3-MP application facilitated the migration and proliferation of microvascular endothelial cells by increasing Akt activation, whereas silencing of 3-MST inhibited the proliferative phenotype. 3-MP also elicited a concentration-dependent increase in microvessel sprouting in rat aortic rings and exerted a pro-angiogenic effect in vivo by promoting the neovascularization of Matrigel plug implants and wound closure in rats. Besides activation of Akt-dependent cellular proliferation, H_2S fueled cellular proliferation by ATP generation through stimulation of mitochondrial bioenergetics at complex II of the mitochondrial electron transfer chain [80]. Thus activation of the mitochondrial pathway for endogenous H_2S synthesis can provide an additional control point for therapeutic angiogenesis in wound-healing and post-ischemic therapy.

Mitochondrial Calcium

Calcium is a ubiquitous second messenger that is an absolute requirement for most vascular endothelium functions, including the synthesis of nitric oxide and vWF, control of vascular tone, vascular permeability, initiation of the inflammatory response, endothelial motility and proliferation, and angiogenesis [81]. Specifically, extracellular mitogens and pro-angiogenic factors bind to endothelial surface receptors, triggering an increase in cytosolic calcium originating from both intracellular stores and the extracellular milieu. Mitochondria modulate calcium signaling by buffering excess cytosolic calcium while at the same time responding metabolically to these calcium signals to fine-tune their functions and elicit a cellular response [82]. Mitochondrial calcium trafficking occurs by uptake mainly through the mitochondrial calcium uniporter and release via the Na^+/Ca^{2+} and H^+/Ca^{2+} exchangers and the mitochondrial permeability transition pore (mPTP) [83] (see Chap. 2 for details).

Cyclophilin D (CypD) Cyclophilin D (CypD) is a peptidyl-prolyl *cis-trans* isomerase localized to the mitochondrial matrix, and a regulatory element of calcium-induced mitochondrial permeability transition pore opening [84]. In vivo, CypD deletion coincides with elevated mitochondrial matrix calcium levels and causes reduced cardiac metabolic flexibility and propensity for heart failure, but is protective against ischemia/reperfusion-induced cell death [85, 86]. Our laboratories were the first to demonstrate a role for CypD in angiogenesis through a complex mechanism involving mitochondrial calcium dependent changes in NAD^+/NADH homeostasis and a SIRT1/Akt signaling axis [87]. We identified a similar mitochondrial calcium-directed and NAD^+/NADH-dependent SIRT1 regulatory mechanism that coordinated endothelial inflammation and leukocyte adhesion [88]. NAD^+/NADH balance regulates intracellular redox state and controls enzyme and transcription factor activities in discrete cellular compartments, due to a substantial mitochondrial localization for NAD^+/NADH and mitochondrial membrane impermeability towards NAD^+ and NADH [89]. We showed in cultured primary endothelial cells that genetic inactivation of CypD was accompanied by an increase in mitochondrial matrix calcium and bioenergetics without net ATP generation, which led to mitochondrial accumulation of NADH and decreased NAD^+/NADH ratios in the mitochondrial and cytosolic/nuclear compartments [87]. These changes commanded a reduction in SIRT1 expression and SIRT1-dependent PTEN inactivation, ultimately leading to Akt phosphorylation. This translated as increased endothelial proliferation and sprouting angiogenesis that sped revascularization and wound closure in CypD-deficient mice [87]. While genetic SIRT1 deletion inhibits sprouting angiogenesis and neovessel formation [90] our results demonstrate a negative relationship between SIRT1 and angiogenesis when SIRT1 levels and activity are only diminished by the absence of CypD. Thus this study proposes a new level of angiogenic control involving mitochondrial calcium and NAD^+/NADH metabolism that does not engage mitochondrial ROS signaling.

Mitochondrial Dynamics

For many years considered static structures, mitochondria are now appreciated for their dynamic ability to move within a cell, as well as divide and fuse with neighboring mitochondria. This dynamic nature of mitochondria allows them to adapt to changing intracellular and environmental demands, translocate to areas of energetic need, alter mitochondrial metabolism, and remove damaged or dysfunctional mitochondria [91]. Disruption in these mitochondrial dynamics impairs cell function and has been linked with neurodegenerative diseases [92]. Physiologically, mitochondrial movement along microtubules allows recruitment and clustering of mitochondria at specific subcellular locations, such as in close contact with the endoplasmic reticulum to ensure efficient calcium release and transfer (see Chap. 10). In endothelial cells mitochondrial mobility plays an important role in hypoxia-directed VEGF transcriptional activation and thus might have important consequences for angiogenic initiation. During hypoxia, endothelial mitochondria clustered around the nucleus through a dynein motor-driven and microtubule-dependent mechanism that increases ROS levels within the nucleus [29]. Disruption of mitochondrial clustering suppressed this nuclear ROS increase without affecting HIF-1α accumulation. However, despite accumulation within the nucleus, HIF-1α failed to associate with the hypoxia response element sequences required for VEGF transcription. The observation that antioxidant ROS suppression attenuated mitochondrial nuclear clustering supports the idea that mitochondrial ROS are responsible for initiating mitochondrial movements around the nucleus (see Chap. 1). As such, mitochondrial motility ensures localized ROS delivery and a targeted cellular response.

Mitochondrial fusion and fission, besides defining mitochondria morphology, are important regulators of mitochondrial apoptosis, with fission facilitating apoptosis and fusion playing a protective role against it, possibly by modulating mitochondrial outer membrane permeabilization [93]. Mitochondrial fusion and fission events also serve physiological roles in regulating mitochondrial respiration and cell growth, as perturbations in both fusion and fission impact cell division and therefore are important modulators of angiogenic processes.

Mitofusins (Mfn1 and Mfn2) Mfn1 and Mfn2 are mitochondrial membrane GTPases that mediate fusion events in outer membrane. Experimentally, deletion of either of the two isoforms of Mfn (Mfn1 and Mfn2) impairs mitochondrial fusion and results in exaggerated fission events that present as small, fragmented mitochondria throughout the cell. Global deletion of either Mfn1 or Mfn2 is embryonically lethal in mice [94]. However, in cultured cells and select tissues, mitochondrial fusion defects negatively influence electron transport and respiration, resulting in impaired cell growth [95]. In endothelial cells, the silencing of Mfn1 or Mfn2 decreased VEGF-mediated endothelial migration and the ability of endothelial cells to form vascular network structures on Matrigel. This angiogenic defect was attributed to decreased cellular viability and augmented apoptosis, and was supported by impaired mitochondrial functions and fragmentation of mitochondrial networks. Despite similar angiogenic phenotypes, the two mitofusins rely on different signaling mechanisms: Mfn1 ablation reduces VEGF-stimulated Akt-eNOS signaling, while Mfn2 knockdown significantly decreased ROS generation and reduced mitochondrial respiratory chain and oxidative metabolism gene expression [96]. This data supports an important role for mitochondrial dynamics in endothelial cell viability and angiogenic potential.

Dynamin-related protein 1 (Drp1) In contrast to mitochondrial fission, deletion of the master regulator of mitochondrial fission dynamin-related protein 1 (Drp1) results in a highly interconnected and reticulated mitochondrial network [97]. Drp1 is essential for brain and embryonic development, and like the Mitofusins, Drp1 deletion is embryonically lethal in mice [98]. At the mitochondrial level, Drp1 mutants exhibit defects in endoplasmic reticulum calcium signaling that sensitizes cells to mitochondrial calcium-dependent apoptosis [99], with important implications for cellular proliferation. In cultured pulmonary artery endothelial cells, hypoxia increased Drp1 expression and mitochondrial fragmentation, which promoted migration and proliferation of endothelial cells while inhibiting apoptosis in response to mitochondrial calcium overload [100]. Conversely, Drp1 silencing inhibited hypoxic-mediated angiogenesis and manifested as reduced tube formation and migration, decreased cellular proliferation, and increased apoptosis due to mitochondrial calcium accumulation. Proliferation of endothelial cells plays a key role in excessive pulmonary angiogenesis and endothelial dysfunction. In a murine model of pulmonary arterial hypertension, treatment with the Drp1 inhibitor Midivi-1 reversed the elevated right ventricular systemic pressure and reduced the increase in microvessel density associated with this disease [100]. Hence targeting mitochondrial fragmentation could become a viable alternative for the inhibition of pathologic angiogenesis.

Mitochondrial Apoptosis

The balance between endothelial angiogenesis and vascular regression is important for the development, maintenance and remodeling of the circulatory system. In endothelial cells, apoptosis is a physiological necessity for vessel regression and can significantly limit angiogenesis by counteracting proliferation. As a result, the selective stimulation of endothelial apoptosis may constitute a promising candidate for anti-angiogenic therapy [101]. Mitochondria actively participate in the apoptotic program through the intrinsic apoptotic pathway by releasing pro-apoptotic molecules (cytochrome c, Smac/DIABLO, Omi/HtrA2) that activate downstream caspases and initiate cellular degradation [102].

B cell lymphoma-2 (Bcl-2) Bcl-2 is an anti-apoptotic member of the Bcl-2 family (together with BCL-X_L, BCL-W, MCL1, BCL-B, and pro-apoptotic BAX, BAK, BOK) that localizes to the outer mitochondrial membrane and regulates apoptosis by controlling cytochrome c release [103]. In agreement with an anti-apoptotic role, pharmacological inhibition and siRNA knockdown of Bcl-2 reduced migration and tube formation in endothelial cells and enhanced growth-factor deprivation-induced apoptosis, while Bcl-2 overexpression had the opposite effect [104]. Similarly, Bcl-2 transfection in senescent endothelial cells that displayed markedly diminished Bcl-2 levels reduced mitochondrial oxidative stress by stimulating glutathione translocation into mitochondria, which improved mitochondrial function and amended their impaired angiogenic capacity. Thus, Bcl-2 plays a crucial role in the regulation of the mitochondrial redox state and Bcl-2 loss exacerbates the impaired angiogenesis by augmenting oxidative stress. In vivo, Bcl-2 adenoviral gene transfer significantly improved angiogenesis in Matrigel plugs implanted into aged mice, whereas the Bcl-2 inhibition reduced angiogenesis in Matrigel plugs implanted into young animals [104]. Accordingly, pharmacological control of mitochondrial apoptotic balance is a viable strategy to manipulate angiogenesis, by ensuring either pro-apoptotic or anti-apoptotic outcomes. An overview of the mitochondrial-targeting compounds with pro-apoptotic and anti-angiogenic properties will be presented in a subsequent section.

Mitochondrial Biogenesis

The dynamic regulation of mitochondrial mass to meet changes in cellular energetic demand is under the transcriptional control of DNA-binding factors that include ERRα, ERRβ, NRF-1, NRF-2, CREB, c-MYC, and PPARγ, and several coregulators modulate transcription factor activity, such as PGC-1α, PGC-1β, PRC and RIP140 [105]. Given that mitochondrial energy production is not required for angiogenic growth, whether the increased glycolytic rate in proliferating endothelial cells is influenced by mitochondrial content is not explicit. Nonetheless, the limited evidence available does support a role for mitochondrial biogenesis during angiogenesis. Firstly, the master regulator of mitochondrial biogenesis, PGC-1α was found to be a strong regulator of VEGF expression and angiogenesis via an ERR-α dependent but HIF-1α-independent pathway [106]. Additionally, PGC-1α played a protective role against oxidative stress in the vascular endothelium via positive regulation of the endothelial mitochondrial antioxidant defense system [107]. Secondly, gene array analysis revealed that VEGF stimulation facilitated the upregulation of several nuclear-encoded mitochondrial genes [108]. Mechanistically, the upregulation of mitochondrial genes was dependent upon the expression of Akt3, silencing of which reduced mitochondrial content as detected by decreased mitochondrial DNA levels, gene expression, and respiratory function. Akt3 further directed mitochondrial biogenesis by modulating the subcellular localization of PGC-1α through control of the major nuclear export receptor CRM-1 [109]. Finally, the angiogenic phenotype of Akt3-null and heterozygous mice supports a role for mitochondrial biogenesis in angiogenesis. Akt3 null animals failed to launch an angiogenic response to growth factor challenge and exhibited a dose-dependent decrease in angiogenesis [109]. In addition, Akt3 knockout animals were unable to form mature, stabilized vasculature compared to WT littermates as assessed by Matrigel plug assay, and cells within Matrigel plugs had reduced mitochondrial content during angiogenesis. Similarly, Akt3 deletion evoked larger, less narrow tubules in Matrigel plugs in vitro, suggesting defects in the cell-cell interactions required for angiogenesis [109]. Taken together, these studies demonstrate that the Akt3/ PGC-1α pathway provides a means to coordinate mitochondrial gene expression and function with growth factor-induced angiogenesis. Whether the increase in mitochondrial mass is necessary to support a surge in cellular energy demand, or to provide enhanced oxidative or metabolic signaling to support angiogenic growth requires further investigation.

Mitochondrial-Targeted Compounds with Pro-angiogenic and Anti-angiogenic Properties

Current therapies aimed at manipulating angiogenesis in cancer patients target growth factors and their receptors, but over time are often met with compensatory cellular adaptations and therapeutic resistance [110]. In theory, endothelial mitochondria would be attractive therapeutic targets for modulating angiogenesis to counter drug resistance because of their ability to regulate both endothelial proliferation and death. For instance, mitochondrial agents that block endothelial proliferation or stimulate apoptosis would be desirable therapies to limit cancer progression. On the other hand, mitochondrial-targeted compounds that stimulate endothelial growth could be employed to facilitate new vessel growth to counteract hypertension or support the recovery of organs damaged by ischemia/reperfusion injury. As discussed earlier in this chapter, endothelial mitochondrial function can be manipulated to activate both anti- and pro-angiogenic signaling mechanisms. Below we present an overview of compounds with known effect on mitochondrial functions and implications in angiogenesis control, based on their ability to target the generation/scavenging of mitochondrial ROS, activate/inhibit apoptosis, increase mitochondrial membrane permeability, or disrupt the flow of electrons through the respiratory chain (Table 21.1).

Table 21.1 Mitochondrial-targeted compounds with pro-angiogenic and anti-angiogenic properties

Compound	Target/mechanism	Effect on angiogenesis	Ref.
MitoQ	Antioxidant	Anti-angiogenic	[111, 114]
		Anti-apoptotic	
SKQ1	Antioxidant	Anti-angiogenic	[115, 116–118]
		Pro-angiogenic	
MitoVit-E	Antioxidant	Anti-apoptotic	[114]
Cyclosporin A	Pro-oxidant	Pro-angiogenic	[119]
Matairesinol	Antioxidant	Anti-angiogenic	[112]
Terpestacin	UQCRB	Anti-angiogenic	[55]
HDNT	UQCRB	Anti-angiogenic	[113, 120]
Red wine polyphenols		Pro-angiogenic (low dose)	[121]
Auraptene	Complex I	Anti-angiogenic	[122]
Embelin		Anti-angiogenic	[17]
GSAO	mPTP/ANT	Anti-angiogenic, pro-apoptotic	[123, 124]
PENAO	mPTP	Anti-angiogenic, pro-apoptotic	[124]
α-TOS	Complex II	Anti-angiogenic, pro-apoptotic	[125]
α-TEA	Complex II	Anti-angiogenic, pro-apoptotic	[125]
MitoVES	Complex II	Anti-angiogenic, pro-apoptotic	[126]
Canstatin	$\alpha_v\beta_3$ and $\alpha_v\beta_5$	Anti-angiogenic, pro-apoptotic	[127]
Endostatin	HK2/VDAC1	Anti-angiogenic, pro-apoptotic	[128]
Survivin inhibitor	IAP	Anti-angiogenic, pro-apoptotic	[129]
Paclitaxel	Microtubules	Anti-angiogenic, pro-apoptotic	[130]
Silibilin		Anti-angiogenic, pro-apoptotic	[131]
Neovastat		Anti-angiogenic, pro-apoptotic	[132]
Angiostatin	MDH2 , ATP synthase	Anti-angiogenic, pro-apoptotic	[133, 134, 134, 135]
		Anti-proliferative	
Magnolol		Anti-angiogenic, pro-apoptotic	[136]
Bet-CA		Anti-angiogenic, pro-apoptotic	[137]

Modulators of Mitochondrial ROS Scavenging and ROS Production

The potential therapeutic benefits of enhancing mitochondrial ROS scavenging capacity were evident from the deleterious effects exerted by augmented ROS generation in genetic models lacking mitochondrial antioxidant proteins [31, 42, 46]. In experimental models, compounds that suppress mitochondrial ROS production effectively destabilize HIF-1α and inhibit the downstream expression of pro-angiogenic genes such as VEGF, resulting in strong anti-angiogenic and anti-proliferative effects not only on endothelial cells, but also on the tumor cells that secrete angiogenic growth factors (MitoQ, Matairesinol, Terpestacin, HDNT) [55, 111–113].

On the other hand, compounds that enhance mitochondrial pro-oxidant load facilitate endothelial apoptosis and suppress angiogenesis (α-TOS, α-TEA, MitoVES) [125, 126], and will be reviewed in the section dedicated to pro-apoptotic mitochondrial compounds. However, initial efforts aimed at manipulating/scavenging mitochondrial ROS to prevent and treat degenerative diseases have been largely disappointing, possibly due to poor uptake and delivery of antioxidants specifically to mitochondria [138]. Next generation mitochondrial-targeted antioxidants more effectively target mitochondria by conjugating the lipophilic triphenylphosphonium cation to an antioxidant moiety. Owing to their net positive electrical charge, these compounds permeate through biological membranes and accumulate in high concentrations within the negatively charged mitochondrial matrix [139]. However, despite proven antioxidant properties, mitochondrial-targeted compounds can participate in redox cycling at high concentrations, resulting in electron leaks and increased ROS generation. This pro-oxidant behavior, in particular at high compound concentrations, has detrimental effects on mitochondrial functions and triggers apoptosis, limiting practical in vivo utilization [140, 141]. As a result, despite proven benefits in scavenging mitochondrial ROS, the effects of these compounds are highly dependent upon the overall oxidant load within the endothelium.

MitoQ and MitoVit-E MitoQ is a mitochondrial-targeted antioxidant synthesized by covalent attachment of ubiquinone to a lipophilic triphenylphosphonium cation. The compound accumulates inside energized mitochondria in a membrane-potential dependent manner and shows recyclable antioxidant activity, being regenerated by the respiratory chain [142]. In cultured cells MitoQ prevented lipid oxidation and protected mitochondria against oxidative damage and hydrogen peroxide-induced apoptosis [142]. The antioxidant properties resulted from repressed ROS production during reverse electron transport at mitochondrial complex I, but a pro-oxidant effect was also noted corresponding to forward electron transport [143]. However, the antioxidant effect was prevalent in porcine aorta endothelial cells where MitoQ treatment scavenged the increase in mitochondrial ROS stimulated during hypoxia, and also inhibited the hypoxic- and ROS-dependent endothelial proliferation via MEK/ERK inactivation [111]. MitoQ efficiently buffered glucose/glucose oxidase-induced oxidative stress in bovine aortic endothelial cells and prevented oxidative stress-induced apoptosis by restraining cytochrome-c release, caspase 3 activation and DNA fragmentation [114]. A similar antioxidant and anti-apoptotic effect in endothelial cells was provided through enhanced scavenging of mitochondrial ROS by supplementation with **MitoVit-E** [114], the mitochondrial compound obtained by conjugating the active antioxidant moiety of vitamin E to the lipophilic triphenylphosphonium cation [144].

SkQ1 SkQ1 (10-(6′-plastoquinonyl) decyltriphenylphosphonium) is a potent, membrane permeable, mitochondria-targeted antioxidant containing a plastoquinone attached by a decane linker to positively charged phosphonium. This derivative exerted enhanced antioxidant activity compared to MitoQ along with much lower pro-oxidant tendencies [145]. SkQ1 showed a strong inhibitory effect on tumor growth with a concomitant anti-angiogenic activity on tumor blood vessels [115]. Dietary supplementation of SkQ1 was found to suppress spontaneous tumor development in p53−/− mice and to inhibit growth of human colon carcinoma xenografts by restraining tumor cell proliferation. In vivo angiogenesis in subcutaneous Matrigel implants was also reduced by SkQ1 that decreased the content of small blood vessels in the implants. The anti-angiogenic and anti-tumor properties of SkQ1 possibly involved scavenging of mitochondrial ROS, which stabilize HIF-1α and stimulate of Ras signaling to promote endothelial growth and phenotypic transformation. However, while these results appear promising, SkQ1 does not provide universal benefit at retarding tumor growth in all cancer models. Specifically, SkQ1 proved ineffective at reducing tumor growth and improving survival in a murine pancreatic ductal adenocarcinoma that manifest with increased levels of pro-angiogenic factors both at tumor level (b-cellulin, VEGF) and systemically (angiopoietin, endolgin) [116]. Aside from cancer, SkQ1 increased transcription and translation of VEGF in a rat model of age-related macular degeneration [117]. A pro-angiogenic effect for SkQ1 is also supported by improved dermal wound healing in old mice that consistently showed poor wound recovery [118]. Mechanistically, dietary SkQ1 supplementation resolved inflammation and increased vessel content in the wound to speed dermal healing. In vitro, while SkQ1 did not have a direct effect on endothelial migration in vitro, treatment of endothelial cells with conditioned media from SkQ1-treated fibroblasts promoted migration and tubulogenesis on Matrigel due to elevated TGFb1 [118]. SkQ1 also prevented TNF-induced disorganization of the actin cytoskeleton and VE-Cadherin contacts in the endothelial monolayer, thus preventing the excessive endothelial response to pro-inflammatory cytokines [118].

UQCRB inhibitors Compounds binding the UQCRB subunit of mitochondrial complex III, including terpestacin and HDNT, show strong anti-angiogenic properties based on suppression of hypoxia-induced VEGF signaling [55, 113]. During hypoxia, mitochondria act as cellular oxygen sensors and generate H_2O_2 at complex III to stabilize HIF-1α [146]. UQCRB inhibitors block hypoxia-induced ROS generation at complex III and prevent HIF-1α stabilization, hindering tumor secretion of VEGF and vascular growth. **Terpestacin**, a small molecule identified in a screen of microbial extracts, functions as an angiogenic inhibitor by reducing in vitro and in vivo angiogenesis, as evidenced by compromised vascular tube formation and decreased blood vessel density in tumor xenografts [55]. Mechanistically, terpestacin disrupted mitochondrial membrane potential and decreased ROS generation without affecting mitochondrial respiration [55]. The small molecule **HDNT** (6-((1-hydroxynaphthalen-4-ylamino)dioxysulfone)-2H–naphtho[1,8-bc]thiophen-2-one) was identified by pharmacological screenings as a synthetic inhibitor of UQCRB with anti-angiogenic properties [113]. HDNT selectively inhibited endothelial cell growth and suppressed VEGF-induced endothelial invasion and tube formation without affecting cell viability. Similar to terpestacin, the mechanism involved a dose-dependent decrease in mitochondrial ROS production, which subsequently inhibited HIF-1α accumulation and VEGF expression during hypoxia [113]. Derivatives of HDNT obtained by functionalization with a sulfonylamide backbone were non-toxic and suppressed mitochondrial ROS-mediated hypoxic signaling and inhibited angiogenesis without toxicity [120]. These compounds effectively hindered neovascularization of the chorioallantoic membrane and during xenograft tumor growth by reductions in HIF-1α and VEGF. The antitumor and anti-angiogenic activity of these derivatives was several-fold more pronounced than of the natural UQCRB inhibitor terpestacin [120].

Matairesinol The lignan matairesinol ((αR,βR)-α,β-bis (4-hydroxy-3-methoxybenzyl) butyrolactone) is a natural antioxidant molecule extracted from *Cedrus deodara* [147]. The anti-angiogenic properties of matairesinol are based on in vitro and in vivo suppression of angiogenesis via disruption of hypoxic mitochondrial ROS production [112]. At non-toxic concentrations, matairesinol inhibited endothelial proliferation, chemoinvasion, tube formation, and angiogenesis of the chorioallantoic membrane, which were mechanistically linked to decreased hypoxia-induced HIF-1α expression and VEGF secretion [112].

Cyclosporin A (CsA) CsA is an immunosuppressant drug used for organ transplantation that binds to cyclophilin A and inhibits calcineurin-dependent NFAT activation and immune response. However, long term use of CsA increases cancer incidence, which cannot be fully explained based on the calcineurin-dependent mechanism [148] CsA also binds the mitochondrial cyclophilin D, a modulator of mPTP opening, and protects against calcium- and oxidative stress-induced necrotic cell death [149]. Recent evidence identified a pro-angiogenic role for CsA, independent of calcineurin signaling, which promoted tumor growth due to mitochondrial ROS accumulation [119]. In mice, CsA treatment increased xenograft and carcinogen-induced tumor growth with pronounced microvessel density. Mechanistically, CsA-directed endothelial migration and proliferation were supported by ERK1/2 activation due to increased mitochondrial ROS [119], which may be due to the inhibitory effect of CsA on mPTP opening and resultant ROS accumulation within mitochondria. Both endothelial proliferation and tumor growth were blocked by antioxidant supplementation, further confirming the involvement of mitochondrial ROS in CsA-directed angiogenesis. Alternatively, independent of ROS generation, the pro-angiogenic effect of mPTP inhibition can be mechanistically linked to mitochondrial calcium accumulation and NAD^+/NADH metabolism-directed Akt activation and endothelial proliferation [87].

Compounds Targeting Mitochondrial Respiratory Capacity

Based on the preferential utilization of glycolysis, and despite a close proximity to oxygenated blood, mitochondrial respiration is lower in endothelial cells than most other oxidative cell types [11] where oxidative phosphorylation subsidizes the cellular energetic demand. Nevertheless, reduced endothelial oxygen utilization is advantageous to preserve oxygen for the underlying tissue, minimize endothelial oxidative damage, and sustain angiogenic ability during hypoxia. Interestingly, endothelial mitochondria retain a significant respiratory reserve that can be accessed to cope with oxidative stress [15]. Targeting mitochondrial respiration and respiratory reserve support a key role for mitochondrial function in establishing angiogenic phenotypes. Compounds that fuel endothelial respiratory capacity are pro-angiogenic (red wine polyphenols) [121] while compounds with inhibitory effects on either endothelial respiration or respiratory reserve prevent neovascularization (embelin, auraptene) [17, 122].

Red wine polyphenols Red wine polyphenolic compounds (RWPC) are red wine extracts that exert protective effects on the cardiovascular system and modulate angiogenesis. However, the effects of RWPC on vascular growth are often paradoxal and concentration-dependent: a low dose (0.2 mg/kg body weight) was pro-angiogenic while a high dose (20 mg/kg body weight) was anti-angiogenic in a rat model of post-ischemic neovascularization [150]. When probing the underlying mechanisms of these findings, it is likely that the beneficial effects of low RWPC doses increased mitochondrial respiratory capacity via an upregulated expression of mitochondrial biogenesis-related genes and increased mitochondrial DNA content [121]. Phenotypically, low doses of RWPC further promoted in vitro capillary tube formation and elongation. High doses of RWPC had no effect on mitochondrial capacity, and inhibition of mitochondrial protein synthesis abolished the pro-angiogenic phenotype observed at low RWPC doses [121]. This data advocates the requirement for increased mitochondrial mass and respiratory function during vessel growth to support the synthesis of either metabolic or oxidative intermediates to promote endothelial proliferation.

Embelin The importance of endothelial respiratory reserve for angiogenic growth is demonstrated through the mitochondrial inhibition by embelin. Embelin is a naturally occurring hydrobenzoquinone from the fruit of Embelia ribes, with anti-tumor and anti-inflammatory properties [151]. Embelin acts as an anti-angiogenic agent that impairs neovascularization associated with tumor growth and wound healing without adversely effecting pre-existent blood vessels [17]. Specifically, embelin treatment reduced tumor microvessel density, which limited tumor growth in xenograft mouse models and attenuated wound closure in mice. Mechanistically, embelin preferentially induced cell death in proliferating but not quiescent endothelial cells independent of ROS production by uncoupling mitochondria and depleting the low respiratory reserve of proliferating cells functioning close to their bioenergetic limit [17]. Unlike quiescent cells, proliferating endothelial cells engage in processes that require increased energy demand and metabolic adaptation and therefore they are more susceptible to energetic impairment. Consequently, depletion of the respiratory reserve would make these cells more susceptible to secondary stresses and cell death [15], a feature that can be exploited therapeutically to inhibit pathologic angiogenesis.

Auraptene Suppression of mitochondrial respiration by auraptene is also inhibitory for pathologic angiogenesis. Auraptene is a natural coumarin derivative derived from citrus fruits that negatively targets mitochondrial complex I to limit tumor cell proliferation [152]. The anti-proliferative effect of auraptene on tumor cells was complemented by an inhibitory effect on tumor neovascularization, as witnessed by reduced density and blood vessels infiltration in Matrigel plugs and impaired vascularization of tumor xenografts [122]. Mechanistically, the anti-angiogenic effect of auraptene correlated with reduced VEGF levels due to inhibition of HIF-1α translation [122], although a direct link between suppressed respiratory function and HIF-1α was not established and requires further study.

Pro-apoptotic Mitochondrial Compounds

Several groups of compounds initiate apoptosis either by targeting directly mitochondria (arsenical compounds, vitamin E analogs) or indirectly by engaging cytosolic signaling pathways that terminate at mitochondrial apoptosis (survivin inhibitors, paclitaxel) [153]. Anti-cancer agents that target and destabilize mitochondria have been termed "mitocans" [154], and negatively influence mitochondrial function through multiple mechanisms. Induction of endothelial apoptosis is an efficient way to tamper angiogenesis by counteracting proliferation and preventing vessel growth, to limit tumor progression and metastasis. Both cytotoxic and anti-angiogenic cancer therapies depend on endothelial apoptosis, the first because endothelial apoptosis precedes cancer cell death due to immediate contact with the chemotherapeutic drugs, and the second because direct targeting of the endothelium triggers endothelial apoptosis and reduces tumor vascularization and growth [155]. The mechanisms of endothelial apoptosis include mPTP activation by targeting of regulatory pore components, increased generation of mitochondrial ROS, and suppression of anti-apoptotic protein functions.

Arsenical derivative compounds The trivalent arsenical compound phenylarsine oxide (PAO) is a potent activator of the mPTP but presents high cellular toxicity [156]. In response, a membrane-impermeable, hydrophilic derivative of PAO, **4-(N-(S- glutathionylacetyl)amino) phenylarsonous acid** (**GSAO**), was formulated by conjugation with a cysteine thiol of reduced glutathione, to facilitate less toxic metabolism at the cell surface [123]. The product resulting from the extracellular metabolism of GSAO contains the arsenical moiety and concentrates in the mitochondria to trigger cell death [123, 157]. GSAO has a high selectivity towards proliferating endothelial cells compared to both quiescent endothelial and tumor cells, and strongly inhibits tumor angiogenesis via mPTP-dependent cell death [123]. The anti-angiogenic effect of GSAO was shown in the chick chorioallantoic membrane and solid subcutaneous tumors that presented with reduced microvessel density and tumor growth without organ toxicity [123]. Mechanistically, GSAO targeted the mPTP modulator adenine nucleotide translocator (ANT) to impair mitochondrial and cellular function via superoxide accumulation, loss of mitochondrial membrane potential, mPTP opening, ATP depletion, caspase activation and surface presentation of phosphatidylserine [123]. A GSAO analog in which the arsenical moiety was localized at the *ortho* position on the phenyl ring (***o*-GSAO**) was found to be a more potent inhibitor of endothelial proliferation and tumor growth, but showed toxic side effects due increased cellular accumulation [124]. Similarly, a cysteine mimetic analog of GSAO, **4-(N-(S-penicillaminylacetyl)amino) phenylarsonous acid** (**PENAO**), exhibited faster endothelial accumulation and stronger anti-proliferative and antitumor efficacy following the same apoptotic mechanism of GSAO without obvious signs of cellular and organ toxicity [158].

Vitamin E analogs Vitamin E analogs with pro-apoptotic function are potent anti-cancer agents that selectively trigger apoptosis in cancer cells. These compounds do not possess the antioxidant activity of other vitamin E derivatives and are considered "redox-silent" vitamin E analogs [159]. Whether the derivative will display pro-apoptotic rather then antioxidant behavior depends upon modifications both within the structural moieties of the vitamin E functional domain together with conformational requirements that modulate compound activity [160]. **α-Tocopheryl-succinate (α-TOS)** and **α-tocopheryloxyacetic acid (α-TEA)** are redox-silent, pro-apoptotic vitamin E analogs. In addition to a strong anti-tumor activity, α-TOS and α-TEA exhibit markedly anti-angiogenic properties by triggering apoptosis specifically in proliferating endothelial cells [125]. The apoptotic activity of α-TOS and α-TEA revolves around the ability to inhibit the binding site of ubiquinone at complex II, resulting in the accumulation of mitochondrial ROS [125]. Phenotypically, endothelial apoptosis was responsible for decreased vascular tube formation in vitro, and compromised wound healing and reduced tumor volume and vascularization in vivo [125]. **Mitochondrial-targeted vitamin E succinate** (**MitoVES**) is a potent pro-apoptotic agent obtained upon conjugation of vitamin E succinate with the positively charged triphenylphosphonium group. MitoVES blocks mitochondrial electron transfer from complex II to complex III and interacts with the UbQ-binding site of complex II, enhancing ROS production and triggering ROS-dependent apoptosis [161]. MitoVES showed a strong anti-angiogenic activity via inhibition of endothelial migration, tube formation and neovascularization in models of breast carcinoma and chronic ischemic wound [126]. Noteworthy, the pro-apoptotic effect was specific for proliferating endothelial cells, since contact-arrested cells exhibited limited mitochondrial uptake of MitoVES due to reduced mitochondrial membrane potential.

Collagen-derived compounds Remodeling of the extracellular matrix (ECM) facilitates endothelial migration and proliferation, and proteolytic fragments of ECM known as matrikines are notorious anti-cancer agents and inhibitors of angiogenesis [162]. Collagen-derived matrikines inhibit tumor growth by targeting both tumor and endothelial cells, restraining their migration and proliferation, while triggering apoptosis [163]. **Canstatin** is a 24-kDa peptide derived from the NC1 domain of the $\alpha 2$ chain of type IV collagen. Canstatin inhibited vascular proliferation both in vitro and in vivo by triggering endothelial death via the combined activation of extrinsic (non-mitochondrial) and intrinsic mitochondrial apoptotic pathways [164]. Mechanistically, extrinsic apoptosis was initiated through membrane death receptors and involved Akt-dependent FLIP downregulation, induction of Fas-ligand expression and caspase-8 activation. Intrinsic apoptosis was triggered by canstatin binding to $\alpha_v\beta_3$ and $\alpha_v\beta_5$ integrins on the cell membrane of both endothelial and tumor cells, followed by disruption of mitochondrial membrane potential and caspase-9 activation [127]. **Endostatin** is a 20 kDa C-terminal fragment of collagen XVIII and one of the most potent inhibitors of angiogenesis. First reports on the pro-apoptotic activity of endostatin showed a marked reduction in the expression of anti-apoptotic proteins Bcl-2 and Bcl-X_L that favored endothelial apoptosis [165]. Another mechanism involved mPTP-dependent apoptosis facilitated by upregulation of voltage-dependent anion channel 1 (VDAC1) as a result of endostatin-directed decrease in the glycolytic enzyme hexokinase 2 (HK2) expression [128]. mPTP activation by endostatin triggered endothelial cell death via loss of mitochondrial membrane potential, mitochondrial membrane permeabilization, cytochrome c release and caspase 9 activation [128].

Survivin inhibitors Survivin is a member of the inhibitor of apoptosis (IAPs) family of proteins. While survivin is largely absent in adult tissues its expression is markedly upregulated in most cancers and angiogenic endothelial cells [166]. Because of this cancer- and angiogenic endothelial-selective expression, the survivin pathway is a viable target for anti-tumor and anti-angiogenic treatments. In tumor cells specifically, survivin expression positively regulates angiogenesis by stimulating the synthesis and secretion of pro-angiogenic VEGF [167]. Thus targeting survivin may have a dual anti-cancer effect by inducing tumor cell apoptosis and suppressing angiogenesis via the localized release of angiogenic growth factors. Indeed, inhibition of the survivin pathway by adenoviral expression of a phosphorylation-defective protein mutant exhibited strong anti-cancer activity by triggering both cancer and endothelial cell apoptosis that subsequently blocked tumor growth and reduced tumor blood vessel density in human breast cancer xenograft models [129]. A DNA vaccine targeting survivin delivered against non-small cell lung carcinoma and deriving metastasis induced a strong T-cell mediated anti-tumor response that triggered tumor cell apoptosis and suppression of tumor neovasculature [168]. On the other hand, endothelial survivin expression was pro-angiogenic and induced capillary network formation, supporting a role for survivin in the angiogenic sustenance of tumor growth [129].

Paclitaxel (Taxol) Paclitaxel is a strong chemotherapeutic drug used for the treatment of breast, ovarian and lung cancer, based on its ability to induce mitochondrial apoptosis upon stabilization of microtubules, mitotic check-point activation and mitotic arrest [169]. Paclitaxel also exhibits anti-angiogenic activity that is similarly mediated by mitochondrial apoptotic signaling. Dependent upon concentration, paclitaxel elicited either a cytotoxic effect or a cytostatic effect on endothelial cells by activating apoptosis or inhibiting endothelial proliferation, respectively, that reduced capillary tube formation [130]. The anti-angiogenic activity was exhibited at both cytotoxic and cytostatic concentrations, involving reduced capillary tubes formation. Mechanistically, cytotoxic concentrations involved destabilization of microtubules, G_2-M arrest, mitochondrial membrane potential collapse, Bcl-2 downregulation and apoptosis. Cytostatic concentrations on the other hand, slowed the cell cycle with an early initiation, but overall suppression of apoptosis upstream of mitochondrial permeabilization [130].

Silibilin Silibinin is a flavonolignan isolated form the seeds of *Silybum marianum* (milk thistle). Silibinin has broad anti-cancer efficacy by targeting tumor cell proliferation and apoptosis, and anti-angiogenic activity by targeting of VEGF, VEGF receptors and iNOS [170]. Explicitly, silibinin compromised endothelial capillary tube formation and disrupted capillary networks due to inhibition of endothelial invasion and migration, and decreased matrix metalloproteinase-2 secretion [131]. The mechanism of endothelial growth inhibition involved cell cycle arrest and apoptosis through mitochondrial membrane potential loss, cytochrome c release, increased Bax expression, and the resultant activation of caspase 3/7 [131].

Neovastat (Æ-941) Neovastat is an anti-angiogenic drug derived from marine cartilages. Neovastat inhibits vascular growth by interfering with multiple signaling pathways involved in angiogenesis, including endothelial apoptosis, matrix metalloproteinases signaling and VEGF-mediated signaling [171]. The pro-apoptotic function of neovastat appeared to be specific to endothelial cells and involved mitochondrial cytochrome c release, activation of caspase-3, caspase-8, and caspase-9, chromatin condensation and DNA fragmentation [132]. The activation of both caspase-8 and caspase-9 indicate a possible crosstalk between the extrinsic and intrinsic pathways of apoptosis.

Angiostatin Angiostatin is a 38 kDa proteolytic degradation fragment of plasminogen with strong anti-angiogenic properties. The anti-angiogenic effect of angiostatin results from suppression of tumor cell VEGF secretion [172], and inhibition of endothelial proliferation and tumor neovascularization [173]. Apoptosis contributes significantly to the anti-angiogenic effects of angiostatin with activation of both mitochondrial and surface receptor apoptotic pathways [133]. Downregulation of anti-apoptotic Bcl-2 also provided to decreased viability of angiostatin-treated endothelial cells and tumors [134]. Interestingly, the targeting of endothelial energetic metabolism by angiostatin contributed to its anti-proliferative effect. On the cell surface, angiostatin bound to the plasma membrane-associated α/β-subunits of ATP synthase [135], and upon internalization, localized to mitochondrial malate dehydrogenase (MDH2) and ATP synthase enzymes, reducing intracellular ATP content [134] and limiting the energetic support for angiogenic functions.

Magnolol Magnolol is a bioactive lignan from bark of *Magnolia officinalis* that exerts anti-tumor activity by inhibiting tumor cell proliferation and activating apoptosis [174]. Mechanistically, magnolol activated ROS-mediated mitochondrial apoptosis via enhanced expression of pro-apoptotic Bax, and suppressed the PI3K/AKT/mTOR pro-survival signaling [136]. Magnolol also negatively influences angiogenesis by inhibiting the endothelial expression of PECAM, which is a cell surface protein required for vascular tube formation and sprouting [136].

Bet-CA Bet-CA is a codrug synthetized by the conjugation of the potent anticancer compounds betulinic acid (3β, hydroxyl-lup-20(29)-en-28-oic acid) and dichloroacetate (DCA). The codrug showed synergistic antitumor properties by selectively triggering mitochondrial apoptosis in cancer cells without apparent cytotoxic manifestations on normal tissues in vitro and in vivo [175]. The antitumor mechanism of Bet-CA involved increased ROS production, mitochondrial membrane potential collapse, mPTP opening and mitochondrial swelling, followed by cytochrome c release, caspase activation and apoptosis. The anti-tumor effect of Bet-CA also suppressed angiogenesis, which significantly attenuated tumor growth and metastatic dissemination in a breast cancer model [137]. Diminished tumor vascular development was associated with arrested endothelial proliferation and migration, which also manifested as decreased angiogenic sprouting in mouse aortic rings and restrictive tube formation in the presence of cancerous effluent [137]. Mechanistically, the anti-angiogenic effect of Bet-CA involved the downregulation of MMP-2 and MMP-9 production by tumor cells and inhibition of VEGFR2 expression and VEGF secretion [137].

Conclusion

Proper mitochondrial functioning and enhanced mitochondrial biogenesis are prerequisites for the accurate execution of endothelial angiogenesis. With reduced energetic support towards vascular growth, mitochondria provide the signaling network required to integrate environmental cues such as growth factors, oxygen and substrate availability with endothelial proliferative, migratory and invasive traits. Mitochondrial functions and mitochondrial proteins are actively involved in coordinating endothelial angiogenesis in multiple ways. Mitochondrial proteins involved in ROS homeostasis ensure the proper balance of pro-oxidants and antioxidants required to support endothelial proliferation and migration, and mitochondrial calcium is employed as a proliferative signal. Metabolic intermediates from mitochondrial fatty acids oxidation fuel DNA synthesis during proliferation, while glucose oxidation and aldehyde metabolism intermediates modulate HIF-1α pro-angiogenic signaling. Mitochondrial dynamics are indispensable to angiogenesis by maintaining the appropriate equilibrium between mitochondrial ROS and calcium-dependent endothelial proliferation and apoptosis, and mitochondrial pro-apoptotic and anti-apoptotic machinery set the threshold for endothelial survival. As such, compounds that target or destabilize mitochondrial functions, particularly with respect to mitochondrial ROS and apoptotic signaling, are attractive opportunities for therapeutic control of angiogenesis, both as positive and negative regulators of vascular growth.

References

1. Drake CJ, Fleming PA. Vasculogenesis in the day 6.5 to 9.5 mouse embryo. Blood. 2000;95:1671–9.
2. Betz C, Lenard A, Belting HG, Affolter M. Cell behaviors and dynamics during angiogenesis. Development. 2016;143:2249–60.
3. Chung AS, Lee J, Ferrara N. Targeting the tumour vasculature: insights from physiological angiogenesis. Nat Rev Cancer. 2010;10:505–14.
4. Bloor CM. Angiogenesis during exercise and training. Angiogenesis. 2005;8:263–71.
5. Timar J, Dome B, Fazekas K, Janovics A, Paku S. Angiogenesis-dependent diseases and angiogenesis therapy. Pathol Oncol Res. 2001;7:85–94.
6. Santulli G, Basilicata MF, De Simone M, Del Giudice C, Anastasio A, Sorriento D, Saviano M, Del Gatto A, Trimarco B, Pedone C, Zaccaro L, Iaccarino G. Evaluation of the anti-angiogenic properties of the new selective alphaVbeta3 integrin antagonist RGDechiHCit. J Transl Med. 2011;9:7.
7. Hellstrom M, Phng LK, Hofmann JJ, Wallgard E, Coultas L, Lindblom P, Alva J, Nilsson AK, Karlsson L, Gaiano N, Yoon K, Rossant J, Iruela-Arispe ML, Kalen M, Gerhardt H, Betsholtz C. Dll4 signalling through Notch1 regulates formation of tip cells during angiogenesis. Nature. 2007;445:776–80.
8. Phng LK, Potente M, Leslie JD, Babbage J, Nyqvist D, Lobov I, Ondr JK, Rao S, Lang RA, Thurston G, Gerhardt H. Nrarp coordinates endothelial Notch and Wnt signaling to control vessel density in angiogenesis. Dev Cell. 2009;16:70–82.
9. Adams RH, Alitalo K. Molecular regulation of angiogenesis and lymphangiogenesis. Nat Rev Mol Cell Biol. 2007;8:464–78.
10. Santulli G, Ciccarelli M, Palumbo G, Campanile A, Galasso G, Ziaco B, Altobelli GG, Cimini V, Piscione F, D'Andrea LD, Pedone C, Trimarco B, Iaccarino G. In vivo properties of the proangiogenic peptide QK. J Transl Med. 2009;7:41.
11. De Bock K, Georgiadou M, Schoors S, Kuchnio A, Wong BW, Cantelmo AR, Quaegebeur A, Ghesquiere B, Cauwenberghs S, Eelen G, Phng LK, Betz I, Tembuyser B, Brepoels K, Welti J, Geudens I, Segura I, Cruys B, Bifari F, Decimo I, Blanco R, Wyns S, Vangindertael J, Rocha S, Collins RT, Munck S, Daelemans D, Imamura H, Devlieger R, Rider M, Van Veldhoven PP, Schuit F, Bartrons R, Hofkens J, Fraisl P, Telang S, Deberardinis RJ, Schoonjans L, Vinckier S, Chesney J, Gerhardt H, Dewerchin M, Carmeliet P. Role of PFKFB3-driven glycolysis in vessel sprouting. Cell. 2013;154:651–63.
12. Oldendorf WH, Cornford ME, Brown WJ. The large apparent work capability of the blood-brain barrier: a study of the mitochondrial content of capillary endothelial cells in brain and other tissues of the rat. Ann Neurol. 1977;1:409–17.
13. Barth E, Stammler G, Speiser B, Schaper J. Ultrastructural quantitation of mitochondria and myofilaments in cardiac muscle from 10 different animal species including man. J Mol Cell Cardiol. 1992;24:669–81.
14. Schoors S, De Bock K, Cantelmo AR, Georgiadou M, Ghesquiere B, Cauwenberghs S, Kuchnio A, Wong BW, Quaegebeur A, Goveia J, Bifari F, Wang X, Blanco R, Tembuyser B, Cornelissen I, Bouche A, Vinckier S, Diaz-Moralli S, Gerhardt H, Telang S, Cascante M, Chesney J, Dewerchin M, Carmeliet P. Partial and transient reduction of glycolysis by PFKFB3 blockade reduces pathological angiogenesis. Cell Metab. 2014;19:37–48.
15. Dranka BP, Hill BG, Darley-Usmar VM. Mitochondrial reserve capacity in endothelial cells: the impact of nitric oxide and reactive oxygen species. Free Radic Biol Med. 2010;48:905–14.
16. Krutzfeldt A, Spahr R, Mertens S, Siegmund B, Piper HM. Metabolism of exogenous substrates by coronary endothelial cells in culture. J Mol Cell Cardiol. 1990;22:1393–404.
17. Coutelle O, Hornig-Do HT, Witt A, Andree M, Schiffmann LM, Piekarek M, Brinkmann K, Seeger JM, Liwschitz M, Miwa S, Hallek M, Kronke M, Trifunovic A, Eming SA, Wiesner RJ, Hacker UT, Kashkar H. Embelin inhibits endothelial mitochondrial respiration and

impairs neoangiogenesis during tumor growth and wound healing. EMBO Mol Med. 2014;6:624–39.
18. Fukumura D, Gohongi T, Kadambi A, Izumi Y, Ang J, Yun CO, Buerk DG, Huang PL, Jain RK. Predominant role of endothelial nitric oxide synthase in vascular endothelial growth factor-induced angiogenesis and vascular permeability. Proc Natl Acad Sci U S A. 2001;98:2604–9.
19. Cleeter MW, Cooper JM, Darley-Usmar VM, Moncada S, Schapira AH. Reversible inhibition of cytochrome c oxidase, the terminal enzyme of the mitochondrial respiratory chain, by nitric oxide. Implications for neurodegenerative diseases. FEBS Lett. 1994;345:50–4.
20. Clementi E, Brown GC, Foxwell N, Moncada S. On the mechanism by which vascular endothelial cells regulate their oxygen consumption. Proc Natl Acad Sci U S A. 1999;96:1559–62.
21. Quintero M, Colombo SL, Godfrey A, Moncada S. Mitochondria as signaling organelles in the vascular endothelium. Proc Natl Acad Sci U S A. 2006;103:5379–84.
22. Davidson SM, Duchen MR. Endothelial mitochondria: contributing to vascular function and disease. Circ Res. 2007;100:1128–41.
23. Xie W, Santulli G, Reiken SR, Yuan Q, Osborne BW, Chen BX, Marks AR. Mitochondrial oxidative stress promotes atrial fibrillation. Sci Rep. 2015;5:11427.
24. Zhang DX, Gutterman DD. Mitochondrial reactive oxygen species-mediated signaling in endothelial cells. Am J Physiol Heart Circ Physiol. 2007;292:H2023–31.
25. Hawkins BJ, Solt LA, Chowdhury I, Kazi AS, Abid MR, Aird WC, May MJ, Foskett JK, Madesh M. G protein-coupled receptor Ca2+-linked mitochondrial reactive oxygen species are essential for endothelial/leukocyte adherence. Mol Cell Biol. 2007;27:7582–93.
26. Murphy MP. How mitochondria produce reactive oxygen species. Biochem J. 2009;417:1–13.
27. Giulivi C, Boveris A, Cadenas E. Hydroxyl radical generation during mitochondrial electron transfer and the formation of 8-hydroxydesoxyguanosine in mitochondrial DNA. Arch Biochem Biophys. 1995;316:909–16.
28. Wang Y, Zang QS, Liu Z, Wu Q, Maass D, Dulan G, Shaul PW, Melito L, Frantz DE, Kilgore JA, Williams NS, Terada LS, Nwariaku FE. Regulation of VEGF-induced endothelial cell migration by mitochondrial reactive oxygen species. Am J Phys Cell Phys. 2011;301:C695–704.
29. Al-Mehdi AB, Pastukh VM, Swiger BM, Reed DJ, Patel MR, Bardwell GC, Pastukh VV, Alexeyev MF, Gillespie MN. Perinuclear mitochondrial clustering creates an oxidant-rich nuclear domain required for hypoxia-induced transcription. Sci Signal. 2012;5:ra47.
30. Chandel NS, McClintock DS, Feliciano CE, Wood TM, Melendez JA, Rodriguez AM, Schumacker PT. Reactive oxygen species generated at mitochondrial complex III stabilize hypoxia-inducible factor-1alpha during hypoxia: a mechanism of O2 sensing. J Biol Chem. 2000;275:25130–8.
31. Li Y, Huang TT, Carlson EJ, Melov S, Ursell PC, Olson JL, Noble LJ, Yoshimura MP, Berger C, Chan PH, Wallace DC, Epstein CJ. Dilated cardiomyopathy and neonatal lethality in mutant mice lacking manganese superoxide dismutase. Nat Genet. 1995;11:376–81.
32. Abid MR, Tsai JC, Spokes KC, Deshpande SS, Irani K, Aird WC. Vascular endothelial growth factor induces manganese-superoxide dismutase expression in endothelial cells by a Rac1-regulated NADPH oxidase-dependent mechanism. FASEB J. 2001;15:2548–50.
33. Suzuki K, Tatsumi H, Satoh S, Senda T, Nakata T, Fujii J, Taniguchi N. Manganese-superoxide dismutase in endothelial cells: localization and mechanism of induction. Am J Phys. 1993;265:H1173–8.
34. Dikalova AE, Bikineyeva AT, Budzyn K, Nazarewicz RR, McCann L, Lewis W, Harrison DG, Dikalov SI. Therapeutic targeting of mitochondrial superoxide in hypertension. Circ Res. 2010;107:106–16.
35. Fang X, Weintraub NL, Rios CD, Chappell DA, Zwacka RM, Engelhardt JF, Oberley LW, Yan T, Heistad DD, Spector AA. Overexpression of human superoxide dismutase inhibits oxidation of low-density lipoprotein by endothelial cells. Circ Res. 1998;82:1289–97.

36. Madsen-Bouterse SA, Zhong Q, Mohammad G, Ho YS, Kowluru RA. Oxidative damage of mitochondrial DNA in diabetes and its protection by manganese superoxide dismutase. Free Radic Res. 2010;44:313–21.
37. Afolayan AJ, Eis A, Teng RJ, Bakhutashvili I, Kaul S, Davis JM, Konduri GG. Decreases in manganese superoxide dismutase expression and activity contribute to oxidative stress in persistent pulmonary hypertension of the newborn. Am J Phys Lung Cell Mol Phys. 2012;303:L870–9.
38. Connor KM, Subbaram S, Regan KJ, Nelson KK, Mazurkiewicz JE, Bartholomew PJ, Aplin AE, Tai YT, Aguirre-Ghiso J, Flores SC, Melendez JA. Mitochondrial H2O2 regulates the angiogenic phenotype via PTEN oxidation. J Biol Chem. 2005;280:16916–24.
39. Zhou Q, Gensch C, Keller C, Schmitt H, Esser J, Moser M, Liao JK. MnTBAP stimulates angiogenic functions in endothelial cells through mitofusin-1. Vasc Pharmacol. 2015;72:163–71.
40. Marrotte EJ, Chen DD, Hakim JS, Chen AF. Manganese superoxide dismutase expression in endothelial progenitor cells accelerates wound healing in diabetic mice. J Clin Invest. 2010;120:4207–19.
41. Wang XR, Zhang MW, Chen DD, Zhang Y, Chen AF. AMP-activated protein kinase rescues the angiogenic functions of endothelial progenitor cells via manganese superoxide dismutase induction in type 1 diabetes. Am J Physiol Endocrinol Metab. 2011;300:E1135–45.
42. Nonn L, Williams RR, Erickson RP, Powis G. The absence of mitochondrial thioredoxin 2 causes massive apoptosis, exencephaly, and early embryonic lethality in homozygous mice. Mol Cell Biol. 2003;23:916–22.
43. Zhang H, Luo Y, Zhang W, He Y, Dai S, Zhang R, Huang Y, Bernatchez P, Giordano FJ, Shadel G, Sessa WC, Min W. Endothelial-specific expression of mitochondrial thioredoxin improves endothelial cell function and reduces atherosclerotic lesions. Am J Pathol. 2007;170:1108–20.
44. Dai S, He Y, Zhang H, Yu L, Wan T, Xu Z, Jones D, Chen H, Min W. Endothelial-specific expression of mitochondrial thioredoxin promotes ischemia-mediated arteriogenesis and angiogenesis. Arterioscler Thromb Vasc Biol. 2009;29:495–502.
45. Price DT, Vita JA, Keaney Jr JF. Redox control of vascular nitric oxide bioavailability. Antioxid Redox Signal. 2000;2:919–35.
46. Conrad M, Jakupoglu C, Moreno SG, Lippl S, Banjac A, Schneider M, Beck H, Hatzopoulos AK, Just U, Sinowatz F, Schmahl W, Chien KR, Wurst W, Bornkamm GW, Brielmeier M. Essential role for mitochondrial thioredoxin reductase in hematopoiesis, heart development, and heart function. Mol Cell Biol. 2004;24:9414–23.
47. Sugiyama T, Michel T. Thiol-metabolizing proteins and endothelial redox state: differential modulation of eNOS and biopterin pathways. Am J Physiol Heart Circ Physiol. 2010;298:H194–201.
48. Hellfritsch J, Kirsch J, Schneider M, Fluege T, Wortmann M, Frijhoff J, Dagnell M, Fey T, Esposito I, Kolle P, Pogoda K, Angeli JP, Ingold I, Kuhlencordt P, Ostman A, Pohl U, Conrad M, Beck H. Knockout of mitochondrial thioredoxin reductase stabilizes prolyl hydroxylase 2 and inhibits tumor growth and tumor-derived angiogenesis. Antioxid Redox Signal. 2015;22:938–50.
49. Brand MD, Esteves TC. Physiological functions of the mitochondrial uncoupling proteins UCP2 and UCP3. Cell Metab. 2005;2:85–93.
50. Mattiasson G, Sullivan PG. The emerging functions of UCP2 in health, disease, and therapeutics. Antioxid Redox Signal. 2006;8:1–38.
51. Shimasaki Y, Pan N, Messina LM, Li C, Chen K, Liu L, Cooper MP, Vita JA, Keaney Jr JF. Uncoupling protein 2 impacts endothelial phenotype via p53-mediated control of mitochondrial dynamics. Circ Res. 2013;113:891–901.
52. Xu MJ, Song P, Shirwany N, Liang B, Xing J, Viollet B, Wang X, Zhu Y, Zou MH. Impaired expression of uncoupling protein 2 causes defective postischemic angiogenesis in mice deficient in AMP-activated protein kinase alpha subunits. Arterioscler Thromb Vasc Biol. 2011;31:1757–65.

53. Nagata D, Mogi M, Walsh K. AMP-activated protein kinase (AMPK) signaling in endothelial cells is essential for angiogenesis in response to hypoxic stress. J Biol Chem. 2003;278:31000–6.
54. Haut S, Brivet M, Touati G, Rustin P, Lebon S, Garcia-Cazorla A, Saudubray JM, Boutron A, Legrand A, Slama A. A deletion in the human QP-C gene causes a complex III deficiency resulting in hypoglycaemia and lactic acidosis. Hum Genet. 2003;113:118–22.
55. Jung HJ, Shim JS, Lee J, Song YM, Park KC, Choi SH, Kim ND, Yoon JH, Mungai PT, Schumacker PT, Kwon HJ. Terpestacin inhibits tumor angiogenesis by targeting UQCRB of mitochondrial complex III and suppressing hypoxia-induced reactive oxygen species production and cellular oxygen sensing. J Biol Chem. 2010;285:11584–95.
56. Jung HJ, Kim Y, Chang J, Kang SW, Kim JH, Kwon HJ. Mitochondrial UQCRB regulates VEGFR2 signaling in endothelial cells. J Mol Med (Berlin). 2013;91:1117–28.
57. Cho YS, Jung HJ, Seok SH, Payumo AY, Chen JK, Kwon HJ. Functional inhibition of UQCRB suppresses angiogenesis in zebrafish. Biochem Biophys Res Commun. 2013;433:396–400.
58. Chang J, Jung HJ, Jeong SH, Kim HK, Han J, Kwon HJ. A mutation in the mitochondrial protein UQCRB promotes angiogenesis through the generation of mitochondrial reactive oxygen species. Biochem Biophys Res Commun. 2014;455:290–7.
59. Chang J, Jung HJ, Park HJ, Cho SW, Lee SK, Kwon HJ. Cell-permeable mitochondrial ubiquinol-cytochrome c reductase binding protein induces angiogenesis in vitro and in vivo. Cancer Lett. 2015;366:52–60.
60. Bleier L, Drose S. Superoxide generation by complex III: from mechanistic rationales to functional consequences. Biochim Biophys Acta. 2013;1827:1320–31.
61. Tatsuta T, Model K, Langer T. Formation of membrane-bound ring complexes by prohibitins in mitochondria. Mol Biol Cell. 2005;16:248–59.
62. Artal-Sanz M, Tavernarakis N. Prohibitin and mitochondrial biology. Trends Endocrinol Metab. 2009;20:394–401.
63. Schleicher M, Shepherd BR, Suarez Y, Fernandez-Hernando C, Yu J, Pan Y, Acevedo LM, Shadel GS, Sessa WC. Prohibitin-1 maintains the angiogenic capacity of endothelial cells by regulating mitochondrial function and senescence. J Cell Biol. 2008;180:101–12.
64. Bourges I, Ramus C, Mousson de Camaret B, Beugnot R, Remacle C, Cardol P, Hofhaus G, Issartel JP. Structural organization of mitochondrial human complex I: role of the ND4 and ND5 mitochondria-encoded subunits and interaction with prohibitin. Biochem J. 2004;383:491–9.
65. Nijtmans LG, de Jong L, Artal Sanz M, Coates PJ, Berden JA, Back JW, Muijsers AO, van der Spek H, Grivell LA. Prohibitins act as a membrane-bound chaperone for the stabilization of mitochondrial proteins. EMBO J. 2000;19:2444–51.
66. Goveia J, Stapor P, Carmeliet P. Principles of targeting endothelial cell metabolism to treat angiogenesis and endothelial cell dysfunction in disease. EMBO Mol Med. 2014;6:1105–20.
67. Harjes U, Kalucka J, Carmeliet P. Targeting fatty acid metabolism in cancer and endothelial cells. Crit Rev Oncol Hematol. 2016;97:15–21.
68. Sherratt HS, Spurway TD. Regulation of fatty acid oxidation in cells. Biochem Soc Trans. 1994;22:423–7.
69. Bonnefont JP, Djouadi F, Prip-Buus C, Gobin S, Munnich A, Bastin J. Carnitine palmitoyltransferases 1 and 2: biochemical, molecular and medical aspects. Mol Asp Med. 2004;25:495–520.
70. Schoors S, Bruning U, Missiaen R, Queiroz KC, Borgers G, Elia I, Zecchin A, Cantelmo AR, Christen S, Goveia J, Heggermont W, Godde L, Vinckier S, Van Veldhoven PP, Eelen G, Schoonjans L, Gerhardt H, Dewerchin M, Baes M, De Bock K, Ghesquiere B, Lunt SY, Fendt SM, Carmeliet P. Fatty acid carbon is essential for dNTP synthesis in endothelial cells. Nature. 2015;520:192–7.
71. Michelakis ED, Webster L, Mackey JR. Dichloroacetate (DCA) as a potential metabolic-targeting therapy for cancer. Br J Cancer. 2008;99:989–94.

72. Sutendra G, Dromparis P, Kinnaird A, Stenson TH, Haromy A, Parker JM, McMurtry MS, Michelakis ED. Mitochondrial activation by inhibition of PDKII suppresses HIF1a signaling and angiogenesis in cancer. Oncogene. 2013;32:1638–50.
73. Klyosov AA, Rashkovetsky LG, Tahir MK, Keung WM. Possible role of liver cytosolic and mitochondrial aldehyde dehydrogenases in acetaldehyde metabolism. Biochemistry. 1996;35:4445–56.
74. Chen CH, Ferreira JC, Gross ER, Mochly-Rosen D. Targeting aldehyde dehydrogenase 2: new therapeutic opportunities. Physiol Rev. 2014;94:1–34.
75. Chen CH, Budas GR, Churchill EN, Disatnik MH, Hurley TD, Mochly-Rosen D. Activation of aldehyde dehydrogenase-2 reduces ischemic damage to the heart. Science. 2008;321:1493–5.
76. Gomes KM, Campos JC, Bechara LR, Queliconi B, Lima VM, Disatnik MH, Magno P, Chen CH, Brum PC, Kowaltowski AJ, Mochly-Rosen D, Ferreira JC. Aldehyde dehydrogenase 2 activation in heart failure restores mitochondrial function and improves ventricular function and remodelling. Cardiovasc Res. 2014;103:498–508.
77. Solito R, Corti F, Chen CH, Mochly-Rosen D, Giachetti A, Ziche M, Donnini S. Mitochondrial aldehyde dehydrogenase-2 activation prevents beta-amyloid-induced endothelial cell dysfunction and restores angiogenesis. J Cell Sci. 2013;126:1952–61.
78. Liu X, Sun X, Liao H, Dong Z, Zhao J, Zhu H, Wang P, Shen L, Xu L, Ma X, Shen C, Fan F, Wang C, Hu K, Zou Y, Ge J, Ren J, Sun A. Mitochondrial aldehyde dehydrogenase 2 regulates revascularization in chronic ischemia: potential impact on the development of coronary collateral circulation. Arterioscler Thromb Vasc Biol. 2015;35:2196–206.
79. Holwerda KM, Karumanchi SA, Lely AT. Hydrogen sulfide: role in vascular physiology and pathology. Curr Opin Nephrol Hypertens. 2015;24:170–6.
80. Coletta C, Modis K, Szczesny B, Brunyanszki A, Olah G, Rios EC, Yanagi K, Ahmad A, Papapetropoulos A, Szabo C. Regulation of vascular tone, angiogenesis and cellular bioenergetics by the 3-mercaptopyruvate sulfurtransferase/H2S pathway: functional impairment by hyperglycemia and restoration by DL-alpha-lipoic acid. Mol Med. 2015;21:1–14.
81. Tran QK, Ohashi K, Watanabe H. Calcium signalling in endothelial cells. Cardiovasc Res. 2000;48:13–22.
82. Santulli G, Xie W, Reiken SR, Marks AR. Mitochondrial calcium overload is a key determinant in heart failure. Proc Natl Acad Sci USA. 2015;112:11389–94.
83. Rizzuto R, Marchi S, Bonora M, Aguiari P, Bononi A, De Stefani D, Giorgi C, Leo S, Rimessi A, Siviero R, Zecchini E, Pinton P. Ca(2+) transfer from the ER to mitochondria: when, how and why. Biochim Biophys Acta. 2009;1787:1342–51.
84. Basso E, Fante L, Fowlkes J, Petronilli V, Forte MA, Bernardi P. Properties of the permeability transition pore in mitochondria devoid of Cyclophilin D. J Biol Chem. 2005;280:18558–61.
85. Elrod JW, Wong R, Mishra S, Vagnozzi RJ, Sakthievel B, Goonasekera SA, Karch J, Gabel S, Farber J, Force T, Brown JH, Murphy E, Molkentin JD. Cyclophilin D controls mitochondrial pore-dependent Ca(2+) exchange, metabolic flexibility, and propensity for heart failure in mice. J Clin Invest. 2010;120:3680–7.
86. Alam MR, Baetz D, Ovize M. Cyclophilin D and myocardial ischemia-reperfusion injury: a fresh perspective. J Mol Cell Cardiol. 2015;78:80–9.
87. Marcu R, Kotha S, Zhi Z, Qin W, Neeley CK, Wang RK, Zheng Y, Hawkins BJ. The mitochondrial permeability transition pore regulates endothelial bioenergetics and angiogenesis. Circ Res. 2015;116:1336–45.
88. Marcu R, Wiczer BM, Neeley CK, Hawkins BJ. Mitochondrial matrix Ca(2)(+) accumulation regulates cytosolic NAD(+)/NADH metabolism, protein acetylation, and sirtuin expression. Mol Cell Biol. 2014;34:2890–902.
89. Stein LR, Imai S. The dynamic regulation of NAD metabolism in mitochondria. Trends Endocrinol Metab. 2012;23:420–8.
90. Potente M, Ghaeni L, Baldessari D, Mostoslavsky R, Rossig L, Dequiedt F, Haendeler J, Mione M, Dejana E, Alt FW, Zeiher AM, Dimmeler S. SIRT1 controls endothelial angiogenic functions during vascular growth. Genes Dev. 2007;21:2644–58.

91. Chen H, Chan DC. Mitochondrial dynamics – fusion, fission, movement, and mitophagy – in neurodegenerative diseases. Hum Mol Genet. 2009;18:R169–76.
92. Itoh K, Nakamura K, Iijima M, Sesaki H. Mitochondrial dynamics in neurodegeneration. Trends Cell Biol. 2013;23:64–71.
93. Chen H, Chan DC. Emerging functions of mammalian mitochondrial fusion and fission. Hum Mol Genet. 2005;14(2):R283–9.
94. Chen H, Detmer SA, Ewald AJ, Griffin EE, Fraser SE, Chan DC. Mitofusins Mfn1 and Mfn2 coordinately regulate mitochondrial fusion and are essential for embryonic development. J Cell Biol. 2003;160:189–200.
95. Chen H, Chomyn A, Chan DC. Disruption of fusion results in mitochondrial heterogeneity and dysfunction. J Biol Chem. 2005;280:26185–92.
96. Lugus JJ, Ngoh GA, Bachschmid MM, Walsh K. Mitofusins are required for angiogenic function and modulate different signaling pathways in cultured endothelial cells. J Mol Cell Cardiol. 2011;51:885–93.
97. Mopert K, Hajek P, Frank S, Chen C, Kaufmann J, Santel A. Loss of Drp1 function alters OPA1 processing and changes mitochondrial membrane organization. Exp Cell Res. 2009;315:2165–80.
98. Ishihara N, Nomura M, Jofuku A, Kato H, Suzuki SO, Masuda K, Otera H, Nakanishi Y, Nonaka I, Goto Y, Taguchi N, Morinaga H, Maeda M, Takayanagi R, Yokota S, Mihara K. Mitochondrial fission factor Drp1 is essential for embryonic development and synapse formation in mice. Nat Cell Biol. 2009;11:958–66.
99. Elgass K, Pakay J, Ryan MT, Palmer CS. Recent advances into the understanding of mitochondrial fission. Biochim Biophys Acta. 1833;2013:150–61.
100. Shen T, Wang N, Yu X, Shi J, Li Q, Zhang C, Fu L, Wang S, Xing Y, Zheng X, Yu L, Zhu D. The critical role of dynamin-related protein 1 in hypoxia-induced pulmonary vascular angiogenesis. J Cell Biochem. 2015;116:1993–2007.
101. Dimmeler S, Zeiher AM. Endothelial cell apoptosis in angiogenesis and vessel regression. Circ Res. 2000;87:434–9.
102. Wang C, Youle RJ. The role of mitochondria in apoptosis*. Annu Rev Genet. 2009;43:95–118.
103. Czabotar PE, Lessene G, Strasser A, Adams JM. Control of apoptosis by the BCL-2 protein family: implications for physiology and therapy. Nat Rev Mol Cell Biol. 2014;15:49–63.
104. Uraoka M, Ikeda K, Kurimoto-Nakano R, Nakagawa Y, Koide M, Akakabe Y, Kitamura Y, Ueyama T, Matoba S, Yamada H, Okigaki M, Matsubara H. Loss of bcl-2 during the senescence exacerbates the impaired angiogenic functions in endothelial cells by deteriorating the mitochondrial redox state. Hypertension. 2011;58:254–63.
105. Hock MB, Kralli A. Transcriptional control of mitochondrial biogenesis and function. Annu Rev Physiol. 2009;71:177–203.
106. Arany Z, Foo SY, Ma Y, Ruas JL, Bommi-Reddy A, Girnun G, Cooper M, Laznik D, Chinsomboon J, Rangwala SM, Baek KH, Rosenzweig A, Spiegelman BM. HIF-independent regulation of VEGF and angiogenesis by the transcriptional coactivator PGC-1alpha. Nature. 2008;451:1008–12.
107. Valle I, Alvarez-Barrientos A, Arza E, Lamas S, Monsalve M. PGC-1alpha regulates the mitochondrial antioxidant defense system in vascular endothelial cells. Cardiovasc Res. 2005;66:562–73.
108. Wright GL, Maroulakou IG, Eldridge J, Liby TL, Sridharan V, Tsichlis PN, Muise-Helmericks RC. VEGF stimulation of mitochondrial biogenesis: requirement of AKT3 kinase. FASEB J. 2008;22:3264–75.
109. Corum DG, Tsichlis PN, Muise-Helmericks RC. AKT3 controls mitochondrial biogenesis and autophagy via regulation of the major nuclear export protein CRM-1. FASEB J. 2014;28:395–407.
110. Dey N, De P, Brian LJ. Evading anti-angiogenic therapy: resistance to anti-angiogenic therapy in solid tumors. Am J Transl Res. 2015;7:1675–98.
111. Schafer M, Schafer C, Ewald N, Piper HM, Noll T. Role of redox signaling in the autonomous proliferative response of endothelial cells to hypoxia. Circ Res. 2003;92:1010–5.

112. Lee B, Kim KH, Jung HJ, Kwon HJ. Matairesinol inhibits angiogenesis via suppression of mitochondrial reactive oxygen species. Biochem Biophys Res Commun. 2012;421:76–80.
113. Jung HJ, Kim KH, Kim ND, Han G, Kwon HJ. Identification of a novel small molecule targeting UQCRB of mitochondrial complex III and its anti-angiogenic activity. Bioorg Med Chem Lett. 2011;21:1052–6.
114. Dhanasekaran A, Kotamraju S, Kalivendi SV, Matsunaga T, Shang T, Keszler A, Joseph J, Kalyanaraman B. Supplementation of endothelial cells with mitochondria-targeted antioxidants inhibit peroxide-induced mitochondrial iron uptake, oxidative damage, and apoptosis. J Biol Chem. 2004;279:37575–87.
115. Agapova LS, Chernyak BV, Domnina LV, Dugina VB, Efimenko AY, Fetisova EK, Ivanova OY, Kalinina NI, Khromova NV, Kopnin BP, Kopnin PB, Korotetskaya MV, Lichinitser MR, Lukashev AL, Pletjushkina OY, Popova EN, Skulachev MV, Shagieva GS, Stepanova EV, Titova EV, Tkachuk VA, Vasiliev JM, Skulachev VP. Mitochondria-targeted plastoquinone derivatives as tools to interrupt execution of the aging program. 3. Inhibitory effect of SkQ1 on tumor development from p53-deficient cells. Biochemistry (Mosc). 2008;73:1300–16.
116. Bazhin AV, Yang Y, D'Haese JG, Werner J, Philippov PP, Karakhanova S. The novel mitochondria-targeted antioxidant SkQ1 modulates angiogenesis and inflammatory micromilieu in a murine orthotopic model of pancreatic cancer. Int J Cancer. 2016;139:130–9.
117. Markovets AM, Fursova AZ, Kolosova NG. Therapeutic action of the mitochondria-targeted antioxidant SkQ1 on retinopathy in OXYS rats linked with improvement of VEGF and PEDF gene expression. PLoS One. 2011;6:e21682.
118. Demyanenko IA, Popova EN, Zakharova VV, Ilyinskaya OP, Vasilieva TV, Romashchenko VP, Fedorov AV, Manskikh VN, Skulachev MV, Zinovkin RA, Pletjushkina OY, Skulachev VP, Chernyak BV. Mitochondria-targeted antioxidant SkQ1 improves impaired dermal wound healing in old mice. Aging (Albany NY). 2015;7:475–85.
119. Zhou AY, Ryeom S. Cyclosporin A promotes tumor angiogenesis in a calcineurin-independent manner by increasing mitochondrial reactive oxygen species. Mol Cancer Res. 2014;12:1663–76.
120. Jung HJ, Cho M, Kim Y, Han G, Kwon HJ. Development of a novel class of mitochondrial ubiquinol-cytochrome c reductase binding protein (UQCRB) modulators as promising anti-angiogenic leads. J Med Chem. 2014;57:7990–8.
121. Duluc L, Jacques C, Soleti R, Iacobazzi F, Simard G, Andriantsitohaina R. Modulation of mitochondrial capacity and angiogenesis by red wine polyphenols via estrogen receptor, NADPH oxidase and nitric oxide synthase pathways. Int J Biochem Cell Biol. 2013;45:783–91.
122. Jang Y, Han J, Kim SJ, Kim J, Lee MJ, Jeong S, Ryu MJ, Seo KS, Choi SY, Shong M, Lim K, Heo JY, Kweon GR. Suppression of mitochondrial respiration with auraptene inhibits the progression of renal cell carcinoma: involvement of HIF-1alpha degradation. Oncotarget. 2015;6:38127–38.
123. Don AS, Kisker O, Dilda P, Donoghue N, Zhao X, Decollogne S, Creighton B, Flynn E, Folkman J, Hogg PJ. A peptide trivalent arsenical inhibits tumor angiogenesis by perturbing mitochondrial function in angiogenic endothelial cells. Cancer Cell. 2003;3:497–509.
124. Dilda PJ, Decollogne S, Rossiter-Thornton M, Hogg PJ. Para to ortho repositioning of the arsenical moiety of the angiogenesis inhibitor 4-(N-(S-glutathionylacetyl)amino)phenylarsenoxide results in a markedly increased cellular accumulation and antiproliferative activity. Cancer Res. 2005;65:11729–34.
125. Dong LF, Swettenham E, Eliasson J, Wang XF, Gold M, Medunic Y, Stantic M, Low P, Prochazka L, Witting PK, Turanek J, Akporiaye ET, Ralph SJ, Neuzil J. Vitamin E analogues inhibit angiogenesis by selective induction of apoptosis in proliferating endothelial cells: the role of oxidative stress. Cancer Res. 2007;67:11906–13.
126. Rohlena J, Dong LF, Kluckova K, Zobalova R, Goodwin J, Tilly D, Stursa J, Pecinova A, Philimonenko A, Hozak P, Banerjee J, Ledvina M, Sen CK, Houstek J, Coster MJ, Neuzil J. Mitochondrially targeted alpha-tocopheryl succinate is antiangiogenic: potential benefit against tumor angiogenesis but caution against wound healing. Antioxid Redox Signal. 2011;15:2923–35.

127. Magnon C, Galaup A, Mullan B, Rouffiac V, Bouquet C, Bidart JM, Griscelli F, Opolon P, Perricaudet M. Canstatin acts on endothelial and tumor cells via mitochondrial damage initiated through interaction with alphavbeta3 and alphavbeta5 integrins. Cancer Res. 2005;65:4353–61.
128. Yuan S, Fu Y, Wang X, Shi H, Huang Y, Song X, Li L, Song N, Luo Y. Voltage-dependent anion channel 1 is involved in endostatin-induced endothelial cell apoptosis. FASEB J. 2008;22:2809–20.
129. Blanc-Brude OP, Mesri M, Wall NR, Plescia J, Dohi T, Altieri DC. Therapeutic targeting of the survivin pathway in cancer: initiation of mitochondrial apoptosis and suppression of tumor-associated angiogenesis. Clin Cancer Res. 2003;9:2683–92.
130. Pasquier E, Carre M, Pourroy B, Camoin L, Rebai O, Briand C, Braguer D. Antiangiogenic activity of paclitaxel is associated with its cytostatic effect, mediated by the initiation but not completion of a mitochondrial apoptotic signaling pathway. Mol Cancer Ther. 2004;3:1301–10.
131. Singh RP, Dhanalakshmi S, Agarwal C, Agarwal R. Silibinin strongly inhibits growth and survival of human endothelial cells via cell cycle arrest and downregulation of survivin, Akt and NF-kappaB: implications for angioprevention and antiangiogenic therapy. Oncogene. 2005;24:1188–202.
132. Boivin D, Gendron S, Beaulieu E, Gingras D, Beliveau R. The antiangiogenic agent Neovastat (AE-941) induces endothelial cell apoptosis. Mol Cancer Ther. 2002;1:795–802.
133. Hanford HA, Wong CA, Kassan H, Cundiff DL, Chandel N, Underwood S, Mitchell CA, Soff GA. Angiostatin(4.5)-mediated apoptosis of vascular endothelial cells. Cancer Res. 2003;63:4275–80.
134. Lee TY, Muschal S, Pravda EA, Folkman J, Abdollahi A, Javaherian K. Angiostatin regulates the expression of antiangiogenic and proapoptotic pathways via targeted inhibition of mitochondrial proteins. Blood. 2009;114:1987–98.
135. Moser TL, Stack MS, Asplin I, Enghild JJ, Hojrup P, Everitt L, Hubchak S, Schnaper HW, Pizzo SV. Angiostatin binds ATP synthase on the surface of human endothelial cells. Proc Natl Acad Sci U S A. 1999;96:2811–6.
136. Kim GD, Oh J, Park HJ, Bae K, Lee SK. Magnolol inhibits angiogenesis by regulating ROS-mediated apoptosis and the PI3K/AKT/mTOR signaling pathway in mES/EB-derived endothelial-like cells. Int J Oncol. 2013;43:600–10.
137. Saha S, Ghosh M, Dutta SK. Role of metabolic modulator Bet-CA in altering mitochondrial hyperpolarization to suppress cancer associated angiogenesis and metastasis. Sci Rep. 2016;6:23552.
138. Murphy MP, Smith RA. Drug delivery to mitochondria: the key to mitochondrial medicine. Adv Drug Deliv Rev. 2000;41:235–50.
139. Murphy MP, Smith RA. Targeting antioxidants to mitochondria by conjugation to lipophilic cations. Annu Rev Pharmacol Toxicol. 2007;47:629–56.
140. Fink BD, Herlein JA, Yorek MA, Fenner AM, Kerns RJ, Sivitz WI. Bioenergetic effects of mitochondrial-targeted coenzyme Q analogs in endothelial cells. J Pharmacol Exp Ther. 2012;342:709–19.
141. Doughan AK, Dikalov SI. Mitochondrial redox cycling of mitoquinone leads to superoxide production and cellular apoptosis. Antioxid Redox Signal. 2007;9:1825–36.
142. Kelso GF, Porteous CM, Coulter CV, Hughes G, Porteous WK, Ledgerwood EC, Smith RA, Murphy MP. Selective targeting of a redox-active ubiquinone to mitochondria within cells: antioxidant and antiapoptotic properties. J Biol Chem. 2001;276:4588–96.
143. O'Malley Y, Fink BD, Ross NC, Prisinzano TE, Sivitz WI. Reactive oxygen and targeted antioxidant administration in endothelial cell mitochondria. J Biol Chem. 2006;281:39766–75.
144. Smith RA, Porteous CM, Coulter CV, Murphy MP. Selective targeting of an antioxidant to mitochondria. Eur J Biochem. 1999;263:709–16.
145. Antonenko YN, Avetisyan AV, Bakeeva LE, Chernyak BV, Chertkov VA, Domnina LV, Ivanova OY, Izyumov DS, Khailova LS, Klishin SS, Korshunova GA, Lyamzaev KG,

Muntyan MS, Nepryakhina OK, Pashkovskaya AA, Pletjushkina OY, Pustovidko AV, Roginsky VA, Rokitskaya TI, Ruuge EK, Saprunova VB, Severina II SRA, Skulachev IV, Skulachev MV, Sumbatyan NV, Sviryaeva IV, Tashlitsky VN, Vassiliev JM, Vyssokikh MY, Yaguzhinsky LS, Zamyatnin Jr AA, Skulachev VP. Mitochondria-targeted plastoquinone derivatives as tools to interrupt execution of the aging program. 1. Cationic plastoquinone derivatives: synthesis and in vitro studies. Biochemistry (Mosc). 2008;73:1273–87.
146. Guzy RD, Hoyos B, Robin E, Chen H, Liu L, Mansfield KD, Simon MC, Hammerling U, Schumacker PT. Mitochondrial complex III is required for hypoxia-induced ROS production and cellular oxygen sensing. Cell Metab. 2005;1:401–8.
147. Tiwari AK, Srinivas PV, Kumar SP, Rao JM. Free radical scavenging active components from *Cedrus deodara*. J Agric Food Chem. 2001;49:4642–5.
148. Mascarell L, Truffa-Bachi P. New aspects of cyclosporin a mode of action: from gene silencing to gene up-regulation. Mini-Rev Med Chem. 2003;3:205–14.
149. Yuan Q, Yang J, Santulli G, Reiken SR, Wronska A, Kim MM, Osborne BW, Lacampagne A, Yin Y, Marks AR. Maintenance of normal blood pressure is dependent on IP3R1-mediated regulation of eNOS. Proc Natl Acad Sci U S A. 2016;113:8532–7.
150. Baron-Menguy C, Bocquet A, Guihot AL, Chappard D, Amiot MJ, Andriantsitohaina R, Loufrani L, Henrion D. Effects of red wine polyphenols on postischemic neovascularization model in rats: low doses are proangiogenic, high doses anti-angiogenic. FASEB J. 2007;21:3511–21.
151. Poojari R. Embelin – a drug of antiquity: shifting the paradigm towards modern medicine. Expert Opin Investig Drugs. 2014;23:427–44.
152. Li J, Mahdi F, Du L, Jekabsons MB, Zhou YD, Nagle DG. Semisynthetic studies identify mitochondria poisons from botanical dietary supplements – geranyloxycoumarins from *Aegle marmelos*. Bioorg Med Chem. 2013;21:1795–803.
153. Park D, Dilda PJ. Mitochondria as targets in angiogenesis inhibition. Mol Asp Med. 2010;31:113–31.
154. Neuzil J, Dong LF, Rohlena J, Truksa J, Ralph SJ. Classification of mitocans, anti-cancer drugs acting on mitochondria. Mitochondrion. 2013;13:199–208.
155. Folkman J. Angiogenesis and apoptosis. Semin Cancer Biol. 2003;13:159–67.
156. Lenartowicz E, Bernardi P, Azzone GF. Phenylarsine oxide induces the cyclosporin A-sensitive membrane permeability transition in rat liver mitochondria. J Bioenerg Biomembr. 1991;23:679–88.
157. Dilda PJ, Ramsay EE, Corti A, Pompella A, Hogg PJ. Metabolism of the tumor angiogenesis inhibitor 4-(N-(S-Glutathionylacetyl)amino)phenylarsonous acid. J Biol Chem. 2008;283:35428–34.
158. Dilda PJ, Decollogne S, Weerakoon L, Norris MD, Haber M, Allen JD, Hogg PJ. Optimization of the antitumor efficacy of a synthetic mitochondrial toxin by increasing the residence time in the cytosol. J Med Chem. 2009;52:6209–16.
159. Neuzil J, Tomasetti M, Zhao Y, Dong LF, Birringer M, Wang XF, Low P, Wu K, Salvatore BA, Ralph SJ. Vitamin E analogs, a novel group of "mitocans," as anticancer agents: the importance of being redox-silent. Mol Pharmacol. 2007;71:1185–99.
160. Neuzil J, Tomasetti M, Mellick AS, Alleva R, Salvatore BA, Birringer M, Fariss MW. Vitamin E analogues: a new class of inducers of apoptosis with selective anti-cancer effects. Curr Cancer Drug Targets. 2004;4:355–72.
161. Dong LF, Jameson VJ, Tilly D, Cerny J, Mahdavian E, Marin-Hernandez A, Hernandez-Esquivel L, Rodriguez-Enriquez S, Stursa J, Witting PK, Stantic B, Rohlena J, Truksa J, Kluckova K, Dyason JC, Ledvina M, Salvatore BA, Moreno-Sanchez R, Coster MJ, Ralph SJ, Smith RA, Neuzil J. Mitochondrial targeting of vitamin E succinate enhances its pro-apoptotic and anti-cancer activity via mitochondrial complex II. J Biol Chem. 2011;286:3717–28.
162. Nyberg P, Xie L, Kalluri R. Endogenous inhibitors of angiogenesis. Cancer Res. 2005;65:3967–79.

163. Monboisse JC, Oudart JB, Ramont L, Brassart-Pasco S, Maquart FX. Matrikines from basement membrane collagens: a new anti-cancer strategy. Biochim Biophys Acta. 2014; 1840:2589–98.
164. Panka DJ, Mier JW. Canstatin inhibits Akt activation and induces fas-dependent apoptosis in endothelial cells. J Biol Chem. 2003;278:37632–6.
165. Dhanabal M, Ramchandran R, Waterman MJ, Lu H, Knebelmann B, Segal M, Sukhatme VP. Endostatin induces endothelial cell apoptosis. J Biol Chem. 1999;274:11721–6.
166. Fukuda S, Pelus LM. Survivin, a cancer target with an emerging role in normal adult tissues. Mol Cancer Ther. 2006;5:1087–98.
167. Sanhueza C, Wehinger S, Castillo Bennett J, Valenzuela M, Owen GI, Quest AF. The twisted survivin connection to angiogenesis. Mol Cancer. 2015;14:198.
168. Xiang R, Mizutani N, Luo Y, Chiodoni C, Zhou H, Mizutani M, Ba Y, Becker JC, Reisfeld RA. A DNA vaccine targeting survivin combines apoptosis with suppression of angiogenesis in lung tumor eradication. Cancer Res. 2005;65:553–61.
169. Weaver BA. How taxol/paclitaxel kills cancer cells. Mol Biol Cell. 2014;25:2677–81.
170. Deep G, Agarwal R. Antimetastatic efficacy of silibinin: molecular mechanisms and therapeutic potential against cancer. Cancer Metastasis Rev. 2010;29:447–63.
171. Falardeau P, Champagne P, Poyet P, Hariton C, Dupont E. Neovastat, a naturally occurring multifunctional antiangiogenic drug, in phase III clinical trials. Semin Oncol. 2001;28:620–5.
172. Hajitou A, Grignet C, Devy L, Berndt S, Blacher S, Deroanne CF, Bajou K, Fong T, Chiang Y, Foidart JM, Noel A. The antitumoral effect of endostatin and angiostatin is associated with a down-regulation of vascular endothelial growth factor expression in tumor cells. FASEB J. 2002;16:1802–4.
173. Griscelli F, Li H, Bennaceur-Griscelli A, Soria J, Opolon P, Soria C, Perricaudet M, Yeh P, Lu H. Angiostatin gene transfer: inhibition of tumor growth in vivo by blockage of endothelial cell proliferation associated with a mitosis arrest. Proc Natl Acad Sci U S A. 1998;95:6367–72.
174. Lin SY, Liu JD, Chang HC, Yeh SD, Lin CH, Lee WS. Magnolol suppresses proliferation of cultured human colon and liver cancer cells by inhibiting DNA synthesis and activating apoptosis. J Cell Biochem. 2002;84:532–44.
175. Saha S, Ghosh M, Dutta SK. A potent tumoricidal co-drug 'Bet-CA' – an ester derivative of betulinic acid and dichloroacetate selectively and synergistically kills cancer cells. Sci Rep. 2015;5:7762.

Chapter 22
High-Density Lipoprotein Regulation of Mitochondrial Function

C. Roger White, Geeta Datta, and Samantha Giordano

HDL serves a prominent anti-atherogenic function by mediating RCT. RCT is initiated by the association of apoA-I in lipid-poor HDL particles with the ATP-binding cassette transporter A1 (ABCA1) on macrophages and other target cells [1]. This interaction allows HDL to act as an acceptor for cholesterol. Lipid-poor HDL can subsequently be converted to a "mature" HDL particle via the action of the HDL-associated enzyme lecithin-cholesterol acyltransferase (LCAT). LCAT activation converts HDL from a nascent discoidal to a mature spherical form and thus increases the cholesterol carrying capacity of the particle. Mature HDL is thought to mediate cholesterol efflux via an interaction with the ABCG1 transporter [1]. The subsequent binding of apoA-I containing HDL particles to the scavenger receptor B1 (SRB1) on hepatocytes permits the unloading of cholesterol which is ultimately secreted as bile [2, 3]. By this mechanism, HDL is thought to attenuate inflammatory injury and reduce atheroma formation [4].

In addition to apoA-I, HDL also serves as a carrier for other exchangeable apolipoproteins, regulatory proteins and anti-oxidant enzymes. Recent proteomic analyses have revealed that more than 85 proteins may associate with HDL particles [5]. This observation suggests that HDL subspecies exist that subserve a variety of cellular functions [5]. Indeed, it is now appreciated that HDL possesses prominent anti-inflammatory and anti-oxidant properties that are independent of its ability to efflux cholesterol [6, 7]. Paraoxonase 1 (PON1) and platelet-activating factor acetyl-hydrolase (PAF-AH) are esterases that bind to helical regions of apoA-I [8]. These enzymes are important anti-oxidant enzymes that catalyze the hydrolysis of oxidized phospholipids [9]. Data also suggest that the HDL-associated lysosphingolipid sphingosine 1-phosphate exerts anti-inflammatory effects. Thus, both protein

C. Roger White • G. Datta • S. Giordano (✉)
University of Alabama at Birmingham, Department of Medicine, Division of Cardiovascular Disease, Birmingham, AL, USA
e-mail: sgiordan@uab.edu

G. Santulli (ed.), *Mitochondrial Dynamics in Cardiovascular Medicine*,
Advances in Experimental Medicine and Biology 982,
DOI 10.1007/978-3-319-55330-6_22

and lipid components of HDL may help to maintain cellular homeostasis. The observation that circulating levels of HDL inversely correlate with mitochondrial DNA damage in humans suggest a specific role for HDL in maintaining mitochondrial integrity [10].

Hypercholesterolemia is a major, cardiovascular risk factor which contributes to the pathogenesis of atherosclerosis, metabolic syndrome and diabetes. Chylomicrons, VLDL and LDL are apoB-containing lipoproteins that serve an important function by delivering cholesterol and triglycerides to peripheral cells. These particles, however, are susceptible to oxidative modification and may adopt pro-atherogenic properties. The uptake of oxidized LDL (oxLDL) by intimal macrophages is a well-characterized component of atherosclerotic plaque formation [11, 12]. Mitochondria are principal sites of ROS formation in the cell and are also targets for redox injury [10, 13–16]. An increase in plasma cholesterol and triglycerides has been implicated in the development of mitochondrial dysfunction. Lysophosphatidylcholine (lysoPC), a pro-inflammatory lipid associated with oxLDL, induces mitochondrial ROS formation and increases permeability in intact cells and isolated mitochondria [17, 18]. These responses are also induced by I-R injury. An increase in oxidant formation thus damages mitochondrial structural elements and impairs respiration by reducing oxidative phosphorylation and ATP formation [19]. Irreversible damage occurs upon dissipation of mitochondrial membrane potential ($\Delta\Psi m$) and opening of the mitochondrial permeability transition pore (mPTP). HDL possesses anti-inflammatory and anti-oxidant properties that protect mitochondria from injury. The goal of this chapter will be to discuss mechanisms by which HDL-associated proteins and lipids modulate mitochondrial function and improve cell survival.

Mitochondrial Structure and Function

The mitochondrion is an energy producing organelle that generates ROS and heat as byproducts of oxidative phosphorylation [20]. Mitochondria are found in all tissues but are most abundant in those with high metabolic requirements, such as cardiac and skeletal muscle [21]. The mitochondrion is a double membraned organelle that contains its own maternally-inherited DNA [22, 23]. The oxidative phosphorylation complexes; Complex I *(NADH: ubiquinone oxidoreductase)*, Complex II *(succinate dehydrogenase)*, Complex III *(CoQH2-cytochrome c reductase)*, Complex IV *(cytochrome c oxidase)*, and Complex V *(ATP synthase)*, are encoded by both the nuclear and mitochondrial genomes. Mitochondria utilize both electron and proton gradients in order to produce energy, in the form of ATP, while consuming oxygen. Electrons enter the electron transport chain at Complexes I and II and are then passed to Complex III via the Q cycle on coenzyme Q [24]. Electrons are shuttled to Complex IV where cytochrome c is reduced. As oxygen is consumed at Complex IV, cytochrome c is reoxidized, and water is formed. As the electrons are shuttled through the complexes, hydrogen ions are pumped from the mitochondrial matrix by Complexes I, III, and IV into the intramembrane space, thus establishing a protomotive force. As the protons flow down the gradient from the intermembrane space back into the mitochondrial matrix, ATP is produced at Complex V [23, 25].

The measurement of mitochondrial respiration can be performed using an oxygen electrode in isolated organelles. Chance and Williams defined the different states of mitochondrial respiration by measuring oxygen consumption in the presence of various mitochondrial substrates [26]. Under control conditions, mitochondria remain in state 3.5 respiration, where there is a generous supply of substrates for the electron transport chain and ADP for ATP production. Using the oxygen electrode, State 3 respiration was defined as the amount of oxygen consumed in the presence of pyruvate, succinate and ADP. State 4 respiration represents oxygen consumption after ADP has been fully converted to ATP. Currently, extracellular flux analysis represents state-of-the-art technology for the measurement of mitochondrial function in whole cells rather than isolated mitochondria. Endogenous substrates required for respiration are found within the cell. Under these conditions, inhibitors of the mitochondrial complexes can be administered to cells in order to determine different indices of mitochondrial function including: basal and maximal mitochondrial respiration; oxygen consumption required for ATP production; proton leak; and non-mitochondrial oxygen consumption [27]. Development of this technology has revolutionized the field of mitochondrial biology [27, 28].

Under normal physiological conditions, some electrons may spill off of the electron transport chain and react with oxygen to form superoxide anion. Complexes I and III are the principal sources of ROS during oxidative phosphorylation [29]. At low levels of production, ROS can act as signaling molecules. When produced in excess, however, ROS induce peroxidation of cellular lipids and proteins, damage mitochondrial and nuclear DNA and impair cell division. It follows that mitochondrial oxidative phosphorylation is significantly impaired. Bioenergetic dysfunction ensues and is associated with dissipation of $\Delta\Psi$m and opening of the mPTP [30]. mPTP induction subsequently leads to mitochondrial swelling and cell death [20, 29]. The antioxidant enzyme manganese superoxide dismutase is present in mitochondria and reduces superoxide to hydrogen peroxide, thus minimizing injury. The ability of the mitochondria to clear ROS thus increases cell survival. Schriner and colleagues reported that overexpression of mitochondrially targeted catalase in cardiac and skeletal muscle resulted in an increase in murine longevity [31]. Mitochondrial DNA damage in myocytes, as assessed by 8-hydroxyguanosine (8-OhdG) formation, increased with age but was attenuated by catalase overexpression in these mice [31]. Atherosclerosis and I-R injury are common cardiovascular disorders that are associated with the development of mitochondrial dysfunction. Defects in electron transport and oxidative phosphorylation enhance ROS formation resulting in mitochondrial injury and the induction of apoptosis via the cleavage of caspase 9. In a mouse model of myocardial infarction, ROS production damages both mitochondrial DNA and proteins, resulting in decreased enzymatic activities of mitochondrial Complexes I, III and IV. Further, coronary I-R injury is associated with opening of the mPTP leading to cell death. Ongoing studies are assessing the effects of mitochondrially-targeted anti-oxidant therapies on the development and progression of cardiovascular disease.

Mitochondrial Clearance

Cells utilize a mitochondrial clearance mechanism termed autophagy in order to avoid cell death secondary to injurious stimuli. The term autophagy was coined by Christian De Duve in the 1960s and is derived from the Greek words auto (self) and phagy (eating) [20, 32, 33]. Today, we know that there are three main types of autophagy: (1) macroautophagy (usually termed autophagy); (2) chaperone mediated autophagy; and (3) microautophagy [34]. Autophagy was first characterized as a response to nutrient starvation and was found to be associated with inactivation of the mammalian target of rapamycin (mTOR) [20, 32, 33]. Autophagy breaks down macromolecules and recycles their components not only to preserve cellular energy but also to clear damaged proteins and mitochondria [20, 32, 33]. More than 30 AuTophagy-related Genes (ATG) are conserved from yeast to mammals and participate in autophagy at different steps throughout the process [35, 36].

Mitochondria can be cleared through macroautophagy, but have also been shown to utilize different ATG proteins to selectively remove damaged mitochondria via mitophagy. As noted above, mitochondria contain their own genome which lacks compact chromatin structure and, compared to nuclear DNA, have a less effective replication and repair system [37]. Further, since mitochondria are the primary cellular producer of ROS, and the mitochondria contains highly reactive iron sulfur clusters, mitochondria are 10–20 times more likely to accumulate DNA damage compared to nuclear DNA. Mitophagy is controlled either in conjunction with general macroautophagy or selectively through specific mitophagy genes. AMPK is activated in response to decreased intracellular ATP, and then phosphorylates the Atg1 homologs ULK1 and ULK2 to activate both general macroautophagy and mitophagy [38]. AMPK or ULK1 deficient hepatocytes exhibit accumulation of p62, ubiquitin aggregates and abnormal mitochondria [38]. In the context of atherosclerosis and I-R injury, mitochondria are damaged and are more likely to produce ROS and less ATP. Mitophagy plays an important role in attenuating apoptosis or necrosis by clearing damaged mitochondria. This prevents the release of cytochrome c, apoptosis-inducing factor (AIF) and other apoptotic factors that can cause the cell to swell and burst. Mitochondrial turnover is thus essential for cell survival.

Mitochondria isolated from rodent hearts have an average half-life of 16–18 days [39]. In rodents exposed to hypoxic conditions, it was shown that mitochondrial half-life is reduced and is associated with a decrease in cardiac cytochrome c content. Further, there was an increase in mitochondrial biogenesis suggesting that formation of new mitochondria is an important response to I-R injury [40]. It is well known that starvation can induce both macroautophagy and mitophagy, but more recent data suggest that autophagy and mitophagy can be initiated by ischemia [41]. The induction of autophagy is essential for the protective effects of ischemic pre- and post- conditioning in the heart [42, 43]. Parkin, pink1 and related proteins play an important role in mitophagy and mitochondrial turnover [44]. In parkin- and pink1-deficient mice, there is an exacerbation of I-R injury, suggesting that the clearance of damaged mitochondria is an essential component of ischemic preconditioning [45]. Since mitochondrial biogenesis and turnover are tightly regulated and have major effects on outcomes, further treatments for heart disease should utilize these pathways in drug discovery.

Mitochondrial and Cell Survival Mechanisms

Cardiac ischemia arises in response to unstable angina and acute myocardial infarction. Upon reperfusion, significant injury occurs at the level of the mitochondrion [46]. Due to the high energy demands of the heart, I-R injury significantly impairs mitochondrial respiration. This is associated with increased ROS formation, uncoupling of oxidative phosphorylation and opening of the mPTP [47]. mPTP opening is associated with dissipation of the $\Delta\Psi m$, calcium influx and the release of pro-apoptotic factors [46]. Ischemic pre-conditioning and post-conditioning have been shown to preserve mitochondrial function thus reducing myocardial reperfusion injury [48, 49]. These conditioning protocols are characterized by brief disruption of blood flow prior to sustained ischemia and reperfusion and result in the activation of two major cell survival pathways [48, 49]. The Reperfusion Injury Salvage Kinase (RISK) pathway is comprised of the pro-survival kinases phosphatidylinositol 3-kinase (PI3K), protein kinase B (*Akt*) and extracellular regulated kinase 1/2 (ERK1/2) [50, 51]. These enzymes phosphorylate multiple substrates in the cell that converge to inhibit opening of mPTP [52–54]. Glycogen synthase kinase 3 beta (GSK3β) appears to be an important target for RISK-dependent cardiomyocyte survival [55]. Phosphorylation of GSK3β reduces its enzymatic activity. Inactive GSK3β attenuates mitochondrial injury by increasing Bcl-2 anti-apoptotic activity, inhibiting the translocation of pro-apoptotic Bax to the outer mitochondrial membrane and preventing mPTP induction via stabilization of the mPTP regulatory protein cyclophilin D [52, 56]. The Survivor Activating Factor Enhancement (SAFE) cascade is an alternate pathway for mitochondrial preservation [57]. TNFα activates the SAFE pathway by binding to TNF receptor type 2 (TNFR2) [58]. TNFR2 engagement results in the activation of Janus kinase (JAK) which phosphorylates the signal transducer and activator of transcription 3 (STAT3). The SAFE pathway has been shown to reduce infarct size in animals undergoing I-R injury by a mechanism involving STAT3 phosphorylation and inhibition of mPTP opening [59]. While the RISK and SAFE pathways preserve mitochondrial function by increasing the activity of specific signaling intermediates, data suggest that crosstalk between these pathways occurs, with stabilization of the mPTP as a final common outcome [48, 57].

HDL and ApoA-I Preserve Mitochondrial Function

Lipid peroxides present in oxLDL have been shown to stimulate ROS formation and impair oxygen consumption at Complexes I, II/III, and IV of the respiratory chain, resulting in mitochondrial dysfunction [60, 61]. The HDL-associated protein PON1 performs an important antioxidant function by hydrolyzing cholesteryl esters and phospholipids in oxidized lipoproteins [61]. PON1 may thus preserve mitochondrial function due to its ability to degrade oxidized lipid species. In contrast to this indirect effect of HDL-associated PON1, a direct role for apoA-I and the lysosphingolipid sphingosine 1-phosphate (S1P) in mediating cardiomyocyte survival has been reported [60–62]. Infarct size in apoA-I$^{-/-}$ mice undergoing coronary artery ligation/reperfusion is significantly increased compared to lesions in wildtype mice [63]. This correlated with a reduction in Coenzyme Q (CoQ) in mitochondria isolated from apoA-I$^{-/-}$ mice. CoQ deficiency was associated with a significant reduction in electron transfer from Complex II to Complex III [63]. In related studies, it was shown that exogenous administration of CoQ restored mitochondrial CoQ levels and attenuated infarct size in apoA-I$^{-/-}$ mice [63]. These results suggested that apoA-I plays a key role in maintaining the coupling of electron transport proteins.

Data suggest that HDL, similar to ischemic pre-conditioning and post-conditioning, prevents mitochondrial injury via activation of RISK and SAFE survival cascades. The SAFE pathway influences mitochondrial function in several ways. First, STAT3-mediated nuclear transcription results in up-regulation of anti-apoptotic Bcl-2 and the antioxidant genes manganese superoxide dismutase and metallothionein while inhibiting pro-apoptotic Bax/Bad expression [48, 64]. STAT3 has also been shown to regulate the electron transport chain and mitochondrial respiration [64, 65]. STAT3 gains access to the mitochondrion with the assistance of the chaperone protein GRIM-19 [66]. At this locus, STAT3 inhibits Complex I and II respiration and the release of cytochrome c [64]. It is proposed that STAT3 ultimately protects mitochondria by reversibly uncoupling electron flow between respiratory complexes, reducing ROS formation and preventing mPTP induction [64, 67–69]. Administration of apoA-I to rodents prior to coronary artery occlusion attenuates morphologic changes associated with ischemic injury and reduces infarct size. Consistent with known targets of the RISK and SAFE signaling cascades, apoA-I treatment increased the phosphorylation of *Akt* and GSK3β. The infarct-sparing response to apoA-I was significantly reduced in animals treated with inhibitors of *Akt*, ERK1/2 and JAK/STAT. These data suggested that the inhibitory effect of apoA-I on myocardial infarct size was due to activation of both RISK and SAFE survival pathways, resulting in attenuation of mitochondrial injury.

HDL-Associated Sphingosine 1-Phosphate: Influences Mitochondrial Function

The lipid composition of HDL plays an important role in determining the function of the lipoprotein particle [70, 71]. Lipid species maintain the structural integrity of HDL and regulate the activities of HDL-associated proteins [72]. Data suggest that the lysosphingolipid S1P of HDL plays an important role in cardioprotection. While S1P is synthesized in hematopoietic and endothelial cells, HDL serves as its principal carrier in plasma [73, 74]. S1P has been shown to act as an inducer of both the

RISK and SAFE pathways [75, 76]. Cardiomyocyte responses to S1P actions are mediated by multiple receptor isoforms (S1P1, S1P2 and S1P3) [77–79]. Addition of HDL or purified S1P to neonatal rat cardiac cardiomyocytes activates S1P2 receptors resulting in STAT3 phosphorylation. In contrast, HDL that is depleted of S1P fails to support STAT3 phosphorylation [75]. HDL treatment was shown to activate *Akt* and ERK1/2 pathways via distinct S1P receptors in mouse cardiomyocytes exposed to hypoxia-reoxygenation [77]. S1P1 binding resulted in activation of ERK1/2, while S1P3 induced *Akt* activation. Under these conditions, levels of phosphorylated GSK3β, a known inhibitor of mPTP opening, were increased [77]. This response was inhibited by S1P receptor blockers and the PI3K inhibitor wortmannin, suggesting that S1P reduces cardiomyocyte injury by activating the RISK pathway [77]. Numerous in vivo studies support a role for S1P in the activation of RISK and SAFE signaling cascades in the context of I-R injury and heart failure [77, 80–82].

Cardioprotective responses to HDL and S1P are ultimately mediated at the level of the mitochondrion. Administration of HDL to mice undergoing I-R injury reduces infarct size in a concentration-dependent manner [83]. This response was associated with STAT3 phosphorylation and inhibition of mPTP opening in isolated cardiomyocytes. Protective effects at the level of the mitochondrion were abolished in $TNF\alpha^{-/-}$ and cardiomyocyte-specific $STAT3^{-/-}$ mice, suggesting that STAT3 mediates the activation of the SAFE pathway [84]. Forkhead box O-1 (FOXO-1) is a transcription factor that is known to increase ROS formation and apoptosis in the non-phosphorylated state [85]. While the RISK pathway and PI3K/*Akt* activation are associated with the inactivation of FOXO-1, data suggest that the SAFE survival pathway also modulates FOXO-1 activity [79, 85]. S1P stimulated the nuclear phosphorylation/inactivation of FOXO-1 in a manner that was blocked by both a JAK/STAT3 and PI3K inhibitor [79]. These observations suggest that the S1P-dependent phosphorylation of FOXO-1 may represent a point of convergence for the RISK and SAFE survival cascades.

Finally, data show that S1P regulates Complex IV assembly and cellular respiration in mitochondria through an interaction with mitochondrial prohibitin-2 (PHB2) [86, 87]. PHB2 is a scaffold protein which functions to stabilize the structure of the inner mitochondrial membrane [88]. Disruption of S1P-PHB2 binding abolishes the cardioprotective response of cardiomyocytes to ischemic pre-conditioning. Under this condition, a reduction in oxidative phosphorylation was associated with opening of mPTP and mitochondrial injury [46, 54]. These data, therefore, suggest that S1P and PHB2 stabilize Complex IV and reduce ROS formation while also supporting oxidative phosphorylation [86].

HDL and Autophagy Induction

Mitochondrial damage induced by I-R injury releases apoptotic factors that damage neighboring mitochondria. It has been proposed that autophagy serves a cytoprotective role by clearing damaged mitochondria and limiting potentially deleterious effects on neighboring organelles [89]. Ischemia initiates autophagy by inducing the de-phosphorylation and inactivation of mTOR which normally acts as a suppressor of autophagy [90]. An increase in the ratio of AMP/ATP concurrently induces AMP-activated protein kinase (AMPK) and stimulates autophagy via multiple mechanisms. The vacuolar protein sorting-34 (Vps34) is a class III PI-3 kinase that initiates phagophore formation through its association with beclin1 [90]. This pre-autophagosomal structure engulfs cytoplasmic components, including damaged mitochondria. Phosphatidylethanolamine (PE) and microtubule-associated protein light chain-3 (LC3 I) interact to form the conjugated product LC3 II. LC3 II and the adaptor protein p62 are recruited to yield the mature autophagosome. The autophagosome then fuses with lysosomes where ingested products are digested. Inhibitors of autophagy promote ROS formation and aggravate mitochondrial injury in response to I-R [91]. They have also been shown to negate cell survival mechanisms associated with ischemic pre-conditioning. It follows that autophagy contributes to the cytoprotection associated with pre-conditioning [42, 92]. A role for S1P in the inhibition of mTOR and activation of autophagy has been reviewed [93, 94]. It follows that inhibitors of S1P formation also prevent the induction of autophagy [95]. As the principal carrier of S1P, HDL may induce autophagy as a cell survival mechanism. In support of this, it was shown that HDL inhibits mTOR activity, stimulates the expression of LC3 II and stimulates the formation of autophagosomes [96].

Alternate Apolipoproteins and Mitochondrial Function

In addition to apoA-I and apoB, other apolipoproteins (i.e., apoJ, apoM, apoC and apoO) serve as regulatory molecules for cholesterol homeostasis [97]. Recent data also suggest a role for these apolipoproteins in the regulation of mitochondrial function. ApoJ, also known as clusterin, is a glycoprotein expressed ubiquitously during development and in adults and is associated with small, dense HDL3 particles [98–100]. Due to the presence of disulfide bonds, apoJ possesses antioxidant properties that inhibit ROS-dependent injury and preserve mitochondrial function [20]. In the H9c2 cardiomyocyte cell line, it was shown that apoJ protects against ROS-induced apoptosis by activating *Akt* and GSK-3β, thus suggesting a role in the activation of the RISK survival cascade [101].

ApoM is apolipoprotein that accounts for approximately 5 % of the protein content of HDL. A major function of apoM is to attenuate atherosclerosis by stimulating the formation of small, dense preβ-HDL particles that play an important role in RCT [102–105]. ApoM protects mitochondrial function by virtue of its ability to bind to S1P and facilitate its incorporation into HDL [80, 104, 106, 107]. The relationship between apoM and S1P-mediated cardioprotection has been evaluated in a murine model of I-R injury [108]. Over-expression of apoM in mice was accompanied by a significant reduction in myocardial infarct size and leukocyte accumulation [108]. The underlying mechanism of apoM-S1P-induced cardioprotection was due to inhibition of cardiomyocyte cell-cell coupling [108]. The passage of death signals through gap junctions was reduced, resulting in attenuation of I-R-induced cardiomyocyte injury [108].

ApoC is an exchangeable apolipoprotein associated with HDL as well as apoB-containing lipoproteins. Three structurally distinct isoforms (apoC-I, apoC-II and apoC-III) have been identified that exert differential effects on lipid metabolism, ROS formation, mitochondrial function and cell death. [109]. Panin and colleagues have tested effects of these apoC isoforms on oxidative phosphorylation in rat liver mitochondria [110]. Using palmitoyl carnitine as a substrate for oxidative phosphorylation, it was shown that apoC-III, but neither apoC-I nor apoC-II, inhibited State 3 respiration [110]. At higher concentrations of apoC-III, the rate of oxygen consumption was reduced ~70 %, and oxidative phosphorylation was completely blocked [110]. These data were the first to show an inhibitory effect of ApoC-III on mitochondrial function. Other data support a role for apoC-I in the development of mitochondrial injury in human aortic smooth muscle cells [111]. Enrichment of HDL with apoC-I was shown to induce cell death in a neutral sphingomyelinase (N-SMAS) dependent manner, with the production of ceramide from N-SMAS stimulating the release of cytochrome c from the mitochondria, the cleavage of caspase 3 and apoptosis. The increased cell death of the aortic smooth muscle cells incubated with ApoC-I may account for the unstable plaque formation seen in patients with hypercholesterolemia [111]. Interestingly, apoC-I enrichment of HDL was associated with a reduction in HDL-associated apoA-I, suggesting that loss of apoA-I and its cytoprotective effects is a component of apoC-I mediated apoptosis [111].

Apoliprotein O is a 198 amino acid protein which is found in association with HDL, LDL and VLDL particles [112]. ApoO is up-regulated by metabolic stress in the diabetic heart where it is thought to play a role in reducing myocardial injury by preventing macrophage lipid accumulation [112]. ApoO and the related apoO-like protein have also been shown to stabilize the inner mitochondrial membrane and the cristea [113–115]. In contrast to this report, over-expression of apoO was shown to impair cardiac function and degrade mitochondrial structure in cardiomyocytes of hypercholesteroleic mice [116]. This observation suggested a pathogenic role for apoO in obesity. To further understand the pathogenic mechanism of apoO, Turkeih and colleagues developed a stable cardiac myoblast cell line that overexpressed apoO. These cells were characterized by an increase in mitochondrial respiration, fatty acid uptake and metabolism, ROS formation and apoptosis compared to control cells [116]. It was suggested that an increase in apoO expression promotes the transition from mitochondrial dysfunction to the development of overt cardiomyopathy [116]. Further research is required to delineate the role of apoO in diverse cardiovascular pathologies.

Conclusion

While HDL plays an important anti-atherogenic role by mediating RCT, it is also an active signaling particle that possesses anti-inflammatory, anti-oxidant and anti-apoptotic properties. HDL-associated proteins and lipids play an important role in the preservation of mitochondrial function. The anti-oxidant enzyme PON1 prevents damage to respiratory complexes by degrading oxidized lipids such as lysoPC [117]. ApoA-I also protects mitochondria by multiple mechanisms. Through an interaction with CoQ, apoA-I stabilizes complex II and inhibits ROS-mediated damage to respiratory complexes [63]. ApoA-I and the HDL-associated lipid S1P also protect mitochondria via the activation of RISK and SAFE survival pathways [118]. *Akt*, ERK1/2 and JAK/STAT3 are critical mediators of these cell survival cascades. The cytoprotective response to S1P is due to an interaction with cellular S1P receptors, resulting in activation of JAK/STAT3 signaling. Binding of apoA-I to ABCA1 is also reported to activate STAT3 signaling and attenuate inflammatory injury [119]. It has been proposed that phosphorylated FOXO-1 and GSK3β are final, common effectors of these pathways and improve cell survival by attenuating mitochondrial ROS formation, mPTP induction and apoptosis [79, 120]. Autophagy represents an additional survival mechanism activated by apoA-I. Under these conditions, ROS-dependent injury may be reduced via the effective clearance of damaged mitochondria.

The ability of HDL to preserve mitochondrial function may be attenuated under pathological conditions. A reduction in circulating HDL concentration is associated with a number of inflammatory disorders [121, 122]. Under these conditions, depletion of apoA-I, PON1, apoM and apoJ from HDL particles may result in a loss of function [121, 123–126]. A decrease in apoM may ablate the protective effects of S1P on mitochondrial function. This response may augment the vascular leakage observed in patients with sepsis [127]. Lipoprotein oxidation is also associated with a reduction in S1P levels and accumulation of pro-inflammatory lysoPC in HDL particles [128]. With respect to apoJ, a decrease in its association with HDL has been implicated in the development of insulin resistance and an increase in apoptosis [129]. Finally, an increase in the incorporation of apoC in the HDL particle may increase mitochondrial injury and apoptosis. Under these conditions, HDL function may be further degraded by the displacement of apoA-I by apoC. These observations suggest that raising circulating HDL concentration as well as its functional properties represents an important therapeutic strategy to minimize mitochondrial injury.

References

1. Yvan-Charvet L, Wang N, Tall AR. Role of HDL, ABCA1, and ABCG1 transporters in cholesterol efflux and immune responses. Arterioscler Thromb Vasc Biol. 2010;30:139–43.
2. Fitzgerald ML, Mujawar Z, Tamehiro N. ABC transporters, atherosclerosis and inflammation. Atherosclerosis. 2010;211:361–70.
3. Zannis VI, Chroni A, Krieger M. Role of apoA-I, ABCA1, LCAT, and SR-BI in the biogenesis of HDL. J Mol Med (Berl). 2006;84:276–94.
4. Säemann MD, Poglitsch M, Kopecky C, Haidinger M, Hörl WH, Weichhart T. The versatility of HDL: a crucial anti-inflammatory regulator. Eur J Clin Investig. 2010;40:1131–43.

5. Shah AS, Tan L, Long JL, Davidson WS. Proteomic diversity of high density lipoproteins: our emerging understanding of its importance in lipid transport and beyond. J Lipid Res. 2013;54:2575–85.
6. Assmann G, Nofer JR. Atheroprotective effects of high-density lipoproteins. Annu Rev Med. 2003;54:321–41.
7. Dunbar RL, Rader DJ. Current drug options for raising HDL cholesterol. Curr Treat Options Cardiovasc Med. 2005;7:15–23.
8. Bashtovyy D, Jones MK, Anantharamaiah GM, Segrest JP. Sequence conservation of apolipoprotein A-I affords novel insights into HDL structure-function. J Lipid Res. 2011;52:435–50.
9. Gaidukov L, Tawfik DS. High affinity, stability, and lactonase activity of serum paraoxonase PON1 anchored on HDL with ApoA-I. Biochemistry. 2005;44:11843–54.
10. Parikh NI, Gerschenson M, Bennett K, Gangcuangco LM, Lopez MS, Mehta NN, Playford MP, Nakamoto BK, Seto TB, Chow DC, Shikuma CM. Lipoprotein concentration, particle number, size and cholesterol efflux capacity are associated with mitochondrial oxidative stress and function in an HIV positive cohort. Atherosclerosis. 2015;239:50–4.
11. Gambardella J, Sardu C, Sacra C, Del Giudice C, Santulli G. Quit smoking to outsmart atherogenesis: molecular mechanisms underlying clinical evidence. Atherosclerosis. 2017;257:242–5.
12. Rocha VZ, Libby P. Obesity, inflammation, and atherosclerosis. Nat Rev Cardiol. 2009;6:399–409.
13. Oliveira HC, Cosso RG, Alberici LC, Maciel EN, Salerno AG, Dorighello GG, Velho JA, de Faria EC, Vercesi AE. Oxidative stress in atherosclerosis-prone mouse is due to low antioxidant capacity of mitochondria. FASEB J. 2005;19:278–80.
14. Xie W, Santulli G, Reiken SR, Yuan Q, Osborne BW, Chen BX, Marks AR. Mitochondrial oxidative stress promotes atrial fibrillation. Sci Rep. 2015;5:11427.
15. Yu E, Mercer J, Bennett M. Mitochondria in vascular disease. Cardiovasc Res. 2012;95:173–82.
16. Honda HM, Korge P, Weiss JN. Mitochondria and ischemia/reperfusion injury. Ann N Y Acad Sci. 2005;1047:248–58.
17. Li X, Fang P, Li Y, Kuo YM, Andrews AJ, Nanayakkara G, Johnson C, Fu H, Shan H, Du F, Hoffman NE, Yu D, Eguchi S, Madesh M, Koch WJ, Sun J, Jiang X, Wang H, Yang X. Mitochondrial reactive oxygen species mediate lysophosphatidylcholine-induced endothelial cell activation. Arterioscler Thromb Vasc Biol. 2016;36:1090–100.
18. Hollie NI, Cash JG, Matlib MA, Wortman M, Basford JE, Abplanalp W, Hui DY. Micromolar changes in lysophosphatidylcholine concentration cause minor effects on mitochondrial permeability but major alterations in function. Biochim Biophys Acta. 1841;2014:888–95.
19. Hulsmans M, Van Dooren E, Holvoet P. Mitochondrial reactive oxygen species and risk of atherosclerosis. Curr Atheroscler Rep. 2012;14:264–76.
20. Lee J, Giordano S, Zhang J. Autophagy, mitochondria and oxidative stress: cross-talk and redox signalling. Biochem J. 2012;441:523–40.
21. Forner F, Foster LJ, Campanaro S, Valle G, Mann M. Quantitative proteomic comparison of rat mitochondria from muscle, heart, and liver. Mol Cell Proteomics. 2006;5:608–19.
22. Fetterman JL, Zelickson BR, Johnson LW, Moellering DR, Westbrook DG, Pompilius M, Sammy MJ, Johnson M, Dunham-Snary KJ, Cao X, Bradley WE, Zhang J, Wei CC, Chacko B, Schurr TG, Kesterson RA, Dell'italia LJ, Darley-Usmar VM, Welch DR, Ballinger SW. Mitochondrial genetic background modulates bioenergetics and susceptibility to acute cardiac volume – overload. Biochem J. 2013;455:157.
23. Amo T, Yadava N, Oh R, Nicholls DG, Brand MD. Experimental assessment of bioenergetic differences caused by the common European mitochondrial DNA haplogroups H and T. Gene. 2008;411:69–76.
24. Sazanov LA. A giant molecular proton pump: structure and mechanism of respiratory complex I. Nat Rev Mol Cell Biol. 2015;16:375–88.

25. Chen Q, Yin G, Stewart S, Hu Y, Lesnefsky EJ. Isolating the segment of the mitochondrial electron transport chain responsible for mitochondrial damage during cardiac ischemia. Biochem Biophys Res Commun. 2010;397:656–60.
26. Chance B, Williams GR. A simple and rapid assay of oxidative phosphorylation. Nature. 1955;175:1120–1.
27. Dranka BP, Benavides GA, Diers AR, Giordano S, Zelickson BR, Reily C, Zou L, Chatham JC, Hill BG, Zhang J, Landar A, Darley-Usmar VM. Assessing bioenergetic function in response to oxidative stress by metabolic profiling. Free Radic Biol Med. 2011;51:1621–35.
28. Nicholls DG, Darley-Usmar VM, Wu M, Jensen PB, Rogers GW, Ferrick DA. Bioenergetic profile experiment using C2C12 myoblast cells. J Vis Exp. 2010;6:2511.
29. Ott M, Gogvadze V, Orrenius S, Zhivotovsky B. Mitochondria, oxidative stress and cell death. Apoptosis. 2007;12:913–22.
30. Wong R, Steenbergen C, Murphy E. Mitochondrial permeability transition pore and calcium handling. Methods Mol Biol. 2012;810:235–42.
31. Schriner SE, Linford NJ, Martin GM, Treuting P, Ogburn CE, Emond M, Coskun PE, Ladiges W, Wolf N, Van Remmen H, Wallace DC, Rabinovitch PS. Extension of murine life span by overexpression of catalase targeted to mitochondria. Science. 2005;308:1909–11.
32. Dodson M, Darley-Usmar V, Zhang J. Cellular metabolic and autophagic pathways: traffic control by redox signaling. Free Radic Biol Med. 2013;63:207–21.
33. Zhang J. Autophagy and mitophagy in cellular damage control. Redox Biol. 2013;1:19–23.
34. Klionsky DJ, Abdelmohsen K, Abe A, Abedin MJ, Abeliovich H, Acevedo Arozena A, Adachi H, Adams CM, Adams PD, Adeli K, Adhihetty PJ, Adler SG, Agam G, Agarwal R, Aghi MK, Agnello M, Agostinis P, Aguilar PV, Aguirre-Ghiso J, Airoldi EM, Ait-Si-Ali S, Akematsu T, Akporiaye ET, Al-Rubeai M, Albaiceta GM, Albanese C, Albani D, Albert ML, Aldudo J, Algül H, Alirezaei M, Alloza I, Almasan A, Almonte-Beceril M, Alnemri ES, Alonso C, Altan-Bonnet N, Altieri DC, Alvarez S, Alvarez-Erviti L, Alves S, Amadoro G, Amano A, Amantini C, Ambrosio S, Amelio I, Amer AO, Amessou M, Amon A, An Z, Anania FA, Andersen SU, Andley UP, Andreadi CK, Andrieu-Abadie N, Anel A, Ann DK, Anoopkumar-Dukie S, Antonioli M, Aoki H, Apostolova N, Aquila S, Aquilano K, Araki K, Arama E, Aranda A, Araya J, Arcaro A, Arias E, Arimoto H, Ariosa AR, Armstrong JL, Arnould T, Arsov I, Asanuma K, Askanas V, Asselin E, Atarashi R, Atherton SS, Atkin JD, Attardi LD, Auberger P, Auburger G, Aurelian L, Autelli R, Avagliano L, Avantaggiati ML, Avrahami L, Awale S, Azad N, Bachetti T, Backer JM, Bae DH, Bae JS, Bae ON, Bae SH, Baehrecke EH, Baek SH, Baghdiguian S, Bagniewska-Zadworna A, Bai H, Bai J, Bai XY, Bailly Y, Balaji KN, Balduini W, Ballabio A, Balzan R, Banerjee R, Bánhegyi G, Bao H, Barbeau B, Barrachina MD, Barreiro E, Bartel B, Bartolomé A, Bassham DC, Bassi MT, Bast RC, Basu A, Batista MT, Batoko H, Battino M, Bauckman K, Baumgarner BL, Bayer KU, Beale R, Beaulieu JF, Beck GR, Becker C, Beckham JD, Bédard PA, Bednarski PJ, Begley TJ, Behl C, Behrends C, Behrens GM, Behrns KE, Bejarano E, Belaid A, Belleudi F, Bénard G, Berchem G, Bergamaschi D, Bergami M, Berkhout B, Berliocchi L, Bernard A, Bernard M, Bernassola F, Bertolotti A, Bess AS, Besteiro S, Bettuzzi S, Bhalla S, Bhattacharyya S, Bhutia SK, Biagosch C, Bianchi MW, Biard-Piechaczyk M, Billes V, Bincoletto C, Bingol B, Bird SW, Bitoun M, Bjedov I, Blackstone C, Blanc L, Blanco GA, Blomhoff HK, Boada-Romero E, Böckler S, Boes M, Boesze-Battaglia K, Boise LH, Bolino A, Boman A, Bonaldo P, Bordi M, Bosch J, Botana LM, Botti J, Bou G, Bouché M, Bouchecareilh M, Boucher MJ, Boulton ME, Bouret SG, Boya P, Boyer-Guittaut M, Bozhkov PV, Brady N, Braga VM, Brancolini C, Braus GH, Bravo-San Pedro JM, Brennan LA, Bresnick EH, Brest P, Bridges D, Bringer MA, Brini M, Brito GC, Brodin B, Brookes PS, Brown EJ, Brown K, Broxmeyer HE, Bruhat A, Brum PC, Brumell JH, Brunetti-Pierri N, Bryson-Richardson RJ, Buch S, Buchan AM, Budak H, Bulavin DV, Bultman SJ, Bultynck G, Bumbasirevic V, Burelle Y, Burke RE, Burmeister M, Bütikofer P, Caberlotto L, Cadwell K, Cahova M, Cai D, Cai J, Cai Q, Calatayud S, Camougrand N, Campanella M, Campbell GR, Campbell M, Campello S, Candau R, Caniggia I, Cantoni L, Cao L, Caplan AB, Caraglia M, Cardinali C, Cardoso SM,

Carew JS, Carleton LA, Carlin CR, Carloni S, Carlsson SR, Carmona-Gutierrez D, Carneiro LA, Carnevali O, Carra S, Carrier A, Carroll B, Casas C, Casas J, Cassinelli G, Castets P, Castro-Obregon S, Cavallini G, Ceccherini I, Cecconi F, Cederbaum AI, Ceña V, Cenci S, Cerella C, Cervia D, Cetrullo S, Chaachouay H, Chae HJ, Chagin AS, Chai CY, Chakrabarti G, Chamilos G, Chan EY, Chan MT, Chandra D, Chandra P, Chang CP, Chang RC, Chang TY, Chatham JC, Chatterjee S, Chauhan S, Che Y, Cheetham ME, Cheluvappa R, Chen CJ, Chen G, Chen GC, Chen H, Chen JW, Chen JK, Chen M, Chen P, Chen Q, Chen SD, Chen S, Chen SS, Chen W, Chen WJ, Chen WQ, Chen X, Chen YH, Chen YG, Chen Y, Chen YJ, Chen YQ, Chen Z, Cheng A, Cheng CH, Cheng H, Cheong H, Cherry S, Chesney J, Cheung CH, Chevet E, Chi HC, Chi SG, Chiacchiera F, Chiang HL, Chiarelli R, Chiariello M, Chieppa M, Chin LS, Chiong M, Chiu GN, Cho DH, Cho SG, Cho WC, Cho YY, Cho YS, Choi AM, Choi EJ, Choi EK, Choi J, Choi ME, Choi SI, Chou TF, Chouaib S, Choubey D, Choubey V, Chow KC, Chowdhury K, Chu CT, Chuang TH, Chun T, Chung H, Chung T, Chung YL, Chwae YJ, Cianfanelli V, Ciarcia R, Ciechomska IA, Ciriolo MR, Cirone M, Claerhout S, Clague MJ, Clària J, Clarke PG, Clarke R, Clementi E, Cleyrat C, Cnop M, Coccia EM, Cocco T, Codogno P, Coers J, Cohen EE, Colecchia D, Coletto L, Coll NS, Colucci-Guyon E, Comincini S, Condello M, Cook KL, Coombs GH, Cooper CD, Cooper JM, Coppens I, Corasaniti MT, Corazzari M, Corbalan R, Corcelle-Termeau E, Cordero MD, Corral-Ramos C, Corti O, Cossarizza A, Costelli P, Costes S, Cotman SL, Coto-Montes A, Cottet S, Couve E, Covey LR, Cowart LA, Cox JS, Coxon FP, Coyne CB, Cragg MS, Craven RJ, Crepaldi T, Crespo JL, Criollo A, Crippa V, Cruz MT, Cuervo AM, Cuezva JM, Cui T, Cutillas PR, Czaja MJ, Czyzyk-Krzeska MF, Dagda RK, Dahmen U, Dai C, Dai W, Dai Y, Dalby KN, Dalla Valle L, Dalmasso G, D'Amelio M, Damme M, Darfeuille-Michaud A, Dargemont C, Darley-Usmar VM, Dasarathy S, Dasgupta B, Dash S, Dass CR, Davey HM, Davids LM, Dávila D, Davis RJ, Dawson TM, Dawson VL, Daza P, de Belleroche J, de Figueiredo P, de Figueiredo RC, de la Fuente J, De Martino L, De Matteis A, De Meyer GR, De Milito A, De Santi M, de Souza W, De Tata V, De Zio D, Debnath J, Dechant R, Decuypere JP, Deegan S, Dehay B, Del Bello B, Del Re DP, Delage-Mourroux R, Delbridge LM, Deldicque L, Delorme-Axford E, Deng Y, Dengjel J, Denizot M, Dent P, Der CJ, Deretic V, Derrien B, Deutsch E, Devarenne TP, Devenish RJ, Di Bartolomeo S, Di Daniele N, Di Domenico F, Di Nardo A, Di Paola S, Di Pietro A, Di Renzo L, Di Antonio A, Díaz-Araya G, Díaz-Laviada I, Diaz-Meco MT, Diaz-Nido J, Dickey CA, Dickson RC, Diederich M, Digard P, Dikic I, Dinesh-Kumar SP, Ding C, Ding WX, Ding Z, Dini L, Distler JH, Diwan A, Djavaheri-Mergny M, Dmytruk K, Dobson RC, Doetsch V, Dokladny K, Dokudovskaya S, Donadelli M, Dong XC, Dong X, Dong Z, Donohue TM, Doran KS, D'Orazi G, Dorn GW, Dosenko V, Dridi S, Drucker L, Du J, Du LL, Du L, du Toit A, Dua P, Duan L, Duann P, Dubey VK, Duchen MR, Duchosal MA, Duez H, Dugail I, Dumit VI, Duncan MC, Dunlop EA, Dunn WA, Dupont N, Dupuis L, Durán RV, Durcan TM, Duvezin-Caubet S, Duvvuri U, Eapen V, Ebrahimi-Fakhari D, Echard A, Eckhart L, Edelstein CL, Edinger AL, Eichinger L, Eisenberg T, Eisenberg-Lerner A, Eissa NT, El-Deiry WS, El-Khoury V, Elazar Z, Eldar-Finkelman H, Elliott CJ, Emanuele E, Emmenegger U, Engedal N, Engelbrecht AM, Engelender S, Enserink JM, Erdmann R, Erenpreisa J, Eri R, Eriksen JL, Erman A, Escalante R, Eskelinen EL, Espert L, Esteban-Martínez L, Evans TJ, Fabri M, Fabrias G, Fabrizi C, Facchiano A, Færgeman NJ, Faggioni A, Fairlie WD, Fan C, Fan D, Fan J, Fang S, Fanto M, Fanzani A, Farkas T, Faure M, Favier FB, Fearnhead H, Federici M, Fei E, Felizardo TC, Feng H, Feng Y, Ferguson TA, Fernández A, Fernandez-Barrena MG, Fernandez-Checa JC, Fernández-López A, Fernandez-Zapico ME, Feron O, Ferraro E, Ferreira-Halder CV, Fesus L, Feuer R, Fiesel FC, Filippi-Chiela EC, Filomeni G, Fimia GM, Fingert JH, Finkbeiner S, Finkel T, Fiorito F, Fisher PB, Flajolet M, Flamigni F, Florey O, Florio S, Floto RA, Folini M, Follo C, Fon EA, Fornai F, Fortunato F, Fraldi A, Franco R, Francois A, François A, Frankel LB, Fraser ID, Frey N, Freyssenet DG, Frezza C, Friedman SL, Frigo DE, Fu D, Fuentes JM, Fueyo J, Fujitani Y, Fujiwara Y, Fujiya M, Fukuda M, Fulda S, Fusco C, Gabryel B, Gaestel M, Gailly P, Gajewska M, Galadari S, Galili G, Galindo I, Galindo MF, Galliciotti

G, Galluzzi L, Galy V, Gammoh N, Gandy S, Ganesan AK, Ganesan S, Ganley IG, Gannagé M, Gao FB, Gao F, Gao JX, García Nannig L, García Véscovi E, Garcia-Macía M, Garcia-Ruiz C, Garg AD, Garg PK, Gargini R, Gassen NC, Gatica D, Gatti E, Gavard J, Gavathiotis E, Ge L, Ge P, Ge S, Gean PW, Gelmetti V, Genazzani AA, Geng J, Genschik P, Gerner L, Gestwicki JE, Gewirtz DA, Ghavami S, Ghigo E, Ghosh D, Giammarioli AM, Giampieri F, Giampietri C, Giatromanolaki A, Gibbings DJ, Gibellini L, Gibson SB, Ginet V, Giordano A, Giorgini F, Giovannetti E, Girardin SE, Gispert S, Giuliano S, Gladson CL, Glavic A, Gleave M, Godefroy N, Gogal RM, Gokulan K, Goldman GH, Goletti D, Goligorsky MS, Gomes AV, Gomes LC, Gomez H, Gomez-Manzano C, Gómez-Sánchez R, Gonçalves DA, Goncu E, Gong Q, Gongora C, Gonzalez CB, Gonzalez-Alegre P, Gonzalez-Cabo P, González-Polo RA, Goping IS, Gorbea C, Gorbunov NV, Goring DR, Gorman AM, Gorski SM, Goruppi S, Goto-Yamada S, Gotor C, Gottlieb RA, Gozes I, Gozuacik D, Graba Y, Graef M, Granato GE, Grant GD, Grant S, Gravina GL, Green DR, Greenhough A, Greenwood MT, Grimaldi B, Gros F, Grose C, Groulx JF, Gruber F, Grumati P, Grune T, Guan JL, Guan KL, Guerra B, Guillen C, Gulshan K, Gunst J, Guo C, Guo L, Guo M, Guo W, Guo XG, Gust AA, Gustafsson A, Gutierrez E, Gutierrez MG, Gwak HS, Haas A, Haber JE, Hadano S, Hagedorn M, Hahn DR, Halayko AJ, Hamacher-Brady A, Hamada K, Hamai A, Hamann A, Hamasaki M, Hamer I, Hamid Q, Hammond EM, Han F, Han W, Handa JT, Hanover JA, Hansen M, Harada M, Harhaji-Trajkovic L, Harper JW, Harrath AH, Harris AL, Harris J, Hasler U, Hasselblatt P, Hasui K, Hawley RG, Hawley TS, He C, He CY, He F, He G, He RR, He XH, He YW, He YY, Heath JK, Hébert MJ, Heinzen RA, Helgason GV, Hensel M, Henske EP, Her C, Herman PK, Hernández A, Hernandez C, Hernández-Tiedra S, Hetz C, Hiesinger PR, Higaki K, Hilfiker S, Hill BG, Hill JA, Hill WD, Hino K, Hofius D, Hofman P, Höglinger GU, Höhfeld J, Holz MK, Hong Y, Hood DA, Hoozemans JJ, Hoppe T, Hsu C, Hsu CY, Hsu LC, Hu D, Hu G, Hu HM, Hu H, Hu MC, Hu YC, Hu ZW, Hua F, Hua Y, Huang C, Huang HL, Huang KH, Huang KY, Huang S, Huang WP, Huang YR, Huang Y, Huber TB, Huebbe P, Huh WK, Hulmi JJ, Hur GM, Hurley JH, Husak Z, Hussain SN, Hussain S, Hwang JJ, Hwang S, Hwang TI, Ichihara A, Imai Y, Imbriano C, Inomata M, Into T, Iovane V, Iovanna JL, Iozzo RV, Ip NY, Irazoqui JE, Iribarren P, Isaka Y, Isakovic AJ, Ischiropoulos H, Isenberg JS, Ishaq M, Ishida H, Ishii I, Ishmael JE, Isidoro C, Isobe K, Isono E, Issazadeh-Navikas S, Itahana K, Itakura E, Ivanov AI, Iyer AK, Izquierdo JM, Izumi Y, Izzo V, Jäättelä M, Jaber N, Jackson DJ, Jackson WT, Jacob TG, Jacques TS, Jagannath C, Jain A, Jana NR, Jang BK, Jani A, Janji B, Jannig PR, Jansson PJ, Jean S, Jendrach M, Jeon JH, Jessen N, Jeung EB, Jia K, Jia L, Jiang H, Jiang L, Jiang T, Jiang X, Jiang Y, Jiménez A, Jin C, Jin H, Jin L, Jin M, Jin S, Jinwal UK, Jo EK, Johansen T, Johnson DE, Johnson GV, Johnson JD, Jonasch E, Jones C, Joosten LA, Jordan J, Joseph AM, Joseph B, Joubert AM, Ju D, Ju J, Juan HF, Juenemann K, Juhász G, Jung HS, Jung JU, Jung YK, Jungbluth H, Justice MJ, Jutten B, Kaakoush NO, Kaarniranta K, Kaasik A, Kabuta T, Kaeffer B, Kågedal K, Kahana A, Kajimura S, Kakhlon O, Kalia M, Kalvakolanu DV, Kamada Y, Kambas K, Kaminskyy VO, Kampinga HH, Kandouz M, Kang C, Kang R, Kang TC, Kanki T, Kanneganti TD, Kanno H, Kanthasamy AG, Kantorow M, Kaparakis-Liaskos M, Kapuy O, Karantza V, Karim MR, Karmakar P, Kaser A, Kaushik S, Kawula T, Kaynar AM, Ke PY, Ke ZJ, Kehrl JH, Keller KE, Kemper JK, Kenworthy AK, Kepp O, Kern A, Kesari S, Kessel D, Ketteler R, Kettelhut IC, Khambu B, Khan MM, Khandelwal VK, Khare S, Kiang JG, Kiger AA, Kihara A, Kim AL, Kim CH, Kim DR, Kim DH, Kim EK, Kim HY, Kim HR, Kim JS, Kim JH, Kim JC, Kim KW, Kim MD, Kim MM, Kim PK, Kim SW, Kim SY, Kim YS, Kim Y, Kimchi A, Kimmelman AC, Kimura T, King JS, Kirkegaard K, Kirkin V, Kirshenbaum LA, Kishi S, Kitajima Y, Kitamoto K, Kitaoka Y, Kitazato K, Kley RA, Klimecki WT, Klinkenberg M, Klucken J, Knævelsrud H, Knecht E, Knuppertz L, Ko JL, Kobayashi S, Koch JC, Koechlin-Ramonatxo C, Koenig U, Koh YH, Köhler K, Kohlwein SD, Koike M, Komatsu M, Kominami E, Kong D, Kong HJ, Konstantakou EG, Kopp BT, Korcsmaros T, Korhonen L, Korolchuk VI, Koshkina NV, Kou Y, Koukourakis MI, Koumenis C, Kovács AL, Kovács T, Kovacs WJ, Koya D, Kraft C, Krainc D, Kramer H, Kravic-Stevovic T, Krek W, Kretz-Remy C, Krick R, Krishnamurthy M, Kriston-Vizi J,

Kroemer G, Kruer MC, Kruger R, Ktistakis NT, Kuchitsu K, Kuhn C, Kumar AP, Kumar A, Kumar D, Kumar R, Kumar S, Kundu M, Kung HJ, Kuno A, Kuo SH, Kuret J, Kurz T, Kwok T, Kwon TK, Kwon YT, Kyrmizi I, La Spada AR, Lafont F, Lahm T, Lakkaraju A, Lam T, Lamark T, Lancel S, Landowski TH, Lane DJ, Lane JD, Lanzi C, Lapaquette P, Lapierre LR, Laporte J, Laukkarinen J, Laurie GW, Lavandero S, Lavie L, LaVoie MJ, Law BY, Law HK, Law KB, Layfield R, Lazo PA, Le Cam L, Le Roch KG, Le Stunff H, Leardkamolkarn V, Lecuit M, Lee BH, Lee CH, Lee EF, Lee GM, Lee HJ, Lee H, Lee JK, Lee J, Lee JH, Lee M, Lee MS, Lee PJ, Lee SW, Lee SJ, Lee SY, Lee SH, Lee SS, Lee S, Lee YR, Lee YJ, Lee YH, Leeuwenburgh C, Lefort S, Legouis R, Lei J, Lei QY, Leib DA, Leibowitz G, Lekli I, Lemaire SD, Lemasters JJ, Lemberg MK, Lemoine A, Leng S, Lenz G, Lenzi P, Lerman LO, Lettieri Barbato D, Leu JI, Leung HY, Levine B, Lewis PA, Lezoualc'h F, Li C, Li F, Li FJ, Li J, Li K, Li L, Li M, Li Q, Li R, Li S, Li W, Li X, Li Y, Lian J, Liang C, Liang Q, Liao Y, Liberal J, Liberski PP, Lie P, Lieberman AP, Lim HJ, Lim KL, Lim K, Lima RT, Lin CS, Lin CF, Lin F, Lin FC, Lin K, Lin KH, Lin PH, Lin T, Lin WW, Lin YS, Lin Y, Linden R, Lindholm D, Lindqvist LM, Lingor P, Linkermann A, Liotta LA, Lipinski MM, Lira VA, Lisanti MP, Liton PB, Liu B, Liu C, Liu CF, Liu F, Liu HJ, Liu J, Liu JJ, Liu JL, Liu K, Liu L, Liu Q, Liu RY, Liu S, Liu W, Liu XD, Liu X, Liu XH, Liu Y, Liu Z, Liuzzi JP, Lizard G, Ljujic M, Lodhi IJ, Logue SE, Lokeshwar BL, Long YC, Lonial S, Loos B, López-Otín C, López-Vicario C, Lorente M, Lorenzi PL, Lõrincz P, Los M, Lotze MT, Lovat PE, Lu B, Lu J, Lu Q, Lu SM, Lu S, Lu Y, Luciano F, Luckhart S, Lucocq JM, Ludovico P, Lugea A, Lukacs NW, Lum JJ, Lund AH, Luo H, Luo J, Luo S, Luparello C, Lyons T, Ma J, Ma Y, Ma Z, Machado J, Machado-Santelli GM, Macian F, MacIntosh GC, MacKeigan JP, MacLeod KF, MacMicking JD, MacMillan-Crow LA, Madeo F, Madesh M, Madrigal-Matute J, Maeda A, Maeda T, Maegawa G, Maellaro E, Maes H, Magariños M, Maiese K, Maiti TK, Maiuri L, Maiuri MC, Maki CG, Malli R, Malorni W, Maloyan A, Mami-Chouaib F, Man N, Mancias JD, Mandelkow EM, Mandell MA, Manfredi AA, Manié SN, Manzoni C, Mao K, Mao Z, Mao ZW, Marambaud P, Marconi AM, Marelja Z, Marfe G, Margeta M, Margittai E, Mari M, Mariani FV, Marin C, Marinelli S, Mariño G, Markovic I, Marquez R, Martelli AM, Martens S, Martin KR, Martin SJ, Martin S, Martin-Acebes MA, Martín-Sanz P, Martinand-Mari C, Martinet W, Martinez J, Martinez-Lopez N, Martinez-Outschoorn U, Martínez-Velázquez M, Martinez-Vicente M, Martins WK, Mashima H, Mastrianni JA, Matarese G, Matarrese P, Mateo R, Matoba S, Matsumoto N, Matsushita T, Matsuura A, Matsuzawa T, Mattson MP, Matus S, Maugeri N, Mauvezin C, Mayer A, Maysinger D, Mazzolini GD, McBrayer MK, McCall K, McCormick C, McInerney GM, McIver SC, McKenna S, McMahon JJ, McNeish IA, Mechta-Grigoriou F, Medema JP, Medina DL, Megyeri K, Mehrpour M, Mehta JL, Mei Y, Meier UC, Meijer AJ, Meléndez A, Melino G, Melino S, de Melo EJ, Mena MA, Meneghini MD, Menendez JA, Menezes R, Meng L, Meng LH, Meng S, Menghini R, Menko AS, Menna-Barreto RF, Menon MB, Meraz-Ríos MA, Merla G, Merlini L, Merlot AM, Meryk A, Meschini S, Meyer JN, Mi MT, Miao CY, Micale L, Michaeli S, Michiels C, Migliaccio AR, Mihailidou AS, Mijaljica D, Mikoshiba K, Milan E, Miller-Fleming L, Mills GB, Mills IG, Minakaki G, Minassian BA, Ming XF, Minibayeva F, Minina EA, Mintern JD, Minucci S, Miranda-Vizuete A, Mitchell CH, Miyamoto S, Miyazawa K, Mizushima N, Mnich K, Mograbi B, Mohseni S, Moita LF, Molinari M, Møller AB, Mollereau B, Mollinedo F, Mongillo M, Monick MM, Montagnaro S, Montell C, Moore DJ, Moore MN, Mora-Rodriguez R, Moreira PI, Morel E, Morelli MB, Moreno S, Morgan MJ, Moris A, Moriyasu Y, Morrison JL, Morrison LA, Morselli E, Moscat J, Moseley PL, Mostowy S, Motori E, Mottet D, Mottram JC, Moussa CE, Mpakou VE, Mukhtar H, Mulcahy Levy JM, Muller S, Muñoz-Moreno R, Muñoz-Pinedo C, Münz C, Murphy ME, Murray JT, Murthy A, Mysorekar IU, Nabi IR, Nabissi M, Nader GA, Nagahara Y, Nagai Y, Nagata K, Nagelkerke A, Nagy P, Naidu SR, Nair S, Nakano H, Nakatogawa H, Nanjundan M, Napolitano G, Naqvi NI, Nardacci R, Narendra DP, Narita M, Nascimbeni AC, Natarajan R, Navegantes LC, Nawrocki ST, Nazarko TY, Nazarko VY, Neill T, Neri LM, Netea MG, Netea-Maier RT, Neves BM, Ney PA, Nezis IP, Nguyen HT, Nguyen HP, Nicot AS, Nilsen H, Nilsson P, Nishimura M, Nishino

I, Niso-Santano M, Niu H, Nixon RA, Njar VC, Noda T, Noegel AA, Nolte EM, Norberg E, Norga KK, Noureini SK, Notomi S, Notterpek L, Nowikovsky K, Nukina N, Nürnberger T, O'Donnell VB, O'Donovan T, O'Dwyer PJ, Oehme I, Oeste CL, Ogawa M, Ogretmen B, Ogura Y, Oh YJ, Ohmuraya M, Ohshima T, Ojha R, Okamoto K, Okazaki T, Oliver FJ, Ollinger K, Olsson S, Orban DP, Ordonez P, Orhon I, Orosz L, Orourke EJ, Orozco H, Ortega AL, Ortona E, Osellame LD, Oshima J, Oshima S, Osiewacz HD, Otomo T, Otsu K, Ou JH, Outeiro TF, Ouyang DY, Ouyang H, Overholtzer M, Ozbun MA, Ozdinler PH, Ozpolat B, Pacelli C, Paganetti P, Page G, Pages G, Pagnini U, Pajak B, Pak SC, Pakos-Zebrucka K, Pakpour N, Palková Z, Palladino F, Pallauf K, Pallet N, Palmieri M, Paludan SR, Palumbo C, Palumbo S, Pampliega O, Pan H, Pan W, Panaretakis T, Pandey A, Pantazopoulou A, Papackova Z, Papademetrio DL, Papassideri I, Papini A, Parajuli N, Pardo J, Parekh VV, Parenti G, Park JI, Park J, Park OK, Parker R, Parlato R, Parys JB, Parzych KR, Pasquet JM, Pasquier B, Pasumarthi KB, Patschan D, Patterson C, Pattingre S, Pattison S, Pause A, Pavenstädt H, Pavone F, Pedrozo Z, Peña FJ, Peñalva MA, Pende M, Peng J, Penna F, Penninger JM, Pensalfini A, Pepe S, Pereira GJ, Pereira PC, Pérez-de la Cruz V, Pérez-Pérez ME, Pérez-Rodríguez D, Pérez-Sala D, Perier C, Perl A, Perlmutter DH, Perrotta I, Pervaiz S, Pesonen M, Pessin JE, Peters GJ, Petersen M, Petrache I, Petrof BJ, Petrovski G, Phang JM, Piacentini M, Pierdominici M, Pierre P, Pierrefite-Carle V, Pietrocola F, Pimentel-Muiños FX, Pinar M, Pineda B, Pinkas-Kramarski R, Pinti M, Pinton P, Piperdi B, Piret JM, Platanias LC, Platta HW, Plowey ED, Pöggeler S, Poirot M, Polčic P, Poletti A, Poon AH, Popelka H, Popova B, Poprawa I, Poulose SM, Poulton J, Powers SK, Powers T, Pozuelo-Rubio M, Prak K, Prange R, Prescott M, Priault M, Prince S, Proia RL, Proikas-Cezanne T, Prokisch H, Promponas VJ, Przyklenk K, Puertollano R, Pugazhenthi S, Puglielli L, Pujol A, Puyal J, Pyeon D, Qi X, Qian WB, Qin ZH, Qiu Y, Qu Z, Quadrilatero J, Quinn F, Raben N, Rabinowich H, Radogna F, Ragusa MJ, Rahmani M, Raina K, Ramanadham S, Ramesh R, Rami A, Randall-Demllo S, Randow F, Rao H, Rao VA, Rasmussen BB, Rasse TM, Ratovitski EA, Rautou PE, Ray SK, Razani B, Reed BH, Reggiori F, Rehm M, Reichert AS, Rein T, Reiner DJ, Reits E, Ren J, Ren X, Renna M, Reusch JE, Revuelta JL, Reyes L, Rezaie AR, Richards RI, Richardson DR, Richetta C, Riehle MA, Rihn BH, Rikihisa Y, Riley BE, Rimbach G, Rippo MR, Ritis K, Rizzi F, Rizzo E, Roach PJ, Robbins J, Roberge M, Roca G, Roccheri MC, Rocha S, Rodrigues CM, Rodríguez CI, de Cordoba SR, Rodriguez-Muela N, Roelofs J, Rogov VV, Rohn TT, Rohrer B, Romanelli D, Romani L, Romano PS, Roncero MI, Rosa JL, Rosello A, Rosen KV, Rosenstiel P, Rost-Roszkowska M, Roth KA, Roué G, Rouis M, Rouschop KM, Ruan DT, Ruano D, Rubinsztein DC, Rucker EB, Rudich A, Rudolf E, Rudolf R, Ruegg MA, Ruiz-Roldan C, Ruparelia AA, Rusmini P, Russ DW, Russo GL, Russo G, Russo R, Rusten TE, Ryabovol V, Ryan KM, Ryter SW, Sabatini DM, Sacher M, Sachse C, Sack MN, Sadoshima J, Saftig P, Sagi-Eisenberg R, Sahni S, Saikumar P, Saito T, Saitoh T, Sakakura K, Sakoh-Nakatogawa M, Sakuraba Y, Salazar-Roa M, Salomoni P, Saluja AK, Salvaterra PM, Salvioli R, Samali A, Sanchez AM, Sánchez-Alcázar JA, Sanchez-Prieto R, Sandri M, Sanjuan MA, Santaguida S, Santambrogio L, Santoni G, Dos Santos CN, Saran S, Sardiello M, Sargent G, Sarkar P, Sarkar S, Sarrias MR, Sarwal MM, Sasakawa C, Sasaki M, Sass M, Sato K, Sato M, Satriano J, Savaraj N, Saveljeva S, Schaefer L, Schaible UE, Scharl M, Schatzl HM, Schekman R, Scheper W, Schiavi A, Schipper HM, Schmeisser H, Schmidt J, Schmitz I, Schneider BE, Schneider EM, Schneider JL, Schon EA, Schönenberger MJ, Schönthal AH, Schorderet DF, Schröder B, Schuck S, Schulze RJ, Schwarten M, Schwarz TL, Sciarretta S, Scotto K, Scovassi AI, Screaton RA, Screen M, Seca H, Sedej S, Segatori L, Segev N, Seglen PO, Seguí-Simarro JM, Segura-Aguilar J, Seki E, Sell C, Seiliez I, Semenkovich CF, Semenza GL, Sen U, Serra AL, Serrano-Puebla A, Sesaki H, Setoguchi T, Settembre C, Shacka JJ, Shajahan-Haq AN, Shapiro IM, Sharma S, She H, Shen CK, Shen CC, Shen HM, Shen S, Shen W, Sheng R, Sheng X, Sheng ZH, Shepherd TG, Shi J, Shi Q, Shi Y, Shibutani S, Shibuya K, Shidoji Y, Shieh JJ, Shih CM, Shimada Y, Shimizu S, Shin DW, Shinohara ML, Shintani M, Shintani T, Shioi T, Shirabe K, Shiri-Sverdlov R, Shirihai O, Shore GC, Shu CW, Shukla D, Sibirny AA, Sica V, Sigurdson CJ, Sigurdsson EM, Sijwali PS,

Sikorska B, Silveira WA, Silvente-Poirot S, Silverman GA, Simak J, Simmet T, Simon AK, Simon HU, Simone C, Simons M, Simonsen A, Singh R, Singh SV, Singh SK, Sinha D, Sinha S, Sinicrope FA, Sirko A, Sirohi K, Sishi BJ, Sittler A, Siu PM, Sivridis E, Skwarska A, Slack R, Slaninová I, Slavov N, Smaili SS, Smalley KS, Smith DR, Soenen SJ, Soleimanpour SA, Solhaug A, Somasundaram K, Son JH, Sonawane A, Song C, Song F, Song HK, Song JX, Song W, Soo KY, Sood AK, Soong TW, Soontornniyomkij V, Sorice M, Sotgia F, Soto-Pantoja DR, Sotthibundhu A, Sousa MJ, Spaink HP, Span PN, Spang A, Sparks JD, Speck PG, Spector SA, Spies CD, Springer W, Clair DS, Stacchiotti A, Staels B, Stang MT, Starczynowski DT, Starokadomskyy P, Steegborn C, Steele JW, Stefanis L, Steffan J, Stellrecht CM, Stenmark H, Stepkowski TM, Stern ST, Stevens C, Stockwell BR, Stoka V, Storchova Z, Stork B, Stratoulias V, Stravopodis DJ, Strnad P, Strohecker AM, Ström AL, Stromhaug P, Stulik J, Su YX, Su Z, Subauste CS, Subramaniam S, Sue CM, Suh SW, Sui X, Sukseree S, Sulzer D, Sun FL, Sun J, Sun SY, Sun Y, Sundaramoorthy V, Sung J, Suzuki H, Suzuki K, Suzuki N, Suzuki T, Suzuki YJ, Swanson MS, Swanton C, Swärd K, Swarup G, Sweeney ST, Sylvester PW, Szatmari Z, Szegezdi E, Szlosarek PW, Taegtmeyer H, Tafani M, Taillebourg E, Tait SW, Takacs-Vellai K, Takahashi Y, Takáts S, Takemura G, Takigawa N, Talbot NJ, Tamagno E, Tamburini J, Tan CP, Tan L, Tan ML, Tan M, Tan YJ, Tanaka K, Tanaka M, Tang D, Tang G, Tanida I, Tanji K, Tannous BA, Tapia JA, Tasset-Cuevas I, Tatar M, Tavassoly I, Tavernarakis N, Taylor A, Taylor GS, Taylor GA, Taylor JP, Taylor MJ, Tchetina EV, Tee AR, Teixeira-Clerc F, Telang S, Tencomnao T, Teng BB, Teng RJ, Terro F, Tettamanti G, Theiss AL, Theron AE, Thomas KJ, Thomé MP, Thomes PG, Thorburn A, Thorner J, Thum T, Thumm M, Thurston TL, Tian L, Till A, Ting JP, Titorenko VI, Toker L, Toldo S, Tooze SA, Topisirovic I, Torgersen ML, Torosantucci L, Torriglia A, Torrisi MR, Tournier C, Towns R, Trajkovic V, Travassos LH, Triola G, Tripathi DN, Trisciuoglio D, Troncoso R, Trougakos IP, Truttmann AC, Tsai KJ, Tschan MP, Tseng YH, Tsukuba T, Tsung A, Tsvetkov AS, Tu S, Tuan HY, Tucci M, Tumbarello DA, Turk B, Turk V, Turner RF, Tveita AA, Tyagi SC, Ubukata M, Uchiyama Y, Udelnow A, Ueno T, Umekawa M, Umemiya-Shirafuji R, Underwood BR, Ungermann C, Ureshino RP, Ushioda R, Uversky VN, Uzcátegui NL, Vaccari T, Vaccaro MI, Váchová L, Vakifahmetoglu-Norberg H, Valdor R, Valente EM, Vallette F, Valverde AM, Van den Berghe G, Van Den Bosch L, van den Brink GR, van der Goot FG, van der Klei IJ, van der Laan LJ, van Doorn WG, van Egmond M, van Golen KL, Van Kaer L, van Lookeren Campagne M, Vandenabeele P, Vandenberghe W, Vanhorebeek I, Varela-Nieto I, Vasconcelos MH, Vasko R, Vavvas DG, Vega-Naredo I, Velasco G, Velentzas AD, Velentzas PD, Vellai T, Vellenga E, Vendelbo MH, Venkatachalam K, Ventura N, Ventura S, Veras PS, Verdier M, Vertessy BG, Viale A, Vidal M, Vieira HL, Vierstra RD, Vigneswaran N, Vij N, Vila M, Villar M, Villar VH, Villarroya J, Vindis C, Viola G, Viscomi MT, Vitale G, Vogl DT, Voitsekhovskaja OV, von Haefen C, von Schwarzenberg K, Voth DE, Vouret-Craviari V, Vuori K, Vyas JM, Waeber C, Walker CL, Walker MJ, Walter J, Wan L, Wan X, Wang B, Wang C, Wang CY, Wang D, Wang F, Wang G, Wang HJ, Wang H, Wang HG, Wang HD, Wang J, Wang M, Wang MQ, Wang PY, Wang P, Wang RC, Wang S, Wang TF, Wang X, Wang XJ, Wang XW, Wang Y, Wang YJ, Wang YT, Wang ZN, Wappner P, Ward C, Ward DM, Warnes G, Watada H, Watanabe Y, Watase K, Weaver TE, Weekes CD, Wei J, Weide T, Weihl CC, Weindl G, Weis SN, Wen L, Wen X, Wen Y, Westermann B, Weyand CM, White AR, White E, Whitton JL, Whitworth AJ, Wiels J, Wild F, Wildenberg ME, Wileman T, Wilkinson DS, Wilkinson S, Willbold D, Williams C, Williams K, Williamson PR, Winklhofer KF, Witkin SS, Wohlgemuth SE, Wollert T, Wolvetang EJ, Wong E, Wong GW, Wong RW, Wong VK, Woodcock EA, Wright KL, Wu C, Wu D, Wu GS, Wu J, Wu M, Wu S, Wu WK, Wu Y, Wu Z, Xavier CP, Xavier RJ, Xia GX, Xia T, Xia W, Xia Y, Xiao H, Xiao J, Xiao S, Xiao W, Xie CM, Xie Z, Xilouri M, Xiong Y, Xu C, Xu F, Xu H, Xu J, Xu L, Xu X, Xu Y, Xu ZX, Xu Z, Xue Y, Yamada T, Yamamoto A, Yamanaka K, Yamashina S, Yamashiro S, Yan B, Yan X, Yan Z, Yanagi Y, Yang DS, Yang JM, Yang L, Yang M, Yang PM, Yang P, Yang Q, Yang W, Yang WY, Yang X, Yang Y, Yang Z, Yao MC, Yao PJ, Yao X, Yao Z, Yasui LS, Ye M, Yedvobnick B, Yeganeh B, Yeh ES, Yeyati PL, Yi F, Yi L, Yin XM, Yip CK, Yoo YM, Yoo YH, Yoon SY,

Yoshida K, Yoshimori T, Young KH, Yu H, Yu JJ, Yu JT, Yu J, Yu L, Yu WH, Yu XF, Yu Z, Yuan J, Yuan ZM, Yue BY, Yue J, Yue Z, Zacks DN, Zacksenhaus E, Zaffaroni N, Zaglia T, Zakeri Z, Zecchini V, Zeng J, Zeng M, Zeng Q, Zervos AS, Zhang DD, Zhang F, Zhang G, Zhang GC, Zhang H, Zhang J, Zhang JP, Zhang L, Zhang MY, Zhang X, Zhang XD, Zhang Y, Zhao M, Zhao WL, Zhao X, Zhao YG, Zhao Y, Zhao YX, Zhao Z, Zhao ZJ, Zheng D, Zheng XL, Zheng X, Zhivotovsky B, Zhong Q, Zhou GZ, Zhou G, Zhou H, Zhou SF, Zhou XJ, Zhu H, Zhu WG, Zhu W, Zhu XF, Zhu Y, Zhuang SM, Zhuang X, Ziparo E, Zois CE, Zoladek T, Zong WX, Zorzano A, Zughaier SM. Guidelines for the use and interpretation of assays for monitoring autophagy (3rd edition). Autophagy. 2016;12:1–222.

35. He C, Klionsky DJ. Regulation mechanisms and signaling pathways of autophagy. Annu Rev Genet. 2009;43:67–93.
36. Chen Y, Klionsky DJ. The regulation of autophagy – unanswered questions. J Cell Sci. 2011;124:161–70.
37. Dunham-Snary KJ, Ballinger SW. Mitochondrial genetics and obesity: evolutionary adaptation and contemporary disease susceptibility. Free Radic Biol Med. 2013;65:1229–37.
38. Egan DF, Shackelford DB, Mihaylova MM, Gelino S, Kohnz RA, Mair W, Vasquez DS, Joshi A, Gwinn DM, Taylor R, Asara JM, Fitzpatrick J, Dillin A, Viollet B, Kundu M, Hansen M, Shaw RJ. Phosphorylation of ULK1 (hATG1) by AMP-activated protein kinase connects energy sensing to mitophagy. Science. 2011;331:456–61.
39. Menzies RA, Gold PH. The turnover of mitochondria in a variety of tissues of young adult and aged rats. J Biol Chem. 1971;246:2425–9.
40. Aschenbrenner V, Zak R, Cutilletta AF, Rabinowitz M. Effect of hypoxia on degradation of mitochondrial components in rat cardiac muscle. Am J Phys. 1971;221:1418–25.
41. Dengjel J, Kristensen AR, Andersen JS. Ordered bulk degradation via autophagy. Autophagy. 2008;4:1057–9.
42. Huang C, Yitzhaki S, Perry CN, Liu W, Giricz Z, Mentzer RM, Gottlieb RA. Autophagy induced by ischemic preconditioning is essential for cardioprotection. J Cardiovasc Transl Res. 2010;3:365–73.
43. Wei C, Li H, Han L, Zhang L, Yang X. Activation of autophagy in ischemic postconditioning contributes to cardioprotective effects against ischemia/reperfusion injury in rat hearts. J Cardiovasc Pharmacol. 2013;61:416–22.
44. Andres AM, Stotland A, Queliconi BB, Gottlieb RA. A time to reap, a time to sow: mitophagy and biogenesis in cardiac pathophysiology. J Mol Cell Cardiol. 2015;78:62–72.
45. Kubli DA, Zhang X, Lee Y, Hanna RA, Quinsay MN, Nguyen CK, Jimenez R, Petrosyan S, Murphy AN, Gustafsson AB. Parkin protein deficiency exacerbates cardiac injury and reduces survival following myocardial infarction. J Biol Chem. 2013;288:915–26.
46. Perrelli MG, Pagliaro P, Penna C. Ischemia/reperfusion injury and cardioprotective mechanisms: role of mitochondria and reactive oxygen species. World J Cardiol. 2011;3:186–200.
47. Hill BG, Darley-Usmar VM. S-nitrosation and thiol switching in the mitochondrion: a new paradigm for cardioprotection in ischaemic preconditioning. Biochem J. 2008;412:e11–3.
48. Lecour S. Multiple protective pathways against reperfusion injury: a SAFE path without Aktion? J Mol Cell Cardiol. 2009;46:607–9.
49. Vinten-Johansen J, Zhao ZQ, Jiang R, Zatta AJ, Dobson GP. Preconditioning and postconditioning: innate cardioprotection from ischemia-reperfusion injury. J Appl Physiol (1985). 2007;103:1441–8.
50. Hausenloy DJ, Mocanu MM, Yellon DM. Cross-talk between the survival kinases during early reperfusion: its contribution to ischemic preconditioning. Cardiovasc Res. 2004;63:305–12.
51. Hausenloy DJ, Yellon DM. Reperfusion injury salvage kinase signalling: taking a RISK for cardioprotection. Heart Fail Rev. 2007;12:217–34.
52. Juhaszova M, Wang S, Zorov DB, Nuss HB, Gleichmann M, Mattson MP, Sollott SJ. The identity and regulation of the mitochondrial permeability transition pore: where the known meets the unknown. Ann N Y Acad Sci. 2008;1123:197–212.

53. Hausenloy DJ, Yellon DM. The mitochondrial permeability transition pore: its fundamental role in mediating cell death during ischaemia and reperfusion. J Mol Cell Cardiol. 2003;35:339–41.
54. Hausenloy DJ, Duchen MR, Yellon DM. Inhibiting mitochondrial permeability transition pore opening at reperfusion protects against ischaemia-reperfusion injury. Cardiovasc Res. 2003;60:617–25.
55. Li J, Xuan W, Yan R, Tropak MB, Jean-St-Michel E, Liang W, Gladstone R, Backx PH, Kharbanda RK, Redington AN. Remote preconditioning provides potent cardioprotection via PI3K/Akt activation and is associated with nuclear accumulation of β-catenin. Clin Sci (Lond). 2011;120:451–62.
56. Miura T, Tanno M, Sato T. Mitochondrial kinase signalling pathways in myocardial protection from ischaemia/reperfusion-induced necrosis. Cardiovasc Res. 2010;88:7–15.
57. Lacerda L, Somers S, Opie LH, Lecour S. Ischaemic postconditioning protects against reperfusion injury via the SAFE pathway. Cardiovasc Res. 2009;84:201–8.
58. Kleinbongard P, Schulz R, Heusch G. TNFα in myocardial ischemia/reperfusion, remodeling and heart failure. Heart Fail Rev. 2011;16:49–69.
59. Heusch G, Boengler K, Schulz R. Inhibition of mitochondrial permeability transition pore opening: the Holy Grail of cardioprotection. Basic Res Cardiol. 2010;105:151–4.
60. García-Heredia A, Marsillach J, Rull A, Triguero I, Fort I, Mackness B, Mackness M, Shih DM, Joven J, Camps J. Paraoxonase-1 inhibits oxidized low-density lipoprotein-induced metabolic alterations and apoptosis in endothelial cells: a nondirected metabolomic study. Mediat Inflamm. 2013;2013:156053.
61. Sangle GV, Chowdhury SK, Xie X, Stelmack GL, Halayko AJ, Shen GX. Impairment of mitochondrial respiratory chain activity in aortic endothelial cells induced by glycated low-density lipoprotein. Free Radic Biol Med. 2010;48:781–90.
62. Cockerill GW, McDonald MC, Mota-Filipe H, Cuzzocrea S, Miller NE, Thiemermann C. High density lipoproteins reduce organ injury and organ dysfunction in a rat model of hemorrhagic shock. FASEB J. 2001;15:1941–52.
63. Dadabayev AR, Yin G, Latchoumycandane C, McIntyre TM, Lesnefsky EJ, Penn MS. Apolipoprotein A1 regulates coenzyme Q10 absorption, mitochondrial function, and infarct size in a mouse model of myocardial infarction. J Nutr. 2014;144:1030–6.
64. Szczepanek K, Chen Q, Derecka M, Salloum FN, Zhang Q, Szelag M, Cichy J, Kukreja RC, Dulak J, Lesnefsky EJ, Larner AC. Mitochondrial-targeted Signal transducer and activator of transcription 3 (STAT3) protects against ischemia-induced changes in the electron transport chain and the generation of reactive oxygen species. J Biol Chem. 2011;286:29610–20.
65. Wegrzyn J, Potla R, Chwae YJ, Sepuri NB, Zhang Q, Koeck T, Derecka M, Szczepanek K, Szelag M, Gornicka A, Moh A, Moghaddas S, Chen Q, Bobbili S, Cichy J, Dulak J, Baker DP, Wolfman A, Stuehr D, Hassan MO, Fu XY, Avadhani N, Drake JI, Fawcett P, Lesnefsky EJ, Larner AC. Function of mitochondrial Stat3 in cellular respiration. Science. 2009;323:793–7.
66. Tammineni P, Anugula C, Mohammed F, Anjaneyulu M, Larner AC, Sepuri NB. The import of the transcription factor STAT3 into mitochondria depends on GRIM-19, a component of the electron transport chain. J Biol Chem. 2013;288:4723–32.
67. Boengler K, Hilfiker-Kleiner D, Heusch G, Schulz R. Inhibition of permeability transition pore opening by mitochondrial STAT3 and its role in myocardial ischemia/reperfusion. Basic Res Cardiol. 2010;105:771–85.
68. Boengler K. Ischemia/reperfusion injury: the benefit of having STAT3 in the heart. J Mol Cell Cardiol. 2011;50:587–8.
69. Boengler K, Ungefug E, Heusch G, Schulz R. The STAT3 inhibitor stattic impairs cardiomyocyte mitochondrial function through increased reactive oxygen species formation. Curr Pharm Des. 2013;19:6890–5.
70. Ashby DT, Rye KA, Clay MA, Vadas MA, Gamble JR, Barter PJ. Factors influencing the ability of HDL to inhibit expression of vascular cell adhesion molecule-1 in endothelial cells. Arterioscler Thromb Vasc Biol. 1998;18:1450–5.

71. Baker PW, Rye KA, Gamble JR, Vadas MA, Barter PJ. Ability of reconstituted high density lipoproteins to inhibit cytokine-induced expression of vascular cell adhesion molecule-1 in human umbilical vein endothelial cells. J Lipid Res. 1999;40:345–53.
72. Wiesner P, Leidl K, Boettcher A, Schmitz G, Liebisch G. Lipid profiling of FPLC-separated lipoprotein fractions by electrospray ionization tandem mass spectrometry. J Lipid Res. 2009;50:574–85.
73. Yatomi Y, Ozaki Y, Ohmori T, Igarashi Y. Sphingosine 1-phosphate: synthesis and release. Prostaglandins Other Lipid Mediat. 2001;64:107–22.
74. Kontush A, Therond P, Zerrad A, Couturier M, Négre-Salvayre A, de Souza JA, Chantepie S, Chapman MJ. Preferential sphingosine-1-phosphate enrichment and sphingomyelin depletion are key features of small dense HDL3 particles: relevance to antiapoptotic and antioxidative activities. Arterioscler Thromb Vasc Biol. 2007;27:1843–9.
75. Frias MA, Lecour S, James RW, Pedretti S. High density lipoprotein/sphingosine-1-phosphate-induced cardioprotection: role of STAT3 as part of the SAFE pathway. JAKSTAT. 2012;1:92–100.
76. Novák J, Olejníčková V, Tkáčová N, Santulli G. Mechanistic role of microRNAs in coupling lipid metabolism and atherosclerosis. Adv Exp Med Biol. 2015;887:79–100.
77. Tao R, Hoover HE, Honbo N, Kalinowski M, Alano CC, Karliner JS, Raffai R. High-density lipoprotein determines adult mouse cardiomyocyte fate after hypoxia-reoxygenation through lipoprotein-associated sphingosine 1-phosphate. Am J Physiol Heart Circ Physiol. 2010;298:H1022–8.
78. Kimura T, Sato K, Malchinkhuu E, Tomura H, Tamama K, Kuwabara A, Murakami M, Okajima F. High-density lipoprotein stimulates endothelial cell migration and survival through sphingosine 1-phosphate and its receptors. Arterioscler Thromb Vasc Biol. 2003;23:1283–8.
79. Somers SJ, Frias M, Lacerda L, Opie LH, Lecour S. Interplay between SAFE and RISK pathways in sphingosine-1-phosphate-induced cardioprotection. Cardiovasc Drugs Ther. 2012;26:227–37.
80. Sattler K, Levkau B. Sphingosine-1-phosphate as a mediator of high-density lipoprotein effects in cardiovascular protection. Cardiovasc Res. 2009;82:201–11.
81. Deshpande GP, McCarthy J, Mardikar H, Lecour S, Opie L. Effects of sphingosine-1-phosphate on acute contractile heart failure (ACHF). Cardiovasc Drugs Ther. 2010;24:459–60.
82. Jin ZQ, Zhou HZ, Zhu P, Honbo N, Mochly-Rosen D, Messing RO, Goetzl EJ, Karliner JS, Gray MO. Cardioprotection mediated by sphingosine-1-phosphate and ganglioside GM-1 in wild-type and PKC epsilon knockout mouse hearts. Am J Physiol Heart Circ Physiol. 2002;282:H1970–7.
83. Frias MA, Pedretti S, Hacking D, Somers S, Lacerda L, Opie LH, James RW, Lecour S. HDL protects against ischemia reperfusion injury by preserving mitochondrial integrity. Atherosclerosis. 2013;228:110–6.
84. Frias MA, James RW, Gerber-Wicht C, Lang U. Native and reconstituted HDL activate Stat3 in ventricular cardiomyocytes via ERK1/2: role of sphingosine-1-phosphate. Cardiovasc Res. 2009;82:313–23.
85. Zhang J, Honbo N, Goetzl EJ, Chatterjee K, Karliner JS, Gray MO. Signals from type 1 sphingosine 1-phosphate receptors enhance adult mouse cardiac myocyte survival during hypoxia. Am J Physiol Heart Circ Physiol. 2007;293:H3150–8.
86. Gomez L, Paillard M, Price M, Chen Q, Teixeira G, Spiegel S, Lesnefsky EJ. A novel role for mitochondrial sphingosine-1-phosphate produced by sphingosine kinase-2 in PTP-mediated cell survival during cardioprotection. Basic Res Cardiol. 2011;106:1341–53.
87. Strub GM, Paillard M, Liang J, Gomez L, Allegood JC, Hait NC, Maceyka M, Price MM, Chen Q, Simpson DC, Kordula T, Milstien S, Lesnefsky EJ, Spiegel S. Sphingosine-1-phosphate produced by sphingosine kinase 2 in mitochondria interacts with prohibitin 2 to regulate complex IV assembly and respiration. FASEB J. 2011;25:600–12.

88. Merkwirth C, Langer T. Prohibitin function within mitochondria: essential roles for cell proliferation and cristae morphogenesis. Biochim Biophys Acta. 1793;2009:27–32.
89. Gustafsson AB. Bnip3 as a dual regulator of mitochondrial turnover and cell death in the myocardium. Pediatr Cardiol. 2011;32:267–74.
90. Dutta D, Xu J, Kim JS, Dunn WA, Leeuwenburgh C. Upregulated autophagy protects cardiomyocytes from oxidative stress-induced toxicity. Autophagy. 2013;9:328–44.
91. Sun K, Xie X, Liu Y, Han Z, Zhao X, Cai N, Zhang S, Song J, Wei L. Autophagy lessens ischemic liver injury by reducing oxidative damage. Cell Biosci. 2013;3:26.
92. Yan WJ, Dong HL, Xiong LZ. The protective roles of autophagy in ischemic preconditioning. Acta Pharmacol Sin. 2013;34:636–43.
93. Li Y, Li S, Qin X, Hou W, Dong H, Yao L, Xiong L. The pleiotropic roles of sphingolipid signaling in autophagy. Cell Death Dis. 2014;5:e1245.
94. Young MM, Kester M, Wang HG. Sphingolipids: regulators of crosstalk between apoptosis and autophagy. J Lipid Res. 2013;54:5–19.
95. Lavieu G, Scarlatti F, Sala G, Carpentier S, Levade T, Ghidoni R, Botti J, Codogno P. Regulation of autophagy by sphingosine kinase 1 and its role in cell survival during nutrient starvation. J Biol Chem. 2006;281:8518–27.
96. Gerster R, Eloranta JJ, Hausmann M, Ruiz PA, Cosin-Roger J, Terhalle A, Ziegler U, Kullak-Ublick GA, von Eckardstein A, Rogler G. Anti-inflammatory function of high-density lipoproteins via autophagy of IkB kinase. Cell Mol Gastroenterol Hepatol. 2015;1:171–87.
97. Dominiczak MH, Caslake MJ. Apolipoproteins: metabolic role and clinical biochemistry applications. Ann Clin Biochem. 2011;48:498–515.
98. de Silva HV, Stuart WD, Duvic CR, Wetterau JR, Ray MJ, Ferguson DG, Albers HW, Smith WR, Harmony JA. A 70-kDa apolipoprotein designated ApoJ is a marker for subclasses of human plasma high density lipoproteins. J Biol Chem. 1990;265:13240–7.
99. Trougakos IP. The molecular chaperone apolipoprotein J/clusterin as a sensor of oxidative stress: implications in therapeutic approaches – a mini-review. Gerontology. 2013;59:514–23.
100. Park S, Mathis KW, Lee IK. The physiological roles of apolipoprotein J/clusterin in metabolic and cardiovascular diseases. Rev Endocr Metab Disord. 2014;15:45–53.
101. Jun HO, Kim DH, Lee SW, Lee HS, Seo JH, Kim JH, Yu YS, Min BH, Kim KW. Clusterin protects H9c2 cardiomyocytes from oxidative stress-induced apoptosis via Akt/GSK-3β signaling pathway. Exp Mol Med. 2011;43:53–61.
102. Christoffersen C, Ahnström J, Axler O, Christensen EI, Dahlbäck B, Nielsen LB. The signal peptide anchors apolipoprotein M in plasma lipoproteins and prevents rapid clearance of apolipoprotein M from plasma. J Biol Chem. 2008;283:18765–72.
103. Dahlbäck B, Nielsen LB. Apolipoprotein M affecting lipid metabolism or just catching a ride with lipoproteins in the circulation? Cell Mol Life Sci. 2009;66:559–64.
104. Blaho VA, Hla T. An update on the biology of sphingosine 1-phosphate receptors. J Lipid Res. 2014;55:1596–608.
105. Elsøe S, Christoffersen C, Luchoomun J, Turner S, Nielsen LB. Apolipoprotein M promotes mobilization of cellular cholesterol in vivo. Biochim Biophys Acta. 1831;2013:1287–92.
106. Duan J, Dahlbäck B, Villoutreix BO. Proposed lipocalin fold for apolipoprotein M based on bioinformatics and site-directed mutagenesis. FEBS Lett. 2001;499:127–32.
107. Christoffersen C, Obinata H, Kumaraswamy SB, Galvani S, Ahnström J, Sevvana M, Egerer-Sieber C, Muller YA, Hla T, Nielsen LB, Dahlbäck B. Endothelium-protective sphingosine-1-phosphate provided by HDL-associated apolipoprotein M. Proc Natl Acad Sci U S A. 2011;108:9613–8.
108. Morel S, Christoffersen C, Axelsen LN, Montecucco F, Rochemont V, Frias MA, Mach F, James RW, Naus CC, Chanson M, Lampe PD, Nielsen MS, Nielsen LB, Kwak BR. Sphingosine-1-phosphate reduces ischaemia-reperfusion injury by phosphorylating the gap junction protein Connexin43. Cardiovasc Res. 2016;109:385–96.
109. Ooi EM, Barrett PH, Chan DC, Watts GF. Apolipoprotein C-III: understanding an emerging cardiovascular risk factor. Clin Sci (Lond). 2008;114:611–24.

110. Panin LE, Shalbueva NI, Polyakov LM. Effects of apolipoproteins C on oxidative phosphorylation in rat liver mitochondria. Bull Exp Biol Med. 2000;130:769–71.
111. Kolmakova A, Kwiterovich P, Virgil D, Alaupovic P, Knight-Gibson C, Martin SF, Chatterjee S. Apolipoprotein C-I induces apoptosis in human aortic smooth muscle cells via recruiting neutral sphingomyelinase. Arterioscler Thromb Vasc Biol. 2004;24:264–9.
112. Lamant M, Smih F, Harmancey R, Philip-Couderc P, Pathak A, Roncalli J, Galinier M, Collet X, Massabuau P, Senard JM, Rouet P. ApoO, a novel apolipoprotein, is an original glycoprotein up-regulated by diabetes in human heart. J Biol Chem. 2006;281:36289–302.
113. Koob S, Reichert AS. Novel intracellular functions of apolipoproteins: the ApoO protein family as constituents of the Mitofilin/MINOS complex determines cristae morphology in mitochondria. Biol Chem. 2014;395:285–96.
114. Hoppins S, Collins SR, Cassidy-Stone A, Hummel E, Devay RM, Lackner LL, Westermann B, Schuldiner M, Weissman JS, Nunnari J. A mitochondrial-focused genetic interaction map reveals a scaffold-like complex required for inner membrane organization in mitochondria. J Cell Biol. 2011;195:323–40.
115. Head BP, Zulaika M, Ryazantsev S, van der Bliek AM. A novel mitochondrial outer membrane protein, MOMA-1, that affects cristae morphology in *Caenorhabditis elegans*. Mol Biol Cell. 2011;22:831–41.
116. Turkieh A, Caubère C, Barutaut M, Desmoulin F, Harmancey R, Galinier M, Berry M, Dambrin C, Polidori C, Casteilla L, Koukoui F, Rouet P, Smih F. Apolipoprotein O is mitochondrial and promotes lipotoxicity in heart. J Clin Invest. 2014;124:2277–86.
117. Huang Y, Wu Z, Riwanto M, Gao S, Levison BS, Gu X, Fu X, Wagner MA, Besler C, Gerstenecker G, Zhang R, Li XM, DiDonato AJ, Gogonea V, Tang WH, Smith JD, Plow EF, Fox PL, Shih DM, Lusis AJ, Fisher EA, DiDonato JA, Landmesser U, Hazen SL. Myeloperoxidase, paraoxonase-1, and HDL form a functional ternary complex. J Clin Invest. 2013;123:3815–28.
118. Kalakech H, Hibert P, Prunier-Mirebeau D, Tamareille S, Letournel F, Macchi L, Pinet F, Furber A, Prunier F. RISK and SAFE signaling pathway involvement in apolipoprotein A-I-induced cardioprotection. PLoS One. 2014;9:e107950.
119. Tang C, Liu Y, Kessler PS, Vaughan AM, Oram JF. The macrophage cholesterol exporter ABCA1 functions as an anti-inflammatory receptor. J Biol Chem. 2009;284:32336–43.
120. Zhang Z, Datta G, Zhang Y, Miller AP, Mochon P, Chen YF, Chatham J, Anantharamaiah GM, White CR. Apolipoprotein A-I mimetic peptide treatment inhibits inflammatory responses and improves survival in septic rats. Am J Physiol Heart Circ Physiol. 2009;297:H866–73.
121. Boden WE, Pearson TA. Raising low levels of high-density lipoprotein cholesterol is an important target of therapy. Am J Cardiol. 2000;85:645–50. A10
122. Boden WE. High-density lipoprotein cholesterol as an independent risk factor in cardiovascular disease: assessing the data from Framingham to the veterans affairs high – density lipoprotein intervention trial. Am J Cardiol. 2000;86:19L–22L.
123. Wu A, Hinds CJ, Thiemermann C. High-density lipoproteins in sepsis and septic shock: metabolism, actions, and therapeutic applications. Shock. 2004;21:210–21.
124. Wu BJ, Ong KL, Shrestha S, Chen K, Tabet F, Barter PJ, Rye KA. Inhibition of arthritis in the Lewis rat by apolipoprotein A-I and reconstituted high-density lipoproteins. Arterioscler Thromb Vasc Biol. 2014;34:543–51.
125. van Leeuwen HJ, Heezius EC, Dallinga GM, van Strijp JA, Verhoef J, van Kessel KP. Lipoprotein metabolism in patients with severe sepsis. Crit Care Med. 2003;31:1359–66.
126. Navab M, Anantharamaiah GM, Reddy ST, Hama S, Hough G, Grijalva VR, Yu N, Ansell BJ, Datta G, Garber DW, Fogelman AM. Apolipoprotein A-I mimetic peptides. Arterioscler Thromb Vasc Biol. 2005;25:1325–31.
127. Maceyka M, Spiegel S. Sphingolipid metabolites in inflammatory disease. Nature. 2014;510:58–67.

128. Kimura T, Sato K, Kuwabara A, Tomura H, Ishiwara M, Kobayashi I, Ui M, Okajima F. Sphingosine 1-phosphate may be a major component of plasma lipoproteins responsible for the cytoprotective actions in human umbilical vein endothelial cells. J Biol Chem. 2001;276:31780–5.
129. Birner-Gruenberger R, Schittmayer M, Holzer M, Marsche G. Understanding high-density lipoprotein function in disease: recent advances in proteomics unravel the complexity of its composition and biology. Prog Lipid Res. 2014;56:36–46.

Chapter 23
MitomiRs Keep the Heart Beating

Samarjit Das, Hannah R. Vasanthi, and Ramesh Parjapath

Introduction

The vital importance of the heart is obvious and diseases of the heart are soaring. Cardiovascular disease (CVD), the leading cause of death globally, can be precipitated by a series of cardiometabolic risks. Several initiatives are in progress to identify the disease at an early stage and manage the disease with utmost care. In this process several scientists are working on microRNAs (miRNAs) as a novel tool in diagnosis, prognosis, and therapeutics.

MiRNAs, a group of non-coding RNAs constituted by 19–25 nucleotides, bind to protein-coding regions of the 3′-untranslated region (UTR) of mRNAs and inhibit the stability and/or translation of that mRNA. They have been used as biomarkers [1] for different kinds of heart disease [2], metabolic disorders [3], and neurodegenerative diseases [4]. Circulating miRNAs can also be used as therapeutic agents [5]. miRNAs that are present in heart tissue have been found to play a powerful role in various cardiovascular diseases [2]. In cases where a certain cardiovascular disease is characterized by the over-expression or loss of a particular gene, the miRNA(s) that targets this gene can contribute to, if not driving, disease progression.

MiRNAs are encoded by the genome of all eukaryotic organisms, and some DNA viruses. In mammalian cells non-coding intronic regions are the major encoding regions for most miRNAs, but some are encoded within the exonic regions. By acting on specific complementary target sequences of mRNA (formation of the RNA induced silencing complex, RISC), miRNAs silence a particular gene [6, 7].

S. Das, PhD (✉)
Department of Pathology, Cardiovascular Division, Johns Hopkins University, Baltimore, MD 21205, USA
e-mail: sdas11@jhmi.edu

H.R. Vasanthi • R. Parjapath
Department of Biotechnology, Pondicherry University, Puducherry, India

G. Santulli (ed.), *Mitochondrial Dynamics in Cardiovascular Medicine*, Advances in Experimental Medicine and Biology 982,
DOI 10.1007/978-3-319-55330-6_23

Several miRNA loci are in close proximity to each other, constituting a polycistonic transcript. Thus, single promoter activation can transcribe the clustered miRNAs together as a single unit. The 5′-end of miRNAs that spans from the 2 to 9 nucleotide position is crucial for target recognition (3′-UTR end of mRNA) and has been termed the "seed"-sequence. The downstream nucleotides of miRNA contain an additional sequence that controls their post-transcriptional properties, including their subcellular translocation patterns [8]. miRNA sequences are highly conserved across organisms. Approximately 60 % of human mRNAs contain at least one conserved miRNA binding site [9]. The expression levels of these miRNAs vary among individuals, and in conditions ranging from normal to diseased.

The present chapter will examine in detail the basics of cardiovascular system with emphasis on the mitochondria as a key organelle in cardiac, anatomy, physiology, biochemistry and pathology which off shoots to cardiovascular medicine.

Development of the Functional Heart

The heart is one of the earliest differentiating and functioning organs [10]. This emphasizes the critical nature of the heart in distributing blood through the vessels and the exchange of nutrients, oxygen, and wastes both to and from the developing baby. It grows into a functional beating heart in the seventh week of gestation which beats throughout life. Cessation of its activity for more than few minutes is detrimental to life.

The fetal heart, which functions in a relatively hypoxic environment, uses glucose and lactate as the predominant fuel substrates. Endogenous glycogen in the fetal and, to a lesser extent, in the neonatal heart is a significant source of glucose and myocardial ATP. However, in conditions of oxygen deprivation, glycogenolysis may be important and may allow the fetal heart to be more resistant to the effects of hypoxia and ischemia as compared to the adult heart [11].

Likewise, fetal hearts have lower mitochondrial content and therefore lower levels of both respiratory complex and TCA cycle activities. Catabolism of fatty acids by beta oxidation provides a small part of the overall ATP production because of low levels of circulating fatty acids and also due to the inhibition of fatty acid oxidation by high lactate levels present in the fetal heart.

The miRNAs which are nothing but non coded RNAs which are less than 22 nucleotides have been identified in cardiac tissue at all stages of development and those highly expressed in the fetal heart include miR-21, miR-29a, miR-129, miR-210, miR-211, miR-320, miR-423, and let-7c [12]. These cardiac miRNAs play a central role in cardiogenesis, heart function and pathology. miR-1 and miR-133a predominantly control early stages of cardiogenesis supporting commitment of cardiac-specific muscle lineage from embryonic stem cells and mesodermal precursors whereas, miR-208 and miR-499 are involved in the late cardiogenic stages mediating differentiation of cardioblasts to cardiomyocytes and fast/slow muscle fiber specification [13]. In the healthy adult heart, based on a large sequencing project, along with other studies, a number of miRNAs that are highly expressed in non-diseased cardiac tissue have been identified and thus likely to play a key role in both normal cardiac maintenance and disease. These include miR-1, miR-16, miR-27b, miR-30d, miR-126, miR-133, miR-143, miR-208 and the let-7 family [14].

Cellular Function of the Beating Heart

The four-chambered heart is made up of different types of cells which contributes to its structural, biochemical, mechanical and electrical properties of the functional heart. The myocardium is made up of the atrial and ventricular cardiomyocytes which forms the muscular walls of the heart aiding in the rhythmic pumping of the blood. Other cell types are fibroblasts, which, by number, represent more than 50 % of cardiac cells, endothelial cells, which form the endocardium, the inner lining of blood vessels and cardiac valves, and smooth muscle cells, which contribute to the coronary arteries aiding the inflow and outflow vasculature [15]. Specialized cardiomyocytes namely pacemaker cells and Purkinje fibers in the conduction system generate and conduct electrical impulses whereas the epicardium gives rise to the precursors of the coronary vasculature and fibroblasts. It is interesting to note that the human heart contains approximately two to three billion cardiac muscle cells however, they account for fewer than a third of the total number of cells in the heart which includes the smooth muscle cells, endothelial cells, fibroblast cells, connective tissue cells and mast cells. These distinct cell pools are not isolated from one another within the heart but instead interact physically and via a variety of soluble paracrine, autocrine, and endocrine factors [16, 17].

The rhythmic contraction of the heart assures constant oxygenation of the various organs and their cells. The central dogma of cardiac contraction-relaxation cycle is orchestrated by the increase and decrease of Ca^{2+} release in the cytoplasm. Further, the sarcoplasmic reticulum, a specialized myocyte organelle is critical for Ca^{2+} ion concentrations, which is an on-off switch for contraction. Intracellular Ca^{2+} handling is regulated by at least three other second messengers: cAMP, cGMP and inositol 1,4,5 trisphosphate in association with SR [18]. The contraction-relaxation cycle starts when an electrical action potential is initiated at the cell membrane of myocyte. Subsequently, an inward current flows depolarizing the cell membrane which in turn opens the voltage-gated Na^+ channels and causes influx of Na^+. Concomitantly, the membrane potential reaches to a permissive level for opening voltage-gated Ca^{2+} channels. These Ca^{2+} ions trigger the release of more Ca^{2+} from the sarcoplasmic reticulum by opening ryanodine receptors (RyRs). This Ca^{2+} induced Ca^{2+} release increases intracellular Ca^{2+} concentration which initiates the contraction cycle by activating contractile protein apparatus myosin and actin [18]. The Ca^{2+} released from sarcoplasmic reticulum diffuses to adjacent myofibril and binds to troponin C of troponin-tropomyosin complex of actin containing thin filament in sarcomeres, which displaces tropomyosin facilitating cross-bridge formation between actin and myosin to enable contraction of the myocardium. Each cycle of the cross-bridge consumes one molecule of ATP which is provided at the myosin ATPase catalytic domain, which is the major site of ATP consumption in the beating heart. Relaxation occurs when intracellular Ca^{2+} is restored to sarcoplasmic reticulum and voltage-gated K^{+} channels opens to permit action potential repolarization, making suitable conditions for relaxation. Withdrawal of intracellular Ca^{2+} takes place via four mechanisms consisting of SR Ca^{2+}-ATPase, sarcolemmal Na^{+}/Ca^{2+} exchanger, sarcolemmal Ca^{2+}-ATPase and mitochondrial Ca^{2+} uptake [18]. Thus ATP, the energy currency of the cell, is the key player for the beating heart. However, other signaling mechanisms aid in the cross talk and help the myocytes to be functional.

Mitochondria the Key Organelle for a Functional Heart

Effective myocardial function depends primarily on oxidative energy production. In humans, at a heart rate of 60–70 beats per minute, the oxygen consumption normalized per gram of myocardium is 20-fold higher than that of skeletal muscle at rest. The human heart consumes 30 kg of ATP per day, fueling basal metabolism and normal contraction that is essential to sustain systemic and pulmonary blood pressure. It is interesting to note that almost all ATP is produced in cardiomyocyte mitochondria.

Mitochondria are abundant in energy-demanding cardiac tissue constituting 20–40 % of cellular volume. The beating heart demands continuous supply of chemical energy not only to perform its mechanical pumping of blood to various organs but also to modulate intracellular and trans-sarcolemmal ionic movements and concentration gradients. Most of the energy required for these laborious processes is provided by mitochondrial bioenergetics. Many interacting bioenergetic pathways add to mitochondrial energetics including pyruvate oxidation, the tricarboxylic acid (TCA) cycle, β-oxidation of fatty acids and the final common pathway oxidative phosphorylation (OXPHOS). About 80–90 % of myocyte energy comes from OXPHOS and to a lesser extent from oxidation of glucose, lactose and other substrates [19]. In addition to energy from the mitochondria, cardiomyocytes also maintain high-energy stores in the form of phosphocreatine (pC) produced by mitochondrial creatine kinase using ATP.

Post-natally, a switch occurs from the carbohydrate metabolism, so that fatty acids become the primary energy substrate in the heart [20]. Likewise, the level of cardiac mitochondrial creatine kinase (mtCK) which is undetectable in early fetal heart, is upregulated during neonatal development. The increased expression of mtCK within the first several weeks after birth is coupled to both the availability of ADP to mitochondria as well as to the structural re-organization of cardiac mitochondria from a random arrangement at day 1 to an organized network by 3 weeks postnatally [21]. Subsequently, the maturation of the phosphocreatine shuttle is coordinated with the development of the contractile properties of the myocardium [22].

Reactive Oxygen Species (ROS), including superoxide and hydroxyl radicals and hydrogen peroxide, are by-products of mitochondrial bioenergetic side reactions of the mitochondrial ETC with molecular oxygen directly generate the superoxide anion radical. The main sites for mitochondrial ROS generation are complex I and III activities of the respiratory chain [23]; either excessive or diminished electron flux at these sites can stimulate the auto-oxidation of flavins and quinones (including coenzyme Q) producing superoxide radicals. The superoxide radicals can be converted to H_2O_2 (in the presence of the enzyme superoxide dismutase) which can further react to form the hydroxyl radical [19]. Generally these cell-damaging oxidants are neutralized by anti-oxidant enzymes, some of which are located in the mitochondria while others are cytosolic. Imbalance in the oxidant and the antioxidant status due to increased ROS generation results in oxidative stress. This may cause profound effects on cardiac cells including increased lipid peroxidation effecting primarily membrane phospholipids and proteins. Oxidative damage also occurs with nucleic acids (particularly targeting mtDNA) including induction of strand breaks, a high incidence of base modification (including the formation of 8-oxoguanosine) and subsequent point mutations and deletions [24].

Mitochondria are the primary site of ROS generation and therefore the critical target of their damaging effects; the mitochondrial respiratory chain located in the inner membrane is often damaged resulting in a further increase in ROS generation leading to a vicious cycle of diminished mitochondrial function. In addition to the well characterized role of ROS in cell damage, recent evidence has provided its role as an important regulatory event [25]. Oxidative species (e.g. H_2O_2) can also function as a potent signal sent from mitochondria to other cellular sites rapidly, and reversibly eliciting an array of intracellular cascades leading to different physiological end-points for the cardiomyocyte (e.g. apoptosis, necrosis, cardio-protection, cell proliferation).

Abnormalities in the mitochondrial organelle structure and function have been found with increasing frequency in association with cardiovascular diseases such as dilated (DCM) and hypertrophic cardiomyopathy (HCM), cardiac conduction defects and sudden death, ischemic and alcoholic cardiomyopathy, as well as myocarditis. Some of the mitochondrial abnormalities may have a genetic basis (e.g. mitochondrial DNA [mtDNA] changes leading to oxidative phosphorylation [OXPHOS] dysfunction, fatty acid oxidation defects due to specific nuclear DNA mutations) while other abnormalities appear to be due to a more sporadic or environmental cardiotoxic insult [19]. Figure 23.1 depicts the influence of various risk factors on the cardiomyocyte mitochondria which lead to biochemical and molecular changes resulting in different cardiovascular diseases.

Mitochondrial Dynamics

Mitochondria are highly dynamic, which allows them to constantly change their morphology. Mitochondria constantly divide (fission) and fuse (fusion); an unsuccessful mitochondrial repair process by fission–fusion activates mitophagy to eliminate the dysfunctional mitochondria from the cell. Recent findings, mainly in

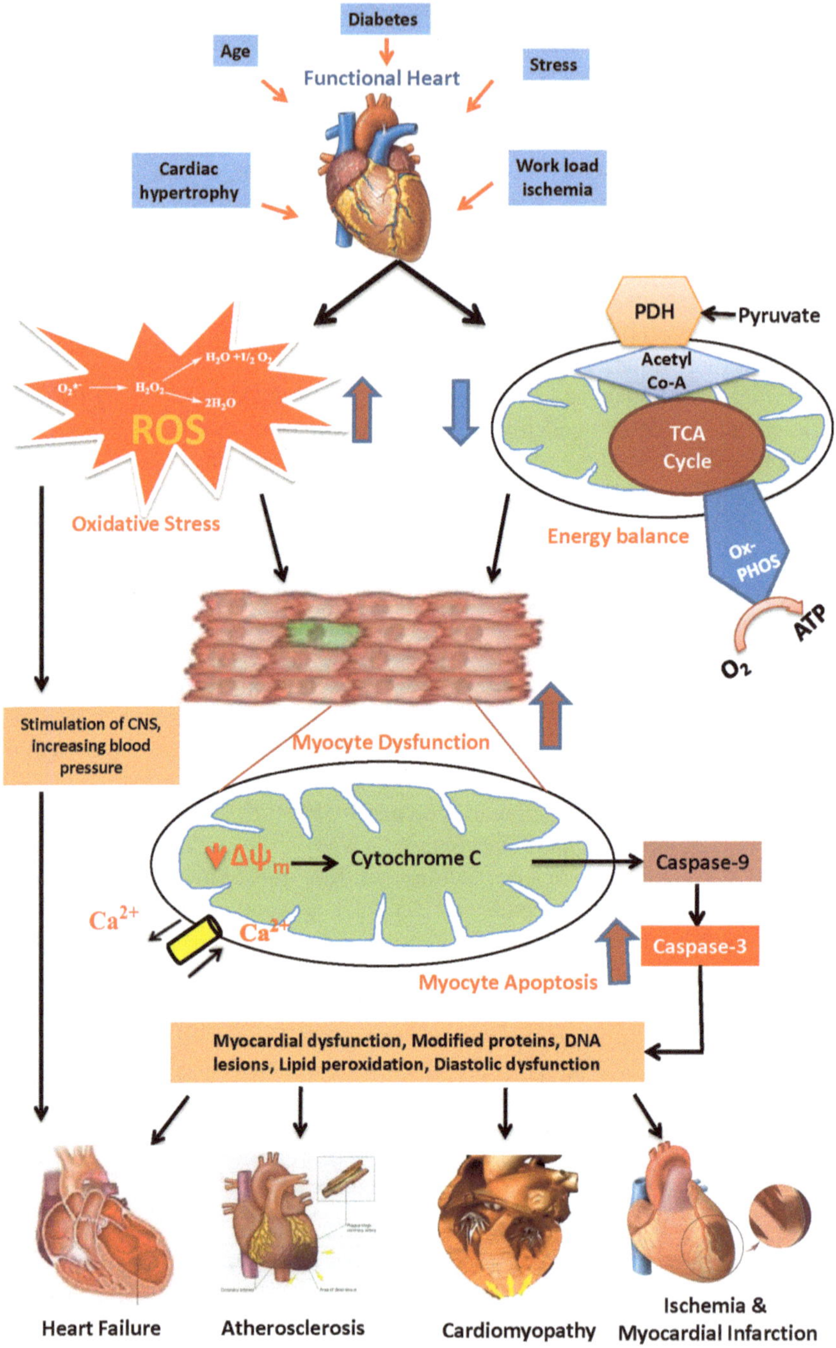

Fig. 23.1 The pivotal role of mitochondria in cardiovascular health

drosophila and mouse models, have been focusing on important role of mitochondrial fission and fusion in cardiac myocytes. The imbalance between mitochondrial fission and fusion affects cellular homeostasis, leading to several disease states, including cancer, neurodegenerative diseases, impaired cardiac development, and cardiovascular diseases. Mitochondrial dynamics also play a critical role in apoptosis, autophagy, and contractile dysfunction [26].

Recent studies focus on the role of miRNAs in the alteration of myocyte function through mitochondrial fission-fusion mechanism. The fission machinery within the mitochondrion is regulated by miR-484, which targets Fis1, a necessary protein for mitochondrial fission and apoptosis [26]. The expression of p53 and dynamin-related protein-1 (Drp1) are down-regulated by miR-30, and thus inhibit mitochondrial fission [27]. Interestingly, dephosphorylation of Drp-1 is suppressed by miR-499, resulting in the inhibition of apoptosis in cardiomyocytes [26]. miR-761 can suppress mitochondrial fission by down-regulating mitochondrial fission factor (MFF) [28]. Subsequent studies have provided additional evidence between mitochondrial morphology and miRNAs; for instance, miR-140 reduce mitochondrial fission by targeting mitofusin-1 [29].

Mitochondrial miRNAs

The mode of action of miRNAs within the cytoplasm has been well studied [6, 7]; however, evidence suggests that miRNAs and Ago2 play an important role also within the mitochondrial compartment [30]. There are studies that suggest the presence of miRNAs in the mitochondrial fraction [30–36], but their functional consequences were largely unknown. In 2012, we demonstrated that miR-181c, a product of the nuclear genome, translocates to the mitochondria, and regulates mitochondrial gene expression. In turn, this alters mitochondrial function by binding to the 3′-end of cytochrome c oxidase subunit 1 (mt-COX1). miR-378 has been shown to target and bind to the mitochondrial transcriptome at the ATP6 locus, causing down-regulation of the protein in the type 1 diabetic heart [34]. ATP6 is a subunit of the F0 complex of the ATP synthase, and its repression impacts ATP generating capacity. In cardiac mitochondria from spontaneous hypertensive rats (SHR) miR-21 has been found to be overexpressed; miR-21 can target mitochondrial genomic product, Cytb [37]. During myogenesis miR-1 enters the mitochondria where it alter the translation of specific mitochondrial genome-encoded transcripts, mt-COX1 and ND-1 [36].

Zhang and colleagues showed that the muscle-specific miR-1 stimulates mitochondrial translation by Ago2-mediated translocation into the mitochondria. The cytoplasmic functions of Ago2 are dependent on its binding partner GW182. In contrast, GW182 has not been detected in mitochondria [36]. While a number of studies (Fig. 23.2) have shown that miRNAs can translocate into the mitochondria when bound to Ago2 (either at pre-RISC or mature-RISC conformation), the underlying mechanism is not well understood. It has been shown that Ago2 acts as a carrier

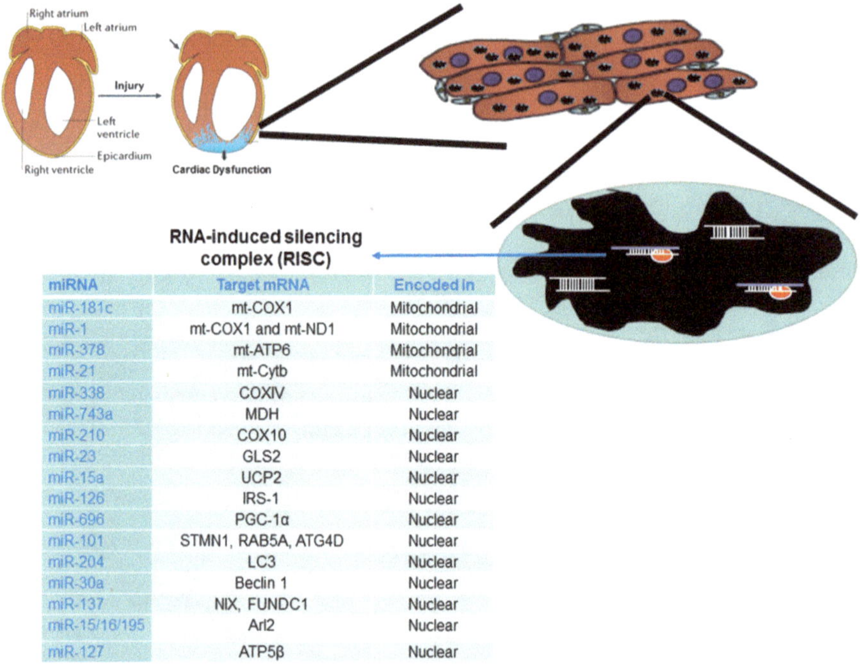

miRNA	Target mRNA	Encoded in
miR-181c	mt-COX1	Mitochondrial
miR-1	mt-COX1 and mt-ND1	Mitochondrial
miR-378	mt-ATP6	Mitochondrial
miR-21	mt-Cytb	Mitochondrial
miR-338	COXIV	Nuclear
miR-743a	MDH	Nuclear
miR-210	COX10	Nuclear
miR-23	GLS2	Nuclear
miR-15a	UCP2	Nuclear
miR-126	IRS-1	Nuclear
miR-696	PGC-1α	Nuclear
miR-101	STMN1, RAB5A, ATG4D	Nuclear
miR-204	LC3	Nuclear
miR-30a	Beclin 1	Nuclear
miR-137	NIX, FUNDC1	Nuclear
miR-15/16/195	Arl2	Nuclear
miR-127	ATP5β	Nuclear

Fig. 23.2 The mechanistic insights of MitomiRs in the pathophysiology of various heart defects

protein, as it can shuttle mitochondrial proteins (heat-shock proteins) [38], ribosomal proteins [39], and mitochondrial genomic products (a non-coding RNA, tRNAMet) [40] into the cytoplasm. This evidence suggests a pivotal role for Ago2 in transporting proteins or non-coding RNAs into and out from the mitochondrial compartment to the cytosolic fraction via mitochondrial membrane channels: TOM at the outer mitochondrial membrane, and TIM situated within the inner membrane of the mitochondrion. In addition, Barrey and colleagues have suggested that both miRNAs and pre-miRNAs are encoded within the mitochondrial genome [31]. However, the biogenesis and the functional aspects of these mitochondrial miRNAs have not been fully elucidated.

Heart Disease

Mitochondria play an important role in the pathophysiology of multiple cardiovascular disease conditions. Thus, preservation of mitochondrial function during the progression of heart defects is one of the primary focuses in cardiovascular therapeutics. MitomiRs can influence various mitochondrial pathways, such as the electron transport chain (TCA), lipid metabolism, and amino acid metabolism. Below we are discussing the role of miRNA in the pathological cardiac phenotypes involving mitochondria.

Congenital Anomalies

The heart is one of the notable organs which is susceptible to congenital defects, but whether miRNAs play a direct role to these defects is unknown. However, a number of reports describing the alteration of miRNAs expression in cardiogenesis is documented. A miRNA cluster, miR-17–92, has been reported to play an important role in septal defects [41]. Human chromosome21-derived miRNAs were overexpressed in the heart with trisomy 21 (Down syndrome), the most common genetic cause of congenital heart defects [42]. Mice lacking of miR-1-2 have thickened walls due to persistent proliferation; septal defects were common and often fatal in this mouse model [43]. Interestingly, miR-1 has been shown to translocate into the mitochondria during muscle development where targets mitochondrial genes, such as ND-1 and mt-COX1 [36]. It has also been reported that in infants with hypoglycemia, progressive lactic acidosis, an increased serum lactate/pyruvate ratio, and elevated plasma alanine are due to defects in the complex I of the mitochondrial respiratory chain [44]. ND-1 is one of the mitochondrial gene derived complex I subunits which plays an important role in the formation of mitochondrial iron-sulfur clusters. Thus, it is an open question that mitochondrial miRNA can play an important role in congenital anomalies of the heart.

Atherosclerosis

Atherosclerosis along with its associated complications are one of the major causes of morbidity and mortality in the United States of America. The inflammatory mediators and shear stress are the two important biochemical and biomechanical stimuli which lead to atherosclerotic lesions. Recently, miRNAs have served as important regulators of multiple pathophysiological cellular effects and molecular signaling pathways involved in atherosclerosis [45, 46]. miR-33a/b regulates HDL synthesis and reverse cholesterol transport by directly binding to the 3′-UTR of ATP-binding cassette transporter A1 (ABCA1). Inhibition of miR-33 in mice on a western-type diet results in elevated plasma HDL. Thus, miR-33 has been suggested to represent a therapeutic target for ameliorating cardiometabolic diseases [47]. ABCA1 can promote foam cell formation by inhibiting macrophage cholesterol efflux. Among other miRNAs, such as miR-758 [48], miR-106b [49], miR-144 [50, 51], miR-128-1 [52], miR-130b [52], miR-148a [52], miR-301b [52], miR-302a [53], miR-33 has been shown to modulate macrophage cholesterol efflux to maintain cholesterol homeostasis [47, 54, 55]. The mitochondrial role of miR-33 has been well established. miR-33 has been found to target both CROT and CPT1A, which eventually affects fatty acid β oxidation (FAO) [56]. miR-33a-5p and miR-33b-5p is encoded from the intronic region of SREBP2 [57], and SREBP1 [58] genes, respectively. By targeting important enzymes like CROT, CPT1A, and 3-Ketoacyl-CoA thiolase, the two family members, miR-33a and miR-33b, affect fatty acid metabolism [1].

Interestingly, the complementary strand of miR- 33a-5p is miR-33a-3p, which has been found to play an important role in FAO by targeting the same targets as miR-33a: CROT and CPT1A [59]. Additionally, miR-148a/b has been shown to regulate mitochondrial function by regulating TCA cycle by directly binding to the 3′-UTR of citrate synthase mRNA [60].

Ischemia and Myocardial Infarction

In 1986, Murry and colleagues described a phenomenon that produces an energy-sparing effect known as ischemic preconditioning during a sustained ischemic insult of the heart [61]. Later, this group and others [62, 63] recognized that ATP utilization of the preconditioned myocardium was much slower than non-preconditioned myocardium. Slower ATP utilization translates into lower stimulation of anaerobic glycolysis, attenuated fall of intracellular pH, and minimization of the ionic alterations that are typical of ischemia. Heart-specific miR-1 and miR-133a may play an important role in myocardial ischemic post-conditioning protection by regulating pro-apoptotic genes, such as CASP9 [64]. Furthermore, miR-1 protects the myocardium from ischemia-reperfusion injury by targeting Hsp90aa1 [65]. As discussed above miR-1 can regulate mitochondrial function by regulating mitochondrial genes [36]. Mitochondrial role in cardiac-ischemia reperfusion injury is well established [63, 66, 67].

An up-regulation of miR-210 expression in the placenta has been reported earlier [68]. Moreover, miR-210 modulates endothelial cell response to hypoxia [69]. Apart from hypoxia, acute myocardial infarction could also upregulate miR-210 [70]. miR-210 also elicits effects on cardiomyocytes and the vasculature. Overexpression of miR-210 in vitro or by intramyocardial injection (in vivo) reduced cell death and improved cardiac function and angiogenesis after myocardial infarction [70]. Importantly, miR-210 regulates electron transport chain constituents by interacting with iron–sulfur cluster scaffold homolog (ISCU), determining the levels of ROS in the mitochondria [71]. miR-210 has also been reported to modulate complex IV function of the ETC by targeting the nuclear encoded mRNA COX10 [72, 73]. Although the specific isoform/component of the complexes has not been validated, it has been reported that miR-210 and miR-141 target complex III and complex V of the ETC, respectively [74].

In cerebral ischemic-injury models, a down-regulation of miR-30a expression has been identified. It has been proposed that the overexpression of beclin1, direct target of miR-30a in the brain, induces autophagy due to the loss of miR-30a during ischemia/reperfusion injury [75]. Furthermore, the expression of p53 and dynamin-related protein-1 (Drp1) are down-regulated by miR-30, resulting in the inhibition of mitochondrial fission [27]. Li and colleagues showed that miR-137 is another hypoxia-responsive miRNA in the brain, which inhibit mitophagy through targeting two mitophagy receptors, FUNDC1 and NIX [76]. Ischemia-induced up-regulation of miR-140 expression in cardiomyocytes [29]. In an in vivo mouse ischemia-reperfusion model, it has been shown that the knock-down of miR-140 attenuates myocardial infarction by inhibition of mitochondrial fission mechanism [29].

A miRNA cluster, miR-23/24/27, has been shown to offer cardiomyocyte-protective effects in vitro by targeting the proapoptotic Bcl-2-like protein 11 (also known as Bim) [77, 78]. Additionally, overexpression of miR-24 has been shown to attenuate infarct size and improve cardiac function in rodents following myocardial infarction [78]. Conversely, it has been reported that the cardiomyocyte-specific overexpression of miR-24 is embryonically lethal in mice [79]. Additionally, on one side it has been shown that miR-24 is downregulated in cardiomyocytes and fibroblasts after myocardial infarction [78], but on the other hand in cultured neonatal ventricular cardiomyocytes (NRVM) miR-24 expression was increased by hypoxia [80]. It has also been demonstrated that miR-23 enhances mitochondrial glutaminase expression and glutamine metabolism by targeting c-Myc [81]. To make the matter more complicated, miR-24 exhibited antiangiogenic effects [82, 83], and impaired excitation–contraction coupling [77], after myocardial infarction. Through these mechanisms, miR-23/24/27 inhibition improved neovascularization and cardiac function [77, 83], thereby raising some serious concerns about the use of this miRNA-cluster overexpression as a therapeutic strategy against ischemia-reperfusion injury.

Arrhythmias

Mitochondria are essential to provide the energy demand of the continuous electrical activity and contractile function of cardiac muscle. Emerging evidence indicates that mitochondrial dysfunction can deteriorate cardiac electrical functioning by impairing the intracellular ion homeostasis, reduced ATP, and excess ROS production, resulting in increased tendency to cardiac arrhythmias [84].

The electron transport chain (ETC) is a redox pathway carried out by five complexes – I, II, III, IV and V (ATP synthase) – which play a pivotal role in ATP and ROS production. The ETC complexes I, III and IV generate a proton gradient by oxidizing NADH/NADPH. This proton gradient is then utilized by ATP synthase to produce ATP.

Oxidative stress has been implicated in the etiology of diabetic cardiomyopathy. Mitochondrial ETC complexes I and III generate ROS as a by-product of their respiratory function. Generation of ROS has been found to have deleterious effects on mitochondria and eventually leads to mitochondrial dysfunction [30, 32].

Iron-sulfur clusters (Fe-S) are essential cofactors for the transfer of electrons in oxidative phosphorylation (OXPHOS) [85]. Thus, the function of Complex I and Complex IV is highly dependent on (Fe-S). The Fe-S assembly enzyme (ISCU) plays an important role in the synthesis of these Fe-S clusters [85]. Under hypoxic conditions, it has been demonstrated that miR-210-5p can directly target ISCU [72, 86]. Succinate dehydrogenase subunit D (SDHD), a subunit of Complex II, has been identified as a target of miR-210 [87]. The authors concluded that miR-210 can ultimately alter complex II activity [87]. COX10, another nuclear encoded subunit of complex IV has been reported as a target of miR-210-5p [72]. Cytochrome c oxidase subunit IV (COXIV), one of the complex IV subunits, plays a vital role during the assembling process of complex IV and in its respiratory function [88]. In vitro studies on neuronal cells showed miR-338-5p alters complex IV activity by targeting the 3′-UTR of COXIV mRNA [89]. In independent studies, it has been shown that the mitochondrial genomic subunit of complex IV, mt-COX1, can be targeted by two different miRNAs, miR-181c [30, 32], and miR-1 [36]. Mitochondrial transcripts are polycistronic in nature, and thus, the effects of miRNA binding to the 3′-UTR of mt-mRNA are still not fully understood. On the one hand, it has been shown that by binding to a miRNA, mt-mRNA expression is downregulated [30, 32, 34]. Conversely, miRNA acts as an activator of post-translational processes when it binds to a mitochondrial encoded-mRNA [36, 37]. Chronic overexpression of a miRNA that targets mitochondrial mRNA confirmed this observation [32, 37].

Zheng and colleagues reported that miR-101-3p negatively regulates the expression of ATP synthase subunit β (ATP5B) [90]. In another study, miR-127-5p was also shown to target the 3′-UTR of the ATP5B transcript, and alters protein content [91]. Additionally, miR-338-5p has been found to target ATP5G1, which ultimately augments ATP synthase activity [89]. Finally, miR-378 has been shown to target and bind to the mitochondrial transcriptome at the ATP6 locus in the heart [34]. ATP6 is a subunit of the F0 complex of the ATP synthase, and its repression impacts ATP generating capacity.

Hypertension

Patients with high blood pressure frequently show abnormalities in cardiac morphology and function, including left ventricular hypertrophy, systolic/diastolic dysfunction, which ultimately leads to heart failure. Once again, excessive ROS production in mitochondria can also be associated with hypertensive cardiomyopathy. The ETC complexes are well known to be the major source of mitochondrial ROS production [32, 66]. Several studies have focused on nuclear-encoded subunits of the ETC and their contributions to mitochondrial ROS production in the spontaneous hypertensive rats (SHR) model. Meng and colleagues proposed that down-regulation of the two mitochondrial enzymes, trifunctional enzyme alpha subunit (Hadha) and NADH dehydrogenase 1 alpha subcomplex 10 (Ndufa10), is associated with the development of cardiac hypertrophy in SHR [92]. Furthermore, in SHR cardiac mitochondria the upregulation of multiple subunits of the ETC complexes, such as complexes I, III and IV, and down-regulation of complexes II and V have been reported [93]. Emerging data suggest that mutations in mitochondrial DNA (mtDNA) can play a role in essential hypertension [94]. Recently, it has been shown that downregulation of mtDNA-encoded cytochrome b (mt-Cytb) in SHR can activate mitochondrial ROS production [37]. The authors also demonstrated that miR-21, a key miRNA induced in SHR, can regulate mitochondrial ROS production by directly binding to the 3′-END of mt-Cytb mRNA [37]. Besides, exogenous miR-21 delivered by recombinant adeno-associated virus (rAAV) was sufficient to lower blood pressure in the SHR model, suggesting a new therapeutic strategy against hypertension [37].

Cardiomyopathy

There are multiple types of cardiomyopathies, including dilated cardiomyopathy (DCM), restrictive cardiomyopathy, hypertrophic cardiomyopathy (HCM), arrhythmogenic right ventricular dysplasia, and unclassified cardiomyopathy. Cardiomyopathy can be acquired or inherited. mtDNA mutation can cause mitochondrial cardiomyopathy, which is estimated about 1 in 5,000 births [95]. Mitochondrial cardiomyopathy can only be inherited from the mother.

Mutations in mitochondrial-encoded genes like Cytb, mt-COX1, mt-COX2 and mt-COX3 all result in the development of DCM [96, 97]. DCM represents the leading cause of mortality in patients with muscular dystrophy. Cardio-specific overexpression of miR-30c mice developed DCM after 6 week of age [98]. These mice showed a significant downregulation of mitochondrial complexes III and IV protein content, altering mitochondrial function in the heart [98]. The expression of miR-30c in the heart during DCM is controversial: some studies have found up-regulation of miR-30c, others have found down-regulation [99–101]. Notably, miR-30c is found in cardiomyocytes as well as in fibroblasts [99]. In vitro, miR-30c has been demonstrated in the regulation of cardiac fibrosis and cardiomyocyte hypertrophy through the alteration of mitochondrial function, apoptosis, and cell proliferation [99]. On the other hand, miR-30c has been proposed to have therapeutic potential for the treatment of hyperlipidemia, as miR-30c can decrease lipid synthesis and lipoprotein secretion [102].

HCM is an autosomal dominant disease characterized by variable expressivity, age penetrance, and a high heterogeneity. The transcriptional profile – miRNAs expression, epigenetic modifications, and posttranslational modifications – seems to be highly relevant for the onset of the disease. Once again, energy deficiency due to HCM ultimately compromises the relaxation capacity of the cardiomyocyte [103, 104]. Thus, mitochondrial role is pivotal in the pathophysiology of HCM. There are several miRNAs that have been found to alter mitochondrial function in HCM. Both in human [105] and mouse models [106], miR-21 has been shown to be upregulated during HCM. On the other hand, miR-1 is down-regulated during HCM. As we have discussed in this chapter both miR-1 and miR-21 can modulate mitochondrial function in the muscle cells by targeting mitochondrial-encoded genes [36, 37]. miR-132 [106] in murine heart and miR-184 in human heart have been found to be up-regulated in HCM. Both miR-132 and miR-184 have been shown to alter mitochondrial transport by targeting CACT for miR-132 [107] and Slc25a22 for miR-184 [108].

Dysfunctional mitochondria are central in the pathogenesis of diabetic cardiomyopathy. MitomiRs can influence diabetic cardiomyopathy by altering multiple pathways: influencing the mitochondrial ETC, altering FAO, amino acid metabolism, nucleotide metabolism, altering mitochondrial transport [109]. Barath and colleagues summarized the possible mechanisms of all the MitomiRs which play an important role in diabetic cardiomyopathy [109].

Conclusions

In conclusion, several data suggest that MitomiRs critically control the pathophysiology of multiple heart diseases. Several MitomiRs might be attractive candidates or targets to improve cardiac health. However, it is evident that further studies are required to identify the mechanisms and binding proteins responsible for the import of miRNAs into mitochondria. More studies are also needed to address the ongoing controversy that when a miRNA binds to mitochondrial-encoded genes, it either increases mt-mRNA activation or mt-mRNA repression. It is clear that by targeting mt-mRNAs, miRNAs are able to influence various aspects of mitochondrial metabolism, resulting in the alteration of mitochondrial function. This alteration ultimately leads to mitochondrial dysfunction, which plays an important role in the development of various cardiac disorders. The use of MitomiRs can offer new therapeutic opportunities for cardiovascular diseases.

Acknowledgments This work was supported by a Scientist Development Grant from the American Heart Association 14SDG18890049 (S.D.).

References

1. Rottiers V, Naar AM. MicroRNAs in metabolism and metabolic disorders. Nat Rev Mol Cell Biol. 2012;13:239–50.
2. Wronska A, Kurkowska-Jastrzebska I, Santulli G. Application of microRNAs in diagnosis and treatment of cardiovascular disease. Acta Physiol (Oxford). 2015;213:60–83.
3. Hulsmans M, Holvoet P. MicroRNAs as early biomarkers in obesity and related metabolic and cardiovascular diseases. Curr Pharm Des. 2013;19:5704–17.
4. Novak J, Kruzliak P, Bienertova-Vasku J, Slaby O, Novak M. MicroRNA-206: a promising theranostic marker. Theranostics. 2014;4:119–33.
5. Sayed D, Rane S, Abdellatif M. MicroRNAs challenge the status quo of therapeutic targeting. J Cardiovasc Transl Res. 2009;2:100–7.
6. Ambros V. The functions of animal microRNAs. Nature. 2004;431:350–5.
7. Bartel DP. MicroRNAs: genomics, biogenesis, mechanism, and function. Cell. 2004;116:281–97.
8. Hwang HW, Wentzel EA, Mendell JT. A hexanucleotide element directs microRNA nuclear import. Science. 2007;315:97–100.
9. Ha M, Kim VN. Regulation of microRNA biogenesis. Nat Rev Mol Cell Biol. 2014;15:509–24.

10. Bruneau BG. Signaling and transcriptional networks in heart development and regeneration. Cold Spring Harb Perspect Biol. 2013;5:a008292.
11. Lopaschuk GD, Collins-Nakai RL, Itoi T. Developmental changes in energy substrate use by the heart. Cardiovasc Res. 1992;26:1172–80.
12. Thum T, Galuppo P, Wolf C, Fiedler J, Kneitz S, van Laake LW, Doevendans PA, Mummery CL, Borlak J, Haverich A, Gross C, Engelhardt S, Ertl G, Bauersachs J. MicroRNAs in the human heart. A clue to fetal gene reprogramming in heart failure. Circulation. 2007;116:258–67.
13. Chistiakov DA, Orekhov AN, Bobryshev YV. Cardiac-specific miRNA in cardiogenesis, heart function, and cardiac pathology (with focus on myocardial infarction). J Mol Cell Cardiol. 2016;94:107–21.
14. Romaine SP, Tomaszewski M, Condorelli G, Samani NJ. MicroRNAs in cardiovascular disease: an introduction for clinicians. Heart. 2015;101:921–8.
15. Xin M, Olson EN, Bassel-Duby R. Mending broken hearts: cardiac development as a basis for adult heart regeneration and repair. Nat Rev Mol Cell Biol. 2013;14:529–41.
16. Tirziu D, Giordano FJ, Simons M. Cell communications in the heart. Circulation. 2010;122:928–37.
17. Santulli G. MicroRNAs and endothelial (dys)function. J Cell Physiol. 2016;231:1638–44.
18. Santulli G, Nakashima R, Yuan Q, Marks AR. Intracellular calcium release channels: an update. J Physiol. 2017 (in press). doi: 10.1113/JP272781.
19. Marín-García J, Goldenthal MJ. The mitochondrial organelle and the heart. Rev Esp Cardiol. 2002;55:1293–310.
20. Sack MN, Harrington LS, Jonassen AK, Mjøs OD, Yellon DM. Coordinate regulation of metabolic enzyme encoding genes during cardiac development and following carvedilol therapy in spontaneously hypertensive rats. Cardiovasc Drugs Ther. 2000;14:31–9.
21. Tiivel T, Kadaya L, Kuznetsov A, Käämbre T, Peet N, Sikk P, Braun U, Ventura-Clapier R, Saks V, Seppet EK. Developmental changes in regulation of mitochondrial respiration by ADP and creatine in rat heart in vivo. Mol Cell Biochem. 2000;208:119–28.
22. Hoerter JA, Kuznetsov A, Ventura-Clapier R. Functional development of the creatine kinase system in perinatal rabbit heart. Circ Res. 1991;69:665–76.
23. Dröse S, Brandt U. Molecular mechanisms of superoxide production by the mitochondrial respiratory chain. Adv Exp Med Biol. Mitochondrial Oxidative Phosphorylation: Springer. 2012;748:145–69.
24. Marín-García J. Heart failure: bench to bedside. New York: Humana Press; 2010.
25. Nulton-Persson AC, Szweda LI. Modulation of mitochondrial function by hydrogen peroxide. J Biol Chem. 2001;276:23357–61.
26. Wang JX, Jiao JQ, Li Q, Long B, Wang K, Liu JP, Li YR, Li PF. miR-499 regulates mitochondrial dynamics by targeting calcineurin and dynamin-related protein-1. Nat Med. 2011;17:71–8.
27. Li J, Donath S, Li Y, Qin D, Prabhakar BS, Li P. miR-30 regulates mitochondrial fission through targeting p53 and the dynamin-related protein-1 pathway. PLoS Genet. 2010;6: e1000795.
28. Long B, Wang K, Li N, Murtaza I, Xiao JY, Fan YY, Liu CY, Li WH, Cheng Z, Li P. miR-761 regulates the mitochondrial network by targeting mitochondrial fission factor. Free Radic Biol Med. 2013;65:371–9.
29. Li J, Li Y, Jiao J, Wang J, Li Y, Qin D, Li P. Mitofusin 1 is negatively regulated by microRNA 140 in cardiomyocyte apoptosis. Mol Cell Biol. 2014;34:1788–99.
30. Das S, Ferlito M, Kent OA, Fox-Talbot K, Wang R, Liu D, Raghavachari N, Yang Y, Wheelan SJ, Murphy E, Steenbergen C. Nuclear miRNA regulates the mitochondrial genome in the heart. Circ Res. 2012;110:1596–603.
31. Barrey E, Saint-Auret G, Bonnamy B, Damas D, Boyer O, Gidrol X. Pre-microRNA and mature microRNA in human mitochondria. PLoS ONE. 2011;6:e20220.
32. Das S, Bedja D, Campbell N, Dunkerly B, Chenna V, Maitra A, Steenbergen C. miR-181c regulates the mitochondrial genome, bioenergetics, and propensity for heart failure in vivo. PLoS ONE. 2014;9:e96820.

33. Dasgupta N, Peng Y, Tan Z, Ciraolo G, Wang D, Li R. miRNAs in mtDNA-less cell mitochondria. Cell Death Discov. 2015;1:15004.
34. Jagannathan R, Thapa D, Nichols CE, Shepherd DL, Stricker JC, Croston TL, Baseler WA, Lewis SE, Martinez I, Hollander JM. Translational regulation of the mitochondrial genome following redistribution of mitochondrial microRNA in the diabetic heart. Circ Cardiovasc Genet. 2015;8:785–802.
35. Srinivasan H, Das S. Mitochondrial miRNA (MitomiR): a new player in cardiovascular health. Can J Physiol Pharmacol. 2015;93:855–61.
36. Zhang X, Zuo X, Yang B, Li Z, Xue Y, Zhou Y, Huang J, Zhao X, Zhou J, Yan Y, Zhang H, Guo P, Sun H, Guo L, Zhang Y, Fu XD. MicroRNA directly enhances mitochondrial translation during muscle differentiation. Cell. 2014;158:607–19.
37. Li H, Zhang X, Wang F, Zhou L, Yin Z, Fan J, Nie X, Wang P, Fu XD, Chen C, Wang DW. MicroRNA-21 lowers blood pressure in spontaneous hypertensive rats by upregulating mitochondrial translation. Circulation. 2016;134:734–51.
38. Beitzinger M, Peters L, Zhu JY, Kremmer E, Meister G. Identification of human microRNA targets from isolated argonaute protein complexes. RNA Biol. 2007;4:76–84.
39. Hock J, Weinmann L, Ender C, Rudel S, Kremmer E, Raabe M, Urlaub H, Meister G. Proteomic and functional analysis of argonaute-containing mRNA-protein complexes in human cells. EMBO Rep. 2007;8:1052–60.
40. Maniataki E, Mourelatos Z. Human mitochondrial tRNAMet is exported to the cytoplasm and associates with the argonaute 2 protein. RNA. 2005;11:849–52.
41. Ventura A, Young AG, Winslow MM, Lintault L, Meissner A, Erkeland SJ, Newman J, Bronson RT, Crowley D, Stone JR, Jaenisch R, Sharp PA, Jacks T. Targeted deletion reveals essential and overlapping functions of the miR-17 through 92 family of miRNA clusters. Cell. 2008;132:875–86.
42. Kuhn DE, Nuovo GJ, Martin MM, Malana GE, Pleister AP, Jiang J, Schmittgen TD, Terry Jr AV, Gardiner K, Head E, Feldman DS, Elton TS. Human chromosome 21-derived miRNAs are overexpressed in down syndrome brains and hearts. Biochem Biophys Res Commun. 2008;370:473–7.
43. Zhao Y, Ransom JF, Li A, Vedantham V, von Drehle M, Muth AN, Tsuchihashi T, McManus MT, Schwartz RJ, Srivastava D. Dysregulation of cardiogenesis, cardiac conduction, and cell cycle in mice lacking miRNA-1-2. Cell. 2007;129:303–17.
44. Moreadith RW, Batshaw ML, Ohnishi T, Kerr D, Knox B, Jackson D, Hruban R, Olson J, Reynafarje B, Lehninger AL. Deficiency of the iron-sulfur clusters of mitochondrial reduced nicotinamide-adenine dinucleotide-ubiquinone oxidoreductase (complex I) in an infant with congenital lactic acidosis. J Clin Invest. 1984;74:685–97.
45. Feinberg MW, Moore KJ. MicroRNA regulation of atherosclerosis. Circ Res. 2016;118:703–20.
46. Novak J, Olejnickova V, Tkacova N, Santulli G. Mechanistic role of microRNAs in coupling lipid metabolism and atherosclerosis. Adv Exp Med Biol. 2015;887:79–100.
47. Najafi-Shoushtari SH, Kristo F, Li Y, Shioda T, Cohen DE, Gerszten RE, Naar AM. MicroRNA-33 and the SREBP host genes cooperate to control cholesterol homeostasis. Science. 2010;328:1566–9.
48. Ramirez CM, Davalos A, Goedeke L, Salerno AG, Warrier N, Cirera-Salinas D, Suarez Y, Fernandez-Hernando C. MicroRNA-758 regulates cholesterol efflux through posttranscriptional repression of ATP-binding cassette transporter A1. Arterioscler Thromb Vasc Biol. 2011;31:2707–14.
49. Kim J, Yoon H, Ramirez CM, Lee SM, Hoe HS, Fernandez-Hernando C, Kim J. MiR-106b impairs cholesterol efflux and increases Abeta levels by repressing ABCA1 expression. Exp Neurol. 2012;235:476–83.
50. de Aguiar Vallim TQ, Tarling EJ, Kim T, Civelek M, Baldan A, Esau C, Edwards PA. MicroRNA-144 regulates hepatic ATP binding cassette transporter A1 and plasma high-density lipoprotein after activation of the nuclear receptor farnesoid X receptor. Circ Res. 2013;112:1602–12.

51. Ramirez CM, Goedeke L, Rotllan N, Yoon JH, Cirera-Salinas D, Mattison JA, Suarez Y, de Cabo R, Gorospe M, Fernandez-Hernando C. MicroRNA 33 regulates glucose metabolism. Mol Cell Biol. 2013;33:2891–902.
52. Goedeke L, Rotllan N, Canfran-Duque A, Aranda JF, Ramirez CM, Araldi E, Lin CS, Anderson NN, Wagschal A, de Cabo R, Horton JD, Lasuncion MA, Naar AM, Suarez Y, Fernandez-Hernando C. MicroRNA-148a regulates LDL receptor and ABCA1 expression to control circulating lipoprotein levels. Nat Med. 2015;21:1280–9.
53. Meiler S, Baumer Y, Toulmin E, Seng K, Boisvert WA. MicroRNA 302a is a novel modulator of cholesterol homeostasis and atherosclerosis. Arterioscler Thromb Vasc Biol. 2015;35:323–31.
54. Rayner KJ, Suarez Y, Davalos A, Parathath S, Fitzgerald ML, Tamehiro N, Fisher EA, Moore KJ, Fernandez-Hernando C. MiR-33 contributes to the regulation of cholesterol homeostasis. Science. 2010;328:1570–3.
55. Marquart TJ, Allen RM, Ory DS, Baldan A. miR-33 links SREBP-2 induction to repression of sterol transporters. Proc Natl Acad Sci U S A. 2010;107:12228–32.
56. Gerin I, Clerbaux LA, Haumont O, Lanthier N, Das AK, Burant CF, Leclercq IA, MacDougald OA, Bommer GT. Expression of miR-33 from an SREBP2 intron inhibits cholesterol export and fatty acid oxidation. J Biol Chem. 2010;285:33652–61.
57. Bommer GT, MacDougald OA. Regulation of lipid homeostasis by the bifunctional SREBF2-miR33a locus. Cell Metab. 2011;13:241–7.
58. Xu X, So JS, Park JG, Lee AH. Transcriptional control of hepatic lipid metabolism by SREBP and ChREBP. Semin Liver Dis. 2013;33:301–11.
59. Goedeke L, Vales-Lara FM, Fenstermaker M, Cirera-Salinas D, Chamorro-Jorganes A, Ramirez CM, Mattison JA, de Cabo R, Suarez Y, Fernandez-Hernando C. A regulatory role for microRNA 33* in controlling lipid metabolism gene expression. Mol Cell Biol. 2013;33:2339–52.
60. Tibiche C, Wang E. MicroRNA regulatory patterns on the human metabolic network. Open Syst Biol J. 2008;1:1–8.
61. Murry CE, Jennings RB, Reimer KA. Preconditioning with ischemia: a delay of lethal cell injury in ischemic myocardium. Circulation. 1986;74:1124–36.
62. Murry CE, Richard VJ, Reimer KA, Jennings RB. Ischemic preconditioning slows energy metabolism and delays ultrastructural damage during a sustained ischemic episode. Circ Res. 1990;66:913–31.
63. Steenbergen C, Perlman ME, London RE, Murphy E. Mechanism of preconditioning. Ionic alterations. Circ Res. 1993;72:112–25.
64. He B, Xiao J, Ren AJ, Zhang YF, Zhang H, Chen M, Xie B, Gao XG, Wang YW. Role of miR-1 and miR-133a in myocardial ischemic postconditioning. J Biomed Sci. 2011;18:22.
65. Zhu WS, Guo W, Zhu JN, Tang CM, Fu YH, Lin QX, Tan N, Shan ZX. Hsp90aa1: a novel target gene of miR-1 in cardiac ischemia/reperfusion injury. Sci Rep. 2016;6:24498.
66. Murphy E, Steenbergen C. Mechanisms underlying acute protection from cardiac ischemia-reperfusion injury. Physiol Rev. 2008;88:581–609.
67. Steenbergen C, Das S, Su J, Wong R, Murphy E. Cardioprotection and altered mitochondrial adenine nucleotide transport. Basic Res Cardiol. 2009;104:149–56.
68. Luo R, Wang Y, Xu P, Cao G, Zhao Y, Shao X, Li YX, Chang C, Peng C, Wang YL. Hypoxia-inducible miR-210 contributes to preeclampsia via targeting thrombospondin type I domain containing 7A. Sci Rep. 2016;6:19588.
69. Fasanaro P, D'Alessandra Y, Di Stefano V, Melchionna R, Romani S, Pompilio G, Capogrossi MC, Martelli F. MicroRNA-210 modulates endothelial cell response to hypoxia and inhibits the receptor tyrosine kinase ligand Ephrin-A3. J Biol Chem. 2008;283:15878–83.
70. Hu S, Huang M, Li Z, Jia F, Ghosh Z, Lijkwan MA, Fasanaro P, Sun N, Wang X, Martelli F, Robbins RC, Wu JC. MicroRNA-210 as a novel therapy for treatment of ischemic heart disease. Circulation. 2010;122:S124–31.
71. Mutharasan RK, Nagpal V, Ichikawa Y, Ardehali H. microRNA-210 is upregulated in hypoxic cardiomyocytes through Akt- and p53-dependent pathways and exerts cytoprotective effects. Am J Physiol Heart Circ Physiol. 2011;301:H1519–30.

72. Chen Z, Li Y, Zhang H, Huang P, Luthra R. Hypoxia-regulated microRNA-210 modulates mitochondrial function and decreases ISCU and COX10 expression. Oncogene. 2010;29: 4362–8.
73. Colleoni F, Padmanabhan N, Yung HW, Watson ED, Cetin I, Tissot van Patot MC, Burton GJ, Murray AJ. Suppression of mitochondrial electron transport chain function in the hypoxic human placenta: a role for miRNA-210 and protein synthesis inhibition. PLoS ONE. 2013;8:e55194.
74. Bienertova-Vasku J, Sana J, Slaby O. The role of microRNAs in mitochondria in cancer. Cancer Lett. 2013;336:1–7.
75. Wang P, Liang J, Li Y, Li J, Yang X, Zhang X, Han S, Li S, Li J. Down-regulation of miRNA-30a alleviates cerebral ischemic injury through enhancing beclin 1-mediated autophagy. Neurochem Res. 2014;39:1279–91.
76. Li W, Zhang X, Zhuang H, Chen HG, Chen Y, Tian W, Wu W, Li Y, Wang S, Zhang L, Chen Y, Li L, Zhao B, Sui S, Hu Z, Feng D. MicroRNA-137 is a novel hypoxia-responsive microRNA that inhibits mitophagy via regulation of two mitophagy receptors FUNDC1 and NIX. J Biol Chem. 2014;289:10691–701.
77. Li RC, Tao J, Guo YB, Wu HD, Liu RF, Bai Y, Lv ZZ, Luo GZ, Li LL, Wang M, Yang HQ, Gao W, Han QD, Zhang YY, Wang XJ, Xu M, Wang SQ. In vivo suppression of microRNA-24 prevents the transition toward decompensated hypertrophy in aortic-constricted mice. Circ Res. 2013;112:601–5.
78. Qian L, Van Laake LW, Huang Y, Liu S, Wendland MF, Srivastava D. miR-24 inhibits apoptosis and represses Bim in mouse cardiomyocytes. J Exp Med. 2011;208:549–60.
79. van Rooij E, Sutherland LB, Liu N, Williams AH, McAnally J, Gerard RD, Richardson JA, Olson EN. A signature pattern of stress-responsive microRNAs that can evoke cardiac hypertrophy and heart failure. Proc Natl Acad Sci U S A. 2006;103:18255–60.
80. Li DF, Tian J, Guo X, Huang LM, Xu Y, Wang CC, Wang JF, Ren AJ, Yuan WJ, Lin L. Induction of microRNA-24 by HIF-1 protects against ischemic injury in rat cardiomyocytes. Physiol Res. 2012;61:555–65.
81. Gao P, Tchernyshyov I, Chang TC, Lee YS, Kita K, Ochi T, Zeller KI, De Marzo AM, Van Eyk JE, Mendell JT, Dang CV. c-Myc suppression of miR-23a/b enhances mitochondrial glutaminase expression and glutamine metabolism. Nature. 2009;458:762–5.
82. Meloni M, Marchetti M, Garner K, Littlejohns B, Sala-Newby G, Xenophontos N, Floris I, Suleiman MS, Madeddu P, Caporali A, Emanueli C. Local inhibition of microRNA-24 improves reparative angiogenesis and left ventricle remodeling and function in mice with myocardial infarction. Mol Ther. 2013;21:1390–402.
83. Fiedler J, Jazbutyte V, Kirchmaier BC, Gupta SK, Lorenzen J, Hartmann D, Galuppo P, Kneitz S, Pena JT, Sohn-Lee C, Loyer X, Soutschek J, Brand T, Tuschl T, Heineke J, Martin U, Schulte-Merker S, Ertl G, Engelhardt S, Bauersachs J, Thum T. MicroRNA-24 regulates vascularity after myocardial infarction. Circulation. 2011;124:720–30.
84. Yang KC, Bonini MG, Dudley Jr SC. Mitochondria and arrhythmias. Free Radic Biol Med. 2014;71:351–61.
85. Tong WH, Rouault T. Distinct iron-sulfur cluster assembly complexes exist in the cytosol and mitochondria of human cells. EMBO J. 2000;19:5692–700.
86. Chan SY, Zhang YY, Hemann C, Mahoney CE, Zweier JL, Loscalzo J. MicroRNA-210 controls mitochondrial metabolism during hypoxia by repressing the iron-sulfur cluster assembly proteins ISCU1/2. Cell Metab. 2009;10:273–84.
87. Puissegur MP, Mazure NM, Bertero T, Pradelli L, Grosso S, Robbe-Sermesant K, Maurin T, Lebrigand K, Cardinaud B, Hofman V, Fourre S, Magnone V, Ricci JE, Pouyssegur J, Gounon P, Hofman P, Barbry P, Mari B. miR-210 is overexpressed in late stages of lung cancer and mediates mitochondrial alterations associated with modulation of HIF-1 activity. Cell Death Differ. 2011;18:465–78.
88. Li Y, Park JS, Deng JH, Bai Y. Cytochrome c oxidase subunit IV is essential for assembly and respiratory function of the enzyme complex. J Bioenerg Biomembr. 2006;38:283–91.

89. Aschrafi A, Schwechter AD, Mameza MG, Natera-Naranjo O, Gioio AE, Kaplan BB. MicroRNA-338 regulates local cytochrome c oxidase IV mRNA levels and oxidative phosphorylation in the axons of sympathetic neurons. J Neurosci. 2008;28:12581–90.
90. Zheng SQ, Li YX, Zhang Y, Li X, Tang H. MiR-101 regulates HSV-1 replication by targeting ATP5B. Antivir Res. 2011;89:219–26.
91. Willers IM, Martinez-Reyes I, Martinez-Diez M, Cuezva JM. miR-127-5p targets the 3′UTR of human beta-F1-ATPase mRNA and inhibits its translation. Biochim Biophys Acta. 2012;1817:838–48.
92. Meng C, Jin X, Xia L, Shen SM, Wang XL, Cai J, Chen GQ, Wang LS, Fang NY. Alterations of mitochondrial enzymes contribute to cardiac hypertrophy before hypertension development in spontaneously hypertensive rats. J Proteome Res. 2009;8:2463–75.
93. Jullig M, Hickey AJ, Chai CC, Skea GL, Middleditch MJ, Costa S, Choong SY, Philips AR, Cooper GJ. Is the failing heart out of fuel or a worn engine running rich? A study of mitochondria in old spontaneously hypertensive rats. Proteomics. 2008;8:2556–72.
94. Ding Y, Xia B, Yu J, Leng J, Huang J. Mitochondrial DNA mutations and essential hypertension (review). Int J Mol Med. 2013;32:768–74.
95. Meyers DE, Basha HI, Koenig MK. Mitochondrial cardiomyopathy: pathophysiology, diagnosis, and management. Tex Heart Inst J. 2013;40:385–94.
96. Casademont J, Miro O. Electron transport chain defects in heart failure. Heart Fail Rev. 2002;7:131–9.
97. Marin-Garcia J, Goldenthal MJ. Understanding the impact of mitochondrial defects in cardiovascular disease: a review. J Card Fail. 2002;8:347–61.
98. Wijnen WJ, van der Made I, van den Oever S, Hiller M, de Boer BA, Picavet DI, Chatzispyrou IA, Houtkooper RH, Tijsen AJ, Hagoort J, van Veen H, Everts V, Ruijter JM, Pinto YM, Creemers EE. Cardiomyocyte-specific miRNA-30c over-expression causes dilated cardiomyopathy. PLoS ONE. 2014;9:e96290.
99. Duisters RF, Tijsen AJ, Schroen B, Leenders JJ, Lentink V, van der Made I, Herias V, van Leeuwen RE, Schellings MW, Barenbrug P, Maessen JG, Heymans S, Pinto YM, Creemers EE. miR-133 and miR-30 regulate connective tissue growth factor: implications for a role of microRNAs in myocardial matrix remodeling. Circ Res. 2009;104:170–8. 6p following 178
100. Leptidis S, El Azzouzi H, Lok SI, de Weger R, Olieslagers S, Kisters N, Silva GJ, Heymans S, Cuppen E, Berezikov E, De Windt LJ, da Costa Martins P. A deep sequencing approach to uncover the miRNOME in the human heart. PLoS ONE. 2013;8:e57800.
101. Thum T, Galuppo P, Wolf C, Fiedler J, Kneitz S, van Laake LW, Doevendans PA, Mummery CL, Borlak J, Haverich A, Gross C, Engelhardt S, Ertl G, Bauersachs J. MicroRNAs in the human heart: a clue to fetal gene reprogramming in heart failure. Circulation. 2007;116:258–67.
102. Soh J, Iqbal J, Queiroz J, Fernandez-Hernando C, Hussain MM. MicroRNA-30c reduces hyperlipidemia and atherosclerosis in mice by decreasing lipid synthesis and lipoprotein secretion. Nat Med. 2013;19:892–900.
103. Poggesi C, Ho CY. Muscle dysfunction in hypertrophic cardiomyopathy: what is needed to move to translation? J Muscle Res Cell Motil. 2014;35:37–45.
104. Guclu A, Germans T, Witjas-Paalberends ER, Stienen GJ, Brouwer WP, Harms HJ, Marcus JT, Vonk AB, Stooker W, Yilmaz A, Klein P, Ten Berg JM, Kluin J, Asselbergs FW, Lammertsma AA, Knaapen P, van Rossum AC, van der Velden J. ENerGetIcs in hypertrophic cardiomyopathy: traNslation between MRI, PET and cardiac myofilament function (ENGINE study). Neth Heart J. 2013;21:567–71.
105. Song L, Su M, Wang S, Zou Y, Wang X, Wang Y, Cui H, Zhao P, Hui R, Wang J. MiR-451 is decreased in hypertrophic cardiomyopathy and regulates autophagy by targeting TSC1. J Cell Mol Med. 2014;18:2266–74.
106. Bagnall RD, Tsoutsman T, Shephard RE, Ritchie W, Semsarian C. Global microRNA profiling of the mouse ventricles during development of severe hypertrophic cardiomyopathy and heart failure. PLoS ONE. 2012;7:e44744.

107. Soni MS, Rabaglia ME, Bhatnagar S, Shang J, Ilkayeva O, Mynatt R, Zhou YP, Schadt EE, Thornberry NA, Muoio DM, Keller MP, Attie AD. Downregulation of carnitine acyl-carnitine translocase by miRNAs 132 and 212 amplifies glucose-stimulated insulin secretion. Diabetes. 2014;63:3805–14.
108. Morita S, Horii T, Kimura M, Hatada I. MiR-184 regulates insulin secretion through repression of Slc25a22. Peer J. 2013;1:e162.
109. Barath R HJ, Hollander JM, Das S. Mitochondrial miRNAs in diabetes: just the tip of the iceberg. Can J Physiol Pharmacol. 2017. (in press).

Chapter 24
Mitochondrial Dysfunction in Cardiovascular Aging

Charles L. Hoppel, Edward J. Lesnefsky, Qun Chen, and Bernard Tandler

Mitochondria are the primary source of energy metabolism in the heart [1]. Previous studies have shown that as a concomitant of the aging process cardiac mitochondria undergo detrimental changes [2]. It turns out that these changes do not occur through the entire mitochondrial population in given cardiomyocytes, but occur predominantly

---

Disclosures
The authors state they have no conflict of interest or financial interests to disclose.

C.L. Hoppel (✉)
Department of Pharmacology, Case Western Reserve University, School of Medicine, Cleveland, OH 44106, USA

Department of Medicine, Case Western Reserve University, School of Medicine, Cleveland, OH 44106, USA

Center for Mitochondrial Disease, Case Western Reserve University, School of Medicine, Wood Bldg. Rm. W147C, 10900 Euclid Avenue, Cleveland, OH 44106, USA
e-mail: charles.hoppel@case.edu

E.J. Lesnefsky
Department of Medicine, Division of Cardiology, Pauley Heart Center, Virginia Commonwealth University, Richmond, VA 23298, USA

Cardiology Section, Medical Service, McGuire Veterans Affairs Medical Center, 1201 Broad Rock Road, Richmond, VA 23249, USA
e-mail: edward.lesnefsky@vcuhealth.org

Q. Chen
Department of Medicine, Division of Cardiology, Pauley Heart Center, Virginia Commonwealth University, Richmond, VA 23298, USA
e-mail: qun.chen@vcuhealth.org

B. Tandler
Department of Biological Sciences, CWRU School of Dental Medicine, Cleveland, OH 44106, USA
e-mail: bernard.tandler@case.edu

G. Santulli (ed.), *Mitochondrial Dynamics in Cardiovascular Medicine*, Advances in Experimental Medicine and Biology 982,
DOI 10.1007/978-3-319-55330-6_24

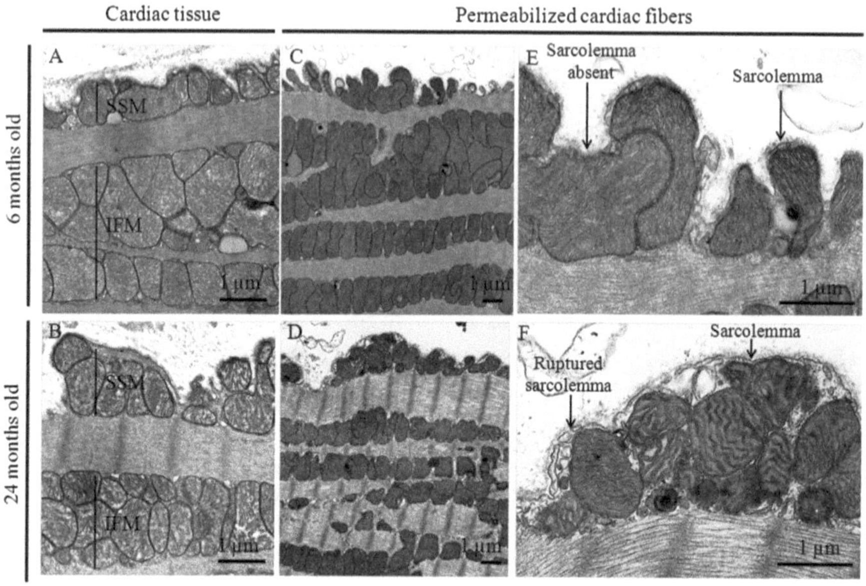

Fig. 24.1 Electron micrographs of the cardiac tissue and permeabilized fibers from young and old Fisher 344 rats: (**a**) 6-month, (**b**) 24-month. The subsarcolemmal (SSM) and interfibrillar (IFM) mitochondria are identified. The ultrastructure of the permeabilized fibers is shown in (**c**) 6-month at low magnification, (**d**) 24-month at low magnification, (**e**) 6-month at high magnification, (**f**) 24-month at high magnification. In the higher magnification micrographs, some segments of the sarcolemma are intact, but other are fragmented; examples are identified with *arrows*. Dimensional bar = 1 µm (Adapted from Lemieux et al. [11] with permission from the Journals of Gerontology, Part A)

in those mitochondria situated between the myofibrils (interfibrillar mitochondria = IFM) [3–5]. In contrast, the subsarcolemmal (SSM) mitochondria are unaffected [3]. These observations are based on isolated mitochondria derived by a technique that separates the two populations.

Ultrastructure of cardiac mitochondria utilizing transmission electron microscopy (TEM) has not revealed obvious differences in structure between the two populations of mitochondria [3, 6, 7]. However, examination by high-resolution scanning electron microscopy (HRSEM) of osmium-extracted, otherwise intact heart tissue revealed differences in the organization of cristae in the two subsets of organelles [8, 9]. This being the case, we proceeded to examine cardiomyocytes that had their plasma membrane disrupted, so-called permeabilized fibers (Fig. 24.1). TEM showed that although remnants of the plasma membrane were still present, there was extensive discontinuities [10].

These permeabilized fibers proved to be excellent simulacra of intact cardiomyocytes in that they use oxygen and elaborate ATP at a level that matches isolated mitochondria [11, 12]. Using such cells we were able to compare the metabolic performance of fibers isolated from three different rat strains that have significantly different life spans [11]. Of the three rat strains used in this study, the Fischer 344 strain has the

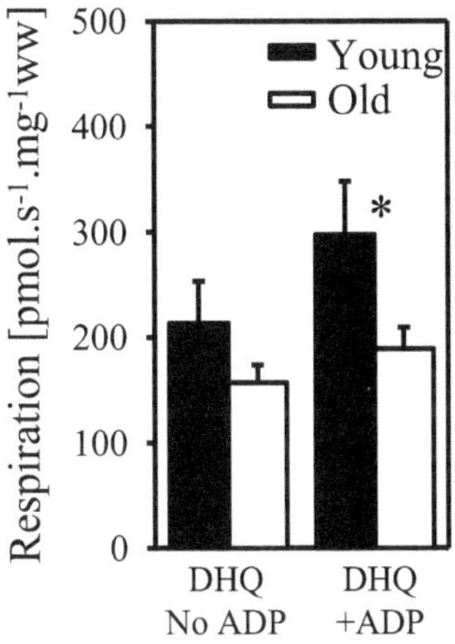

Fig. 24.2 Duroquinol (DHQ) + rotenone oxidation in permeabilized cardiac fibers from young and old Fisher 344. Resting respiration is measured in the presence of DHQ + rotenone without ADP (DHQ, no ADP). Coupled respiration is measured after the addition of saturating ADP to the DHQ (DHQ, +ADP). Data for the different age groups. The *asterisk* indicates significant differences between age groups for the same state and strain (Adapted from Lemieux et al. [11] with permission from the Journals of Gerontology, Part A)

shortest life and exhibits the greatest decrease in mitochondrial function. Brown Norway rats have a longer life span and show less severe alterations in key mitochondrial functions. Hybrids of the two rat strains live significantly longer lives and avert the mitochondrial alterations that occur in their parent strains [11]. From these observations, it can be concluded that cardiac mitochondrial performance is correlated with life span; in other words, the more profound the mitochondrial damage in the heart, the shorter the lifespan, probably because of production of reactive oxygen species and activation of mitochondria-based programs leading to cell death [13–18]. Figure 24.2 shows the decrease in oxidative phosphorylation starting at complex III in cardiac fibers from aged Fischer 334 rats Lemieux et al. [11].

The use of permeabilized isolated cardiac muscle fibers has been touted to be superior to the use of isolated mitochondria [11, 12]. However, although permeabilized fibers do in fact reveal mitochondrial abnormalities it is impossible to determine if these changes involve the entire content of mitochondria or are restricted to a particular subset. In contrast, by separating SSM and IFM it is possible to determine if the damage is ubiquitous or affects only one, thus providing more precise information than the fiber preparations are capable of showing [3]. For this reason, we rely principally on the use of cell fractions to obtain separate sets of mitochondria that are situated in the two discrete venues. The result of these studies is that we determined that mitochondrial damage in the short-lived animals is not ubiquitous for the entire mitochondrial content of these cells, but is restricted to the IFM, whereas the SSM remained unaffected [2, 3, 5]. Integrated mitochondrial function, measured as oxidative phosphorylation, is decreased in IFM because of a defect in

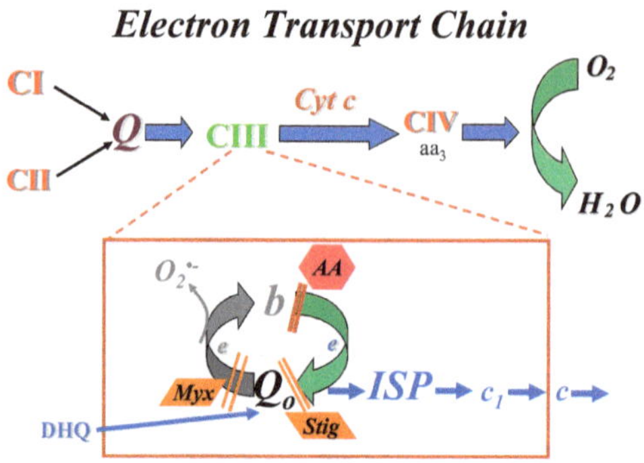

Fig. 24.3 A cartoon of the electron transport chain (*upper portion*) showing the convergence of reducing equivalents from complexes I and II to coenzyme Q. Reduced coenzyme Q is oxidized by complex III to reduce cytochrome c, which is oxidized by complex IV with formation of water. The *lower portion* of the cartoon below the dashed lines shows the Q_o site on cytochrome b with reduction by the coenzyme Q analog duroquinol, the myxothiazol-binding site (Myx) and the stigmatellin-binding site (Stig) in the Q_o site. Superoxide from leakage at the Q_o site is shown. Antimycin A is a traditional complex III inhibitor. The Reiske iron-sulfur protein (ISP) and cytochrome c_1 (c_1) with cytochrome b are the active sites for electron transfer in complex III

oxidation coupled with an increase in net ROS production, again only in the IFM [3, 16]. We showed that the oxidative defect is localized to complexes III and IV, an observation subsequently confirmed by Hagen and colleagues [3, 16, 17, 19]. Further, the site of ROS production was determined to be complex III, locating the defect in complex III at the hub of mitochondrial damage in aging [16]. This defect can be ameliorated by administration of acetylcarnitine to the aged rat [3, 16].

To return to the case of the permeabilized fibers, although the plasmalemma was disrupted, the mitochondria had a normal appearance, both SSM and IFM (Fig. 24.1). Mitochondrial morphology was not altered by age, in that the organelles of both 6- and 24-month old rats were structurally identical by TEM [3, 6, 7]. Thus biochemical alterations that have occurred in the 18-month interval are unrelated to morphology of these mitochondria. It should be noted that the differences in crista architecture between the SSM and IFM that we observed by HRSEM in adult, 6-month rat hearts are effaced by 24 months [10].

As mentioned previously, complexes III and IV are the sites of oxidative defects in IFM in the aged rat heart [3, 4, 16]. Under normal circumstances, complex III oxidizes reduced coenzyme Q and through a series of steps ultimately reduces cytochrome *c* [1, 2] (Fig. 24.3). In the aged rat, the binding site for reduced coenzyme Q (Qo of cytochrome *b*) is altered, resulting in decreased activity of complex III with increased ROS production at that site [3, 4, 16]. Complex IV activity lessens in the IFM of the 24-month rat because of a structural rearrangement of the complex that can be restored by phospholipids [3]. Taken together, the alterations in complexes III and IV lead to an increased susceptibility of the aged cardiomyocyte to stress, including infarction, so that contraction of the cardiomyocyte is interfered with.

This susceptibility in brought about largely by the ROS production at the Qo site in complex III [16]. ROS production is influenced by infarction, which causes a loss of iron from the iron-sulfur protein (ISP) [20] that constitutes a catalyst in complex III, adding a further defect to the aging effect. The defect in the ISP leads to a further increase in ROS production as a retrograde effect, exacerbating the production of ROS, thus worsening the problem [20].

Defects in Electron Transport Induced by Age

Supercomplexes (respirasomes) are assemblies of complex I, the dimer of complex III, and one to four copies of complex IV that facilitate the interaction between components in terms of reducing equivalents [21–28]. In aged animals, changes in the subsidiary complexes III and IV of supercomplexes result in a diminution of the content of the host supercomplex in IFM, leading to a decrease in oxidation [29]. In contrast, complex I activity is unaffected by age in murine hearts [30].

Cytochrome c *Oxidase (Complex IV)*

Cytochrome *c* oxidase plays a significant role in respiration. With age, respiration through cytochrome c oxidase decreases in IFM because of an altered inner membrane, which depends in part on biophysical properties [3, 31, 32]. However, the addition of phospholipid liposomes restored the cytochrome c oxidase respiration and enzyme activity [3, 32]. In IFM (both in situ and isolated) there is a 25% reduction in subunit VIIa, as shown by immunoelectron microscopy [19]. Because analytical immunoblotting fails to reveal such a decrement [19], it may be concluded that the subunit is actually present but is not readily demonstrable by the aforementioned structural study because of morphological alterations in the inner membrane.

The same decrease in complex IV activity and content is present in the heart of aged humans [33]. Decreased cytochrome oxidase activity has an upstream effect on complex III, enhancing the production of mitochondrial ROS as previously noted in a variety of cardiac models [34, 35], eventually favoring activation of cell death programs [36].

Complex III

As in the case for complex IV, complex III activity is decreased only in IFM in aging [3]. It has been found that the Qo site of cytochrome *b* is the critical recipient of the aging effect [16]. This observation is based on the use of two inhibitors, each of which binds to different domains of the Qo site. Stigmatellin inhibits the reduction of cytochrome b in IFM from the aged heart in the same fashion as it does in young hearts. In contrast, myxothiazol only partially inhibits the reduction of cytochrome *b*, resulting in a leak of electrons from the Qo site of IFM from the aged heart (Fig. 24.3). These observations support the notion that the aging defect in complex III is specifically localized to the myxothiazol-binding site of the Qo site of cytochrome b [16].

It has been shown that the major site of ROS production is complex III and that the Qo site is the principal source of ROS [34, 37]. The superoxide so produced is in the intermembrane space and can either pass directly into the cytosol through the voltage-dependent anion channel [38] or be converted to H_2O_2 by Cu, Zn-superoxide dismutase in the intermembrane space [34]. The H_2O_2 subsequently can leave the mitochondria. In both cases, the released H_2O_2 can damage all extra-mitochondrial structures within the myocyte [39, 40]. The ROS in the intermembrane space activates the permeability transition pores [41] leading to cytochrome *c* release into the cytosol triggering programmed cell death. The inner membrane permeability transition pores may consist in part of complex V [42]. The decrease in complex V activity with aging combined with the presence of ROS might increase the propensity of the inner membrane to elaborate pores that could contribute to mitochondrial damage [14].

Cardiolipin

The phospholipids of the mitochondrial inner membrane include a unique lipid, namely, cardiolipin (CL) [43, 44]. It is essential for the activity of complex I, III, IV, and V as well as translocases and transporters of the inner membrane [44]. Albeit others have attributed aging effects in cardiac mitochondria to a reduction in their content of CL [32], in our hands the content of CL in SSM or IFM in aged Fischer 344 rats remains unaltered [45]. The discrepancies that have been noted might be due to variations in analytical techniques used in these studies [32, 45]. As a result of our findings, we conclude that CL is unrelated to aged cardiac mitochondrial defects.

However in ischemia of the aged heart, a change in composition of CL takes place [46, 47]; this change does not occur in ischemic young hearts [48]. It is paralleled by the increased susceptibility to damage that occurs in the elderly [49]. Some have held that the changes in CL during ischemia are due to insertion in CL of a higher molecular weight fatty acid [50, 51], whereas it also has been reported that oxidation of CL elevates its molecular weight [46, 47]. These differences in interpretation have not be reconciled.

Mitochondrial DNA, Aging, and the Heart

Mitochondria contain DNA (mtDNA) that differs from that in the nucleus in that mtDNA is arranged in circles and utilizes a different coding language to direct synthesis of specific mitochondrial proteins, all of which are components of the electron transport chain and are hydrophobic. The mtDNA is susceptible to mutation, which may be brought about by exposure to ROS [52, 53]. An increase in mtDNA mutations in aged heart has been reported [52, 53]. Despite the occurrence of mutations, the copy number of mtDNA is unaffected in the aged heart, neither in SSM nor IFM [54, 55], in contrast to what occurs in skeletal muscle and hepatic mitochondria where the copy number decreases [54].

Mitochondria contain an enzyme – 8-oxodeoxyguandine glycosylase – that repairs damaged mtDNA [56, 57]. When this enzyme is knocked down, mitochondrial mutations increase, an enhancement that rises dramatically with age [55]. This observation supports the idea that aging per se is a cause of mtDNA mutations, again implicating ROS [52, 53]. Not only do the latter affect the production of new proteins they also damage already existing mitochondrial proteins, both enzymatic and structural interfering with the function of these proteins [2].

Dynamic Mitochondria Remodeling in the Aged Heart

Mitochondria in the heart have a limited half-life, but the number of mitochondria per myocyte remains fairly constant. This being the case, there must be a source for the generation of replacement organelles. All evidence indicates that new mitochondria are generated by division of already existing mitochondria [58]. TEM studies have revealed that cardiac mitochondria can follow two distinctly different morphological pathways for division [59]. The first involves the formation of a partition that spans the inner compartment. Ingrowth of the outer membrane in a fashion that resembles a closing iris diaphragm ultimately leads to the separation of the two mitochondrial moieties. Little is known of the biochemical events related to this modality of division. The second pathway, one that has been intensely studied by a spectrum of cell biological methods, involves the formation of a belt-like band around the organelle; its constriction ultimately leads to the fission of the host mitochondrion [59]. These belts consist of dynamin-related protein 1 (DRP-1), which has been attracted to specific outer membrane proteins to form a "noose" around the entire organelle [59]. It is the tightening of this noose that carries out the scission process. As the outer membrane is contracting it impinges on the inner membrane, thus involving the entire covering of the mitochondria. [59].

Based on cell fractionation, it has been reported that a concomitant of aging is that the yield of mitochondria is reduced [3]. In fact, it is only one of the two populations of cardiac mitochondria, namely, the IFM that undergo this reduction, whereas the SSM yield remains unaltered [3]. In order for this reduction in mitochondrial yield to be brought about, either there is a slowing of IFM division or an increase in autophagy (mitophagy) or a combination of the two. In our EM studies we failed to note increased evidence of mitophagy [9, 59], bolstering the notion that it is most likely fission that has been slowed. Even though mitophagy does not seem to increase, its steady occurrence will help to shrink the mitochondrial population. Mitophagy probably involves the elimination of those defective organelles that were set free by organelle division. Even after the two processes (fission and mitophagy) have taken place, the residual mitochondria in aged animals are unable to engage in oxidative phosphorylation at the same rate as mitochondria in younger animals are able [3]. The possibility of enhanced removal of mitochondria in specific regions of the aged myocyte is a provocative notion. Depolarization of mitochondrial membrane potential activates mitophagy [60]. Perhaps IFM, with their defects in oxidative phosphorylation and enhanced susceptibility to permeability transition pore opening, both risks for depolarization of mitochondrial membrane potential, may undergo enhanced rates of removal by mitophagy in comparison to the age-unaltered SSM population [2].

Therapy

Mitochondrial alterations appear to be a major culprit in damage to the aging heart [1, 2]. Age-induced decrements in mitochondrial function are superimposed upon disease-related mitochondrial impairments [20]. Ischemia-reperfusion [61, 62] and congestive heart failure [63] each induce mitochondrial damage that is superimposed upon pre-existing aging defects. Therefore any therapy that is directed towards improving the performance of these organelles should have salubrious effects in hearts that have been exposed to stress, including ischemia and aging [2]. The concept to attenuate disease-induced impairment of the aged heart by addressing pre-existing age-related defects is an appealing therapeutic approach. Using experimental animals, such therapies include behavior modifications such as caloric restriction [64] and exercise [65], nutriceuticals – coenzyme Q10 plus PUFA [66] and acetylcarnitine [67] – and an assortment of pharmacological agents.

Behavioral Modifications

Although caloric restriction in young rats does not improve cardiac mitochondrial function, in aged rats it ameliorates oxidative phosphorylation as noted in permeabilized cardiac fibers derived from these animals [64]. Calorie restriction further improves mitochondrial function related to activation of sirtuins 1 and 3 [68–70]. Cardiac mitochondrial function also improves in elderly rats that have been exercised. This is accompanied by a reduction in H_2O_2 production and MnSOD activity in both SSM and IFM [71].

Nutraceuticals

Among the nutraceuticals administered to elderly animals are resveratrol, Coenzyme Q10 plus poly-unsaturated fatty acids (PUFA), and acetylcarnitine. Resveratrol delays aged-mediated cardiovascular dysfunction by activation of sirtuins 1 and 3 [72–75].

The latter affects mitochondrial function by deacetylation of acetyllysines. In hearts from untreated elderly rats the activity of sirtuin 1 is decreased. This negative effect is exaggerated by the fact that NAD+, an activator of sirtuins, also decreases with age. The result of these changes is decreased activity of complex I-IV and increased content of NADH+ in elderly female Wistar rats [73]; in other words, the consequence is a weakening of mitochondrial performance. A therapy that increases NAD+ content in skeletal muscle and brain, but its impact on heart has not been studied [76–78].

Administration of CoQ10 in PUFA to 24 months old rats extends their life span. This supplementation enhances the activity of cytochrome *c* oxidase in cardiac mitochondria in 24-month old mice, but CoQ10 alone does not affect either ETC activities or ROS production in these mice [79].

Acetylcarnitine is another nutraceutical that has positive effects on cardiac mitochondria in aged rats [80] by increasing mitochondrial oxidative phosphorylation in IFM to values comparable to those of adults, raising complex III and IV activity, and increasing mitochondrial content of cytochromes *b* and aa_3 [67]. It has been noted that elderly hearts do not respond well to stress such as ischemia-reperfusion [49]. The renaissance in mitochondrial activity brought about by acetylcarnitine provides the elderly heart with the ability to counter this type of stress.

Pharmacological agents

Two chemical agents have been studied in terms of their effect on the hearts of experimental animals. SS-31, a cell permeable tetrapeptide, has been shown to protect against the ravages of ischemia-reperfusion in the heart and improves mitochondrial function in heart failure [81, 82]. It is proposed that SS-31 acts through binding to cardiolipin, protecting cytochrome *c*. Clearly these studies need to be extended to the aged heart especially because it has been shown to improve mitochondrial function in aged skeletal muscle [83].

A second chemical agent that has been studied in the context of life span is rapamycin. When administered in late life to mice it extends their life span, probably by slowing metabolism [84] by affecting gene and protein expression in heart and improved cardiac function [85]. Administration of rapamycin to aged female mice for a 10-week period increases mitochondrial biogenesis and energy metabolism, altering the myocardial metabolome. Admittedly studies of the sort have been limited in scope and need to be pursued in the context of cardiac function, pathology, and cardiac mitochondrial function, especially in aging.

Conclusions

Overall, cardiac mitochondrial dysfunction accompanies aging. Rather than affecting the mitochondrial population in toto in the cardiac myocyte, this dysfunction is confined solely to the interfibrillar mitochondria. Integrated mitochondrial function measured as oxidative phosphorylation is decreased in the latter because of a defect in complexes III and IV. At the same time the number of IFM per myocyte decreases, compounding the loss of mitochondrial activity. The fact that aged cardiomyocytes can be restored to near adult levels of resistance to stress by the simple expedient of improving mitochondrial function is compelling evidence that it is the mitochondria that are the leading players in the pathophysiology of the aging heart.

Acknowledgements This work was supported by a R21AG049461 from the National Institute on Aging, and a Scientist Development Grant (11SDG5120011, QC) and Grant-in-aids (15GRNT24480123, QC) and (12GRNT20510024, CLH) from the American Heart Association, VCU's CTSA (UL1TR000058 from the National Institutes of Health's National Center for Advancing Translational Science) and the CCTR Endowment Fund of the Virginia Commonwealth University (QC), the Office of Research and Development, Medical Research Service Merit Review Award (2IO1BX001355-05A2), Department of Veterans Affairs (EJL), and the Pauley Heart Center, Virginia Commonwealth University (QC, EJL).

References

1. Lesnefsky EJ, Moghaddas S, Tandler B, Kerner J, Hoppel CL. Mitochondrial dysfunction in cardiac disease: ischemia-reperfusion, aging, and heart failure. J Mol Cell Cardiol. 2001;33:1065–89.
2. Lesnefsky EJ, Chen Q, Hoppel CL. Mitochondrial metabolism in aging heart. Circ Res. 2016;118:1593–611.
3. Fannin SW, Lesnefsky EJ, Slabe TJ, Hassan MO, Hoppel CL. Aging selectively decreases oxidative capacity in rat heart interfibrillar mitochondria. Arch Biochem Biophys. 1999;372:399–407.
4. Lesnefsky EJ, Gudz TI, Moghaddas S, Migita CT, Ikeda-Saito M, Turkaly PJ, Hoppel CL. Aging decreases electron transport complex III activity in heart interfibrillar mitochondria by alteration of the cytochrome c binding site. J Mol Cell Cardiol. 2001;33:37–47.
5. Lesnefsky EJ, Hoppel CL. Oxidative phosphorylation and aging. Ageing Res Rev. 2006;5:402–33.
6. Palmer JW, Tandler B, Hoppel CL. Biochemical properties of subsarcolemmal and interfibrillar mitochondria isolated from rat cardiac muscle. J Biol Chem. 1977;252:8731–9.
7. Palmer JW, Tandler B, Hoppel CL. Biochemical differences between subsarcolemmal and interfibrillar mitochondria from rat cardiac muscle: effects of procedural manipulations. Arch Biochem Biophys. 1985;236:691–702.
8. Riva A, Tandler B, Lesnefsky EJ, Conti G, Loffredo F, Vazquez E, Hoppel CL. Structure of cristae in cardiac mitochondria of aged rat. Mech Ageing Dev. 2006;127:917–21.
9. Riva A, Tandler B, Loffredo F, Vazquez E, Hoppel C. Structural differences in two biochemically defined populations of cardiac mitochondria. Am J Physiol Heart Circ Physiol. 2005;289:H868–72.
10. El'darov CM, Vays VB, Vangeli IM, Kolosova NG, Bakeeva LE. Morphometric examination of mitochondrial ultrastructure in aging cardiomyocytes. Biochemistry (Mosc). 2015; 80:604–9.
11. Lemieux H, Vazquez EJ, Fujioka H, Hoppel CL. Decrease in mitochondrial function in rat cardiac permeabilized fibers correlates with the aging phenotype. J Gerontol A Biol Sci Med Sci. 2010;65:1157–64.
12. Hughey CC, Hittel DS, Johnsen VL, Shearer J. Respirometric oxidative phosphorylation assessment in saponin-permeabilized cardiac fibers. J Vis Exp. 2011;48:2431, 1–7.

13. Giorgio V, Petronilli V, Ghelli A, Carelli V, Rugolo M, Lenaz G, Bernardi P. The effects of idebenone on mitochondrial bioenergetics. Biochim Biophys Acta. 2012;1817:363–9.
14. Hafner AV, Dai J, Gomes AP, Xiao CY, Palmeira CM, Rosenzweig A, Sinclair DA. Regulation of the mPTP by SIRT3-mediated deacetylation of CypD at lysine 166 suppresses age-related cardiac hypertrophy. Aging (Albany NY). 2010;2:914–23.
15. Halestrap AP, Clarke SJ, Javadov SA. Mitochondrial permeability transition pore opening during myocardial reperfusion – a target for cardioprotection. Cardiovasc Res. 2004;61:372–85.
16. Moghaddas S, Hoppel CL, Lesnefsky EJ. Aging defect at the Qo site of complex III augments oxyradical production in rat heart interfibrillar mitochondria. Arch Biochem Biophys. 2003;414:59–66.
17. Suh JH, Heath SH, Hagen TM. Two subpopulations of mitochondria in the aging rat heart display heterogenous levels of oxidative stress. Free Radic Biol Med. 2003;35:1064–72.
18. Hsieh CC, Papaconstantinou J. Thioredoxin-ASK1 complex levels regulate ROS-mediated p38 MAPK pathway activity in livers of aged and long-lived Snell dwarf mice. FASEB J. 2006;20:259–68.
19. Fujioka H, Moghaddas S, Murdock DG, Lesnefsky EJ, Tandler B, Hoppel CL. Decreased cytochrome c oxidase subunit VIIa in aged rat heart mitochondria: immunocytochemistry. Anat Rec (Hoboken). 2011;294:1825–33.
20. Lesnefsky EJ, Gudz TI, Migita CT, Ikeda-Saito M, Hassan MO, Turkaly PJ, Hoppel CL. Ischemic injury to mitochondrial electron transport in the aging heart: damage to the iron-sulfur protein subunit of electron transport complex III. Arch Biochem Biophys. 2001;385:117–28.
21. Enriquez JA. Supramolecular organization of respiratory complexes. Annu Rev Physiol. 2016;78:533–61.
22. Schagger H, Pfeiffer K. Supercomplexes in the respiratory chains of yeast and mammalian mitochondria. EMBO J. 2000;19:1777–83.
23. Schagger H, de Coo R, Bauer MF, Hofmann S, Godinot C, Brandt U. Significance of respirasomes for the assembly/stability of human respiratory chain complex I. J Biol Chem. 2004;279:36349–53.
24. Enriquez JA, Lenaz G. Coenzyme q and the respiratory chain: coenzyme q pool and mitochondrial supercomplexes. Mol Syndromol. 2014;5:119–40.
25. Acin-Perez R, Enriquez JA. The function of the respiratory supercomplexes: the plasticity model. Biochim Biophys Acta. 2014;1837:444–50.
26. Cogliati S, Frezza C, Soriano ME, Varanita T, Quintana-Cabrera R, Corrado M, Cipolat S, Costa V, Casarin A, Gomes LC, Perales-Clemente E, Salviati L, Fernandez-Silva P, Enriquez JA, Scorrano L. Mitochondrial cristae shape determines respiratory chain supercomplexes assembly and respiratory efficiency. Cell. 2013;155:160–71.
27. Lapuente-Brun E, Moreno-Loshuertos R, Acin-Perez R, Latorre-Pellicer A, Colas C, Balsa E, Perales-Clemente E, Quiros PM, Calvo E, Rodriguez-Hernandez MA, Navas P, Cruz R, Carracedo A, Lopez-Otin C, Perez-Martos A, Fernandez-Silva P, Fernandez-Vizarra E, Enriquez JA. Supercomplex assembly determines electron flux in the mitochondrial electron transport chain. Science. 2013;340:1567–70.
28. Acin-Perez R, Fernandez-Silva P, Peleato ML, Perez-Martos A, Enriquez JA. Respiratory active mitochondrial supercomplexes. Mol Cell. 2008;32:529–39.
29. Gomez LA, Monette JS, Chavez JD, Maier CS, Hagen TM. Supercomplexes of the mitochondrial electron transport chain decline in the aging rat heart. Arch Biochem Biophys. 2009;490:30–5.
30. Kwong LK, Sohal RS. Age-related changes in activities of mitochondrial electron transport complexes in various tissues of the mouse. Arch Biochem Biophys. 2000;373:16–22.
31. Paradies G, Ruggiero FM, Petrosillo G, Gadaleta MN, Quagliariello E. The effect of aging and acetyl-L-carnitine on the function and on the lipid composition of rat heart mitochondria. Ann N Y Acad Sci. 1994;717:233–43.
32. Paradies G, Ruggiero FM, Petrosillo G, Quagliariello E. Age-dependent decline in the cytochrome c oxidase activity in rat heart mitochondria: role of cardiolipin. FEBS Lett. 1997;406:136–8.

33. Muller-Hocker J. Cytochrome-c-oxidase deficient cardiomyocytes in the human heart – an age-related phenomenon. A histochemical ultracytochemical study. Am J Pathol. 1989;134:1167–73.
34. Chen Q, Vazquez EJ, Moghaddas S, Hoppel CL, Lesnefsky EJ. Production of reactive oxygen species by mitochondria: central role of complex III. J Biol Chem. 2003;278:36027–31.
35. Qian T, Nieminen AL, Herman B, Lemasters JJ. Mitochondrial permeability transition in pH-dependent reperfusion injury to rat hepatocytes. Am J Phys. 1997;273:C1783–92.
36. Schull S, Gunther SD, Brodesser S, Seeger JM, Tosetti B, Wiegmann K, Pongratz C, Diaz F, Witt A, Andree M, Brinkmann K, Kronke M, Wiesner RJ, Kashkar H. Cytochrome c oxidase deficiency accelerates mitochondrial apoptosis by activating ceramide synthase 6. Cell Death Dis. 2015;6:e1691.
37. St-Pierre J, Buckingham JA, Roebuck SJ, Brand MD. Topology of superoxide production from different sites in the mitochondrial electron transport chain. J Biol Chem. 2002;277: 44784–90.
38. Han D, Antunes F, Canali R, Rettori D, Cadenas E. Voltage-dependent anion channels control the release of the superoxide anion from mitochondria to cytosol. J Biol Chem. 2003;278:5557–63.
39. Becker LB. New concepts in reactive oxygen species and cardiovascular reperfusion physiology. Cardiovasc Res. 2004;61:461–70.
40. Becker LB, vanden Hoek TL, Shao ZH, Li CQ, Schumacker PT. Generation of superoxide in cardiomyocytes during ischemia before reperfusion. Am J Phys. 1999;277:H2240–6.
41. Chen Q, Paillard M, Gomez L, Li H, Hu Y, Lesnefsky EJ. Postconditioning modulates ischemia-damaged mitochondria during reperfusion. J Cardiovasc Pharmacol. 2012;59:101–8.
42. Giorgio V, von Stockum S, Antoniel M, Fabbro A, Fogolari F, Forte M, Glick GD, Petronilli V, Zoratti M, Szabo I, Lippe G, Bernardi P. Dimers of mitochondrial ATP synthase form the permeability transition pore. Proc Natl Acad Sci U S A. 2013;110:5887–92.
43. Hoch FL. Cardiolipins and biomembrane function. Biochim Biophys Acta. 1992;1113:71–133.
44. Houtkooper RH, Vaz FM. Cardiolipin, the heart of mitochondrial metabolism. Cell Mol Life Sci. 2008;65:2493–506.
45. Moghaddas S, Stoll MS, Minkler PE, Salomon RG, Hoppel CL, Lesnefsky EJ. Preservation of cardiolipin content during aging in rat heart interfibrillar mitochondria. J Gerontol A Biol Sci Med Sci. 2002;57:B22–8.
46. Lesnefsky EJ, Hoppel CL. Cardiolipin as an oxidative target in cardiac mitochondria in the aged rat. Biochim Biophys Acta. 2008;1777:1020–7.
47. Lesnefsky EJ, Minkler P, Hoppel CL. Enhanced modification of cardiolipin during ischemia in the aged heart. J Mol Cell Cardiol. 2009;46:1008–15.
48. Lesnefsky EJ, Moghaddas S, Stoll MS, Minkler P, Hoppel C. Oxidative damage to the mitochondrial phospholipid cardiolipin occurs during ischemia in the senescent, but not the adult heart. FASEB J. 2003;17:A880. (abstract 574.18)
49. Lesnefsky EJ, Gallo DS, Ye J, Whittingham TS, Lust WD. Aging increases ischemia-reperfusion injury in the isolated, buffer-perfused heart. J Lab Clin Med. 1994;124:843–51.
50. Sparagna GC, Chicco AJ, Murphy RC, Bristow MR, Johnson CA, Rees ML, Maxey ML, McCune SA, Moore RL. Loss of cardiac tetralinoleoyl cardiolipin in human and experimental heart failure. J Lipid Res. 2007;48:1559–70.
51. Sparagna GC, Lesnefsky EJ. Cardiolipin remodeling in the heart. J Cardiovasc Pharmacol. 2009;53:290–301.
52. Vermulst M, Bielas JH, Kujoth GC, Ladiges WC, Rabinovitch PS, Prolla TA, Loeb LA. Mitochondrial point mutations do not limit the natural lifespan of mice. Nat Genet. 2007;39:540–3.
53. Vermulst M, Bielas JH, Loeb LA. Quantification of random mutations in the mitochondrial genome. Methods. 2008;46:263–8.
54. Barazzoni R, Short KR, Nair KS. Effects of aging on mitochondrial DNA copy number and cytochrome c oxidase gene expression in rat skeletal muscle, liver, and heart. J Biol Chem. 2000;275:3343–7.

55. Marin-Garcia J, Ananthakrishnan R, Goldenthal MJ. Mitochondrial gene expression in rat heart and liver during growth and development. Biochem Cell Biol. 1997;75:137–42.
56. de Souza-Pinto NC, Eide L, Hogue BA, Thybo T, Stevnsner T, Seeberg E, Klungland A, Bohr VA. Repair of 8-oxodeoxyguanosine lesions in mitochondrial dna depends on the oxoguanine dna glycosylase (OGG1) gene and 8-oxoguanine accumulates in the mitochondrial dna of OGG1-defective mice. Cancer Res. 2001;61:5378–81.
57. Klungland A, Rosewell I, Hollenbach S, Larsen E, Daly G, Epe B, Seeberg E, Lindahl T, Barnes DE. Accumulation of premutagenic DNA lesions in mice defective in removal of oxidative base damage. Proc Natl Acad Sci U S A. 1999;96:13300–5.
58. Ong SB, Hall AR, Hausenloy DJ. Mitochondrial dynamics in cardiovascular health and disease. Antioxid Redox Signal. 2013;19:400–14.
59. Fujioka H, Tandler B, Hoppel CL. Mitochondrial division in rat cardiomyocytes: an electron microscope study. Anat Rec (Hoboken). 2012;295:1455–61.
60. Huang C, Andres AM, Ratliff EP, Hernandez G, Lee P, Gottlieb RA. Preconditioning involves selective mitophagy mediated by Parkin and p62/SQSTM1. PLoS One. 2011;6:e20975.
61. Chen Q, Moghaddas S, Hoppel CL, Lesnefsky EJ. Ischemic defects in the electron transport chain increase the production of reactive oxygen species from isolated rat heart mitochondria. Am J Physiol Cell Physiol. 2008;294:C460–6.
62. Chen Q, Moghaddas S, Hoppel CL, Lesnefsky EJ. Reversible blockade of electron transport during ischemia protects mitochondria and decreases myocardial injury following reperfusion. J Pharmacol Exp Ther. 2006;319:1405–12.
63. Rosca MG, Vazquez EJ, Kerner J, Parland W, Chandler MP, Stanley W, Sabbah HN, Hoppel CL. Cardiac mitochondria in heart failure: decrease in respirasomes and oxidative phosphorylation. Cardiovasc Res. 2008;80:30–9.
64. Abete P, Testa G, Ferrara N, De Santis D, Capaccio P, Viati L, Calabrese C, Cacciatore F, Longobardi G, Condorelli M, Napoli C, Rengo F. Cardioprotective effect of ischemic preconditioning is preserved in food-restricted senescent rats. Am J Physiol Heart Circ Physiol. 2002;282:H1978–87.
65. Abete P, Testa G, Galizia G, Mazzella F, Della Morte D, de Santis D, Calabrese C, Cacciatore F, Gargiulo G, Ferrara N, Rengo G, Sica V, Napoli C, Rengo F. Tandem action of exercise training and food restriction completely preserves ischemic preconditioning in the aging heart. Exp Gerontol. 2005;40:43–50.
66. Quiles JL, Pamplona R, Ramirez-Tortosa MC, Naudi A, Portero-Otin M, Araujo-Nepomuceno E, Lopez-Frias M, Battino M, Ochoa JJ. Coenzyme Q addition to an n-6 PUFA-rich diet resembles benefits on age-related mitochondrial DNA deletion and oxidative stress of a MUFA-rich diet in rat heart. Mech Ageing Dev. 2010;131:38–47.
67. Lesnefsky EJ, He D, Moghaddas S, Hoppel CL. Reversal of mitochondrial defects before ischemia protects the aged heart. FASEB J. 2006;20:1543–5.
68. Cencioni C, Spallotta F, Mai A, Martelli F, Farsetti A, Zeiher AM, Gaetano C. Sirtuin function in aging heart and vessels. J Mol Cell Cardiol. 2015;83:55–61.
69. Sack MN, Finkel T. Mitochondrial metabolism, sirtuins, and aging. Cold Spring Harb Perspect Biol. 2012;4:a013102, 1–10.
70. Winnik S, Auwerx J, Sinclair DA, Matter CM. Protective effects of sirtuins in cardiovascular diseases: from bench to bedside. Eur Heart J. 2015;36:3404–12.
71. Judge S, Jang YM, Smith A, Selman C, Phillips T, Speakman JR, Hagen T, Leeuwenburgh C. Exercise by lifelong voluntary wheel running reduces subsarcolemmal and interfibrillar mitochondrial hydrogen peroxide production in the heart. Am J Physiol Regul Integr Comp Physiol. 2005;289:R1564–72.
72. Ahn BH, Kim HS, Song S, Lee IH, Liu J, Vassilopoulos A, Deng CX, Finkel T. A role for the mitochondrial deacetylase Sirt3 in regulating energy homeostasis. Proc Natl Acad Sci U S A. 2008;105:14447–52.
73. Braidy N, Guillemin GJ, Mansour H, Chan-Ling T, Poljak A, Grant R. Age related changes in NAD+ metabolism oxidative stress and Sirt1 activity in wistar rats. PLoS One. 2011; 6:e19194.

74. Lombard DB, Alt FW, Cheng HL, Bunkenborg J, Streeper RS, Mostoslavsky R, Kim J, Yancopoulos G, Valenzuela D, Murphy A, Yang Y, Chen Y, Hirschey MD, Bronson RT, Haigis M, Guarente LP, Farese Jr RV, Weissman S, Verdin E, Schwer B. Mammalian Sir2 homolog SIRT3 regulates global mitochondrial lysine acetylation. Mol Cell Biol. 2007;27:8807–14.
75. Sack MN. The role of SIRT3 in mitochondrial homeostasis and cardiac adaptation to hypertrophy and aging. J Mol Cell Cardiol. 2012;52:520–5.
76. Gong B, Pan Y, Vempati P, Zhao W, Knable L, Ho L, Wang J, Sastre M, Ono K, Sauve AA, Pasinetti GM. Nicotinamide riboside restores cognition through an upregulation of proliferator-activated receptor-gamma coactivator 1alpha regulated beta-secretase 1 degradation and mitochondrial gene expression in Alzheimer's mouse models. Neurobiol Aging. 2013;34:1581–8.
77. Mendelsohn AR, Larrick JW. Partial reversal of skeletal muscle aging by restoration of normal NAD(+) levels. Rejuvenation Res. 2014;17:62–9.
78. Ruggieri S, Orsomando G, Sorci L, Raffaelli N. Regulation of NAD biosynthetic enzymes modulates NAD-sensing processes to shape mammalian cell physiology under varying biological cues. Biochim Biophys Acta. 2015;1854:1138–49.
79. Ochoa JJ, Quiles JL, Huertas JR, Mataix J. Coenzyme Q10 protects from aging-related oxidative stress and improves mitochondrial function in heart of rats fed a polyunsaturated fatty acid (PUFA)-rich diet. J Gerontol A Biol Sci Med Sci. 2005;60:970–5.
80. Rosca MG, Lemieux H, Hoppel CL. Mitochondria in the elderly: is acetylcarnitine a rejuvenator? Adv Drug Deliv Rev. 2009;61:1332–42.
81. Szeto HH, Birk AV. Serendipity and the discovery of novel compounds that restore mitochondrial plasticity. Clin Pharmacol Ther. 2014;96:672–83.
82. Kloner RA, Hale SL, Dai W, Gorman RC, Shuto T, Koomalsingh KJ, Gorman 3rd JH, Sloan RC, Frasier CR, Watson CA, Bostian PA, Kypson AP, Brown DA. Reduction of ischemia/reperfusion injury with bendavia, a mitochondria-targeting cytoprotective Peptide. J Am Heart Assoc. 2012;1:e001644.
83. Siegel MP, Kruse SE, Percival JM, Goh J, White CC, Hopkins HC, Kavanagh TJ, Szeto HH, Rabinovitch PS, Marcinek DJ. Mitochondrial-targeted peptide rapidly improves mitochondrial energetics and skeletal muscle performance in aged mice. Aging Cell. 2013;12:763–71.
84. Wilkinson JE, Burmeister L, Brooks SV, Chan CC, Friedline S, Harrison DE, Hejtmancik JF, Nadon N, Strong R, Wood LK, Woodward MA, Miller RA. Rapamycin slows aging in mice. Aging Cell. 2012;11:675–82.
85. Chiao YA, Kolwicz SC, Basisty N, Gagnidze A, Zhang J, Gu H, Djukovic D, Beyer RP, Raftery D, MacCoss M, Tian R, Rabinovitch PS. Rapamycin transiently induces mitochondrial remodeling to reprogram energy metabolism in old hearts. Aging (Albany NY). 2016;8:314–27.

Chapter 25
Insulin Resistance and Mitochondrial Dysfunction

Alba Gonzalez-Franquesa and Mary-Elizabeth Patti

The Epidemic of Diabetes

Diabetes is now recognized as a worldwide epidemic. The International Diabetes Federation (www.idf.org) estimated that diabetes was responsible for 5.1 million deaths in 2013, and projected that diabetes will be the seventh leading cause of death in 2030 [1]. The costs for health care related to diabetes are also enormous. American and Caribbean nations spent around 263 billion USD in 2013 (an amount equivalent to that spent by half of the world), while Europe spent 147 million USD. While diabetes is skyrocketing in South-East Asia and Africa, less than 1% of the global healthcare diabetes budget is spent in these regions.

Among the millions of people worldwide suffering from diabetes, type 2 diabetes (T2D) accounts for ~90% of the cases, with type 1 diabetes accounting for about 10%. Monogenic diabetes and other rare forms make up a small percentage [2]. In this review, we will focus exclusively on T2D and its relationship to mitochondrial function.

Pathogenesis of Type 2 Diabetes

T2D is characterized by elevations in plasma glucose. Once diabetes is diagnosed, the key defects which can be observed include (1) ***insulin resistance***, or reduced responsiveness of a tissue to insulin, and (2) ***abnormal insulin secretion*** by

A. Gonzalez-Franquesa • M.-E. Patti (✉)
Research Division, Joslin Diabetes Center, and Harvard Medical School, Boston, MA, USA
e-mail: Alba.Gonzalez-Franquesa@joslin.harvard.edu; mary.Elizabeth.Patti@joslin.harvard.edu

G. Santulli (ed.), *Mitochondrial Dynamics in Cardiovascular Medicine*, Advances in Experimental Medicine and Biology 982,
DOI 10.1007/978-3-319-55330-6_25

pancreatic islet β-cells. In this chapter, we will consider these two defects as largely independent, although there is likely a reciprocal relationship. For example, sustained insulin resistance in multiple tissues (including the β-cell) may contribute to loss of insulin secretion, while altered insulin secretion may also increase glucose levels and induce further insulin resistance. Additional defects contributing to the complexity of diabetes risk include altered intestinal incretin hormone secretion and/or responsiveness, altered glucagon secretion by pancreatic α-cells, impaired oxidative metabolism at rest and during exercise, altered insulin-independent glucose disposal, perturbed lipid and amino acid metabolism, dysregulation of neural control of appetite in the brain, and reduced renal glucose excretion [3].

Which of These Defects Are Primary to Pathogenesis of T2D in Humans?

Elegant studies by DeFronzo and Groop demonstrated that insulin resistance was not only a feature of established T2D, but could also be detected in individuals at risk for T2D. For example, cross-sectional analysis demonstrated reduced insulin-stimulated glucose uptake and glycogen synthesis in normal glucose tolerant humans with diabetes family history [4]. Such insulin resistance was subclinical, as plasma glucose levels were maintained within the normal range by increased insulin secretion by pancreatic β-cells.

DeFronzo and colleagues proposed the concept of the Starling's curve of the pancreas [5], with an inverted U-shaped curve showing the initial increase and subsequent decrease in insulin secretion over time with sustained insulin resistance. Time-dependent loss of insulin secretion would ultimately lead to increased glucose levels and the clinical onset of impaired glucose tolerance (rise in glucose after eating) or overt T2D [6, 7]. These concepts were subsequently extended by the work of Bergman and colleagues, who described a hyperbolic relationship between insulin sensitivity and insulin secretion, with their product defined as disposition index [8]. In this context, reductions in insulin sensitivity are compensated by increased insulin secretion by healthy β-cells, maintaining glucose homeostasis. Once the disposition index is reduced, glucose values rise. From these concepts, it is not surpising that β-cell function may be already reduced by as much as 80% once T2D is diagnosed [9–12], with similar reductions in β-cell mass [13, 14].

Longitudinal studies have also underscored the importance of insulin resistance as not only a feature of diabetes risk, but a predictor of T2D development. For example, metabolic analysis of normal glucose tolerant individuals with T2D family history (both parents with T2D) revealed that insulin resistance at baseline, but not insulin secretion, predicted subsequent T2D development over a mean of 25 years [15]. Moreover, longitudinal population analysis over 13 years demonstrated that defects in insulin secretion emerged relatively late in those who ultimately developed T2D [16].

Molecular Etiology of Insulin Resistance

The burden of insulin resistance and T2D is closely linked to the epidemics of obesity and physical inactivity [17], as both obesity and physical inactivity are typically associated with insulin resistance.

As with many phenotypes in humans, there is substantial interindividual variability in the susceptibility to insulin resistance and T2D, likely related in part to genetic background [18, 19]. For example, rare genetic mutations can result in monogenic forms of T2D [20–22]. Genome-wide association studies (GWAS) in individuals with common forms of T2D have identified many polymorphisms associated with disease risk [23–25]. Interestingly, these effects are modest in size, and collectively explain only ≈10% of disease heritability [26]. Genes proposed to be related to these SNPs have largely been assumed to mediate effects via insulin secretion. However, it remains uncertain whether such SNPs are mediating risk via a nearby or more distant gene or noncoding RNA, and many SNP-related genes are expressed rather broadly, making functional assessment of SNP-mediated mechanisms difficult.

More recent GWAS studies and meta-analysis [27] have identified *loci* associated with diabetes-related continuous metabolic variables (i.e., HOMA-IR, HOMA-B, fasting glucose, fasting insulin). For example, a SNP adjacent to insulin receptor substrate 1 (IRS1) was reported as the first T2D risk *locus* associated with insulin resistance and hyperinsulinemia [28]. Moreover, SNPs within the 1-Mb region flanking the gene MTNR1B are associated with fasting glucose, independently of body mass index (BMI) [29–31]. Moreover, T2D, fasting glucose, and glycated hemoglobin [30] are also associated with a variant near MTNR1B. Recently, two large consortia have performed whole-genome and exome sequencing in a large population, and conclude that lower-frequency variants are common, but do not have a major impact on T2D predisposition [32].

Collectively these data suggest that substantial T2D risk is related to non-genetic factors. These may include epigenetic factors, developmental history and exposures, and postnatal environmental factors such as personal or family lifestyle, dietary patterns, physical activity, and the microbiome.

Tissue-Specific Insulin Resistance

While insulin resistance can be defined by reduced responsiveness to insulin, the metabolic manifestations of insulin differ according to tissue and cell type.

Skeletal muscle accounts for over 80% of insulin-stimulated glucose uptake. Thus, insulin signaling in skeletal muscle is essential for normal insulin-stimulated whole-body glucose uptake. Indeed, analysis of human skeletal muscle biopsy samples demonstrated that insulin stimulation activates not only insulin signaling [33], but also downstream metabolic endpoints such as glycogen synthesis [4] and potent transcriptional responses [34]. In insulin resistant individuals, insulin signaling is disrupted, with increased serine (Ser) phosphorylation of IRS1, decreased AKT phosphorylation, and reduced translocation of the glucose transporter GLUT4 to the sarcolemmal membrane, thus impairing glucose uptake [35]. These signaling defects are also pathway-specific, with preserved activity of the ERK-MAPK pathway despite defects in the AKT pathway [33].

In the ***liver***, insulin resistance results in failure to suppress gluconeogenesis, promoting increased hepatic glucose production (HGP). Indeed, insulin inhibits the transcription of two key gluconeogenic enzymes (phosphoenolpyruvate carboxykinase (PEPCK) and glucose-6 phosphatase (G6Pase) [36]), through AKT and FOXO-dependent mechanisms [37]. Moreover, insulin resistance is often accompanied by hepatic steatosis and nonalcoholic fatty liver disease, reflecting at least in part, increased lipogenesis. In turn, lipid accumulation may also reduce insulin signaling, via activation of PKC-ε and JNK1-dependent pathways [38].

Insulin resistance in ***white adipose tissue*** (WAT) results in not only reduced glucose uptake but also failure to suppress lipolysis, leading to increased release of free fatty acids (FFA) and glycerol from adipose stores, further increasing plasma FFA. This can result in excessive delivery and so-called "ectopic" accumulation of lipids in non-adipose tissues such as skeletal muscle and liver [39]. Furthermore, obesity-related insulin resistance is also associated with altered secretion of adipokines, both pro- and anti-inflammatory, and low-grade systemic inflammation [40].

Since the discovery of human ***brown adipose tissue*** (BAT) [41–43], this tissue has been established as an important endocrine organ which can modulate whole-body glucose metabolism, insulin sensitivity, and thermogenesis [44]. This was demonstrated experimentally in animal models, in which high fat diet (HFD)-fed mice receiving a BAT transplant had reduced body weight, increased insulin sensitivity, and improved glucose tolerance [45]. Whether improvements in systemic metabolism result from BAT secretory products (e.g., circulating norepinephrine, IL6, FGF21), from increased glucose and lipid uptake by brown fat depots, or from other mechanisms, remains uncertain [45]. Metabolically active adipocytes residing in WAT, due to a process called *browning*, may also modulate thermogenesis and whole-body energy expenditure.

Less traditionally insulin-responsive tissues also play key roles in metabolic defects associated with insulin resistance. For example, insulin resistance in the ***brain***, specifically in the satiety centres of the hypothalamus (e.g., arcuate nucleus) [46], can promote hyperphagia and reduce energy expenditure, thus increasing adiposity. Furthermore, brain insulin resistance may also directly regulate metabolism in other organs, e.g., increasing hepatic glucose production and reducing muscle glucose uptake in rodents [47, 48].

The ***gastrointestinal system*** also plays a critical role in regulation of metabolism via secretion of incretin hormones such as GLP-1 and GIP in response to oral intake. These hormones not only increase insulin secretion, but also regulate intestinal motility, appetite, and systemic metabolism [49–51]. GLP-1 and GIP secretion in response to a meal are impaired in normoglycemic insulin resistant humans, in parallel with the degree of insulin resistance [52]. Thus, incretin deficiency is a likely contributor to progressive decreases in insulin secretion in insulin resistant individuals.

Mitochondrial Function and Human Insulin Resistance

Mitochondria are essential for maintaining cellular energy homeostasis and viability. They are often called the cellular powerhouses given their important role in the generation of adenosine triphosphate (ATP). However, mitochondria also have important roles in reactive oxygen species (ROS) generation, detoxification and scavenging, ROS-dependent intracellular signaling [53, 54], apoptosis and autophagy control [55], and regulation of cytoplasmatic calcium [56]. We will first outline key mitochondrial processes, and then consider the evidence supporting links between insulin resistance, T2D risk, and disruption in mitochondrial content or functional capacity in tissues relevant for metabolic regulation.

Mitochondrial Bioenergetics

Mitochondria produce ATP mainly through oxidative phosphorylation (OxPhos) located in the mitochondrial inner membrane *cristae*, foldings which expand the membrane surface area. Electrons generated from nutrient oxidation are carried by the electron donors NADH and $FADH_2$ and enter the electron transport system (ETS) at complex I (CI) and II (CII), respectively. The flow of electrons results in pumping of protons from the mitochondrial matrix to the intermembrane space, creating both a pH gradient and a membrane potential. When protons return to the matrix via complex V (CV or ATP synthase), ATP is synthesized from adenosine diphosphate (ADP) and inorganic phosphate [57]. Protons can also leak across the inner membrane through uncoupler proteins, thus bypassing ATP synthesis and releasing energy as heat instead, the so-called non-shivering thermogenesis.

Mitochondrial Biogenesis

The content, volume and mass of mitochondria are tightly regulated in response to coordinated signals from both nuclear- and mitochondrial-encoded proteins. One key regulator of this process is the PGC1 family of coactivators [58]. PGC1α and β coactivate transcription factors such as PPAR nuclear receptors, NRF, and ERR, promoting coordinated increases in transcription of mitochondrial genes. This response, which ultimately increases mitochondrial DNA (mtDNA) content, mitochondrial mass, and activity, can be initiated by a wide range of stimuli, including exercise, cellular stress, temperature, nutrients, and nitric oxide (NO) [59].

Mitochondrial Dynamics

Tight regulation of mitochondrial dynamics, including mitochondrial fusion, fission, and movement, is important for cellular bioenergetics, senescence and survival, quality control, and intracellular signaling [60, 61]. Fusion results in elongated, connected mitochondria, a response often seen with nutrient restriction; conversely, fission, promoting fragmented, separated mitochondria, is more common in a nutrient-rich environment [62, 63]. Indeed, repression, ablation, or loss-of-function mutation of fusion genes (i.e., MFN1, MFN2 and OPA1) leads to mitochondrial fragmentation or reduced mitochondrial filaments [64–67], while re-expression normalizes length [68–70]. By contrast, experimental reduction of mitochondrial fission protein (i.e., DRP1, FIS1, MTP18) content or activity elongates the mitochondrial network [71–75], while overexpression promotes fragmentation [73, 74, 76].

Collectively, fusion and fission are both normal parts of the mitochondrial life cycle which promote not only a healthy rearrangement of mitochondrial components but also the release of dysfunctional components, thereby helping to maintaining a healthy pool of mitochondria. Ultimately, damaged mitochondria can be eliminated via the process of mitophagy.

Disruptions in T2D and Insulin Resistance

Obesity, insulin resistance, and established T2D have all been shown to alter mitochondrial number and/or function in different tissues, resulting in the presence of the overly broad term "mitochondrial dysfunction" as a key phenotype linked to T2D pathophysiology. Despite abundant data, it remains uncertain whether perturbations in mitochondrial functional capacity are a cause, consequence, or key contributor to insulin resistance and T2D risk [77–79]. Indeed, this controversy has been the source of much debate in the scientific community [80, 81] (Table 25.1). In the sections below, we will examine data from studies assessing mitochondrial function in different tissues in order to dissect this key question for insulin resistance and T2D pathophysiology.

Skeletal Muscle

Skeletal muscle is a major metabolic tissue of the body. While many consider muscle to be relatively homogeneous, it is important to recognize its heterogeneity from a metabolic perspective. Classification by fiber type, based on immunohistochemical or gene expression patterns, can be useful to illustrate differences in metabolism in discrete muscle groups. For example, type I fibers have higher mitochondrial density and oxidative capacity, while type II fibers have high glycogen content and fewer capillaries [105] and can be further subdivided into types IIa, IIb, and IIx. As

Table 25.1 Main studies supporting links between mitochondrial dysfunction and insulin resistance

Authors	Year	Journal	Sample	Observations
Insulin resistance is associated with impaired mitochondrial function (decreased FA oxidation and increased DAGs)				
Vondra et al. [82]	1977	Diabetologia	T2D subjects	Decreased mitochondrial activity
Kelley et al. [83]	1999	Am J Physiol	T2D patients	30–40% lower mitochondrial enzyme activity
Simoneau et al. [84]	1999	FASEB J		
He et al. [85]	2001	Diabetes		
Kelley et al. [86]	2002	Diabetes		
Kelley et al. [86]	2002	Diabetes	Obese individuals	Smaller mitochondrial size, reduced mitochondrial content, and altered mitochondrial performance
Petersen et al. [87]	2003	Science	Healthy lean elderly	Severe insulin resistance in skeletal muscle, higher levels of TG (in muscle and liver), decreased mitochondrial oxidative capapcity, and decreased ATP synthesis
Mootha et al. [88]	2003	Nat Genet	Obese Caucasians with IGT and T2D	Decreased expression of nuclear-encoded genes regulating mito biogenesis (i.e. PGC1α-responsive genes)
Patti et al. [89]	2003	PNAS	Obese and overweight non-diabetic Mexican-Americans	Decreased expression of PGC1α and β and nuclear OxPhos genes
Petersen et al. [90]	2004	N Engl J Med	Young-insulin resistant offspring of parents with T2D	Decreased mitochondrial activity, increased intramyocellular FA, and lower ratio of type 1 (oxidative) to type 2 (glycolytic) fibers in muscle
Morino et al. [91]	2005	J Clin Invest	T2D and insulin resistant individuals	30% reduction in mitochondrial density
Mitochondrial dysfunction is not in causal relationship to insulin resistance				
Miller et al. [92]	1984	J Appl Physiol Respir Environ Exerc Physiol	Rodents with HFD	Increased levels of mitochondrial marker enzymes (beta-hydroxybutyrate DH and CS)
McAinch et al. [93]	2003	Obes Res		

(continued)

Table 25.1 (continued)

Authors	Year	Journal	Sample	Observations
Han et al. [94]	2004	Am J Physiol Endocrinol Metab	Transgenic rodents with mito dysfunction	Low substrate oxidative capacity, increased AMP, activated AMPK, increased GLUT4 (i.e., improving insulin responsiveness)
Wredenberg et al. [95]	2006	Biochem Bioph Res Co		
Han et al. [96]	2011	PLoS ONE		
Han et al. [94]	2004	Am J Physiol Endocrinol Metab	Transgenic rodents with muscle mitochondrial deficiency/ dysfunction	Increased glucose tolerance, increased insulin sensitivity, and increased insulin-stimulated glucose uptake
Wredenberg et al. [95]	2006	Biochem Bioph Res Co		
Pospisilik et al. [97]	2007	Cell		
Zechner et al. [98]	2010	Cell Metab		
Garcia-Roves et al. [99]	2007	PNAS	Rodents with intermittently elevating plasma FA (with insulin resistance)	Increase in mitochondrial biogenesis
Turner et al. [100]	2007	Diabetes	Rodents fed with HFD (with insulin resistance)	Increase in mitochondrial biogenesis (increased marker enzymes, FA oxidation, and mtDNA copy number)
Hancock et al. [101]	2008	PNAS		
Toledo et al. [102]	2008	Diabetes	Diet-induced weight loss in humans	Improved insulin resistance without increasing mitochondrial capacity
Nair et al. [103]	2008	Diabetes	Insulin resistant Asian Indians with T2D	Muscle mitochondrial capacity similar to nondiabetic Indians and higher than Caucasians
Jin et al. [104]	2011	J Clin Invest	Caucasians with insulin resistance	No change in PGC1 or OXPHOS expression in isolated IR in individuals matched for BMI to insulin sensitive individuals

previously noted, skeletal muscle is responsible for an estimated 80% of glucose uptake during intravenous insulin infusion in healthy individuals [106–108]. Reduced muscle glucose uptake contributes to reduced glycogen synthesis, reduced glucose oxidation, whole-body insulin resistance and postprandial hyperglycemia in T2D [109–112]. These defects are also observed in high-risk individuals prior to the onset of T2D, as indicated by 70% reduction in muscle glycogen synthesis in young lean and glucose tolerant offspring of parents with T2D [113]. Notably, these defects are not observed in other tissues such as brain, fat and splanchnic tissue, which maintain their glucose utilization in T2D [5] (Fig. 25.1). Collectively, these data have stimulated focus on skeletal muscle as a dominant contributor to insulin resistance and diabetes risk.

The molecular mechanisms mediating muscle insulin signaling to glucose uptake have been the focus of intense study for many years. Insulin binds and activates the receptor tyrosine (Tyr) kinase, resulting in increased phosphorylation of Tyr residues in the β-subunit of the insulin receptor [114–117] and subsequent recruitment and phosphorylation of IRS1. IRS1 further activates downstream targets such as PI3K and AKT, ultimately resulting in translocation of GLUT4 to the plasma membrane. Additional effects of insulin include activation of NO synthase, leading to arterial vasodilatation [118, 119].

In patients with T2D, multiple steps of insulin signaling are disrupted. Reduced phosphorylation of the insulin receptor and IRS1 has been demonstrated in T2D [33, 114], in insulin resistant offspring of T2D parents [120, 121], and in obese

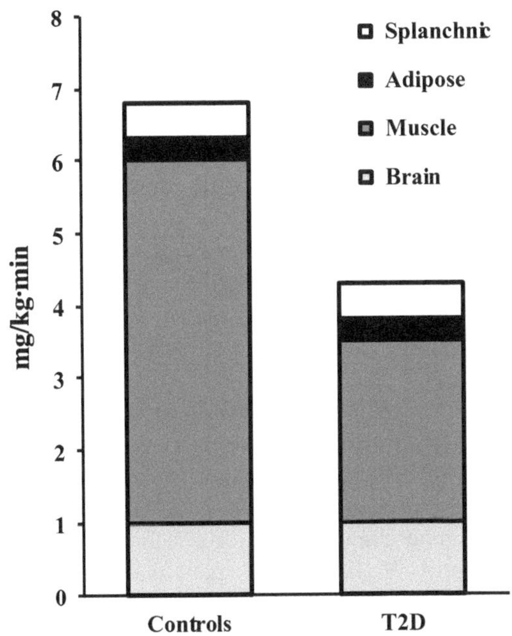

Fig. 25.1 Glucose uptake during euglycemic hyperinsulinemic clamp (100 μU/ml) in healthy control and T2D individuals (Adapted from DeFronzo [5])

individuals with normal glucose tolerance [33]. While IRS1 Tyr phosphorylation is reduced, inhibitory serine (Ser) phosphorylation (Ser307 in humans, Ser302 in rodents) [122] is increased in insulin resistance and T2D. This results in reduced interaction of IRS1 with the insulin receptor [123] and reduced downstream signaling [124].

Mitochondrial Function and Insulin Resistance in Skeletal Muscle

Many studies using a variety of methods have demonstrated abnormalities in mitochondrial function and/or number in muscle from humans with insulin resistance and T2D [86, 90, 125, 126] (Table 25.1). In vivo nuclear magnetic resonance (NMR) spectroscopy studies have revealed decreased mitochondrial ATP production and phosphocreatine recovery in both T2D and lean insulin resistant offspring of individuals with T2D [90, 127]. Reduced ATP production has also been observed in response to exercise in healthy young offspring of parents with T2D, suggesting this phenotype may be consistently related to insulin resistance [128].

Beyond NMR, more traditional assays of mitochondrial enzymatic activity in muscle biopsy samples evaluated ex vivo have also been used to probe functional defects in insulin resistance and T2D. Analysis of OxPhos enzymatic activity has yielded varying results, depending on the methodology and the population under study. Initial studies of Simoneau and colleagues reported decreased activity of citrate synthase (CS), a surrogate for mitochondrial content and activity, in both obesity and T2D [84]. Subsequent analysis of OxPhos activity, as measured by NADH:O_2 oxidoreductase activity (CI activity), demonstrated a 40% reduction in T2D patients [86]. Similarly, lower muscle OxPhos capacity in T2D was demonstrated in studies analyzing either muscle biopsy samples [86, 129–131] or isolated mitochondria [132]. These differences remained even after matching for age, sex and body mass [127, 129]. Moreover, analysis of lean insulin resistant but normoglycemic offspring also revealed a 50% lower cytochrome C oxidase activity` [91]. Elderly insulin resistant subjects have also decreased mitochondrial OxPhos activity [87]. By contrast, other studies have reported normal mitochondrial activity in both T2D [133, 134] and in obese and insulin resistant individuals [135]. Interestingly, studies in South Asian Indians demonstrated increased OxPhos activity despite insulin resistance [103].

The impact of insulin resistance and T2D on mitochondrial content and subcellular localization has also been evaluated using electron microscopy. In general, reduced mitochondrial content is observed in either T2D [86, 91, 136, 137] or insulin resistance [91, 137]. These differences are particularly prominent in the subsarcolemmal compartment, with a threefold reduction in patients with T2D [126]. Collectively, these data suggest that differences in mitochondrial energetics may result, at least in part, from reduced mitochondrial number.

Transcriptional Regulation of Mitochondrial Biogenesis and Function in Skeletal Muscle

Reduced mitochondrial biogenesis may contribute to observed alterations in mitochondrial number and function. Transcriptomic studies revealed a coordinated reduction in expression of genes encoded by PGC1α in patients with T2D in both the fasting [89] and insulin-stimulated states [138]. In addition, expression of PGC1α and β was reduced in obese insulin resistant individuals [89]. However, this reduction has not been consistently observed in lean, insulin resistant offspring of parents with T2D or non-diabetic relatives [91, 104, 139], highlighting the potential impact of obesity as driver of insulin resistance-related phenotypes.

Perturbations in mitochondrial biogenesis can also be associated with altered mitochondrial dynamics contributing to functional derangements. Obese insulin resistant individuals have reduced expression of the mitochondrial fusion gene MFN2 [140], while mutations in the fission protein OPA1 reduce muscle ATP synthesis [141]. Mice with either genetic (*ob/ob*) or diet-induced obesity have increased expression of DRP1 and FIS1 in parallel with smaller, shorter mitochondria [142]. Surprisingly, protein content of MFN2, DRP1, and OPA1 are increased in obesity with hyperglycemia (*db/db* mice) [143]. Conversely, inhibition of mitochondrial fission improves both muscle insulin signaling and whole-body insulin sensitivity [142], suggesting that fission might underlie both mitochondrial derangements and insulin resistance in muscle.

Mitochondrial Alterations and Insulin Resistance in Skeletal Muscle: Interpretation

The variability in results of analysis of mitochondrial function in insulin resistance and T2D suggests not only differences in methodology, but also genotypic and phenotypic differences in populations under study. These include age, BMI, diet, activity and physical fitness, muscle fiber type, early-life environmental exposures, and magnitude of insulin resistance. Among these, obesity appears to be a potential cofounding factor in many studies: a reduction in mitochondrial functional capacity might be related to obesity itself rather than insulin resistance, given that obese patients show small mitochondria. Careful matching for BMI reduces differences in mitochondrial content and activity in some studies [129, 130, 132] but not in others [127, 129]. Moreover, muscle mitochondrial function appears to be dissociated from insulin resistance given that mitochondrial content, activity, and OxPhos capacity do not correlate with insulin sensitivity in some insulin resistant cohorts [103, 135, 144].

Defects in function can be dissociated from mitochondrial mass, a parameter less precisely measured, especially in skeletal muscle. When oxygen consumption measurements in muscle fibers are normalized by mtDNA content or CS activity, differences between T2D and controls are reduced or completely abolished in some studies [130]. In other studies, ADP-stimulated and maximal respiration is reduced in T2D, despite unchanged mitochondrial content [129]. Similarly, reductions in OxPhos activity can be observed even in isolated subsarcolemmal mitochondria [126]. In this setting, differences in fission/fusion/tethering to other organelles, or to local metabolite gradients potentially disrupted during the isolation and assay process, could contribute to variable functional capacity.

Given the variability inherent to human studies, rodent models have also been used to decipher the relationship between insulin resistance, T2D, and mitochondrial dysfunction. Mitochondrial respiratory capacity is increased in glycolytic muscle from *db/db* [143] and *ob/ob* mice [145], but is unaltered [145] or reduced [143] in oxidative muscle, in parallel with decreased mtDNA content and expression of mitochondrial biogenesis regulators [143]. Mitochondrial function in rats is also variable, with increases in Zucker diabetic fatty (ZDF) rats [146] and HFD-fed animals [100, 147], but unchanged capacity in other studies using ZDF rats [148].

Rodent models have also demonstrated that experimentally-induced alterations in mitochondrial function are not consistently linked to insulin resistance. For example, rats made iron deficient have reduced protein content of cytochrome C and cytochrome C oxidase subunits I and IV, but unchanged insulin-stimulated glucose uptake [96]. Furthermore, experimental disruption of OxPhos CI (e.g., muscle- and liver-specific apoptosis-induced factor (AIF)-deficient mice) increases insulin sensitivity [97]. Similarly, deficiency of TFAM [95], the PGC1 family of transcriptional activators [98], or cytochrome C oxidase subunit IV peptide 2a (COX6A2) [149] yields protection against insulin resistance. Thus, rodent studies support an inconsistent relationship, and in some cases, full dissociation between insulin resistance and mitochondrial function.

Potential Factors Beyond Insulin Resistance Contributing to Mitochondrial Functional Defects Associated with Insulin Resistance and T2D

The data above from both humans and rodent models suggest that insulin resistance is often, but not always, associated with disruption in some components of mitochondrial metabolic function. Moreover, experimental disruption of mitochondrial OxPhos, sufficient to reduce energy metabolism, does not produce insulin resistance, at least in animal models. What additional factors contribute to mitochondrial dysfunction but not directly linked to insulin resistance might account for this? Potential mediators (Fig. 25.2) include ectopic lipid accumulation, oxidative stress (e.g., ROS), imbalance betwen fatty acid oxidation (FAO) and TCA cycle/OxPhos capacity, incomplete FAO, all of which may in turn be linked to physical inactivity.

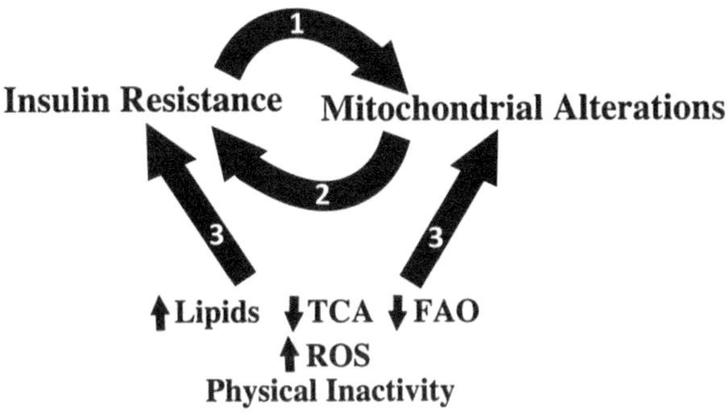

Fig. 25.2 Main hypotheses linking insulin resistance and mitochondrial dysfunction. (*1*) Insulin resistance causes alterations in multiple facets of mitochondrial function. (*2*) Mitochondrial alterations lead to insulin resistance. (*3*) Other factors lead to both insulin resistance and mitochondrial alterations concomitantly, including increased tissue lipid supply, imbalance between tricarboxylic acid cycle and OxPhos capacity, incomplete lipid oxidation, and increased ROS

Ectopic Lipid Accumulation

One phenomenon characteristic of insulin resistance and T2D in skeletal muscle is inappropriate or "ectopic" intramyocellular lipid accumulation [150], reviewed in [151, 152], and in [153]. Tissue accumulation of triglycerides (TG) can also predict insulin resistance in both adults and children [154–157]. Moreover, inhibition or activation of genes regulating muscle lipid uptake (i.e., LPL, FATP, CD36) in rodents yields protection or susceptibility to insulin resistance, respectively [158–162], confirming the important role of ectopic lipid accumulation. One mechanism by which accumulation of excess lipid yields insulin resistance is via disruption of insulin signaling. Experimentally increasing plasma fatty acids (FA) in both humans [163] and rodents [164, 165] blunts insulin activation of IRS1-associated PI3K activity in muscle. The delayed time course of insulin resistance in this protocol (>3 h after lipid infusion) suggests that tissue lipid uptake and metabolism are required to reduce glucose uptake [163].

It remains unclear which molecular lipid species are the dominant mediators of insulin resistance. While initial studies focused on TG accumulation as a common signature, these species are also elevated in muscle of highly trained and insulin-sensitive athletes, and thus not likely to play the dominant role in the pathophysiology [166, 167]. Rather, additional lipids such as diacylglycerols (DAGs) may be more closely linked to insulin resistance [122, 168–170]. DAGs disrupt insulin signaling by activating serine/threonine (Ser/Thr) kinases such as PKC, inhibitory IKKβ and NFκB, reducing post-receptor insulin action [171]. Moreover, FFA may directly activate Toll-like receptors, mediating inflammatory signals, IKK and JNK activation, and downregulation of insulin signaling [172–174]. Ceramides have also been linked to insulin resistance in obese humans [175, 176].

Different studies report increased FAO in insulin resistance and T2D [177–180]. Furthermore, when lipid oxidation is increased as a result of increased lipid delivery, but not matched by appropriate increases in oxidative capacity, products of incomplete FAO may accumulate; this may also contribute to oxidative stress, activation of Ser/Thr kinases, and additional stimulus for insulin resistance [100]. Hence, lipid accumulation can parallel reduced mitochondrial oxidative capacity [87, 90, 181]. Whether lipid excess could either cause or result from mitochondrial dysfunction remains unclear [86, 182, 183]. Experimental lipid infusion in healthy individuals reduces insulin-stimulated ATP synthesis [184]. However, OxPhos activity is not reduced in obesity in the fasting state (when FA substrate utilization is maximal) [185, 186], despite insulin resistance [187, 188]. Not surprisingly, lipid oxidation and mitochondrial content are not consistently correlated in individuals with obesity or T2D, potentially due to reduced expression of PGC1α and dysregulated FA activation of PPAR [189, 190].

Rodent studies have also provided conflicting data. In obese ZDF rats, intramyocellular lipids are increased without altered mitochondrial function [191]. In other studies, elevations in circulating FFA accompanying HFD-induced obesity are associated with increased mitochondrial OxPhos capacity and biogenesis [99, 101, 192], together with increased FAO and expression of mitochondrial genes [100]. Increased lipid supply may actually promote lipid oxidative capacity via activation of PPARδ [99]. Similarly, experimental increases in FAO induced by HFD in mice with reduced malonyl CoA levels (acetyl-CoA carboxylase 2 KO mice) yields protection against diet-induced insulin resistance [193].

The relationship between lipid acccumulation, insulin sensitivity, and mitochondrial function has also been tested during intervention studies. Weight loss consistently increases insulin sensitivity, but alterations in intramyocellular TG content and mitochondrial respiration or enzymatic activity are not always observed [102, 194]. Exercise training in T2D improves both insulin sensitivity and mitochondrial content, but does not consistently decrease intramyocellular lipids [195]. Strategies to pharmacologically alter lipid content have also been used to study this relationship: administration of acipimox, a drug that reduces lipolysis, modestly increases insulin sensitivity and decreases ROS in muscle, but does not improve basal or insulin-stimulated OxPhos capacity in obese individuals with or without T2D [196, 197].

Oxidative Stress

Another potential mechanism contributing to insulin resistance and mitochondrial dysfunction could be oxidative stress (i.e., increased levels of ROS), since mitochondria are one of the major sites of ROS production in the cell [198]. Normal ROS production and signaling are essential for cellular and mitochondrial functionality, a phenomenon known as *mitohormesis* [199, 200]. However, chronic sustained increases in ROS may be deleterious, damaging DNA, protein, and lipids, and altering transcriptional regulation. With chronic nutrient oversupply, excessive reduction of NADH and $FADH_2$ can lead to increased mitochondrial membrane potential; if demand for ATP or uncoupling is not increased in parallel, this may increase mitochondrial ROS production [201] and could contribute to lower intrinsic mitochondrial activity in T2D individuals [126, 129, 132].

Several experimental paradigms support the possibility that excessive ROS could contribute to insulin resistance. Treatment of cells with glucocorticoids or the pro-inflammatory hormone TNFα leads to insulin resistance and a parallel increase in ROS and related transcriptional effectors [202]. Moreover, increased ROS has been associated with insulin resistance in animal models [203]. For example, hypercaloric feeding in rodents increases plasma levels of hydrogen peroxide (H_2O_2) [204], while HFD feeding increases mitochondrial H_2O_2 production, decreasing redox-buffering capacity [201], and decreases oxidative capacity and ATP levels [205]. Studies in T2D individuals failed to show an increase in mitochondrial H_2O_2 [206, 207], while antioxidant capacity was reduced in skeletal muscle [208].

Mechanisms linking ROS to insulin resistance remain uncertain. Mitochondrial ROS have been linked to activation of pro-inflammatory molecules such as IKKs [209] or PKCs [91], both of which can yield disruption of insulin signaling. These and other oxidative stress species activate Ser kinases, ultimately disrupting insulin signaling [115, 201, 210]. Conversely, ROS have been shown to oxidize and eversibly inhibit protein tyrosine phosphatases, increasing insulin signaling in some studies [211].

Since association between mitochondrial-derived oxidative stress and insulin resistance cannot establish causality, several studies have attempted to reduce oxidative stress by treatment with antioxidants or overexpression of antioxidant genes. For example, mitochondrial-targeted antioxidant treatment and overexpression of mitochondrial catalase in HFD-fed rodents preserves insulin sensitivity [201], and reduces DAG concentration [212]. Similarly, overexpression of manganese superoxide dismutase ameliorates insulin resistance [213]. By contrast, antioxidant treatment in another study normalized muscle mitochondrial function but did not normalize insulin sensitivity or glucose tolerance [205].

Physical inactivity is also linked to mitochondrial dysfunction, independent of insulin sensitivity. Exercise training increases mitochondrial biogenesis in skeletal muscle [214], and restores oxidative capacity even in T2D [207, 215]; in some studies, insulin sensitivity and metabolic flexibility also improve in parallel [216]. However, insulin sensitivity and mitochondrial function do not always respond concordantly. For example, insulin sensitivity is improved with weight loss achieved through diet and exercise without measurable changes in mitochondrial capacity [102]. Furthermore, the effectiveness of physical training may differ from individual to individual. For example, offspring of mothers with T2D are not able to improve insulin sensitivity as much as non-diabetic controls after 9 days of intensive training, despite similar skeletal muscle OxPhos capacity [217].

Taken together, these data support two principal hypotheses for mechanisms mediating the link between mitochondrial function and insulin resistance (Fig. 25.3). Firstly, increases in intracellular lipids due to increased delivery and/or altered metabolism can disrupt insulin signaling. Secondly, increased mitochondrial ROS generation, due to imbalance between oxidative demand and mitochondrial capacity,

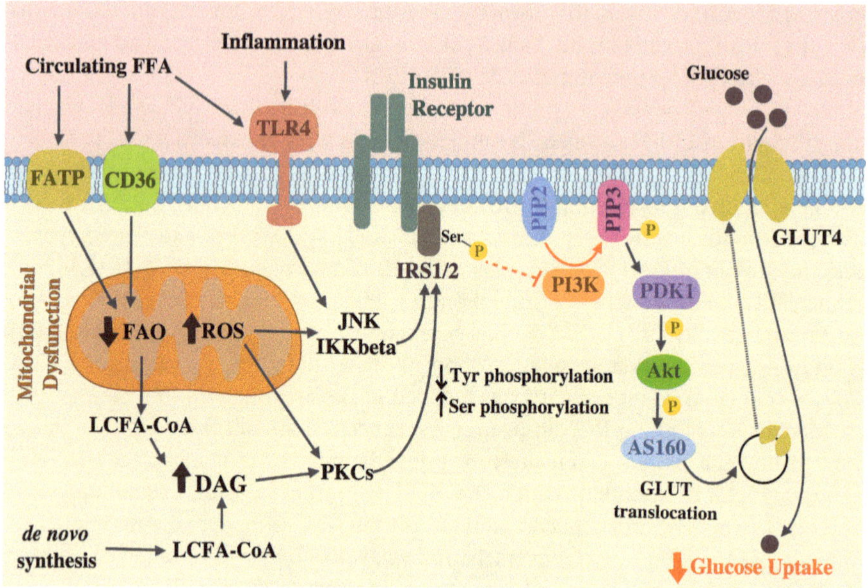

Fig. 25.3 Main hypotheses linking mitochondrial dysfunction to insulin resistance. Low mitochondrial fatty acid oxidation (*FAO*) relative to supply may contribute to incomplete oxidation and/or accumulation of species such as DAG that activate PKCs and disrupt insulin signaling. ROS overproduction within mitochondrial can activate inflammatory pathways ultimately disrupting insulin signaling

can also perturb insulin signaling. Both of these mechanisms may mediate, at least in part, the impact of obesity and physical inactivity on both insulin sensitivity and mitochondrial function.

Liver

The liver plays a key role in maintenance of whole body energy metabolism via functions such as storage of glucose-derived carbons, synthesis of plasma proteins, and hormone production and detoxification. The liver can store glucose as glycogen (glycogenesis) after a meal, while producing glucose in the fasting state and during exercise by depolymerizing glycogen (glycogenolysis), or synthesizing glucose from other sugars, amino acids and glycerol (gluconeogenesis). Apart from carbohydrate metabolism, the liver also has an important role in lipid metabolism, via both oxidation of TG and de novo lipogenesis from excess sugar and amino acids, and in the synthesis of cholesterol and bile acids. The liver can also deaminate amino acids and produce plasma proteins and urea.

Obesity and insulin resistance are associated with abnormalities in nuclear-encoded genes regulating OxPhos in the liver, in parallel with reduced PGC1α expression and increased hepatic lipid accumulation [218]. By contrast, OxPhos complexes genes are increased in parallel with BMI and insulin resistance in Japanese obese T2D individuals [219]. It remains uncertain whether these differences are related to ethnic differences or degree of obesity in the population under study. Given the difficulty in accessing liver tissue sample for analysis in humans, rodent experimental models are required, but have yielded discordant conclusions. Analysis of isolated mitochondria after HFD-feeding in rodents [220–222] revealed reduced respiratory capacity and increased oxidative stress. Similarly, mice with genetic obesity (*db/db* and *ob/ob*) have decreased activity of mitochondrial complexes CI to CV in both isolated mitochondria [223] or liver homogenates [143, 145]. By contrast, other studies revealed unchanged [224] or increased liver mitochondrial oxidative capacity in adult *ob/ob* mice [225] and insulin resistant diabetic GK rats [226]. The reasons for these differences remain unclear, but may be related to different diets, magnitude of obesity, and experimental methods used to assess mitochondrial function (i.e., mitochondrial respiration, mtDNA analysis, etc). Regarding lipid accumulation in the liver, genetic defects in FAO (i.e., lack of LCAD) predispose animals to hepatic steatosis and insulin resistance [148].

White Adipose Tissue

Adipose tissue is now recognized as a key endocrine tissue that participates in the pathogenesis of insulin resistance and T2D. While lipid-storing adipocytes represent the main component of adipose tissue, the stromal vascular fraction (SVF) also contains pre-adipocytes, endothelial cells, fibroblasts and resident macrophages. Full differentiation is critical for lipid accumulation and the metabolic function of adipose tissue. Differentiated WAT acts as a nutrient and hormonal sensor coordinating either FA or glucose oxidation or storage [227]. Beyond lipid storage, adipose tissue can also regulate systemic metabolism by secreting adipokines (i.e., cytokines, chemokines, hormone-like factors, other mediators) that regulate appetite, inflammation, insulin action, and glucose and lipid metabolism [228, 229].

There are two major types of WAT: visceral adipose tissue (VAT) and subcutaneous adipose tissue (SAT). VAT has been considered the *harmful* fat, as compared with *beneficial* SAT [230]. VAT is associated with insulin resistance, increased T2D risk, dyslipidemia, atherosclerosis and mortality [231–233]. Indeed, VAT secretes less adiponectin [234] and more pro-inflammatory cytokines [235] while accumulating more macrophages.

With sustained obesity, insulin resistance, and T2D, adipose tissue is characterized by low-grade chronic inflammation [236], with macrophage infiltration and secretion of pro-inflammatory cytokines such as TNFα, IL6 and MCP1 among others contributing to obesity-associated insulin resistance and hyperglycemia [236–247]. In fact, antibody-mediated neutralization of TNFα improves insulin sensitivity [237].

Obesity-related inflammation has been linked to mitochondria since these organelles can activate the inflammasome, a multiprotein complex that both initiates and controls the inflammatory response to stress conditions, such as oxidative stress [248, 249]. This complex can also be activated by mtDNA [250] and lipids such as ceramides [251]. Interestingly, ceramides have been associated themselves with mitochondrial dysfunction [252].

Mitochondrial Health and Adipocyte Differentiation

Mitochondrial health is intimately linked to adipocyte differentiation. Indeed, some studies reported an increase in mitochondrial biogenesis and mitochondrial protein levels during differentiation [253–256]. During the course of differentiation, there is also a 20 to 30-fold increase in mitochondrial content, together with increases in β-oxidation [254, 255], mtDNA copy number, and transcription of both nuclear and mitochondrial-encoded genes [257]. Conversely, adequate mitochondria are required for regulation of adipocyte differentiation [258]. For example, reduction in expression of the mitochondrial transcription factor TFAM, a highly expressed gene during differentiation, reduces ETS capacity and disrupted GLUT4 translocation in 3T3-L1 adipocytes despite promoting insulin-stimulated AKT phosphorylation [259]; these data suggest that glucose transport in adipocytes might be disrupted by mitochondrial dysfunction via a mechanism downstream of AKT-dependent signaling. Similarly, rotenone-induced inhibition of mitochondrial respiration altered adipocyte differentiation [253]. Moreover, obesity-induced alterations in mitochondrial biogenesis also disrupt differentiation [254, 260], while in mature adipocytes, obesity, age and T2D are associated with reduced mitochondrial content [260].

Mitochondrial Dysfunction in WAT in Obesity

Metabolic function of adipose mitochondria is also altered in obesity. Adipocytes facing nutritional overload have increased lipid storage, increased glycolysis, and reduced mitochondrial biogenesis [261] and function [262–264]. Mitochondrial content and number are reduced in subcutaneous WAT biopsies from humans with obesity and T2D [265]. Mitochondrial number correlated with lipogenesis but not with BMI or insulin sensitivity [266]. Moreover, reduced abundance of mitochondria and expression of mitochondrial genes was reported in patients with insulin resistance, T2D and severe obesity [267]. In these patients, functional studies in WAT indicated a decrease in adipocyte oxygen consumption and ATP production [268], with similar findings in both VAT and SAT from obese individuals [269]. In studies of monozygotic twins, the more obese co-twin was more insulin resistant, in association with reduced expression of mitochondrial OxPhos-encoding genes in abdominal SAT [270]. Surprisingly, VAT adipocytes have higher numbers of mitochondria [271] and higher mitochondrial respiration compared to SAT [272]. However, T2D has been associated with decreased expression of mitochondrial ETS genes in VAT regardless of obesity [273], with potential links to mTORC1-dependent mechanisms [274].

Rodent models of obesity, both genetic (*ob/ob*) and dietary (HFD feeding) obesity, and T2D (*db/db*) have also been used to assess derangements in adipose tissue. WAT from *ob/ob* and from *db/db* mice showed reduced expression of markers of mitochondrial capacity [275–277], with downregulation of PPARα, ERRα and PGC1α in *db/db* and obese HFD-fed mice [278]. At a protein level, mitochondrial proteins are markedly decreased in the adipocytes of *db/db* mice, but not in *ob/ob* mice [276]. Moreover, protein levels of ETS components (i.e., NADH dehydrogenase NDUFB8, succinate dehydrogenase SDHB and COX4I1) are reduced in HFD-fed mice [279]. Electron microscopy also demonstrates autophagosomes near mitochondria in HFD-fed mice; parallele increases in autophagy related proteins such as PINK1 and PARKIN also suggests autophagy-related metabolic remodeling [279].

Oxidative Stress in WAT

As in many tissues, optimal levels of ROS are required for normal function, including differentiation, in adipocytes [258]. Indeed, the adipocyte can tolerate high levels of ROS under basal conditions without damaging the cell or leading to apoptosis [202, 280]. However, most studies indicate that oxidative stress is chronically increased at high levels in WAT from models of dietary and genetic obesity [203, 212, 281]. This stress is a likely consequence of increased ROS production in adipose mitochondria [201] that may be related to chronic increases in non-esterified FA (NEFA) and TNFα [282, 283]. Excessive ROS may have several deleterious consequences, including reduced mitochondrial biogenesis [203], reduced proliferation of adipogenesis precursors [284], substrate and TG accumulation due to incomplete oxidation, and carbonylation of key insulin signaling intermediates such as IRS1 and 2, reducing insulin-stimulated Tyr phosphorylation [124]. Collectively, this can lead to hypertrophic, dysfunctional, and large adipocytes. Conversely, antioxidants can restore mitochondrial function and promote adipogenesis in parallel [285].

Brown and Beige Adipose Tissue

The presence of brown adipose tissue (BAT) in human adults was discovered less than a decade ago [41, 43, 286]. BAT is now considered a key tissue potentially contributing to modulation of whole-body metabolism and thermogenesis [44]. Brown adipocytes have many mitochondria with high oxidative capacity that also contain UCP1 uncoupling protein along the inner membrane [287]. In parallel, white adipose tissue has been shown to be capable of so-called *browning* or *beiging*, increasing mitochondrial number and expression of UCP1, and conferring increased metabolic activity [288, 289]. Indeed, cold exposure can potently increase UCP1 protein levels in WAT undergoing *browning*, with levels similar to those in constitutive BAT; the relative contribution of these tissues to thermogenesis remains uncertain [290, 291]. These data have stimulated an intensive search for pharmacologic agents that could either increase activity of BAT or promote *beiging* of WAT, in hopes that stimulation of systemic energy expenditure could decrease obesity [292].

Mitochondrial Uncoupling and Glucose Metabolism in BAT

The capacity for uncoupling mitochondrial respiration is a defining feature of BAT and beige WAT; in turn, the ability to fine-tune respiration may play an important regulatory role in whole-body glucose and lipid homeostasis. Data suggest that activated BAT may contribute to as much as 75% of glucose clearance in mice; whether BAT similarly contributes to glucose metabolism in humans remains uncertain [293, 294]. Studies of BAT in humans rely on fluorodeoxyglucose uptake during PET scans to assess activity [293]. Indeed, BAT glucose consumption is increased during active thermogenesis, potentially via both insulin-dependent (GLUT4) and insulin-independent (GLUT1) uptake [293, 295], and could contribute to oxidative metabolism of glucose [296, 297] or to glucose-derived lipogenesis [295, 298]. BAT may be also a major site of triglyceride clearance from the circulation [299].

Conversely, loss of uncoupling blunts the increase in glucose uptake and oxidation [300]. Whether this contributes to diabetes risk remains uncertain, but it is interesting that human studies have associated polymorphisms at the UCP1 *locus* with T2D [301, 302]. UCP1 deficiency in mice also increases FGF21 expression and release in this tissue [303]. Interestingly, FGF21 is released from BAT during active thermogenesis [304], promotes BAT uncoupling and glucose oxidation [305], and reduces glucose in rodent models of T2D [306]. Thus, FGF21 may play a key role as an effector maintaining activity of BAT.

Central Nervous System Regulation

Central nervous system regulation is an essential element of control of appetite and energy expenditure, and thus whole-body energy homeostasis. Substantial progress has been made in the last decade in identifying the neural pathways controlling food intake. The dominant circuits regulating metabolism are located in the hypothalamus [307], with the arcuate nucleus (in the mediobasal hypothalamus) having a particularly important role in energy intake and feeding [308]. Neuronal populations in this region express orexigenic (agouti-related protein, AGRP; and neuropeptide Y, NPY) or anorexigenic neuropeptides (pro-opiomelanocortin, POMC; and cocaine and amphetamine-related transcript, CART) [309].

Insulin action in the hypothalamus suppresses appetite in rodents [310–312] via modulation of neuropeptides [313]. Indeed, insulin increases POMC and reduces NPY and AGRP expression, leading to activation of second-order neurons ultimately reducing food intake and increasing energy expenditure. Central action of insulin also regulates systemic glucose metabolism, including inhibition of HGP [314, 315] and glycogen synthesis in skeletal muscle [316], and stimulates lipogenesis in adipose tissue [317]. Leptin can also regulate central insulin signaling and action [317] through a PI3K-dependent mechanism [318]. Thus, both insulin and leptin signaling [319, 320] are essential regulators of hypothalamic control of whole-body energy homeostasis.

Hypothalamic regulation is perturbed in obesity and insulin resistance. For example, obese, insulin resistant individuals have reduced appetite inhibition upon glucose intake [3], despite higher insulin levels, indicating hypothalamic insulin resistance. Similarly, studies of obese rodents indicate central insulin resistance, contributing to increased HGP and reduced muscle glucose uptake [47, 48]. Lipotoxicity and increased ceramides [321–323], inflammation [319, 324] and endoplasmatic reticulum (ER) stress [325] have all been implicated as contributors to pathogenesis of central insulin and leptin resistance [312, 326, 327].

Mitochondrial Function in the Hypothalamus

Analysis of mitochondrial function in the hypothalamus is challenging, given the presence of both orexigenic and anorexigenic neuronal populations. In permeabilized hypothalamus from obese Zucker rats, no differences in respiratory control ratio were observed in comparison to lean rats [328]; by contrast, increased mitochondrial respiratory capacity has been observed in isolated mitochondria from obese rats [329]. Moreover, mitochondrial function in the hypothalamus may be compromised by altered mitochondrial dynamics. For example, ablation of mitofusin2 (Mfn2) in POMC neurons leads to disrupted POMC processing, leptin resistance, reduced energy expenditure and obesity [309], with similar impact in AGRP neurons [309, 330].

Oxidative Stress in the Hypothalamus

ROS have been shown to modulate hypothalamic neuronal function, thus potentially linking perturbed mitochondrial function and central regulation. ROS can acutely activate POMC firing; conversely, suppression of ROS inhibits POMC neuronal activity [331]. Effects of ROS are more modest in NPY/AGRP neuronal populations [332], in which ROS production is decreased with activation of UCP2 [333]. In T2D, increased ROS due to hyperglycemia can inactivate nicotinic acetylcholine receptors on autonomic neurons, suggesting oxidative stress may act as a second messenger to disrupt synaptic transmission [334].

Stem Cell Populations

Stem cells are undifferentiated cells that can both divide into other stem cells or differentiate into specialized cells required for normal development, differentiation, function and repair of tissues. Hence, alterations in stem cell function may impact developmental trajectories and even tissue function, thus potentially contributing to risk of metabolic disease emerging during adult life (e.g., insulin resistance and T2D) [335].

Mitochondrial Function in Insulin Resistant Stem Cells

Analysis of metabolic function in stem cells has been used to decipher relationship between insulin resistance and mitochondrial function, in the absence of complicating differentiation-dependent phenotypes. For example, stem cells obtained from patients with monogenic insulin resistance (i.e., mutations in the insulin receptor) [336, 337] have robust metabolic phenotypes, with decreased spare respiratory capacity, decreased endogenous CS activity, and a 34% increase in ADP/ATP ratio, suggesting reduced energy availability [337]. Moreover, insulin resistant cells have a 22% increase in mitochondrial number and a 16% decrease in mitochondrial area, along with increased expression of the mitochondrial fission genes MFF and INF2 [337]. Given that mitochondrial mass, mtDNA and CS protein remained unchanged, these data collectively support that insulin resistance may promote a switch to mitochondrial fission and energetic stress. Although these studies indicate that insulin resistance in stem cell populations can promote mitochondrial alterations, it remains unclear whether more common forms of insulin resistance or T2D, which typically arise in a polygenic background, can also be a primary driver of mitochondrial dysfunction.

Alterations in mitochondrial metabolism can also affect the function of germ cells, including both oocytes and sperm. This may instigate a vicious cycle whereby alterations in parental metabolism can promote metabolic disease in offspring (reviewed in [338, 339].

Mitochondrion as a Therapeutic Target

Collectively, the data presented above underscore the strong relationships between mitochondrial dysfunction and metabolic diseases such as insulin resistance and T2D. Whether these relationships are essential for the pathogenesis of disease, or are secondary phenomena, remains uncertain. Thus, testing whether improvement in mitochondrial function can improve metabolism is a key element. In this section, we will review the impact of therapeutic strategies to modify, regulate and control mitochondrial bioenergetics, biogenesis and overall functional capacity (Fig. 25.4).

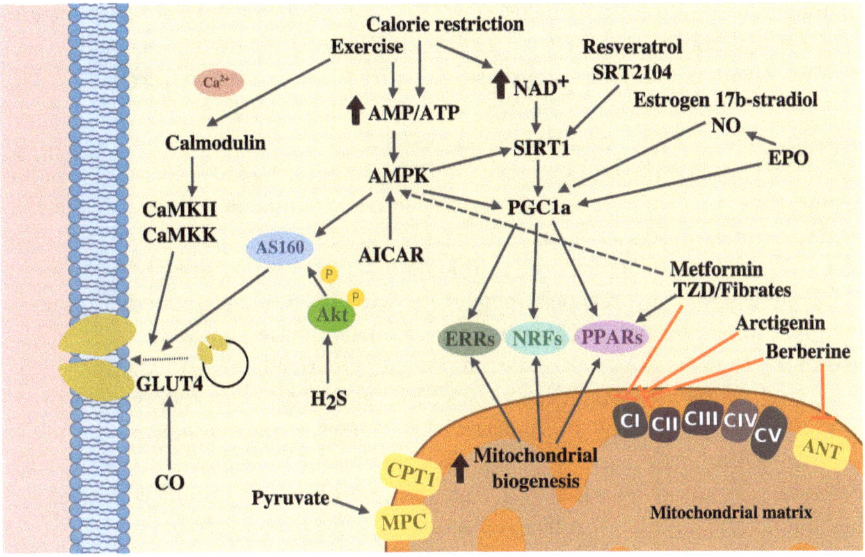

Fig. 25.4 Cellular mechanisms by which drugs, compounds or other interventions may modify, regulate and control mitochondrial health

Non-pharmacological Interventions: Exercise and Weight Loss

Lifestyle interventions such as physical activity and calorie restriction improve insulin sensitivity and glucose tolerance in patients with insulin resistance and T2D, and in parallel, improve mitochondrial health. Multiple studies have reported the benefits of exercise and muscle contraction on insulin sensitivity and glucose homeostasis [340–343], largely via insulin-independent AMPK-dependent pathways (lund 1995 [344, 345] [342]. Exercise also promotes anti-inflammatory responses [346], with release of IL-6 and other myokines [347, 348] as well as cytokines (i.e., IL10, sTNFR) that inhibit the production of pro-inflammatory cytokines (i.e., IL1β, TNFα) [349, 350].

Exercise training can robustly increase OxPhos capacity in rat skeletal muscle [351], in parallel with an increase in mitochondrial number and widening of *cristae*. Endurance training is particularly potent to increase muscle mitochondrial content [352], CS activity [207, 353], cardiolipin [216], and protein content of OxPhos complexes [195]. Muscle mitochondrial biogenesis (i.e., PPARγ, NRF1 and TFAM genes) is also stimulated after endurance exercise [354]. Exercise training also induces neovascularization through pro-angiogenic factors (i.e., VEGF) in an ERRα- and PGC1α-dependent manner [355], thus coordinately regulating oxygen delivery and oxygen consumption. Mitochondrial ROS production is also decreased after physical activity, possibly due to an increase in CI and CIII capacity, the sites of ROS production, or in muscle UCP3 [207] that may limit excessive mitochondrial ROS [356]. However, some but not all studies indicate that responses to exercise training are reduced in individuals with T2D, with no increase in either mitochondrial-related proteins (i.e., PDK4, COX1 and COX4) or mitochondrial content [195, 207] even with improved insulin sensitivity after strength training [357].

Exercise interventions in humans can also impact the liver, yielding reduced hepatic lipids [358]; in rodents, exercise results in increased liver insulin sensitivity and AKT phosphorylation [359], even during continued HFD feeding [360]. In this setting, endurance training reverses the decrease in maximal FCCP-mediated uncoupled respiration observed in sedentary, HFD-fed rats [361]. Interestingly, maximal effects were observed in response to voluntary wheel exercise, irrespective of diet.

Adipose tissue also responds potently to exercise training. Individuals with or without a family history of T2D (first-degree relative) have improvements in VO_{2max} and broad changes in gene expression, including enrichment of the OxPhos pathway [362]. Moreover, exercise decreases expression of inflammatory markers [363–366]. Similar effects have been observed in rodent adipose tissue in response to exercise. Mitochondrial activity, including cytochrome C oxidase and malate dehydrogenase activity, increases in VAT after swimming [367], in parallel with increased mitochondrial biogenesis, mtDNA content and glucose uptake in SAT [368]. In rodent models of insulin resistance and T2D, physical activity normalized VAT mitochondrial content and improved glucose homeostasis [263]. The effects of exercise are often associated with *browning* in SAT, with increased basal oxygen consumption reflecting increased mitochondrial biogenesis and upregulation of PGC1α, PRDM16, CIDEA, and UCP1 [369–375], and increased blood vessel number potentially linked to upregulation of the pro-angiogenic gene VEGFA [374, 375]. NO synthesis may also be required for exercise-dependent effects [368]. Interestingly, the effect of exercise on adipose tissue persists after transplantation of adipose from an exercised animal to a sedentary animal, suggesting that secreted factors may play a role in this response [374].

Calorie restriction sufficient to induce weight loss also improves insulin sensitivity, and in parallel, normalizes respiratory exchange ratio (RER) in both the basal [194] and insulin-stimulated states [376]. Moreover, this dietary intervention in humans (i.e., calorie restriction with weight loss) improves insulin resistance and fasting hyperglycemia, and decreases hepatocellular lipid accumulation [377]. The combination of both exercise and diet can normalize many phenotypes (i.e., body weight, glucose and insulin tolerance, fasting glucose, plasma insulin, TG, cholesterol, NEFA, insulin action) alongside with normalized expression of genes regulating mitochondrial function and biogenesis, decreased ROS, and improved ATP synthesis [378]. By contrast, many studies show that weight loss alone does not increase mtDNA content or maximal aerobic capacity [379], with either no increase [102, 190] or even a decrease in mitochondrial function [194]. Furthermore, weight loss without increased physical activity improved insulin resistance without increased muscle mitochondrial function [102], thus indicating that restoration of insulin sensitivity in muscle is independent of mitochondrial function, at least in some paradigms.

More directed dietary interventions in humans, such as omega-3 FA feeding, can increase PPARα-dependent FAO [380, 381] and reduce steatosis [381]. Omega-3

FA feeding also increases plasma adiponectin [380] and promote a shift to mitochondrial fusion, potentially contributing to reduced oxidative stress [382].

Thus, both exercise and calorie restriction improve aspects of mitochondrial function, largely in parallel with improved insulin sensitivity; both of these interventions often converge on key regulatory points of energy metabolism, such as AMPK or PGC1α [383, 384].

Pharmacological Interventions

Beyond lifestyle interventions, targeting mitochondrial dysfunction has been considered a potential approach to reduce insulin resistance. The goal of such pharmacological interventions could include (a) uncoupling mitochondrial respiration to increase energy expenditure and FAO [385, 386], (b) increasing mitochondrial biogenesis, and (c) inhibiting or decreasing OxPhos-generated oxidative stress.

Inducing Mitochondrial Uncoupling

Uncoupling has been considered as one strategy to increase energy expenditure and promote weight loss. Late in the 1930s the uncoupler 2,4-dinitrophenol (DNP) was used for treatment of obesity [387], but its narrow therapeutic window and serious side effects ultimately resulted in its withdrawal from the market in 1938. Additional uncouplers with reduced activity and improved safety profiles (e.g., Rhodamine 19) have been proposed over the ensuing years [386, 388].

Inducing Mitochondrial Biogenesis and Activity

Different metabolic challenges affect cellular energetics (i.e., exercise, calorie restriction), increasing AMP and NAD^+ and activating the metabolic sensors AMPK and SIRT1, respectively. In turn, AMPK and SIRT1 can stimulate PGC1α phosphorylation and deacetylation, ultimately increasing its activity [389].

Natural antioxidants that are able to scavenge carbon and lipid radicals and ROS have been used in attempt to increase mitochondrial biogenesis and function. The best studied compound is the polyphenol resveratrol (3,5,4′-trihydroxystillbene), a compound found in red wine that mimics the effects of calorie restriction through activation of SIRT1 and AMPK [390, 391], and increases mitochondrial biogenesis via upregulation of PGC1α, NRF1 and TFAM [392]. Of note, resveratrol has been also proposed to exert antiaging effects and extend survival [390, 393, 394]. In rodents, resveratrol protects against diet-induced obesity, activates PGC1α and increases aerobic capacity [395]. Interestingly, resveratrol may also act via SIRT1-independent mechanisms [396], potentially via increased mitochondrial content [390, 395, 397]. However, mitochondrial content increases are not observed in all studies [398, 399]. These relatively modest and inconsistent effects may also explain in part the lack of impact on body weight in humans [400–402].

Other small molecules and vitamins which increase NAD^+, activate SIRT1, or induce antioxidant responses have also been studied [403–405], some specifically targeting mitochondria [406]. The SIRT1-activating compounds, SRT1720 and SRT2104 improve lipid profiles and induce weight loss in obese individuals [407]. Furthermore, SRT2104 increases oxidative capacity in elderly lean individuals, without affecting body weight [408]. Moreover, treatment with NAD^+ or the NAD^+ precursor nicotinamide riboside could also stimulate SIRT1 activity. Indeed, NAD^+ supplementation in HFD-fed mice protects against weight gain, potentially via increased energy expenditure [409] and increased BAT mitochondrial biogenesis and thermogenesis.

Multiple other pharmacologic agents have been evaluated for their impact on mitochondrial biogenesis. **Nitric oxide** (NO) and its derivates can activate mitochondrial biogenesis at the transcriptional level [410] through activation of cGMP and PGC1α [411, 412]. **Hydrogen sulfide** (H_2S) preserves mitochondrial function in injured heart muscle [413, 414] via increased AKT phosphorylation, increased nuclear localization of NRF1/2, and increased mitochondrial biogenesis [414]. Inhalation of low doses of **carbon monoxide** (CO) improves muscle mitochondrial density, capillary density, and translocation of GLUT4 [415]. The estrogen 17beta-estradiol can also activate NRF1 and 2, PGC1α and TFAM [416], while **erythropoietin** activates mitochondrial biogenesis via NRF1, PGC1α and NO [417]. **Adiponectin** increases mitochondrial content via AMPK and SIRT1 [418, 419], and can also activate PGC1α and p38 MAPK via MAPK phosphatase 1 suppression [420]. **Fibrates** activate the nuclear receptor PPARα and have been used to treat dyslipidemias [421], but may also have an effect on mitochondrial function via PGC1α action [422]. **Thiazolidinediones** (TZD) can improve mitochondrial function in human SAT [265] in parallel with stimulation of FA uptake and oxidation [276, 277, 423]. Natural compounds such as the alkaloid **berberine** increase energy expenditure [424] potentially due to increased BAT mass and *browning* in WAT, via AMPK- and PGC1α-dependent mechanisms [425]. Furthermore, animal studies indicate that this compound delays obesity onset [424–428].

AMPK activators have also been studied as potential inducers of mitochondrial biogenesis. 5-Aminoimizadole-4-Carboxamide Ribonucleotide (AICAR) administration to sedentary mice increases endurace performance [429], potentially through PGC1α-induced mitochondrial biogenesis [430] and/or direct modulation of gluconeogenesis and glycogenolysis [431, 432]. Pharmacological manipulation of AMPK [433] may also enhance glucose and lipid oxidation [434, 435].

Ameliorating Insulin Resistance by Inhibiting Mitochondrial Respiration

While mitochondrial dysfunction has been associated with insulin resistance and T2D in many studies, inhibition of specific enzymes within mitochondrial OxPhos can ameliorate insulin resistance and improve overall metabolism. For example, inhibition of certain OxPhos complexes (i.e., CI) could alter mitochondrial substrate load and potentially reduce deleterious levels of ROS production. This concept has also been supported by animal studies demonstrating that disruption of OxPhos in multiple tissues is associated with improved insulin sensitivity [97].

Metformin and TZD are best recognized for their efficacy to improve insulin resistance and T2D; and their effects may be mediated by their impact to reduce mitochondrial function. **Metformin** decreases plasma glucose by decreasing HGP, increasing insulin-stimulated glucose uptake, and in parallel, inhibiting CI of OxPhos in cells, isolated mitochondria and skeletal muscle [436], similar to effects of other biguanides [437, 438]. Metformin inhibition of CI may reduce ATP production and content, thus activating AMPK and stimulating mitochondrial biogenesis and lipid oxidation via PGC1-dependent mechanisms [430, 439] and reducing ROS production [438, 440–442]. A novel anti-diabetic drug that also inhibits CI, activates AMPK, and increases muscle glucose uptake has been shown to be even more potent than metformin [443], with fewer off-target effects [444, 445]. Whether this drug will be useful clinically remains unknown.

Like metformin, **TZD** also inhibit CI. TZD are best known as activating ligands for the PPAR family of nuclear receptors (i.e., PPARγ in adipose, PPARα and PPARδ in muscle), increasing expression of genes related to differentiation, and yielding adipocytes with increased lipogenic capacity, mitochondrial number, and insulin action [257, 446]. Increased capacity for adipose lipid storage ultimately decreases ectopic lipid accumulation in muscle and liver, reduces circulating lipids [447, 448], and thus improves whole-body insulin sensitivity. Whether enhanced mitochondrial biogenesis in this context is simply a feature of increased adipose differentiation, or also contributes independently to metabolic effects in either adipose [265, 449] or muscle [450–452] remains uncertain. TZD can also directly inhibit CI [453], promoting shifts in cellular energetics toward anaerobic fuel utilization [454].

Inhibition of CI may be a common mechanism by which many compounds improve insulin resistance. **Fibrates**, the family of PPARα ligands [455], reduce plasma lipids by stimulating PPARα-dependent FAO in the liver [456]. Interestingly, fibrates also inhibit CI activity [453] and stimulate CS and CIV activity [457]. Other compounds which inhibit CI also improve insulin resistance and reduce plasma glucose levels; these include **berberine** [458] and **arctigenin** [459]. Berberine also inhibits adenine nucleotide translocate (ANT) [460], while its metabolite demethyleneberberine decreases ROS levels [461].

Inhibition of the mitochondrial pyruvate carrier (MPC) may also be a therapeutic approach. While first described for TZD [462] even specific inhibitors of the MPC can increase glucose uptake in both human and rodent myotubes [462]. Similarly, inhibition of carnitine-palmitoyl-transferase 1 (CPT1), another mitochondrial transporter, improves insulin resistance in experimental models and humans [463].

Collectively, these data suggest that limiting nutrient oxidation, ATP production, and/or ROS generation in the setting of chronic nutrient excess may have beneficial metabolic effects in obesity and diabetes. Reducing ATP and NADH levels may stimulate activation of nutrient sensing effector pathways such as AMPK/SIRT1/

PGC1, increasing glucose uptake and mitochondrial biogenesis in parallel. Since there are likely to be important tissue-specific effects of each of these strategies in vivo, additional studies will be required to fully validate these approaches.

Conclusion

Mitochondrial dysfunction is a consistent phenotype associated with insulin resistance in multiple tissues. Deciphering whether this is a critical causal link in the pathogenesis of human insulin resistance and type 2 diabetes remains a challenging goal for several reasons. Firstly, mitochondrial function cannot be defined by a single phenotype, as mitochondria have such a broad impact on multiple aspects of cellular function. Unfortunately, few studies comprehensively and simultaneously analyze multiple phenotypes. Secondly, it is likely that oxidative dysfunction is not simply an "on-off" phenotype but rather a continuous spectrum which progresses over time during the life course of disease risk. By the time clinical disease is recognized, the phenotype represents a mixture of primary effects as well as chronic secondary effects of perturbed metabolism, altered ROS, obesity, and hyperglycemia on regulation of transcriptional and epigenetic mechanisms. Even insulin resistant but non-diabetic individuals at risk for disease have likely had subclinical perturbations for years, as demonstrated by the studies of Tabak and colleagues [16]. Thus, dissecting the primary effects from secondary effects is challenging, especially in human disease.

Recent studies in undifferentiated human stem cells indicate that primary insulin resistance due to monogenic mutations in the insulin receptor, even in the absence of differentiation, induces changes in mitochondrial size, number, and functional capacity. Thus, chronic "primary" insulin resistance can yield secondary changes in mitochondrial metabolism. Whether these results are generalizable to more common forms of diabetes risk resulting from the combined impact of polygenic genetic, epigenetic, and environmental influences, remains an unanswered question currently under study. It remains possible that tissue-specific primary defects which alter mitochondrial metabolism could underlie additional diabetes-related risk patterns, influencing regulation of insulin secretion, exercise capacity, appetite, and sensitivity to calorie excess and inactivity.

Regardless of whether defects in mitochondrial metabolism are primary or secondary culprits in disease pathogenesis, it is likely that these defects can initiate a vicious cycle of reducing energetic capacity, increasing oxidative stress, and further reducing mitochondrial biogenesis and dynamics. Reduced mitochondrial capacity in the setting of chronic nutrient excess and inactivity may further impair exercise capacity, promote weight gain in an obesogenic environment, and contribute to metabolic instability. Thus, both lifestyle and pharmacologic strategies focused on interrupting this vicious cycle to improve mitochondrial metabolism may be an effective approach for disease prevention and treatment.

References

1. Mathers CD, Loncar D. Projections of global mortality and burden of disease from 2002 to 2030. PLoS Med. 2006;3:e442.
2. Organization WH. 2012 [updated 2012]; Available from: http://www.who.int/diabetes/fact/en.
3. DeFronzo RA. Banting lecture. From the triumvirate to the ominous octet: a new paradigm for the treatment of type 2 diabetes mellitus. Diabetes. 2009;58:773–95.
4. Eriksson J, Franssila-Kallunki A, Ekstrand A, Saloranta C, Widen E, Schalin C, Groop L. Early metabolic defects in persons at increased risk for non-insulin-dependent diabetes mellitus. N Engl J Med. 1989;321:337–43.
5. DeFronzo RA. Lilly lecture 1987. The triumvirate: β-cell, muscle, liver: a collusion responsible for NIDDM. Diabetes. 1988;37:667–87.
6. Cersosimo E, Triplitt C, Mandarino LJ, DeFronzo RA. Pathogenesis of type 2 diabetes mellitus. In: De Groot LJ, Beck-Peccoz P, Chrousos G, Dungan K, Grossman A, Hershman JM, et al., editors. Endotext. South Dartmouth: MDText.com; 2000.
7. Bergman RN, Finegood DT, Kahn SE. The evolution of beta-cell dysfunction and insulin resistance in type 2 diabetes. Eur J Clin Investig. 2002;32(Suppl 3):35–45.
8. Bergman RN. Lilly lecture 1989. Toward physiological understanding of glucose tolerance. Minimal model approach. Diabetes. 1989;38:1512–27.
9. Gastaldelli A, Ferrannini E, Miyazaki Y, Matsuda M, DeFronzo RA. Beta-cell dysfunction and glucose intolerance: results from the San Antonio metabolism (SAM) study. Diabetologia. 2004;47:31–9.
10. Ferrannini E, Gastaldelli A, Miyazaki Y, Matsuda M, Mari A, DeFronzo RA. beta-cell function in subjects spanning the range from normal glucose tolerance to overt diabetes: a new analysis. J Clin Endocrinol Metab. 2005;90:493–500.
11. Abdul-Ghani MA, Jenkinson CP, Richardson DK, Tripathy D, DeFronzo RA. Insulin secretion and action in subjects with impaired fasting glucose and impaired glucose tolerance: results from the veterans administration genetic epidemiology study. Diabetes. 2006;55:1430–5.
12. Abdul-Ghani MA, Tripathy D, DeFronzo RA. Contributions of beta-cell dysfunction and insulin resistance to the pathogenesis of impaired glucose tolerance and impaired fasting glucose. Diabetes Care. 2006;29:1130–9.
13. Yoon KH, Ko SH, Cho JH, Lee JM, Ahn YB, Song KH, Yoo SJ, Kang MI, Cha BY, Lee KW, Son HY, Kang SK, Kim HS, Lee IK, Bonner-Weir S. Selective beta-cell loss and alpha-cell expansion in patients with type 2 diabetes mellitus in Korea. J Clin Endocrinol Metab. 2003;88:2300–8.
14. Butler AE, Janson J, Bonner-Weir S, Ritzel R, Rizza RA, Butler PC. Beta-cell deficit and increased beta-cell apoptosis in humans with type 2 diabetes. Diabetes. 2003;52:102–10.
15. Martin BC, Warram JH, Krolewski AS, Bergman RN, Soeldner JS, Kahn CR. Role of glucose and insulin resistance in development of type II diabetes mellitus: results of a 25-year follow-up study. Lancet. 1992;340:925–9.
16. Tabak AG, Jokela M, Akbaraly TN, Brunner EJ, Kivimaki M, Witte DR. Trajectories of glycaemia, insulin sensitivity, and insulin secretion before diagnosis of type 2 diabetes: an analysis from the Whitehall II study. Lancet. 2009;373:2215–21.
17. James WP. The fundamental drivers of the obesity epidemic. Obes Rev Off J Int Assoc Study Obes. 2008;9(Suppl 1):6–13.
18. DeFronzo RA. Pathogenesis of type 2 diabetes: metabolic and molecular implications for identifying diabetes genes. Diabetes Rev. 1997;5:177–269.
19. Groop L, Lyssenko V. Genes and type 2 diabetes mellitus. Curr Diabetes Rep. 2008;8:192–7.
20. Zeggini E, Scott LJ, Saxena R, Voight BF, Marchini JL, Hu T, de Bakker PI, Abecasis GR, Almgren P, Andersen G, Ardlie K, Bostrom KB, Bergman RN, Bonnycastle LL, Borch-Johnsen

K, Burtt NP, Chen H, Chines PS, Daly MJ, Deodhar P, Ding CJ, Doney AS, Duren WL, Elliott KS, Erdos MR, Frayling TM, Freathy RM, Gianniny L, Grallert H, Grarup N, Groves CJ, Guiducci C, Hansen T, Herder C, Hitman GA, Hughes TE, Isomaa B, Jackson AU, Jorgensen T, Kong A, Kubalanza K, Kuruvilla FG, Kuusisto J, Langenberg C, Lango H, Lauritzen T, Li Y, Lindgren CM, Lyssenko V, Marvelle AF, Meisinger C, Midthjell K, Mohlke KL, Morken MA, Morris AD, Narisu N, Nilsson P, Owen KR, Palmer CN, Payne F, Perry JR, Pettersen E, Platou C, Prokopenko I, Qi L, Qin L, Rayner NW, Rees M, Roix JJ, Sandbaek A, Shields B, Sjogren M, Steinthorsdottir V, Stringham HM, Swift AJ, Thorleifsson G, Thorsteinsdottir U, Timpson NJ, Tuomi T, Tuomilehto J, Walker M, Watanabe RM, Weedon MN, Willer CJ, Illig T, Hveem K, Hu FB, Laakso M, Stefansson K, Pedersen O, Wareham NJ, Barroso I, Hattersley AT, Collins FS, Groop L, McCarthy MI, Boehnke M, Altshuler D. Meta-analysis of genome-wide association data and large-scale replication identifies additional susceptibility loci for type 2 diabetes. Nat Genet. 2008;40:638–45.
21. Morris AP, Voight BF, Teslovich TM, Ferreira T, Segre AV, Steinthorsdottir V, Strawbridge RJ, Khan H, Grallert H, Mahajan A, Prokopenko I, Kang HM, Dina C, Esko T, Fraser RM, Kanoni S, Kumar A, Lagou V, Langenberg C, Luan J, Lindgren CM, Muller-Nurasyid M, Pechlivanis S, Rayner NW, Scott LJ, Wiltshire S, Yengo L, Kinnunen L, Rossin EJ, Raychaudhuri S, Johnson AD, Dimas AS, Loos RJ, Vedantam S, Chen H, Florez JC, Fox C, Liu CT, Rybin D, Couper DJ, Kao WH, Li M, Cornelis MC, Kraft P, Sun Q, van Dam RM, Stringham HM, Chines PS, Fischer K, Fontanillas P, Holmen OL, Hunt SE, Jackson AU, Kong A, Lawrence R, Meyer J, Perry JR, Platou CG, Potter S, Rehnberg E, Robertson N, Sivapalaratnam S, Stancakova A, Stirrups K, Thorleifsson G, Tikkanen E, Wood AR, Almgren P, Atalay M, Benediktsson R, Bonnycastle LL, Burtt N, Carey J, Charpentier G, Crenshaw AT, Doney AS, Dorkhan M, Edkins S, Emilsson V, Eury E, Forsen T, Gertow K, Gigante B, Grant GB, Groves CJ, Guiducci C, Herder C, Hreidarsson AB, Hui J, James A, Jonsson A, Rathmann W, Klopp N, Kravic J, Krjutskov K, Langford C, Leander K, Lindholm E, Lobbens S, Mannisto S, Mirza G, Muhleisen TW, Musk B, Parkin M, Rallidis L, Saramies J, Sennblad B, Shah S, Sigurethsson G, Silveira A, Steinbach G, Thorand B, Trakalo J, Veglia F, Wennauer R, Winckler W, Zabaneh D, Campbell H, van Duijn C, Uitterlinden AG, Hofman A, Sijbrands E, Abecasis GR, Owen KR, Zeggini E, Trip MD, Forouhi NG, Syvanen AC, Eriksson JG, Peltonen L, Nothen MM, Balkau B, Palmer CN, Lyssenko V, Tuomi T, Isomaa B, Hunter DJ, Qi L, Shuldiner AR, Roden M, Barroso I, Wilsgaard T, Beilby J, Hovingh K, Price JF, Wilson JF, Rauramaa R, Lakka TA, Lind L, Dedoussis G, Njolstad I, Pedersen NL, Khaw KT, Wareham NJ, Keinanen-Kiukaanniemi SM, Saaristo TE, Korpi-Hyovalti E, Saltevo J, Laakso M, Kuusisto J, Metspalu A, Collins FS, Mohlke KL, Bergman RN, Tuomilehto J, Boehm BO, Gieger C, Hveem K, Cauchi S, Froguel P, Baldassarre D, Tremoli E, Humphries SE, Saleheen D, Danesh J, Ingelsson E, Ripatti S, Salomaa V, Erbel R, Jockel KH, Moebus S, Peters A, Illig T, De Fair U, Hamsten A, Morris AD, Donnelly PJ, Frayling TM, Hattersley AT, Boerwinkle E, Melander O, Kathiresan S, Nilsson PM, Deloukas P, Thorsteinsdottir U, Groop LC, Stefansson K, Hu F, Pankow JS, Dupuis J, Meigs JB, Altshuler D, Boehnke M, McCarthy MI. Large-scale association analysis provides insights into the genetic architecture and pathophysiology of type 2 diabetes. Nat Genet. 2012;44:981–90.
22. Mahajan A, Go MJ, Zhang W, Below JE, Gaulton KJ, Ferreira T, Horikoshi M, Johnson AD, Ng MC, Prokopenko I, Saleheen D, Wang X, Zeggini E, Abecasis GR, Adair LS, Almgren P, Atalay M, Aung T, Baldassarre D, Balkau B, Bao Y, Barnett AH, Barroso I, Basit A, Been LF, Beilby J, Bell GI, Benediktsson R, Bergman RN, Boehm BO, Boerwinkle E, Bonnycastle LL, Burtt N, Cai Q, Campbell H, Carey J, Cauchi S, Caulfield M, Chan JC, Chang LC, Chang TJ, Chang YC, Charpentier G, Chen CH, Chen H, Chen YT, Chia KS, Chidambaram M, Chines PS, Cho NH, Cho YM, Chuang LM, Collins FS, Cornelis MC, Couper DJ, Crenshaw AT, van Dam RM, Danesh J, Das D, De Faire U, Dedoussis G, Deloukas P, Dimas AS, Dina C, Doney AS, Donnelly PJ, Dorkhan M, van Duijn C, Dupuis J, Edkins S, Elliott P, Emilsson V, Erbel R, Eriksson JG, Escobedo J, Esko T, Eury E, Florez JC, Fontanillas P, Forouhi NG, Forsen T, Fox C, Fraser RM, Frayling TM, Froguel P, Frossard P, Gao Y, Gertow K, Gieger C, Gigante B, Grallert H, Grant GB, Grrop LC, Groves CJ, Grundberg E, Guiducci C, Hamsten A, Han BG, Hara K, Hassanali N, Hattersley AT, Hayward C, Hedman AK, Herder

C, Hofman A, Holmen OL, Hovingh K, Hreidarsson AB, Hu C, Hu FB, Hui J, Humphries SE, Hunt SE, Hunter DJ, Hveem K, Hydrie ZI, Ikegami H, Illig T, Ingelsson E, Islam M, Isomaa B, Jackson AU, Jafar T, James A, Jia W, Jockel KH, Jonsson A, Jowett JB, Kadowaki T, Kang HM, Kanoni S, Kao WH, Kathiresan S, Kato N, Katulanda P, Keinanen-Kiukaanniemi KM, Kelly AM, Khan H, Khaw KT, Khor CC, Kim HL, Kim S, Kim YJ, Kinnunen L, Klopp N, Kong A, Korpi-Hyovalti E, Kowlessur S, Kraft P, Kravic J, Kristensen MM, Krithika S, Kumar A, Kumate J, Kuusisto J, Kwak SH, Laakso M, Lagou V, Lakka TA, Langenberg C, Langford C, Lawrence R, Leander K, Lee JM, Lee NR, Li M, Li X, Li Y, Liang J, Liju S, Lim WY, Lind L, Lindgren CM, Lindholm E, Liu CT, Liu JJ, Lobbens S, Long J, Loos RJ, Lu W, Luan J, Lyssenko V, Ma RC, Maeda S, Magi R, Mannisto S, Matthews DR, Meigs JB, Melander O, Metspalu A, Meyer J, Mirza G, Mihailov E, Moebus S, Mohan V, Mohlke KL, Morris AD, Muhleisen TW, Muller-Nurasyid M, Musk B, Nakamura J, Nakashima E, Navarro P, Ng PK, Nica AC, Nilsson PM, Njolstad I, Nothen MM, Ohnaka K, Ong TH, Owen KR, Palmer CN, Pankow JS, Park KS, Parkin M, Pechlivanis S, Pedersen NL, Peltonen L, Perry JR, Peters A, Pinidiyapathirage JM, Platou CG, Potter S, Price JF, Qi L, Radha V, Rallidis L, Rasheed A, Rathman W, Rauramaa R, Raychaudhuri S, Rayner NW, Rees SD, Rehnberg E, Ripatti S, Robertson N, Roden M, Rossin EJ, Rudan I, Rybin D, Saaristo TE, Salomaa V, Saltevo J, Samuel M, Sanghera DK, Saramies J, Scott J, Scott LJ, Scott RA, Segre AV, Sehmi J, Sennblad B, Shah N, Shah S, Shera AS. Genome-wide trans-ancestry meta-analysis provides insight into the genetic architecture of type 2 diabetes susceptibility. Nat Genet. 2014;46:234–44.

23. Groves CJ, Zeggini E, Minton J, Frayling TM, Weedon MN, Rayner NW, Hitman GA, Walker M, Wiltshire S, Hattersley AT, McCarthy MI. Association analysis of 6,736 U.K. subjects provides replication and confirms TCF7L2 as a type 2 diabetes susceptibility gene with a substantial effect on individual risk. Diabetes. 2006;55:2640–4.
24. Sladek R, Rocheleau G, Rung J, Dina C, Shen L, Serre D, Boutin P, Vincent D, Belisle A, Hadjadj S, Balkau B, Heude B, Charpentier G, Hudson TJ, Montpetit A, Pshezhetsky AV, Prentki M, Posner BI, Balding DJ, Meyre D, Polychronakos C, Froguel P. A genome-wide association study identifies novel risk loci for type 2 diabetes. Nature. 2007;445:881–5.
25. Mohlke KL, Boehnke M. Recent advances in understanding the genetic architecture of type 2 diabetes. Hum Mol Genet. 2015;24:R85–92.
26. Billings LK, Florez JC. The genetics of type 2 diabetes: what have we learned from GWAS? Ann N Y Acad Sci. 2010;1212:59–77.
27. Hong KW, Chung M, Cho SB. Meta-analysis of genome-wide association study of homeostasis model assessment beta cell function and insulin resistance in an East Asian population and the European results. Mol Gen Genet MGG. 2014;289:1247–55.
28. Rung J, Cauchi S, Albrechtsen A, Shen L, Rocheleau G, Cavalcanti-Proenca C, Bacot F, Balkau B, Belisle A, Borch-Johnsen K, Charpentier G, Dina C, Durand E, Elliott P, Hadjadj S, Jarvelin MR, Laitinen J, Lauritzen T, Marre M, Mazur A, Meyre D, Montpetit A, Pisinger C, Posner B, Poulsen P, Pouta A, Prentki M, Ribel-Madsen R, Ruokonen A, Sandbaek A, Serre D, Tichet J, Vaxillaire M, Wojtaszewski JF, Vaag A, Hansen T, Polychronakos C, Pedersen O, Froguel P, Sladek R. Genetic variant near IRS1 is associated with type 2 diabetes, insulin resistance and hyperinsulinemia. Nat Genet. 2009;41:1110.
29. Prokopenko I, Langenberg C, Florez JC, Saxena R, Soranzo N, Thorleifsson G, Loos RJ, Manning AK, Jackson AU, Aulchenko Y, Potter SC, Erdos MR, Sanna S, Hottenga JJ, Wheeler E, Kaakinen M, Lyssenko V, Chen WM, Ahmadi K, Beckmann JS, Bergman RN, Bochud M, Bonnycastle LL, Buchanan TA, Cao A, Cervino A, Coin L, Collins FS, Crisponi L, de Geus EJ, Dehghan A, Deloukas P, Doney AS, Elliott P, Freimer N, Gateva V, Herder C, Hofman A, Hughes TE, Hunt S, Illig T, Inouye M, Isomaa B, Johnson T, Kong A, Krestyaninova M, Kuusisto J, Laakso M, Lim N, Lindblad U, Lindgren CM, McCann OT, Mohlke KL, Morris AD, Naitza S, Orru M, Palmer CN, Pouta A, Randall J, Rathmann W, Saramies J, Scheet P, Scott LJ, Scuteri A, Sharp S, Sijbrands E, Smit JH, Song K, Steinthorsdottir V, Stringham HM, Tuomi T, Tuomilehto J, Uitterlinden AG, Voight BF,

Waterworth D, Wichmann HE, Willemsen G, Witteman JC, Yuan X, Zhao JH, Zeggini E, Schlessinger D, Sandhu M, Boomsma DI, Uda M, Spector TD, Penninx BW, Altshuler D, Vollenweider P, Jarvelin MR, Lakatta E, Waeber G, Fox CS, Peltonen L, Groop LC, Mooser V, Cupples LA, Thorsteinsdottir U, Boehnke M, Barroso I, Van Duijn C, Dupuis J, Watanabe RM, Stefansson K, McCarthy MI, Wareham NJ, Meigs JB, Abecasis GR. Variants in MTNR1B influence fasting glucose levels. Nat Genet. 2009;41:77–81.

30. Bouatia-Naji N, Bonnefond A, Cavalcanti-Proenca C, Sparso T, Holmkvist J, Marchand M, Delplanque J, Lobbens S, Rocheleau G, Durand E, De Graeve F, Chevre JC, Borch-Johnsen K, Hartikainen AL, Ruokonen A, Tichet J, Marre M, Weill J, Heude B, Tauber M, Lemaire K, Schuit F, Elliott P, Jorgensen T, Charpentier G, Hadjadj S, Cauchi S, Vaxillaire M, Sladek R, Visvikis-Siest S, Balkau B, Levy-Marchal C, Pattou F, Meyre D, Blakemore AI, Jarvelin MR, Walley AJ, Hansen T, Dina C, Pedersen O, Froguel P. A variant near MTNR1B is associated with increased fasting plasma glucose levels and type 2 diabetes risk. Nat Genet. 2009;41:89–94.

31. Lyssenko V, Nagorny CL, Erdos MR, Wierup N, Jonsson A, Spegel P, Bugliani M, Saxena R, Fex M, Pulizzi N, Isomaa B, Tuomi T, Nilsson P, Kuusisto J, Tuomilehto J, Boehnke M, Altshuler D, Sundler F, Eriksson JG, Jackson AU, Laakso M, Marchetti P, Watanabe RM, Mulder H, Groop L. Common variant in MTNR1B associated with increased risk of type 2 diabetes and impaired early insulin secretion. Nat Genet. 2009;41:82–8.

32. Fuchsberger C, Flannick J, Teslovich TM, Mahajan A, Agarwala V, Gaulton KJ, Ma C, Fontanillas P, Moutsianas L, McCarthy DJ, Rivas MA, Perry JR, Sim X, Blackwell TW, Robertson NR, Rayner NW, Cingolani P, Locke AE, Fernandez Tajes J, Highland HM, Dupuis J, Chines PS, Lindgren CM, Hartl C, Jackson AU, Chen H, Huyghe JR, van de Bunt M, Pearson RD, Kumar A, Muller-Nurasyid M, Grarup N, Stringham HM, Gamazon ER, Lee J, Chen Y, Scott RA, Below JE, Chen P, Huang J, Go MJ, Stitzel ML, Pasko D, Parker SC, Varga TV, Green T, Beer NL, Day-Williams AG, Ferreira T, Fingerlin T, Horikoshi M, Hu C, Huh I, Ikram MK, Kim BJ, Kim Y, Kim YJ, Kwon MS, Lee J, Lee S, Lin KH, Maxwell TJ, Nagai Y, Wang X, Welch RP, Yoon J, Zhang W, Barzilai N, Voight BF, Han BG, Jenkinson CP, Kuulasmaa T, Kuusisto J, Manning A, Ng MC, Palmer ND, Balkau B, Stancakova A, Abboud HE, Boeing H, Giedraitis V, Prabhakaran D, Gottesman O, Scott J, Carey J, Kwan P, Grant G, Smith JD, Neale BM, Purcell S, Butterworth AS, Howson JM, Lee HM, Lu Y, Kwak SH, Zhao W, Danesh J, Lam VK, Park KS, Saleheen D, So WY, Tam CH, Afzal U, Aguilar D, Arya R, Aung T, Chan E, Navarro C, Cheng CY, Palli D, Correa A, Curran JE, Rybin D, Farook VS, Fowler SP, Freedman BI, Griswold M, Hale DE, Hicks PJ, Khor CC, Kumar S, Lehne B, Thuillier D, Lim WY, Liu J, van der Schouw YT, Loh M, Musani SK, Puppala S, Scott WR, Yengo L, Tan ST, Taylor Jr HA, Thameem F, Wilson Sr G, Wong TY, Njolstad PR, Levy JC, Mangino M, Bonnycastle LL, Schwarzmayr T, Fadista J, Surdulescu GL, Herder C, Groves CJ, Wieland T, Bork-Jensen J, Brandslund I, Christensen C, Koistinen HA, Doney AS, Kinnunen L, Esko T, Farmer AJ, Hakaste L, Hodgkiss D, Kravic J, Lyssenko V, Hollensted M, Jorgensen ME, Jorgensen T, Ladenvall C, Justesen JM, Karajamaki A, Kriebel J, Rathmann W, Lannfelt L, Lauritzen T, Narisu N, Linneberg A, Melander O, Milani L, Neville M, Orho-Melander M, Qi L, Qi Q, Roden M, Rolandsson O, Swift A, Rosengren AH, Stirrups K, Wood AR, Mihailov E, Blancher C, Carneiro MO, Maguire J, Poplin R, Shakir K, Fennell T, De Pristo M, Hrabe de Angelis M, Deloukas P, Gjesing AP, Jun G, Nilsson P, Murphy J, Onofrio R, Thorand B, Hansen T, Meisinger C, Hu FB, Isomaa B, Karpe F, Liang L, Peters A, Huth C, O' Rahilly SP, Palmer CN, Pedersen O, Rauramaa R, Tuomilehto J, Salomaa V, Watanabe RM, Syvanen AC, Bergman RN, Bharadwaj D, Bottinger EP, Cho YS, Chandak GR, Chan JC, Chia KS, Daly MJ, Ebrahim SB, Langenberg C, Elliott P, Jablonski KA, Lehman DM, Jia W, Ma RC, Pollin TI, Sandhu M, Tandon N, Froguel P, Barroso I, Teo YY, Zeggini E, Loos RJ, Small KS, Ried JS, DeFronzo RA, Grallert H, Glaser B, Metspalu A, Wareham NJ, Walker M, Banks E, Gieger C, Ingelsson E, Im HK, Illig T, Franks PW, Buck G, Trakalo J, Buck D, Prokopenko I, Magi R, Lind L, Farjoun Y, Owen KR, Gloyn AL, Strauch K, Tuomi T, Kooner JS, Lee JY, Park T, Donnelly P, Morris AD, Hattersley AT, Bowden DW, Collins FS, Atzmon G, Chambers JC, Spector TD, Laakso M, Strom TM, Bell GI, Blangero J,

Duggirala R, Tai ES, McVean G, Hanis CL, Wilson JG, Seielstad M, Frayling TM, Meigs JB, Cox NJ, Sladek R, Lander ES, Gabriel S, Burtt NP, Mohlke KL, Meitinger T, Groop L, Abecasis G, Florez JC, Scott LJ, Morris AP, Kang HM, Boehnke M, Altshuler D, McCarthy MI. The genetic architecture of type 2 diabetes. Nature. 2016;536:41–7.

33. Cusi K, Maezono K, Osman A, Pendergrass M, Patti ME, Mandel G, DeFronzo RA, Kahn CR, Mandarino LJ. Insulin resistance differentially affects the PI 3-kinase-and MAP kinase – mediated signaling in human muscle. J Clin Invest. 2000;105:311–20.
34. Rome S, Clement K, Rabasa-Lhoret R, Loizon E, Poitou C, Barsh GS, Riou JP, Laville M, Vidal H. Microarray profiling of human skeletal muscle reveals that insulin regulates ~ 800 genes during an hyperinsulinemic clamp. J Biol Chem. 2003;278:18063.
35. Zierath JR, He L, Guma A, Odegoard Wahlstrom E, Klip A, Wallberg-Henriksson H. Insulin action on glucose transport and plasma membrane GLUT4 content in skeletal muscle from patients with NIDDM. Diabetologia. 1996;39:1180–9.
36. Liao J, Barthel A, Nakatani K, Roth RA. Activation of protein kinase B/Akt is sufficient to repress the glucocorticoid and cAMP induction of phosphoenolpyruvate carboxykinase gene. J Biol Chem. 1998;273:27320–4.
37. Hall RK, Yamasaki T, Kucera T, O'Brien RM, Granner DK. Regulation of phosphoenolpyruvate carboxykinase and insulin-like growth factor binding protein-1 gene expression by insulin. J Biol Chem. 2000;275:30169–75.
38. Samuel VT, Liu ZX, Qu X, Elder BD, Bilz S, Befroy D, Romanelli AJ, Shulman GI. Mechanism of hepatic insulin resistance in non-alcoholic fatty liver disease. J Biol Chem. 2004;279:32345–53.
39. Boden G. Obesity, insulin resistance and free fatty acids. Curr Opin Endocrinol Diabetes Obes. 2011;18:139–43.
40. Suganami T, Tanaka M, Ogawa Y. Adipose tissue inflammation and ectopic lipid accumulation. Endocr J. 2012;59:849–57.
41. Cypess AM, Lehman S, Williams G, Tal I, Rodman D, Goldfine AB, Kuo FC, Palmer EL, Tseng YH, Doria A, Kolodny GM, Kahn CR. Identification and importance of brown adipose tissue in adult humans. N Engl J Med. 2009;360:1509–17.
42. Nedergaard J, Bengtsson T, Cannon B. Unexpected evidence for active brown adipose tissue in adult humans. Am J Physiol Endocrinol Metab. 2007;293:E444–52.
43. Virtanen KA, Lidell ME, Orava J, Heglind M, Westergren R, Niemi T, Taittonen M, Laine J, Savisto NJ, Enerback S, Nuutila P. Functional brown adipose tissue in healthy adults. N Engl J Med. 2009;360:1518–25.
44. Chondronikola M, Volpi E, Borsheim E, Porter C, Annamalai P, Enerback S, Lidell ME, Saraf MK, Labbe SM, Hurren NM, Yfanti C, Chao T, Andersen CR, Cesani F, Hawkins H, Sidossis LS. Brown adipose tissue improves whole-body glucose homeostasis and insulin sensitivity in humans. Diabetes. 2014;63:4089–99.
45. Stanford KI, Middelbeek RJ, Townsend KL, An D, Nygaard EB, Hitchcox KM, Markan KR, Nakano K, Hirshman MF, Tseng YH, Goodyear LJ. Brown adipose tissue regulates glucose homeostasis and insulin sensitivity. J Clin Invest. 2013;123:215–23.
46. Marino JS, Xu Y, Hill JW. Central insulin and leptin-mediated autonomic control of glucose homeostasis. Trends Endocrinol Metab. 2011;22:275–85.
47. Obici S, Feng Z, Tan J, Liu L, Karkanias G, Rossetti L. Central melanocortin receptors regulate insulin action. J Clin Invest. 2001;108:1079–85.
48. Obici S, Feng Z, Karkanias G, Baskin DG, Rossetti L. Decreasing hypothalamic insulin receptors causes hyperphagia and insulin resistance in rats. Nat Neurosci. 2002;5:566–72.
49. Drucker DJ. The biology of incretin hormones. Cell Metab. 2006;3:153–65.
50. Drucker DJ, Nauck MA. The incretin system: glucagon-like peptide-1 receptor agonists and dipeptidyl peptidase-4 inhibitors in type 2 diabetes. Lancet. 2006;368:1696–705.
51. Meier JJ, Nauck MA. Incretins and the development of type 2 diabetes. Curr Diab Rep. 2006;6:194–201.
52. Rask E, Olsson T, Soderberg S, Johnson O, Seckl J, Holst JJ, Ahren B. Impaired incretin response after a mixed meal is associated with insulin resistance in nondiabetic men. Diabetes Care. 2001;24:1640–5.

53. Starkov AA. The role of mitochondria in reactive oxygen species metabolism and signaling. Ann N Y Acad Sci. 2008;1147:37–52.
54. Murphy MP. How mitochondria produce reactive oxygen species. Biochem J. 2009;417:1–13.
55. Murgia M, Giorgi C, Pinton P, Rizzuto R. Controlling metabolism and cell death: at the heart of mitochondrial calcium signalling. J Mol Cell Cardiol. 2009;46:781–8.
56. Santulli G, Marks AR. Essential roles of intracellular calcium release channels in muscle, brain, metabolism, and aging. Curr Mol Pharmacol. 2015;8:206–22.
57. Mitchell P. Coupling of phosphorylation to electron and hydrogen transfer by a chemi-osmotic type of mechanism. Nature. 1961;191:144–8.
58. Puigserver P, Wu A, Park CW, Graves R, Wright M, Spiegelman BR. A cold-inducible coactivator of nuclear receptors linked to adaptive thermogenesis. Cell. 1998;92:829–39.
59. Scarpulla RC. Transcriptional paradigms in mammalian mitochondrial biogenesis and function. Physiol Rev. 2008;88:611–38.
60. Liesa M, Palacin M, Zorzano A. Mitochondrial dynamics in Mammalian health and disease. Physiol Rev. 2009;89:799–845.
61. Twig G, Hyde B, Shirihai OS. Mitochondrial fusion, fission and autophagy as a quality control axis: the bioenergetic view. Biochim Biophys Acta. 2008;1777:1092–7.
62. Gomes LC, Di Benedetto G, Scorrano L. During autophagy mitochondria elongate, are spared from degradation and sustain cell viability. Nat Cell Biol. 2011;13:589–98.
63. Molina AJ, Wikstrom JD, Stiles L, Las G, Mohamed H, Elorza A, Walzer G, Twig G, Katz S, Corkey BE, Shirihai OS. Mitochondrial networking protects beta-cells from nutrient-induced apoptosis. Diabetes. 2009;58:2303–15.
64. Arnoult D, Grodet A, Lee YJ, Estaquier J, Blackstone C. Release of OPA1 during apoptosis participates in the rapid and complete release of cytochrome c and subsequent mitochondrial fragmentation. J Biol Chem. 2005;280:35742–50.
65. Chen H, Detmer SA, Ewald AJ, Griffin EE, Fraser SE, Chan DC. Mitofusins Mfn1 and Mfn2 coordinately regulate mitochondrial fusion and are essential for embryonic development. J Cell Biol. 2003;160:189–200.
66. Eura Y, Ishihara N, Yokota S, Mihara K. Two mitofusin proteins, mammalian homologues of FZO, with distinct functions are both required for mitochondrial fusion. J Biochem. 2003;134:333–44.
67. Bach D, Pich S, Soriano FX, Vega N, Baumgartner B, Oriola J, Daugaard JR, Lloberas J, Camps M, Zierath JR, Rabasa-Lhoret R, Wallberg-Henriksson H, Laville M, Palacin M, Vidal H, Rivera F, Brand M, Zorzano A. Mitofusin-2 determines mitochondrial network architecture and mitochondrial metabolism. A novel regulatory mechanism altered in obesity. J Biol Chem. 2003;278:17190–7.
68. de Brito OM, Scorrano L. Mitofusin 2 tethers endoplasmic reticulum to mitochondria. Nature. 2008;456:605–10.
69. Cipolat S, Martins de Brito O, Dal Zilio B, Scorrano L. OPA1 requires mitofusin 1 to promote mitochondrial fusion. Proc Natl Acad Sci U S A. 2004;101:15927–32.
70. Song Z, Chen H, Fiket M, Alexander C, Chan DC. OPA1 processing controls mitochondrial fusion and is regulated by mRNA splicing, membrane potential, and Yme1L. J Cell Biol. 2007;178:749–55.
71. Frank S, Gaume B, Bergmann-Leitner ES, Leitner WW, Robert EG, Catez F, Smith CL, Youle RJ. The role of dynamin-related protein 1, a mediator of mitochondrial fission, in apoptosis. Dev Cell. 2001;1:515–25.
72. Pitts KR, Yoon Y, Krueger EW, McNiven MA. The dynamin-like protein DLP1 is essential for normal distribution and morphology of the endoplasmic reticulum and mitochondria in mammalian cells. Mol Biol Cell. 1999;10:4403–17.
73. James DI, Parone PA, Mattenberger Y, Martinou JC. hFis1, a novel component of the mammalian mitochondrial fission machinery. J Biol Chem. 2003;278:36373–9.
74. Stojanovski D, Koutsopoulos OS, Okamoto K, Ryan MT. Levels of human Fis1 at the mitochondrial outer membrane regulate mitochondrial morphology. J Cell Sci. 2004;117:1201–10.

75. Tondera D, Santel A, Schwarzer R, Dames S, Giese K, Klippel A, Kaufmann J. Knockdown of MTP18, a novel phosphatidylinositol 3-kinase-dependent protein, affects mitochondrial morphology and induces apoptosis. J Biol Chem. 2004;279:31544–55.
76. Tondera D, Czauderna F, Paulick K, Schwarzer R, Kaufmann J, Santel A. The mitochondrial protein MTP18 contributes to mitochondrial fission in mammalian cells. J Cell Sci. 2005;118:3049–59.
77. Dumas JF, Simard G, Flamment M, Ducluzeau PH, Ritz P. Is skeletal muscle mitochondrial dysfunction a cause or an indirect consequence of insulin resistance in humans? Diabetes Metab. 2009;35:159–67.
78. Pagel-Langenickel I, Bao J, Pang L, Sack MN. The role of mitochondria in the pathophysiology of skeletal muscle insulin resistance. Endocr Rev. 2010;31:25–51.
79. Hanson M. Understanding the origins of diabetes. JAMA. 2014;311:575–6.
80. Goodpaster BH. Mitochondrial deficiency is associated with insulin resistance. Diabetes. 2013;62:1032–5.
81. Holloszy JO. "Deficiency" of mitochondria in muscle does not cause insulin resistance. Diabetes. 2013;62:1036–40.
82. Vondra K, Rath R, Bass A, Slabochova Z, Teisinger J, Vitek V. Enzyme activities in quadriceps femoris muscle of obese diabetic male patients. Diabetologia. 1977;13:527–9.
83. Kelley DE, Goodpaster B, Wing RR, Simoneau JA. Skeletal muscle fatty acid metabolism in association with insulin resistance, obesity, and weight loss. Am J Phys. 1999;277:E1130–41.
84. Simoneau JA, Veerkamp JH, Turcotte LP, Kelley DE. Markers of capacity to utilize fatty acids in human skeletal muscle: relation to insulin resistance and obesity and effects of weight loss. FASEB J. 1999;13:2051–60.
85. He J, Watkins S, Kelley DE. Skeletal muscle lipid content and oxidative enzyme activity in relation to muscle fiber type in type 2 diabetes and obesity. Diabetes. 2001;50:817–23.
86. Kelley DE, He J, Menshikova EV, Ritov VB. Dysfunction of mitochondria in human skeletal muscle in type 2 diabetes. Diabetes. 2002;51:2944–50.
87. Petersen KF, Befroy D, Dufour S, Dziura J, Ariyan C, Rothman DL, DiPietro L, Cline GW, Shulman GI. Mitochondrial dysfunction in the elderly: possible role in insulin resistance. Science. 2003;300:1140–2.
88. Mootha VK, Lindgren CM, Eriksson KF, Subramanian A, Sihag S, Lehar J, Puigserver P, Carlsson E, Ridderstrale M, Laurila E, Houstis N, Daly MJ, Patterson N, Mesirov JP, Golub TR, Tamayo P, Spiegelman B, Lander ES, Hirschhorn JN, Altshuler D, Groop LC. PGC-1alpha-responsive genes involved in oxidative phosphorylation are coordinately downregulated in human diabetes. Nat Genet. 2003;34:267–73.
89. Patti ME, Butte AJ, Crunkhorn S, Cusi K, Berria R, Kashyap S, Miyazaki Y, Kohane I, Costello M, Saccone R, Landaker EJ, Goldfine AB, Mun E, DeFronzo R, Finlayson J, Kahn CR, Mandarino LJ. Coordinated reduction of genes of oxidative metabolism in humans with insulin resistance and diabetes: potential role of PGC1 and NRF1. Proc Natl Acad Sci USA. 2003;100:8466–71.
90. Petersen KF, Dufour S, Befroy D, Garcia R, Shulman GI. Impaired mitochondrial activity in the insulin-resistant offspring of patients with type 2 diabetes. N Engl J Med. 2004;350:664–71.
91. Morino K, Petersen KF, Dufour S, Befroy D, Frattini J, Shatzkes N, Neschen S, White MF, Bilz S, Sono S, Pypaert M, Shulman GI. Reduced mitochondrial density and increased IRS-1 serine phosphorylation in muscle of insulin-resistant offspring of type 2 diabetic parents. J Clin Invest. 2005;115:3587–93.
92. Miller WC, Bryce GR, Conlee RK. Adaptations to a high-fat diet that increase exercise endurance in male rats. J Appl Physiol Respir Environ Exerc Physiol. 1984;56:78–83.
93. McAinch AJ, Lee JS, Bruce CR, Tunstall RJ, Hawley JA, Cameron-Smith D. Dietary regulation of fat oxidative gene expression in different skeletal muscle fiber types. Obes Res. 2003;11:1471–9.

94. Han DH, Nolte LA, Ju JS, Coleman T, Holloszy JO, Semenkovich CF. UCP-mediated energy depletion in skeletal muscle increases glucose transport despite lipid accumulation and mitochondrial dysfunction. Am J Physiol Endocrinol Metab. 2004;286:E347–53.
95. Wredenberg A, Freyer C, Sandstrom ME, Katz A, Wibom R, Westerblad H, Larsson NG. Respiratory chain dysfunction in skeletal muscle does not cause insulin resistance. Biochem Biophys Res Commun. 2006;350:202–7.
96. Han DH, Hancock CR, Jung SR, Higashida K, Kim SH, Holloszy JO. Deficiency of the mitochondrial electron transport chain in muscle does not cause insulin resistance. PLoS One. 2011;6:e19739.
97. Pospisilik JA, Knauf C, Joza N, Benit P, Orthofer M, Cani PD, Ebersberger I, Nakashima T, Sarao R, Neely G, Esterbauer H, Kozlov A, Kahn CR, Kroemer G, Rustin P, Burcelin R, Penninger JM. Targeted deletion of AIF decreases mitochondrial oxidative phosphorylation and protects from obesity and diabetes. Cell. 2007;131:1–16.
98. Zechner C, Lai L, Zechner JF, Geng T, Yan Z, Rumsey JW, Collia D, Chen Z, Wozniak DF, Leone TC, Kelly DP. Total skeletal muscle PGC-1 deficiency uncouples mitochondrial derangements from fiber type determination and insulin sensitivity. Cell Metab. 2010;12:633–42.
99. Garcia-Roves P, Huss JM, Han DH, Hancock CR, Iglesias-Gutierrez E, Chen M, Holloszy JO. Raising plasma fatty acid concentration induces increased biogenesis of mitochondria in skeletal muscle. Proc Natl Acad Sci USA. 2007;104:10709–13.
100. Turner N, Bruce CR, Beale SM, Hoehn KL, So T, Rolph MS, Cooney GJ. Excess lipid availability increases mitochondrial fatty acid oxidative capacity in muscle: evidence against a role for reduced fatty acid oxidation in lipid-induced insulin resistance in rodents. Diabetes. 2007;56:2085–92.
101. Hancock CR, Han DH, Chen M, Terada S, Yasuda T, Wright DC, Holloszy JO. High-fat diets cause insulin resistance despite an increase in muscle mitochondria. Proc Natl Acad Sci USA. 2008;105:7815–20.
102. Toledo FG, Menshikova EV, Azuma K, Radikova Z, Kelley CA, Ritov VB, Kelley DE. Mitochondrial capacity in skeletal muscle is not stimulated by weight loss despite increases in insulin action and decreases in intramyocellular lipid content. Diabetes. 2008;57:987–94.
103. Nair KS, Bigelow ML, Asmann YW, Chow LS, Coenen-Schimke JM, Klaus KA, Guo ZK, Sreekumar R, Irving BA. Asian Indians have enhanced skeletal muscle mitochondrial capacity to produce ATP in association with severe insulin resistance. Diabetes. 2008;57:1166–75.
104. Jin W, Goldfine AB, Boes T, Henry RR, Ciaraldi TP, Kim EY, Emecan M, Fitzpatrick C, Sen A, Shah A, Mun E, Vokes V, Schroeder J, Tatro E, Jimenez-Chillaron J, Patti ME. Increased SRF transcriptional activity in human and mouse skeletal muscle is a signature of insulin resistance. J Clin Invest. 2011;121:918–29.
105. Schiaffino S, Reggiani C. Fiber types in mammalian skeletal muscles. Physiol Rev. 2011;91:1447–531.
106. DeFronzo RA, Jacot E, Jequier E, Maeder E, Wahren J, Felber JP. The effect of insulin on the disposal of intravenous glucose: results from indirect calorimetry and hepatic and femoral venous catheterization. Diabetes. 1981;30:1000–7.
107. Sasson S, Cerasi E. Substrate regulation of the glucose transport system in skeletal muscle. J Biol Chem. 1986;261:16827–33.
108. DeFronzo RA, Ferrannini E, Hendler R, Wahren J, Felig P. Influence of hyperinsulinemia, hyperglycemia, and the route of glucose administration on splanchnic glucose exchange. Proc Natl Acad Sci U S A. 1978;75:5173–7.
109. DeFronzo RA, Gunnarson R, Biorkman O, Olsson M, Warren J. Effects of insulin on peripheral and splanchnic glucose metabolism in non-insulin dependent (Type II) diabetes mellitus. J Clin Invest. 1985;76:149–55.
110. Mitrakou A, Kelley D, Veneman T, Jenssen T, Pangburn T, Reilly J, Gerich J. Contribution of abnormal muscle and liver glucose metabolism to postprandial hyperglycemia in NIDDM. Diabetes. 1990;39:1381–90.

111. Shulman GI, Rothman DL, Jue R, Stein P, DeFronzo RA, Shulman RG. Quantitation of muscle glycogen synthesis in normal subjects and subjects with non-insulin dependent diabetes by 13C nuclear magnetic resonance spectroscopy. N Engl J Med. 1990;322: 223–8.
112. Groop LC, Bonadonna RC, DelPrato S, Ratheiser K, Zyck K, Ferrannini B, DeFronzo RA. Glucose and free fatty acid metabolism in non-insulin-dependent diabetes mellitus. Evidence for multiple sites of insulin resistance. J Clin Invest. 1989;84:205–13.
113. Rothman DL, Magnusson I, Cline G, Gerard D, Kahn CR, Shulman RG, Shulman GI. Decreased muscle glucose transport/phosphorylation is an early defect in the pathogenesis of non-insulin-dependent diabetes mellitus. Proc Natl Acad Sci USA. 1995;92: 983–7.
114. Bajaj M, Defronzo RA. Metabolic and molecular basis of insulin resistance. J Nucl Cardiol Off Publ Am Soc Nucl Cardiol. 2003;10:311–23.
115. Saltiel AR, Kahn CR. Insulin signalling and the regulation of glucose and lipid metabolism. Nature. 2001;414:799–806.
116. Taniguchi CM, Emanuelli B, Kahn CR. Critical nodes in signalling pathways: insights into insulin action. Nat Rev Mol Cell Biol. 2006;7:85–96.
117. Musi N, Goodyear LJ. Insulin resistance and improvements in signal transduction. Endocrine. 2006;29:73–80.
118. Kashyap SR, Roman LJ, Lamont J, Masters BS, Bajaj M, Suraamornkul S, Belfort R, Berria R, Kellogg Jr DL, Liu Y, DeFronzo RA. Insulin resistance is associated with impaired nitric oxide synthase activity in skeletal muscle of type 2 diabetic subjects. J Clin Endocrinol Metab. 2005;90:1100–5.
119. Montagnani M, Chen H, Barr VA, Quon MJ. Insulin-stimulated activation of eNOS is independent of Ca2+ but requires phosphorylation by Akt at Ser(1179). J Biol Chem. 2001;276:30392–8.
120. Kashyap SR, Belfort R, Berria R, Suraamornkul S, Pratipranawatr T, Finlayson J, Barrentine A, Bajaj M, Mandarino L, DeFronzo R, Cusi K. Discordant effects of a chronic physiological increase in plasma FFA on insulin signaling in healthy subjects with or without a family history of type 2 diabetes. Am J Physiol Endocrinol Metab. 2004;287:537–46.
121. Pratipanawatr W, Pratipanawatr T, Cusi K, Berria R, Adams JM, Jenkinson CP, Maezono K, DeFronzo RA, Mandarino LJ. Skeletal muscle insulin resistance in normoglycemic subjects with a strong family history of type 2 diabetes is associated with decreased insulin-stimulated insulin receptor substrate-1 tyrosine phosphorylation. Diabetes. 2001;50:2572–8.
122. Yu C, Chen Y, Cline GW, Zhang D, Zong H, Wang Y, Bergeron R, Kim JK, Cushman SW, Cooney GJ, Atcheson B, White MF, Kraegen EW, Shulman GI. Mechanism by which fatty acids inhibit insulin activation of insulin receptor substrate-1 (IRS-1)-associated phosphatidylinositol 3-kinase activity in muscle. J Biol Chem. 2002;277:50230–6.
123. Aguirre V, Werner ED, Giraud J, Lee YH, Shoelson SE, White MF. Phosphorylation of Ser307 in insulin receptor substrate-1 blocks interactions with the insulin receptor and inhibits insulin action. J Biol Chem. 2002;277:1531–7.
124. Folli F, Saad MJA, Backer JM, Kahn CR. Regulation of phosphatidylinositol 3-kinase activity in liver and muscle of animal models of insulin-resistant and insulin-deficient diabetes mellitus. J Clin Invest. 1993;92:1787–94.
125. Befroy D, Petersen KF, Dufour S, Mason GF, de Graaf RA, Rothman D, Shulman G. Impaired mitochondrial substrate oxidation in muscle of insulin-resistant offspring of type 2 diabetic patients. Diabetes. 2007;56:1376–81.
126. Ritov VB, Menshikova EV, He J, Ferrell RE, Goodpaster BH, Kelley DE. Deficiency of subsarcolemmal mitochondria in obesity and type 2 diabetes. Diabetes. 2005;54:8–14.
127. Schrauwen-Hinderling VB, Kooi ME, Hesselink MK, Jeneson JA, Backes WH, van Echteld CJ, van Engelshoven JM, Mensink M, Schrauwen P. Impaired in vivo mitochondrial function but similar intramyocellular lipid content in patients with type 2 diabetes mellitus and BMI-matched control subjects. Diabetologia. 2007;50:113–20.

128. Kacerovsky-Bielesz G, Chmelik M, Ling C, Pokan R, Szendroedi J, Farukuoye M, Kacerovsky M, Schmid AI, Gruber S, Wolzt M, Moser E, Pacini G, Smekal G, Groop L, Roden M. Short-term exercise training does not stimulate skeletal muscle ATP synthesis in relatives of humans with type 2 diabetes. Diabetes. 2009;58:1333–41.
129. Phielix E, Schrauwen-Hinderling VB, Mensink M, Lenaers E, Meex R, Hoeks J, Kooi ME, Moonen-Kornips E, Sels JP, Hesselink MK, Schrauwen P. Lower intrinsic ADP-stimulated mitochondrial respiration underlies in vivo mitochondrial dysfunction in muscle of male type 2 diabetic patients. Diabetes. 2008;57:2943–9.
130. Boushel R, Gnaiger E, Schjerling P, Skovbro M, Kraunsoe R, Dela F. Patients with type 2 diabetes have normal mitochondrial function in skeletal muscle. Diabetologia. 2007;50:790–6.
131. Ritov VB, Menshikova EV, Azuma K, Wood R, Toledo FG, Goodpaster BH, Ruderman NB, Kelley DE. Deficiency of electron transport chain in human skeletal muscle mitochondria in type 2 diabetes mellitus and obesity. Am J Physiol Endocrinol Metab. 2010;298:E49–58.
132. Mogensen M, Sahlin K, Fernstrom M, Glintborg D, Vind BF, Beck-Nielsen H, Hojlund K. Mitochondrial respiration is decreased in skeletal muscle of patients with type 2 diabetes. Diabetes. 2007;56:1592–9.
133. Szendroedi J, Schmid AI, Chmelik M, Toth C, Brehm A, Krssak M, Nowotny P, Wolzt M, Waldhausl W, Roden M. Muscle Mitochondrial atp synthesis and glucose transport/phosphorylation in type 2 diabetes. PLoS Med. 2007;4:e154.
134. Gnaiger E. Capacity of oxidative phosphorylation in human skeletal muscle: new perspectives of mitochondrial physiology. Int J Biochem Cell Biol. 2009;41:1837–45.
135. Lefort N, Glancy B, Bowen B, Willis WT, Bailowitz Z, De Filippis EA, Brophy C, Meyer C, Hojlund K, Yi Z, Mandarino LJ. Increased reactive oxygen species production and lower abundance of complex I subunits and carnitine palmitoyltransferase 1B protein despite normal mitochondrial respiration in insulin-resistant human skeletal muscle. Diabetes. 2010;59:2444–52.
136. Chomentowski P, Coen PM, Radikova Z, Goodpaster BH, Toledo FG. Skeletal muscle mitochondria in insulin resistance: differences in intermyofibrillar versus subsarcolemmal subpopulations and relationship to metabolic flexibility. J Clin Endocrinol Metab. 2011;96: 494–503.
137. Heilbronn LK, Gan SK, Turner N, Campbell LV, Chisholm DJ. Markers of mitochondrial biogenesis and metabolism are lower in overweight and obese insulin-resistant subjects. J Clin Endocrinol Metab. 2007;92:1467–73.
138. Mootha VK, Handschin C, Arlow D, Xie X, St Pierre J, Sihag S, Yang W, Altshuler D, Puigserver P, Patterson N, Willy PJ, Schulman IG, Heyman RA, Lander ES, Spiegelman BM. Erralpha and Gabpa/b specify PGC-1alpha-dependent oxidative phosphorylation gene expression that is altered in diabetic muscle. Proc Natl Acad Sci U S A. 2004;101:6570–5.
139. Karlsson HK, Ahlsen M, Zierath JR, Wallberg-Henriksson H, Koistinen HA. Insulin signaling and glucose transport in skeletal muscle from first-degree relatives of type 2 diabetic patients. Diabetes. 2006;55:1283–8.
140. Nisoli E, Carruba MO. Nitric oxide and mitochondrial biogenesis. J Cell Sci. 2006;119: 2855–62.
141. Lodi R, Tonon C, Valentino ML, Iotti S, Clementi V, Malucelli E, Barboni P, Longanesi L, Schimpf S, Wissinger B, Baruzzi A, Barbiroli B, Carelli V. Deficit of in vivo mitochondrial ATP production in OPA1-related dominant optic atrophy. Ann Neurol. 2004;56:719–23.
142. Jheng HF, Tsai PJ, Guo SM, Kuo LH, Chang CS, Su IJ, Chang CR, Tsai YS. Mitochondrial fission contributes to mitochondrial dysfunction and insulin resistance in skeletal muscle. Mol Cell Biol. 2012;32:309–19.
143. Holmstrom MH, Iglesias-Gutierrez E, Zierath JR, Garcia-Roves PM. Tissue-specific control of mitochondrial respiration in obesity-related insulin resistance and diabetes. Am J Physiol Endocrinol Metab. 2012;302:E731–9.
144. Karakelides H, Irving BA, Short KR, O'Brien P, Nair KS. Age, obesity, and sex effects on insulin sensitivity and skeletal muscle mitochondrial function. Diabetes. 2010;59:89–97.

145. Holmstrom MH, Tom RZ, Bjornholm M, Garcia-Roves PM, Zierath JR. Effect of leptin treatment on mitochondrial function in obese leptin-deficient ob/ob mice. Metabolism. 2013;62:1258–67.
146. Lenaers E, De Feyter HM, Hoeks J, Schrauwen P, Schaart G, Nabben M, Nicolay K, Prompers JJ, Hesselink MK. Adaptations in mitochondrial function parallel, but fail to rescue, the transition to severe hyperglycemia and hyperinsulinemia: a study in Zucker diabetic fatty rats. Obesity (Silver Spring). 2010;18:1100–7.
147. van den Broek NM, Ciapaite J, De Feyter HM, Houten SM, Wanders RJ, Jeneson JA, Nicolay K, Prompers JJ. Increased mitochondrial content rescues in vivo muscle oxidative capacity in long-term high-fat-diet-fed rats. FASEB J. 2010;24:1354–64.
148. Zhang D, Liu ZX, Choi CS, Tian L, Kibbey R, Dong J, Cline GW, Wood PA, Shulman GI. Mitochondrial dysfunction due to long-chain Acyl-CoA dehydrogenase deficiency causes hepatic steatosis and hepatic insulin resistance. Proc Natl Acad Sci USA. 2007;104:17075–80.
149. Quintens R, Singh S, Lemaire K, De Bock K, Granvik M, Schraenen A, Vroegrijk IO, Costa V, Van Noten P, Lambrechts D, Lehnert S, Van Lommel L, Thorrez L, De Faudeur G, Romijn JA, Shelton JM, Scorrano L, Lijnen HR, Voshol PJ, Carmeliet P, Mammen PP, Schuit F. Mice deficient in the respiratory chain gene Cox6a2 are protected against high-fat diet-induced obesity and insulin resistance. PLoS One. 2013;8:e56719.
150. Dela F, Helge JW. Insulin resistance and mitochondrial function in skeletal muscle. Int J Biochem Cell Biol. 2013;45:11–5.
151. Yki-Jarvinen H. Ectopic fat accumulation: an important cause of insulin resistance in humans. J R Soc Med. 2002;95(Suppl 42):39–45.
152. Samuel VT, Petersen KF, Shulman GI. Lipid-induced insulin resistance: unravelling the mechanism. Lancet. 2010;375:2267–77.
153. Samuel VT, Shulman GI. Mechanisms for insulin resistance: common threads and missing links. Cell. 2012;148:852–71.
154. Boden G, Shulman GI. Free fatty acids in obesity and type 2 diabetes: defining their role in the development of insulin resistance and beta-cell dysfunction. Eur J Clin Invest. 2002;32(Suppl 3):14–23.
155. Perseghin G, Scifo P, De Cobelli F, Pagliato E, Battezzati A, Arcelloni C, Vanzulli A, Testolin G, Pozza G, Del Maschio A, Luzi L. Intramyocellular triglyceride content is a determinant of in vivo insulin resistance in humans: a 1H-13C nuclear magnetic resonance spectroscopy assessment in offspring of type 2 diabetic parents. Diabetes. 1999;48:1600–6.
156. Szczepaniak LS, Babcock EE, Schick F, Dobbins RL, Garg A, Burns DK, McGarry JD, Stein DT. Measurement of intracellular triglyceride stores by H spectroscopy: validation in vivo. Am J Phys. 1999;276:E977–89.
157. Weiss R, Dufour S, Taksali SE, Tamborlane WV, Petersen KF, Bonadonna RC, Boselli L, Barbetta G, Allen K, Rife F, Savoye M, Dziura J, Sherwin R, Shulman GI, Caprio S. Prediabetes in obese youth: a syndrome of impaired glucose tolerance, severe insulin resistance, and altered myocellular and abdominal fat partitioning. Lancet. 2003;362:951–7.
158. Wang H, Knaub LA, Jensen DR, Young Jung D, Hong EG, Ko HJ, Coates AM, Goldberg IJ, de la Houssaye BA, Janssen RC, McCurdy CE, Rahman SM, Soo Choi C, Shulman GI, Kim JK, Friedman JE, Eckel RH. Skeletal muscle-specific deletion of lipoprotein lipase enhances insulin signaling in skeletal muscle but causes insulin resistance in liver and other tissues. Diabetes. 2009;58:116–24.
159. Kim JK, Gimeno RE, Higashimori T, Kim HJ, Choi H, Punreddy S, Mozell RL, Tan G, Stricker-Krongrad A, Hirsch DJ, Fillmore JJ, Liu ZX, Dong J, Cline G, Stahl A, Lodish HF, Shulman GI. Inactivation of fatty acid transport protein 1 prevents fat-induced insulin resistance in skeletal muscle. J Clin Invest. 2004;113:756–63.
160. Goudriaan JR, Dahlmans VE, Teusink B, Ouwens DM, Febbraio M, Maassen JA, Romijn JA, Havekes LM, Voshol PJ. CD36 deficiency increases insulin sensitivity in muscle, but induces insulin resistance in the liver in mice. J Lipid Res. 2003;44:2270–7.

161. Hajri T, Han XX, Bonen A, Abumrad NA. Defective fatty acid uptake modulates insulin responsiveness and metabolic responses to diet in CD36-null mice. J Clin Invest. 2002;109:1381–9.
162. Kim JK, Fillmore JJ, Chen Y, Yu C, Moore IK, Pypaert M, Lutz EP, Kako Y, Velez-Carrasco W, Goldberg IJ, Breslow JL, Shulman GI. Tissue-specific overexpression of lipoprotein lipase causes tissue-specific insulin resistance. Proc Natl Acad Sci USA. 2001;98:7522–7.
163. Dresner A, Laurent D, Marcucci M, Griffin ME, Dufour S, Cline GW, Slezak LA, Andersen DK, Hundal RS, Rothman DL, Petersen KF, Shulman GI. Effects of free fatty acids on glucose transport and IRS-1-associated phosphatidylinositol 3-kinase activity. J Clin Invest. 1999;103:253–9.
164. Griffin ME, Marcucci MJ, Cline GW, Bell K, Barucci N, Lee D, Goodyear LJ, Kraegen EW, White MF, Shulman GI. Free fatty acid-induced insulin resistance is associated with activation of protein kinase C theta and alterations in the insulin signaling cascade. Diabetes. 1999;48:1270–4.
165. Kim JK, Fillmore JJ, Sunshine MJ, Albrecht B, Higashimori T, Kim DW, Liu ZX, Soos TJ, Cline GW, O'Brien WR, Littman DR, Shulman GI. PKC-theta knockout mice are protected from fat-induced insulin resistance. J Clin Invest. 2004;114:823–7.
166. Krssak M, Petersen KF, Bergeron R, Price T, Laurent D, Rothman DL, Roden M, Shulman GI. Intramuscular glycogen and intramyocellular lipid utilization during prolonged exercise and recovery in man: a 13C and 1H nuclear magnetic resonance spectroscopy study. J Clin Endocrinol Metab. 2000;85:748–54.
167. Goodpaster BH, He J, Watkins S, Kelley DE. Skeletal muscle lipid content and insulin resistance: evidence for a paradox in endurance-trained athletes. J Clin Endocrinol Metab. 2001;86:5755–61.
168. Schmitz-Peiffer C, Browne CL, Oakes ND, Watkinson A, Chisholm DJ, Kraegen EW, Biden TJ. Alterations in the expression and cellular localization of protein kinase C isozymes epsilon and theta are associated with insulin resistance in skeletal muscle of the high-fat-fed rat. Diabetes. 1997;46:169–78.
169. Dube JJ, Amati F, Toledo FG, Stefanovic-Racic M, Rossi A, Coen P, Goodpaster BH. Effects of weight loss and exercise on insulin resistance, and intramyocellular triacylglycerol, diacylglycerol and ceramide. Diabetologia. 2011;54:1147–56.
170. Itani SI, Ruderman NB, Schmieder F, Boden G. Lipid-induced insulin resistance in human muscle is associated with changes in diacylglycerol, protein kinase C, and IkappaB-alpha. Diabetes. 2002;51:2005–11.
171. Schinner S, Scherbaum WA, Bornstein SR, Barthel A. Molecular mechanisms of insulin resistance. Diabet Med. 2005;22:674–82.
172. Shi H, Kokoeva MV, Inouye K, Tzameli I, Yin H, Flier JS. TLR4 links innate immunity and fatty acid-induced insulin resistance. J Clin Invest. 2006;116:3015–25.
173. Kim F, Pham M, Luttrell I, Bannerman DD, Tupper J, Thaler J, Hawn TR, Raines EW, Schwartz MW. Toll-like receptor-4 mediates vascular inflammation and insulin resistance in diet-induced obesity. Circ Res. 2007;100:1589–96.
174. Senn JJ. Toll-like receptor-2 is essential for the development of palmitate-induced insulin resistance in myotubes. J Biol Chem. 2006;281:26865–75.
175. Coen PM, Dube JJ, Amati F, Stefanovic-Racic M, Ferrell RE, Toledo FG, Goodpaster BH. Insulin resistance is associated with higher intramyocellular triglycerides in type I but not type II myocytes concomitant with higher ceramide content. Diabetes. 2010;59:80–8.
176. Adams 2nd JM, Pratipanawatr T, Berria R, Wang E, DeFronzo RA, Sullards MC, Mandarino LJ. Ceramide content is increased in skeletal muscle from obese insulin-resistant humans. Diabetes. 2004;53:25–31.
177. Ara I, Larsen S, Stallknecht B, Guerra B, Morales-Alamo D, Andersen JL, Ponce-Gonzalez JG, Guadalupe-Grau A, Galbo H, Calbet JA, Helge JW. Normal mitochondrial function and increased fat oxidation capacity in leg and arm muscles in obese humans. Int J Obes (Lond). 2011;35:99–108.

178. Goodpaster BH, Wolfe RR, Kelley DE. Effects of obesity on substrate utilization during exercise. Obes Res. 2002;10:575–84.
179. Groop LC, Saloranta C, Shank M, Bonadonna RC, Ferrannini E, DeFronzo RA. The role of free fatty acid metabolism in the pathogenesis of insulin resistance in obesity and noninsulin-dependent diabetes mellitus. J Clin Endocrinol Metab. 1991;72(1):96–107.
180. Boon H, Blaak EE, Saris WH, Keizer HA, Wagenmakers AJ, van Loon LJ. Substrate source utilisation in long-term diagnosed type 2 diabetes patients at rest, and during exercise and subsequent recovery. Diabetologia. 2007;50:103–12.
181. Petersen KF, Dufour S, Shulman GI. Decreased insulin-stimulated ATP synthesis and phosphate transport in muscle of insulin-resistant offspring of type 2 diabetic parents. PLoS Med. 2005;2:e233.
182. Lowell BB, Shulman GI. Mitochondrial dysfunction and type 2 diabetes. Science. 2005;307:384–7.
183. Neschen S, Morino K, Dong J, Wang-Fischer Y, Cline GW, Romanelli AJ, Rossbacher JC, Moore IK, Regittnig W, Munoz DS, Kim JH, Shulman GI. n-3 Fatty acids preserve insulin sensitivity in vivo in a peroxisome proliferator-activated receptor-alpha-dependent manner. Diabetes. 2007;56:1034–41.
184. Brehm A, Krssak M, Schmid AI, Nowotny P, Waldhausl W, Roden M. Increased lipid availability impairs insulin-stimulated ATP synthesis in human skeletal muscle. Diabetes. 2006;55:136–40.
185. Holloway GP, Thrush AB, Heigenhauser GJ, Tandon NN, Dyck DJ, Bonen A, Spriet LL. Skeletal muscle mitochondrial FAT/CD36 content and palmitate oxidation are not decreased in obese women. Am J Physiol Endocrinol Metab. 2007;292:E1782–9.
186. Bandyopadhyay GK, Yu JG, Ofrecio J, Olefsky JM. Increased malonyl-CoA levels in muscle from obese and type 2 diabetic subjects lead to decreased fatty acid oxidation and increased lipogenesis; thiazolidinedione treatment reverses these defects. Diabetes. 2006;55:2277–85.
187. Brehm A, Krssak M, Schmid AI, Nowotny P, Waldhausl W, Roden M. Acute elevation of plasma lipids does not affect ATP synthesis in human skeletal muscle. Am J Physiol Endocrinol Metab. 2010;299:E33–8.
188. Chavez AO, Kamath S, Jani R, Sharma LK, Monroy A, Abdul-Ghani MA, Centonze VE, Sathyanarayana P, Coletta DK, Jenkinson CP, Bai Y, Folli F, Defronzo RA, Tripathy D. Effect of short-term free fatty acids elevation on mitochondrial function in skeletal muscle of healthy individuals. J Clin Endocrinol Metab. 2010;95:422–9.
189. Szendroedi J, Phielix E, Roden M. The role of mitochondria in insulin resistance and type 2 diabetes mellitus. Nat Rev Endocrinol. 2012;8:92–103.
190. van de Weijer T, Sparks LM, Phielix E, Meex RC, van Herpen NA, Hesselink MK, Schrauwen P, Schrauwen-Hinderling VB. Relationships between mitochondrial function and metabolic flexibility in type 2 diabetes mellitus. PLoS One. 2013;8:e51648.
191. De Feyter HM, Lenaers E, Houten SM, Schrauwen P, Hesselink MK, Wanders RJ, Nicolay K, Prompers JJ. Increased intramyocellular lipid content but normal skeletal muscle mitochondrial oxidative capacity throughout the pathogenesis of type 2 diabetes. FASEB J. 2008;22:3947–55.
192. Holloway GP, Benton CR, Mullen KL, Yoshida Y, Snook LA, Han XX, Glatz JF, Luiken JJ, Lally J, Dyck DJ, Bonen A. In obese rat muscle transport of palmitate is increased and is channeled to triacylglycerol storage despite an increase in mitochondrial palmitate oxidation. Am J Physiol Endocrinol Metab. 2009;296:E738–47.
193. Abu-Elheiga L, Oh W, Kordari P, Wakil SJ. Acetyl-CoA carboxylase 2 mutant mice are protected against obesity and diabetes induced by high-fat/high-carbohydrate diets. Proc Natl Acad Sci U S A. 2003;100:10207–12.
194. Rabol R, Svendsen PF, Skovbro M, Boushel R, Haugaard SB, Schjerling P, Schrauwen P, Hesselink MK, Nilas L, Madsbad S, Dela F. Reduced skeletal muscle mitochondrial respiration and improved glucose metabolism in nondiabetic obese women during a very low calorie dietary intervention leading to rapid weight loss. Metabolism. 2009;58:1145–52.

195. Meex RC, Schrauwen-Hinderling VB, Moonen-Kornips E, Schaart G, Mensink M, Phielix E, van de Weijer T, Sels JP, Schrauwen P, Hesselink MK. Restoration of muscle mitochondrial function and metabolic flexibility in type 2 diabetes by exercise training is paralleled by increased myocellular fat storage and improved insulin sensitivity. Diabetes. 2010;59: 572–9.
196. Phielix E, Jelenik T, Nowotny P, Szendroedi J, Roden M. Reduction of non-esterified fatty acids improves insulin sensitivity and lowers oxidative stress, but fails to restore oxidative capacity in type 2 diabetes: a randomised clinical trial. Diabetologia. 2014;57:572–81.
197. Makimura H, Stanley TL, Suresh C, De Sousa-Coelho AL, Frontera WR, Syu S, Braun LR, Looby SE, Feldpausch MN, Torriani M, Lee H, Patti ME, Grinspoon SK. Metabolic effects of long-term reduction in free fatty acids with acipimox in obesity: a randomized trial. J Clin Endocrinol Metab. 2016;101:1123–33.
198. Quinlan CL, Perevoshchikova IV, Hey-Mogensen M, Orr AL, Brand MD. Sites of reactive oxygen species generation by mitochondria oxidizing different substrates. Redox Biol. 2013;1:304–12.
199. Yun J, Finkel T. Mitohormesis. Cell Metab. 2014;19:757–66.
200. Ristow M, Schmeisser K. Mitohormesis: promoting health and lifespan by increased levels of reactive oxygen species (ROS). Dose-Response Publ Int Hormesis Soc. 2014;12:288–341.
201. Anderson EJ, Lustig ME, Boyle KE, Woodlief TL, Kane DA, Lin CT, Price III JW, Kang L, Rabinovitch PS, Szeto HH, Houmard JA, Cortright RN, Wasserman DH, Neufer PD. Mitochondrial H2O2 emission and cellular redox state link excess fat intake to insulin resistance in both rodents and humans. J Clin Invest. 2009;119:573–81.
202. Houstis N, Rosen ED, Lander ES. Reactive oxygen species have a causal role in multiple forms of insulin resistance. Nature. 2006;440:944–8.
203. Curtis JM, Grimsrud PA, Wright WS, Xu X, Foncea RE, Graham DW, Brestoff JR, Wiczer BM, Ilkayeva O, Cianflone K, Muoio DE, Arriaga EA, Bernlohr DA. Downregulation of adipose glutathione S-transferase A4 leads to increased protein carbonylation, oxidative stress, and mitochondrial dysfunction. Diabetes. 2010;59:1132–42.
204. Bonnard C, Durand A, Peyrol S, Chanseaume E, Chauvin MA, Morio B, Vidal H, Rieusset J. Mitochondrial dysfunction results from oxidative stress in the skeletal muscle of diet-induced insulin-resistant mice. J Clin Invest. 2008;118:789.
205. Boudina S, Sena S, Sloan C, Tebbi A, Han YH, O'Neill BT, Cooksey RC, Jones D, Holland WL, McClain DA, Abel ED. Early mitochondrial adaptations in skeletal muscle to diet-induced obesity are strain dependent and determine oxidative stress and energy expenditure but not insulin sensitivity. Endocrinology. 2012;153:2677–88.
206. Abdul-Ghani MA, Jani R, Chavez A, Molina-Carrion M, Tripathy D, Defronzo RA. Mitochondrial reactive oxygen species generation in obese non-diabetic and type 2 diabetic participants. Diabetologia. 2009;52:574–82.
207. Hey-Mogensen M, Hojlund K, Vind BF, Wang L, Dela F, Beck-Nielsen H, Fernstrom M, Sahlin K. Effect of physical training on mitochondrial respiration and reactive oxygen species release in skeletal muscle in patients with obesity and type 2 diabetes. Diabetologia. 2010;53:1976–85.
208. Larsen S, Stride N, Hey-Mogensen M, Hansen CN, Andersen JL, Madsbad S, Worm D, Helge JW, Dela F. Increased mitochondrial substrate sensitivity in skeletal muscle of patients with type 2 diabetes. Diabetologia. 2011;54:1427–36.
209. Nishikawa T, Araki E. Impact of mitochondrial ROS production in the pathogenesis of diabetes mellitus and its complications. Antioxid Redox Signal. 2007;9:343–53.
210. Evans JL, Goldfine ID, Maddux BA, Grodsky GM. Are oxidative stress-activated signaling pathways mediators of insulin resistance and beta-cell dysfunction? Diabetes. 2003;52:1–8.
211. Loh K, Deng H, Fukushima A, Cai X, Boivin B, Galic S, Bruce C, Shields BJ, Skiba B, Ooms LM, Stepto N, Wu B, Mitchell CA, Tonks NK, Watt MJ, Febbraio MA, Crack PJ, Andrikopoulos S, Tiganis T. Reactive oxygen species enhance insulin sensitivity. Cell Metab. 2009;10:260–72.

212. Lee HY, Choi CS, Birkenfeld AL, Alves TC, Jornayvaz FR, Jurczak MJ, Zhang D, Woo DK, Shadel GS, Ladiges W, Rabinovitch PS, Santos JH, Petersen KF, Samuel VT, Shulman GI. Targeted expression of catalase to mitochondria prevents age-associated reductions in mitochondrial function and insulin resistance. Cell Metab. 2010;12:668–74.
213. Boden MJ, Brandon AE, Tid-Ang JD, Preston E, Wilks D, Stuart E, Cleasby ME, Turner N, Cooney GJ, Kraegen EW. Overexpression of manganese superoxide dismutase ameliorates high-fat diet-induced insulin resistance in rat skeletal muscle. Am J Physiol Endocrinol Metab. 2012;303:E798–805.
214. Hoppeler H, Howald H, Conley K, Lindstedt SL, Claassen H, Vock P, Weibel ER. Endurance training in humans: aerobic capacity and structure of skeletal muscle. J Appl Physiol. 1985;59:320–7.
215. Phielix E, Meex R, Moonen-Kornips E, Hesselink MK, Schrauwen P. Exercise training increases mitochondrial content and ex vivo mitochondrial function similarly in patients with type 2 diabetes and in control individuals. Diabetologia. 2010;53:1714–21.
216. Toledo FG, Menshikova EV, Ritov VB, Azuma K, Radikova Z, DeLany J, Kelley DE. Effects of physical activity and weight loss on skeletal muscle mitochondria and relationship with glucose control in type 2 diabetes. Diabetes. 2007;56:2142–7.
217. Irving BA, Short KR, Nair KS, Stump CS. Nine days of intensive exercise training improves mitochondrial function but not insulin action in adult offspring of mothers with type 2 diabetes. J Clin Endocrinol Metab. 2011;96:E1137–41.
218. Pihlajamaki J, Boes T, Kim EY, Dearie F, Kim BW, Schroeder J, Mun E, Nasser I, Park PJ, Bianco AC, Goldfine AB, Patti ME. Thyroid hormone-related regulation of gene expression in human fatty liver. J Clin Endocrinol Metab. 2009;94:3521–9.
219. Takamura T, Misu H, Matsuzawa-Nagata N, Sakurai M, Ota T, Shimizu A, Kurita S, Takeshita Y, Ando H, Honda M, Kaneko S. Obesity upregulates genes involved in oxidative phosphorylation in livers of diabetic patients. Obesity (SilverSpring). 2008;16:2601–9.
220. Iossa S, Lionetti L, Mollica MP, Crescenzo R, Botta M, Barletta A, Liverini G. Effect of high-fat feeding on metabolic efficiency and mitochondrial oxidative capacity in adult rats. Br J Nutr. 2003;90:953–60.
221. Raffaella C, Francesca B, Italia F, Marina P, Giovanna L, Susanna I. Alterations in hepatic mitochondrial compartment in a model of obesity and insulin resistance. Obesity (Silver Spring). 2008;16:958–64.
222. Mantena SK, Vaughn DP, Andringa KK, Eccleston HB, King AL, Abrams GA, Doeller JE, Kraus DW, Darley-Usmar VM, Bailey SM. High fat diet induces dysregulation of hepatic oxygen gradients and mitochondrial function in vivo. Biochem J. 2009;417:183–93.
223. Wang Y, Lam KS, Lam JB, Lam MC, Leung PT, Zhou M, Xu A. Overexpression of angiopoietin-like protein 4 alters mitochondria activities and modulates methionine metabolic cycle in the liver tissues of db/db diabetic mice. Mol Endocrinol. 2007;21:972–86.
224. Flamment M, Arvier M, Gallois Y, Simard G, Malthiery Y, Ritz P, Ducluzeau PH. Fatty liver and insulin resistance in obese Zucker rats: no role for mitochondrial dysfunction. Biochimie. 2008;90:1407–13.
225. Brady LJ, Brady PS, Romsos DR, Hoppel CL. Elevated hepatic mitochondrial and peroxisomal oxidative capacities in fed and starved adult obese (ob/ob) mice. Biochem J. 1985;231:439–44.
226. Ferreira FM, Palmeira CM, Seica R, Santos MS. Alterations of liver mitochondrial bioenergetics in diabetic Goto-Kakizaki rats. Metabolism. 1999;48:1115–9.
227. Sun K, Kusminski CM, Scherer PE. Adipose tissue remodeling and obesity. J Clin Invest. 2011;121:2094–101.
228. Rasouli N, Kern PA. Adipocytokines and the metabolic complications of obesity. J Clin Endocrinol Metab. 2008;93:S64–73.
229. Cao H, Gerhold K, Mayers JR, Wiest MM, Watkins SM, Hotamisligil GS. Identification of a lipokine, a lipid hormone linking adipose tissue to systemic metabolism. Cell. 2008;134:933–44.

230. Tran TT, Yamamoto Y, Gesta S, Kahn CR. Beneficial effects of subcutaneous fat transplantation on metabolism. Cell Metab. 2008;7:410–20.
231. Carey VJ, Walters EE, Colditz GA, Solomon CG, Willett WC, Rosner BA, Speizer FE, Manson JE. Body fat distribution and risk of non-insulin-dependent diabetes mellitus in women. The nurses' health study. Am J Epidemiol. 1997;145:614–9.
232. Wang Y, Rimm EB, Stampfer MJ, Willett WC, Hu FB. Comparison of abdominal adiposity and overall obesity in predicting risk of type 2 diabetes among men. Am J Clin Nutr. 2005;81:555–63.
233. Zhang C, Rexrode KM, van Dam RM, Li TY, Hu FB. Abdominal obesity and the risk of all-cause, cardiovascular, and cancer mortality: sixteen years of follow-up in US women. Circulation. 2008;117:1658–67.
234. Gil A, Olza J, Gil-Campos M, Gomez-Llorente C, Aguilera CM. Is adipose tissue metabolically different at different sites? Int J Pediatr Obes. 2011;6(Suppl 1):13–20.
235. O'Rourke RW, Metcalf MD, White AE, Madala A, Winters BR, Maizlin II, Jobe BA, Roberts Jr CT, Slifka MK, Marks DL. Depot-specific differences in inflammatory mediators and a role for NK cells and IFN-gamma in inflammation in human adipose tissue. Int J Obes. 2009;33:978–90.
236. Xu H, Barnes GT, Yang Q, Tan G, Yang D, Chou CJ, Sole J, Nichols A, Ross JS, Tartaglia LA, Chen H. Chronic inflammation in fat plays a crucial role in the development of obesity-related insulin resistance. J Clin Invest. 2003;112:1821–30.
237. Hotamisligil GS, Shargill NS, Spiegelman BM. Adipose expression of tumor necrosis factor-α: direct role in obesity-linked insulin resistance. Science. 1993;259:87–91.
238. GS H. Inflammation and metabolic disorders. Nature. 2006;444:860–7.
239. Heilbronn LK, Campbell LV. Adipose tissue macrophages, low grade inflammation and insulin resistance in human obesity. Curr Pharm Des. 2008;14:1225–30.
240. Kanda H, Tateya S, Tamori Y, Kotani K, Hiasa K, Kitazawa R, Kitazawa S, Miyachi H, Maeda S, Egashira K, Kasuga M. MCP-1 contributes to macrophage infiltration into adipose tissue, insulin resistance, and hepatic steatosis in obesity. J Clin Invest. 2006;116:1494–505.
241. Schenk S, Saberi M, Olefsky JM. Insulin sensitivity: modulation by nutrients and inflammation. J Clin Invest. 2008;118:2992–3002.
242. Weisberg SP, McCann D, Desai M, Rosenbaum M, Leibel RL, Ferrante Jr AW. Obesity is associated with macrophage accumulation in adipose tissue. J Clin Invest. 2003;112:1796–808.
243. Kern PA, Ranganathan S, Li C, Wood L, Ranganathan G. Adipose tissue tumor necrosis factor and interleukin-6 expression in human obesity and insulin resistance. Am J Physiol Endocrinol Metab. 2001;280:745–51.
244. Wellen KE, Hotamisligil GS. Inflammation, stress, and diabetes. J Clin Invest. 2005;115:1111–9.
245. Galic S, Oakhill JS, Steinberg GR. Adipose tissue as an endocrine organ. Mol Cell Endocrinol. 2010;316:129–39.
246. Feingold KR, Soued M, Staprans I, Gavin LA, Donahue ME, Huang BJ, Moser AH, Gulli R, Grunfeld C. Effect of tumor necrosis factor (TNF) on lipid metabolism in the diabetic rat. Evidence that inhibition of adipose tissue lipoprotein lipase activity is not required for TNF-induced hyperlipidemia. J Clin Invest. 1989;83:1116–21.
247. Grunfeld C, Feingold KR. The metabolic effects of tumor necrosis factor and other cytokines. Biotherapy. 1991;3:143–58.
248. Zhou R, Yazdi AS, Menu P, Tschopp J. A role for mitochondria in NLRP3 inflammasome activation. Nature. 2011;469:221–5.
249. Lane T, Flam B, Lockey R, Kolliputi N. TXNIP shuttling: missing link between oxidative stress and inflammasome activation. Front Physiol. 2013;4:50.
250. Shimada K, Crother TR, Karlin J, Dagvadorj J, Chiba N, Chen S, Ramanujan VK, Wolf AJ, Vergnes L, Ojcius DM, Rentsendorj A, Vargas M, Guerrero C, Wang Y, Fitzgerald KA, Underhill DM, Town T, Arditi M. Oxidized mitochondrial DNA activates the NLRP3 inflammasome during apoptosis. Immunity. 2012;36:401–14.

251. Vandanmagsar B, Youm YH, Ravussin A, Galgani JE, Stadler K, Mynatt RL, Ravussin E, Stephens JM, Dixit VD. The NLRP3 inflammasome instigates obesity-induced inflammation and insulin resistance. Nat Med. 2011;17:179–88.
252. Yu J, Novgorodov SA, Chudakova D, Zhu H, Bielawska A, Bielawski J, Obeid LM, Kindy MS, Gudz TI. JNK3 signaling pathway activates ceramide synthase leading to mitochondrial dysfunction. J Biol Chem. 2007;282:25940–9.
253. Lu RH, Ji H, Chang ZG, Su SS, Yang GS. Mitochondrial development and the influence of its dysfunction during rat adipocyte differentiation. Mol Biol Rep. 2010;37:2173–82.
254. Gao CL, Zhu C, Zhao YP, Chen XH, Ji CB, Zhang CM, Zhu JG, Xia ZK, Tong ML, Guo XR. Mitochondrial dysfunction is induced by high levels of glucose and free fatty acids in 3T3-L1 adipocytes. Mol Cell Endocrinol. 2010;320:25–33.
255. Maclaren R, Cui W, Simard S, Cianflone K. Influence of obesity and insulin sensitivity on insulin signaling genes in human omental and subcutaneous adipose tissue. J Lipid Res. 2008;49:308–23.
256. Wilson-Fritch L, Burkart A, Bell G, Mendelson K, Leszyk J, Nicoloro S, Czech M, Corvera S. Mitochondrial biogenesis and remodeling during adipogenesis and in response to the insulin sensitizer rosiglitazone. Mol Cell Biol. 2003;23:1085–94.
257. Boudina S, Graham TE. Mitochondrial function/dysfunction in white adipose tissue. Exp Physiol. 2014;99:1168–78.
258. Tormos KV, Anso E, Hamanaka RB, Eisenbart J, Joseph J, Kalyanaraman B, Chandel NS. Mitochondrial complex III ROS regulate adipocyte differentiation. Cell Metab. 2011;14:537–44.
259. Shi X, Burkart A, Nicoloro SM, Czech MP, Straubhaar J, Corvera S. Paradoxical effect of mitochondrial respiratory chain impairment on insulin signaling and glucose transport in adipose cells. J Biol Chem. 2008;283:30658–67.
260. Asterholm IW, Mundy DI, Weng J, Anderson RG, Scherer PE. Altered mitochondrial function and metabolic inflexibility associated with loss of caveolin-1. Cell Metab. 2012;15:171–85.
261. Kusminski CM, Scherer PE. Mitochondrial dysfunction in white adipose tissue. Trends Endocrinol Metab. 2012;23:435–43.
262. Valerio A, Cardile A, Cozzi V, Bracale R, Tedesco L, Pisconti A, Palomba L, Cantoni O, Clementi E, Moncada S, Carruba MO, Nisoli E. TNF-alpha downregulates eNOS expression and mitochondrial biogenesis in fat and muscle of obese rodents. J Clin Invest. 2006;116:2791–8.
263. Laye MJ, Rector RS, Warner SO, Naples SP, Perretta AL, Uptergrove GM, Laughlin MH, Thyfault JP, Booth FW, Ibdah JA. Changes in visceral adipose tissue mitochondrial content with type 2 diabetes and daily voluntary wheel running in OLETF rats. J Physiol. 2009;587:3729–39.
264. Sutherland LN, Capozzi LC, Turchinsky NJ, Bell RC, Wright DC. Time course of high-fat diet-induced reductions in adipose tissue mitochondrial proteins: potential mechanisms and the relationship to glucose intolerance. Am J Physiol Endocrinol Metab. 2008;295:E1076–83.
265. Bogacka I, Xie H, Bray GA, Smith SR. Pioglitazone induces mitochondrial biogenesis in human subcutaneous adipose tissue in vivo. Diabetes. 2005;54:1392–9.
266. Kaaman M, Sparks LM, van Harmelen V, Smith SR, Sjolin E, Dahlman I, Arner P. Strong association between mitochondrial DNA copy number and lipogenesis in human white adipose tissue. Diabetologia. 2007;50:2526–33.
267. Krishnan J, Danzer C, Simka T, Ukropec J, Walter KM, Kumpf S, Mirtschink P, Ukropcova B, Gasperikova D, Pedrazzini T, Krek W. Dietary obesity-associated Hif1alpha activation in adipocytes restricts fatty acid oxidation and energy expenditure via suppression of the Sirt2-NAD+ system. Genes Dev. 2012;26:259–70.
268. Bogacka I, Ukropcova B, McNeil M, Gimble JM, Smith SR. Structural and functional consequences of mitochondrial biogenesis in human adipocytes in vitro. J Clin Endocrinol Metab. 2005;90:6650–6.

269. Yin X, Lanza IR, Swain JM, Sarr MG, Nair KS, Jensen MD. Adipocyte mitochondrial function is reduced in human obesity independent of fat cell size. J Clin Endocrinol Metab. 2014;99:E209–16.
270. Mustelin L, Pietilainen KH, Rissanen A, Sovijarvi AR, Piirila P, Naukkarinen J, Peltonen L, Kaprio J, Yki-Jarvinen H. Acquired obesity and poor physical fitness impair expression of genes of mitochondrial oxidative phosphorylation in monozygotic twins discordant for obesity. Am J Physiol Endocrinol Metab. 2008;295:E148.
271. Deveaud C, Beauvoit B, Salin B, Schaeffer J, Rigoulet M. Regional differences in oxidative capacity of rat white adipose tissue are linked to the mitochondrial content of mature adipocytes. Mol Cell Biochem. 2004;267:157–66.
272. Kraunsoe R, Boushel R, Hansen CN, Schjerling P, Qvortrup K, Stockel M, Mikines KJ, Dela F. Mitochondrial respiration in subcutaneous and visceral adipose tissue from patients with morbid obesity. J Physiol. 2010;588:2023–32.
273. Dahlman I, Forsgren M, Sjogren A, Nordstrom EA, Kaaman M, Naslund E, Attersand A, Arner P. Downregulation of electron transport chain genes in visceral adipose tissue in type 2 diabetes independent of obesity and possibly involving tumor necrosis factor-alpha. Diabetes. 2006;55:1792–9.
274. Ost A, Svensson K, Ruishalme I, Brannmark C, Franck N, Krook H, Sandstrom P, Kjolhede P, Stralfors P. Attenuated mTOR signaling and enhanced autophagy in adipocytes from obese patients with type 2 diabetes. Mol Med. 2010;16:235–46.
275. Wilson-Fritch L, Nicoloro S, Chouinard M, Lazar MA, Chui PC, Leszyk J, Straubhaar J, Czech MP, Corvera S. Mitochondrial remodeling in adipose tissue associated with obesity and treatment with rosiglitazone. J Clin Invest. 2004;114:1281–9.
276. Choo HJ, Kim JH, Kwon OB, Lee CS, Mun JY, Han SS, Yoon YS, Yoon G, Choi KM, Ko YG. Mitochondria are impaired in the adipocytes of type 2 diabetic mice. Diabetologia. 2006;49:784–91.
277. Rong JX, Qiu Y, Hansen MK, Zhu L, Zhang V, Xie M, Okamoto Y, Mattie MD, Higashiyama H, Asano S, Strum JC, Ryan TE. Adipose mitochondrial biogenesis is suppressed in db/db and high-fat diet-fed mice and improved by rosiglitazone. Diabetes. 2007;56:1751–60.
278. Keller MP, Attie AD. Physiological insights gained from gene expression analysis in obesity and diabetes. Annu Rev Nutr. 2010;30:341–64.
279. Cummins TD, Holden CR, Sansbury BE, Gibb AA, Shah J, Zafar N, Tang Y, Hellmann J, Rai SN, Spite M, Bhatnagar A, Hill BG. Metabolic remodeling of white adipose tissue in obesity. Am J Physiol Endocrinol Metab. 2014;307:E262–77.
280. De Pauw A, Tejerina S, Raes M, Keijer J, Arnould T. Mitochondrial (dys)function in adipocyte (de)differentiation and systemic metabolic alterations. Am J Pathol. 2009;175:927–39.
281. Chen XH, Zhao YP, Xue M, Ji CB, Gao CL, Zhu JG, Qin DN, Kou CZ, Qin XH, Tong ML, Guo XR. TNF-alpha induces mitochondrial dysfunction in 3T3-L1 adipocytes. Mol Cell Endocrinol. 2010;328:63–9.
282. Wang T, Si Y, Shirihai OS, Si H, Schultz V, Corkey RF, Hu L, Deeney JT, Guo W, Corkey BE. Respiration in adipocytes is inhibited by reactive oxygen species. Obesity (Silver Spring). 2010;18:1493–502.
283. Lin Y, Berg AH, Iyengar P, Lam TK, Giacca A, Combs TP, Rajala MW, Du X, Rollman B, Li W, Hawkins M, Barzilai N, Rhodes CJ, Fantus IG, Brownlee M, Scherer PE. The hyperglycemia-induced inflammatory response in adipocytes: the role of reactive oxygen species. J Biol Chem. 2005;280:4617–26.
284. Mahadev K, Motoshima H, Wu X, Ruddy JM, Arnold RS, Cheng G, Lambeth JD, Goldstein BJ. The NAD(P)H oxidase homolog Nox4 modulates insulin-stimulated generation of H2O2 and plays an integral role in insulin signal transduction. Mol Cell Biol. 2004;24:1844–54.
285. Goldstein BJ, Mahadev K, Wu X, Zhu L, Motoshima H. Role of insulin-induced reactive oxygen species in the insulin signaling pathway. Antioxid Redox Signal. 2005;7:1021–31.
286. van Marken Lichtenbelt WD, Vanhommerig JW, Smulders NM, Drossaerts JM, Kemerink GJ, Bouvy ND, Schrauwen P, Teule GJ. Cold-activated brown adipose tissue in healthy men. N Engl J Med. 2009;360:1500–8.

287. Ricquier D, Bouillaud F. The uncoupling protein homologues: UCP1, UCP2, UCP3, StUCP and AtUCP. Biochem J. 2000;345:161–79.
288. Giralt M, Villarroya F. White, brown, beige/brite: different adipose cells for different functions? Endocrinology. 2013;154:2992–3000.
289. Sidossis L, Kajimura S. Brown and beige fat in humans: thermogenic adipocytes that control energy and glucose homeostasis. J Clin Invest. 2015;125:478–86.
290. Shabalina IG, Petrovic N, de Jong JM, Kalinovich AV, Cannon B, Nedergaard J. UCP1 in brite/beige adipose tissue mitochondria is functionally thermogenic. Cell Rep. 2013;5:1196–203.
291. Okamatsu-Ogura Y, Fukano K, Tsubota A, Uozumi A, Terao A, Kimura K, Saito M. Thermogenic ability of uncoupling protein 1 in beige adipocytes in mice. PLoS One. 2013;8:e84229.
292. Nedergaard J, Cannon B. The browning of white adipose tissue: some burning issues. Cell Metab. 2014;20:396–407.
293. Orava J, Nuutila P, Lidell ME, Oikonen V, Noponen T, Viljanen T, Scheinin M, Taittonen M, Niemi T, Enerback S, Virtanen KA. Different metabolic responses of human brown adipose tissue to activation by cold and insulin. Cell Metab. 2011;14:272–9.
294. Ouellet V, Labbe SM, Blondin DP, Phoenix S, Guerin B, Haman F, Turcotte EE, Richard D, Carpentier AC. Brown adipose tissue oxidative metabolism contributes to energy expenditure during acute cold exposure in humans. J Clin Invest. 2012;122:545–52.
295. Cannon B, Nedergaard J. Brown adipose tissue: function and physiological significance. Physiol Rev. 2004;84:277–359.
296. Lopez-Soriano FJ, Fernandez-Lopez JA, Mampel T, Villarroya F, Iglesias R, Alemany M. Amino acid and glucose uptake by rat brown adipose tissue. Effect of cold-exposure and acclimation. Biochem J. 1988;252:843–9.
297. Cannon B, Nedergaard J. The physiological role of pyruvate carboxylation in hamster brown adipose tissue. Eur J Biochem. 1979;94:419–26.
298. Buckley MG, Rath EA. Regulation of fatty acid synthesis and malonyl-CoA content in mouse brown adipose tissue in response to cold-exposure, starvation or re-feeding. Biochem J. 1987;243:437–42.
299. Bartelt A, Bruns OT, Reimer R, Hohenberg H, Ittrich H, Peldschus K, Kaul MG, Tromsdorf UI, Weller H, Waurisch C, Eychmuller A, Gordts PL, Rinninger F, Bruegelmann K, Freund B, Nielsen P, Merkel M, Heeren J. Brown adipose tissue activity controls triglyceride clearance. Nat Med. 2011;17:200–5.
300. Inokuma K, Ogura-Okamatsu Y, Toda C, Kimura K, Yamashita H, Saito M. Uncoupling protein 1 is necessary for norepinephrine-induced glucose utilization in brown adipose tissue. Diabetes. 2005;54:1385–91.
301. Jia JJ, Tian YB, Cao ZH, Tao LL, Zhang X, Gao SZ, Ge CR, Lin QY, Jois M. The polymorphisms of UCP1 genes associated with fat metabolism, obesity and diabetes. Mol Biol Rep. 2010;37:1513–22.
302. Del Mar Gonzalez-Barroso M, Ricquier D, Cassard-Doulcier AM. The human uncoupling protein-1 gene (UCP1): present status and perspectives in obesity research. Obes Rev Off J Int Assoc Study Obes. 2000;1:61–72.
303. Keipert S, Kutschke M, Lamp D, Brachthauser L, Neff F, Meyer CW, Oelkrug R, Kharitonenkov A, Jastroch M. Genetic disruption of uncoupling protein 1 in mice renders brown adipose tissue a significant source of FGF21 secretion. Mol Metab. 2015;4:537–42.
304. Chartoumpekis DV, Habeos IG, Ziros PG, Psyrogiannis AI, Kyriazopoulou VE, Papavassiliou AG. Brown adipose tissue responds to cold and adrenergic stimulation by induction of FGF21. Mol Med. 2011;17:736–40.
305. Hondares E, Iglesias R, Giralt A, Gonzalez FJ, Giralt M, Mampel T, Villarroya F. Thermogenic activation induces FGF21 expression and release in brown adipose tissue. J Biol Chem. 2011;286:12983–90.
306. Kharitonenkov A, Shiyanova TL, Koester A, Ford AM, Micanovic R, Galbreath EJ, Sandusky GE, Hammond LJ, Moyers JS, Owens RA, Gromada J, Brozinick JT, Hawkins ED,

Wroblewski VJ, Li DS, Mehrbod F, Jaskunas SR, Shanafelt AB. FGF-21 as a novel metabolic regulator. J Clin Invest. 2005;115:1627–35.
307. Rachid B, van de Sande-Lee S, Rodovalho S, Folli F, Beltramini GC, Morari J, Amorim BJ, Pedro T, Ramalho AF, Bombassaro B, Tincani AJ, Chaim E, Pareja JC, Geloneze B, Ramos CD, Cendes F, Saad MJ, Velloso LA. Distinct regulation of hypothalamic and brown/beige adipose tissue activities in human obesity. Int J Obes (Lond). 2015;39:1515–22.
308. Williams KW, Elmquist JK. From neuroanatomy to behavior: central integration of peripheral signals regulating feeding behavior. Nat Neurosci. 2012;15:1350–5.
309. Schneeberger M, Dietrich MO, Sebastian D, Imbernon M, Castano C, Garcia A, Esteban Y, Gonzalez-Franquesa A, Rodriguez IC, Bortolozzi A, Garcia-Roves PM, Gomis R, Nogueiras R, Horvath TL, Zorzano A, Claret M. Mitofusin 2 in POMC neurons connects ER stress with leptin resistance and energy imbalance. Cell. 2013;155:172–87.
310. Plum L, Belgardt BF, Bruning JC. Central insulin action in energy and glucose homeostasis. J Clin Invest. 2006;116:1761–6.
311. Bruning JC, Gautam D, Burks DJ, Gillette J, Schubert M, Orban PC, Klein R, Krone W, Muller-Wieland D, Kahn CR. Role of brain insulin receptor in control of body weight and reproduction. Science. 2000;289:2122–5.
312. Schwartz MW, Woods SC, Porte Jr D, Baskin RJ, Baskin DG. Cental nervous system control of food intake. Nature. 2000;404:661–71.
313. Plum L, Schubert M, Bruning JC. The role of insulin receptor signaling in the brain. Trends Endocrinol Metab. 2005;16:59–65.
314. Obici S, Zhang BB, Karkanias G, Rossetti L. Hypothalamic insulin signaling is required for inhibition of glucose production. Nat Med. 2002;8:1376–82.
315. Gelling RW, Morton GJ, Morrison CD, Niswender KD, Myers Jr MG, Rhodes CJ, Schwartz MW. Insulin action in the brain contributes to glucose lowering during insulin treatment of diabetes. Cell Metab. 2006;3:67–73.
316. Perrin C, Knauf C, Burcelin R. Intracerebroventricular infusion of glucose, insulin, and the adenosine monophosphate-activated kinase activator, 5-aminoimidazole-4-carboxamide-1-beta-D-ribofuranoside, controls muscle glycogen synthesis. Endocrinology. 2004;145: 4025–33.
317. Koch L, Wunderlich FT, Seibler J, Konner AC, Hampel B, Irlenbusch S, Brabant G, Kahn CR, Schwenk F, Bruning JC. Central insulin action regulates peripheral glucose and fat metabolism in mice. J Clin Invest. 2008;118:2132–47.
318. Morton GJ, Gelling RW, Niswender KD, Morrison CD, Rhodes CJ, Schwartz MW. Leptin regulates insulin sensitivity via phosphatidylinositol-3-OH kinase signaling in mediobasal hypothalamic neurons. Cell Metab. 2005;2:411–20.
319. De Souza CT, Araujo EP, Bordin S, Ashimine R, Zollner RL, Boschero AC, Saad MJ, Velloso LA. Consumption of a fat-rich diet activates a proinflammatory response and induces insulin resistance in the hypothalamus. Endocrinology. 2005;146:4192–9.
320. Munzberg H, Flier JS, Bjorbaek C. Region-specific leptin resistance within the hypothalamus of diet-induced obese mice. Endocrinology. 2004;145:4880–9.
321. Obici S, Feng Z, Morgan K, Stein D, Karkanias G, Rossetti L. Central administration of oleic acid inhibits glucose production and food intake. Diabetes. 2002;51:271–5.
322. Morgan K, Obici S, Rossetti L. Hypothalamic responses to long-chain fatty acids are nutritionally regulated. J Biol Chem. 2004;279:31139–48.
323. Contreras C, Gonzalez-Garcia I, Martinez-Sanchez N, Seoane-Collazo P, Jacas J, Morgan DA, Serra D, Gallego R, Gonzalez F, Casals N, Nogueiras R, Rahmouni K, Dieguez C, Lopez M. Central ceramide-induced hypothalamic lipotoxicity and ER stress regulate energy balance. Cell Rep. 2014;9:366–77.
324. Johnson AM, Olefsky JM. The origins and drivers of insulin resistance. Cell. 2013;152: 673–84.
325. Ozcan L, Ergin AS, Lu A, Chung J, Sarkar S, Nie D, Myers Jr MG, Ozcan U. Endoplasmic reticulum stress plays a central role in development of leptin resistance. Cell Metab. 2009;9:35–51.

326. Elmquist JK, Elias CF, Saper CB. From lesions to leptin: hypothalamic control of food intake and body weight. Neuron. 1999;22:221–32.
327. Quaresma PG, Reencober N, Zanotto TM, Santos AC, Weissmann L, de Matos AH, Lopes-Cendes I, Folli F, Saad MJ, Prada PO. Pioglitazone treatment increases food intake and decreases energy expenditure partially via hypothalamic adiponectin/adipoR1/AMPK pathway. Int J Obes. 2016;40:138–46.
328. Benani A, Barquissau V, Carneiro L, Salin B, Colombani AL, Leloup C, Casteilla L, Rigoulet M, Penicaud L. Method for functional study of mitochondria in rat hypothalamus. J Neurosci Methods. 2009;178:301–7.
329. Colombani AL, Carneiro L, Benani A, Galinier A, Jaillard T, Duparc T, Offer G, Lorsignol A, Magnan C, Casteilla L, Penicaud L, Leloup C. Enhanced hypothalamic glucose sensing in obesity: alteration of redox signaling. Diabetes. 2009;58:2189–97.
330. Dietrich MO, Liu ZW, Horvath TL. Mitochondrial dynamics controlled by mitofusins regulate Agrp neuronal activity and diet-induced obesity. Cell. 2013;155:188–99.
331. Diano S, Liu ZW, Jeong JK, Dietrich MO, Ruan HB, Kim E, Suyama S, Kelly K, Gyengesi E, Arbiser JL, Belsham DD, Sarruf DA, Schwartz MW, Bennett AM, Shanabrough M, Mobbs CV, Yang X, Gao XB, Horvath TL. Peroxisome proliferation-associated control of reactive oxygen species sets melanocortin tone and feeding in diet-induced obesity. Nat Med. 2011;17:1121–7.
332. Horvath TL, Andrews ZB, Diano S. Fuel utilization by hypothalamic neurons: roles for ROS. Trends Endocrinol Metab. 2009;20:78–87.
333. Andrews ZB, Liu ZW, Walllingford N, Erion DM, Borok E, Friedman JM, Tschop MH, Shanabrough M, Cline G, Shulman GI, Coppola A, Gao XB, Horvath TL, Diano S. UCP2 mediates ghrelin's action on NPY/AgRP neurons by lowering free radicals. Nature. 2008;454:846–51.
334. Campanucci V, Krishnaswamy A, Cooper E. Diabetes depresses synaptic transmission in sympathetic ganglia by inactivating nAChRs through a conserved intracellular cysteine residue. Neuron. 2010;66:827–34.
335. Woo M, Isganaitis E, Cerletti M, Fitzpatrick C, Wagers A, Jimenez-Chillaron J, Patti ME. Early life nutrition modulates muscle stem cell number: implications for muscle mass and repair. Stem Cells Dev. 2011;20:1763.
336. Iovino S, Burkart AM, Kriauciunas K, Warren L, Hughes KJ, Molla M, Lee YK, Patti ME, Kahn CR. Genetic insulin resistance is a potent regulator of gene expression and proliferation in human iPS cells. Diabetes. 2014;63:4130–42.
337. Burkart AM, Tan K, Warren L, Iovino S, Hughes KJ, Kahn CR, Patti ME. Insulin resistance in human iPS cells reduces mitochondrial size and function. Sci Rep. 2016;6:22788.
338. Turner N, Robker RL. Developmental programming of obesity and insulin resistance: does mitochondrial dysfunction in oocytes play a role? Mol Hum Reprod. 2015;21:23–30.
339. Rando OJ, Simmons RA. I'm eating for two: parental dietary effects on offspring metabolism. Cell. 2015;161:93–105.
340. Richter EA, Garetto LP, Goodman MN, Ruderman NB. Muscle glucose metabolism following exercise in the rat: increased sensitivity to insulin. J Clin Invest. 1982;69:785–93.
341. Ren JM, Semenkovich CF, Gulve EA, Gao J, Holloszy JO. Exercise induces rapid increases in GLUT4 expression, glucose transport capacity, and insulin-stimulated glycogen storage in muscle. J Biol Chem. 1994;269:14396–401.
342. Hansen PA, Nolte LA, Chen MM, Holloszy JO. Increased GLUT-4 translocation mediates enhanced insulin sensitivity of muscle glucose transport after exercise. J Appl Physiol (1985). 1998;85:1218–22.
343. Wallberg-Henriksson H, Holloszy JO. Activation of glucose transport in diabetic muscle: responses to contraction and insulin. Am J Phys. 1985;249:C233–7.
344. Ojuka EO, Nolte LA, Holloszy JO. Increased expression of GLUT-4 and hexokinase in rat epitrochlearis muscles exposed to AICAR in vitro. J Appl Physiol (1985). 2000;88:1072–5.
345. Hevener AL, Reichart D, Olefsky J. Exercise and thiazolidinedione therapy normalize insulin action in the obese Zucker fatty rat. Diabetes. 2000;49:2154–9.

346. Starkie R, Ostrowski SR, Jauffred S, Febbraio M, Pedersen BK. Exercise and IL-6 infusion inhibit endotoxin-induced TNF-alpha production in humans. FASEB J. 2003;17:884–6.
347. Pedersen BK, Febbraio MA. Muscles, exercise and obesity: skeletal muscle as a secretory organ. Nat Rev Endocrinol. 2012;8:457–65.
348. Pedersen BK, Akerstrom TC, Nielsen AR, Fischer CP. Role of myokines in exercise and metabolism. J Appl Physiol (1985). 2007;103:1093–8.
349. Pedersen BK, Febbraio MA. Muscle as an endocrine organ: focus on muscle-derived interleukin-6. Physiol Rev. 2008;88:1379–406.
350. Petersen AM, Pedersen BK. The anti-inflammatory effect of exercise. J Appl Physiol (1985). 2005;98:1154–62.
351. Holloszy JO. Biochemical adaptations in muscle. Effects of exercise on mitochondrial oxygen uptake and respiratory enzyme activity in skeletal muscle. J Biol Chem. 1967;242:2278–82.
352. Larsen S, Nielsen J, Hansen CN, Nielsen LB, Wibrand F, Stride N, Schroder HD, Boushel R, Helge JW, Dela F, Hey-Mogensen M. Biomarkers of mitochondrial content in skeletal muscle of healthy young human subjects. J Physiol. 2012;590:3349–60.
353. Dela F, Larsen JJ, Mikines KJ, Ploug T, Petersen LN, Galbo H. Insulin-stimulated muscle glucose clearance in patients with NIDDM. Effects of one-legged physical training. Diabetes. 1995;44:1010–20.
354. Hawley JA, Gibala MJ. What's new since Hippocrates? Preventing type 2 diabetes by physical exercise and diet. Diabetologia. 2012;55:535–9.
355. Asterholm IW, Scherer PE. Enhanced metabolic flexibility associated with elevated adiponectin levels. Am J Pathol. 2010;176:1364–76.
356. Nabben M, Hoeks J, Briede JJ, Glatz JF, Moonen-Kornips E, Hesselink MK, Schrauwen P. The effect of UCP3 overexpression on mitochondrial ROS production in skeletal muscle of young versus aged mice. FEBS Lett. 2008;582:4147–52.
357. Holten MK, Zacho M, Gaster M, Juel C, Wojtaszewski JF, Dela F. Strength training increases insulin-mediated glucose uptake, GLUT4 content, and insulin signaling in skeletal muscle in patients with type 2 diabetes. Diabetes. 2004;53:294–305.
358. Senechal M, Rempel M, Duhamel TA, MacIntosh AC, Hay J, Wicklow B, Wittmeier K, Shen GX, McGavock JM. Fitness is a determinant of the metabolic response to endurance training in adolescents at risk of type 2 diabetes mellitus. Obesity (Silver Spring). 2015;23:823–32.
359. Rao X, Zhong J, Xu X, Jordan B, Maurya S, Braunstein Z, Wang TY, Huang W, Aggarwal S, Periasamy M, Rajagopalan S, Mehta K, Sun Q. Exercise protects against diet-induced insulin resistance through downregulation of protein kinase Cbeta in mice. PLoS One. 2013;8:e81364.
360. Tsuzuki T, Shinozaki S, Nakamoto H, Kaneki M, Goto S, Shimokado K, Kobayashi H, Naito H. Voluntary exercise can ameliorate insulin resistance by reducing iNOS-mediated S-nitrosylation of Akt in the liver in obese rats. PLoS One. 2015;10:e0132029.
361. Goncalves IO, Maciel E, Passos E, Torrella JR, Rizo D, Viscor G, Rocha-Rodrigues S, Santos-Alves E, Domingues MR, Oliveira PJ, Ascensao A, Magalhaes J. Exercise alters liver mitochondria phospholipidomic profile and mitochondrial activity in non-alcoholic steatohepatitis. Int J Biochem Cell Biol. 2014;54:163–73.
362. Ronn T, Volkov P, Tornberg A, Elgzyri T, Hansson O, Eriksson KF, Groop L, Ling C. Extensive changes in the transcriptional profile of human adipose tissue including genes involved in oxidative phosphorylation after a 6-month exercise intervention. Acta Physiol (Oxford). 2014;211:188–200.
363. King DE, Carek P, Mainous 3rd AG, Pearson WS. Inflammatory markers and exercise: differences related to exercise type. Med Sci Sports Exerc. 2003;35:575–81.
364. Abramson JL, Vaccarino V. Relationship between physical activity and inflammation among apparently healthy middle-aged and older US adults. Arch Intern Med. 2002;162:1286–92.
365. Geffken DF, Cushman M, Burke GL, Polak JF, Sakkinen PA, Tracy RP. Association between physical activity and markers of inflammation in a healthy elderly population. Am J Epidemiol. 2001;153:242–50.

366. Mattusch F, Dufaux B, Heine O, Mertens I, Rost R. Reduction of the plasma concentration of C-reactive protein following nine months of endurance training. Int J Sports Med. 2000;21:21–4.
367. Stallknecht B, Vinten J, Ploug T, Galbo H. Increased activities of mitochondrial enzymes in white adipose tissue in trained rats. Am J Phys. 1991;261:E410–4.
368. Trevellin E, Scorzeto M, Olivieri M, Granzotto M, Valerio A, Tedesco L, Fabris R, Serra R, Quarta M, Reggiani C, Nisoli E, Vettor R. Exercise training induces mitochondrial biogenesis and glucose uptake in subcutaneous adipose tissue through eNOS-dependent mechanisms. Diabetes. 2014;63:2800–11.
369. Stanford KI, Middelbeek RJ, Goodyear LJ. Exercise effects on white adipose tissue: beiging and metabolic adaptations. Diabetes. 2015;64:2361–8.
370. Mahoney DJ, Parise G, Melov S, Safdar A, Tarnopolsky MA. Analysis of global mRNA expression in human skeletal muscle during recovery from endurance exercise. FASEB J. 2005;19:1498–500.
371. Gollisch KS, Brandauer J, Jessen N, Toyoda T, Nayer A, Hirshman MF, Goodyear LJ. Effects of exercise training on subcutaneous and visceral adipose tissue in normal- and high-fat diet-fed rats. Am J Physiol Endocrinol Metab. 2009;297:E495–504.
372. Craig BW, Hammons GT, Garthwaite SM, Jarett J, Holloszy JO. Adaptation of fat cells to exercise: response of glucose uptake and oxidation to insulin. J Appl Physiol. 1981;51:1500–6.
373. Sutherland LN, Bomhof MR, Capozzi LC, Basaraba SA, Wright DC. Exercise and adrenaline increase PGC-1{alpha} mRNA expression in rat adipose tissue. J Physiol. 2009;587:1607–17.
374. Stanford KI, Middelbeek RJ, Townsend KL, Lee MY, Takahashi H, So K, Hitchcox KM, Markan KR, Hellbach K, Hirshman MF, Tseng YH, Goodyear LJ. A novel role for subcutaneous adipose tissue in exercise-induced improvements in glucose homeostasis. Diabetes. 2015;64:2002–14.
375. Vernochet C, Mourier A, Bezy O, Macotela Y, Boucher J, Rardin MJ, An D, Lee KY, Ilkayeva OR, Zingaretti CM, Emanuelli B, Smyth G, Cinti S, Newgard CB, Gibson BW, Larsson NG, Kahn CR. Adipose-specific deletion of TFAM increases mitochondrial oxidation and protects mice against obesity and insulin resistance. Cell Metab. 2012;16:765–76.
376. Kelley DE, Mandarino LJ. Fuel selection in human skeletal muscle in insulin resistance: a reexamination. Diabetes. 2000;49:677–83.
377. Petersen KF, Dufour S, Befroy D, Lehrke M, Hendler RE, Shulman GI. Reversal of nonalcoholic hepatic steatosis, hepatic insulin resistance, and hyperglycemia by moderate weight reduction in patients with type 2 diabetes. Diabetes. 2005;54:603–8.
378. Suga T, Kinugawa S, Takada S, Kadoguchi T, Fukushima A, Homma T, Masaki Y, Furihata T, Takahashi M, Sobirin MA, Ono T, Hirabayashi K, Yokota T, Tanaka S, Okita K, Tsutsui H. Combination of exercise training and diet restriction normalizes limited exercise capacity and impaired skeletal muscle function in diet-induced diabetic mice. Endocrinology. 2014;155:68–80.
379. Snel M, Gastaldelli A, Ouwens DM, Hesselink MK, Schaart G, Buzzigoli E, Frolich M, Romijn JA, Pijl H, Meinders AE, Jazet IM. Effects of adding exercise to a 16-week very low-calorie diet in obese, insulin-dependent type 2 diabetes mellitus patients. J Clin Endocrinol Metab. 2012;97:2512–20.
380. Neschen S, Morino K, Rossbacher JC, Pongratz RL, Cline GW, Sono S, Gillum M, Shulman GI. Fish oil regulates adiponectin secretion by a peroxisome proliferator-activated receptor-gamma-dependent mechanism in mice. Diabetes. 2006;55:924–8.
381. Nakatani T, Kim HJ, Kaburagi Y, Yasuda K, Ezaki O. A low fish oil inhibits SREBP-1 proteolytic cascade, while a high-fish-oil feeding decreases SREBP-1 mRNA in mice liver: relationship to anti-obesity. J Lipid Res. 2003;44:369–79.
382. Lionetti L, Mollica MP, Donizzetti I, Gifuni G, Sica R, Pignalosa A, Cavaliere G, Gaita M, De Filippo C, Zorzano A, Putti R. High-lard and high-fish-oil diets differ in their effects on function and dynamic behaviour of rat hepatic mitochondria. PLoS One. 2014;9:e92753.

383. Reznick RM, Shulman GI. The role of AMP-activated protein kinase in mitochondrial biogenesis. J Physiol. 2006;574:33–9.
384. Rockl KS, Witczak CA, Goodyear LJ. Signaling mechanisms in skeletal muscle: acute responses and chronic adaptations to exercise. IUBMB Life. 2008;60:145–53.
385. Perry RJ, Zhang D, Zhang XM, Boyer JL, Shulman GI. Controlled-release mitochondrial protonophore reverses diabetes and steatohepatitis in rats. Science. 2015;347:1253–6.
386. Lou PH, Hansen BS, Olsen PH, Tullin S, Murphy MP, Brand MD. Mitochondrial uncouplers with an extraordinary dynamic range. Biochem J. 2007;407:129–40.
387. Koch RA, Lee RC, Tainter ML. Dinitrophenol on liver function. Cal West Med. 1935;43:337–9.
388. Antonenko YN, Avetisyan AV, Cherepanov DA, Knorre DA, Korshunova GA, Markova OV, Ojovan SM, Perevoshchikova IV, Pustovidko AV, Rokitskaya TI, Severina II, Simonyan RA, Smirnova EA, Sobko AA, Sumbatyan NV, Severin FF, Skulachev VP. Derivatives of rhodamine 19 as mild mitochondria-targeted cationic uncouplers. J Biol Chem. 2011;286:17831–40.
389. Canto C, Auwerx J. PGC-1alpha, SIRT1 and AMPK, an energy sensing network that controls energy expenditure. Curr Opin Lipidol. 2009;20:98–105.
390. Baur JA, Pearson KJ, Price NL, Jamieson HA, Lerin C, Kalra A, Prabhu VV, Allard JS, Lopez-Lluch G, Lewis K, Pistell PJ, Poosala S, Becker KG, Boss O, Gwinn D, Wang M, Ramaswamy S, Fishbein KW, Spencer RG, Lakatta EG, Le Couteur D, Shaw RJ, Navas P, Puigserver P, Ingram DK, de Cabo R, Sinclair DA. Resveratrol improves health and survival of mice on a high-calorie diet. Nature. 2006;444:337–42.
391. Towler MC, Hardie DG. AMP-activated protein kinase in metabolic control and insulin signaling. Circ Res. 2007;100:328–41.
392. Biala A, Tauriainen E, Siltanen A, Shi J, Merasto S, Louhelainen M, Martonen E, Finckenberg P, Muller DN, Mervaala E. Resveratrol induces mitochondrial biogenesis and ameliorates Ang II-induced cardiac remodeling in transgenic rats harboring human renin and angiotensinogen genes. Blood Press. 2010;19:196–205.
393. Lin SJ, Defossez PA, Guarente L. Requirement of NAD and SIR2 for life-span extension by calorie restriction in *Saccharomyces cerevisiae*. Science. 2000;289:2126–8.
394. Cohen HY, Miller C, Bitterman KJ, Wall NR, Hekking B, Kessler B, Howitz KT, Gorospe M, de Cabo R, Sinclair DA. Calorie restriction promotes mammalian cell survival by inducing the SIRT1 deacetylase. Science. 2004;305:390–2.
395. Lagouge M, Argmann C, Gerhart-Hines Z, Meziane H, Lerin C, Daussin F, Messadeq N, Milne J, Lambert P, Elliott P, Geny B, Laakso M, Puigserver P, Auwerx J. Resveratrol improves mitochondrial function and protects against metabolic disease by activating SIRT1 and PGC-1alpha. Cell. 2006;127:1109–22.
396. Pacholec M, Bleasdale JE, Chrunyk B, Cunningham D, Flynn D, Garofalo RS, Griffith D, Griffor M, Loulakis P, Pabst B, Qiu X, Stockman B, Thanabal V, Varghese A, Ward J, Withka J, Ahn K. SRT1720, SRT2183, SRT1460, and resveratrol are not direct activators of SIRT1. J Biol Chem. 2010;285:8340–51.
397. Csiszar A, Labinskyy N, Pinto JT, Ballabh P, Zhang H, Losonczy G, Pearson K, de Cabo R, Pacher P, Zhang C, Ungvari Z. Resveratrol induces mitochondrial biogenesis in endothelial cells. Am J Physiol Heart Circ Physiol. 2009;297:H13–20.
398. Timmers S, Konings E, Bilet L, Houtkooper RH, van de Weijer T, Goossens GH, Hoeks J, van der Krieken S, Ryu D, Kersten S, Moonen-Kornips E, Hesselink MK, Kunz I, Schrauwen-Hinderling VB, Blaak EE, Auwerx J, Schrauwen P. Calorie restriction-like effects of 30 days of resveratrol supplementation on energy metabolism and metabolic profile in obese humans. Cell Metab. 2011;14:612–22.
399. Jackson JR, Ryan MJ, Alway SE. Long-term supplementation with resveratrol alleviates oxidative stress but does not attenuate sarcopenia in aged mice. J Gerontol A Biol Sci Med Sci. 2011;66:751–64.
400. Bhatt JK, Thomas S, Nanjan MJ. Resveratrol supplementation improves glycemic control in type 2 diabetes mellitus. Nutr Res (New York, NY). 2012;32:537–41.

401. Crandall JP, Oram V, Trandafirescu G, Reid M, Kishore P, Hawkins M, Cohen HW, Barzilai N. Pilot study of resveratrol in older adults with impaired glucose tolerance. J Gerontol A Biol Sci Med Sci. 2012;67:1307–12.
402. Yoshino J, Conte C, Fontana L, Mittendorfer B, Imai S, Schechtman KB, Gu C, Kunz I, Rossi Fanelli F, Patterson BW, Klein S. Resveratrol supplementation does not improve metabolic function in nonobese women with normal glucose tolerance. Cell Metab. 2012;16:658–64.
403. Milne JC, Lambert PD, Schenk S, Carney DP, Smith JJ, Gagne DJ, Jin L, Boss O, Perni RB, Vu CB, Bemis JE, Xie R, Disch JS, Ng PY, Nunes JJ, Lynch AV, Yang H, Galonek H, Israelian K, Choy W, Iffland A, Lavu S, Medvedik O, Sinclair DA, Olefsky JM, Jirousek MR, Elliott PJ, Westphal CH. Small molecule activators of SIRT1 as therapeutics for the treatment of type 2 diabetes. Nature. 2007;450:712–6.
404. Maguire JJ, Wilson DS, Packer L. Mitochondrial electron transport-linked tocopheroxyl radical reduction. J Biol Chem. 1989;264:21462–5.
405. Yi X, Maeda N. alpha-Lipoic acid prevents the increase in atherosclerosis induced by diabetes in apolipoprotein E-deficient mice fed high-fat/low-cholesterol diet. Diabetes. 2006;55:2238–44.
406. Kelso GF, Porteous CM, Coulter CV, Hughes G, Porteous WK, Ledgerwood EC, Smith RA, Murphy MP. Selective targeting of a redox-active ubiquinone to mitochondria within cells: antioxidant and antiapoptotic properties. J Biol Chem. 2001;276:4588–96.
407. Baksi A, Kraydashenko O, Zalevkaya A, Stets R, Elliott P, Haddad J, Hoffmann E, Vlasuk GP, Jacobson EW. A phase II, randomized, placebo-controlled, double-blind, multi-dose study of SRT2104, a SIRT1 activator, in subjects with type 2 diabetes. Br J Clin Pharmacol. 2014;78:69–77.
408. Libri V, Brown AP, Gambarota G, Haddad J, Shields GS, Dawes H, Pinato DJ, Hoffman E, Elliot PJ, Vlasuk GP, Jacobson E, Wilkins MR, Matthews PM. A pilot randomized, placebo controlled, double blind phase I trial of the novel SIRT1 activator SRT2104 in elderly volunteers. PLoS One. 2012;7:e51395.
409. Canto C, Houtkooper RH, Pirinen E, Youn DY, Oosterveer MH, Cen Y, Fernandez-Marcos PJ, Yamamoto H, Andreux PA, Cettour-Rose P, Gademann K, Rinsch C, Schoonjans K, Sauve AA, Auwerx J. The NAD(+) precursor nicotinamide riboside enhances oxidative metabolism and protects against high-fat diet-induced obesity. Cell Metab. 2012;15:838–47.
410. Nisoli E, Clementi E, Moncada S, Carruba MO. Mitochondrial biogenesis as a cellular signaling framework. Biochem Pharmacol. 2004;67:1–15.
411. Nisoli E, Clementi E, Paolucci C, Cozzi V, Tonello C, Sciorati C, Bracale R, Valerio A, Francolini M, Moncada S, Carruba MO. Mitochondrial biogenesis in mammals: the role of endogenous nitric oxide. Science. 2003;299:896–9.
412. Gao Y. The multiple actions of NO. Pflugers Arch. 2010;459:829–39.
413. Elrod JW, Calvert JW, Morrison J, Doeller JE, Kraus DW, Tao L, Jiao X, Scalia R, Kiss L, Szabo C, Kimura H, Chow CW, Lefer DJ. Hydrogen sulfide attenuates myocardial ischemia-reperfusion injury by preservation of mitochondrial function. Proc Natl Acad Sci U S A. 2007;104:15560–5.
414. Calvert JW, Elston M, Nicholson CK, Gundewar S, Jha S, Elrod JW, Ramachandran A, Lefer DJ. Genetic and pharmacologic hydrogen sulfide therapy attenuates ischemia-induced heart failure in mice. Circulation. 2010;122:11–9.
415. Pecorella SR, Potter JV, Cherry AD, Peacher DF, Welty-Wolf KE, Moon RE, Piantadosi CA, Suliman HB. The HO-1/CO system regulates mitochondrial-capillary density relationships in human skeletal muscle. Am J Phys Lung Cell Mol Physiol. 2015;309:L857–71.
416. Mattingly KA, Ivanova MM, Riggs KA, Wickramasinghe NS, Barch MJ, Klinge CM. Estradiol stimulates transcription of nuclear respiratory factor-1 and increases mitochondrial biogenesis. Mol Endocrinol. 2008;22:609–22.
417. Carraway MS, Suliman HB, Jones WS, Chen CW, Babiker A, Piantadosi CA. Erythropoietin activates mitochondrial biogenesis and couples red cell mass to mitochondrial mass in the heart. Circ Res. 2010;106:1722–30.

418. Iwabu M, Yamauchi T, Okada-Iwabu M, Sato K, Nakagawa T, Funata M, Yamaguchi M, Namiki S, Nakayama R, Tabata M, Ogata H, Kubota N, Takamoto I, Hayashi YK, Yamauchi N, Waki H, Fukayama M, Nishino I, Tokuyama K, Ueki K, Oike Y, Ishii S, Hirose K, Shimizu T, Touhara K, Kadowaki T. Adiponectin and AdipoR1 regulate PGC-1alpha and mitochondria by Ca(2+) and AMPK/SIRT1. Nature. 2010;464:1313–9.
419. Civitarese AE, Ukropcova B, Carling S, Hulver M, DeFronzo RA, Mandarino L, Ravussin E, Smith SR. Role of adiponectin in human skeletal muscle bioenergetics. Cell Metab. 2006;4:75–87.
420. Qiao L, Kinney B, Yoo HS, Lee B, Schaack J, Shao J. Adiponectin increases skeletal muscle mitochondrial biogenesis by suppressing mitogen-activated protein kinase phosphatase-1. Diabetes. 2012;61:1463–70.
421. Tenenbaum A, Motro M, Fisman EZ. Dual and pan-peroxisome proliferator-activated receptors (PPAR) co-agonism: the bezafibrate lessons. Cardiovasc Diabetol. 2005;4:14.
422. Hondares E, Rosell M, Diaz-Delfin J, Olmos Y, Monsalve M, Iglesias R, Villarroya F, Giralt M. Peroxisome proliferator-activated receptor alpha (PPARalpha) induces PPARgamma coactivator 1alpha (PGC-1alpha) gene expression and contributes to thermogenic activation of brown fat: involvement of PRDM16. J Biol Chem. 2011;286:43112–22.
423. Boden G, Homko C, Mozzoli M, Showe LC, Nichols C, Cheung P. Thiazolidinediones upregulate fatty acid uptake and oxidation in adipose tissue of diabetic patients. Diabetes. 2005;54:880–5.
424. Lee YS, Kim WS, Kim KH, Yoon MJ, Cho HJ, Shen Y, Ye JM, Lee CH, Oh WK, Kim CT, Hohnen-Behrens C, Gosby A, Kraegen EW, James DE, Kim JB. Berberine, a natural plant product, activates AMP-activated protein kinase with beneficial metabolic effects in diabetic and insulin-resistant states. Diabetes. 2006;55:2256–64.
425. Zhang Z, Zhang H, Li B, Meng X, Wang J, Zhang Y, Yao S, Ma Q, Jin L, Yang J, Wang W, Ning G. Berberine activates thermogenesis in white and brown adipose tissue. Nat Commun. 2014;5:5493.
426. Gomes AP, Duarte FV, Nunes P, Hubbard BP, Teodoro JS, Varela AT, Jones JG, Sinclair DA, Palmeira CM, Rolo AP. Berberine protects against high fat diet-induced dysfunction in muscle mitochondria by inducing SIRT1-dependent mitochondrial biogenesis. Biochim Biophys Acta. 2012;1822:185–95.
427. Xia X, Yan J, Shen Y, Tang K, Yin J, Zhang Y, Yang D, Liang H, Ye J, Weng J. Berberine improves glucose metabolism in diabetic rats by inhibition of hepatic gluconeogenesis. PLoS One. 2011;6:e16556.
428. Kim WS, Lee YS, Cha SH, Jeong HW, Choe SS, Lee MR, Oh GT, Park HS, Lee KU, Lane MD, Kim JB. Berberine improves lipid dysregulation in obesity by controlling central and peripheral AMPK activity. Am J Physiol Endocrinol Metab. 2009;296:E812–9.
429. Narkar VA, Downes M, Yu RT, Embler E, Wang YX, Banayo E, Mihaylova MM, Nelson MC, Zou Y, Juguilon H, Kang H, Shaw RJ, Evans RM. AMPK and PPARdelta agonists are exercise mimetics. Cell. 2008;134:405–15.
430. Jager S, Handschin C, St Pierre J, Spiegelman BM. AMP-activated protein kinase (AMPK) action in skeletal muscle via direct phosphorylation of PGC-1alpha. Proc Natl Acad Sci USA. 2007;104:12017–22.
431. Vincent MF, Marangos PJ, Gruber HE, Van den Berghe G. Inhibition by AICA riboside of gluconeogenesis in isolated rat hepatocytes. Diabetes. 1991;40:1259–66.
432. Young ME, Radda GK, Leighton B. Activation of glycogen phosphorylase and glycogenolysis in rat skeletal muscle by AICAR – an activator of AMP-activated protein kinase. FEBS Lett. 1996;382:43–7.
433. Winder WW, Holmes BF, Rubink DS, Jensen EB, Chen M, Holloszy JO. Activation of AMP-activated protein kinase increases mitochondrial enzymes in skeletal muscle. J Appl Physiol. 2000;88:2219–26.
434. Fujii N, Jessen N, Goodyear LJ. AMP-activated protein kinase and the regulation of glucose transport. Am J Physiol Endocrinol Metab. 2006;291:E867–77.

435. Ruderman NB, Saha AK, Kraegen EW. Minireview: malonyl CoA, AMP-activated protein kinase, and adiposity. Endocrinology. 2003;144:5166–71.
436. Brunmair B, Staniek K, Gras F, Scharf N, Althaym A, Clara R, Roden M, Gnaiger E, Nohl H, Waldhausl W, Furnsinn C. Thiazolidinediones, like metformin, inhibit respiratory complex I: a common mechanism contributing to their antidiabetic actions? Diabetes. 2004;53:1052–9.
437. Owen MR, Doran E, Halestrap AP. Evidence that metformin exerts its anti-diabetic effects through inhibition of complex 1 of the mitochondrial respiratory chain. Biochem J. 2000;348(Pt 3):607–14.
438. Matsuzaki S, Humphries KM. Selective inhibition of deactivated mitochondrial complex I by biguanides. Biochemistry. 2015;54:2011–21.
439. Hardie DG. Sensing of energy and nutrients by AMP-activated protein kinase. Am J Clin Nutr. 2011;93:891s–6.
440. Hou X, Song J, Li XN, Zhang L, Wang X, Chen L, Shen YH. Metformin reduces intracellular reactive oxygen species levels by upregulating expression of the antioxidant thioredoxin via the AMPK-FOXO3 pathway. Biochem Biophys Res Commun. 2010;396:199–205.
441. Cahova M, Palenickova E, Dankova H, Sticova E, Burian M, Drahota Z, Cervinkova Z, Kucera O, Gladkova C, Stopka P, Krizova J, Papackova Z, Oliyarnyk O, Kazdova L. Metformin prevents ischemia reperfusion-induced oxidative stress in the fatty liver by attenuation of reactive oxygen species formation. Am J Physiol Gastrointest Liver Physiol. 2015;309:G100–11.
442. Algire C, Moiseeva O, Deschenes-Simard X, Amrein L, Petruccelli L, Birman E, Viollet B, Ferbeyre G, Pollak MN. Metformin reduces endogenous reactive oxygen species and associated DNA damage. Cancer Prev Res (Phila). 2012;5:536–43.
443. Martin DS, Leonard S, Devine R, Redondo C, Kinsella GK, Breen CJ, McEneaney V, Rooney MF, Munsey TS, Porter RK, Sivaprasadarao A, Stephens JC, Findlay JB. Novel mitochondrial complex I inhibitors restore glucose-handling abilities of high-fat fed mice. J Mol Endocrinol. 2016;56:261–71.
444. Florez H, Luo J, Castillo-Florez S, Mitsi G, Hanna J, Tamariz L, Palacio A, Nagendran S, Hagan M. Impact of metformin-induced gastrointestinal symptoms on quality of life and adherence in patients with type 2 diabetes. Postgrad Med. 2010;122:112–20.
445. Kirpichnikov D, McFarlane SI, Sowers JR. Metformin: an update. Ann Intern Med. 2002;137:25–33.
446. Harrington WW, S Britt C, G Wilson J, O Milliken N, G Binz J, C Lobe D, R Oliver W, C Lewis M, M Ignar D. The effect of PPARalpha, PPARdelta, PPARgamma, and PPARpan agonists on body weight, body mass, and serum lipid profiles in diet-induced obese AKR/J mice. PPAR Res. 2007;2007:97125.
447. Hauner H. The mode of action of thiazolidinediones. Diabetes Metab Res Rev. 2002;18(Suppl 2):S10–5.
448. Day C. Thiazolidinediones: a new class of antidiabetic drugs. Diabet Med. 1999;16:179–92.
449. Bogacka I, Xie H, Bray GA, Smith SR. The effect of pioglitazone on peroxisome proliferator-activated receptor-gamma target genes related to lipid storage in vivo. Diabetes Care. 2004;27:1660–7.
450. Mensink M, Hesselink MK, Russell AP, Schaart G, Sels JP, Schrauwen P. Improved skeletal muscle oxidative enzyme activity and restoration of PGC-1alpha and PPARbeta/delta gene expression upon rosiglitazone treatment in obese patients with type 2 diabetes mellitus. Int J Obes (Lond). 2007;31:1302.
451. Schrauwen-Hinderling VB, Mensink M, Hesselink MK, Sels JP, Kooi ME, Schrauwen P. The insulin-sensitizing effect of rosiglitazone in type 2 diabetes mellitus patients does not require improved in vivo muscle mitochondrial function. J Clin Endocrinol Metab. 2008;93:2917–21.
452. Skov V, Glintborg D, Knudsen S, Tan Q, Jensen T, Kruse TA, Beck-Nielsen H, Hojlund K. Pioglitazone enhances mitochondrial biogenesis and ribosomal protein biosynthesis in skeletal muscle in polycystic ovary syndrome. PLoS ONE. 2008;3:e2466.
453. Brunmair B, Lest A, Staniek K, Gras F, Scharf N, Roden M, Nohl H, Waldhausl W, Furnsinn C. Fenofibrate impairs rat mitochondrial function by inhibition of respiratory complex I. J Pharmacol Exp Ther. 2004;311:109–14.

454. Brunmair B, Gras F, Neschen S, Roden M, Wagner L, Waldhausl W, Furnsinn C. Direct thiazolidinedione action on isolated rat skeletal muscle fuel handling is independent of peroxisome proliferator-activated receptor-gamma-mediated changes in gene expression. Diabetes. 2001;50:2309–15.
455. Schoonjans K, Staels B, Auwerx J. Role of the peroxisome proliferator-activated receptor (PPAR) in mediating the effects of fibrates and fatty acids on gene expression. J Lipid Res. 1996;37:907–25.
456. Steiner G. Atherosclerosis in type 2 diabetes: a role for fibrate therapy? Diab Vasc Dis Res. 2007;4:368–74.
457. Yao CX, Li WY, Zhang SF, Zhang SF, Zhang HF, Zang MX. Effects of Doxorubicin and Fenofibrate on the activities of NADH oxidase and citrate synthase in mice. Basic Clin Pharmacol Toxicol. 2011;109:452–6.
458. Turner N, Li JY, Gosby A, To SW, Cheng Z, Miyoshi H, Taketo MM, Cooney GJ, Kraegen EW, James DE, Hu LH, Li J, Ye JM. Berberine and its more biologically available derivative, dihydroberberine, inhibit mitochondrial respiratory complex I: a mechanism for the action of berberine to activate AMP-activated protein kinase and improve insulin action. Diabetes. 2008;57:1414–8.
459. Duan S, Huang S, Gong J, Shen Y, Zeng L, Feng Y, Ren W, Leng Y, Hu Y. Design and synthesis of novel arctigenin analogues for the amelioration of metabolic disorders. ACS Med Chem Lett. 2015;6:386–91.
460. Pereira CV, Machado NG, Oliveira PJ. Mechanisms of berberine (natural yellow 18)-induced mitochondrial dysfunction: interaction with the adenine nucleotide translocator. Toxicol Sci Off J Soc Toxicol. 2008;105:408–17.
461. Zhang P, Qiang X, Zhang M, Ma D, Zhao Z, Zhou C, Liu X, Li R, Chen H, Zhang Y. Demethyleneberberine, a natural mitochondria-targeted antioxidant, inhibits mitochondrial dysfunction, oxidative stress, and steatosis in alcoholic liver disease mouse model. J Pharmacol Exp Ther. 2015;352:139–47.
462. Divakaruni AS, Wiley SE, Rogers GW, Andreyev AY, Petrosyan S, Loviscach M, Wall EA, Yadava N, Heuck AP, Ferrick DA, Henry RR, McDonald WG, Colca JR, Simon MI, Ciaraldi TP, Murphy AN. Thiazolidinediones are acute, specific inhibitors of the mitochondrial pyruvate carrier. Proc Natl Acad Sci U S A. 2013;110:5422–7.
463. Timmers S, Nabben M, Bosma M, van Bree B, Lenaers E, van Beurden D, Schaart G, Westerterp-Plantenga MS, Langhans W, Hesselink MK, Schrauwen-Hinderling VB, Schrauwen P. Augmenting muscle diacylglycerol and triacylglycerol content by blocking fatty acid oxidation does not impede insulin sensitivity. Proc Natl Acad Sci U S A. 2012;109:11711–6.

Chapter 26
Mechanistic Role of Kinases in the Regulation of Mitochondrial Fitness

Daniela Sorriento, Jessica Gambardella, Antonella Fiordelisi, Bruno Trimarco, Michele Ciccarelli, Guido Iaccarino, and Gaetano Santulli

Mounting evidence indicates that mitochondria contain multiple phosphorylation substrates and that protein kinases translocate into mitochondria, suggesting that protein phosphorylation in this organelle could be fundamental for the regulation of its own function. Here we examine the mechanistic role of cellular kinases in the fine regulation of key mitochondrial activities, including mitochondrial quality control, fission/fusion processes, metabolism, and mitophagy.

Mitochondrial Respiration

Mitochondria are cytoplasmic organelles that are involved in oxidative energy metabolism, producing the most of our cellular energy by oxidative phosphorylation [1, 2], through the metabolization of nutrients and production of ATP. Mitochondrial ATP production relies on the electron transport chain (ETC), composed of respiratory chain complexes I–IV.

D. Sorriento • A. Fiordelisi • B. Trimarco
Department of Advanced Biomedical Sciences, "Federico II" University, Naples, Italy

J. Gambardella
Department of Medicine, Surgery and Dentistry, University of Salerno, Baronissi, Italy

Columbia University Medical Center, New York, NY, USA

M. Ciccarelli • G. Iaccarino
Department of Medicine, Surgery and Dentistry, University of Salerno, Baronissi, Italy

G. Santulli (✉)
Dept. of Biomedical Advanced Sciences, Federico II University, Naples, Italy

Columbia University Medical Center, New York Presbyterian Hospital—Manhattan, New York, NY, USA
e-mail: gsantulli001@gmail.com

G. Santulli (ed.), *Mitochondrial Dynamics in Cardiovascular Medicine*, Advances in Experimental Medicine and Biology 982,
DOI 10.1007/978-3-319-55330-6_26

Table 26.1 Substrates of SRC in the ETC e the effects of substrate phosphorylation

SRC substrates	Phosphorylation sites	Effect
ANT1	Tyr^{190} and Tyr^{194}	Protection of cardiac mitochondria against ischemic-reperfusion injury
Cytochrome c oxidase	unknown	Increase of the enzymatic activity of complex IV
NDUFV2	Tyr^{193}	Increase of NADH dehydrogenase activity
SDHA	Tyr^{215}	Regulation of reactive oxygen species production

The Regulation of Proteins in the ETC

SRC family kinases (SFKs) are the most involved in tyrosine phosphorylation of mitochondrial proteins to regulate mitochondrial respiration. Src kinases affect the enzymatic activities of complexes I, III–V; importantly, SRC phosphorylates different substrates in the ETC (Table 26.1). Indeed, Src is present inside mitochondria where it phosphorylates the subunit II of cytochrome c oxidase [3], increasing the enzymatic activity of complex IV. Moreover, adenine nucleotide translocase 1 (ANT1), which is able to transport ADP from cytosol within mitochondria in exchange of ATP, is known to be phosphorylated on Tyr^{190} and Tyr^{194} by c-Src kinase [4–6].

Src is also known to phosphorylate NDUFV2 (NADH dehydrogenase [ubiquinone] flavoprotein 2) of complex I at Tyr^{193}, and SDHA (succinate dehydrogenase A) of complex II at Tyr^{215} [7]. NDUFV2 phosphorylation is required for NADH dehydrogenase activity, affecting respiration activity and cellular ATP content, while SDHA phosphorylation has no effect on enzyme activity, but affects reactive oxygen species production. Furthermore, the tyrosine protein kinase Fgr phosphorylates flavoprotein of succinate dehydrogenase at Tyr^{535} and Tyr^{596} and aconitase at Tyr^{71}, Tyr^{544}, and Tyr^{665} [8, 9].

Mitochondrial Biogenesis

Mitochondrial biogenesis can be defined as the growth and division of pre-existing mitochondria leading to a variation in number, size, and mass. It is dependent on different signaling cascades and transcriptional complexes that promote the formation and assembly of mitochondria.

The Regulation of PCG1α

The peroxisome proliferator-activated receptor γ coactivator 1 (PGC1) family of transcriptional coactivators has recently emerged as central regulator of metabolism being a positive modulator of mitochondrial biogenesis and respiration [10]. PGC1α is a co-transcriptional regulation factor that induces mitochondrial biogenesis by activating different transcription factors, including NRF-1 and NRF-2. These latters on turn activate Tfam to increase replication of mtDNA and to induce the transcription of key mitochondrial enzymes [11]. AMPK, p38 MAPK, and GSK3β are the best-characterized protein kinases known to target PGC1α. It has been shown that PGC1α is increased in response to activation of AMPK and is reduced in both AMPK null [12] and dominant negative mice [13]. Indeed, AMPK directly phosphorylates PGC1α on Thr^{177} and Ser^{538} [14] thereby enhancing its co-transcriptional activity and consequently mitochondrial gene expression. p38 MAPK phosphorylates PGC1α at Thr^{262}, Ser^{265}, and Thr^{298} in response to cytokine stimulation in muscle cells [15]. Moreover, it enhances the activity of PGC1α by increasing its stability and by disrupting the inactivating interaction between PGC1α and the co-repressor p160MBP in myoblasts [16]. Finally, PGC-1α is also phosphorylated by glycogen synthase kinase 3β (GSK3β), which inhibits PGC1α by enhancing its proteasomal degradation in the nucleus during oxidative stress [17].

Mitochondrial Quality Control

Mitochondria have an important role in the regulation of cell survival, cell death and metabolic homeostasis. They continuously fuse or divide to maintain their functions and damaged mitochondria after fission are removed through mitophagy. Thus, several mechanisms are involved in the regulation of mitochondrial quality control including mitochondrial fission and fusion [18], Parkin-dependent pathways [19] and degradation of damaged mitochondria by lysosomes [20] and autophagosomes [21].

The Regulation of Mitochondrial Fission

Mitochondria fission and fusion are mainly mediated by highly conserved guanosine triphosphatases (GTPases) [22, 23]. Among them, dynamin-related protein 1 (DRP1) is the GTPase that regulates mitochondrial fission [24, 25]. It is a cytosolic protein that once activated translocates to the outer mitochondrial membrane where multimerizes in order to create a ring-like structure that constricts and divides the organelle [26, 27]. Drp1 activity is mainly regulated through phosphorylation in different sites by several protein kinases (Table 26.2). However, data on the effects of DRP1 phosphorylation are rather puzzling since different kinases phosphorylates the same site in DRP1 resulting in opposing effects, as described below.

Table 26.2 Protein kinases that phosphorylate DRP1 and their effects on DRP1 activation

Protein kinases	DRP1 phosphorylation sites	Effect on DRP1
PKA	Ser^{656}	Inhibition
	Ser^{637}	Inhibition
CaMK1α	Ser^{637}	Activation
ERK	Ser^{616}	Activation
CDK	Ser^{616}	Inhibition

PKA has a key role in the regulation of DRP1 activity by preventing its translocation to the mitochondria and inhibiting the process of mitochondrial fission. Indeed, Cribbs & Strack demonstrated that PKA phosphorylates Drp1 at Ser^{656} and this attenuates the GTPase activity of Drp1 promoting cell survival [28]. Chang & Blackstone discovered that PKA phosphorylates DRP1 also at Ser^{637} inhibiting its GTPase activity [29]. Accordingly, the phospho-mimetic substitution Ser637Asp blocks mitochondrial fission and apoptotic cell death [29]. Thus, PKA exerts an inhibitory effect on DRP1 activation by phosphorylation of both Ser^{656} and Ser^{637}, even if no data are available to understand whether phosphorylation at these sites could have different physiological implications. On the contrary, phosphorylation of Ser^{637}, which is inhibitory in PKA signaling, induced mitochondrial fission when is due to Ca^{2+}/calmodulin-dependent protein kinase Iα (CaMKIα) [30]. Similarly, phosphorylation of DRP1 at Ser^{616} by ERK2 activates DRP1 and promotes mitochondrial fission [31], whereas phosphorylation at the same residue by CDK5 exerts opposite effects [32]. Thus, the effects of DRP1 phosphorylation on mitochondrial fission depend on both the type of kinase and the specific phosphorylated residues. However, further studies are needed to clarify such effects.

Regulation of Mitochondrial Fusion

Mitofusins (MFNs) 1 and 2 are a class of conserved GTPases of the mitochondrial outer membrane that are essential for mitochondrial fusion and consequently to maintain normal mitochondrial morphology. MFNs are essential for normal cardiac function. Indeed, the combined deletion of MFN1 and MFN2 in murine hearts induces mitochondrial and cardiomyocyte dysfunction which rapidly leads to progressive and lethal dilated cardiomyopathy. MFN1 and 2 can be phosphorylated and such phosphorylation affects their ability to modulate mitochondrial fusion (Table 26.3).

Indeed, MFN2 was found to be phosphorylated by PINK1 at Thr^{111} and S^{442} to become a mitochondrial receptor for Parkin and eventually promote mitophagy. Moreover, MFN2 can be also phosphorylated by JNK at Ser^{27} causing its degradation through the ubiquitin-proteasome pathway which in turn affects both mitochondrial dynamics and apoptosis [33]. Also MFN1 can be phosphorylated at Thr^{562} by ERK to modulate apoptotic responses [34]. Indeed, this inhibitory phosphorylation of MFN1 induces its association with BAK facilitating its oligomerization and inducing cytochrome *c* release and cell death [34].

Table 26.3 Mitofusins phosphorylation sites and effects of phosphorylation on mitophagy

Protein kinases	MFN phosphorylation sites	Effect on mitophagy
PINK1	T^{111} of MFN2	Activation
PINK1	S^{442} of MFN2	Activation
JNK	Ser^{27} of MFN2	Inhibition
ERK	T^{562} of MFN1	Inhibition

Parkin-Dependent Mechanisms

PINK1 is a kinase associated with mitochondria: the loss of this kinase expression causes mitochondrial dysfunction and mitophagy [35–37]. Parkin is an E3 ubiquitin ligase suggested to be downstream of PINK1 to regulate the removal of damaged mitochondria. Indeed, PINK1 is activated by mitochondria membrane potential depolarization and is imported into mitochondria to activate Parkin [38–41]. This latter causes proteasomal degradation of outer mitochondrial membrane proteins [42, 43] and selective autophagy of damaged mitochondria [44], suggesting that PINK1 and Parkin mediate a mitochondrial quality control pathway. The removal of damaged mitochondria was thought to be mainly attributable to the activation of PINK1-Parkin-Ubiquitin cascade: PINK1 directly phosphorylates Parkin at Ser^{65} which on turn activates Ubiquitin [45]. Actually, recent findings changed this view: PINK1 has been shown to recruit Parkin to mitochondria also in presence of mutation of Ser^{65} to Alanine suggesting the ability of PINK1 to regulate Parkin in a phosphorylation independent manner. Moreover, PINK1 directly phosphorylates Ubiquitin at Ser^{65} which on turn activates Parkin [46]. Thus, PINK1 phosphorylates at Ser^{65} both Parkin and Ubiquitin to induce the full activation of Parkin, as summarized in Fig. 26.1.

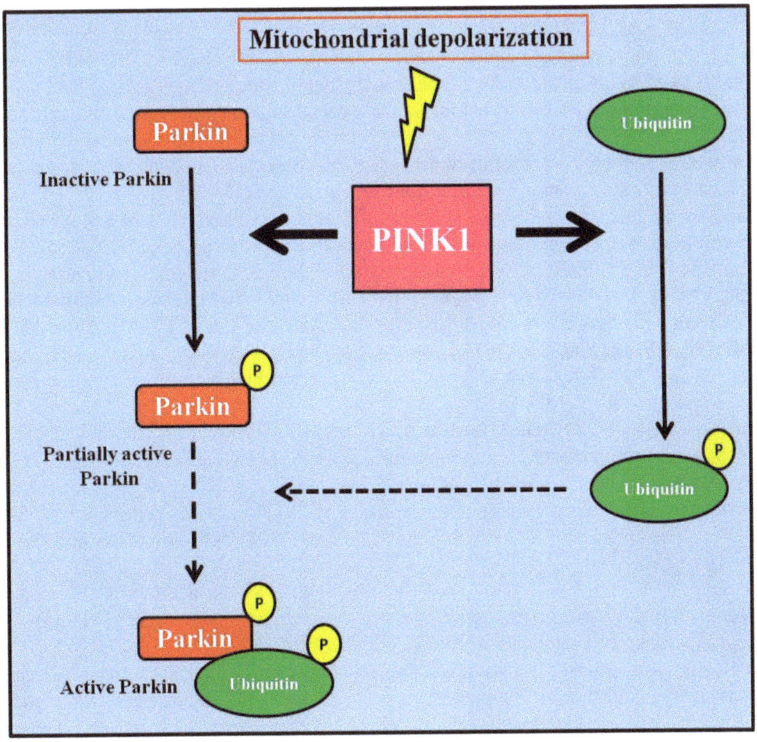

Fig. 26.1 The full activation of Parkin is dependent on PINK1-mediated phosphorylation of both Parkin and ubiquitin at Ser^{65}

Acknowledgements Dr. Gaetano Santulli, M.D., Ph.D. is supported by the National Institutes of Health (NIH, Grant NIDDK107895).

References

1. Chance B, Sies H, Boveris A. Hydroperoxide metabolism in mammalian organs. Physiol Rev. 1979;59:527–605.
2. Sorriento D, Ciccarelli M, Santulli G, Illario M, Trimarco B, Iaccarino G. Trafficking GRK2: cellular and metabolic consequences of GRK2 subcellular localization. Transl Med UniSa. 2014;10:3–7.
3. Miyazaki T, Neff L, Tanaka S, Horne WC, Baron R. Regulation of cytochrome c oxidase activity by c-Src in osteoclasts. J Cell Biol. 2003;160:709–18.
4. Lewandrowski U, Sickmann A, Cesaro L, Brunati AM, Toninello A, Salvi M. Identification of new tyrosine phosphorylated proteins in rat brain mitochondria. FEBS Lett. 2008;582:1104–10.
5. Feng J, Zhu M, Schaub MC, Gehrig P, Roschitzki B, Lucchinetti E, Zaugg M. Phosphoproteome analysis of isoflurane-protected heart mitochondria: phosphorylation of adenine nucleotide translocator-1 on Tyr194 regulates mitochondrial function. Cardiovasc Res. 2008;80:20–9.
6. Feng J, Lucchinetti E, Enkavi G, Wang Y, Gehrig P, Roschitzki B, Schaub MC, Tajkhorshid E, Zaugg K, Zaugg M. Tyrosine phosphorylation by Src within the cavity of the adenine nucleotide translocase 1 regulates ADP/ATP exchange in mitochondria. Am J Phys Cell Phys. 2010;298:C740–8.
7. Ogura M, Yamaki J, Homma MK, Homma Y. Mitochondrial c-Src regulates cell survival through phosphorylation of respiratory chain components. Biochem J. 2012;447:281–9.
8. Salvi M, Morrice NA, Brunati AM, Toninello A. Identification of the flavoprotein of succinate dehydrogenase and aconitase as in vitro mitochondrial substrates of Fgr tyrosine kinase. FEBS Lett. 2007;581:5579–85.
9. Thomson M. Evidence of undiscovered cell regulatory mechanisms: phosphoproteins and protein kinases in mitochondria. Cellular Mol Life Sci: CMLS. 2002;59:213–9.
10. Handschin C, Spiegelman BM. Peroxisome proliferator-activated receptor gamma coactivator 1 coactivators, energy homeostasis, and metabolism. Endocr Rev. 2006;27:728–35.
11. Virbasius JV, Scarpulla RC. Activation of the human mitochondrial transcription factor A gene by nuclear respiratory factors: a potential regulatory link between nuclear and mitochondrial gene expression in organelle biogenesis. Proc Natl Acad Sci U S A. 1994;91:1309–13.
12. Iglesias MA, Furler SM, Cooney GJ, Kraegen EW, Ye JM. AMP-activated protein kinase activation by AICAR increases both muscle fatty acid and glucose uptake in white muscle of insulin-resistant rats in vivo. Diabetes. 2004;53:1649–54.
13. Watt MJ, Steinberg GR, Chan S, Garnham A, Kemp BE, Febbraio MA. Beta-adrenergic stimulation of skeletal muscle HSL can be overridden by AMPK signaling. FASEB J: Off Publ Feder Am Soc Exp Biol. 2004;18:1445–6.
14. Jager S, Handschin C, St-Pierre J, Spiegelman BM. AMP-activated protein kinase (AMPK) action in skeletal muscle via direct phosphorylation of PGC-1alpha. Proc Natl Acad Sci U S A. 2007;104:12017–22.
15. Puigserver P, Rhee J, Lin J, Wu Z, Yoon JC, Zhang CY, Krauss S, Mootha VK, Lowell BB, Spiegelman BM. Cytokine stimulation of energy expenditure through p38 MAP kinase activation of PPARgamma coactivator-1. Mol Cell. 2001;8:971–82.
16. Fan M, Rhee J, St-Pierre J, Handschin C, Puigserver P, Lin J, Jaeger S, Erdjument-Bromage H, Tempst P, Spiegelman BM. Suppression of mitochondrial respiration through recruitment of p160 myb binding protein to PGC-1alpha: modulation by p38 MAPK. Genes Dev. 2004;18:278–89.
17. Anderson RM, Barger JL, Edwards MG, Braun KH, O'Connor CE, Prolla TA, Weindruch R. Dynamic regulation of PGC-1alpha localization and turnover implicates mitochondrial adaptation in calorie restriction and the stress response. Aging Cell. 2008;7:101–11.

18. van der Bliek AM, Shen Q, Kawajiri S. Mechanisms of mitochondrial fission and fusion. Cold Spring Harb Perspect Biol. 2013;5.
19. Ding WX, Yin XM. Mitophagy: mechanisms, pathophysiological roles, and analysis. Biol Chem. 2012;393:547–64.
20. Soubannier V, Rippstein P, Kaufman BA, Shoubridge EA, McBride HM. Reconstitution of mitochondria derived vesicle formation demonstrates selective enrichment of oxidized cargo. PLoS ONE. 2012;7:e52830.
21. Youle RJ, Narendra DP. Mechanisms of mitophagy. Nat Rev Mol Cell Biol. 2011;12:9–14.
22. Youle RJ, van der Bliek AM. Mitochondrial fission, fusion, and stress. Science. 2012;337: 1062–5.
23. Ishihara N, Otera H, Oka T, Mihara K. Regulation and physiologic functions of GTPases in mitochondrial fusion and fission in mammals. Antioxid Redox Signal. 2013;19:389–99.
24. Cribbs JT, Strack S. Functional characterization of phosphorylation sites in dynamin-related protein 1. Methods Enzymol. 2009;457:231–53.
25. Chen H, Detmer SA, Ewald AJ, Griffin EE, Fraser SE, Chan DC. Mitofusins Mfn1 and Mfn2 coordinately regulate mitochondrial fusion and are essential for embryonic development. J Cell Biol. 2003;160:189–200.
26. Zhu PP, Patterson A, Stadler J, Seeburg DP, Sheng M, Blackstone C. Intra- and intermolecular domain interactions of the C-terminal GTPase effector domain of the multimeric dynamin-like GTPase Drp1. J Biol Chem. 2004;279:35967–74.
27. Lee YJ, Jeong SY, Karbowski M, Smith CL, Youle RJ. Roles of the mammalian mitochondrial fission and fusion mediators Fis1, Drp1, and Opa1 in apoptosis. Mol Biol Cell. 2004;15:5001–11.
28. Cribbs JT, Strack S. Reversible phosphorylation of Drp1 by cyclic AMP-dependent protein kinase and calcineurin regulates mitochondrial fission and cell death. EMBO Rep. 2007;8:939–44.
29. Chang CR, Blackstone C. Cyclic AMP-dependent protein kinase phosphorylation of Drp1 regulates its GTPase activity and mitochondrial morphology. J Biol Chem. 2007;282: 21583–7.
30. Han XJ, Lu YF, Li SA, Kaitsuka T, Sato Y, Tomizawa K, Nairn AC, Takei K, Matsui H, Matsushita M. CaM kinase I alpha-induced phosphorylation of Drp1 regulates mitochondrial morphology. J Cell Biol. 2008;182:573–85.
31. Kashatus JA, Nascimento A, Myers LJ, Sher A, Byrne FL, Hoehn KL, Counter CM, Kashatus DF. Erk2 phosphorylation of Drp1 promotes mitochondrial fission and MAPK-driven tumor growth. Mol Cell. 2015;57:537–51.
32. Cho B, Cho HM, Kim HJ, Jeong J, Park SK, Hwang EM, Park JY, Kim WR, Kim H, Sun W. CDK5-dependent inhibitory phosphorylation of Drp1 during neuronal maturation. Exp Mol Med. 2014;46:e105.
33. Leboucher GP, Tsai YC, Yang M, Shaw KC, Zhou M, Veenstra TD, Glickman MH, Weissman AM. Stress-induced phosphorylation and proteasomal degradation of mitofusin 2 facilitates mitochondrial fragmentation and apoptosis. Mol Cell. 2012;47:547–57.
34. Pyakurel A, Savoia C, Hess D, Scorrano L. Extracellular regulated kinase phosphorylates mitofusin 1 to control mitochondrial morphology and apoptosis. Mol Cell. 2015;58:244–54.
35. Geisler S, Holmstrom KM, Skujat D, Fiesel FC, Rothfuss OC, Kahle PJ, Springer W. PINK1/Parkin-mediated mitophagy is dependent on VDAC1 and p62/SQSTM1. Nat Cell Biol. 2010;12:119–31.
36. Narendra DP, Jin SM, Tanaka A, Suen DF, Gautier CA, Shen J, Cookson MR, Youle RJ. PINK1 is selectively stabilized on impaired mitochondria to activate Parkin. PLoS Biol. 2010;8:e1000298.
37. Matsuda N, Sato S, Shiba K, Okatsu K, Saisho K, Gautier CA, Sou YS, Saiki S, Kawajiri S, Sato F, Kimura M, Komatsu M, Hattori N, Tanaka K. PINK1 stabilized by mitochondrial depolarization recruits Parkin to damaged mitochondria and activates latent Parkin for mitophagy. J Cell Biol. 2010;189:211–21.
38. Lin W, Kang UJ. Characterization of PINK1 processing, stability, and subcellular localization. J Neurochem. 2008;106:464–74.

39. Jin SM, Lazarou M, Wang C, Kane LA, Narendra DP, Youle RJ. Mitochondrial membrane potential regulates PINK1 import and proteolytic destabilization by PARL. J Cell Biol. 2010;191:933–42.
40. Deas E, Wood NW, Plun-Favreau H. Mitophagy and Parkinson's disease: the PINK1-parkin link. Biochim Biophys Acta. 1813;2011:623–33.
41. Yamano K, Youle RJ. PINK1 is degraded through the N-end rule pathway. Autophagy. 2013;9:1758–69.
42. Tanaka K, Matsuda N, Okatsu K. Mechanisms underling the cause of Parkinson's disease: the functions of Parkin/PINK1. Rinsho Shinkeigaku = Clin Neurol. 2010;50:867.
43. Yoshii SR, Kishi C, Ishihara N, Mizushima N. Parkin mediates proteasome-dependent protein degradation and rupture of the outer mitochondrial membrane. J Biol Chem. 2011; 286:19630–40.
44. Narendra D, Tanaka A, Suen DF, Youle RJ. Parkin is recruited selectively to impaired mitochondria and promotes their autophagy. J Cell Biol. 2008;183:795–803.
45. Kondapalli C, Kazlauskaite A, Zhang N, Woodroof HI, Campbell DG, Gourlay R, Burchell L, Walden H, Macartney TJ, Deak M, Knebel A, Alessi DR, Muqit MM. PINK1 is activated by mitochondrial membrane potential depolarization and stimulates Parkin E3 ligase activity by phosphorylating Serine 65. Open Biol. 2012;2:120080.
46. Kane LA, Lazarou M, Fogel AI, Li Y, Yamano K, Sarraf SA, Banerjee S, Youle RJ. PINK1 phosphorylates ubiquitin to activate Parkin E3 ubiquitin ligase activity. J Cell Biol. 2014;205:143–53.

Chapter 27
Mitochondria Damage and Kidney Disease

Pu Duann and Pei-Hui Lin

Abbreviations

AKI	Acute kidney injury
AMPK	AMP-activated protein kinase
ATP	Adenosine triphosphate
Ca^{2+}	Calcium ion
CKD	Chronic kidney disease
coQ	Coenzyme Q
CsA	Cyclosporin A
cyt C	Cytochrome C
DN	Diabetic nephropathy
dNTPs	Deoxynucleotides triphosphates
DRP1	Dynamin related protein 1
ESRD	End-stage renal disease
ETC	Electron transport chain
FSGS	Focal and segmental glomerulosclerosis
GN	Glomerulonephritis
GSK	Glycogen synthase kinase
H^+	Proton
I/R injury or IRI	Ischemic reperfusion injury

P. Duann (✉)
Department of Surgery, Baylor College of Medicine, Houston, TX 77030, USA
e-mail: Pu.Duann@bcm.edu

P.-H. Lin (✉)
Davis Heart and Lung Research Institute, The Ohio State University, Columbus, OH 43210, USA

Department of Surgery, The Ohio State University, Columbus, OH 43210, USA
e-mail: Pei-Hui.Lin@osumc.edu

G. Santulli (ed.), *Mitochondrial Dynamics in Cardiovascular Medicine*,
Advances in Experimental Medicine and Biology 982,
DOI 10.1007/978-3-319-55330-6_27

IMM	Inner mitochondrial membrane
KLF-6	Krüppel-like factor 6
MELAS	Myopathy encephalopathy lactic acidosis and stroke-like episodes
Mfn1 and 2	Mitofusins 1 and 2
MPT	Mitochondrial permeability transition
mPTP	Mitochondrial permeability transition pore
mtDNA	Mitochondrial DNA
MA-5	Mitochonic acid 5
NAD^+	Nicotinamide adenine dinucleotide
nDNA	Nuclear DNA
NRF	Nuclear respiratory factors
OMM	Outer mitochondrial membrane
OPA1	Optical atrophy 1
OXPHOS	Oxidative phosphorylation
PGC-1α	PPARγ-coactivator -1α
PPAR	Peroxisome proliferator-activated receptor
ROS	Reactive oxygen species
SIRT	Sirtuin
TLR	Toll-like receptor
$\Delta\Psi m$	Mitochondrial inner membrane potential

In human body, the kidney receives 20% of cardiac output and consumes 10% of the body's oxygen to perform its functions. The kidney is composed of multiple cell populations which are involved in several vital functions that maintain body's homeostasis such as acid-base and electrolytes balance, blood pressure regulation, nutrients reabsorption and hormone secretion [1]. The human adult kidney is composed with about 3 million functional units, or "nephrons", and their primary functions and structures are subdivided into a glomerular filtration unit and several tubular segments. The glomerular filtration is a process of filtering blood circulation, retaining circulating cells and useful macromolecules. The tubular segments are involved in active transport processes to reabsorb water, electrolytes, and nutrients to maintain fluid and osmolarity homeostasis. The coordination of both processes maintains body homeostasis and results in metabolic waste excretion with only trace amount of proteins in the urine output [2]. Hence, kidney dysfunction often results in kidney disease with systemic complications.

Kidney diseases can be classified into two categories- acute kidney injury (AKI) or chronic kidney disease (CKD). Renal disease and complications are the current global health concerns due to the vital roles of kidney in body homeostasis [3–5]. AKI is characterized by rapid renal function decline along with accompanying electrolytes abnormalities, fluid overload, severe acidosis, hematologic abnormalities (e.g. anemia, uremic platelet dysfunction), and possibly multi-organs failure and poor clinical outcomes [6]. CKD is linked to gradual loss of renal function over time. CKD could derive from defects in glomerular filtration units or chronic tubular injuries which could lead to multiple complications including cardiovascular disease, high blood pressure, bone loss, and malnutrition. Both AKI and CKD are closely integrated and could serve as a risk factor for one another which often linked with increase cardiovascular risk and uprising mortality and morbidity rates [7, 8]. Additionally, kidney functions are susceptible to certain inherited genetic diseases and aging processes [9, 10]. All different forms of renal injuries could ultimately progress into severe end-stage renal disease (ESRD) with renal replacement as the only therapeutic option. Therefore, therapeutic strategies are in dire need to prevent renal diseases progression.

Mitochondria is the cell's energy-producing organelle that maintains cellular redox and energy homeostasis, and therefore a major source of intracellular oxidative stress [11]. In addition to its role in adenosine triphosphate (ATP) generation through oxidative phosphorylation (OXPHOS), the mitochondria also play an essential role in metabolic signaling such as pyrimidine, heme biosynthesis, TCA cycle and fatty acid β-oxidation pathways, in calcium ion (Ca^{2+}) homeostasis, thermogenesis, proliferation and regulating intrinsic apoptotic pathway. Mitochondria are heterogeneous and dynamic organelles. The mitochondria populations can be different in size, mass, metabolic activity, and membrane potential within a cell. Additionally, mitochondria constantly change shape, dynamics and turnover to maintain cellular homeostasis. Cellular stress can induce compromised mitochondria membrane integrity or dysfunctions, leading to cell death via different mechanisms. This could involve the release of apoptotic molecules such as cytochrome C (cyt C), and apoptosomes from mitochondrial inter-membrane spaces. Alternatively, activation of mitochondrial permeability transition (MPT) can trigger mitochondrial inner membrane potential ($\Delta\Psi m$) dissipations and subsequent loss of energy production [12].

Mitochodrial damages and dysfunction are recognized as a leading factor to many chronic and acute renal diseases [11]. Therefore, it is remarkably important to understand mitochondrial biology and pathophysiology for effective therapeutics discoveries in renal diseases. In this chapter, we discuss evidence supporting mitochondrial dysfunction in the pathogenesis of kidney disease, and summarize the recent development of mitochondria-targeted therapies which hold high promise alleviating renal injury.

Mitochondria in Kidney Health

Mitochondria are especially important in metabolically active organs such as brain, heart, kidney and muscle. Kidney consumes roughly 7% of the body's daily ATP energy expenditure [13, 14]. Due to various energy demands, different nephron segments have different mitochondria densities and distributions. It is generally accepted that renal tubule cells are rich in mitochondria, with the S1 segment containing the highest mitochondria density. These renal tubule cells require this large amount of mitochondria due to their high-energy demand for reabsorption and secretion against chemical gradients, which heavily rely on normal mitochondrial oxidative phosphorylation to supply ATP as an energy source [15]. Nevertheless, the high-energy requirement of the podocytes was only recently highlighted for possible mechanisms in structural stability, organization of cytoskeletal and extracellular matrix proteins, motility, remodeling of foot process, uptake of filtered proteins, and some other more obscure mechanisms [16]. Bioenergetic profiles studies of mitochondrial function confirmed that podocytes are very susceptible to dysfunction in energy supply during stress condtions [17]. Several factors such as mitochondrial biogenesis and turnover, bioenergetics, dynamics, and autophagy (mitophagy) regulate the conditions of mitochondria.

Mitochondrial Biogenesis

Mitochondrial turnover is an exquisitely coordinated process to maintain mitochondrial homeostasis. In this regard, dysfunctional mitochondria are selectively eliminated and replaced through biogenesis, a coordinated process between nuclear and mitochondrial genomes, to increase the mass and number of functional mitochondria in compensation for the lost or damaged mitochondria [18]. The mitochondrial DNA (mtDNA) encodes 37 genes which include 13 structural subunits of the mitochondrial respiratory chain. The molecular mechanism of mitochondrial biogenesis during tissue injury and repair are actively studied topics [19, 20] (Fig. 27.1a, b). The mitochondrial biogenesis program involves mito-nuclear communication and several key regulators have been identified such as: peroxisome proliferator-activated receptor (PPAR), PPARγ coactivator 1α (PGC-1α), sirtuin-1 and 3 (SIRT 1/3) family deacetylase, AMP-activated protein kinase (AMPK), and nuclear respiratory factors 1 and 2 (NRF1 and NRF2) [21]. AMPK and SIRT1 positively regulate PGC-1α, the master regulator of mitochondria biogenesis, through post-translational modifications with phosphorylation or deacetylation respectively. In addition, AMPK and SIRT1 are important modulators of energy metabolism. In kidney, the PGC-1α transcriptional coactivator is predominantly expressed in proximal tubules, and therefore is pivotal to tubular homeostasis. PGC-1α regulates the expression of NRF1 and NRF2 and increases the biosynthesis of nicotinamide adenine dinucleotide (NAD^+), a central metabolic coenzyme/cosubstrate involved in cellular energy metabolism, to link oxidative metabolism to renal protection [22–24]. Both NRF1 and NRF2 are nuclear DNA (nDNA)-encoded transcription factors that activate genes coding for the OXPHOS system and genes involved in mtDNA transcription and replication [25, 26].

Mitochondrial Bioenergetics

The mitochondrial OXPHOS respiratory chain is composed of four protein complexes which are known as complexes I–IV. These complexes transfer electrons and protons (H^+) across inner mitochondrial membrane (IMM) to generate an electrochemical gradient for ATP synthesis in a fifth protein complex, known as

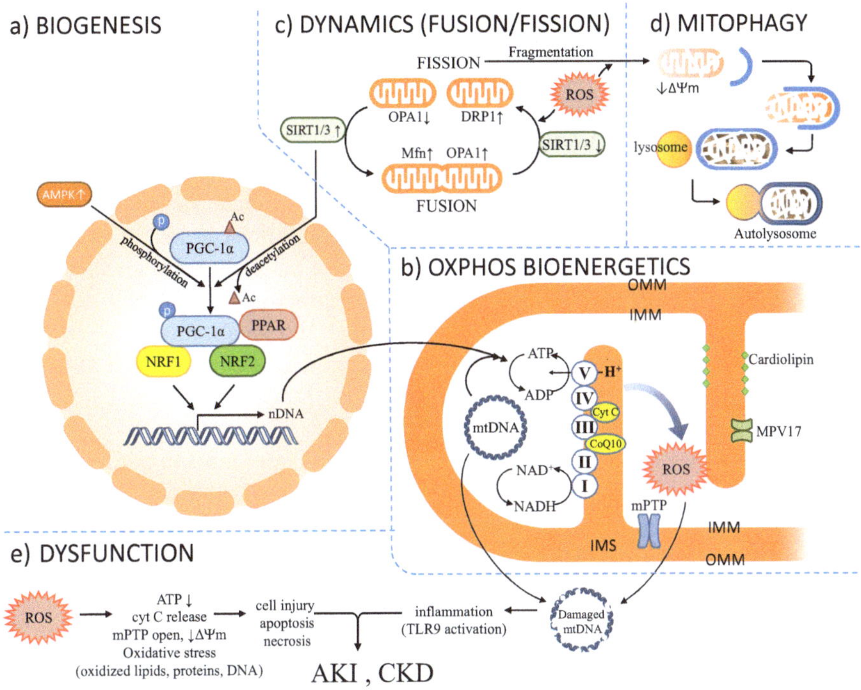

Fig. 27.1 The health and disease of kidney are regulated by the pathophysiological conditions of mitochondria. Several processes regulate the health and disease states of mitochondria which include (panel **a**) mitochondrial biogenesis – a process requires coordinated expression of both nDNA and mtDNA and several transcriptional factors and co-activators; (panel **b**) OXPHOS bioenergetics – an electrochemical gradient and ATP synthesis process which depends on respiratory electron transfer chain (complexes I–V), cofactors (MPV17 and others), and cardiolipin to maintain normal oxidative phosphorylation; (panel **c**) mitochondrial dynamics – a process requires functional balance between fusion and fission; and (panel **d**) mitochondrial turnover by autophagy (or mitophagy) – an important process to clear the dysfunctional mitochondria. (Panel **e**) Disturbance in any steps of the processes in panels **a**–**d** will cause ROS overproduction, decrease in ATP generation, loss of IMM potential (ΔΨm), mPTP opening and cyt C release. The oxidative stress could lead to cell apoptosis or cell injury and the damaged mtDNA, nDNA could stimulate innate immune response through activation of Toll-like receptor (TLR) and inflammation which, all together, lead to AKI or CKD. See Text and Abbreviations for details

complex V (ATP synthase) (Fig. 27.1b). Some important factors were identified in this process. For example, NAD concentrations determine the rate-limiting process [27, 28] and Coenzyme Q10 (coQ10) shuttles electrons in the respiratory chain [29]. Additionally, depending on its interaction with cardiolipin, the IMM resident protein cytochrome C (cyt C) regulates an intricate balance between mitochondrial respiration and apoptosis. Cardiolipin was identified as a phospholipid that was exclusively expressed on the IMM, where it forms microdomains and plays a central structural role in cristae formation, a property of cardiolipin on membrane curvature, organization of the electron transport chain (ETC) complexes into supra-complex for OXPHOS activity, and as a platform for initiation of apoptosis [30, 31]. The interaction of cyt C and cardiolipin determines whether or not cyt C functions as an electron carrier or a peroxidase.

During normal electron transfer, some (<4%) of the consumed oxygen is converted into superoxide radicals such as reactive oxygen free radical species (ROS) and reactive nitrogen free radical species (RNS) via electron leakage, which constitutes different forms of oxidative stress [32]. However, under ischemic conditions, Ca^{2+} overflows into mitochondria during rapid loss of ATP and aggravates ROS production. These reactive radicals cause modification of biomolecules of DNAs (nDNAs and mtDNAs) and proteins, lead to lipid peroxidation, and may impair their bioactivities which eventually lead to the opening of the mitochondrial permeability transition pore (mPTP) and loss of mitochondrial membrane potential ($\Delta\Psi$m) [12].

The importance of the IMM-resident MPV17 protein in mitochondrial physiology has come to light recently. MPV17 is encoded by nDNA and functions as a $\Delta\Psi$m-modulating channel that contributes to mitochondrial homeostasis under different conditions. MPV17 forms weak cation-selective channesl and has shown several subconductance states in vitro [33]. Interestingly, MPV17 channel protein allows for the passage of small molecules such as deoxynucleotides triphosphates (dNTPs). It is possible that MPV17 maintains mitochondrial dNTP homeostasis, and therefore the perturbation of dNTP pools is a recognized cause of mitochondrial genomic instability [34]. Mutations of MPV17 are associated with mtDNA depletion syndrome, an inherited autosomal recessive disease in humans. Additionally, MPV-like proteins are identified as epithelial and neuronal restricted protein and implicated in ROS metabolism and apoptosis regulation through its binding and functional interaction with mitochondrial serine protease, HTRA2 [35]. Altogether, excess oxidative stress accumulation, mtDNA instability, and imbalance of bioenergetics will lead to mitochondrial dysfunction which could be important in the pathogenesis of kidney diseases [36].

Mitochondrial Dynamics

Mitochondria are highly dynamic organelles – their number, size and locations change constantly in response to energy demands. They constantly move, fuse and divide – switching between elongated interconnected networks or fragmented discrete morphologies via coordinated fusion and fission. These dynamics are essential to their size, morphology, energy biogenesis, function, and maintenance of cellular homeostasis and viability [37–39]. Perturbation of mitochondrial dynamics results in mitochondrial dysfunction and is linked to aging, end organ injury, and human diseases [40, 41]. Mitochondrial dynamics are regulated by a complex relationship between fission proteins (the large GTPase, dynamin related protein 1 (DRP1), and mitochondrial fission 1 (Fis1)) and fusion proteins (mitofusins 1 and 2 (Mfn1, Mfn2) and optical atrophy (OPA1)) [5, 42] (Fig. 27.1c). The NAD^+-dependent SIRT family deacetylases, especially the mitochondrial matrix-resided SIRT3 protein, play an important role in regulating mitochondrial dynamics and function. In this role, the SIRT3 deacetylates and activates the mitochondrial fusion protein OPA1 to improve mitochondrial function under stress conditions [43] or acute kidney injury [44].

Mitochondrial Turnover: Autophagy and Mitophagy

Autophagy is a non-selective process where cytoplasmic contents, e.g. damaged organelles and aged protein aggregates, are sequestered into autophagosomes for lysosome delivery and bulk degradation within the cell. Autophagy occurs during nutrient stress for ATP preservation, amino acid recycling, and anabolic protein synthesis. In certain stress conditions, selective autophagy occurs to efficiently remove toxic cellular materials such as damaged organelles. Mitophagy is one of the selective autophagy processes to remove accumulated dysfunctional mitochondria. Mitophagy clearance encompasses several steps – the damaged mitochondria are first marked by a loss of mitochondrial inner membrane potential ($\Delta\Psi m$) via intricate protein interactions involving kinases, E3 ubiquitin ligase, and proteins regulating mitochondrial dynamics and transportation [45] (Fig. 27.1d). The damaged mitochondria are eventually encapsulated into autophagosomes and degraded in autolysosomes. Autophagy (and mitophagy) is ROS-dependent, which is heavily regulated by cellular redox activity. Growing evidences support the notion that disturbance of autophagy/mitophagy is associated with the pathogenesis of renal diseases such as AKI [5], diabetic nephropathy [45] and glomerulosclerosis [46].

Mitochondria Damage in Kidney Diseases

Renal diseases encompassing both acute and chronic conditions of kidneys injury with declined renal functions are current global health concerns with tremendous medical burdens [3–5]. A common link between all forms of acute and chronic kidney injuries is the generation of toxic ROS and RNS when the disease manifests. This oxidative stress injury could be derived from ischemia/reperfusion, energy shortage from impaired mitochondrial biogenesis or ATP energetics, or defective clearance of damaged mitochondria (mitophagy). All of the listed conditions could link to recruitment of immune cells, inflammatory cytokines accumulation, apoptosis, and tissue injury. The renal phenotypes of mitochondrial dysfunction may manifest as proximal tubular dysfunction, tubularinterstitial disease, cystic kidney disease, podocytopathy, and nephrotic syndromes. All different forms of renal injuries might increase cardiovascular risk with uprising mortality and morbidity rates.

Acute Kidney Injury (AKI)

AKI has been a global concern with worldwide prevalence of 13 million people and death toll roughly 1.7 million annually [47]. AKI is a clinical condition that is commonly related to an acute episode of systemic injuries that occur in cases such as: kidney transplantation and septic and trauma patients. In the USA, AKI is the source of 1–2% of all hospital admissions and the mortality rates in hospitalized ICU patients with AKI could reach 50–70% [48]. Several etiologies such as ischemic, toxic, septic, and hypertensive injuries cause AKI with symptoms of abrupt reduction in kidney functions. AKI symptoms are manifested as acute decline in glomerular filtration rate, concomitant decreased urinary output, tubular necrosis, tubular interstitial inflammation and vascular permeability changes [5, 49].

Interestingly, in a polymicrobial septic AKI model, Tsuji and colleagues recently identified circulating mtDNA that stimulate systemic inflammation via *Toll*-like receptor 9 (TLR9) signaling as a novel pathogenic mechanism for AKI (Fig. 27.1e). The source of circulating mtDNA has been speculated to be derived from immune cells upon septic clearance or from spleen immune response [50, 51]. This study suggests that timely protection and removal of damaged mitochondria would reduce inflammation and could be an effective therapeutic strategy against AKI.

Ischemic-Reperfusion (I/R) Induced AKI

Renal ischemia-reperfusion injury (IRI) is a common cause of AKI and post-transplantation kidney allograft dysfunction resulting from an acute decline in general or localized renal oxygen and nutrients supplies to the affected tissue, and impairment of timely removal of metabolite wastes in kidney cells. Extensive oxidative stress generated during this process injures the tubular epithelial cells, causing histologically characteristic inflammation and necrotic cell death (necrosis) which was referred as acute tubule necrosis previously [52]. Emerging evidence supports the notion that multiple signaling pathways are involved in the pathobiology of AKI [52, 53]. Of these, damaged mitochondria with disrupted matrix cristae and mitochondrial permeability-mediated necrosis are crucial to ischemic AKI. Proper functional autophagy (mitophagy) is reno-protective during renal tissue regeneration if the injury is under a certain threshold, whereas defective mitophagy impedes tissue regeneration [5].

Nephrotoxic AKI

Exposure to nephrotoxins, such as some chemotherapy drugs, medications, intravascular contrast media, trace heavy metals, drug abuse, or certain chemicals/drugs combinations could induce AKI, especially in the elderly, young children, and high-risk patients [54, 55]. Drug nephrotoxicity was found to be responsible for 19% of all AKI cases on critically ill patients. The kidney is susceptible to intrinsic renal toxin injury which could result in various clinical syndromes ranging from minimal changes in tubular function to fulminant kidney failure. Injury may also occur to altered hemodynamic AKI. Some examples include: NSAIDs and renin-angiotensin system (RAS) antagonism, tubular epithelium direct cell injury from chemotherapy drugs like cisplatin, tubular obstruction of urinary flow due to precipitates of toxins or their metabolites, and angiopathogenesis from vascular injury and interstitial inflammation. Several research efforts have been dedicated to identify biomarkers of nephrotoxic AKI [56]. Cisplatin-induced renal mitochondrial structural, functional and homeostatic changes include swollen mitochondria, ultra-structurally disrupted IMM cristae, reduced mitochondrial mass, mtDNA depletion, and diminished cyt C oxidase activity in animal model [57]. Permeability of mitochondrial OMM upregulates the expression of mitochondrial dynamin related protein 1 (DRP1), an OMM resident fission protein, which induces mitochondrial fragmentation, cyt C release and apoptosis [58]. Reduction of mitochondria biogenesis regulators such as PGC-1α and SIRT3 were also documented [44]. Moreover, kidney specific overexpression of SIRT1 was protective against cisplatin-induced AKI [59]. The nephrotoxic effect and its pathophysiology of cisplatin-induced AKI had been a seminary topic of a recent review [60].

Septic AKI

Septic shock is one of the most frequent injury which could account for nearly half of the cause of AKI in critical ill patients [61]. Mitochondrial damage is an important contributing factor in the pathogenesis of septic AKI [62]. Clinical biopsies obtained from intensive care of human sepsis subjects also support this notion [63]. Apparent mitochondrial ultrastructure changes and biochemical studies revealed mitochondrial pathogenesis includes loss of mitochondrial mass, swelling, fragmentation due to aberrant mitochondrial dynamics, PGC-1α reduction, extensive mitochondrial cristae remodeling, and cyt C release during apoptosis. As aforementioned, mitochondrial damage and the release of mtDNA is linked to systemic inflammation and tissue injury. The intricate interplays between mitochondrial energy precursor NAD/NADH, mitochondrial health and kidney function were shown in several studies of septic AKI animal model [23, 24]. In summary, mitochondrial NAD concentration regulates the rate-limiting process of mitochondrial energy production and, hence, PGC-1α causes an increase of NAD biogenesis, mitochondrial numbers and subsequent renoprotection.

Chronic Kidney Disease (CKD)

CKD is characterized by persistent renal dysfunction for more than 3 months. The causes of CKD have been commonly associated with a number of comorbid conditions such as type 2 diabetes and hypertension [64]. The prevalence of CKD is estimated to be roughly 8–16% globally. According to the United States Renal Data System report, more than 13% of adults develop CKD and 30% of the population at the age of 65 and above experience some form of kidney failure in the USA [65]. CKD is staged based on estimated glomerular filtration rate (eGFR) and albuminuria with conditions ranging from asymptomatic stage to end-stage renal disease (ESRD) [66]. Different from AKI, CKD is recognized as a separate, irreversible and progressive pathologic condition which often inevitably leads to ESRD [7, 8, 49]. Individuals with diabetes or hypertension have high risk to develop CKD. Recent findings support the close relation between AKI and CKD and confirm, clinically, that one could serve as the risk factor to the other in maladaptive recovery conditions [7, 8].

Diabetic Nephropathy (DN)

Diabetic Nephropathy (DN), a progressive microvascular complication that arises from diabetes mellitus (DM) which affects individuals with both type 1 or type 2 DM, is the leading cause of CKD. DN is a chronic disease characterized by decreased glomerular filtration, proteinuria, glomerular hypertrophy, and renal fibrosis with gradual decline in renal function which contributes to 40% kidney failure and eventually requires renal replacement therapy [67–69].

The pathogenesis of DN is complex due to all four renal compartments-glomeruli, tubules, interstitium and blood vessel being involved. Moreover, etiologies of DN originated from hyperglycemia mediated microvasculate abnormalities which affect several signaling pathways. These include: increased glucose metabolite flux, advanced glycation end-products formation, endoplasmic reticulum stress, ROS overproduction, pro-inflammation and apoptotoc cell death in podocytes [68, 69]. It is well documented that aberrant mitochondrial homeostasis plays pivotal roles in DN progression [11, 68–73]. Hyperglycemia induced increased oxidative stress derived from renal ROS over-production in conditions like impaired mito-glutathione transport [74], increased mitochondrial activation of an ubiquitous NAD(P)H oxidase (Nox4) in respiratory chain [75], decreased mtDNA stability to against ROS production [76] and podocytes apoptosis [77] all contribute to cell damage and lethal effects linked to diabetic complications. Interestingly, hyperglycemia triggers mitochondrial fission and is mediated by Rho-associated coiled coil-containing protein kinase 1 (ROCK1)-dependent phosphorylation of dynamin related protein 1 (DRP1) and its recruitment to mitochondria [78]. These results suggest that mitochondrial dynamics play an important role in the pathogenesis of DN. Additionally, defects in mitochondrial biogenesis with downregulated expression or activity of PGC-1α is recognized to contribute to DN [71, 72]. It appears that the protective role of PGC-1α is associated with the inhibition of DRP1-mediated mitochondrial dynamics remodeling and ROS production [79]. Growing evidence also suggests that autophagy or mitophagy is altered in DN [45]. It is plausible that impairment in autophagic flux would lead to the DN complication associated with extracellular matrix deposition and fibrosis.

Glomerulonephritis (GN)

Chronic glomerulonephritis (GN), characterized by dysfunction in the glomerular filtration barrier, is a major feature of CKD. GN accounts for roughly 10% of CKD. GN is the inflammation of the glomeruli with clinical symptoms which may include hematuria, proteinuria, edema, and hypertension. Congenital GN can occur as genetic mutations on mitochondrial proteins (see section on "Genetic Mitochondrial Disease Affecting Kidney Functions"). Immune diseases may also cause chronic GN. Abnormal-shaped mitochondria are often found before the disease progress into secondary (or acquired) focal segmental glomerulosclerosis (FSGS), a severe form of GN [46]. Therefore, these evidence support the notion that impaired autophagic mitochondrial turnover is responsible for FSGS phenotype in mice.

Recently, Kruppel-like factors 6 (KLF6), a subfamily of DNA-binding zinc finger protein, was identified as an inducible early injury response gene that encodes a crucial transcriptional regulator of mitochondrial function in podocytes under stress response [80–83]. Podocyte-specific loss of *Klf6* reduced mitochondrial function and increased the susceptibility to FSGS in mice. Likewise, podocyte-specific KLF6 expression was significantly reduced in FSGS patients when compared to healthy individuals. Mechanistic studies indicated KLF6 prevents mitochondrial injury via enhancing cyt C assembly [82].

The function roles of MPV17 were explored further as discussed earlier. Growing body of evidence indicated that the glomerulosclerosis gene *mpv17* and its encoded MPV17 protein as important factors in regulating peroxisomal metabolism of ROS [84]. MPV17 was identified as an IMM- localized protein in podocyte. Loss of *Mpv17* was found to be associated with glomerulosclerosis and knockout mice developed renal failure later on in their lifespan (9–12 months). The glomerular lesions were caused by toxic oxidative stress and lipid peroxidation adducts. Interestingly, expression of MPV17 is diminished in several glomerular injury models and also in human FSGS subjects [85]. Altogether, the evidence further confirmed that MPV17 is important for mitochondrial homeostasis in podocytes.

Genetic Mitochondrial Disease Affecting Kidney Functions

Genetic disorders that affect mitochondrial function can be mutations related to either the nuclear or mitochondrial genomes which encode mitochondrial proteins. They represent genetically and clinically heterogeneous disorders which affect mostly metabolic-active organs. Genetic mitochondrial mutations could be derived from primary (inherited) or secondary to environmental predisposition to drugs or oxidative stress injury. Kidney phenotypes are often manifested as tubular dysfunction, interstitial nephritis, glomerular dysfunction, and renal tumors [9, 86].

MELAS and tRNA-Leu (m.A3243G) Transition

MELAS (mitochondrial encephalopathy, lactic acidosis, and stroke like symptoms) is a mitochondrial syndrome manifested in patients that are 40-years old or younger with symptoms like seizure and stroke-like episodes, muscle myopathy, lactic acidosis, and maternally inherited diabetes with deafness. A base transition (A3243G) in mitochondrial gene tRNA-Leu accounts for 80% of MELAS cases with occasional renal phenotypes of FSGS, or less commonly with tubular interstitial nephritis, which results in progressive renal insufficiency [87, 88]. Electron microscopic images from renal biopsies of MELAS patient revealed abnormal mitochondria in podocytes and tubular cells, indicating dysfunctional mitochondria is central to disease pathogenesis.

Mitochondrial DNA Deletion

Different from the mtDNA point mutations, few mitochondrial disease syndromes which are often associated with large mtDNA deletion at different genome locations and affect tubular pathologies, are Fanconi [89], Kearns-Sayre, and Pearson syndromes [90]. The clinical features of Pearson syndrome are manifested with pancreatic fibrosis with insulin-dependent diabetes, whereas the kidney phenotype of Kearns-Sayre syndrome includes renal tubular acidosis (proximal or distal) which occasionally progress to ESRD. Although occurrences may be rare, these syndromes involve renal tubulopathy (proximal or distal) manifested as electrolyte abnormalities and fibrosis. Renal biopsies of patients with Kearns-Sayre syndrome revealed impaired cyt C oxidase activity, defect in energy metabolism, and mitochondrial ultrastructures alterations [90].

Mutations in Nuclear-Encoded Genes for Mitochondrial Proteins

The mitochondria proteome is composed with about 1,000 proteins which are mostly encoded by nDNA. nDNA mutations with prominent tubular dysfunction include mutations that affect energy metabolism, often affecting complex III and IV, and mutations that affect fatty acid metabolism. Podocyte dysfunction is often correlated with multi-mutations in coenzyme Q10 biosynthesis pathways and the pathogenic mechanism was confirmed in *pdss2* gene encoding prenyl diphosphate synthase subunit 2, the first enzyme of coQ10 pathway [29, 91].

Aging in Kidney Disease

Older age is associated with an elevated risk of kidney disease and increased population of older renal transplant recipients and poor graft survival. The renal aging phenotype includes profound functional alterations leading to increased renal vascular resistance and reduced repair capacity. Aging kidney is often associated with histological changes such as tubular atrophy, interstitial fibrosis and glomerulosclerosis. Clinically, renal aging is associated with an increased susceptibility to AKI [92], and CKD may share accelerated vascular disease phenotype similar to the condition of premature aging [93]. The kidney suffers from increasing oxidative stress and disturbance in autophagy with aging [94, 95]. The molecular mechanism of renal cell senescence has been explored to certain extent. Downregulation of PGC-1α may link telomere-dysfunction to compromised renal mitochondrial function through p53 activation and alteration of transcription [96]. The renin–angiotensin–aldosterone system (RAAS) is a key regulator of blood pressure, fluid homeostasis and cardiovascular physiology. Angiotensin II and the age-dependent switch of mitochondrial expression of the two types of cell surface angiotensin receptors (AT1R *vs* AT2R) are important factor in determining respiratory activity during aging [97]. Further studies are needed to explore the role of the telomere–mitochondrial axis in renal physiology, aging and disease.

Mitochondria-Targeted Therapeutics

As protection of mitochondria could serve as a potential effective therapeutic strategy, it has been research endeavors focused on this development toward kidney diseases [49, 98–101]. These agents include antagonizing mitochondria oxidants, promoting mitochondrial biogenesis and ATP synthesis, regulating ROS metabolism, cardiolipin protection, inhibitors of mPTP, or inhibitors of mitochondrial fragmentation as detailed below (Table 27.1).

Mitochondria-Targeted Antioxidants

Mitochondria-targeted antioxidants (e.g. MitoQ, MitoTEMPO, MitoE, Mito-CP, SkQ1 and SkQR1) have been a part of the efforts to reduce mitochondrial oxidative stress. These agents are based on the mechanism of delivery of known redox agents to mitochondrial matrix through conjugation with TPP+ (triphenylalkylphosphonium

Table 27.1 Mitochondria-targeted therapeutics

Therapeutics	Chemicals	Action mechanisms	Experimental model
Antioxidants			
MitoQ[a]	Mitoquinone	Anti-oxidant concentrate at matrix in a $\Delta\Psi$m-dependent manner; ROS scavenger	IRI-AKI, cisplatin-AKI, sepsis-AKI
Biogenesis activator			
Resveratrol (SRT501)	Small peptides	AMPK / SIRT-1 / PGC-1α axis activator	Hemorrhagic shock, sepsis-AKI
AICAR		SIRT3 activator, AMPK agonist	Cisplatin-AKI, IRI-AKI
Formoterol		β2-adrenoceptor agonist, increase PGC-1α synthesis	IRI-AKI
ROS metabolism and bioenergetics			
Mitochonic acid (MA-5)	Synthetic compound	OXPHOS-independent increase of ATP synthesis, reduce ROS	Sepsis-AKI
Cardiolipin protection			
Bendavia (SS-31)[a]	Szeto-Schiller tetrapeptide	Protect cardiolipin from peroxidation; increase ATP and reduce ROS	AKI and CKD
		Prevent cyt C transformation into peroxidase	
mPTP inhibitor			
cyclosporin A (CsA)[a]	Small molecule	Interact with cyclophilin D (a mPTP structural protein)	Adriamycin-induced podocyte injury
TDZD-8		mPTP inhibitors, GSK3β inhibitor	Podocyte injury or drug induced AKI
Fission inhibitor			
Mdivi1	Small molecule	Inhibitor of mitochondrial division; induce mitochondria fusion	IRI-AKI
		Block DRP1 assembly and translocation	
K(ATP) channel opener			
Simdax (Levosimendan)[a]	Small molecule	Calcium sensitizer, K(ATP) channel opener	IRI-AKI
cyt C assembly			
KLF6	DNA-binding zinc finger transcriptional regulator	enhance cyt C assembly	Podocytes FSGS (animal study)

[a]Clinical trials. https://clinicaltrials.gov

cation) moiety [100]. These ROS scavenger compounds could cross the membrane lipid bilayer and concentrate at the matrix in a membrane potential ($\Delta\Psi m$) – dependent manner. MitoQ was shown to be safe in clinical trial with Parkinson's disease, fatty acid disease, and is currently under clinical trial (NCT02364648) for CKD. In summary, several animal models have confirmed that mitochondrial-targeted antioxidants represent an effective strategy to prevent or attenuate different forms of kidney diseases.

Activator of Mitochondria Biogenesis

Activation of Mitochondria biogenesis is required for the increased metabolic and energy demands during the recovery phase of acute organ injury. The AMPK/SIRT/PGC-1α axis plays crucial roles in mitochondrial biogenesis. PGC-1α is a pivotal factor that coordinately regulates NAD biogenesiss, reduces oxidative stress, and facilitates recovery from AKI [24]. SIRT1 is a NAD-dependent deacetylase that positively regulates PGC-1α expression and activity. Agents in this category include resveratrol, AICAR, and formoterol. Treatment with resveratrol, a natural plant phytoalexin, in a rat hemorrhagic shock model increases mitochondria mass, restores mitochondrial respiratory capacity, and reduces oxidative stress [102]. Moreover, resveratrol modulates immune response by suppressing inflammation driven by macrophage in a lipopolysaccharide -induced septic AKI model [103]. Pretreatment with AICAR, an AMPK activator, attenuated injury and tubular necrosis, decreased nitrosative stress, and ameliorated renal function in a rat I/R induced AKI [104]. Formoterol, a potent agonist of β2-adrenoreceptor, induces mitochondrial biogenesis by increasing mtDNA copy numbers, oxygen consumption rate, and PGC-1α synthesis. In doing so it restores renal function, rescues renal tubules from injury, and diminishes necrosis in an I/R-induced AKI animal model [105]. In summary, agents affect mitochondrial biogenesis and NAD modulation hold great promise in treating kidney disease, but their clinical translation still awaits further investigation.

Modulation of ROS Metabolism and ATP Synthesis

Mitochonic acid 5 (MA-5), a synthetic derivative of indole acetic acid plant hormone, was identified as a compound to enhance ATP production. MA-5 was found to improve the survival of fibroblasts collected from patients with mitochondrial diseases, including Leigh syndrome and MELAS [106]. Evidence supports that MA-5 enhances ATP production by promoting assembly and oligomerization of complex V at mitochondrial crista junction, and therefore preserving the mitochondria dynamics and preventing its fragmentation. These new findings provide new mechanisms in renal and cardiac protection [106, 107].

Cardiolipin Protections

As cardiolipin regulates IMM structural and functional plasticity, peroxidation or loss of cardiolipin has been associated with aging and several forms of AKI and CKD. Development of cardiolipin-target compound to optimize efficiency of the ETC and thereby restore cellular bioenergetics has been an innovative discovery. SS-31 peptide (Szeto-Schiller peptide, also named MTP-131 or Bendavia) is a cardiolipin-targeted tetrapeptide (D-Arg-dimethylTyr-Lys-Phe-NH2) that stabilizes cardiolipin, scavenges mitochondrial ROS, regulates cyt C activity, and inhibits the mitochondrial permeability transition (MPT) pore in AKI and CKD models [108–110]. SS-31 treatment demonstrated mitochondrial protective effects on swine stenotic-kidney injury and improved renal hemodynamics and function [111]. SS-31 has entered several clinical trials for treatment with AKI renal microvascular dysfunction in hypertension, and I/R injury of myocardial infarction [108].

mPTP Inhibitors

Opening of the IMM localized mitochondrial permeability transition pore (mPTP), a Ca^{2+}-dependent nonselective pore, under certain pathological conditions such as Ca^{2+} overload or oxidative stress leads to the IMM permeability to small molecules (<1,500 Da in molecular weight), loss of proton motive force, uncoupling of OXPHOS, and mitochondrial swelling which often followed by cell death. Opening of mPTP appears to be a pivotal event in AKI [112]. Cyclosporin A (CsA) is a potent inhibitor of mPTP through interaction with cyclophilin D, a structural component and mediator of mPTP which critically regulates Ca^{2+}- and ROS-dependent opening. Low dose of CsA (at submicromolar concentration) suppress mPTP opening and mitochondria swelling, and is currently under clinical trial for its effect on reperfusion injury on acute myocardial infarction (NCT01595958). CsA reduced podocyte damage through mitochondrial protection in an adriamycin-induced podocyte injury animal model [113]. A clinical trial (NCT02397213) is underway to test CsA renal protection in cardiac surgery (potential AKI condition). High dose of CsA is renotoxic and thus limits its clinical use.

Evidence suggests that glycogen synthase kinase (GSK)-3β resides at the nexus of multiple signaling pathways implicated in the regulation of mitochondrial permeability transition (MPT). The selective GSK-3β inhibitor TDZD-8, a thiadiazolidinone derivative, works as a non-ATP competitive inhibitor to prevent MPT and improve cell viability during AKI [112, 114].

Fission Inhibitors

Mitochondrial fusion is required for homogenous distribution of mtDNA, matrix proteins, and lipids across all fused mitochondria. However, mitochondria fission is important in mitochondrial proliferation following mitosis and is also involved in damaged (fragmented) mitochondrial removal by mitophagy. Mdivi-1 (mitochondrial fission inhibitor-1) was identified as a small molecule that selectively and reversibly inhibits the assembly and the GTPase activity of fission protein DRP1 [115]. Mdivi-1 induces rapid and reversible formation of fused mitochondria in wide varieties of cells and several animal disease models including I/R-injury in the heart, liver, kidney, and brain [116]. Mdivi-1 ameliorates rhabdomyolysis-induced AKI in rat by reducing ROS stress and tubular epithelial cell apoposis with simultaneous increase of ATP production [117] and protection from mPTP opening in acute cardiorenal syndrome [118]. Despite its therapeutic potential, Mdivi-1 exhibits divergent functions which could be cytoprotective or cytotoxic and the renoprotective effect of Mdivi-1 in humans is still lacking [116].

K(ATP) Channel Opener

Levosimendan is a Ca^{2+} sensitizing smooth muscle vasodilator clinically used for the treatment of heart failure. Levosimendan exerts pleiotropic beneficial effects including mitochondria protection in cardiomyocytes during ischemic heart disease [119]. Possible molecular mechanisms of Levosimendan might involve its function as a mitochondrial ATP-sensitive potassium channel to favorably conserve mitochondrial energy in cardiomyocytes. Pre-clinical and clinical data also indicated positive circulatory effects in the brain, lung, and renoprotection under potentially lethal stress conditions. Currently, a clinical trial (NCT01720030) with Levosimendan in an AKI study is ongoing.

Concluding Remarks and Future Perspectives

Mitochondria are the "power house" of the high-energy demanding kidney cells. The mitochondrial networks also affect the cross-talk with other cellular organelles such as the endoplasmic reticulum, nucleus, peroxisomes, and thus impact many cellular functions. Renal mitochondrial dysfunction could derive from defects on mitochondrial biogenesis, bioenergetics, dynamics, turnover and genetic mutations which may contribute to respiratory chain-derived oxidative stress, abnormal ultrastructures, increased sensitivity to apoptosis, and accumulation of damaged mitochondria with unstable mitochondrial DNA. Disturbance of renal mitochondrial homeostasis could lead to damages to microvasculatures, inflammation, fibrosis, and kidney failure. Although kidney mitochondrial dysfunctions have been exquisitely demonstrated to be tightly linked to pathobiology in many preclinical animal studies of CKD, AKI, aging and genetic defected mitochondrial disease, many direct cause-effect mechanisms remain elusive before development of clinical translations.

The mitochondria-targeting therapeutics as effective interventions to preserve mitochondria structures and functions have been demonstrated in several animal models of kidney injuries. Among these, MitoQ and SS-31 have shown promising potential in clinical trials towards certain kidney injuries and diseases. With better understandings of mitochondrial genome and its physiological functions, advances in in vivo animal studies, and human clinical trials to confirm efficacy and safety of mitochondria-targeting therapeutics, we have high hope to translate these discoveries into therapeutic strategies to ameliorate kidney injuries and diseases in the future.

Acknowledgments PHL is supported, in part, by an OSU intramural Lockwood Fund.

Conflict of Interest No competing interest.

References

1. Skorecki K, Chertow GM, Marsden PA, Taal MW, Yu ASL. Brenner and Rector's the kidney. 10th ed. Philadelphia: Saunders; 2015. 2748 p
2. Scott RP, Quaggin SE. Review series: the cell biology of renal filtration. J Cell Biol. 2015;209(2):199–210. Epub 2015/04/29
3. Lameire NH, Bagga A, Cruz D, De Maeseneer J, Endre Z, Kellum JA, et al. Acute kidney injury: an increasing global concern. Lancet. 2013;382(9887):170–9. Epub 2013/06/04
4. Hill NR, Fatoba ST, Oke JL, Hirst JA, O'Callaghan CA, Lasserson DS, et al. Global prevalence of chronic kidney disease – a systematic review and meta-analysis. PLoS One. 2016;11(7):e0158765. Epub 2016/07/08
5. Duann P, Lianos EA, Ma J, Lin PH. Autophagy, innate immunity and tissue repair in acute kidney injury. Int J Mol Sci. 2016;17(5):E662, p1–19. Epub 2016/05/07
6. Weisbord SD, Palevsky PM. Design of clinical trials in acute kidney injury: lessons from the past and future directions. Semin Nephrol. 2016;36(1):42–52. Epub 2016/04/18
7. Ferenbach DA, Bonventre JV. Acute kidney injury and chronic kidney disease: from the laboratory to the clinic. Nephrol Ther. 2016;12(Suppl 1):S41–8. Epub 2016/03/15
8. Ferenbach DA, Bonventre JV. Mechanisms of maladaptive repair after AKI leading to accelerated kidney ageing and CKD. Nat Rev Nephrol. 2015;11(5):264–76. Epub 2015/02/04
9. Emma F, Montini G, Parikh SM, Salviati L. Mitochondrial dysfunction in inherited renal disease and acute kidney injury. Nat Rev Nephrol. 2016;12(5):267–80.
10. Divers J, Freedman BI. Genetics in kidney disease in 2013: susceptibility genes for renal and urological disorders. Nat Rev Nephrol. 2014;10(2):69–70. Epub 2013/12/04
11. Ratliff BB, Abdulmahdi W, Pawar R, Wolin MS. Oxidant mechanisms in renal injury and disease. Antioxid Redox Signal. 2016;25:119. Epub 2016/02/26
12. Kroemer G, Galluzzi L, Brenner C. Mitochondrial membrane permeabilization in cell death. Physiol Rev. 2007;87(1):99–163. Epub 2007/01/24
13. Hoenig MP, Zeidel ML. Homeostasis, the milieu interieur, and the wisdom of the nephron. Clin J Am Soc Nephrol. 2014;9(7):1272–81. Epub 2014/05/03
14. Forbes JM. Mitochondria-power players in kidney function? Trends Endocrinol Metab. 2016;27(7):441–2. Epub 2016/05/25
15. Hall AM, Unwin RJ, Parker N, Duchen MR. Multiphoton imaging reveals differences in mitochondrial function between nephron segments. J Am Soc Nephrol. 2009;20(6):1293–302. Epub 2009/05/28
16. Muller-Deile J, Schiffer M. The podocyte power-plant disaster and its contribution to glomerulopathy. Front Endocrinol (Lausanne). 2014;5:209. Epub 2015/01/08
17. Stieger N, Worthmann K, Teng B, Engeli S, Das AM, Haller H, et al. Impact of high glucose and transforming growth factor-beta on bioenergetic profiles in podocytes. Metabolism. 2012;61(8):1073–86. Epub 2012/03/01

18. Weinberg JM. Mitochondrial biogenesis in kidney disease. J Am Soc Nephrol. 2011;22(3):431–6. Epub 2011/03/01
19. Funk JA, Schnellmann RG. Accelerated recovery of renal mitochondrial and tubule homeostasis with SIRT1/PGC-1alpha activation following ischemia-reperfusion injury. Toxicol Appl Pharmacol. 2013;273(2):345–54. Epub 2013/10/08
20. Stallons LJ, Funk JA, Schnellmann RG. Mitochondrial homeostasis in acute organ failure. Curr Pathobiol Rep. 2013;1(3):169–77. Epub 2014/01/05
21. Quiros PM, Mottis A, Auwerx J. Mitonuclear communication in homeostasis and stress. Nat Rev Mol Cell Biol. 2016;17(4):213–26. Epub 2016/03/10
22. Tran M, Parikh SM. Mitochondrial biogenesis in the acutely injured kidney. Nephron Clin Pract. 2014;127(1–4):42–5. Epub 2014/10/25
23. Tran M, Tam D, Bardia A, Bhasin M, Rowe GC, Kher A, et al. PGC-1alpha promotes recovery after acute kidney injury during systemic inflammation in mice. J Clin Invest. 2011;121(10):4003–14. Epub 2011/09/02
24. Tran MT, Zsengeller ZK, Berg AH, Khankin EV, Bhasin MK, Kim W, et al. PGC1alpha drives NAD biosynthesis linking oxidative metabolism to renal protection. Nature. 2016;531(7595):528–32. Epub 2016/03/17
25. Dinkova-Kostova AT, Abramov AY. The emerging role of Nrf2 in mitochondrial function. Free Radic Biol Med. 2015;88(Pt B):179–88. Epub 2015/05/16
26. Zoja C, Benigni A, Remuzzi G. The Nrf2 pathway in the progression of renal disease. Nephrol Dial Transplant. 2014;29(Suppl 1):i19–24. Epub 2013/06/14
27. Verdin E. NAD(+) in aging, metabolism, and neurodegeneration. Science. 2015;350(6265):1208–13. Epub 2016/01/20
28. Canto C, Menzies KJ, Auwerx J. NAD(+) metabolism and the control of energy homeostasis: a balancing act between mitochondria and the nucleus. Cell Metab. 2015;22(1):31–53. Epub 2015/06/30
29. Saha SP, Whayne Jr TF. Coenzyme Q-10 in human health: supporting evidence? South Med J. 2016;109(1):17–21. Epub 2016/01/08
30. Ascenzi P, Coletta M, Wilson MT, Fiorucci L, Marino M, Polticelli F, et al. Cardiolipin-cytochrome c complex: switching cytochrome c from an electron-transfer shuttle to a myoglobin- and a peroxidase-like heme-protein. IUBMB Life. 2015;67(2):98–109. Epub 2015/04/11
31. O'Brien ES, Nucci NV, Fuglestad B, Tommos C, Wand AJ. Defining the apoptotic trigger: the interaction of cytochrome c and cardiolipin. J Biol Chem. 2015;290(52):30879–87. Epub 2015/10/22
32. Drose S, Brandt U. Molecular mechanisms of superoxide production by the mitochondrial respiratory chain. Adv Exp Med Biol. 2012;748:145–69. Epub 2012/06/26
33. Antonenkov VD, Isomursu A, Mennerich D, Vapola MH, Weiher H, Kietzmann T, et al. The human mitochondrial DNA depletion syndrome gene MPV17 encodes a non-selective channel that modulates membrane potential. J Biol Chem. 2015;290(22):13840–61. Epub 2015/04/12
34. Dalla Rosa I, Camara Y, Durigon R, Moss CF, Vidoni S, Akman G, et al. MPV17 loss causes deoxynucleotide insufficiency and slow DNA replication in mitochondria. PLoS Genet. 2016;12(1):e1005779. Epub 2016/01/14
35. Krick S, Shi S, Ju W, Faul C, Tsai SY, Mundel P, et al. Mpv17l protects against mitochondrial oxidative stress and apoptosis by activation of Omi/HtrA2 protease. Proc Natl Acad Sci U S A. 2008;105(37):14106–11. Epub 2008/09/06
36. Che R, Yuan Y, Huang S, Zhang A. Mitochondrial dysfunction in the pathophysiology of renal diseases. Am J Phys Renal Phys. 2014;306(4):F367–78. Epub 2013/12/07
37. Putti R, Sica R, Migliaccio V, Lionetti L. Diet impact on mitochondrial bioenergetics and dynamics. Front Physiol. 2015;6:109. Epub 2015/04/24
38. Liesa M, Shirihai OS. Mitochondrial dynamics in the regulation of nutrient utilization and energy expenditure. Cell Metab. 2013;17(4):491–506. Epub 2013/04/09

39. Pernas L, Scorrano L. Mito-morphosis: mitochondrial fusion, fission, and cristae remodeling as key mediators of cellular function. Annu Rev Physiol. 2016;78:505–31. Epub 2015/12/17
40. McDonnell E, Peterson BS, Bomze HM, Hirschey MD. SIRT3 regulates progression and development of diseases of aging. Trends Endocrinol Metab. 2015;26(9):486–92. Epub 2015/07/04
41. Benigni A, Perico L, Macconi D. Mitochondrial dynamics is linked to longevity and protects from end-organ injury: the emerging role of sirtuin 3. Antioxid Redox Signal. 2016;25(4):185–99. Epub 2016/03/15
42. Zhan M, Brooks C, Liu F, Sun L, Dong Z. Mitochondrial dynamics: regulatory mechanisms and emerging role in renal pathophysiology. Kidney Int. 2013;83(4):568–81. Epub 2013/01/18
43. Samant SA, Zhang HJ, Hong Z, Pillai VB, Sundaresan NR, Wolfgeher D, et al. SIRT3 deacetylates and activates OPA1 to regulate mitochondrial dynamics during stress. Mol Cell Biol. 2014;34(5):807–19. Epub 2013/12/18
44. Morigi M, Perico L, Rota C, Longaretti L, Conti S, Rottoli D, et al. Sirtuin 3-dependent mitochondrial dynamic improvements protect against acute kidney injury. J Clin Invest. 2015;125(2):715–26. Epub 2015/01/22
45. Higgins GC, Coughlan MT. Mitochondrial dysfunction and mitophagy: the beginning and end to diabetic nephropathy? Br J Pharmacol. 2014;171(8):1917–42. Epub 2014/04/12
46. Kawakami T, Gomez IG, Ren S, Hudkins K, Roach A, Alpers CE, et al. Deficient autophagy results in mitochondrial dysfunction and FSGS. J Am Soc Nephrol. 2015;26(5):1040–52. Epub 2014/11/20
47. Mehta RL, Cerda J, Burdmann EA, Tonelli M, Garcia-Garcia G, Jha V, et al. International Society of Nephrology's 0by25 initiative for acute kidney injury (zero preventable deaths by 2025): a human rights case for nephrology. Lancet. 2015;385(9987):2616–43. Epub 2015/03/18
48. Bellomo R, Kellum JA, Ronco C. Acute kidney injury. Lancet. 2012;380(9843):756–66. Epub 2012/05/24
49. Eirin A, Lerman A, Lerman LO. The Emerging role of mitochondrial targeting in kidney disease. Handbook experimental pharmacology. Berlin/Heidelberg: Springer; 2016. p. 1–22. Copyright Holder Springer International Publishing Switzerland. Epub 2016/06/19
50. Tang C, Dong Z. Mitochondria in kidney injury: when the power plant fails. J Am Soc Nephrol. 2016;27(7):1869–72. Epub 2016/01/09
51. Tsuji N, Tsuji T, Ohashi N, Kato A, Fujigaki Y, Yasuda H. Role of mitochondrial DNA in septic AKI via toll-like receptor 9. J Am Soc Nephrol. 2016;27(7):2009–20. Epub 2015/11/18
52. Zuk A, Bonventre JV. Acute kidney injury. Annu Rev Med. 2016;67:293–307. Epub 2016/01/16
53. Kers J, Leemans JC, Linkermann A. An overview of pathways of regulated necrosis in acute kidney injury. Semin Nephrol. 2016;36(3):139–52. Epub 2016/06/25
54. Perazella MA, Luciano RL. Review of select causes of drug-induced AKI. Expert Rev Clin Pharmacol. 2015;8(4):367–71. Epub 2015/06/26
55. Pendergraft 3rd WF, Herlitz LC, Thornley-Brown D, Rosner M, Niles JL. Nephrotoxic effects of common and emerging drugs of abuse. Clin J Am Soc Nephrol. 2014;9(11):1996–2005. Epub 2014/07/19
56. Ferguson MA, Vaidya VS, Bonventre JV. Biomarkers of nephrotoxic acute kidney injury. Toxicology. 2008;245(3):182–93. Epub 2008/02/26
57. Mukhopadhyay P, Horvath B, Zsengeller Z, Zielonka J, Tanchian G, Holovac E, et al. Mitochondrial-targeted antioxidants represent a promising approach for prevention of cisplatin-induced nephropathy. Free Radic Biol Med. 2012;52(2):497–506. Epub 2011/11/29
58. Brooks C, Wei Q, Cho SG, Dong Z. Regulation of mitochondrial dynamics in acute kidney injury in cell culture and rodent models. J Clin Invest. 2009;119(5):1275–85. Epub 2009/04/08
59. Hasegawa K, Wakino S, Yoshioka K, Tatematsu S, Hara Y, Minakuchi H, et al. Kidney-specific overexpression of Sirt1 protects against acute kidney injury by retaining peroxisome function. J Biol Chem. 2010;285(17):13045–56. Epub 2010/02/09

60. Ozkok A, Edelstein CL. Pathophysiology of cisplatin-induced acute kidney injury. Biomed Res Int. 2014;2014:967826. Epub 2014/08/29
61. Uchino S, Kellum JA, Bellomo R, Doig GS, Morimatsu H, Morgera S, et al. Acute renal failure in critically ill patients: a multinational, multicenter study. JAMA. 2005;294(7):813–8. Epub 2005/08/18
62. Parikh SM, Yang Y, He L, Tang C, Zhan M, Dong Z. Mitochondrial function and disturbances in the septic kidney. Semin Nephrol. 2015;35(1):108–19. Epub 2015/03/22
63. Takasu O, Gaut JP, Watanabe E, To K, Fagley RE, Sato B, et al. Mechanisms of cardiac and renal dysfunction in patients dying of sepsis. Am J Respir Crit Care Med. 2013;187(5):509–17. Epub 2013/01/26
64. Jha V, Garcia-Garcia G, Iseki K, Li Z, Naicker S, Plattner B, et al. Chronic kidney disease: global dimension and perspectives. Lancet. 2013;382(9888):260–72. Epub 2013/06/04
65. SYSTEM USRD. USRDS annual data report: epidemiology of kidney disease in the United States. Bethesda: USRDS- United States Renal Data System; 2015.
66. KDIGO 2012, CKD Work Group. KDIGO 2012 clinical practice guideline for the evaluation and management of chronic kidney disease. Kidney Int. 2013;2012 Suppl:1–150.
67. D'Agati VD, Chagnac A, de Vries AP, Levi M, Porrini E, Herman-Edelstein M, et al. Obesity-related glomerulopathy: clinical and pathologic characteristics and pathogenesis. Nat Rev Nephrol. 2016;12(8):453–71. Epub 2016/06/07
68. Badal SS, Danesh FR. New insights into molecular mechanisms of diabetic kidney disease. Am J Kidney Dis. 2014;63(2 Suppl 2):S63–83. Epub 2014/01/28
69. Forbes JM, Cooper ME. Mechanisms of diabetic complications. Physiol Rev. 2013;93(1):137–88. Epub 2013/01/11
70. Blake R, Trounce IA. Mitochondrial dysfunction and complications associated with diabetes. Biochim Biophys Acta. 2014;1840(4):1404–12. Epub 2013/11/20
71. Sharma K. Obesity, oxidative stress, and fibrosis in chronic kidney disease. Kidney Int Suppl (2011). 2014;4(1):113–7. Epub 2014/11/18
72. Sharma K. Obesity and diabetic kidney disease: role of oxidant stress and redox balance. Antioxid Redox Signal. 2016;25(4):208–16. Epub 2016/03/18
73. Hallan S, Sharma K. The role of mitochondria in diabetic kidney disease. Curr Diabetes Rep. 2016;16(7):61. Epub 2016/05/10
74. Lash LH. Mitochondrial glutathione in diabetic nephropathy. J Clin Med. 2015;4(7):1428–47. Epub 2015/08/05
75. Block K, Gorin Y, Abboud HE. Subcellular localization of Nox4 and regulation in diabetes. Proc Natl Acad Sci U S A. 2009;106(34):14385–90. Epub 2009/08/27
76. Savu O, Sunkari VG, Botusan IR, Grunler J, Nikoshkov A, Catrina SB. Stability of mitochondrial DNA against reactive oxygen species (ROS) generated in diabetes. Diabetes Metab Res Rev. 2011;27(5):470–9. Epub 2011/04/13
77. Susztak K, Raff AC, Schiffer M, Bottinger EP. Glucose-induced reactive oxygen species cause apoptosis of podocytes and podocyte depletion at the onset of diabetic nephropathy. Diabetes. 2006;55(1):225–33. Epub 2005/12/29
78. Wang W, Wang Y, Long J, Wang J, Haudek SB, Overbeek P, et al. Mitochondrial fission triggered by hyperglycemia is mediated by ROCK1 activation in podocytes and endothelial cells. Cell Metab. 2012;15(2):186–200. Epub 2012/02/14
79. Guo K, Lu J, Huang Y, Wu M, Zhang L, Yu H, et al. Protective role of PGC-1alpha in diabetic nephropathy is associated with the inhibition of ROS through mitochondrial dynamic remodeling. PLoS One. 2015;10(4):e0125176. Epub 2015/04/09
80. Mallipattu SK, He JC. KLF 6: a mitochondrial regulator in the kidney. Oncotarget. 2015;6(18):15720–1. Epub 2015/07/15
81. Mallipattu SK, He JC. The podocyte as a direct target for treatment of glomerular disease? Am J Phys Renal Phys. 2016;311(1):F46–51. Epub 2016/04/22
82. Mallipattu SK, Horne SJ, D'Agati V, Narla G, Liu R, Frohman MA, et al. Kruppel-like factor 6 regulates mitochondrial function in the kidney. J Clin Invest. 2015;125(3):1347–61. Epub 2015/02/18

83. Kopp JB. Loss of Kruppel-like factor 6 cripples podocyte mitochondrial function. J Clin Invest. 2015;125(3):968–71. Epub 2015/02/18
84. Zwacka RM, Reuter A, Pfaff E, Moll J, Gorgas K, Karasawa M, et al. The glomerulosclerosis gene Mpv17 encodes a peroxisomal protein producing reactive oxygen species. EMBO J. 1994;13(21):5129–34. Epub 1994/11/01
85. Casalena G, Krick S, Daehn I, Yu L, Ju W, Shi S, et al. Mpv17 in mitochondria protects podocytes against mitochondrial dysfunction and apoptosis in vivo and in vitro. Am J Phys Renal Phys. 2014;306(11):F1372–80. Epub 2014/03/07
86. O'Toole JF. Renal manifestations of genetic mitochondrial disease. Int J Nephrol Renov Dis. 2014;7:57–67. Epub 2014/02/12
87. Guery B, Choukroun G, Noel LH, Clavel P, Rotig A, Lebon S, et al. The spectrum of systemic involvement in adults presenting with renal lesion and mitochondrial tRNA(Leu) gene mutation. J Am Soc Nephrol. 2003;14(8):2099–108. Epub 2003/07/23
88. Tzen CY, Tsai JD, Wu TY, Chen BF, Chen ML, Lin SP, et al. Tubulointerstitial nephritis associated with a novel mitochondrial point mutation. Kidney Int. 2001;59(3):846–54. Epub 2001/03/07
89. Carney EF. Tubular disease: mistargeted protein disrupts mitochondrial metabolism in inherited Fanconi syndrome. Nat Rev Nephrol. 2014;10(3):125. Epub 2014/01/22
90. Goto Y, Itami N, Kajii N, Tochimaru H, Endo M, Horai S. Renal tubular involvement mimicking Bartter syndrome in a patient with Kearns-Sayre syndrome. J Pediatr. 1990;116(6):904–10. Epub 1990/06/01
91. Peng M, Falk MJ, Haase VH, King R, Polyak E, Selak M, et al. Primary coenzyme Q deficiency in Pdss2 mutant mice causes isolated renal disease. PLoS Genet. 2008;4(4):e1000061. Epub 2008/04/26
92. Xue JL, Daniels F, Star RA, Kimmel PL, Eggers PW, Molitoris BA, et al. Incidence and mortality of acute renal failure in Medicare beneficiaries, 1992 to 2001. J Am Soc Nephrol. 2006;17(4):1135–42. Epub 2006/02/24
93. Kooman JP, Kotanko P, Schols AM, Shiels PG, Stenvinkel P. Chronic kidney disease and premature ageing. Nat Rev Nephrol. 2014;10(12):732–42. Epub 2014/10/08
94. Romano AD, Serviddio G, de Matthaeis A, Bellanti F, Vendemiale G. Oxidative stress and aging. J Nephrol. 2010;23(Suppl 15):S29–36. Epub 2010/11/27
95. Fougeray S, Pallet N. Mechanisms and biological functions of autophagy in diseased and ageing kidneys. Nat Rev Nephrol. 2015;11(1):34–45. Epub 2014/11/12
96. Schmitt R, Susnik N, Melk A. Molecular aspects of renal senescence. Curr Opin Organ Transplant. 2015;20(4):412–6. Epub 2015/07/01
97. Abadir PM, Foster DB, Crow M, Cooke CA, Rucker JJ, Jain A, et al. Identification and characterization of a functional mitochondrial angiotensin system. Proc Natl Acad Sci U S A. 2011;108(36):14849–54. Epub 2011/08/20
98. Tabara LC, Poveda J, Martin-Cleary C, Selgas R, Ortiz A, Sanchez-Nino MD. Mitochondria-targeted therapies for acute kidney injury. Expert Rev Mol Med. 2014;16:e13. Epub 2014/08/12
99. Granata S, Dalla Gassa A, Tomei P, Lupo A, Zaza G. Mitochondria: a new therapeutic target in chronic kidney disease. Nutr Metab (Lond). 2015;12:49. Epub 2015/11/28
100. Kezic A, Spasojevic I, Lezaic V, Bajcetic M. Mitochondria-targeted antioxidants: future perspectives in kidney ischemia reperfusion injury. Oxidative Med Cell Longev. 2016;2016:2950503. Epub 2016/06/18
101. Ishimoto Y, Inagi R. Mitochondria: a therapeutic target in acute kidney injury. Nephrol Dial Transplant. 2016;31(7):1062–9. Epub 2015/09/04
102. Wang H, Guan Y, Karamercan MA, Ye L, Bhatti T, Becker LB, et al. Resveratrol rescues kidney mitochondrial function following hemorrhagic shock. Shock. 2015;44(2):173–80. Epub 2015/04/22
103. Chen L, Yang S, Zumbrun EE, Guan H, Nagarkatti PS, Nagarkatti M. Resveratrol attenuates lipopolysaccharide-induced acute kidney injury by suppressing inflammation driven by macrophages. Mol Nutr Food Res. 2015;59(5):853–64. Epub 2015/02/04

104. Lempiainen J, Finckenberg P, Levijoki J, Mervaala E. AMPK activator AICAR ameliorates ischaemia reperfusion injury in the rat kidney. Br J Pharmacol. 2012;166(6):1905–15. Epub 2012/02/14
105. Jesinkey SR, Funk JA, Stallons LJ, Wills LP, Megyesi JK, Beeson CC, et al. Formoterol restores mitochondrial and renal function after ischemia-reperfusion injury. J Am Soc Nephrol. 2014;25(6):1157–62. Epub 2014/02/11
106. Suzuki T, Yamaguchi H, Kikusato M, Matsuhashi T, Matsuo A, Sato T, et al. Mitochonic Acid 5 (MA-5), a derivative of the plant hormone Indole-3-acetic acid, improves survival of fibroblasts from patients with mitochondrial diseases. Tohoku J Exp Med. 2015;236(3):225–32. Epub 2015/06/30
107. Suzuki T, Yamaguchi H, Kikusato M, Hashizume O, Nagatoishi S, Matsuo A, et al. Mitochonic acid 5 binds mitochondria and ameliorates renal tubular and cardiac myocyte damage. J Am Soc Nephrol. 2016;27(7):1925–32. Epub 2015/11/27
108. Szeto HH. First-in-class cardiolipin-protective compound as a therapeutic agent to restore mitochondrial bioenergetics. Br J Pharmacol. 2014;171(8):2029–50. Epub 2013/10/15
109. Szeto HH, Birk AV. Serendipity and the discovery of novel compounds that restore mitochondrial plasticity. Clin Pharmacol Ther. 2014;96(6):672–83. Epub 2014/09/05
110. Birk AV, Liu S, Soong Y, Mills W, Singh P, Warren JD, et al. The mitochondrial-targeted compound SS-31 re-energizes ischemic mitochondria by interacting with cardiolipin. J Am Soc Nephrol. 2013;24(8):1250–61. Epub 2013/07/03
111. Eirin A, Ebrahimi B, Zhang X, Zhu XY, Woollard JR, He Q, et al. Mitochondrial protection restores renal function in swine atherosclerotic renovascular disease. Cardiovasc Res. 2014;103(4):461–72. Epub 2014/06/21
112. Wang Z, Ge Y, Bao H, Dworkin L, Peng A, Gong R. Redox-sensitive glycogen synthase kinase 3beta-directed control of mitochondrial permeability transition: rheostatic regulation of acute kidney injury. Free Radic Biol Med. 2013;65:849–58. Epub 2013/08/27
113. Guan N, Ren YL, Liu XY, Zhang Y, Pei P, Zhu SN, et al. Protective role of cyclosporine A and minocycline on mitochondrial disequilibrium-related podocyte injury and proteinuria occurrence induced by adriamycin. Nephrol Dial Transplant. 2015;30(6):957–69. Epub 2015/02/04
114. Wang Z, Bao H, Ge Y, Zhuang S, Peng A, Gong R. Pharmacological targeting of GSK3beta confers protection against podocytopathy and proteinuria by desensitizing mitochondrial permeability transition. Br J Pharmacol. 2015;172(3):895–909. Epub 2014/09/30
115. Cassidy-Stone A, Chipuk JE, Ingerman E, Song C, Yoo C, Kuwana T, et al. Chemical inhibition of the mitochondrial division dynamin reveals its role in Bax/Bak-dependent mitochondrial outer membrane permeabilization. Dev Cell. 2008;14(2):193–204. Epub 2008/02/13
116. Rosdah AA, Holien JK, Delbridge LM, Dusting GJ, Lim SY. Mitochondrial fission – a drug target for cytoprotection or cytodestruction? Pharmacol Res Perspect. 2016;4(3):e00235. Epub 2016/07/20
117. Tang WX, Wu WH, Qiu HY, Bo H, Huang SM. Amelioration of rhabdomyolysis-induced renal mitochondrial injury and apoptosis through suppression of Drp-1 translocation. J Nephrol. 2013;26(6):1073–82. Epub 2013/04/05
118. Sumida M, Doi K, Ogasawara E, Yamashita T, Hamasaki Y, Kariya T, et al. Regulation of mitochondrial dynamics by dynamin-related protein-1 in acute cardiorenal syndrome. J Am Soc Nephrol. 2015;26(10):2378–87. Epub 2015/02/04
119. Papp Z, Edes I, Fruhwald S, De Hert SG, Salmenpera M, Leppikangas H, et al. Levosimendan: molecular mechanisms and clinical implications: consensus of experts on the mechanisms of action of levosimendan. Int J Cardiol. 2012;159(2):82–7. Epub 2011/07/26

Chapter 28
Mitochondrial Dysfunction in the Diabetic Kidney

Kumar Sharma

The pathobiology of diabetic kidney disease has primarily focused on a variety of metabolic pathways linked to glucose metabolism and inflammatory mediators [1]. However, many of the pathways implicated including the poly-ol pathway, hexosamine formation, glycosylation, protein kinase C have not translated to interventions showing clinical utility [1–3]. Beginning with the work from Brownlee et al., the concept that mitochondrial superoxide production played an upstream role in regulating a variety of pathways has held sway [4, 5]. Unfortunately, clinical approaches to reduce superoxide production have not had promising results [6, 7]. However, recent work has not found evidence of an increase in mitochondrial superoxide production in the diabetic kidney [8, 9]. In contrast, an opposing theory stating that reduced mitochondrial function (or dysfunction) is a major upstream consequence of hyperglycemia and diabetes has been gaining traction [10]. In support of this newer view, an approach to augment mitochondrial function may provide novel therapeutics to delay and arrest diabetic complications.

K. Sharma (✉)
Institute of Metabolomic Medicine, Center for Renal Translational Medicine, University of California San Diego/Veterans Affairs San Diego Healthcare System,
Stein Clinical Research Building, 4th Floor, 9500 Gilman Drive, La Jolla, CA 92093-0711, USA

Division of Nephrology-Hypertension, Veterans Affairs San Diego Healthcare System,
La Jolla, CA 92093, USA
e-mail: kumarsharma@ucsd.edu

G. Santulli (ed.), *Mitochondrial Dynamics in Cardiovascular Medicine*,
Advances in Experimental Medicine and Biology 982,
DOI 10.1007/978-3-319-55330-6_28

Mitochondrial Function in the Kidney

Apart from the heart, the kidney is considered to contain the greatest density of mitochondria per mass of tissue [11]. An explanation for this observation may be that apart from the heart, the kidney is constantly required to be functioning to maintain adequate homeostasis of water, electrolytes, minerals and many other organic and inorganic ions within the organism. Therefore, a loss of mitochondrial function or content could be the basis or consequence of reduced organ function. The role of mitochondria in organ dysfunction may be best understood in the setting of ischemia-reperfusion syndrome in the heart and kidney. The normal heart has to continuously perform a coordinated function linking electrical activity to diastolic and systolic activity of the right and left ventricles. The energy demands amount to about 30 kg of ATP production on a daily basis and much of the ATP generation is considered to be derived from ox-phos activity. However, it is exceedingly difficult to quantitate the amount and function of mitochondria in an organ. One approach is to quantify the supramolecular complexes or respirasomes [12]. Recently, respirasomes were found to be reduced in heart failure [12] setting up a paradigm that reduced mitochondrial energy output could contribute to the progressive failure of heart function in CHF [13].

Similarly, the kidney tubular cells are considered to require ATP for the enormous requirements of the proximal tubular cell to retain water, sodium and electrolytes. The TAL cell also requires a large amount of ATP to pump sodium, potassium and chloride into the interstitium and generate the counter-current gradient. In models of ischemia-reperfusion injury and septic kidney disease there appears to be a marked depletion of mitochondrial function [11, 14], which may underlie the basis of reduced renal function despite normal appearing glomeruli and restoration of renal blood flow. Whether mitochondrial dysfunction plays a role in human diabetic kidney disease was not comprehensively evaluated until quite recently.

Metabolomic Studies Implicating Mitochondrial Dysfunction in Diabetes

Recently, several groups have applied both urine and serum metabolomics to characterize biochemical disturbances in diabetic kidney disease. Blood studies revealed a role for pseudouridine as a biomarker for diabetic kidney disease [15, 16], however the role of pseudouridine remains unclear. Other metabolites that have been identified include indoxylsulfate [17] which may be a marker of renal dysfunction and has been recognized as a uremic metabolite and partly controlled by gut bacteria [16]. Kynurenine has also been found to be elevated in patients and may predispose to kidney disease [18]. Additionally, plasma sphingomyelin has been found to be elevated in patients with diabetes and may pre-date kidney disease [19].

Urine metabolomics studies have been plagued by lack of quality control in collecting timed samples of urine [3]. A well conducted study using both GC-MS and

LC-MS based analysis of urine [20] identified that acyl-carnitines, acyl-glycines and tryptophan metabolites discriminated patients with DKD. In our study of subjects with diabetes and evidence of kidney disease, characterized by reduced eGFR (<75 ml/min/1.73 m^2) using 24 h urine collections, there was a consistent reduction in a panel of 13 metabolites [21]. These 13 metabolites were found to be reduced in subjects from many parts of the USA and Europe with diabetes and CKD. Using a network analysis approach it was found that the 13 metabolites were all linked in a large network and that 12/13 metabolites were controlled by enzymes localized to mitochondria. As there was a significant reduction of the metabolites it was concluded that overall mitochondrial function was reduced in patients with diabetic kidney disease. Support from this conclusion came from studies of kidney biopsies which indicated a reduction in mitochondrial protein (complex IV) and reduction in PGC1α the master regulator of biogenesis.

A reduction in mitochondria in patients with diabetic kidney disease was also demonstrated by a reduction of urine exosomal mtDNA [21] and a reduction in blood levels of mtDNA [22]. These studies demonstrate from a variety of approaches that diabetic kidney disease is marked by reduced mitochondrial content and function. Of major interest is that non-diabetic kidney disease is also marked by reduction of circulating mtDNA [23] and reduced PGC1α in kidney tissues [24].

Mitochondrial Content and Function in Experimental DKD

Mitochondrial function has been considered to be at the heart of understanding diabetic complications as glucose oxidation fundamentally involves the mitochondrial TCA cycle and oxidative phosphorylation. However, the response of mitochondria to excess glucose in many cells and organs still remain unclear. Using a targeted metabolomics approach, Mootha's group studied the metabolic response to glucose [25]. In healthy volunteers the ingestion of an oral glucose challenge resulted in a remarkable increase in glycolytic intermediates and an increase in bile acids in the blood. Surprisingly, there was a marked reduction in many amino acids, including the branched chain amino acids, leucine, isoleucine and valine. There was also a marked reduction in the ketone body, β-OH butyrate. The overall response was considered a switch from catabolism to anabolism. Once mitochondrial function analysis was developed initial studies focused on the skeletal muscle response in patients with diabetes. As skeletal muscle is considered to contain the greatest content of mitochondria, it is logical to assess the skeletal muscle mitochondrial response in the diabetic milieu. A series of studies found that mitochondrial content and function were reduced in the skeletal muscle [26, 27]. However, therapeutic approaches to improve mitochondrial function in patients with diabetes have not been pursued until very recently.

The kidney is considered to have the greatest amount of mitochondria per tissue mass, and is only second to the heart [28]. With hyperglycemia from type 1 diabetes, the kidney response was found to reduce the content of mitochondria per mass of tissue [8]. Electron chain complex activity was measured with classical methods and found to be generally reduced across all of the complexes [8]. There may be a reduction of pyruvate entry into the mitochondria as the pyruvate dehydrogenase complex was phosphorylated [8]. Despite the reduction in mitochondrial electron chain complex activity and PDH, several studies have found an increase in TCA metabolites in the urine of mouse models of type 1 diabetes. Recent studies have also addressed mitochondrial function and content with type 1 diabetic animal models. With STZ-induced diabetes in the rat, mitochondrial fragmentation was found at 4 weeks of diabetes in association with reduced ATP content and these disturbances preceded albuminuria and elevation in urinary KIM1 levels [9]. Reduced mitochondrial electron chain complex III activity was found in the diabetic rat kidney by a separate group [29]. Mitochondrial DNA content was found to be increased with diabetes in this model. Mitochondrial DNA was found to be elevated in mesangial cells exposed to high glucose in the first 4 days before reduction was noted with longer durations [22].

Several groups have also found marked differences in mitochondria with type 2 diabetes. In the *db/db* mouse model, there is a fragmentation and reduced function of mitochondria in the podocytes [30]. There was an increase in mitochondrial fission which was mediated via Rho-associated coiled coil-containing protein kinase 1 (ROCK1) activation. The role of dynamin related protein (Drp1) was recently implicated as mice with podocyte specific inducible reduction of Drp1 had improved mitochondrial structure and function [31]. Rap1b, a small GTPase, was also found to improve mitochondrial function in renal tubular cells treated with high glucose and Rap1b was recently found to be reduced in tubular cells in biopsy tissue from patients with diabetic nephropathy [32]. Using electron microscopy, there was evidence of reduction of elongated mitochondria in tubular cells of diabetic nephropathy specimens [32].

In patients with diabetic kidney disease our group initially found a reduction in mitochondrial DNA content in urine exosomes [21]. As urine exosomes are largely derived from podocytes and tubular epithelial cells, the reduction in urine exosomal mtDNA was considered to reflect a reduction in renal epithelial mitochondrial content. Support for this hypothesis was found with reduction of mitochondrial complex protein levels in kidney biopsies from patients with diabetic nephropathy. The reduction of mitochondria is likely due in large part due to reduction of PGC1α, the master regulator of mitochondrial biogenesis.

Role of PGC1α in CKD and DKD

The dependence of mitochondrial function for recovery of kidney function may be exemplified by the role of PGC1α. As the master regulator of mitochondrial biogenesis, PGC1α has been recognized to be critically important in a variety of tissues including muscle tissue [33]. With diabetes, it was recognized that PGC1α was reduced in the skeletal muscle of patients with type 2 diabetes. However, the role of PGC1α in the muscle remains unclear as overexpression of PGC1α leads to insulin resistance (see Chap. 25). A recent study found that upregulation of PGC1α was associated with stimulation of a specific metabolite, 3-hydroxyisobutyrate (3-HIB), a catabolic intermediate of valine [34]. 3-HIB mediated enhanced trans-endothelial fatty acid transport and may play a role in the link with branched chain amino acids and development of insulin resistance and type 2 diabetes [34]. Interestingly, one of the 13 metabolites that were reduced in the urine of patients with DKD is 3-HIB and we were the first to recognize that PGC1α was reduced in kidney disease [21].

The role of PGC1α was found to play a major role initially in acute kidney injury. In studies by Parikh and colleagues, tubular reduction of PGC1α made mice susceptible to ischemic and sepsis related injury [35] whereas over expression of PGC1α in tubular cells conferred protection [14]. The benefit of upregulating PGC1α in ischemic renal injury may partly be due to regulation of NAD. In a model of chronic kidney disease and renal fibrosis with folic acid-induced nephropathy, tubular upregulation of PGC1α was found to be remarkably protective [24].

In animal models of diabetic kidney disease, PGC1α has been found to be reduced and with stimulation of AMPK, PGC1α protein levels were restored along with markers of mitochondrial content and electron transport chain activity [8]. Several groups have now demonstrated a similar relationship in diabetic kidney disease [36, 37]. The regulation of PGC1α and mitochondrial biogenesis in DKD appears to be primarily be regulated by AMPK.

AMPK, Mitochondrial Function and ROS in DKD

PGC1α can be affected by numerous signaling pathways as well as potentially by epigenetic regulation of the PGC1α promoter. In the context of diabetes, a major upstream pathway appears to be AMPK, the master energy sensor. 5′-adenosine monophosphate (AMP)-activated protein kinase (AMPK) is an evolutionarily conserved serine/threonine kinase that is a key player in maintaining cellular energy homeostasis. AMPK is highly regulated by the AMP/ATP ratio and is activated under conditions of low ATP levels. In this context activation of AMPK serves to improve energy efficiency by reducing protein synthesis and enhancing glucose entry into cells and stimulation of PGC1α and mitochondrial biogenesis. AMPK is also now recognized to have a variety of functional roles and is also regulated by a variety of kinases such as liver kinase B1 (LKB1), calcium−/calmodulin-dependent kinase kinase 2 (CAMKK2), TGF-β activated kinase 1 (TAK1) as well as a variety of phosphatases, such as protein phosphatase 2A (PP2A), protein phosphatase 2C (PP2C) and Mg^{2+}/Mn^{2+} dependent protein phosphatase 1E (PPM1E) [38]. Several groups have identified that AMPK is reduced in models of type 1 and type 2 diabetic kidney disease as well as in the diet-induced obesity model of kidney disease [8, 39–41].

The regulation of AMPK in diabetic kidney disease remains unclear. Studies by Kasinath's group identified that LKB1 activity was reduced by high glucose due to acetylation of LKB1 and contributed to reduced AMPK activity [42]. Similar results were found by Susztak's group as deletion of LKB1 in tubular cells led to AMPK reduction and apoptosis whereas stimulation of AMPK was able to rescue the phenotype [43]. Recently, our group evaluated the regulation of ATP, ADP and AMP in the diabetic kidney using mass spec imaging [44]. As ATP is easily metabolized in harvested tissue and AMP is of very low levels, accurate measurements are difficult to perform. With the advent of mass spec imaging on frozen tissues, the aspect of tissue processing is limited and the likelihood that ATP levels are similar to the in vivo condition is higher. Using MALDI-TOF approach with targeted mass spec imaging, we found that glomerular levels of ATP/AMP or ATP/ADP were elevated in diabetic kidneys. Thus, it is likely that at the glomerular level, the reduction of AMPK is at least in part due to an elevation in the ATP/AMP ratio.

Stimulation of AMPK in diabetic kidney disease was found to have a remarkable effect to reduce albuminuria, mesangial matrix expansion and the stimulation of matrix molecules and TGF-β. AMPK activation also reduced a key producer of reactive oxygen species in the kidney, NADPH oxidase. The benefit of AMPK activation with AICAR was found to be dependent on AMPK, as mice deficient in AMPKα2 had increased albuminuria but no beneficial effect with AICAR. AMPK activation also markedly reduced mTOR activation in the diabetic kidney and restored PGC1α levels in the diabetic kidney independent of lowering blood glucose or changes in body weights in type 1 models of diabetic kidney disease.

A key insight was the effect of AMPK activation on markers of mitochondrial biogenesis and electron chain activity. The diabetic kidney was found to have a reduction of mitochondrial content and reduced activity of the electron chain complexes. This was completely restored with AMPK activation, likely via stimulation of PGC1α. Therefore, this study provided compelling evidence that stimulation of mitochondrial biogenesis and mitochondrial electron transport chain activity led to beneficial outcomes in diabetic kidney disease. This study also provided new insight into the role of mitochondrial ROS in the context of diabetic kidney disease.

It has been widely believed that overproduction of superoxide from mitochondria played a key role in causing diabetic complications and diabetic kidney disease. However, this theory had not been supported by direct in vivo measurement of mitochondrial superoxide. We developed a DHE-based in vivo assessment of superoxide in mice and found that the normal kidney had a high level of superoxide production [8]. Surprisingly, with type 1 diabetes there was a marked reduction of renal superoxide production in association with reduced mitochondrial electron transport chain activity. Therefore, we concluded that the reduction of mitochondrial electron transport chain activity was reflected by reduced production of superoxide. As the mitochondrial contribution of superoxide in the baseline state is widely believed to be 80% of basal superoxide production, our DHE-based studies support the idea that mitochondrial superoxide is reduced in the diabetic kidney. Additional studies using electron paramagnetic resonance with kidney extracts also found a reduced level of superoxide from the diabetic kidney. Finally, isolated mitochondria from the diabetic kidney also exhibited reduced levels of ROS production.

To further evaluate the role of mitochondrial superoxide, studies were also performed in the $SOD2^{+/-}$ mouse. The reduced levels of mitochondrial SOD were associated with a high level of DHE-based superoxide production in the kidney. However, despite this high level of superoxide the $SOD2^{+/-}$ mouse has no evidence of kidney disease [8]. Even with diabetes, this mouse does not develop any worsening of diabetic kidney disease. These results indicate that mitochondrial superoxide are neither necessary nor sufficient for development of diabetic kidney disease. The major detrimental source of ROS in the diabetic kidney is likely to be non-mitochondrial and possibly arising from NADPH oxidase. Indeed, stimulation of mitochondrial electron transport chain activity is associated with an enhancement of superoxide production and amelioration of renal inflammation and fibrosis. This concept is captured in the theory of mitochondrial hormesis in the context of diabetic complications. At present, several pharmaceutical companies are pursuing strategies to improve mitochondrial function for many chronic conditions, including diabetic nephropathy.

New Concepts Linking Reduced Mitochondrial Function to Diabetic Kidney Disease

A major impediment to advancing novel treatments for diabetic kidney disease is the lack of insightful biomarkers. Recently an inhibitor of Nox4 was found to benefit diabetic kidney disease by several groups, however the downstream target of Nox4 activity has not been identified. Using urine metabolomics, we found that inhibition of Nox4 was found to dose-dependently reduce levels of urine fumarate in type 1 diabetic micei [45]. Tissue levels of fumarate were also increased in the diabetic kidney and reduced with the Nox4 inhibitor.

Fumarate is a key metabolite within the TCA cycle and arises from conversion of succinate to fumarate by succinate dehydrogenase (SDH). Fumarate itself is converted to malate by fumarate hydratase. As fumarate was increased in the diabetic tissue, we hypothesized that FH activity would be reduced in the diabetic kidney. Indeed there was a marked reduction of FH activity and FH protein in the diabetic kidney. FH appears to be a key target of Nox4, as overexpression of Nox4 led to marked reduction of FH levels and Nox4 inhibition restored FH levels. Importantly, FH was also found to be reduced in the human diabetic kidney.

Thus, there is now supportive data that a key mitochondrial enzyme, FH, is affected in diabetic kidney disease and likely contributes to an upregulation of tissue and urine fumarate levels. A recent study using patients with diabetes found that urine levels of fumarate were increased in patients with diabetes who had progressive renal disease. Further studies are ongoing with larger prospective cohorts to determine the role of urine fumarate as a biomarker of progressive diabetic kidney disease.

Fumarate itself has been identified as a "fibrogenic metabolite". Addition of fumarate to renal cells leads to stimulation of TGF-β and gene expression of extracellular matrix molecules. Fumarate is also a potent stimulator of HIF1α accumulation by inhibiting prolyl-hydroxylase activity. The combination of increased HIF1α and TGF-β may work in concert to suppress mitochondrial ox-phos activity, suppress mitochondrial biogenesis and promote matrix accumulation and fibrosis.

Another major insight has been the recognition that reduced fatty acid oxidation and impaired cholesterol export could contribute to progressive kidney disease, including diabetic kidney disease. With reduced mitochondrial biogenesis, there could be reduced fatty acid oxidation leading to accumulation of fatty acids. In addition, reduced mitochondrial biogenesis may contribute to reduced cholesterol export pathways and lipotoxicity. Stimulation of fatty acid oxidation, possibly via PPARα stimulation is an attractive target for future therapeutics.

Concluding Remarks

In the past 5 years there has been a remarkable resurgence of interest in the role of mitochondrial function in diabetic kidney disease. In contrast to the prevailing theory, accumulating evidence suggests that mitochondrial dysfunction and suppression are consequence of diabetes and not causative for disease progression. By restoring mitochondrial biogenesis, reducing mitochondrial fragmentation and improving mitochondrial activity there is marked improvement in disease parameters. Future studies to specifically target pathways of mitochondrial biogenesis via ROCK1 inhibitors, AMPK activation and PGC1α stimulation would provide novel approaches. Newer targets such as fumarate hydratase activators and pathways to enhance fatty acid oxidation may also provide potent therapeutic approaches. With the parallel improvements in metabolomics, a coordinated companion diagnostic approach would be of added value to provide informative biomarkers to assure that disease pathways are being engaged by the therapeutic and efficacy will hopefully be predicted.

References

1. Filla LA, Edwards JL. Metabolomics in diabetic complications. Mol BioSyst. 2016;12: 1090–105.
2. Hallan S, Sharma K. The role of mitochondria in diabetic kidney disease. Curr Diabetes Rep. 2016;16:61.
3. Darshi M, Van Espen B, Sharma K. Metabolomics in diabetic kidney disease: unraveling the biochemistry of a silent killer. Am J Nephrol. 2016;44:92–103.
4. Nishikawa T, Edelstein D, Du XL, Yamagishi S, Matsumura T, Kaneda Y, et al. Normalizing mitochondrial superoxide production blocks three pathways of hyperglycaemic damage. Nature. 2000;404:787–90.

5. Nishikawa T, Brownlee M, Araki E. Mitochondrial reactive oxygen species in the pathogenesis of early diabetic nephropathy. J Diabetes Investig. 2015;6:137–9.
6. de Zeeuw D, Akizawa T, Audhya P, Bakris GL, Chin M, Christ-Schmidt H, et al. Bardoxolone methyl in type 2 diabetes and stage 4 chronic kidney disease. N Engl J Med. 2013;369:2492.
7. Alkhalaf A, Kleefstra N, Groenier KH, Bilo HJ, Gans RO, Heeringa P, et al. Effect of benfotiamine on advanced glycation end products and markers of endothelial dysfunction and inflammation in diabetic nephropathy. PLoS One. 2012;7:e40427.
8. Dugan LL, You YH, Ali SS, Diamond-Stanic M, Miyamoto S, DeCleves AE, et al. AMPK dysregulation promotes diabetes-related reduction of superoxide and mitochondrial function. J Clin Invest. 2013;123:4888–99.
9. Coughlan MT, Nguyen TV, Penfold SA, Higgins GC, Thallas-Bonke V, Tan SM, et al. Mapping time-course mitochondrial adaptations in the kidney in experimental diabetes. Clin Sci (Lond). 2016;130:711–20.
10. Sharma K. Mitochondrial hormesis and diabetic complications. Diabetes. 2015;64:663–72.
11. Parikh SM, Yang Y, He L, Tang C, Zhan M, Dong Z. Mitochondrial function and disturbances in the septic kidney. Semin Nephrol. 2015;35:108–19.
12. Rosca MG, Vazquez EJ, Kerner J, Parland W, Chandler MP, Stanley W, et al. Cardiac mitochondria in heart failure: decrease in respirasomes and oxidative phosphorylation. Cardiovasc Res. 2008;80:30–9.
13. Santulli G, Xie W, Reiken SR, Marks AR. Mitochondrial calcium overload is a key determinant in heart failure. Proc Natl Acad Sci USA. 2015;112:11389–94.
14. Tran MT, Zsengeller ZK, Berg AH, Khankin EV, Bhasin MK, Kim W, et al. PGC1alpha drives NAD biosynthesis linking oxidative metabolism to renal protection. Nature. 2016;531:528–32.
15. Solini A, Manca ML, Penno G, Pugliese G, Cobb JE, Ferrannini E. Prediction of declining renal function and albuminuria in patients with type 2 diabetes by metabolomics. J Clin Endocrinol Metab. 2016;101:696–704.
16. Niewczas MA, Sirich TL, Mathew AV, Skupien J, Mohney RP, Warram JH, et al. Uremic solutes and risk of end-stage renal disease in type 2 diabetes: metabolomic study. Kidney Int. 2014;85:1214–24.
17. Atoh K, Itoh H, Haneda M. Serum indoxyl sulfate levels in patients with diabetic nephropathy: relation to renal function. Diabetes Res Clin Pract. 2009;83:220–6.
18. Rhee EP, Thadhani R. New insights into uremia-induced alterations in metabolic pathways. Curr Opin Nephrol Hypertens. 2011;20:593–8.
19. Makinen VP, Tynkkynen T, Soininen P, Forsblom C, Peltola T, Kangas AJ, et al. Sphingomyelin is associated with kidney disease in type 1 diabetes (The FinnDiane Study). Metabolomics. 2012;8:369–75.
20. van der Kloet FM, Tempels FW, Ismail N, van der Heijden R, Kasper PT, Rojas-Cherto M, et al. Discovery of early-stage biomarkers for diabetic kidney disease using ms-based metabolomics (FinnDiane study). Metabolomics. 2012;8:109–19.
21. Sharma K, Karl B, Mathew AV, Gangoiti JA, Wassel CL, Saito R, et al. Metabolomics reveals signature of mitochondrial dysfunction in diabetic kidney disease. J Am Soc Nephrol. 2013;24:1901–12.
22. Czajka A, Ajaz S, Gnudi L, Parsade CK, Jones P, Reid F, et al. Altered mitochondrial function, mitochondrial DNA and reduced metabolic flexibility in patients with diabetic nephropathy. EBioMedicine. 2015;2:499–512.
23. Tin A, Grams ME, Ashar FN, Lane JA, Rosenberg AZ, Grove ML, et al. Association between mitochondrial DNA copy number in peripheral blood and incident CKD in the atherosclerosis risk in communities study. J Am Soc Nephrol. 2016;27:2467–73.
24. Kang HM, Ahn SH, Choi P, Ko YA, Han SH, Chinga F, et al. Defective fatty acid oxidation in renal tubular epithelial cells has a key role in kidney fibrosis development. Nat Med. 2015;21:37–46.
25. Shaham O, Wei R, Wang TJ, Ricciardi C, Lewis GD, Vasan RS, et al. Metabolic profiling of the human response to a glucose challenge reveals distinct axes of insulin sensitivity. Mol Syst Biol. 2008;4:214.

26. Kelley DE, He J, Menshikova EV, Ritov VB. Dysfunction of mitochondria in human skeletal muscle in type 2 diabetes. Diabetes. 2002;51:2944–50.
27. Dobson AW, Kelley MR, Wilson GL, LeDoux SP. Targeting DNA repair proteins to mitochondria. Methods Mol Biol. 2002;197:351–62.
28. Pagliarini DJ, Calvo SE, Chang B, Sheth SA, Vafai SB, Ong SE, et al. A mitochondrial protein compendium elucidates complex I disease biology. Cell. 2008;134:112–23.
29. Mustata GT, Rosca M, Biemel KM, Reihl O, Smith MA, Viswanathan A, et al. Paradoxical effects of green tea (*Camellia sinensis*) and antioxidant vitamins in diabetic rats: improved retinopathy and renal mitochondrial defects but deterioration of collagen matrix glycoxidation and cross-linking. Diabetes. 2005;54:517–26.
30. Wang W, Wang Y, Long J, Wang J, Haudek SB, Overbeek P, et al. Mitochondrial fission triggered by hyperglycemia is mediated by ROCK1 activation in podocytes and endothelial cells. Cell Metab. 2012;15:186–200.
31. Ayanga BA, Badal SS, Wang Y, Galvan DL, Chang BH, Schumacker PT, et al. Dynamin-related protein 1 deficiency improves mitochondrial fitness and protects against progression of diabetic nephropathy. J Am Soc Nephrol. 2016;27:2733–47.
32. Xiao L, Zhu X, Yang S, Liu F, Zhou Z, Zhan M, et al. Rap1 ameliorates renal tubular injury in diabetic nephropathy. Diabetes. 2014;63:1366–80.
33. Leone TC, Lehman JJ, Finck BN, Schaeffer PJ, Wende AR, Boudina S, et al. PGC-1alpha deficiency causes multi-system energy metabolic derangements: muscle dysfunction, abnormal weight control and hepatic steatosis. PLoS Biol. 2005;3:e101.
34. Jang C, Oh SF, Wada S, Rowe GC, Liu L, Chan MC, et al. A branched-chain amino acid metabolite drives vascular fatty acid transport and causes insulin resistance. Nat Med. 2016;22:421–6.
35. Tran M, Tam D, Bardia A, Bhasin M, Rowe GC, Kher A, et al. PGC-1alpha promotes recovery after acute kidney injury during systemic inflammation in mice. J Clin Invest. 2011;121: 4003–14.
36. Hong YA, Lim JH, Kim MY, Kim TW, Kim Y, Yang KS, et al. Fenofibrate improves renal lipotoxicity through activation of AMPK-PGC-1alpha in db/db mice. PLoS One. 2014;9: e96147.
37. Wu L, Wang Q, Guo F, Zhou Y, Ji H, Liu F, et al. Activation of FoxO1/PGC-1alpha prevents mitochondrial dysfunction and ameliorates mesangial cell injury in diabetic rats. Mol Cell Endocrinol. 2015;413:1–12.
38. Jeon SM. Regulation and function of AMPK in physiology and diseases. Exp Mol Med. 2016;48:e245.
39. Eid AA, Ford BM, Block K, Kasinath BS, Gorin Y, Ghosh-Choudhury G, et al. AMP-activated protein kinase (AMPK) negatively regulates Nox4-dependent activation of p53 and epithelial cell apoptosis in diabetes. J Biol Chem. 2010;285:37503–12.
40. Lee MJ, Feliers D, Mariappan MM, Sataranatarajan K, Mahimainathan L, Musi N, et al. A role for AMP-activated protein kinase in diabetes-induced renal hypertrophy. Am J Physiol Renal Physiol. 2007;292:F617–27.
41. Decleves AE, Mathew AV, Cunard R, Sharma K. AMPK mediates the initiation of kidney disease induced by a high-fat diet. J Am Soc Nephrol. 2011;22:1846–55.
42. Lee MJ, Feliers D, Sataranatarajan K, Mariappan MM, Li M, Barnes JL, et al. Resveratrol ameliorates high glucose-induced protein synthesis in glomerular epithelial cells. Cell Signal. 2010;22:65–70.
43. Han SH, Malaga-Dieguez L, Chinga F, Kang HM, Tao J, Reidy K, et al. Deletion of Lkb1 in renal tubular epithelial cells leads to CKD by altering metabolism. J Am Soc Nephrol. 2016;27:439–53.
44. Miyamoto S, Hsu CC, Hamm G, Darshi M, Diamond-Stanic M, Decleves AE, et al. Mass spectrometry imaging reveals elevated glomerular ATP/AMP in diabetes/obesity and identifies sphingomyelin as a possible mediator. EBioMedicine. 2016;7:121–34.
45. You YH, Quach T, Saito R, Pham J, Sharma K. Metabolomics reveals a key role for fumarate in mediating the effects of NADPH oxidase 4 in diabetic kidney disease. J Am Soc Nephrol. 2016;27(2):466–81.

Chapter 29
Prohibitin Signaling at the Kidney Filtration Barrier

Christina Ising and Paul T. Brinkkoetter

Podocytes and Their Importance for the Kidney Filtration Barrier

Each kidney comprises about a million small vascular filtration units termed glomeruli where approximately 160 l of urine are filtered and cleared per day. The kidney filter is characterized by its size and charge selectivity. The glomerular ultrafiltrate, i.e. the primary urine is almost completely free of proteins. The ability of the kidney to retain plasma proteins is essential for life and a defective filtration barrier often becomes apparent due to the increased loss of proteins into the urine and its associated side effects, namely edema [1]. More than 5% of all human beings worldwide suffer from chronic kidney diseases (CKD) with great impact on quality of life and rising socio-economical burdens on our societies.

Anatomically, the glomerular filtration barrier consists of three layers: a fenestrated endothelial cell layer with a highly negatively charged glycocalyx, the glomerular basement membrane and as third and outer layer the glomerular epithelial cells, the podocytes (Fig. 29.1).

C. Ising, PhD
Department II of Internal Medicine and Center for Molecular Medicine Cologne, University of Cologne, Robert-Koch-Street 21, 50931 Cologne, Germany

Department of Neurology, Hope Center for Neurological Disorders, Charles F. and Joanne Knight Alzheimer's Disease Research Center, Washington University School of Medicine, 660 S. Euclid Ave, St. Louis, MO 63110, USA
e-mail: isingc@wustl.edu

P.T. Brinkkoetter, MD (✉)
Department II of Internal Medicine and Center for Molecular Medicine Cologne, University of Cologne, Robert-Koch-Street 21, 50931 Cologne, Germany
e-mail: paul.brinkkoetter@uk-koeln.de

G. Santulli (ed.), *Mitochondrial Dynamics in Cardiovascular Medicine*, Advances in Experimental Medicine and Biology 982,
DOI 10.1007/978-3-319-55330-6_29

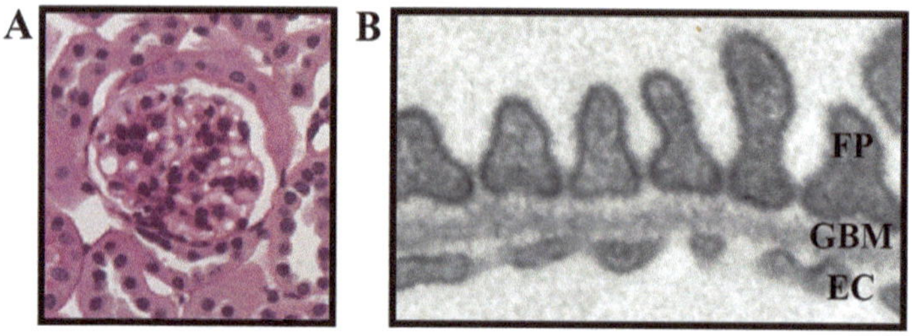

Fig. 29.1 The glomerular filtration barrier, (**a**) PAS staining of a kidney showing a glomerulus with the surrounding tubular system. (**b**) Electron microscopy image depicting the three layers of the glomerular filtration barrier. (*FP* podocyte foot process, *GBM* glomerular basement membrane, *EC* endothelial cell layer)

In the past several years, accumulating evidence suggested that glomerular podocytes are crucial for the function of the kidney filter and critically involved in the development of proteinuria [2, 3]. Podocytes are terminally differentiated and post-mitotic cells resting within the G0 phase of the cell cycle. They elaborate primary and secondary processes with long, regularly spaced foot processes that completely enwrap the glomerular capillaries [4]. The foot processes of neighboring podocytes interdigitate and form a specialized intercellular junction, the slit diaphragm, which contributes to the selective filtration barrier and regulates podocyte function [5–7]. Podocyte function is critical for the filtration barrier. They not only seal the filtration barrier and prevent the loss of albumin into the urine; they also maintain the GBM and secret growth factors that influence endothelial cell function.

Glomerular disorders, with diabetic nephropathy being the leading cause, account for the majority of cases of chronic kidney disease. Although multiple pathways contribute to the glomerular injury, the severity of glomerular diseases of any kind is defined by the degree of podocyte damage. The stereotypical response of podocytes to injury is shape change (effacement), dysfunction, and eventually cell death and/or detachment from the glomerular basement membrane [8]. Given their post-mitotic nature, podocyte damage and eventually loss cannot easily be compensated by proliferation of neighboring cells, subsequently leading to the development of proteinuria and irreversible scarring, i.e. glomerulosclerosis [2, 3, 6, 9, 10]. Importantly, proteinuria and chronic kidney disease are independently associated with an increased risk for end stage renal failure and cardiovascular disease [11–13]. Accordingly, one of the principal goals of kidney research is to understand the pathobiology of proteinuria.

The Slit Diaphragm Controls Signaling Networks for Podocyte Function and Survival

The slit diaphragm is a unique structure in the mammalian organism that consists of a well-defined set of proteins including the IgG-domain proteins NEPH and nephrin, which recruit cytoplasmic adaptor proteins to initiate signal transduction events that regulate podocyte function [14]. Among these proteins is podocin, a stomatin/prohibitin/flotillin/HflK/C (SPFH) domain-containing protein, which orchestrates the lipid environment at the slit diaphragm and by doing so regulates the TRPC6 ion-channel [15]. Podocin (NPHS2) gene mutations lead to childhood onset of proteinuria and steroid-resistant forms of focal-segmental glomerulosclerosis (FSGS) [16]. Interestingly, the podocin homolog MEC-2 in *C. elegans* regulates the activity of the ion-channel MEC-4/-10, which mediates mechanosensation in touch receptor neurons [17]. There is a strong similarity of the mechanosensory machinery in *C. elegans* and the complex organization of the slit diaphragm, including a tight anchoring to the cytoskeleton. This led to the hypothesis that podocin might be involved in glomerular mechanosensation. Although this is still speculative, podocin could play an important role as part of a protein complex that regulates glomerular mechanotransduction by linking the slit diaphragm to the cytoskeleton and regulating membrane bilayer stiffness [15].

The Mitochondrial Contribution to Renal Disease

Mitochondria are the primary cellular source of ATP and essential for all eukaryotic cells [18]. They are highly dynamic structures that constantly divide and fuse to build an intracellular network that assures the distribution within the cells and is required to execute their pleiotropic activities beyond aerobic ATP production. These include intrinsic apoptosis pathways, participation in RNA and DNA production, biosynthesis of heme, certain amino acids, lipids and others [19, 20]. In addition, the inner mitochondrial membrane potential participates in the uptake of cytosolic Ca^{2+} and potentially serves as buffer for locally increased intracellular Ca^{2+} levels [21].

Mitochondrial diseases might occur as a consequence of mutations in either the nuclear or the mitochondrial genome (for review see O'Toole et al. [22] and Che et al. [23]). In general, there is a poor relationship between the mitochondrial gene defect and the clinical presentation challenging clinicians to establish a definitive diagnosis and to understand the pathobiology underlying the clinical symptoms. Presentations during childhood and adolescence are often associated with tubular dysfunction.

Podocytes have the most abundant mitochondrial network among glomerular cells [24]. Several lines of evidence suggest a critical role for a mitochondrial contribution in glomerulopathies. Mutations in enzymes responsible for coenzyme Q generation, i.e. COQ2 [25] and PDSS2 [26], and in the mitochondrial $tRNA^{Leu}$(UUR) [27, 28] are associated with FSGS (Table 29.1).

In addition, the susceptibility gene for renal disease in kd/kd mice, a model for collapsing glomerulopathy, was mapped to encode a prenyltransferase-like mitochondrial protein (PMLP) [30]. In the puromycin aminonucleoside nephrosis (PAN) model of FSGS there is not only a strong correlation between mitochondrial reactive oxygen species (ROS) production and glomerular disease but also reduced Cox

Table 29.1 Glomerular manifestations of genetic mitochondrial disease

Gene	Locus	Inheritance	Protein	Function	Phenotype or syndrome
COQ2	4q21–q22	AR	Polyprenyl-transferase	CoQ10 biosynthesis, which transfers electrons from the mitochondrial respiratory chain	COQ10 deficiency, early-onset SRNS, with or without encephalo-myopathy
PDSS2	6q21	AR	Decaprenyl diphosphate synthase-2	CoQ10 biosynthesis, which transfers electrons from the mitochondrial respiratory chain	COQ10 deficiency, Leigh syndrome and SRNS
MTTL1	mtDNA		tRNA-LEU	Mitochondrial tRNA for leucine	MELAS syndrome. Mitochondrial diabetes, deafness and FSGS, with or without nephrotic syndrome

According to Machuca et al. [29]

I levels, reduced copy numbers of total mtDNA and decreased TFAM/NRF-1 levels [31]. Diabetic rats following STZ administration show accumulation of oxidative damage of mtDNA [32]. Holthofer et al. reported that the expression of the mitochondrially encoded respiratory chain complex is decreased in glomeruli from patients suffering from congenital nephrotic syndrome of the Finnish type [33]. Moreover, activated protein C controls expression of the partially mitochondrial localized redox-regulating protein p66Shc, thus linking the extracellular protease APC to mitochondrial function in the diabetic milieu [34]. Additionally, work performed in a transformed podocyte cell line and primary cells suggested that podocytes primarily depend on mitochondria for energy homeostasis using a range of substrates including glucose, pyruvate, palmitate and lactate [35]. Inhibition of glycolysis or oxidative phosphorylation each reduced ATP levels and may explain the susceptibility of podocytes to oxidative stress as observed in patients with diabetic nephropathy. Additional work by Ozawa et al. [36] further evaluated the roles of mitochondrial OXPHOS and glycolysis in podocytes. While blocking mitochondrial function had only minor effects on cell shape and migratory ability, inhibiting glycolysis significantly reduced the formation of lamellipodia, decreased the cell migratory ability and induced the onset of apoptosis. Consistently, the local ATP production in lamellipodia was predominantly regulated by glycolysis. These data suggest that mitochondria and glycolysis play parallel but distinct roles in podocytes. Last, podocytes employ a number of cellular pathways to inhibit intrinsic apoptotic pathways. Among these is Cdk5 that governs podocyte survival by regulating levels of the pro-survival protein Bcl-2 through the ERK1/2 pathway [37–40].

The Effect of Prohibitin Depletion on Mitochondrial Function in Podocytes

Maintaining a physiological balance between mitochondrial fusion and fission events is essential to prevent neurodegeneration due to damaged post-mitotic neurons. Mitochondrial dynamics are needed as a quality control, as they help to alleviate effects of damaged mitochondria. Fusion leads to redistribution of content from damaged to healthy mitochondria, whereas fission ensures elimination of dysfunctional mitochondria by mitophagy. These processes are necessary not only to maintain a healthy mitochondrial network, but also to protect from mitochondria-induced apoptosis [41]. In this regard, disturbed processing of the central regulator of fusion, OPA1, leads to severe mitochondrial fragmentation and a higher susceptibility to cell death [42–45]. Interestingly, loss of the mitochondrial membrane proteins prohibitins induces selective cleavage of OPA1 to short over long isoforms leading to disorganized and swollen mitochondrial cristae structures [45].

Two prohibitins – Prohibitin-1 and Prohibitin-2 (PHB1 and PHB2) – are involved in this process. Loss of either one leads to loss of the other, as they stabilize each other [46, 47]. This relationship has also been seen in glomerular podocytes, where loss of PHB2 destabilizes PHB1 [48]. In line with results obtained in neurons, a podocyte-specific PHB2-knockout mouse model revealed fragmented mitochondrial cristae structures within podocytes (Fig. 29.2). These mice presented with fast progressing glomerulosclerosis, severe albuminuria and eventually premature death within 4–5 weeks postnatally. Interestingly, there are reports on disturbed mitochondrial cristae structures in podocytes in some cases of FSGS in patients as well as in disease models [31, 49–53].

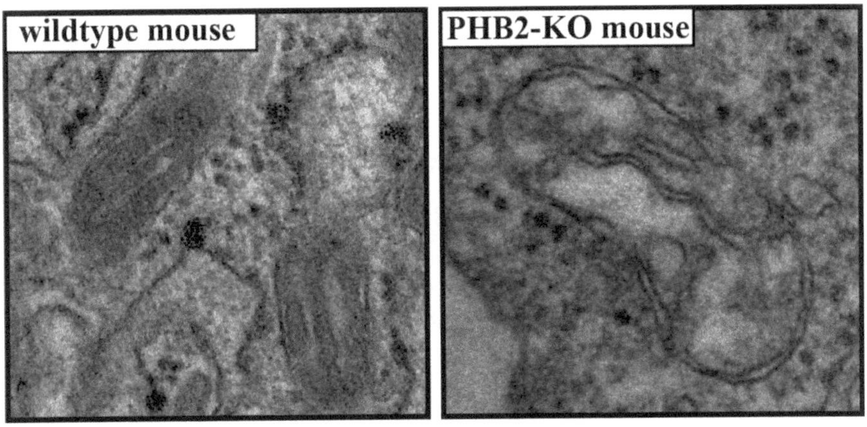

Fig. 29.2 Mitochondrial appearance in podocytes. Podocytes of wildtype mice have healthy mitochondria with normal cristae structures (*left*), whereas podocytes of PHB2 knockout mice contain mitochondria with disorganized and swollen cristae structures (*right*)

Fragmented cristae structures could lead to impaired respiratory function, as the respiratory chain proteins are localized within cristae structures. However, studies investigating the effect of prohibitin loss on ATP levels and the formation of reactive oxygen species (ROS) over the last years revealed conflicting results, with some in vitro studies showing increased ROS production [54–56] and other in vitro and in vivo studies finding no effect on respiratory chain activity [45, 57, 58]. Investigation of the respiratory chain function as well as oxygen consumption specifically in PHB2-knockdown podocytes revealed no impairments in these cells [48]. However, these results did not address the question whether the disruption of the mitochondrial network plays a role in the development of glomerulosclerosis or if it is a secondary effect of podocyte damage.

Prohibitins and Their Relationship to the Insulin Signaling Pathway

In a landmark paper in 2009, Artal-Sanz and Tavernarakis described a so far unknown link between prohibitins and insulin signaling. They were able to show that already long-lived *daf-2* mutant worms – *daf-2* being the *C elegans* homolog of the mammalian insulin receptor - had an even longer lifespan upon loss of *phb-2*. Additionally, *phb-2*-deficiency promoted longevity under dietary restriction in *C. elegans* [59]. In contrast to the nematode, in mammals two receptors mediate the action of growth factors on cell metabolism: the insulin receptor (IR) and the Insulin-like Growth Factor 1 Receptor (IGF1R). Both receptors have been shown to be expressed in podocytes [60, 61].

Intriguingly, by using an IR/IGF1R/PHB2-triple podocyte-specific knockout mouse line, the loss of insulin/IGF-1 signaling in podocytes alleviated not only renal disease and improved kidney function as compared to single PHB2-podocyte knockout mice, but also delayed the onset of kidney failure and thereby prolonged survival of the PHB2-podocyte knockout mice for several weeks [48]. Further analysis of the podocyte-specific knockout mice revealed increased mTORC1 activation in glomeruli of these mice, which is a downstream target of the insulin signaling pathway. This is of particular interest as hyperactivation of mTORC1 can cause glomerular diseases in mice and humans [62]. However, several studies since then have identified a phenotype more closely resembling diabetic nephropathy with thickening of the glomerular basement membrane and mesangial expansion than showing features of FSGS [62–65]. Furthermore, mTORC1 activation in podocytes induced by loss of PHB2 revealed some striking differences between this model and other mouse models that more directly induced mTORC1 hyperactivation. For example, PHB2-knockout mice present with an aggravated course of disease and shortened life span as compared to mice with a podocyte-specific knockout of the negative regulator of mTOR signaling, TSC1 [63]. While treatment with the mTORC1-inhibitor rapamycin alleviated albuminuria in TSC1 knockout mice, it had no effect on levels of albuminuria in PHB2 knockout mice even though it prolonged survival in this model [48, 63]. Of note, hyperactivity of mTORC1, but also suppression of mTORC1 may cause podocyte damage and ultimately glomerular diseases [62, 66, 67]. These reports emphasize the importance of a balanced mTORC1 activation in podocyte biology, which is disturbed upon loss of PHB2 [48].

Prohibitin-2 at the Slit Diaphragm

While prohibitin complexes are mainly considered as an integral part of a functional mitochondrial network, several studies suggest that prohibitins might also localize to other parts of a cell. For example, prohibitins have been shown to be important for the regulation of apoptosis [45, 68–70], to localize to the nucleus where they modulate transcriptional activity [68, 70–73] or modulate signaling at the plasma membrane [74, 75]. With regard to localization in podocytes PHB2 can be found not only in mitochondria but also at the slit diaphragm of podocytes [76] (Fig. 29.3).

Initially, podocyte-specific PHB2 knockout mice develop normal slit diaphragm structures between their foot processes, which are then lost again after a short period of time [76]. The localization of PHB2 to the slit diaphragm could contribute to the onset of albuminuria observed in these mice, as loss of the slit diaphragm structure severely impairs the filtration barrier [77–80]. Intriguingly, PHB2 knockout mice with an additional knockout of the insulin and IGF1 receptor or treated with the mTOR inhibitor rapamycin remained albuminuric despite a delayed onset of kidney failure and prolonged survival [48], further arguing into the direction of distinct roles of prohibitins in mitochondria and at the slit diaphragm in podocytes.

Proteins at the slit diaphragm have also been implicated in signaling processes in healthy podocytes [14]. Given the strong similarity of the mechanosensory machin-

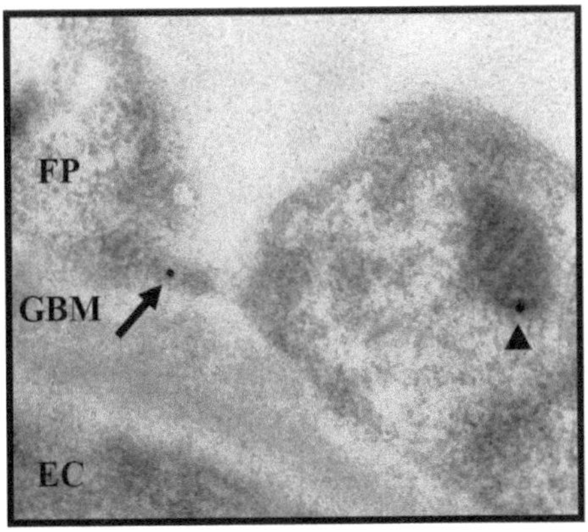

Fig. 29.3 Immunogold labeling for PHB2 in podocytes. Immunogold labeling of kidney sections showed PHB2 not only in mitochondria (*arrowhead*), but also close to the slit diaphragm in podocytes (*arrow*)

ery of the nematode *C. elegans* and the complex organization of the slit diaphragm in mammals, *C. elegans* has gained much attention as model organism to study signaling events at the slit diaphragm [15, 81, 82]. Interestingly, the mechanosensory protein complex in *C. elegans* contains MEC-2 [83, 84] – the homolog of the well-known slit diaphragm protein podocin – but also the worm homolog of PHB2 [76]. The mechanosensory complex converts mechanical forces into electrical signals, measurable by a touch response. This response is reduced upon depletion of either PHB-1 or PHB-2 specifically from the touch neurons, where PHB-2 was shown to co-localize with MEC-2 [76]. Even though there is only little experimental evidence for localization of PHB1 to the slit diaphragm and/or mechanosensory complex, it seems very likely that both, PHB1 and PHB2, localize to these protein complexes as they are usually stabilizing each other [76].

Prohibitins: A Dual Function Within Podocytes

Work on the role of prohibitins in the kidney revealed an unexpected impact of these proteins on podocyte health. On the one hand, PHB2 and possibly PHB1 can be found at the slit diaphragm, where their loss may contribute to the development of albuminuria [76] (Fig. 29.4). However, it remains to be elucidated what role prohibitins play with regard to signaling at this specialized cell-cell contact. On the other hand, loss of prohibitins leads to activation of the insulin/IGF1 signaling

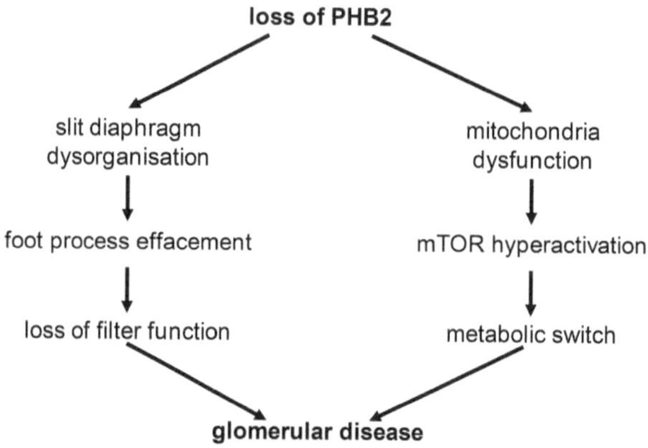

Fig. 29.4 Suggested model for why loss of PHB2 leads to glomerular disease [76]. Loss of PHB2 ultimately leads to glomerular disease potentially via two distinct pathways. On the one hand, PHB2 depletion results in foot process effacement because of its localization to the slit diaphragm, which results in loss of filter function and the onset of albuminuria. On the other hand, PHB2 deficiency leads to mitochondrial dysfunction that enhances mTOR activation, resulting in a detrimental metabolic switch in podocytes leading to podocyte loss. These two pathways together underlie the severe glomerular disease seen in PHB2 knockout mice. (*mTOR* mammalian target of rapamycin, *PHB* prohibitin)

pathway, resulting in downstream mTORC1 activation. Interestingly, treatment with the mTORC1 inhibitor rapamycin can improve kidney function and prolong survival [48], which renders the FDA-approved drug a potential treatment strategy for patients presenting with glomerular diseases involving mitochondrial structural changes.

In 2009, a direct link between insulin signaling and prohibitins has been proposed for the first time [75, 85]. Recently, metabolomic profiling in *C. elegans* has revealed changes in the fatty acid composition after PHB depletion. These modulations lead to a shortened lifespan in wild-type worms, but an increased lifespan of already long-lived *daf2* mutants [86]. Intriguingly, recent work in cardiomyocytes unraveled a metabolic shift from lipid utilization to carbohydrate utilization in a mitochondrial-fusion deficient model system in vitro and in vivo [87]. These findings taken together might be transferable to podocytes, where loss of mitochondrially localized prohibitins leads to podocyte loss, ultimately leading to kidney failure [48]. However, it remains to be studied if the metabolomic changes as reported in other model systems have a similar impact on the phenotype observed in podocyte-specific PHB2 knockout mice.

References

1. Pavenstädt H. The charge for going by foot: modifying the surface of podocytes. Exp Nephrol. 1998;6:98–103.
2. Elger M, Kriz W. Podocytes and the development of segmental glomerulosclerosis. Nephrol Dial Transplant Off Publ Eur Dial Transplant Assoc – Eur Ren Assoc. 1998;13:1368–73.
3. Kriz W, Gretz N, Lemley KV. Progression of glomerular diseases: is the podocyte the culprit? Kidney Int. 1998;54:687–97.
4. Höhne M, Ising C, Hagmann H, et al. Light microscopic visualization of podocyte ultrastructure demonstrates oscillating glomerular contractions. Am J Pathol. 2013;182:332–8.
5. Kreidberg JA. Podocyte differentiation and glomerulogenesis. J Am Soc Nephrol JASN. 2003;14:806–14.
6. Kriz W, Lemley KV. The role of the podocyte in glomerulosclerosis. Curr Opin Nephrol Hypertens. 1999;8:489–97.
7. Pavenstädt H, Kriz W, Kretzler M. Cell biology of the glomerular podocyte. Physiol Rev. 2003;83:253–307.
8. Shankland SJ. The podocyte's response to injury: role in proteinuria and glomerulosclerosis. Kidney Int. 2006;69:2131–47.
9. Kerjaschki D. Dysfunctions of cell biological mechanisms of visceral epithelial cell (podocytes) in glomerular diseases. Kidney Int. 1994;45:300–13.
10. Salant DJ. The structural biology of glomerular epithelial cells in proteinuric diseases. Curr Opin Nephrol Hypertens. 1994;3:569–74.
11. Sarnak MJ, Levey AS, Schoolwerth AC, et al. Kidney disease as a risk factor for development of cardiovascular disease: a statement from the American Heart Association Councils on Kidney in Cardiovascular Disease, High Blood Pressure Research, Clinical Cardiology, and Epidemiology and Prevention. Hypertension. 2003;42:1050–65.
12. Asanuma K, Yanagida-Asanuma E, Takagi M, et al. The role of podocytes in proteinuria. Nephrol Carlton Vic. 2007;12(Suppl 3):S15–20.

13. Foley RN, Parfrey PS, Sarnak MJ. Clinical epidemiology of cardiovascular disease in chronic renal disease. Am J Kidney Dis Off J Natl Kidney Found. 1998;32:S112–9.
14. Benzing T. Signaling at the slit diaphragm. J Am Soc Nephrol JASN. 2004;15:1382–91.
15. Huber TB, Schermer B, Müller RU, et al. Podocin and MEC-2 bind cholesterol to regulate the activity of associated ion channels. Proc Natl Acad Sci U S A. 2006;103:17079–86.
16. Machuca E, Hummel A, Nevo F, et al. Clinical and epidemiological assessment of steroid-resistant nephrotic syndrome associated with the NPHS2 R229Q variant. Kidney Int. 2009; 75:727–35.
17. Goodman MB, Ernstrom GG, Chelur DS, et al. MEC-2 regulates *C. elegans* DEG/ENaC channels needed for mechanosensation. Nature. 2002;415:1039–42.
18. Detmer SA, Chan DC. Functions and dysfunctions of mitochondrial dynamics. Nat Rev Mol Cell Biol. 2007;8:870–9.
19. Schmidt O, Pfanner N, Meisinger C. Mitochondrial protein import: from proteomics to functional mechanisms. Nat Rev Mol Cell Biol. 2010;11:655–67.
20. Osman C, Voelker DR, Langer T. Making heads or tails of phospholipids in mitochondria. J Cell Biol. 2011;192:7–16.
21. Xie W, Santulli G, Reiken SR, Yuan Q, Osborne BW, Chen BX, Marks AR. Mitochondrial oxidative stress promotes atrial fibrillation. Sci Rep. 2015;5:11427.
22. O'Toole JF. Renal manifestations of genetic mitochondrial disease. Int J Nephrol Renov Dis. 2014;7:57–67.
23. Che R, Yuan Y, Huang S, et al. Mitochondrial dysfunction in the pathophysiology of renal diseases. Am J Phys Renal Phys. 2014;306:F367–78.
24. Bartlett P, Keegan J, Schaefer H. Mechanism of aminonucleoside-induced nephrosis in the rat. III Kidney mitochondrial phosphorylation and dephosphorylation activity. Proc Soc Exp Biol Med Soc Exp Biol Med N Y N. 1963;112:96–101.
25. Diomedi-Camassei F, Di Giandomenico S, Santorelli FM, et al. COQ2 nephropathy: a newly described inherited mitochondriopathy with primary renal involvement. J Am Soc Nephrol JASN. 2007;18:2773–80.
26. López LC, Schuelke M, Quinzii CM, et al. Leigh syndrome with nephropathy and CoQ10 deficiency due to decaprenyl diphosphate synthase subunit 2 (PDSS2) mutations. Am J Hum Genet. 2006;79:1125–9.
27. Jansen JJ, Maassen JA, van der Woude FJ, et al. Mutation in mitochondrial tRNA(Leu(UUR)) gene associated with progressive kidney disease. J Am Soc Nephrol JASN. 1997;8:1118–24.
28. Kurogouchi F, Oguchi T, Mawatari E, et al. A case of mitochondrial cytopathy with a typical point mutation for MELAS, presenting with severe focal-segmental glomerulosclerosis as main clinical manifestation. Am J Nephrol. 1998;18:551–6.
29. Machuca E, Benoit G, Antignac C. Genetics of nephrotic syndrome: connecting molecular genetics to podocyte physiology. Hum Mol Genet. 2009;18:R185–94.
30. Peng M, Jarett L, Meade R, et al. Mutant prenyltransferase-like mitochondrial protein (PLMP) and mitochondrial abnormalities in kd/kd mice. Kidney Int. 2004;66:20–8.
31. Hagiwara M, Yamagata K, Capaldi RA, et al. Mitochondrial dysfunction in focal segmental glomerulosclerosis of puromycin aminonucleoside nephrosis. Kidney Int. 2006;69:1146–52.
32. Kakimoto M, Inoguchi T, Sonta T, et al. Accumulation of 8-hydroxy-2′-deoxyguanosine and mitochondrial DNA deletion in kidney of diabetic rats. Diabetes. 2002;51:1588–95.
33. Holthöfer H, Kretzler M, Haltia A, et al. Altered gene expression and functions of mitochondria in human nephrotic syndrome. FASEB J Off Publ Fed Am Soc Exp Biol. 1999;13:523–32.
34. Bock F, Shahzad K, Wang H, et al. Activated protein C ameliorates diabetic nephropathy by epigenetically inhibiting the redox enzyme p66Shc. Proc Natl Acad Sci U S A. 2013;110: 648–53.
35. Abe Y, Sakairi T, Kajiyama H, et al. Bioenergetic characterization of mouse podocytes. Am J Phys Cell Phys. 2010;299:C464–76.
36. Ozawa S, Ueda S, Imamura H, et al. Glycolysis, but not Mitochondria, responsible for intracellular ATP distribution in cortical area of podocytes. Sci Rep. 2015;5:18575.

37. Brinkkoetter PT, Olivier P, Wu JS, et al. Cyclin I activates Cdk5 and regulates expression of Bcl-2 and Bcl-XL in postmitotic mouse cells. J Clin Invest. 2009;119:3089–101.
38. Brinkkoetter PT, Pippin JW, Shankland SJ. Cyclin I-Cdk5 governs survival in post-mitotic cells. Cell Cycle Georget Tex. 2010;9:1729–31.
39. Brinkkoetter PT, Wu JS, Ohse T, et al. p35, the non-cyclin activator of Cdk5, protects podocytes against apoptosis in vitro and in vivo. Kidney Int. 2010;77:690–9.
40. Taniguchi Y, Pippin JW, Hagmann H, et al. Both cyclin I and p35 are required for maximal survival benefit of cyclin-dependent kinase 5 in kidney podocytes. Am J Phys Renal Phys. 2012;302:F1161–71.
41. Tatsuta T, Langer T. Quality control of mitochondria: protection against neurodegeneration and ageing. EMBO J. 2008;27:306–14.
42. Chen H, Chomyn A, Chan DC. Disruption of fusion results in mitochondrial heterogeneity and dysfunction. J Biol Chem. 2005;280:26185–92.
43. MacVicar T, Langer T. OPA1 processing in cell death and disease – the long and short of it. J Cell Sci. 2016;129:2297–306.
44. Duvezin-Caubet S, Jagasia R, Wagener J, et al. Proteolytic processing of OPA1 links mitochondrial dysfunction to alterations in mitochondrial morphology. J Biol Chem. 2006; 281:37972–9.
45. Merkwirth C, Dargazanli S, Tatsuta T, et al. Prohibitins control cell proliferation and apoptosis by regulating OPA1-dependent cristae morphogenesis in mitochondria. Genes Dev. 2008;22: 476–88.
46. Berger KH, Yaffe MP. Prohibitin family members interact genetically with mitochondrial inheritance components in *Saccharomyces cerevisiae*. Mol Cell Biol. 1998;18:4043–52.
47. Tatsuta T, Model K, Langer T. Formation of membrane-bound ring complexes by prohibitins in mitochondria. Mol Biol Cell. 2005;16:248–59.
48. Ising C, Koehler S, Brähler S, et al. Inhibition of insulin/IGF-1 receptor signaling protects from mitochondria-mediated kidney failure. EMBO Mol Med. 2015;7:275–87.
49. Barisoni L, Madaio MP, Eraso M, et al. The kd/kd mouse is a model of collapsing glomerulopathy. J Am Soc Nephrol JASN. 2005;16:2847–51.
50. Hotta O, Inoue CN, Miyabayashi S, et al. Clinical and pathologic features of focal segmental glomerulosclerosis with mitochondrial tRNALeu(UUR) gene mutation. Kidney Int. 2001; 59:1236–43.
51. Güçer S, Talim B, Aşan E, et al. Focal segmental glomerulosclerosis associated with mitochondrial cytopathy: report of two cases with special emphasis on podocytes. Pediatr Dev Pathol Off J Soc Pediatr Pathol Paediatr Pathol Soc. 2005;8:710–7.
52. Markowitz GS, Appel GB, Fine PL, et al. Collapsing focal segmental glomerulosclerosis following treatment with high-dose pamidronate. J Am Soc Nephrol JASN. 2001;12:1164–72.
53. Kawakami T, Gomez IG, Ren S, et al. Deficient autophagy results in mitochondrial dysfunction and FSGS. J Am Soc Nephrol JASN. 2015;26:1040–52.
54. Liu D, Lin Y, Kang T, et al. Mitochondrial dysfunction and adipogenic reduction by prohibitin silencing in 3T3-L1 cells. PLoS One. 2012;7:e34315.
55. Schleicher M, Shepherd BR, Suarez Y, et al. Prohibitin-1 maintains the angiogenic capacity of endothelial cells by regulating mitochondrial function and senescence. J Cell Biol. 2008;180:101–12.
56. Kathiria AS, Butcher LD, Feagins LA, et al. Prohibitin 1 modulates mitochondrial stress-related autophagy in human colonic epithelial cells. PLoS One. 2012;7:e31231.
57. Merkwirth C, Martinelli P, Korwitz A, et al. Loss of prohibitin membrane scaffolds impairs mitochondrial architecture and leads to tau hyperphosphorylation and neurodegeneration. PLoS Genet. 2012;8:e1003021.
58. Baris OR, Klose A, Kloepper JE, et al. The mitochondrial electron transport chain is dispensable for proliferation and differentiation of epidermal progenitor cells. Stem Cells Dayt Ohio. 2011;29:1459–68.
59. Artal-Sanz M, Tavernarakis N. Prohibitin couples diapause signalling to mitochondrial metabolism during ageing in *C. elegans*. Nature. 2009;461:793–7.

60. Coward RJM, Welsh GI, Yang J, et al. The human glomerular podocyte is a novel target for insulin action. Diabetes. 2005;54:3095–102.
61. Bridgewater DJ, Ho J, Sauro V, et al. Insulin-like growth factors inhibit podocyte apoptosis through the PI3 kinase pathway. Kidney Int. 2005;67:1308–14.
62. Gödel M, Hartleben B, Herbach N, et al. Role of mTOR in podocyte function and diabetic nephropathy in humans and mice. J Clin Invest. 2011;121:2197–209.
63. Inoki K, Mori H, Wang J, et al. mTORC1 activation in podocytes is a critical step in the development of diabetic nephropathy in mice. J Clin Invest. 2011;121:2181–96.
64. Ito N, Nishibori Y, Ito Y, et al. mTORC1 activation triggers the unfolded protein response in podocytes and leads to nephrotic syndrome. Lab Invest J Tech Methods Pathol. 2011;91:1584–95.
65. Daniel C, Ziswiler R, Frey B, et al. Proinflammatory effects in experimental mesangial proliferative glomerulonephritis of the immunosuppressive agent SDZ RAD, a rapamycin derivative. Exp Nephrol. 2000;8:52–62.
66. Vollenbröker B, George B, Wolfgart M, et al. mTOR regulates expression of slit diaphragm proteins and cytoskeleton structure in podocytes. Am J Phys Renal Phys. 2009;296:F418–26.
67. George B, Vollenbröker B, Saleem MA, et al. GSK3β inactivation in podocytes results in decreased phosphorylation of p70S6K accompanied by cytoskeletal rearrangements and inhibited motility. Am J Phys Renal Phys. 2011;300:F1152–62.
68. Fusaro G, Dasgupta P, Rastogi S, et al. Prohibitin induces the transcriptional activity of p53 and is exported from the nucleus upon apoptotic signaling. J Biol Chem. 2003;278: 47853–61.
69. Supale S, Thorel F, Merkwirth C, et al. Loss of prohibitin induces mitochondrial damages altering β-cell function and survival and is responsible for gradual diabetes development. Diabetes. 2013;62:3488–99.
70. Delage-Mourroux R, Martini PG, Choi I, et al. Analysis of estrogen receptor interaction with a repressor of estrogen receptor activity (REA) and the regulation of estrogen receptor transcriptional activity by REA. J Biol Chem. 2000;275:35848–56.
71. Wang S, Fusaro G, Padmanabhan J, et al. Prohibitin co-localizes with Rb in the nucleus and recruits N-CoR and HDAC1 for transcriptional repression. Oncogene. 2002;21:8388–96.
72. Park S-E, Xu J, Frolova A, et al. Genetic deletion of the repressor of estrogen receptor activity (REA) enhances the response to estrogen in target tissues in vivo. Mol Cell Biol. 2005;25:1989–99.
73. Kurtev V, Margueron R, Kroboth K, et al. Transcriptional regulation by the repressor of estrogen receptor activity via recruitment of histone deacetylases. J Biol Chem. 2004; 279:24834–43.
74. Chiu C-F, Ho M-Y, Peng J-M, et al. Raf activation by Ras and promotion of cellular metastasis require phosphorylation of prohibitin in the raft domain of the plasma membrane. Oncogene. 2013;32:777–87.
75. Ande SR, Mishra S. Prohibitin interacts with phosphatidylinositol 3,4,5-triphosphate (PIP3) and modulates insulin signaling. Biochem Biophys Res Commun. 2009;390:1023–8.
76. Ising C, Bharill P, Brinkkoetter S, et al. Prohibitin-2 depletion unravels extra-mitochondrial functions at the kidney filtration barrier. Am J Pathol. 2016;186:1128–39.
77. Boute N, Gribouval O, Roselli S, et al. NPHS2, encoding the glomerular protein podocin, is mutated in autosomal recessive steroid-resistant nephrotic syndrome. Nat Genet. 2000;24: 349–54.
78. Shih NY, Li J, Karpitskii V, et al. Congenital nephrotic syndrome in mice lacking CD2-associated protein. Science. 1999;286:312–5.
79. Philippe A, Nevo F, Esquivel EL, et al. Nephrin mutations can cause childhood-onset steroid-resistant nephrotic syndrome. J Am Soc Nephrol JASN. 2008;19:1871–8.
80. Winn MP, Conlon PJ, Lynn KL, et al. A mutation in the TRPC6 cation channel causes familial focal segmental glomerulosclerosis. Science. 2005;308:1801–4.
81. Schermer B, Benzing T. Lipid-protein interactions along the slit diaphragm of podocytes. J Am Soc Nephrol JASN. 2009;20:473–8.

82. Müller R-U, Zank S, Fabretti F, et al. *Caenorhabditis elegans*, a model organism for kidney research: from cilia to mechanosensation and longevity. Curr Opin Nephrol Hypertens. 2011;20:400–8.
83. Zhang Y, Ma C, Delohery T, et al. Identification of genes expressed in C elegans touch receptor neurons. Nature. 2002;418:331–5.
84. Zhang S, Arnadottir J, Keller C, et al. MEC-2 is recruited to the putative mechanosensory complex in *C. elegans* touch receptor neurons through its stomatin-like domain. Curr Biol CB. 2004;14:1888–96.
85. Ande SR, Gu Y, Nyomba BLG, et al. Insulin induced phosphorylation of prohibitin at tyrosine 114 recruits Shp1. Biochim Biophys Acta. 1793;2009:1372–8.
86. Lourenço AB, Muñoz-Jiménez C, Venegas-Calerón M, et al. Analysis of the effect of the mitochondrial prohibitin complex, a context-dependent modulator of longevity, on the *C. elegans* metabolome. Biochim Biophys Acta. 2015;1847:1457–68.
87. Wai T, García-Prieto J, Baker MJ, et al. Imbalanced OPA1 processing and mitochondrial fragmentation cause heart failure in mice. Science. 2015;350:aad0116.

Chapter 30
Mitochondrial Heteroplasmy

George B. Stefano, Christina Bjenning, Fuzhou Wang, Nan Wang, and Richard M. Kream

Intriguingly, a small subset of heritable polymorphisms in the haploid mtDNA genome is associated with the phenotypic characteristics of severe genetic disorders, originating from markedly compromised mitochondrial function [1–4]. This observation is supported by a relatively recent body of literature, demonstrating interactive effects of heritable mtDNA polymorphisms on altered expression of select nuclear genes that are associated with several human disease states [5–7]. Arguably, a therapeutically more attractive approach than attempting to target the more elusive heritable genetic polymorphisms that, in concert with accumulated somatic mutations and epigenetic factors, influence the etiology and progression of cardiovascular [8] metastatic [9, 10], autoimmune [11, 12], and neurodegenerative [10, 13, 14] disorders.

G.B. Stefano (✉) • R.M. Kream
International Scientific Information, Inc.,
150 Broadhollow Rd, Ste 114, Melville, NY 11747, USA
e-mail: george.stefano@mitogenetics.com

C. Bjenning
Cardiometabolic Designs LLC, 160 W15th Ave, Suite 303, Sea Cliff, NY 11579, USA

F. Wang, MD, PhD
Division of Neuroscience, Bonoi Academy of Science & Education,
Chapel Hill, NC 27510, USA

N. Wang, MD
Department of Anesthesiology, Affiliated Hospital of OB/GYN, Nanjing Medical University,
Nanjing 210004, China

G. Santulli (ed.), *Mitochondrial Dynamics in Cardiovascular Medicine*,
Advances in Experimental Medicine and Biology 982,
DOI 10.1007/978-3-319-55330-6_30

Biological Significance of Mitochondrial Heteroplasmy

To successfully translate human genetic aberrations to effective therapeutic approaches/therapies, we must seek critical foci to probe the multiple biological roles of mitochondrial heteroplasmy. For example, the dynamically determined co-expression of wildtype (WT) inherited polymorphisms and somatic mutations in varying ratios within individual mtDNA genomes distributed throughout the intra-organelle compartments of individual cells are noteworthy (Fig. 30.1). Human mtDNA is a 16.6-kilo base circular genome, corresponding to subunits ND4 and ND4L including ND1-6, of respiratory complex I, catalytic units CO1-3 of respiratory complex IV, subunits ATP8 and ATP6 of F1F0 ATPase, and CYTB of respiratory complex III. The remaining genes encode 22 tRNAs and 12 and 16S rRNAs [15]. Since the number of mitochondrial genomes, normalized against each diploid nuclear genome, varies according to cell type and total mtDNA copy number, estimates in the biomedical literature range from 100 to 10,000 [2, 15, 16]. Considering the complex sequence heterogeneity within mixed intra-mitochondrial populations of several thousand individual heteroplasmic mtDNA genomes within individual cells; it may be technically too challenging to sort empirically sort presumed biological activities of distinct patterns of mtDNA heteroplasmy based on molecular biological, biochemical, physiological, and bioinformatic criteria [17, 18].

The existence of highly regulated and the habitual presence of heteroplasmic mtDNA genomes, in mitochondrial compartments of separate human cell types which complements normative mitochondrial function, can be postulated based on findings of distinct and repeatable patterns of mtDNA heteroplasmy, varying across different cell types from the same individual [17, 19]. This observation is in concert with an earlier study monitoring intra-mitochondrial organization of heterologous heteroplasmic mtDNA genomes into DNA-protein complexes termed nucleoids [20]. Observing heterologous mtDNAs stably maintained in distinct nucleoid populations while trans-complementation of heteroplasmic nucleoids appeared to be achieved by the diffusion of mtDNA-derived transcripts within the mitochondrial matrix, the authors speculated on the presence of a restorative mechanism of trans-complementation to operationally increase homoplasmic WT mtDNA and mitochondrial bioenergetics. The patterns of tissue- and cell-specific heteroplasmic mtDNA appear mechanistically maintained via intra-mitochondrial trans-complementation of heteroplasmic nucleoids and mtDNA-derived transcripts [20] as does the intercellular exchange of mtDNA [17]. Further, polycistronic nature heteroplasmic mtDNA-encoded transcripts introduce an additional level of intricacy which can be employed to evaluate theoretical facilitative roles of preserved patterns of mtDNA heteroplasmy on homeostatic metabolic processes [15].

A Drosophila genetic model, utilizing a temperature-sensitive lethal mtDNA mutation in functional linkage to the cytochrome c oxidase subunit I (CO1) locus [21], validates of the potential existential role of cell-specific patterns of mtDNA heteroplasmy on normative mitochondrial functions. Importantly, expressing an alternative CO1 oxidase, which specifically rendered restorative mtDNA hetero-

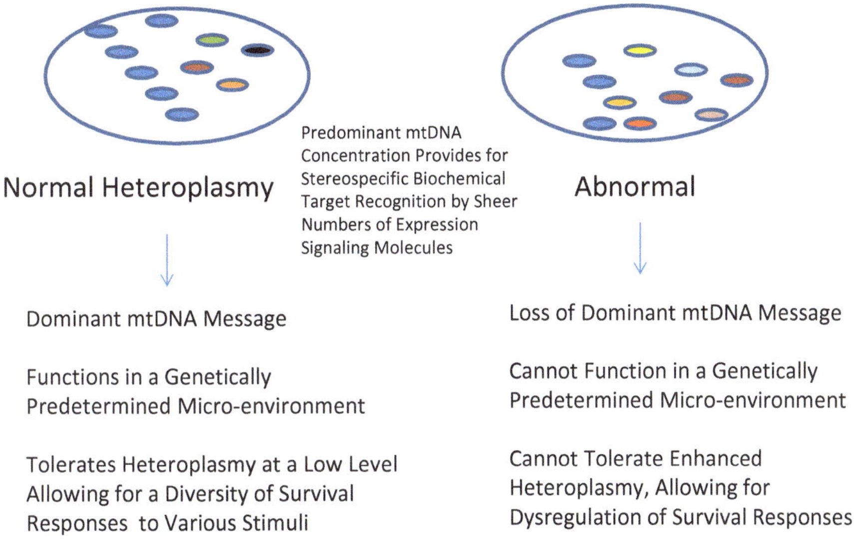

Fig. 30.1 Heteroplasmy: Mitochondrial diversity. Alterations of mitochondrial DNA (mtDNA) are largely associated with post-mitotic neural- and muscular-cell disorders. The pathology may emerge from primary mtDNA mutations and alterations in nuclear-encoded mtDNA maintenance proteins. Our understanding of these phenomena is lacking. However, answers may be present from our understanding of mitochondrial origin from large populations of bacteria. Indeed, vesicular and tunneling nanotubular phenomena may represent a process whereby the vast mtDNA variations are summated, shared, and the dominant ratio emerges. Alterations occur when various modified genomes have micro-environmental survival mechanisms (affinities) or the immediate microbiome (Gut Phenomena) selects a special microbe genome to emerge (as a comparison), bringing together intrinsic selective alteration and those imposed by environmental idiosyncrasies. Under these circumstances, an abnormally expressed genome may become the 'norm', until the conditions or physiological environment changes. Thus, this dynamic mixing system provides functional mitochondria that are dynamically suited for the prevailing circumstances by providing genetic diversity when called upon. In this regard, pathological ratio expression emerges when the chemical signals are misinterpreted or a totally novel "stress" appears. However, under normal circumstances, abundant and normal ratios emerge to meet an appropriate and specific demand. The regulation comes from the population numbers of the functioning mitochondria, much like a summation process. Because a mitochondrion environment is so hazardous, e.g., ROS, they must rely on numbers to survive by buffering the potential hazards

plasmy associated with fully viable and tissue-specific phenotypes, maintained the viability of homoplasmic flies despite restrictive temperatures. On the contrary, restoration of mtDNA homoplasmy in the eye of heteroplasmic flies, using a genetic replacement model to induce tissue-specific mutant CO1 homoplasmy, resulted in severe neurodegeneration at restrictive temperatures.

Findings in this Drosophila model further show a decrease in the frequency of the mutant allele in heteroplasmic flies in the germ line and over multiple generations [22]. We surmise, in reviewing the studies noted above, that selection against potentially disadvantageous mtDNA heteroplasmic mutations in oogenesis may be in contrast to developmentally determined, cell-specific patterns of mtDNA heteroplasmy generated during various stages of tissue differentiation (non-random gene flow rather than drift). Concurrently, cellular processes that regulate physiologically compatible patterns of heteroplasmic mtDNA may undergo state-dependent dysregulation, resulting in the altered metabolically-compromised phenotype characteristic of congestive heart failure [23] and other pathophysiologically altered states [24]. Of further relevance is a recent study linking allele-specific expression of nuclear DNA and mtDNA heteroplasmy with functional impairment in bioenergetics and a favorable selection for impaired mitochondrial function [25].

The mtDNA genome represents a self-contained genetic system, preserved through evolution, encoding vital catalytic and regulatory subunits of the respiratory complexes noted earlier [26–28]. Ultimately, the maintenance of redox potential by electron transport through respiratory complexes represents a major mechanism responsible for reciprocal regulation of energy production. Thus, the transformation of cell-specific mitochondria into powerful bio-engines shows dependency on the accuracy of ongoing gene expression within insular metabolic or physiological demands. A recent, pivotal review depicts a binary regulatory system responsible for selective expression of genes contained within mitochondrial and chloroplast genomes [27–29]. Functional mitochondria are crucial in maintaining, differentiating and reprogramming induced pluripotent stem cells (iPSCs) [30]. The critical regulatory roles of cell-specific patterns of mtDNA heteroplasmy are fundamental in maintaining essential metabolic capacity during habitual aging processes [17].

For instance, in Barth syndrome, an inherited dilated cardiomyopathy stemming from a mutation in TAZ, the gene encoding the mitochondrial protein tafazzin, decreased basal mitochondrial oxygen consumption associated with structural changes in mitochondrial respiratory chain super complexes, results in increased output of putatively harmful ROS. Consequently, human iPSCs from Barth syndrome patients demonstrate structural and functional abnormalities, including increased ROS, which intriguingly reverse by the reintroduction of WT TAZ gene, indicating a therapeutic potential of human stem cell-derived cardiomyocytes [8, 31–33].

Furthermore, hypoxia-inducible factor one alpha (HIF1α) signaling pathways, may strongly regulate the transition from oxidative to glycolytic metabolic processes in cardiomyocytes providing a novel insight into survival in a hypoxic microenvironment [34, 35]. A shift from somatic mitochondrial oxidative metabolism to glycolytic metabolism, highly reminiscent of cancer cells, is requisite for successful reprogramming of iPSCs [35]. Consequently, somatic mitochondria and affiliated oxidative bioenergetics are profoundly altered with the induction of an iPSC-like phenotype. The authors conclude that early induction of HIF1α target genes may be required for iPSC derivation via the activation of a glycolytic program highly reminiscent of undifferentiated cancer cells. This process in all likelihood may occur in cardiomyocytes given the vulnerabilities of cells to hypoxia.

Examples of Heteroplasmy in Cardiomyopathy

Cardiomyopathies refer to a heterogeneous group of diseases characterized by the impaired function of heart muscle. Although the causes remain largely unclear, recent studies have recognized a significant role for genetic factors [36]. Although the true prevalence of mtDNA-related cardiomyopathy is unknown, at least 1 in 10,000–15,000 of the general population will be affected based on the prevalence of mtDNA disease and also the frequency of cardiac involvement.

Large-Scale Rearrangements of mtDNA

Large-scale rearrangements of mtDNA including single deletions and duplications are always heteroplasmic. A variety of clinical presentations exists, such as the sporadic chronic progressive external ophthalmoplegia (CPEO), Kearns-Sayre syndrome (KSS), and other rare syndromes [36]. KSS is commonly caused by single deletions in mtDNA and probably the most characteristic clinical syndrome with an onset before 20 years of age, pigmentary retinopathy, and external ophthalmoplegia. Cardiomyopathy is rarely observed in this syndrome and if present, has a late clinical onset [36].

Point Mutations (PM) of mtDNA

tRNA$^{Leu(UUR)}$ Gene

A3243G Mutation

This point mutation in the mitochondrial DNA is the most prevalent mtDNA mutation and results in the nucleotide substitution of Guanine for Adenine at nucleotide 3243 in the leucine tRNA gene. The A3243G mutation is associated with 80% of patients with MELAS syndrome (mitochondrial encephalomyopathy, lactic acidosis, and stroke-like episodes) [37], even though this syndrome is also associated with at least 12 other mtDNA gene mutations reported in the literature [37]. MELAS syndrome is clinically characterized by abnormal development, lactic acidosis, intermittent vomiting, seizures, stroke-like episodes and hemiparesis or cortical blindness [36]. About 20–38% patients with MELAS may show cardiac involvement [36, 37]. In a study of 17 patients with mitochondrial disease and cardiac involvement, Anan et al. [38] found that five patients with MELAS all had the A3243G point mutation, and two of them showed left ventricular hypertrophy on echocardiogram. Likewise, Majamaa-Voltti et al. [39] found that the risk of left ventricular hypertrophy was significantly increased in patients with the A3243G mutation. Recently, Lindroos et al. [40] reported that myocardial glucose uptake is decreased in patients with the A3243G mutation, adding to the impaired cardiac energetics and likely contributing to the progression of the mitochondrial cardiomyopathy. The A3243G mutation may be a part of a multisystem involvement but rarely present as the sole clinical manifestation of cardiomyopathy [41]. Other phenotypes as CPEO, deafness, and maternally inherited diabetes mellitus, have also been described [41].

A3260G Mutation

Zeviani et al. first described a large pedigree with a clinical presentations including proximal muscle weakness, increased blood lactate, exercise intolerance, both during exercise and at rest, and reduced cardiac ejection fraction. Hypertrophic cardiomyopathy was recognized in three of the probands by echocardiography. Affected individuals were heteroplasmic for the A3260G mutation. This syndrome was categorized under the category "Maternally Inherited Myopathy and Cardiomyopathies" by Zeviani et al. [42].

C3303T Mutation

This mutation in the $tRNA^{Leu(UUR)}$ gene was first reported by Silvestri et al. [43]. Then, Bruno et al. [44] described the same mutation in eight patients from different families and tissues from the individuals appeared to be homoplasmic or near homoplasmic for the C3303T mutation in all families. Finally, Ueki et al. [45] studied 44 patients with lactic acidosis. Among them, one patient had the clinical manifestation of isolated cardiomyopathy and then tested positive for the C3303T mutation. It is interesting to note that the A3302G mutation, adjacent to the C3303T mutation, has also been described. In a study discussing two case reports with the A3302G mutation [46], one of them died of cardiorespiratory failure. Since the two genes are adjacent, we raise the question of whether the A3302G mutation could also lead to cardiomyopathy, with a higher level of mutant mtDNA resulting in a more severe manifestation of the heart [46].

$tRNA^{Leu(CUN)}$ Gene

T12297C Mutation

The mutation was initially reported by Tessa et al. in 1999 [47]. They studied ten children of eight unrelated families with dilated cardiomyopathy. The T12297C mutation was identified in two siblings of unrelated parents with dilated cardiomyopathy and endocardial fibroelastosis, and the novel mutation was abundant and heteroplasmic in both patients. The same mutation was also reported in a 36-year-old man who presented with congestive heart failure and dilated ventricles by echocardiography. The ultrastructural studies of tissues showed proliferation of mitochondria, with a wide range of changes in shape and size. The mutation was heteroplasmic, and 88% mutant was in the heart DNA [48]. Recently, Wang et al. [49] reported a family with multiple members who had the T12297C mutation and T14487C mutation, which is found in Leigh syndrome. All subjects tested were homoplasmic for the T12297C mutation, but no individuals had dilated cardiomyopathy.

tRNA^{-Ile} Gene

A4317 Mutation

The mutation was first reported in a 1-year-old boy with metabolic acidosis and general weakness. The boy developed hypertrophic cardiomyopathy and then died of heart failure on the seventh day of his presentation due to fatal infantile cardiomyopathy [50]. Tanaka et al. [50] and subsequently Tomari et al. suggested that the A4317G mutation caused structural rearrangement of the T-stem of the tRNA^{Ile}, leading to the instability in the mitochondria, and hence may result in mitochondrial dysfunction [51].

A4269G Mutation

The first patient with this mutation was an 18-year-old boy with short stature, who finally died as a result of heart failure. The boy initially presented at the age of 4, epilepsy, multiorgan dysfunction, and later developed dilated cardiomyopathy. In the study, the A4269G mutation was found in the index case with no mutation found in the controls [52]. Then, it was found that the A4269G point mutation inhibits protein synthesis completely in mitochondria, and therefore results in a reduction of the respiratory activity [53, 54]. Hino et al. studied the molecular mechanism by which the A4269Gmutation affects the translational activity of the mutant tRNA^{Ile}. Interestingly, they found that the point mutation destabilizes the T-stem of the tRNA and also decreases the affinity of binding of the elongation factor EF-Tu, and finally can't protect the mutant tRNA^{Ile} from degradation [53].

A4295G Mutation

This mutation was initially reported in a 7-month-old girl. The child had a sudden onset of cyanotic spells and finally died of complications of hypertrophic cardiomyopathy [55]. Ultrastructural findings showed massive mitochondrial proliferation in heart and liver but not in skeletal muscle fibers. The brother of the patient was also found to have left ventricular hypertrophic cardiomyopathy and underwent heart transplantation because of a sudden decline in cardiac function.

A4300G Mutation

The mutation was detected in a large Italian family with maternally inherited cardiomyopathy. The proband and ten other family members all developed isolated hypertrophic cardiomyopathy. Importantly, the heart appeared to be the only organ affected, unlike other point mutations in the tRNA (Ile) gene [36].

m.4322dupC Mutation

Mahjoub et al. [56] identified a novel heteroplasmic mtDNA (m.4322dupC) mutation in the $tRNA^{Ile}$ gene associated with isolated dilated cardiomyopathy as a maternal trait. This new mutation was heteroplasmic in muscle and blood, was not present in healthy controls, and was accompanied by slightly reduced the level of activity of respiratory chain enzyme.

tRNA^{Lys} Gene

A8344G Mutation

The A8344G mutation in the $tRNA^{Lys}$ gene has typically been associated with MERRF syndrome (Myoclonic Epilepsy with Ragged Red Fibers). The multisystem disorder is characterized by myoclonus, generalized epilepsy, weakness, ataxia, and ragged red fibers. Patients diagnosed with MERRF usually present in childhood or early adulthood. Recently, a review of [53, 57] patients showed that 22% of them developed cardiomyopathy. The A8344G mutation is also found in other clinical phenotypes, such as Leigh's syndrome, myoclonus or myopathy with truncal lipomas, and proximal myopathy [58].

Vallance et al. [59] reported a case of infantile histiocytoid cardiomyopathy caused by the A8344G (MERRF) mutation, which is another rare cardiomyopathy characterized by pathognomonic histiocyte-like cells within the subendocardium.

G8363A Mutation

This mutation was first described in two families with maternally inherited hypertrophic cardiomyopathy [60]. The proband in the first family was a boy from Spain who presented with heart failure and cognitive regression at the age of 8 and died of cardiorespiratory arrest at 17 years old. The patient's sibling presented with mild developmental delay, weakness of pelvic girdle, and hypertrophic cardiomyopathy at the age of 4 years. In the second family, the proband was a 44-year-old American, who developed progressive hearing loss at the age of 35 years. Later, there were a variety of clinical features, which included gait difficulties, slurred speech, shortness of breath, and chest pain. Echocardiography suggested cardiomyopathy. The proband's daughter had congenital hypertrophic cardiomyopathy and severe mental retardation [60]. A comprehensive systematic review regarding the clinical and molecular data of G8363A mutation in 6 studies showed that 4 of 41 patients with a varying level of the mutational load in the blood developed cardiomyopathy [61].

tRNA^{Val} Gene

G1644A Mutation

The novel pathological mtDNA mutation in the tRNA^{Val} was found in a family. The proband was diagnosed with MELAS and hypertrophic cardiomyopathy. Meanwhile, her mother was diagnosed with mild hypertrophy and both mother and grandmother manifested sensorineural deafness [62].

tRNA^{glycine} Gene

T9997C Mutation

Merante et al. [63] reported a unique heteroplasmic T-to-C transition at nucleotide 9997 in the mitochondrial tRNA^{glycine} gene in a multiplex family who showed non-obstructive cardiomyopathy and the severity of the symptoms correlated with the degree of mtDNA heteroplasmy.

12S rRNA Gene

A1555G Mutation

This mutation in the 12S rRNA gene was reported in a family with maternally inherited cardiomyopathy and was associated with restrictive cardiomyopathy, unlike other mtDNA mutations associated with cardiomyopathy. The mutation was found to be heteroplasmic in several tissues including the heart muscle [64].

Polypeptide-encoding mtDNA

Leigh Syndrome

Leigh syndrome, a necrotizing encephalomyelopathy, is characterized by progressive neurodegenerative disorder, affecting infants and adults. One of the main diagnostic criteria is the presence of brainstem degeneration with or without basal ganglia disease [65]. It constitutes one of the progressive neurodegenerative diseases associated with an abnormal generation of mitochondrial energy. The T8993G and T8993C mtDNA mutations account for 10–20% of Leigh syndrome mutations and other mtDNA mutations account for another 10–20% of Leigh syndrome mutations [66]. Extraneurologic symptoms might include cardiac involvement and particularly hypertrophic cardiomyopathy [67].

T8528C Mutation

Four unrelated infants with infantile hypertrophic cardiomyopathy and multisystem disease were reported by Ware et al. [68]. Sequencing of the entire mitochondrial genome revealed the novel T8528C mitochondrial mutation. This mutation leads to alteration of both ATPases 6 and 8 and results in cardiomyopathy when nearly homoplasmic. Position 8528 is located at the overlapping region encoding the ATPase 6 and ATPase 8 subunits, and the nucleotide alteration leads to the change of the initiation of methionine to threonine in the ATPase 6 subunit, abrogating the start of translation. Recently, Imai et al. [69] report a case of a Japanese patient with the T8528C mutation who died of a rapidly progressive cardiomyopathy within 5.5 months of age, demonstrating that the mutation is highly heteroplasmic in the heart (90%), and protein synthesis of both ATPase 6 and 8 was impaired in the heart. They further confirm the T8528C mutation in blood using Sanger Sequencing, and quantitative PCR showed a high heteroplasmy rate in blood DNA (88%), demonstrating that a high heteroplasmy rate of a causative mitochondrial DNA mutation in blood DNA extracted from a Guthrie card is similar to that in heart tissue in severe infantile cardiomyopathy [70].

A8701G Mutation

Zhu et al. [71] report a Han Chinese family in which several members were diagnosed with maternally transmitted hypertension and dilated cardiomyopathy. The study found a significantly associated mitochondrial variant, i.e., A8701G mutation, which was enriched in this family, may be due to the consanguineous marriage in the first generation. This mutation was heteroplasmic in blood and was accompanied by slightly reduced deficiency of mitochondrial respiratory chain enzymes.

Mitochondrial ND1 *Gene*

G3337A Mutation

Recently, Zifa et al. [72] reported a novel G3337A mutation in mitochondrial NADH dehydrogenase-1 gene (*ND1*) in two unrelated patients. The first patient was a newborn female who required resuscitation at birth. She had respiratory dysfunction and was hypotonic. Cardiac findings included aneurysm of the intra-auricular septum, and narrowing of the pulmonary artery, and some features suggestive of cardiomyopathy. The second patient was a 65-year old woman with diabetes mellitus, presenting with chest pain and dyspnea. The echocardiogram showed left ventricular hypertrophy. In the study, the mutation was 100% homoplasmic in both of the patients' blood, and in the blood of the mother and grandmother of the first patient. Surprisingly, the first patient's mother only complained of mild muscle hypotonia, exercise intolerance and weakness. Her maternal grandmother was asymptomatic at 65-years-old, as well as her father. The G3337A mutation was not detected in 150 control individuals [73].

T3398C Mutation

A recent study reported an association between an m.3398T.C MTND1 variant and left ventricular non-compaction [73]. Left ventricular non-compaction is caused by abnormal compaction of myofibrils during cardiac development and leads to progressive ventricular dilatation, and systolic and has been recognized as a cardiac symptom of mtDNA disease, particularly in pediatric populations, and most commonly as part of multisystem disease [74].

Mitochondrial DNA D-Loop

Boles et al. [75] demonstrated that mtDNA control region point heteroplasmy, particularly of the D-loop, is associated with a protean clinical disorder in children "at risk" for mitochondrial disease, being found in 15 of 75 patients. Four among the 15 children were noted to show severe cardiomyopathy with congestive heart failure in infancy, which completely or essentially resolved with supportive therapy only.

mtDNA Lesions Transmitted in a Mendelian Manner

This group of mitochondrial diseases is associated with mtDNA abnormalities due to nuclear DNA (nDNA) mutations and are inherited as Mendelian genetic traits. Then the nDNA mutation secondarily affects mtDNA leading to multiple mtDNA deletions or mtDNA depletion syndrome [36, 76].

Multiple mtDNA Deletions

An Italian family with adult-onset mitochondrial encephalomyopathy was described by Zeviani et al. [76]. The disease is characterized by progressive external ophthalmoplegia (PEO) and inherited as an autosomal dominant (AD) trait. There were different mtDNA deletions in the proband's muscle mtDNA by Southern blot analysis. Maternal inheritance was excluded as the disease was inherited through both maternal and paternal lineages in the subsequent generations. Multiple mtDNA deletions in muscle specimens from two siblings with optic atrophy, PEO, peripheral neuropathy and muscle weakness have been reported. The siblings were born to consanguineous healthy parents [44, 66, 76]. Then, an autosomal recessive syndrome associated with multiple mtDNA deletions in muscle was shown in six patients from two unrelated families. The patients presented with childhood-onset, PEO and severe cardiomyopathy requiring cardiac transplantation. The study has found that cardiac abnormalities are found in 37% of patients with AD-PEO with multiple mtDNA deletions [76–78].

mtDNA Depletion Syndrome

Mitochondrial DNA depletion syndrome represents an autosomal recessive disorder characterized by decreased mitochondrial DNA copy number in affected tissues, which has been reported in hepatic failure and skeletal muscle myopathies [79]. Thus, mtDNA depletion is tissue-specific occurring in either liver or muscle. The first report of depleted mtDNA with hypertrophic cardiomyopathy was a 5-month-old child with a history of loss of muscle tone, progressive weakness, and developmental delay from 3 months of age. Left ventricular hypertrophy without outflow obstruction was shown by echocardiogram. The patient died from circulatory shock and respiratory distress [79].

In sum, mitochondrial heteroplasmy plays a critical role in contributing to the pathogenesis of some cardiomyopathy, in which it functions as an etiological aspect. As among the above-mentioned mtDNA-associated cardiomyopathies, heteroplasmy resulted from congenital, or acquired mutation determines the eventual outcomes. Of them, some are from hereditary, and some are from sporadic. For this emerging field, a lot of questions must be examined. We speculate that, in addition to the mutation-related heteroplasmy, most have congenital characteristic. The common cardiac diseases like idiopathic hypertension, hypertrophic cardiomyopathy, and congestive heart failure etc. have a predilection in particular populations within whom the cardiac mtDNAs are prone to acquire mutation under chronic mutation-evoking stresses (Fig. 30.2). If this could be verified, corresponding novel therapeutic treatments can be implemented to conquer, or at least to delay the progression, of the disease through focusing on the heteroplasmy.

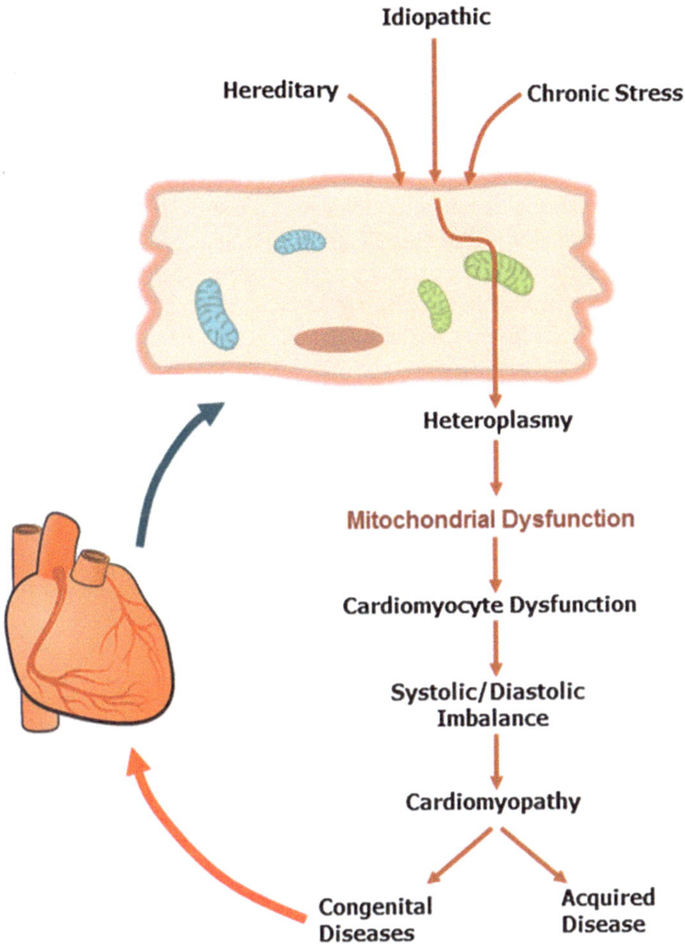

Fig. 30.2 Proposed cardiomyopathy related to mitochondrial heteroplasmy. Mitochondrial genetic materials of cardiomyocytes undergoing hereditary, idiopathic, and chronic stress (hypertension, hypoxia, homeostatic imbalance, etc.) respond correspondingly, and the heteroplasmy now initiated results in mitochondrial dysfunction. Subsequently, cardiomyocyte function is undermined and the overall systolic/diastolic balance is overturned, and then the eventual outcome like congenital or acquired diseases occur. This proposed etiological contribution of heteroplasmy to various common or uncommon cardiac diseases needs to be tested further, and corresponding therapeutic targets implemented

Conclusions

We postulate that biological significance of mtDNA heteroplasmy is demonstrated by the ability of cellular mitochondria to effectively modulate energy state-dependent changes by concerted transcriptional and translational mechanisms [17, 80, 81]. Altered homeostatic regulation of mtDNA heteroplasmy patterns can be amplified in the initiation and progression of biomedical pathophysiological processes linked with human disease states [16, 17, 82–85]. Moreover, bidirectional intra-cellular communication involving mitochondrial signaling pathways allow for co-ordinate regulation of nuclear DNA- and mtDNA-derived gene expression. This phenomenon occurs within the dynamic of a constantly changing physiological environment, which is designed to promote molecular switching of the metabolic machinery, e.g., anabolic to catabolic demands [27, 86]. Since technological transplantation of functionally viable mitochondria envisages a significant restoration of normative cellular function, a comprehensive understanding and mapping of cell-specific mosaic patterns of heteroplasmic mtDNA expression are essential to the therapeutic potential of future translational studies.

Acknowledgements The authors would like to thank Ms. Danielle Benz, for her assistance in the preparation and formatting of the content of this chapter.

References

1. Taylor RW, Turnbull DM. Mitochondrial DNA mutations in human disease. Nat Rev Genet. 2005;6:389–402.
2. Song S, Pursell ZF, Copeland WC, Longley MJ, Kunkel TA, Mathews CK. DNA precursor asymmetries in mammalian tissue mitochondria and possible contribution to mutagenesis through reduced replication fidelity. Proc Natl Acad Sci U S A. 2005;102:4990–5.
3. Crimi M, O'Hearn SF, Wallace DC, Comi GP. Molecular research technologies in mitochondrial diseases: the microarray approach. IUBMB Life. 2005;57:811–8.
4. Wallace DC. The mitochondrial genome in human adaptive radiation and disease: on the road to therapeutics and performance enhancement. Gene. 2005;354:169–80.
5. Kenney MC, Chwa M, Atilano SR, Falatoonzadeh P, Ramirez C, Malik D, Tarek M, Caceres-del-Carpio J, Nesburn AB, Boyer DS, Kuppermann BD, Vawter M, Jazwinski SM, Miceli M, Wallace DC, Udar N. Inherited mitochondrial DNA variants can affect complement, inflammation and apoptosis pathways: insights into mitochondrial-nuclear interactions. Hum Mol Genet. 2014;23:3537–51.
6. Kenney MC, Chwa M, Atilano SR, Falatoonzadeh P, Ramirez C, Malik D, Tarek M, Del Carpio JC, Nesburn AB, Boyer DS, Kuppermann BD, Vawter MP, Jazwinski SM, Miceli MV, Wallace DC, Udar N. Molecular and bioenergetic differences between cells with African versus European inherited mitochondrial DNA haplogroups: implications for population susceptibility to diseases. Biochim Biophys Acta. 1842;2014:208–19.
7. Atilano SR, Malik D, Chwa M, Caceres-Del-Carpio J, Nesburn AB, Boyer DS, Kuppermann BD, Jazwinski SM, Miceli MV, Wallace DC, Udar N, Kenney MC. Mitochondrial DNA variants can mediate methylation status of inflammation, angiogenesis and signaling genes. Hum Mol Genet. 2015;24:4491–503.
8. Wang G, McCain ML, Yang L, He A, Pasqualini FS, Agarwal A, Yuan H, Jiang D, Zhang D, Zangi L, Geva J, Roberts AE, Ma Q, Ding J, Chen J, Wang DZ, Li K, Wang J, Wanders RJ, Kulik W, Vaz FM, Laflamme MA, Murry CE, Chien KR, Kelley RI, Church GM, Parker KK, Pu WT. Modeling the mitochondrial cardiomyopathy of Barth syndrome with induced pluripotent stem cell and heart-on-chip technologies. Nat Med. 2014;20:616–23.
9. Vogelstein B, Papadopoulos N, Velculescu VE, Zhou S, Diaz Jr LA, Kinzler KW. Cancer genome landscapes. Science. 2013;339:1546–58.
10. Frank SA. Somatic mosaicism and disease. Curr Biol. 2014;24:R577–81.

11. Joseph CG, Darrah E, Shah AA, Skora AD, Casciola-Rosen LA, Wigley FM, Boin F, Fava A, Thoburn C, Kinde I, Jiao Y, Papadopoulos N, Kinzler KW, Vogelstein B, Rosen A. Association of the autoimmune disease scleroderma with an immunologic response to cancer. Science. 2014;343:152–7.
12. Ross KA. Coherent somatic mutation in autoimmune disease. PLoS One. 2014;9:e101093.
13. Poduri A, Evrony GD, Cai X, Walsh CA. Somatic mutation, genomic variation, and neurological disease. Science. 2013;341:1237758.
14. Jamuar SS, Lam AT, Kircher M, D'Gama AM, Wang J, Barry BJ, Zhang X, Hill RS, Partlow JN, Rozzo A, Servattalab S, Mehta BK, Topcu M, Amrom D, Andermann E, Dan B, Parrini E, Guerrini R, Scheffer IE, Berkovic SF, Leventer RJ, Shen Y, Wu BL, Barkovich AJ, Sahin M, Chang BS, Bamshad M, Nickerson DA, Shendure J, Poduri A, Yu TW, Walsh CA. Somatic mutations in cerebral cortical malformations. N Engl J Med. 2014;371:733–43.
15. Anderson S, Bankier AT, Barrell BG, de Bruijn MH, Coulson AR, Drouin J, Eperon IC, Nierlich DP, Roe BA, Sanger F, Schreier PH, Smith AJ, Staden R, Young IG. Sequence and organization of the human mitochondrial genome. Nature. 1981;290:457–65.
16. Valero T. Mitochondrial biogenesis: pharmacological approaches. Curr Pharm Des. 2014;20:5507–9.
17. Jayaprakash AD, Benson EK, Gone S, Liang R, Shim J, Lambertini L, Toloue MM, Wigler M, Aaronson SA, Sachidanandam R. Stable heteroplasmy at the single-cell level is facilitated by intercellular exchange of mtDNA. Nucleic Acids Res. 2015;43:2177–87.
18. Stefano GB, Kream RM. Cancer: mitochondrial origins. Med Sci Monit. 2015;21:3736–9.
19. He Y, Wu J, Dressman DC, Iacobuzio-Donahue C, Markowitz SD, Velculescu VE, Diaz Jr LA, Kinzler KW, Vogelstein B, Papadopoulos N. Heteroplasmic mitochondrial DNA mutations in normal and tumour cells. Nature. 2010;464:610–4.
20. Schon EA, Gilkerson RW. Functional complementation of mitochondrial DNAs: mobilizing mitochondrial genetics against dysfunction. Biochim Biophys Acta. 1800;2010:245–9.
21. Chen Z, Qi Y, French S, Zhang G, Covian Garcia R, Balaban R, Xu H. Genetic mosaic analysis of a deleterious mitochondrial DNA mutation in Drosophila reveals novel aspects of mitochondrial regulation and function. Mol Biol Cell. 2015;26:674–84.
22. Hill JH, Chen Z, Xu H. Selective propagation of functional mitochondrial DNA during oogenesis restricts the transmission of a deleterious mitochondrial variant. Nat Genet. 2014;46:389–92.
23. Meyers DE, Basha HI, Koenig MK. Mitochondrial cardiomyopathy: pathophysiology, diagnosis, and management. Tex Heart Inst J. 2013;40:385–94.
24. Greaves LC, Reeve AK, Taylor RW, Turnbull DM. Mitochondrial DNA and disease. J Pathol. 2012;226:274–86.
25. Li M, Schroder R, Ni S, Madea B, Stoneking M. Extensive tissue-related and allele-related mtDNA heteroplasmy suggests positive selection for somatic mutations. Proc Natl Acad Sci U S A. 2015;112:2491–6.
26. Stefano GB, Mantione KJ, Casares FM, Kream RM. Anaerobically functioning mitochondria: Evolutionary perspective on modulation of energy metabolism in *Mytilus edulis*. Invertebr Surviv J. 2015;12:22–8.
27. Snyder C, Stefano GB. Mitochondria and chloroplasts shared in animal and plant tissues: significance of communication. Med Sci Monit. 2015;21:1507–11.
28. Stefano GB, Snyder C, Kream RM. Mitochondria, chloroplasts in animal and plant cells: significance of conformational matching. Med Sci Monit. 2015;21:REV2064–9.
29. Allen JF. Why chloroplasts and mitochondria retain their own genomes and genetic systems: colocation for redox regulation of gene expression. PNAS. 2015;112:10231–8.
30. Xu X, Duan S, Yi F, Ocampo A, Liu GH, Izpisua Belmonte JC. Mitochondrial regulation in pluripotent stem cells. Cell Metab. 2013;18:325–32.
31. Jung G, Bernstein D. hiPSC modeling of inherited cardiomyopathies. Curr Treat Options Cardiovasc Med. 2014;16:320.

32. Ellen Kreipke R, Wang Y, Miklas JW, Mathieu J, Ruohola-Baker H. Metabolic remodeling in early development and cardiomyocyte maturation. Semin Cell Dev Biol. 2016;52:84–92.
33. Schlame M, Ren M. Barth syndrome, a human disorder of cardiolipin metabolism. FEBS Lett. 2006;580:5450–5.
34. Prigione A, Rohwer N, Hoffmann S, Mlody B, Drews K, Bukowiecki R, Blumlein K, Wanker EE, Ralser M, Cramer T, Adjaye J. HIF1alpha modulates cell fate reprogramming through early glycolytic shift and upregulation of PDK1-3 and PKM2. Stem Cells. 2014;32:364–76.
35. Stefano GB, Kream RM. Hypoxia defined as a common culprit/initiation factor in mitochondrial-mediated proinflammatory processes. Med Sci Monit. 2015;21:1478–84.
36. Santorelli FM, Tessa A, D'Amati G, Casali C. The emerging concept of mitochondrial cardiomyopathies. Am Heart J. 2001;141:E1.
37. Fayssoil A. Heart diseases in mitochondrial encephalomyopathy, lactic acidosis, and stroke syndrome. Congest Heart Fail. 2009;15:284–7.
38. Anan R, Nakagawa M, Miyata M, Higuchi I, Nakao S, Suehara M, Osame M, Tanaka H. Cardiac involvement in mitochondrial diseases. A study on 17 patients with documented mitochondrial DNA defects. Circulation. 1995;91:955–61.
39. Majamaa-Voltti K, Peuhkurinen K, Kortelainen ML, Hassinen IE, Majamaa K. Cardiac abnormalities in patients with mitochondrial DNA mutation 3243A>G. BMC Cardiovasc Disord. 2002;2:12.
40. Lindroos MM, Parkka JP, Taittonen MT, Iozzo P, Karppa M, Hassinen IE, Knuuti J, Nuutila P, Majamaa K. Myocardial glucose uptake in patients with the m.3243A > G mutation in mitochondrial DNA. J Inherit Metab Dis. 2016;39:67–74.
41. Vilarinho L, Santorelli FM, Rosas MJ, Tavares C, Melo-Pires M, DiMauro S. The mitochondrial A3243G mutation presenting as severe cardiomyopathy. J Med Genet. 1997;34:607–9.
42. Zeviani M, Gellera C, Antozzi C, Rimoldi M, Morandi L, Villani F, Tiranti V, DiDonato S. Maternally inherited myopathy and cardiomyopathy: association with mutation in mitochondrial DNA tRNA(Leu)(UUR). Lancet. 1991;338:143–7.
43. Silvestri G, Santorelli FM, Shanske S, Whitley CB, Schimmenti LA, Smith SA, DiMauro S. A new mtDNA mutation in the tRNA(Leu(UUR)) gene associated with maternally inherited cardiomyopathy. Hum Mutat. 1994;3:37–43.
44. Bruno C, Kirby DM, Koga Y, Garavaglia B, Duran G, Santorelli FM, Shield LK, Xia W, Shanske S, Goldstein JD, Iwanaga R, Akita Y, Carrara F, Davis A, Zeviani M, Thorburn DR, DiMauro S. The mitochondrial DNA C3303T mutation can cause cardiomyopathy and/or skeletal myopathy. J Pediatr. 1999;135:197–202.
45. Ueki I, Koga Y, Povalko N, Akita Y, Nishioka J, Yatsuga S, Fukiyama R, Matsuishi T. Mitochondrial tRNA gene mutations in patients having mitochondrial disease with lactic acidosis. Mitochondrion. 2006;6:29–36.
46. van den Bosch BJ, de Coo IF, Hendrickx AT, Busch HF, de Jong G, Scholte HR, Smeets HJ. Increased risk for cardiorespiratory failure associated with the A3302G mutation in the mitochondrial DNA encoded tRNALeu(UUR) gene. Neuromuscul Disord. 2004;14:683–8.
47. Tessa A, Vilarinho L, Casali C, Santorelli FM. MtDNA-related idiopathic dilated cardiomyopathy. Eur J Hum Genet. 1999;7:847–8.
48. Grasso M, Diegoli M, Brega A, Campana C, Tavazzi L, Arbustini E. The mitochondrial DNA mutation T12297C affects a highly conserved nucleotide of tRNA(Leu(CUN)) and is associated with dilated cardiomyopathy. Eur J Hum Genet. 2001;9:311–5.
49. Wang J, Brautbar A, Chan AK, Dzwiniel T, Li FY, Waters PJ, Graham BH, Wong LJ. Two mtDNA mutations 14487T>C (M63V, ND6) and 12297T>C (tRNA Leu) in a Leigh syndrome family. Mol Genet Metab. 2009;96:59–65.
50. Tanaka M, Ino H, Ohno K, Hattori K, Sato W, Ozawa T, Tanaka T, Itoyama S. Mitochondrial mutation in fatal infantile cardiomyopathy. Lancet. 1990;336:1452.
51. Tomari Y, Hino N, Nagaike T, Suzuki T, Ueda T. Decreased CCA-addition in human mitochondrial tRNAs bearing a pathogenic A4317G or A10044G mutation. J Biol Chem. 2003;278:16828–33.

52. Taniike M, Fukushima H, Yanagihara I, Tsukamoto H, Tanaka J, Fujimura H, Nagai T, Sano T, Yamaoka K, Inui K, et al. Mitochondrial tRNA(Ile) mutation in fatal cardiomyopathy. Biochem Biophys Res Commun. 1992;186:47–53.
53. Hino N, Suzuki T, Yasukawa T, Seio K, Watanabe K, Ueda T. The pathogenic A4269G mutation in human mitochondrial tRNA(Ile) alters the T-stem structure and decreases the binding affinity for elongation factor Tu. Genes Cells. 2004;9:243–52.
54. Hayashi J, Ohta S, Kagawa Y, Takai D, Miyabayashi S, Tada K, Fukushima H, Inui K, Okada S, Goto Y, et al. Functional and morphological abnormalities of mitochondria in human cells containing mitochondrial DNA with pathogenic point mutations in tRNA genes. J Biol Chem. 1994;269:19060–6.
55. Merante F, Myint T, Tein I, Benson L, Robinson BH. An additional mitochondrial tRNA(Ile) point mutation (A-to-G at nucleotide 4295) causing hypertrophic cardiomyopathy. Hum Mutat. 1996;8:216–22.
56. Mahjoub S, Sternberg D, Boussaada R, Filaut S, Gmira F, Mechmech R, Jardel C, Arab SB. A novel mitochondrial DNA tRNAIle (m.4322dupC) mutation associated with idiopathic dilated cardiomyopathy. Diagn Mol Pathol. 2007;16:238–42.
57. Wahbi K, Larue S, Jardel C, Meune C, Stojkovic T, Ziegler F, Lombes A, Eymard B, Duboc D, Laforet P. Cardiac involvement is frequent in patients with the m.8344A>G mutation of mitochondrial DNA. Neurology. 2010;74:674–7.
58. Silvestri G, Ciafaloni E, Santorelli FM, Shanske S, Servidei S, Graf WD, Sumi M, DiMauro S. Clinical features associated with the A – >G transition at nucleotide 8344 of mtDNA ("MERRF mutation"). Neurology. 1993;43:1200–6.
59. Vallance HD, Jeven G, Wallace DC, Brown MD. A case of sporadic infantile histiocytoid cardiomyopathy caused by the A8344G (MERRF) mitochondrial DNA mutation. Pediatr Cardiol. 2004;25:538–40.
60. Santorelli FM, Mak SC, El-Schahawi M, Casali C, Shanske S, Baram TZ, Madrid RE, DiMauro S. Maternally inherited cardiomyopathy and hearing loss associated with a novel mutation in the mitochondrial tRNA(Lys) gene (G8363A). Am J Hum Genet. 1996;58:933–9.
61. Virgilio R, Ronchi D, Bordoni A, Fassone E, Bonato S, Donadoni C, Torgano G, Moggio M, Corti S, Bresolin N, Comi GP. Mitochondrial DNA G8363A mutation in the tRNA Lys gene: clinical, biochemical and pathological study. J Neurol Sci. 2009;281:85–92.
62. Menotti F, Brega A, Diegoli M, Grasso M, Modena MG, Arbustini E. A novel mtDNA point mutation in tRNA(Val) is associated with hypertrophic cardiomyopathy and MELAS. Ital Heart J. 2004;5:460–5.
63. Merante F, Tein I, Benson L, Robinson BH. Maternally inherited hypertrophic cardiomyopathy due to a novel T-to-C transition at nucleotide 9997 in the mitochondrial tRNA(glycine) gene. Am J Hum Genet. 1994;55:437–46.
64. Santorelli FM, Tanji K, Manta P, Casali C, Krishna S, Hays AP, Mancini DM, DiMauro S, Hirano M. Maternally inherited cardiomyopathy: an atypical presentation of the mtDNA 12S rRNA gene A1555G mutation. Am J Hum Genet. 1999;64:295–300.
65. Wang SB, Weng WC, Lee NC, Hwu WL, Fan PC, Lee WT. Mutation of mitochondrial DNA G13513A presenting with Leigh syndrome, Wolff-Parkinson-White syndrome and cardiomyopathy. Pediatr Neonatol. 2008;49:145–9.
66. Thorburn DR, Rahman S. Mitochondrial DNA-associated leigh syndrome and NARP. In: Pagon RA, Adam MP, Ardinger HH, Wallace SE, Amemiya A, LJH B, et al., editors. GeneReviews®. Seattle: University of Washington; 2003.
67. Agapitos E, Pavlopoulos PM, Patsouris E, Davaris P. Subacute necrotizing encephalomyelopathy (Leigh's disease): a clinicopathologic study of ten cases. Gen Diagn Pathol. 1997;142:335–41.
68. Ware SM, El-Hassan N, Kahler SG, Zhang Q, Ma YW, Miller E, Wong B, Spicer RL, Craigen WJ, Kozel BA, Grange DK, Wong LJ. Infantile cardiomyopathy caused by a mutation in the overlapping region of mitochondrial ATPase 6 and 8 genes. J Med Genet. 2009;46:308–14.

69. Imai A, Fujita S, Kishita Y, Kohda M, Tokuzawa Y, Hirata T, Mizuno Y, Harashima H, Nakaya A, Sakata Y, Takeda A, Mori M, Murayama K, Ohtake A, Okazaki Y. Rapidly progressive infantile cardiomyopathy with mitochondrial respiratory chain complex V deficiency due to loss of ATPase 6 and 8 protein. Int J Cardiol. 2016;207:203–5.
70. Imai A, Kishita Y, Nakayama Y, Fujita S, Futatani T, Kohda M, Yatsuka Y, Nakaya A, Sakata Y, Murayama K, Ohtake A, Okazaki Y. Dried blood spots for newborn screening allows easy determination of a high heteroplasmy rate in severe infantile cardiomyopathy. Int J Cardiol. 2016;221:446–9.
71. Zhu Y, Gu X, Xu C. A mitochondrial DNA A8701G mutation associated with maternally inherited hypertension and dilated cardiomyopathy in a Chinese pedigree of a consanguineous marriage. Chin Med J. 2016;129:259–66.
72. Zifa E, Theotokis P, Kaminari A, Maridaki H, Leze H, Petsiava E, Mamuris Z, Stathopoulos C. A novel G3337A mitochondrial ND1 mutation related to cardiomyopathy co-segregates with tRNALeu(CUN) A12308G and tRNAThr C15946T mutations. Mitochondrion. 2008;8:229–36.
73. Tang S, Batra A, Zhang Y, Ebenroth ES, Huang T. Left ventricular noncompaction is associated with mutations in the mitochondrial genome. Mitochondrion. 2010;10:350–7.
74. Finsterer J. Cardiogenetics, neurogenetics, and pathogenetics of left ventricular hypertrabeculation/noncompaction. Pediatr Cardiol. 2009;30:659–81.
75. Boles RG, Luna C, Ito M. Severe reversible cardiomyopathy in four unrelated infants associated with mitochondrial DNA D-loop heteroplasmy. Pediatr Cardiol. 2003;24:484–7.
76. Zeviani M, Petruzzella V, Carrozzo R. Disorders of nuclear-mitochondrial intergenomic signalling. J Bioenerg Biomembr. 1997;29:121–30.
77. Zeviani M, Servidei S, Gellera C, Bertini E, DiMauro S, DiDonato S. An autosomal dominant disorder with multiple deletions of mitochondrial DNA starting at the D-loop region. Nature. 1989;339:309–11.
78. Bohlega S, Tanji K, Santorelli FM, Hirano M, al-Jishi A, DiMauro S. Multiple mitochondrial DNA deletions associated with autosomal recessive ophthalmoplegia and severe cardiomyopathy. Neurology. 1996;46:1329–34.
79. Marin-Garcia J, Ananthakrishnan R, Goldenthal MJ. Hypertrophic cardiomyopathy with mitochondrial DNA depletion and respiratory enzyme defects. Pediatr Cardiol. 1998;19:266–8.
80. Campello S, Lacalle RA, Bettella M, Manes S, Scorrano L, Viola A. Orchestration of lymphocyte chemotaxis by mitochondrial dynamics. J Exp Med. 2006;203:2879–86.
81. Campello S, Scorrano L. Mitochondrial shape changes: orchestrating cell pathophysiology. EMBO Rep. 2010;11:678–84.
82. Wallace DC. A mitochondrial paradigm of metabolic and degenerative diseases, aging, and cancer: a dawn for evolutionary medicine. Annu Rev Genet. 2005;39:359–407.
83. Calloway CD, Reynolds RL, Herrin Jr GL, Anderson WW. The frequency of heteroplasmy in the HVII region of mtDNA differs across tissue types and increases with age. Am J Hum Genet. 2000;66:1384–97.
84. Stefano GB, Kream RM. Dysregulated mitochondrial and chloroplast bioenergetics from a translational medical perspective (Review). Int J Mol Med. 2016;37:547–55.
85. Stefano GB, Kream RM. Mitochondrial DNA heteroplasmy in human health and disease. Biomed Rep. 2016;4:259–62.
86. Stefano GB, Challenger S, Kream RM. Hyperglycemia-associated alterations in cellular signaling and dysregulated mitochondrial bioenergetics in human metabolic disorders. Eur J Nutr. 2016;55:2339–45.

Chapter 31
Mitochondrial Transplantation in Myocardial Ischemia and Reperfusion Injury

Borami Shin, Douglas B. Cowan, Sitaram M. Emani, Pedro J. del Nido, and James D. McCully

Ischemic heart disease remains the leading cause of death worldwide and ischemia and reperfusion injury hallmark the central pathophysiology that determines clinical outcomes [1]. Robust investigations have elucidated the mechanisms of myocardial ischemia [2–4] and reperfusion injury [5] as well as the intrinsic adaptive mechanisms of myocardial stunning [6], hibernation [7], preconditioning [8, 9] and postconditioning [10–12]. They enabled the current knowledge of cardiac metabolism, unraveling the complex biochemical and cellular changes resulting from myocardial ischemia and reperfusion including disruptions of cellular energy, ionic homeostasis and oxidative stress which all converge on mitochondrial dysfunction to culminate in cellular apoptosis and necrosis [13–16].

B. Shin, MD
Department of Cardiac Surgery, Boston Children's Hospital, Boston, MA, USA
e-mail: borami.shin@childrens.harvard.edu

D.B. Cowan, PhD
Department of Anesthesiology, Division of Cardiac Anesthesia Research, Harvard Medical School, Boston Children's Hospital, Boston, MA, USA
e-mail: douglas.cowan@childrens.harvard.edu

S.M. Emani, MD
Division of Cardiovascular Critical Care, Harvard Medical School, Boston Children's Hospital, Boston, MA, USA
e-mail: sitaram.emani@childrens.harvard.edu

P.J. del Nido, MD
Department of Cardiac Surgery, William E. Ladd Professor of Child Surgery, Harvard Medical School, Boston Children's Hospital, Boston, MA, USA
e-mail: pedro.delnido@cardiac.chboston.org

J.D. McCully, PhD (✉)
Department of Cardiac Surgery, Harvard Medical School, Boston Children's Hospital, Boston, USA
e-mail: james_mccully@hms.harvard.edu

G. Santulli (ed.), *Mitochondrial Dynamics in Cardiovascular Medicine*, Advances in Experimental Medicine and Biology 982,
DOI 10.1007/978-3-319-55330-6_31

The functional and structural derangements governing the pathophysiology of ischemia-reperfusion injury as well as changes in myocardial survival pathways inspired many pharmacological, genetic and procedural therapies in a variety of models including energy substrates, antioxidants, inhibitors of calcium (Ca^{2+}) and sodium accumulation, mitochondrial components, immunomodulation as well as procedural techniques. However, a major dichotomy exists between positive results of preclinical studies and equivocal or negative results of the same therapies on limiting infarct size in human clinical trials [17]. Meanwhile, multi-disciplinary approaches in both experimental and clinical research targeting myocardial ischemia continue to highlight mitochondrial malfunction as the common denominator of cardiomyocyte's ischemic response, emphasizing that the preservation of mitochondrial function and energy production may be the essential gateway to cardioprotection.

Rather than ameliorating one mediator or a step within the vast array of mitochondrial responses to ischemia and reperfusion, a newly emerging method of replacing damaged cardiomyocyte mitochondria with healthy mitochondria isolated from non-ischemic tissue of the same patient is being explored in various in vitro and in vivo models with promising results. In this chapter, we will explore (1) the basic physiology of myocardial ischemia and reperfusion injury, (2) the key features of mitochondrial metabolism and changes in ischemia and reperfusion injury, (3) current methodologies of mitochondrial transplantation, and (4) potential mechanisms of cardioprotection conferred by mitochondrial transplantation based on the most current experimental developments.

Mitochondria in Cardiac Tissue

The heart is a syncytial, muscular network of cardiomyocytes specialized to propagate electrical impulses across communicating cells allowing the varied components of the heart to act as a single contractile apparatus. The incessant contractility of the myocardium and cellular homeostasis is almost exclusively powered by the production of ATP by the mitochondria which constitute about 30% of the myocardial cell volume and provide for the daily turnover rate of ATP up to 15–20 times the weight of the heart in humans [18]. Thus, tight coupling of metabolic proficiency and myocardial contraction is essential for normal cardiac function and govern the intricate set of metabolic pathways resulting in both ATP producing and non-ATP producing endpoints for different classes of substrates under normal physiological conditions and during severe energy deprivation.

ATP is produced by the process of oxidative phosphorylation carried out by a series of four multi-subunit complexes embedded in the inner mitochondrial membrane as the Electron Transport Chain (ETC). The components of the electron transport chain include: Complex I (NADH dehydrogenase), II (succinate ubiquinone oxidoreductase), III (ubiquinol: cytochrome c oxidoreductase), IV (cytochrome c oxidase), V (ATP synthase), and two main electron carriers ubiquinone and cytochrome *c*. Thirteen of these 76 protein subunits making up the five complexes of the electron transport chain are encoded by the mitochondrial DNA and the remaining are encoded by the nuclear DNA. The various catabolic pathways of energy-yielding substrates (fatty acid, glucose, amino acids and ketones) converge on Acetyl-CoA production which mainly enter the tricarboxylic acid cycle (TCA) in the mitochondrial matrix to generate reducing equivalents NADH and $FADH_2$. As electrons from NADH and $FADH_2$ flow through the ETC to reduce oxygen to water (by complex IV), the resultant release of potential energy is used to pump protons into the mitochondrial intermembrane space (by complex I, III and IV) that generates an electrochemical gradient (−200 mV) across the mitochondrial inner membrane which is harnessed by complex V (ATP synthase) to couple P_i and ADP to ATP. The ATP synthesized in the matrix is transported into the intermembrane space by an inner membrane channel adenine nucleotide translocase (ANT), where mitochondrial creatine kinase (miCK) transfers the high energy phosphate of ATP to create phosphocreatine which is then stored in the cytosol as energy reserves. Additional ATP is transported out to the cytosol via a complex formed by ANT, miCK and the voltage-dependent anion channel (VDAC) located in the outer membrane, where miCK bridges the space between ANT and VDAC, allowing for low conductance transfer of ATP out into the cytosol. In addition to the maintenance of cellular homeostasis, ATP consuming reactions vital to cardiomyocytes are exemplified by three main reactions which include (1) Myosin-ATPase which provides energy for myosin filament sliding for muscle contraction, (2) Ca^{2+}/Mg^{2+}-ATPase which stores Ca^{2+} in the sarcoplasmic reticulum for activation of myocardial contraction upon release, and (3) Na^{+}/K^{+}-ATPase used to maintain cellular excitability and Ca^{2+} efflux [19].

Cardiac mitochondria consist of two spatially distinct subpopulations, namely subsarcolemmal mitochondria (SSM), located directly beneath the cellular membrane, and interfibrillar mitochondria (IFM) which reside in rows between the myofibrils [20, 21]. Scanning and transmission electron microscopy of left ventricular papillary muscle sections of Japanese Monkeys and humans have revealed many structural differences between SSM and IFM [22, 23]. IFM are generally elongated in shape (1.5–2.0 μm), and occupy the space between Z-lines flanked by the junctional sarcoplasmic reticulum, and have predominantly tubular cristae. SSM located beneath the sarcolemma have more variable lengths and sizes (0.4–3.0 μm), possessing more lamelliform and closely-packed cristae [23, 24]. Studies using three dimensional modeling approach coupled with MitoTracker Red staining also revealed that IFM were arranged in a more orderly pattern in contrast to SSM which had a more random arrangement [25, 26]. Numerous reports have also indicated that SSM and IFM possess distinct functional differences such as substrate utilizations [20, 27, 28], enzymatic rates [21] and ATP synthesis [29, 30] which have been found to be generally higher in IFM.

Mitochondrial subpopulations appear to be differentially influenced during different pathological insults. However, many inconsistencies exist across different animal models. For example, cardiac ischemia has been reported to primarily effect SSM in rats [31] and rabbit models [32, 33] but both SSM and IFM populations in canines [21, 34]. In contrast to some of the ischemic models, ischemia and reperfusion injury have been reported to effect both populations in the mouse, rat and rabbit models [35–37]. The clinical significance of spatial distinctions of cardiac mitochondria with regards to pathological conditions and therapeutic interventions thereof, remain to be investigated.

Mitochondria in Myocardial Ischemia and Reperfusion

Ischemia-reperfusion injury occurs as a result of attenuation or cessation of coronary blood flow such that oxygen delivery to the myocardium is insufficient to meet the myocardial oxygen requirements to preserve cellular function and contraction. The energy stores generated by the mitochondria are only sufficient to sustain myocardial contractions for a few seconds, and this explains the need for the robustness of the cardiac metabolism, consuming the highest rate of oxygen per unit weight basis. Under normal conditions, 75% of the oxygen supplied by the coronary vasculature is extracted by the myocardium during a single passage through the heart. Therefore, the heart is highly susceptible to alterations in oxygen delivery where an increase in myocardial oxygen consumption (MVO_2) can only be met by an increase in coronary blood flow, a phenomena unique to the cardiac muscle. The loss of myocardial oxygen delivery has been shown to exhaust myocardial oxygen reserves within 8 seconds after the onset of normothermic global ischemia with subsequent decrease in energy reserves [38]. In addition, studies measuring energy content by phosphorous-31 NMR spectroscopy have demonstrated that decreases in myocardial energy reserves are directly associated with decreases in myocardial function and viability which in turn are associated with detrimental clinical pathologies [39–43].

Reversible Myocardial Ischemia

With the onset of coronary occlusion, myocardial injury begins in approximately 20 min, being first evident in the subendocardium and papillary muscle, then extending into the mid-myocardial bed-at-risk by about 60–90 min, such that the wavefront of irreversible transmural infarction is complete in 3–4 hrs [44–47]. The infarcted myocardium subsequently exhibits features of inflammation and necrosis. Depending on the extent and duration of the ischemic event, myocardial injury may be reversible where the myocardium activates physiologic adaptations of survival manifested by myocardial stunning, hibernation and pre- and post-ischemic conditioning. These conditions result from changes in gene expressions that sustain cell survival under oxygen deprivation and during the stresses of reperfusion.

Myocardial stunning, first described by Heyndrickx and colleagues in 1975, refers to a state of contractile dysfunction which persists after reperfusion despite the absence of irreversible myocellular damage and return of normal or near-normal perfusion. Heyndrickx demonstrated the reversibility of ischemic myocardium following a 5 min or 15 min episode of ischemia and reperfusion in dogs, which resulted in a prolonged regional contractile deficit that required 6 h to recover from a 5 min ischemic episode and greater than 24 h to recover from a 15 min ischemic episode [48]. Stress of reperfusion can be exemplified in cardiac surgery patients where, despite modern methods of cardioprotection, the heart weaning from cardiopulmonary bypass often exhibit varying degrees of myocardial stunning and requires inotropes hours to days after surgery without objective evidence of myocardial infarction [49, 50].

In contrast to the stunned myocardium, the hibernating myocardium refers to a myocardium with chronically reduced contractility associated with local reduction in myocardial perfusion [51, 52]. Although the hypo-perfused cardiomyocytes can develop degenerative changes over time, the contractile function of the hibernating myocardium can be restored by reperfusion [53–55], allowing the myocardium to be called dysfunctional but 'viable'. As in line with the 'smart heart hypothesis', hibernation is considered an adaptive mechanism of myocardial metabolism in which myocardial contractility and cellular activity is deliberately reduced to minimize metabolic demands and recruit coronary flow reserves in the face of severely depressed local perfusion [51]. However, hibernating myocardium cannot maintain viability indefinitely and delayed revascularization is associated with worse outcomes [56, 57]. It is now widely believed that repetitive episodes of stunning (i.e. ischemic dysfunction) create a sustained depression of contractile function as in hibernation and that stunning and hibernation are likely to be part of a continuous disease spectrum [58, 59]. In a swine model where the heart was subjected to repetitive episodes of 90 min of coronary occlusion followed by 12 h of reperfusion cycled up to six times, the regional myocardial function was depressed despite normalization of perfusion, reflecting myocardial stunning. However, additional episodes of coronary occlusion did not result in further decreases in ventricular function, reflecting the mismatch between perfusion and contractile function which characterizes hibernation [59]. This model along with several older studies demonstrated the hallmarks of chronic dysfunction in the human hibernating myocardium including loss of myofibrils, higher dependence on glucose and glycogen accumulation [60–62].

Irreversible Myocardial Ischemia and Mitochondrial Dysfunction

Despite regional differences in myocardial hypoxic tolerance and intrinsic adaptive mechanisms, irreversible myocardial injury inevitably follows prolonged ischemia. Countless studies have identified the pathological modulators of irreversible myocardial ischemic injury and the intricate linkages that converge on mitochondrial damage as the central culprit of cascade to myocardial cell death and tissue damage.

With the cessation of blood flow and oxygen supply, declines in the rate of oxygen consumption by oxidative phosphorylation and the synthesis of high energy phosphate products are immediately detectable. Mitochondrial function assessed by oximetry in many animal models have shown that ischemia significantly decreases state three oxygen consumption and respiratory control index in malate (complex I substrate) and succinate (complex III substrate) in energized mitochondria [63, 64]. Similar findings were measured for complex III and cytochrome oxidase in buffer-perfused rat hearts [32] and rabbit hearts [65] where global ischemia decreases their Vmax, mRNA levels as well as the content of cytochrome *c*. These declines were also shown to be paralleled with significant decreases in phosphocreatine, inorganic phosphate, ATP, and high energy stores, up to 36 ± 3% following 30 min of ischemia which remained decreased at 15 min post reperfusion via ^{31}P nuclear magnetic resonance (NMR) [43].

With ischemic depletion of ATP synthesis, multiple biochemical and ultrastructural changes occur in the mitochondria that are hallmarked by several key cellular events which are cellular acidosis and accumulation of intracellular, mitochondrial and even nuclear Ca^{2+} ($[Ca^{+2}]_n$). With depletion of oxygen, the heart switches from fatty acid oxidation to anaerobic glycolysis resulting in a build-up of lactate and a decrease in the cellular pH. This rapid metabolic acidosis can be quantified by the measurement of tissue PCO_2 and ^{21}P NMR spectroscopy in the lab as well as with a pH probe in the operating room [66, 67]. Accumulation of intracellular hydrogen ion (H^+) opens the sodium-hydrogen (Na^+/H^+) exchanger (NHE), resulting in the transport of H^+ into the extracellular space and the movement of Na^+ into the cytosol [68]. The excess Na^+ is then extruded in exchange for the intake of Ca^{2+} through the reverse action of the plasma membrane sodium-Ca^{2+} (Na^+/Ca^{2+}) exchanger (NCX), increasing the cytosolic Ca^{2+}concentration ($[Ca^{2+}])_i$. The increase in $[Ca^{2+}]_i$ is further augmented by depolarization of the plasma membrane potential, which leads to the opening of the L-type Ca^{2+}channels causing further Ca^{2+} influx, as well as the cessation of uptake of Ca^{2+} into the sarcoplasmic reticulum by the sarco/endoplasmic reticulum Ca^{2+}-ATPase (SERCA) due to the decline in ATP [69]. Cellular Ca^{2+}-dependent phospholipases and proteases are in turn activated inducing membrane injury and further Ca^{2+} entry into the cell.

The increase in cytosolic Ca^{2+} leads to the augmented mitochondrial Ca^{2+} that has deleterious effects on their structure and function [70]. In a normal state, mitochondrial inner membrane pump H^+ out to the cytosol creating a voltage gradient that provides the passive energy for Ca^{2+} influx into the matrix by the Ca^{2+}-uniporter. During ischemia, increases in the intracellular Ca^{2+} ($[Ca^{2+}]_i$) lead to increased Ca^{2+} uptake by the mitochondria which destabilizes the inner membrane potential. As Ca^{2+} accumulates, mitochondria use ATP to transport excess Ca^{2+} out of the mitochondria against the electrochemical gradient which wastes more ATP required to maintain cellular viability, a mechanism called 'futile Ca^{2+} cycling' [70–72]. As Ca^{2+} overloads in the mitochondria, water passively follow and the mitochondria swell. This has been visualized via ECM and light scattering technique on isolated perfused heart and in vivo pig and sheep models, where greater than 88% of myocardial mitochondria are electron translucent and swollen following 20–30 min of global ischemia and exhibit increased intermembrane space, enlarged cristae and disrupted matrix [63, 64, 73, 74]. These alterations in mitochondrial structure are noted to be damaging to ATP transfer from the mitochondria via the adenine nucleotide translocase (ANT)-mitochondrial creatine kinase (miCK)-voltage-dependent anion carrier (VDAC) complex. It has been proposed that mitochondrial swelling increases intermembrane distance and dissociate the functional interaction between ANT, miCK and VDAC that compromise efficient energy transfer during reperfusion and decrease high energy synthesis in the myocardium post reperfusion [75].

The role of Ca^{2+} overload has been demonstrated by studies that showed warm global ischemia results in rapid accumulation of intracellular and mitochondrial Ca^{2+} which is associated with the depletion of ATP, and precedes the onset of ischemic contracture in various animal models [76–78]. Conversely, amelioration of this Ca^{2+} overload has been shown to correlate with maintenance of high energy phosphates in the myocardium, results of which formed the conceptual basis for optimization of many cardioplegic solutions in cardiac surgery. For example, Hearse et al. and many subsequent investigators showed that magnesium, when added to potassium cardioplegic solutions, significantly decreases cytosolic and nuclear Ca^{2+} accumulation during global ischemia [79, 80], enhances preservation and re-synthesis of high-energy phosphates, and significantly decreases myocardial injury upon reperfusion in perfused rabbit heart [43, 78, 81] and in situ blood-perfused sheep and pig heart models [47, 82, 83]. Furthermore, the addition of pharmacological agents such as diazoxide to the K^+/Mg^{2+} cardioplegia solution, was shown to ameliorate mitochondrial Ca^{2+} accumulation by opening of the mitochondrial ATP-sensitive K^+ channels which correlated with prevention of mitochondrial swelling, decrease in apoptosis and improved post-ischemic function [82, 84–87].

Myocardial Reperfusion and Mitochondrial Injury

While restoration of blood flow is the absolute therapeutic aim to salvage ischemic cell death, it is proposed that reperfusion itself independently activates a cascade of cellular injuries which inflate the damages in excess of those produced by ischemia alone. First described by Jennings et al. in 1960 in their description of histological features of reperfused ischemic canine myocardium, they reported that features of cellular injury after 3 h of ischemia followed by 1 h of reperfusion were far worse than changes observed after 4 h of ischemia alone; the reperfused myocardium had significantly more cell swelling, contracture of myofibrils and mitochondrial and sarcolemmal disruption [88]. Clinically, reperfusion injury leads to arrhythmias, conversion of myocardial necrosis into a hemorrhagic infarct with prominent contraction band necrosis, disruption of the microvasculature leading to the 'no-reflow' phenomenon, which refers to impairment of blood flow to an ischemic region in spite of opening the infarct-related artery [17, 89–91]. In addition, it is associated with activation of inflammatory reactions which can be extensive enough to cause inflammatory responses on distant, non-ischemic organs [92].

Extensive experimental efforts have identified the potential mechanisms of reperfusion injury that build on the existing elements of the damages sustained from ischemia, described by distinct biochemical changes that occur from reperfusion in the mitochondria. These changes include (1) rapid normalization of pH, (2) exacerbation of existing mitochondrial and intracellular Ca^{2+} overload, and (3) generation of reactive oxygen species.

Reperfusion and re-oxygenation results in rapid restoration of intracellular pH by lactic acid washout, clearance of H^+ by the Na^+/H^+ exchanger (NHE) and Na^+/HCO_3^- exchanger, and the restoration of the mitochondrial membrane potential. As the mitochondrial membrane potential normalizes, the mitochondrial Ca^{2+}-uniporter (MCU) opens causing an even more influx of Ca^{2+} into an already Ca^{2+}-overloaded mitochondria [93]. Cytosolic Ca^{2+} is also proposed to be further increased by the damaged sarcolemmal membrane and oxidative stress-induced dysfunction of the sarcoplasmic reticulum, all of which can overwhelm the cardiomyocyte's mechanisms for ionic homeostasis.

Many experimental studies have also implicated Reactive Oxygen Species (ROS) as partial mediators of reperfusion injury and the damaged mitochondria as the potential primary source of ROS production. Various modalities including electron spin resonance spectrometer [94] have detected bursts of ROS during myocardial reperfusion in different animal models and patients undergoing coronary revascularization [95]. Examples include H_2O_2, HO^-, HOCl, and O_2^-, the superoxide anion radial which is considered to be the most toxic ROS generated primarily by the mitochondria. In addition, it has been recently proposed that mitochondrial alterations at complex I of the electron transport chain and the accumulation of succinate during ischemia may be related to mitochondrial oxidative injury during reperfusion. Succinate is the substrate of complex II, and the sole mitochondrial component to the three metabolites that accumulate during ischemia: succinate, xanthine and hypoxanthine [96]. Accumulation succinate is detected in the mitochondria when the electron transport chain comes to a halt during ischemia. It is proposed that upon reperfusion, succinate is abruptly hydrolyzed by complex II [97, 98] at a maximal rate that overwhelms the speed of ATP synthesis generating an excess pool of electrons that start to flow backwards by a process termed Reverse Electron Transport (RET) [99]. As the excess electrons are forced back through complex I, the reverse reaction of NADH dehydrogenase generates large amounts of superoxide. This has been demonstrated by various animal experiments which showed that inhibition of complex I slows reactivation of mitochondria during the early phase of reperfusion, reduce ROS and protects against ischemic-reperfusion injury [100–103]. Excessive ROS generated during reperfusion cause tissue damage by a myriad of mechanisms including peroxidation of lipids, inactivation of proteins/enzymes, nuclear and mitochondrial DNA strand breaks, oxidation of Cardiolipin, SER dysfunction and activate inflammation, all of which serve to activate apoptotic and necrotic signaling pathways. ROS also reduce the viability of the vasodilator nitric oxide (NO), removing its cardioprotective effects such as inactivation of superoxide radicals, inhibition of neutrophil accumulation and improvement of coronary flow [104].

Multitude of the acute fluctuations described above are proposed to act in concert to lead to the opening of the voltage dependent, nonspecific pore in the inner mitochondrial membrane known as the Mitochondrial Permeability Transition Pore (mTPT), which is thought to be the most noxious step marking the transition from reversible to irreversible reperfusion injury. The mTPT is closed during normal conditions and ischemia. Opening of mTPT pore at reperfusion allows the free passage of any molecule smaller than 1.5 kDa, causing a buildup of colloidal osmotic pressure leading to severe mitochondrial swelling, complete collapse of the mitochondrial membrane potential, uncoupling of oxidative phosphorylation and disruption of the mitochondrial membrane and eventual rupture. This process leads to release of the cytochrome c from the ETC, which is a potent activator of apoptotic signaling cascades leading to cardiomyocyte death [105, 106]. Systemically, activation of corresponding immune responses leads to leukocyte aggregation, leukocyte-mediated tissue destruction and endothelial dysfunction leading to platelet aggregation causing damage to the surrounding microvasculature.

Ischemia Versus Reperfusion Injury

The concept of reperfusion injury as an independent mediator of cardiomyocyte death distinct from ischemic injury has been controversial. The uncertainty relates to the inability to accurately assess the exact progression of necrosis during the transition from myocardial ischemia to reperfusion in situ [17]. Moreover, understanding the sequential evolution of mitochondrial damage in relation to the timeframe along myocardial ischemia and reperfusion is not straightforward. In spite of these challenges, studies demonstrate that critical events of mitochondrial damage occur at the time of ischemia and persist through reperfusion.

Transcriptomic and proteomic enrichment analyses in both Langendorff-perfused and in situ blood-perfused rabbit hearts have demonstrated that global ischemia downregulates genes/proteins associated with mitochondrial structure, function, energy production, cofactor metabolism, generation of precursor metabolites of energy, carbohydrate metabolism and regulation of biosynthesis and transcription [107]. In contrast, the same analyses in globally ischemic rabbit hearts treated with cardioplegia showed that cardioprotection was associated with significant increases in differentially expressed transcripts/proteins associated with mitochondrial function, energy production, fatty acid beta oxidation, cellular lipid metabolism as well as processes of muscle contraction [107]. These results indicate that (1) the mitochondrion plays a significant role in both global ischemia and cardioprotective pathways, (2) the mechanisms of mitochondrial damage resulting from global ischemia as well as pathways of cardioprotection are modulated, at least in part, by RNA- and protein-dependent mechanisms, and (3) these critical events occur at the time of ischemia rather than reperfusion.

The challenges of elucidating reperfusion injury are also demonstrated by the equivocal results of studies targeting the accepted mediators of reperfusion injury.

While attractive targets of intervention such as the accumulation of Ca^{2+}, reactive oxygen species, components of the electron transport chain and inflammation have yielded promising preclinical results, their clinical translations have been largely disappointing. For example, efforts to decrease Ca^{2+} overload with antagonists of the sarcolemmal Ca^{2+} ion channels, mitochondrial Ca^{2+} uniporter, or the Na^+/H^+ exchanger have shown to decrease myocardial infarct sizes up to 50% in animal models [108–110], but have been unsuccessful in the corresponding clinical studies [111–113]. The same trend is present in animal studies where these interventions were applied just at reperfusion to yield reductions in the final infarct sizes up to 50%, their applications in patients have been largely negative or inconclusive [17]. A host of antioxidants such as superoxide dismutase, trimetazimide, vitamin C and E with positive results in animal models failed to show meaningful improvement in infarct size, post-ischemic cardiac function nor mortality benefit in human clinical trials [114–116]. Similarly, attempts at targeting neutrophils such as leukocyte depleted blood [117], antibodies against the cell adhesion molecule P selection [118] as well as inhibitors of complement activation [119] have shown significant cardioprotection in animals with negative results in their corresponding clinical studies [120–124].

Several reasons may exist for the discrepancy between the experimental animal studies and human trials. First, although the existence of ischemia and reperfusion injury and the essential role of the mitochondria is well established in humans, variabilities may exist in the detailed biochemical mechanisms underlying cell death pathways amongst different species. Second, heterogeneities in a clinical setting in the varying degrees of ischemia as well as timing and context of intervention may contribute to the inconclusive results of clinical studies. Third, the interventions examined so far may have been of questionable benefit in preclinical studies or lacking in adequate knowledge of their pharmacological actions and kinetics in clinical subjects. Although the unequivocal role of mitochondrial dysfunction in myocardial ischemia and reperfusion is increasingly evident, the low rate of clinical success in the vast number of therapeutic targets investigated thus far calls for a need to evaluate our current understanding of the sequential mechanism of mitochondrial dysfunction during ischemia and reperfusion, as well as the way we devise our remedies.

Mitochondrial Transplantation

Mitochondrial dysfunction is the prominent feature of myocardial ischemic injury which are compounded by the process of reperfusion to have lasting effects on post-ischemic function and cellular viability. A myriad of experimental studies confirms that loss of mitochondrial function during ischemia significantly decreases cardiomyocyte viability and post-ischemic recovery. Despite this clear trend, there is currently no meaningful therapy for mitochondrial rescue from ischemia-reperfusion injury with growing discrepancies between preclinical and clinical studies. This calls for a review and a shift in the current paradigm and exploration of new approaches in therapeutic design. Instead of targeting a single step or one mediator of a complex, interconnected reactions of mitochondrial injury, a new therapy of mitochondrial transplantation is being investigated with the hypothesis that replacement of damaged mitochondria with viable, respiration-competent mitochondria isolated from non-ischemic tissue would overcome a host of deleterious effects of native mitochondrial dysfunction resulting in cardioprotection.

Mitochondrial Transplantation for Cardioprotection

Based on the hypothesis that myocardial mitochondrial injury occurs during ischemia and persists during reperfusion to significantly compromise post-ischemic myocardial function and viability, the concept of mitochondrial transplantation at the time of reperfusion was proposed in the recent decade. The first validation of mitochondrial transplantation in the ischemic heart model was demonstrated in the Langendorff-perfused ischemic rabbit heart in 2009 [125]. In this study, viable, respiration-competent mitochondria isolated from healthy left ventricular tissue of donor rabbits were directly injected into the ischemic zone of isolated hearts after 30 min of regional ischemia just prior to 120 min of reperfusion. Ischemic hearts that were transplanted with healthy mitochondria were found to have significantly enhanced cardioprotection by increased ATP content, reduction in infarct size, decreased cardiomyocyte loss, and improved post-ischemic myocardial function. Confocal microscopy showed that injected mitochondria were present and viable after 120 min of reperfusion and were distributed from the epicardium to the subendocardium. These findings heightened new interests in mitochondrial therapy calling for a deeper understanding of the mechanisms behind mitochondrial transfer, uptake, cardioprotection as well as the optimization of methods for enhancing clinical applicability.

Mitochondrial Source and Viability

In efforts to enhance clinical practicality and reduce immunogenic complications, rather than transplanting mitochondria from an allogenic source such as a donor organism, an approach of isolating mitochondria from a remote, nonischemic tissue in the same organism (autologous) to be transplanted into an ischemic region of the myocardium have been explored with promising results [126, 127]. A recent study in an in-vivo rabbit model of regional myocardial ischemia and reperfusion, autologous mitochondria from the pectoralis major muscle of the same organism were isolated via the newest isolation technique of standard homogenization and filtration, and injected in the myocardium following 29 min of ischemia immediately prior to reperfusion [126]. Results showed that transplantation of mitochondria isolated this way significantly decreased infarct size, improved cardiomyocyte viability and myocardial function at 4 weeks of recovery. Furthermore, additional studies showed that transplantation of autogenic mitochondria did not induce any autoimmunity via multiplex and ELISA assays which did not show any increases in inflammatory cytokines associated with immune responses seen in patients with heart transplant rejection such as C-Reactive Protein (CRP), Tumor Necrosis Factor-alpha (TNFα), Interleukins -1, -4, -6, -12, -18, Macrophage Inflammatory Protein -1 alpha and -1 beta (MIP-1α, -1β) [126].

In addition to the benefits of the autologous origin of mitochondria, efficacy of mitochondrial isolation from various sources of tissues has been investigated. As expected, different sources of tissues provide different absolute number of mitochondria, with liver being the highest, followed by skeletal muscle followed by cardiac muscle [127]. However, no differences were detected in the cardioprotective effects of mitochondria derived from different tissue types [125, 126], nor their subpopulations (subsarcolemmal or interfibrillar) [125]. On the other hand, mitochondrial viability is found to be a prime determinant of efficacy of mitochondrial transplantation for cardioprotection. Studies have shown that Nonviable mitochondria isolated from frozen tissue with confirmed absence of oxygen consumption failed to provide cardioprotection [125]. Similarly, mitochondrial proteins, mitochondrial complexes and mitochondrial RNA and DNA have not been found to provide cardioprotection [125]. Injection of exogenous ATP, ATP synthesis promoters, and exogenous ADP have also not shown to result in any protection against myocardial ischemia [125, 128]. These findings together indicate the imperative requirement of transplanted mitochondria to be viable, freshly-isolated and respiration-competent to be efficacious.

Mitochondrial Isolation and Purification

To assure the viability of the transplanted mitochondria and the clinical practicality of mitochondrial transplantation, it is critical that the isolation process of mitochondria is simple, cost-effective and feasible within the clinical timeframe of intervention for myocardial ischemia. For example, in the case of ST-elevation myocardial infarction (STEMI), the goal time from the onset of evaluation to revascularization via PCI (door-to-balloon time) is less than 90 min. For cardiac surgery, most operations last from 40 min to several hours in duration. However, the time from the onset of actual incision to revascularization or to the time to wean from bypass are often less than 1 h. This makes the standard technique of mitochondrial isolation which takes over 90 min at best, impractical and unreliable for yielding viable mitochondria. In 2014, a new method of rapid isolation and purification of mitochondria was developed by Preble and colleagues, who employ standardized tissue dissociator and differential filtration rather than centrifugation, cutting the isolation time to less than 30 min [129]. The purity and viability of the mitochondria isolated from this method have been validated by many different quality control parameters involving oxygen consumption rates via the Clarke-type electrode, ATP assays, fluorescent probe analysis, as well as light microscopy using fluorescent mitochondrial labels (MitoTracker CMXros) and transmission electron microscopy [125, 126, 129, 130].

Mitochondrial Delivery, Mode of Uptake and Distribution

To further enhance the therapeutic potential of mitochondrial transplantation, different modes of delivery of exogenous mitochondria have been investigated which include direct myocardial injection and intracoronary vascular infusion. From initial studies in both Langendorff-perfused and in situ-blood perfused rabbit hearts, respiration competent, autologous mitochondria were injected directly into eight to ten sites in the ischemic left ventricular free wall which showed significant improvement in infarct size and myocardial function. In a recent study using Langendorff-perfused rabbit heart model of both regional and global ischemia, the mode of vascular delivery via intracoronary infusion was validated and compared to the method of direct injection with regards to myocardial distribution using ^{18}F-rhodamine 6G and iron oxide nanoparticles labeling followed by visualization with positron emission tomography, microcomputed tomography and magnetic resonance imaging [127]. Intracoronary delivery of mitochondria was shown to result in a rapid and a more widespread distribution throughout the myocardium as compared to direct injection, which permitted higher concentrations of mitochondria in the targeted region. The efficacy of cardioprotection via intracoronary delivery of autologous mitochondria paralleled that of direct injection in another experiment using Langendorff-perfused rabbit hearts of regional ischemia which showed similar decrease in myocardial infarct size and improvement in both global and regional myocardial function [127]. While additional in vivo studies are needed, the ability to deliver mitochondria by vascular infusion expands the therapeutic potential such as global delivery to the ischemic heart just prior to reperfusion after cardiopulmonary bypass or delivery into the coronary artery at time of percutaneous coronary interventions (PCI) for revascularization.

On a histological level, both in situ and in isolated heart models, transplanted mitochondria have been shown to be initially present in the interstitial spaces surrounding cardiomyocytes coinciding with the time of increase in myocardial function, approximately 10 min post injection. Within 1–2 h post-delivery, the transplanted mitochondria were detectable within cardiomyocytes residing near the sarcolemma between Z-lines of the sarcomeres, in clusters around endogenous damaged mitochondria as well as near the nucleus. This internalization has been confirmed using human mitochondria isolated from HeLA cells in a rabbit heart. The use of human mitochondria in the rabbit model allows for the differentiation between native rabbit mitochondria and transplanted human mitochondria based on immune reactivity to a monoclonal anti-human mitochondrion, and internalization of the mitochondria into cardiomyocytes were visualized via transmission electron microscopy using immune-gold staining [125, 126]. These studies along with other in vitro studies demonstrated that the uptake of mitochondria is linear and time dependent, and remain viable for at least 24 h, significantly enhancing rate of oxygen consumption and high-energy synthesis in both neonatal and adult cardiomyocytes [125, 126, 131].

Many hypotheses have been proposed as to the mechanism of mitochondrial internalization by cardiomyocytes such as actin-meditated endocytosis, Caveolae-dependent-clathrin dependent endocytosis [132], tunneling nanotubes [133] and macro-pinocytosis [134] based on prior experiments in cell types other than cardiomyocytes. A recent in vitro study of cardiomyocytes using specific inhibitors of these mechanisms showed that only Cyctochalasin D, a specific inhibitor of actin polymerization, decreased internalization of mitochondria into cardiomyocytes and deceased ATP content, suggesting actin-mediated endocytosis as a potential mechanism of mitochondrial internalization by cardiomyocytes [131]. Additionally, the internalized mitochondrial did not co-localize with lysosomal or autophagocytosis markers raising another mechanism by which the internalized mitochondria escape endosomes. Actin-mediated endocytosis of mitochondria is in agreement with the widely accepted theory of the endosymbiotic origin of mitochondria. While externally presented mitochondria may be internalized via actin-mediated endocytosis in cardiomyocytes, intercellular transfer of mitochondria between cells may involve tunneling nanotubes within the syncytial architecture of cardiomyocytes in vivo [135].

Mechanisms of Cardioprotection

The mechanism through which transplanted mitochondria provide cardioprotection reflect the mediators of mitochondrial damage accrued during ischemic injury. Based on the established observations, at the least, cardioprotection via autologous mitochondrial transplantation are considered to involve (1) increased myocardial ATP content and metabolic recovery (2) change in proteomic and transcriptomics (3) upregulations of cytokine and chemokines that enhance post-infarct myocardial function and (4) replacement of damaged mitochondrial DNA.

To date, all studies of mitochondrial transplantation in cardiac tissue have shown that transplanted mitochondria increase cellular ATP content and oxygen consumption. In the in vivo rabbit model of regional ischemia/reperfusion injury, total tissue ATP content in the area of risk in mitochondria-transplanted hearts were significantly increased starting as early as 15 min of mitochondrial delivery and present even at 21 days of recovery compared to their controls. On the other hand, the use of exogenous ATP or ADP have not shown to restore high-energy phosphate stores nor have beneficial effects on post-ischemic functional recovery [126, 128, 136]. This is believed to be due, in part, to the lability of ATP and its short half-life in vivo. In line with this observation, cardioprotection by mitochondrial transplantation requires mitochondria that are intact, viable and respiration competent. Use of non-viable mitochondria or mitochondrial components such as mitochondrial RNA and DNA have not shown to confer improvement in postischemic function or cellular viability, although some mitochondrial complexes such as cytochrome c [137] and coenzyme Q_{10} have been shown to have modest benefit at very high doses [138].

The increases in the energy level and metabolic recovery is also reflected by the changes in myocardial proteomics and increases in differentially expressed proteins post mitochondrial transplantation. Study using proteomic analysis and functional annotation clustering in post-ischemic rabbit myocardial tissue have shown that proteome clusters of the mitochondrion, the generations of precursor metabolites for energy, and cellular respiration were significantly increased in hearts treated with mitochondrial transplantation compared with their controls [126].

Mitochondrial transplantation has also been shown to result in downregulation of inflammatory cytokines and upregulation of chemokines that play key roles in angiogenesis, ateriogenesis, progenitor cell migration, prevention and protection against cardiomyocyte apoptosis and enhanced cardiac functional recovery. Rabbit hearts with regional ischemia treated with autologous mitochondrial transplantation were shown in vivo to have significantly decreased inflammatory markers when compared to untreated regional ischemic hearts, including tumor necrosis factor alpha (TNFα), interleukin -6 and -10 (IL-6, -10), monocyte chemoattractant protein-1 (MCP-1) and high-sensitivity C-reactive protein (hsCRP). Multiplex analysis indicated that autologous mitochondria significantly upregulated expression of epidermal growth factor (EGF), growth-related oncogene (GRO), interleukin-6 (IL-6) and monocyte chemoattractant protein-3 (MCP-3). EGF been shown to play a key protective role in myocardial ischemic injury by stimulating cell growth, proliferation and migration [139, 140]. GRO and IL-6 have been implicated in reconstitution of cardiac tissue mass post myocardial infarction by acting as chemoattractants for vascularization and protection against cardiomyocyte apoptosis [141]. These chemokines all act with MCP-3 to enhance post-infarct myocardial function and improve myocardial remodeling. It has been proposed that together with increased tissue ATP levels in the area of ischemic injury, induction of cardioprotective cytokines augment the role of differentially expressed proteins and mitochondrial pathways to enhance metabolic recovery and post ischemic cardiac function.

The effect on reactive oxygen species (ROS) by mitochondrial transplantation remains to be delineated, although results are gearing more towards the idea that ROS may not to play a major role in cardioprotection provided by mitochondrial transplantation. In Langendorff-perfused rabbit hearts of regional ischemia, ROS in form of thiobarbituric acid-reactive substances (TBARS) were found to be significantly decreased in hearts that received mitochondrial transplantation when compared to untreated hearts. However, ROS scavenger 2-Mercaptopropionylglycine (MPG) failed to block cardioprotection afforded by mitochondrial mitochondria when used throughout reperfusion, or added to the transplanted mitochondria, suggesting that cardioprotection by transplanted mitochondria occurs through mechanism that are not significantly modified by ROS [126]. Whether these observations represent primary or secondary effects along with more clear role of ROS in the spectrum of mitochondrial injury and recue remain open to continued investigation.

Consistent with studies in many other fields, recent studies have also shown that mitochondrial transplantation not only rescues cellular energy and function, but also replaces damaged mitochondrial DNA [131]. In vitro studies using human HeLa cells depleted of mitochondrial DNA (HeLa p^0) incapable of oxygen consumption were rescued by co-incubation with mitochondria containing intact mtDNA with significantly increased ATP content and oxygen consumption rates detected even at 2 weeks post co-incubation. Quantitative real-time RT-PCR analysis confirmed replacement of mtDNA in HeLa p^0 cells following mitochondrial transplantation. Whether the mechanism of this type of mitochondrial rescue is related to the replication of mtDNA in the transplanted mitochondria or expansion of transplanted mitochondrial population requires further characterization.

Mitochondrial Transplantation in Human Subjects

First clinical application of mitochondrial transplantation therapy has been done in 2016 in pediatric patients who suffered myocardial ischemia-reperfusion injury at our institution. Five pediatric patients in critical condition who were unable to be weaned off extracorporeal membrane oxygenation (ECMO) support due to myocardial dysfunction related to ischemia and reperfusion were treated with autologous mitochondria isolated from the patients' rectus abdominis muscle and were injected at ten different places in the most hypokinetic regions of the myocardium [142]. All five patients had significant improvement in their myocardial systolic function and all but one patient were successfully weaned off ECMO support by the second day post mitochondrial transplantation. The single patient who was unable to wean off ECMO support suffered irreversible multi-organ failure despite the recovery of myocardial function following mitochondrial transplantation. As shown in animal models, mitochondrial auto-transplantation successfully improved post-ischemic myocardial function and was not associated with adverse short-term complications related to mitochondrial delivery such as arrhythmia, intramyocardial hematoma or scarring. This case demonstrates the potential role of the novel therapy of mitochondrial transplantation to improve ventricular dysfunction following ischemia-reperfusion injury in humans. Additional experimental investigations and larger scale trials are necessary to evaluate details of safety, efficacy, dosing and modes of transplantation with paralleled improvements in our understanding of ischemia and reperfusion injury and the role of mitochondria.

Conclusion

Mitochondrial dysfunction is the principal feature of myocardial ischemic injury which prevails through reperfusion to severely compromise cardiomyocyte survival and post-ischemic ventricular function. These include alterations in mitochondrial structure, function, ionic accumulations, mitochondrial enzyme and complex activity, high energy synthesis, mitochondrial DNA structure, transcriptomics, proteomics, and mitochondria-mediated apoptotic pathways. Mitochondrial auto-transplantation involves replacing damaged mitochondria in the ischemic heart with fresh, viable, and respiration competent mitochondria isolated from a nonischemic area of the patient's own body to circumvent damaged mitochondrial function accrued during ischemia and rescue myocardial function. This novel method has been shown to improve myocardial infarct severity, cellular viability, and ventricular function in many different animal models and preliminarily, in humans. The known mechanisms of cardioprotection by mitochondrial transplantation include increase in myocardial ATP content, changes in mitochondrial proteomics and transcriptomics, upregulation of cytokine and chemokines that enhance post-infarct myocardial function, as well as replacement of damaged mitochondrial DNA. Moreover, the advent of mitochondrial isolation to a simple, rapid and highly pure technique, permits application of this method within the clinical timeframe of most medical and surgical procedures.

Possible applications of mitochondrial transplantation abound, as mitochondrial damage underlie a vast number of disorders including Alzheimer's disease, Parkinson's disease, ischemia-reperfusion injury of the liver, kidney, brain and muscle compartments, as well as rare mitochondrial genetic disorders. Recently, mitochondrial transplantation has been shown to restore damaged mitochondrial function in Parkinson's' disease brain in rats [143], decrease cell death from ischemia-reperfusion injury in rat liver [144] and attenuate hypoxic pulmonary hypertension in rats [145]. A recent study by Hayakawa et al. described the transfer of healthy mitochondria from astrocytes to surrounding neuronal cells during transient focal cerebral ischemia to induce cell survival pathways [146].

From natural phenomena to constructed therapies, the diversified transmissibility of mitochondria and their exertive roles on cellular transformation and survival are difficult to underestimate. Much remains in question with regards to the mechanism of mitochondrial influence on cell survival as well as their specific roles in different disease models; and optimizations of safety and long-term outcomes provoke future experimental tasks and innovation. Along with many varied fields of mitochondrial disorders, mitochondrial transplantation offers a valuable strategy for cardioprotection in ischemic heart disease and opens the gateway to deeper understanding of mitochondrial function and therapeutic potential.

Acknowledgments This work was supported by: The Richard A. and Susan F. Smith Foundation, President's Innovation Award, Boston Children's Hospital, Michael B. Klein and Family, the Sidman Family Foundation, the Bulens/Capozzi Foundation, the Boston Children's Hospital Anesthesia Foundation, the Kenneth C. Griffin Charitable Research Fund and the Boston Investment Council. This work in part was funded by NIH 5T32HL007734 fellowship to Dr. Shin.

Disclosures Dr. McCully, Dr. Cowan and Dr. del Nido have patents pending for the isolation and usage of mitochondria. There are no other conflicts of interest by any of the authors. The authors attest they had full freedom to explore the data, analyze the results independent from any sponsor and that they had sole authority to make the final decision to submit the material for publication.

References

1. Benjamin EJ, Blaha MJ, Chiuve SE, Cushman M, Das SR, Deo R, de Ferranti SD, Floyd J, Fornage M, Gillespie C, Isasi CR, Jiménez MC, Jordan LC, Judd SE, Lackland D, Lichtman JH, Lisabeth L, Liu S, Longenecker CT, Mackey RH, Matsushita K, Mozaffarian D, Mussolino ME, Nasir K, Neumar RW, Palaniappan L, Pandey DK, Thiagarajan RR, Reeves MJ, Ritchey M, Rodriguez CJ, Roth GA, Rosamond WD, Sasson C, Towfighi A, Tsao CW, Turner MB, Virani SS, Voeks JH, Willey JZ, Wilkins JT, Wu JH, Alger HM, Wong SS, Muntner P, American Heart Association Statistics Committee and Stroke Statistics Subcommittee. Heart disease and stroke statistics-2017 update: a report from the American Heart Association. Circulation. 2017;135:e146–e603.
2. Jennings RB, Reimer KA. Lethal myocardial ischemic injury. Am J Pathol. 1981;102: 241–55.
3. Buja LM, Willerson JT. Abnormalities of volume regulation and membrane integrity in myocardial tissue slides after early ischemic injury in the dog: effects of mannitol, polyethylene glycol, and propranolol. Am J Pathol. 1981;103:79–95.
4. Santulli G, Nakashima R, Yuan Q, Marks AR. Intracellular calcium release channels: an update. J Physiol. 2017 (in press). doi: 10.1113/JP272781.
5. Braunwald E, Kloner RA. Myocardial reperfusion: a double-edge sword? J Clin Invest. 1985;76:1713–9.
6. Fassina L, Rozzi G, Rossi S, Scacchi S, Galetti M, Lo Muzio FP, Del Bianco F, Colli Franzone P, Petrilli G, Faggian G, Miragoli M. Cardiac kinematic parameters computed from video of in situ beating heart. Sci Rep. 2017;7:46143 doi:10.1038/srep46143.
7. Kloner RA, Przyklenk K, Patel B. Altered myocardial states: the stunned and hibernating myocardium. Am J Med. 1989;86:14–22.
8. Murry CE, Jennings RB, Reimer KA. Preconditioning with ischemia: a delay of lethal cell injury in ischemic myocardium. Circulation. 1986;74:1124–36.
9. Murry CE, Richard VJ, Reimer KA, Jennings RB. Ischemic preconditioning slows energy metabolism and delays ultrastructural damage during a sustained ischemic episode. Circ Res. 1990;66:913–31.
10. Zhao ZQ, Corvera JS, Halkos ME, Kerendi F, Wang NP, Guyton RA, Vinten-Johansen J. Inhibition of myocardial injury by ischemic postconditioning during reperfusion: comparison with ischemic preconditioning. Am J Physiol Heart Circ Physiol. 2003;285:H579–88.
11. Zhao ZQ, Vinten-Johansen J. Postconditioning: reduction of reperfusion-induced injury. Cardiovasc Res. 2006;70:200–11.
12. Cohen MV, Yang XM, Downey JM. The pH hypothesis of postconditioning: staccato reperfusion reintroduces oxygen and perpetuates myocardial acidosis. Circulation. 2007;115:1895–903.
13. Kolwicz Jr SC, Purohit S, Tian R. Cardiac metabolism and its interactions with contraction, growth and survival of cardiomyocytes. Circ Res. 2013;113:603–16.
14. Carden DL, Granger DN. Pathophysiology of ischemia-reperfusion injury. J Pathol. 2000;190:255–66.
15. Collard CD, Gelman S. Pathophysiology, clinical manifestations, and prevention of ischemia-reperfusion injury. Anesthesiology. 2001;94:1133–8.
16. Turer AT, Hill JA. Pathogenesis of myocardial ischemia-reperfusion injury and rationale for therapy. Am J Cardiol. 2010;106:360–8.
17. Yellon DM, Hausenloy DJ. Myocardial reperfusion injury. N Engl J Med. 2007;357:1121–35.
18. Meerson FZ, Zaletayeva TA, Lagutchev SS, Pshennikova MG. Structure and mass of mitochondria in the process of compensatory hyperfunction and hypertrophy of the heart. Exp Cell Res. 1964;36:568–78.
19. Feinberg H, Levitsky S. Biochemical rationale of cardioplegia. In: Engelman RM, Levitsky S, editors. A textbook of clinical cardioplegia. Mt Kisco, New York: Futura; 1982. p. 131–9.

20. Palmer JW, Tandler B, Hoppel CL. Biochemical properties of subsarcolemmal and interfibrillar mitochondria isolated from rat cardiac muscle. J Biol Chem. 1977;252:8731–9.
21. Weinstein ES, Benson DW, Fry ED. Subpopulations in human heart mitochondria. J Surg Res. 1986;40:495–8.
22. Dalen H, Odegarden S, Saetersdal T. The application of various electron microscopic techniques for ultrastructural characterization of the human papillary heart muscle cell in biopsy material. Virchows Arch A Pathol Anat Histopathol. 1987;410:265–79.
23. Shimada T, Horita K, Murakami M, Ogura R. Morphological studies of different mitochondrial populations in monkey myocardial cells. Cell Tissue Res. 1984;238:577–82.
24. Lukyanenko V, Chikando A, Lederer WJ. Mitochondria in cardiomyocyte Ca^{2+} signaling. Int J Biochem Cell Biol. 2009;41:1957–71.
25. Birkedal R, Shiels HA, Vendelin M. Three-dimensional mitochondrial arrangement in ventricular myocytes: from chaos to order. Am J Physiol Cell Physiol. 2006;291:C1148–58.
26. Hollander JM, Thapa D, Shepherd D. Physiological and structural differences in spatially distinct subpopulations of cardiac mitochondria. Am J Physiol Heart Circ Physiol. 2014;307:H1–14.
27. Muller W. Subsarcolemmal mitochondria and capillarization of soleus muscle fibers in young rats subjected to an endurance training. A morphometric study of semithin sections. Cell Tissue Res. 1976;174:367–89.
28. Rosca MG, Hoppel CL. Mitochondrial dysfunction in heart failure. Heart Fail Rev. 2013;18:607–22.
29. Baseler WA, Thapa D, Jagannathan R, Dabkowski ER, Crsoton TL, Hollander JM. miR-141 as a regulator of the mitochondrial phosphate carrier (Slc25a3) in the type 1 diabetic heart. Am J Physiol Cell Physiol. 2012;303:C1244–51.
30. Croston TL, Shepherd DL, Thapa D, Nichols CE, Lewis SE, Dabkowski ER, Jagannathan R, Baseler WA, Hollander JM. Evaluation of the cardiolipin biosynthetic pathway and its interactions in the diabetic heart. Life Sci. 2013;93:313–22.
31. Chen Q, Moghaddas S, Hoppel CL, Lesnefsky EJ. Ischemic defects in the electron transport chain increase the production of reactive oxygen species from isolated rat heart mitochondria. Am J Physiol Cell Physiol. 2008;294:C460–6.
32. Lesnefsky EJ, Gudz TI, Migita CT, Ikeda-Saito M, Hassan MO, Turkaly PJ, Hoppel CL. Ischemic injury to mitochondrial electron transport in the aging heart: damage to the iron-sulfur protein subunit of electron transport complex III. Arch Biochem Biophys. 2001;385:117–28.
33. Lesnefsky EJ, Chen Q, Moghaddas S, Hassan MO, Tandler B, Hoppel CL. Blockade of electron transport chain during ischemia protects cardiac mitochondria. J Biol Chem. 2004;279:47961–7.
34. Rosca MG, Vazquez EJ, Kerner J, Parland W, Chandler MP, Stanely W, Sabbah HN, Hoppel CL. (2008)Cardiac mitochondria in heart failure: decrease in respirasomes and oxidative phosphorylation. Cardiovasc Res. 2008;80:30–9.
35. Dabkowski ER, Williamson CL, Hollander JM. Mitochondria-specific transgenic overexpression of phospholipid hydroperoxide glutathione peroxidase (GPx4) attenuates ischemia/reperfusion-associated cardiac dysfunction. Free Radic Biol Med. 2008;45:855–65.
36. Duan J, Karmazyn M. Relationship between oxidative phosphorylation and adenine nucleotide translocase activity of two populations of cardiac mitochondria and mechanical recovery of ischemic hearts following reperfusion. Can J Physiol Pharmacol. 1989;67:704–9.
37. Lesnefsky EJ, Chen Q, Slabe TJ, Stoll MS, Minkler PE, Hassan MO, Tandler B, Hoppel CL. Ischemia, rather than reperfusion, inhibits respiration through cytochrome oxidase in the isolated, perfused rabbit heart: role of cardiolipin. Am J Physiol Heart Circ Physiol. 2004;287:H258–67.
38. Kubler W, Spieckermann PG. Regulation of glycolysis in the ischemic and anoxic myocardium. J Mol Cell Cardiol. 1970;1:351–77.
39. Conway MA, Allis J, Ouwerkerk R, Niioka T, Rajagopalan B, Radda GK. Detection of low phosphocreatine to ATP ratio in failing hypertrophied human myocardium by 31P magnetic resonance spectroscopy. Lancet. 1991;338:973–6.

40. Hardy CJ, Weiss RG, Bottomley PA, Gerstenblith G. Altered myocardial high-energy phosphate metabolites in patients with dilated cardiomyopathy. Am Heart J. 1991;112:795–801.
41. Neubauer S, Krahe T, Schindler R, Horn M, Hillenbrand H, Entzeroth C, Mader H, Kromer EP, Riegger GA, Lackner K. 31P magnetic resonance spectroscopy in dilated cardiomyopathy and coronary artery disease. Altered cardiac high-energy phosphate metabolism in heart failure. Circulation. 1992;86:1810–8.
42. Neubauer S, Horn M, Cramer M, Harre K, Newell JB, Peters W, Pabst T, Ertl G, Hahn D, Ingwall JS, Kochsiek K. Myocardial phosphocreatine-to-ATP ratio is a predictor of mortality in patients with dilated cardiomyopathy. Circulation. 1997;96:2190–6.
43. Tsukube T, McCully JD, Metz KR, Cook CU, Levitsky S. Amelioration of ischemic calcium overload correlates with high-energy phosphates in the senescent myocardium. Am J Phys. 1997;273:H418–25.
44. Reimer KA, Lowe JE, Rasmussen MM, Jennings RB. The wavefront phenomenon of ischemic cell death: 1. Myocardial infarct size vs duration of coronary occlusion in dogs. Circulation. 1977;56:786–94.
45. Reimer KA, Jennings RB. The "wavefront phenomenon" of myocardial ischemic cell death: II. Transmural progression of necrosis within the framework of ischemic bed size (myocardium at risk) and collateral flow. Lab Investig. 1979;40:633–44.
46. Reimer KA, Jennings RB. Myocardial ischemia, hypoxia, and infarction. In: Fozzard HA, Haber E, Jennings RB, Katz AM, Morgan HE, editors. The heart and cardiovascular system: scientific foundations, vol. II. 2nd ed. New York: Raven Press; 1992. p. 1875–973.
47. McCully JD, Levitsky S. Mitochondrial ATP-sensitive potassium channels in surgical cardioprotection. Arch Biochem Biophys. 2003;420:237–45.
48. Heyndrickx GR, Millard RW, McRitchie RJ, Maroko PR, Vatner SF. Regional myocardial functional and electrophysiological alterations after brief coronary artery occlusion in conscious dogs. J Clin Invest. 1975;56:978–85.
49. Gray R, Maddahi J, Bernan D, Raymond M, Waxman A, Ganz W, Matloff J, Swan HJ. Scintigraphic and hemodynamic demonstration of transient left ventricular dysfunction immediately after uncomplicated coronary artery bypass grafting. J Thorac Cardiovasc Surg. 1979;77:504–10.
50. Kloner RA, Przyklenk K, Kay GL. Clinical evidence for stunned myocardium after coronary artery bypass surgery. J Card Surg. 1994;9:397–402.
51. Rahimtoola SH. A perspective on the three large multicenter randomized clinical trials of coronary bypass surgery for chronic stable angina. Circulation. 1985;72:V123–35.
52. Buja LM, Vander Heide RS. Pathobiology of ischemic heart disease: past, present and future. Cardiovasc Pathol. 2016;25:214–20.
53. Chen C, Liu J, Hua D, Ma L, Lai T, Fallon JT, Knibbs D, Gillam L, Mangion J, Knight DR, Waters D. Impact of delayed reperfusion of myocardial hibernation on myocardial ultrastructure and function and their recoveries after reperfusion in a pig model of myocardial hibernation. Cardiovasc Pathol. 2000;9:67–84.
54. Lai T, Fallon JT, Liu J, Mangion J, Gillam L, Waters D, Chen C. Reversibility and pathohistological basis of left ventricular remodeling in hibernating myocardium. Cardiovasc Pathol. 2000;9:323–35.
55. Shah BN, Khattar RS, Senior R. The hibernating myocardium: current concepts, diagnostic dilemmas, and clinical challenges in the post-STICH era. Eur Heart J. 2013;34:1323–36.
56. Bax JJ, Schinkel AF, Boersma E, Rizzello V, Elhendy A, Maat A, Roelandt JR, van der Wall EE, Poldermans D. Early versus delayed revascularization in patients with ischemic cardiomyopathy and substantial viability: impact on outcome. Circulation. 2003;108:II39–42.
57. Beanlands RS, Hendry PJ, Masters RG, de Kemp RA, Woodend K, Ruddy TD. Delay in revascularization is associated with increased mortality rate in patients with severe left ventricular dysfunction and viable myocardium on fluorine 18-fluorodeoxyglucose positron emission tomography imaging. Circulation. 1998;98:II51–6.
58. Vanoverschelde JL, Wijns W, Depre C, Essamri B, Heyndrickx GR, Borgers M, Bol A, Melin JA. Mechanisms of chronic regional postischemic dysfunction in humans. New insights from the study of noninfarcted collateral-dependent myocardium. Circulation. 1993;87:1513–23.

59. Kim SJ, Peppas A, Hong SK, Yang G, Huang Y, Diaz G, Sadoshima J, Vatner DE, Vatner SF. Persistent stunning induces myocardial hibernation and protection: flow/function and metabolic mechanisms. Circ Res. 2003;92:1233–9.
60. Borgers M, Ausma J. Structural aspects of the chronic hibernating myocardium in man. Basic Res Cardiol. 1995;90:44–6.
61. Arai AE, Pantely GA, Anselone CG, Bristow J, Bristow JD. Active downregulation of myocardial energy requirements during prolonged moderate ischemia in swine. Circ Res. 1991;69:1458–69.
62. Schulz R, Rose J, Martin C, Brodde OE, Heusch G. Development of short-term myocardial hibernation. Its limitation by the severity of ischemia and inotropic stimulation. Circulation. 1993;88:684–95.
63. Rousou AJ, Ericsson M, Federman M, Levitsky S, McCully JD. Opening of mitochondrial KATP channels enhances cardioprotection through the modulation of mitochondrial matrix volume, calcium accumulation and respiration. Am J Physiol Heart Circ Physiol. 2004;287:H1967–76.
64. McCully JD, Rousou AJ, Parker RA, Levitsky S. Age and gender-related differences in mitochondrial oxygen consumption and calcium with cardioplegia and diazoxide. Ann Thorac Surg. 2007;83:1102–9.
65. Faulk EA, McCully JD, Hadlow NC, Tsukube T, Krukenkamp IB, Federman M, Levitsky S. Magnesium cardioplegia enhances mRNA levels and the maximal velocity of cytochrome oxidase I in the senescent myocardium during global ischemia. Circulation. 1995;92:II405–12.
66. Walker CA, Crawford Jr FA, Spinale FG. Myoctye contractile dysfunction with hypertrophy and failure: relevance to cardiac surgery. J Thorac Cardiovasc Surg. 2000;119:388–400.
67. Khabbaz KR, Zankoul F, Warner KG. Operative metabolic monitoring of the heart II. Online measurement of myocardial tissue pH. Ann Thorac Surg. 2001;72:S2227–33.
68. Karmazyn M. The role of the myocardial sodium-hydrogen exchanger (NHE) and its role in mediating ischemic and reperfusion injury. Keio J Med. 1998;47:65–72.
69. Kaplan P, Hendrikx M, Mattheussen M, Mubagwa K, Flameng W. Effect of ischemia and reperfusion on sarcoplasmic reticulum calcium uptake. Circ Res. 1992;71:1123–30.
70. Santulli G, Xie W, Reiken SR, Marks AR. Mitochondrial calcium overload is a key determinant in heart failure. Proc Natl Acad Sci U S A. 2015;112:11389–94.
71. Frolkis VV, Frolkis RA, Mkhitarian LS, Shevchuk VG, Fraifeld VE, Vakulenko LG, Syrovy I. Contractile function and Ca^{2+} transport system of myocardium in ageing. Gerontology. 1988;34:64–74.
72. Peng CF, Kane JJ, Murphy ML, Straub KD. Abnormal mitochondrial oxidative phosphorylation of ischemic myocardium by calcium chelating agents. J Mol Cell Cardiol. 1977;9:897–908.
73. Suleiman MS, Halestrap AP, Griffiths EJ. Mitochondria: a target for myocardial protection. Pharmacol Ther. 2001;89:29–46.
74. Ozcan C, Holmuhamedov EL, Jahangir A, Terzic A. Diazoxide protects mitochondria from anoxic injury: implication for myopreservation. J Thorac Cardiovasc Surg. 2001;121:298–306.
75. Garlid KD. Opening mitochondrial K(ATP) in the heart-what happens, and what does not happen. Basic Res Cardiol. 2000;95:275–9.
76. Steenbergen C, Murphy E, Watts JA, London RE. Correlation between cytosolic free calcium, contracture, ATP, and irreversible ischemic injury in perfused rat heart. Circ Res. 1990;66:135–46.
77. Ataka K, Chen D, Levitsky S, Jimenez E, Feinberg H. Effect of aging on intracellular Ca^{2+}, pH_i, and contractility during ischemia and reperfusion. Circulation. 1992;86:II371–6.
78. Tsukube T, McCully JD, Faulk EA, Federman M, LoCicero 3rd J, Krukenkamp IB, Levitsky S. Magnesium cardioplegia reduces cytosolic and nuclear calcium and DNA fragmentation in the senescent myocardium. Ann Thorac Surg. 1994;58:1005–11.

79. Hearse DJ, Garlick PB, Humphrey SM. Ischemic contracture of the myocardium: mechanism and prevention. Am J Cardiol. 1977;39:986–93.
80. Hearse DJ, Stewart DA, Braimbridge MV. Myocardial protection during ischemic cardiac arrest: the importance of magnesium in cardioplegic infusates. J Thorac Cardiovasc Surg. 1978;75:877–85.
81. Faulk EA, McCully JD, Tsukube T, Hadlow NC, Krukenkamp IB, Levitsky S. Myocardial mitochondrial calcium accumulation modulates nuclear calcium accumulation and DNA fragmentation. Ann Thorac Surg. 1995;60:338–44.
82. McCully JD, Wakiyama H, Cowan DB, Federman M, Levitsky S. Diazoxide amelioration of myocardial injury and mitochondrial damage during cardiac surgery. Ann Thorac Surg. 2002;74:2138–45.
83. Wakiyama H, Cowan DB, Toyoda Y, Federman M, Levitsky S, McCully JD. Selective opening of mitochondrial ATP-sensitive potassium channels during surgically induced myocardial ischemia increases necrosis and apoptosis. Eur J Cardiothorac Surg. 2002;21:424–33.
84. Halestrap AP, Kerr PM, Javadov S, Suleiman S. The mitochondrial permeability transition: role in ischemia/reperfusion injury. Sepsis. 1998;2:313–25.
85. Holmuhamedov EL, Jovanovic S, Dzeja PP, Jovanovic A, Terzic A. Mitochondrial ATP-sensitive K^+ channels modulate cardiac mitochondrial function. Am J Phys. 1998;275:H1567–76.
86. Holmuhamedov EL, Wang L, Terzic A. ATP-sensitive K^+ channel openers prevent CA^{2+} overload in rat cardiac mitochondria. J Physiol. 1999;519:347–60.
87. Murata M, Akao M, O'Rourke B, Marban E. Mitochondrial ATP-sensitive potassium channels attenuate matrix CA^{2+} overload during stimulated ischemia and reperfusion: possible mechanism of cardioprotection. Circ Res. 2001;89:891–8.
88. Jennings RB, Sommers HM, Smyth GA, Flack HA, Linn H. Myocardial necrosis induced by temporary occlusion of a coronary artery in the dog. Arch Pathol. 1960;70:68–78.
89. Ito H. No-reflow phenomenon and prognosis in patients with acute myocardial infarction. Nat Clin Pract Cardiovasc Med. 2006;499:506.
90. Buja LM. Myocardial ischemia and reperfusion injury. Cardiovasc Pathol. 2005;14:170–5.
91. Gross GJ, Auchampach JA. Reperfusion injury: does it exist? J Mol Cell Cardiol. 2007;42:12–8.
92. Barry MC, Kelly C, Burke P, Sheehan S, Redmond HP, Bouchier-Hayes D. Immunological and physiological responses to aortic surgery: effect of reperfusion on neutrophil and monocyte activation and pulmonary function. Br J Surg. 1997;84:513–9.
93. Pan X, Liu J, Nguyen T, Liu C, Sun J, Teng Y, Fergusson MM, Rovira II, Allen M, Springer DA, Aponte AM, Gucek M, Balaban RS, Murphy E, Finkel T. The physiological role of mitochondrial calcium revealed by mice lacking the mitochondrial calcium uniporter. Nat Cell Biol. 2013;15:1464–72.
94. Arroyo CM, Kramer JH, Dickens BF, Weglicki WB. Identification of free radicals in myocardial ischemia/reperfusion by spin trapping with nitrone DMPO. FEBS Lett. 1987;221:101–4.
95. Ferrari R, Alfieri O, Curello S, Ceconi C, Cargnoni A, Marzollo P, Pardini A, Caradonna E, Visioli O. Occurrence of oxidative stress during reperfusion of the human heart. Circulation. 1990;81:201–11.
96. Pacher P, Nivorozhkin A, Szabó C. (2006)Therapeutic effects of xanthine oxidase inhibitors: renaissance half a century after the discovery of allopurinol. Pharmacol Rev. 2006;58:87–114.
97. Chouchani ET, Pell VR, Guade E, Aksentijevic D, Sundier SY, Robb EL, Logan A, Nadtochiy SM, Ord EN, Smith AC, Eyassu F, Shirley R, Hu CH, Dare AJ, James AM, Rogatti S, Hartley RC, Eaton S, Costa AS, Brookes PS, Davidson SM, Duchen MR, Saeb-Parsy K, Shattock MJ, Robinson AJ, Work LM, Frezza C, Krieg T, Murphy MP. Ischaemic accumulation of succinate controls reperfusion injury through mitochondrial ROS. Nature. 2014;515:431–5.
98. Ashrafian H, Czibik G, Bellahcene M, Aksentijevic D, Smith AC, Mitchell SJ, Dodd MS, Kirwan J, Byrne JJ, Ludwig C, Isackson H, Yavari A, Stottrup NB, Contractor H, Cahill TJ,

Sahgal N, Ball DR, Birkler RI, Hargreaves I, Tennant DA, Land J, Lygate CA, Johannsen M, Kharbanda RK, Neubauer S, Redwood C, de Cabo R, Ahmet I, Talan M, Günther UL, Robinson AJ, Viant MR, Pollard PJ, Tyler DJ, Watkins H. Fumarate is cardioprotective via activation of the Nrf2 antioxidant pathway. Cell Metab. 2012;15:361–71.
99. Murphy MP. How mitochondria produce reactive oxygen species. Biochem J. 2009;417:1–3.
100. Chouchani ET, Methner C, Nadtochiy SM, Logan A, Pell VR, Ding S, James AM, Cochemé HM, Reinhold J, Lilley KS, Partridge L, Fearnley IM, Robinson AJ, Hartley RC, Smith RA, Krieg T, Brookes PS, Murphy MP. Cardioprotection by S-nitrosation of cysteine switch on mitochondrial complex I. Nat Med. 2013;19:753–9.
101. Stewart S, Lesnefsky EJ, Chen Q. Reversible blockade of electron transport chain with amobarbital at the onset of reperfusion attenuates cardiac injury. Transl Res. 2009;153:224–31.
102. Burwell LS, Nadtochiy SM, Tompkins AJ, Young S, Brookes PS. Direct evidence for S-nitrosation of mitochondrial complex I. Biochem J. 2006;394:627–34.
103. Methner C, Chouchani ET, Buonincontri G, Pell VR, Sawiak SJ, Murphy MP, Krieg T. Mitochondria selective S-nitrosation by mitochondria-targeted S-nitrosothiol protects against post-infarct heart failure in mouse hearts. Eur J Heart Fail. 2014;16:712–7.
104. Zweier JL, Talukder MA. The role of oxidants and free radicals in reperfusion injury. Cardiovasc Res. 2006;70:181–90.
105. Crompton M. The mitochondrial permeability transition pore and its role in cell death. Biochem J. 1999;341:233–49.
106. Halestrap AP, Clarke SJ, Javadov SA. Mitochondrial permeability transition pore opening during myocardial reperfusion- a target for cardioprotection. Cardiovasc Res. 2004;61:372–85.
107. McCully JD, Bhasin MK, Daly C, Guerrero MC, Dillon S, Liberman TA, Cowan DB, Mably JD, McGowan FX, Levitsky S. Transcriptomic and proteomic analysis of global ischemia and cardioprotection in the rabbit heart. Physiol Genomics. 2009;38:125–37.
108. Klein HH, Pich S, Lindert S, Nebendahl K, Warneke G, Kreuzer H. Treatment of reperfusion injury with intracoronary calcium channel antagonists and reduced coronary free calcium concentration in regionally ischemic, reperfused porcine hearts. J Am Coll Cardiol. 1989;13:1395–401.
109. Carry MM, Mrak RE, Murphy ML, Peng CF, Straub KD, Fody EP. Reperfusion injury in ischemic myocardium: protective effects of ruthenium red and of nitroprusside. Am J Cardiovasc Pathol. 1989;2:335–44.
110. Gumina RJ, Buerger E, Eickmeier C, Moore J, Daemmgen J, Gross GJ. Inhibition of the Na^+/H^+ exchanger confers greater cardioprotection against 90 minutes of myocardial ischemia than ischemic preconditioning in dogs. Circulation. 1999;100:2519–26.
111. Boden WE, van Gilst W, Scheldewaert RG, Starkey IR, Carlier MF, Julian DG, Whitehead A, Bertrand ME, Col JJ, Pedersen OL, Lie KI, Santoni JP, Fox KM. Diltiazem in acute myocardial infarction treated with thrombolytic agents: a randomized placebo-controlled trial. Incomplete Infarction Trial of European Research Collaborators Evaluating Prognosis post-Thrombolysis (INTERCEPT). Lancet. 2000;355:1751–6.
112. Théroux P, Chaitman BR, Danchin N, Erhardt L, Meinertz T, Schroeder JS, Tognoni G, White HD, Willerson JT, Jessel A. (2000)Inhibition of the sodium-hydrogen exchanger with cariporide to prevent myocardial infarction in high-risk ischemic situations. Main results of the GUARDIAN trial. Guard during ischemia against necrosis (GUARDIAN) investigators. Circulation. 2000;102:3032–8.
113. Zeymer U, Suryapranata H, Monassier JP, Opolski G, Davies J, Rasmanis G, Linssen G, Tebbe U, Schröder R, Tiemann R, Machnig T, Neuhaus KL, ESCAMI Investigators. The Na(+)/H(+) exchange inhibitor eniporide as an adjunct to early reperfusion therapy for acute myocardial infarction. Results of the evaluation of the safety and cardioprotective effects of eniporide in acute myocardial infarction (ESCAMI) trial. J Am Coll Cardiol. 2001;38:1644–50.

114. Flaherty JT, Pitt B, Gruber JW, Heuser RR, Rothbaum DA, Burwell LR, George BS, Kereiakes DJ, Deitchman D, Gustafson N. Recombinant human superoxide dismutase (h-SOD) fails to improve recovery of ventricular function in patients undergoing coronary angioplasty for acute myocardial infarction. Circulation. 1994;89:1982–91.
115. Downey JM. Free radicals and their involvement during long-term myocardial ischemia and reperfusion. Annu Rev Physiol. 1990;52:487–504.
116. Sesso HD, Buring JE, Christen WG, Kurth T, Belanger C, MacFadyen J, Manson JE, Glynn RJ, Gaziano JM. Vitamins E and C in the prevention of cardiovascular disease in men: the Physicians' Health Study II randomized controlled trial. JAMA. 2008;300:2123–33.
117. Litt MR, Jeremy RW, Weisman HF, Winkelstein JA, Becker LC. Neutrophil depletion limited to reperfusion reduces myocardial infarct size after 90 minutes of ischemia: evidence for neutrophil-mediated reperfusion injury. Circulation. 1989;80:1816–27.
118. Hayward R, Campbell B, Shin YK, Scalia R, Lefer AM. Recombinant soluble P-selectin glycoprotein ligand-1 protects against myocardial ischemic reperfusion injury in cats. Cardiovasc Res. 1999;41:65–76.
119. Vakeva AP, Agah A, Rollins SA, Matis LA, Li L, Stahl GL. Myocardial infarction and apoptosis after myocardial ischemia and reperfusion: role of the terminal complement components and inhibition by anti-C5 therapy. Circulation. 1998;97:2259–67.
120. Baran KW, Nguyen M, McKendall GR, Lambrew CT, Dykstra G, Palmeri ST, Gibbons RJ, Borzak S, Sobel BE, Gourlay SG, Rundle AC, Gibson CM, Barron HV, Limitation of Myocardial Infarction Following Thrombolysis in Acute Myocardial Infarction (LIMIT AMI) Study Group. Double-blind, randomized trial of anti-CD18 antibody in conjunction with recombinant tissue plasminogen activator for acute myocardial infarction: limitation of myocardial infarction following thrombolysis in acute myocardial infarction (LIMIT AMI) study. Circulation. 2001;104:2778–83.
121. Faxon DP, Gibbons RJ, Chronos NA, Gurbel PA, Sheehan F, HALT-MI Investigators. The effect of blockade of the CD11/CD18 integrin receptor on infarct size in patients with acute myocardial infarction treated with direct angioplasty: the results of the HALT-MI study. J Am Coll Cardiol. 2002;40:1199–204.
122. Mertens P, Maes A, Nuyts J, Belmans A, Desmet W, Esplugas E, Charlier F, Figueras J, Sambuceti G, Schwaiger M, Mortelmans L, Van de Werf F, PSALM investigators. Recombinant P-selectin glycoprotein ligand-immunoglobulin, a P-selectin antagonist, as an adjunct to thrombolysis in acute myocardial infarction. The P-Selectin Antagonist Limiting Myonecrosis (PSALM) trial. Am Heart J. 2006;152:125.e1–8.
123. Mahaffey KW, Granger CB, Nicolau JC, Ruzyllo W, Weaver WD, Theroux P, Hochman JS, Filloon TG, Mojcik CF, Todaro TG, Armstrong PW, COMPLY Investigators. Effect of pexelizumab, an anti-C5 complement antibody, as adjunctive therapy to fibrinolysis in acute myocardial infarction: the COMPlement inhibition in myocardial infarction treated with thromboLYtics (COMPLY) trial. Circulation. 2003;108:1176–83.
124. Granger CB, Mahaffey KW, Weaver WD, Theroux P, Hochman JS, Filloon TG, Rollins S, Todaro TG, Nicolau JC, Ruzyllo W, Armstrong PW, COMMA Investigators. Pexelizumab, an anti-C5 complement antibody, as adjunctive therapy to primary percutaneous coronary intervention in acute myocardial infarction: the COMplement inhibition in Myocardial infarction treated with Angioplasty (COMMA) trial. Circulation. 2003;108:1184–90.
125. McCully JD, Cowan DB, Pacak CA, Toumpoulis IK, Dayalan H, Levitsky S. Injection of isolated mitochondria during early reperfusion for cardioprotection. Am J Physiol Heart Circ Physiol. 2009;296:H94–H105.
126. Masuzawa A, Black KM, Pacak CA, Ericsson M, Barnett RJ, Drumm C, Seth P, Bloch DB, Levitsky S, Cowan DB, McCully JD. Transplantation of autologously-derived mitochondria protects the heart from ischemia-reperfusion injury. Am J Phys Heart Circ Physiol. 2013;304:H966–82.
127. Cowan DB, Yao R, Akurathi V, Snay ER, Thedsanamoorthy JK, Zurakowski D, Ericsson M, Friehs I, Wu Y, Levitsky S, Del Nido PJ, Packard AB, McCully JD. Intracoronary delivery of mitochondria to the ischemic heart for cardioprotection. PLoS ONE. 2016;11:e0160889.

128. Bolling SF, Le B, Bove EL. Effect of ATP synthesis promoters on postischemic myocardial recovery. J Surg Res. 1990;49:205–11.
129. Preble JM, Pacak CA, Kondo H, MacKay AA, Cowan DB, McCully JD. Rapid isolation and purification of mitochondria for transplantation by tissue dissociation and differential filtration. J Vis Exp. 2014;91:e51682. doi:10.3791/51682.
130. Preble JM, Kondo H, Levitsky S, McCully JD. Quality control parameters for mitochondria transplant in cardiac tissue. JSM Biochem Mol Biol. 2014;2:1008.
131. Pacak AP, Preble JM, Kondo H, Seibel P, Levitsky S, Del Nido PJ, Cowan DB, McCully JD. Actin-dependent mitochondrial internalization in cardiomyocytes: evidence for rescue of mitochondrial function. Biol Open. 2015;4:622–6.
132. Pfeiffer ER, Wright AT, Edwards AG, Stowe JC, McNall K, Tan J, Niesman I, Patel HH, Roth DM, Omens JH, McCulloch AD. Caveolae in ventricular myocytes are required for stretch-dependent conduction slowing. J Mol Cell Cardiol. 2014;76:265–74.
133. Lou E, Fujisawa S, Morozov A, Barlas A, Romin Y, Dogan Y, Gholami S, Moreira AL, Manova-Todorova K, Moore MA. Tunneling nanotubes provide a unique conduit for intercellular transfer of cellular contents in human malignant pleural mesothelioma. PLoS ONE. 2012;7(3):e33093.
134. Kitani T, Kami D, Matoba S, Gojo S. Internalization of isolated functional mitochondria: involvement of micropinocytosis. J Cell Mol Med. 2014;18:1694–703.
135. Huang X, Sun L, Ji S, Zhao T, Zhang W, Xu J, Zhang J, Wang Y, Wang X, Franzini-Armstrong C, Zheng M, Cheng H. Kissing and nanotunneling mediate intermitochondrial communication in the heart. Proc Natl Acad Sci U S A. 2013;110(8):2846–51.
136. Stark G, Domanowits H, Sterz F, Startk U, Bachernegg M, Kickenweiz E, Decrinis M, Laggner AN, Tritthart HA. Action of ATP on ventricular automaticity. J Cardiovasc Pharmacol. 1994;24:740–4.
137. Piel DA, Gruber PJ, Weinheimer CJ, Courtois MR, Robertson CM, Coopersmith CM, Deutschman CS, Levy RJ. Mitochondrial resuscitation with exogenous cytochrome c in the septic heart. Crit Care Med. 2007;35:2120–7.
138. Pepe S, Marasco SF, Haaas SJ, Sheeran FL, Krum H, Rosenfeldt FL. Coenzyme Q_{10} in cardiovascular disease. Mitochondrion. 2007;7(Suppl):S154–67.
139. Lorita J, Soley M, Ramirez I. Epidermal growth factor protects the heart against low-flow ischemia-induced injury. J Phys Biochem. 2012;66:55–62.
140. Roche S, D'lppolito G, Gomez LA, Bouckenooghe T, Lehmann S, Monteromenei CN, Schiller PC. Comparative analysis of protein expression of three stem cell populations: models of cytokine delivery system in vivo. Int J Pharm. 2012; doi:10.1016/j.ijpharm.2011.12.041.
141. Kocher AA, Schuster MD, Bonaros N, Lietz K, Xiang G, Martens TP, Kurlansky PA, Sondermeijer H, Witkowski P, Boyle A, Homma S, Wang SF, Itescu S. Myocardial homing and neovascularization by human bone marrow blasts is regulated by IL-8/Gro CXC chemochines. J Mol Cell Cardiol. 2006;40:455–64.
142. Emani SM, Piekarski BL, Baird C, Kaza A, Harrild D, del Nido PJ, McCully JD. Autologous mitochondria transplantation by ventricular dysfunction following myocardial ischemia-reperfusion injury. (personal communication). n.d.
143. Chang JC, Wu SL, Liu KH, Chen YH, Chuang CS, Cheng FC, Su HL, Wei YH, Kuo SJ, Liu CS. Allogenic/xenogenic transplantation of peptide-labeled mitochondria in Parkinson's disease: restoration of mitochondria functions and attenuation of 6-hydroxydopamine-induced neurotoxicity. Transl Res. 2016;170:40–56.
144. Lin HC, Liu SY, Lai HS, Lai IR. Isolated mitochondria infusion mitigates ischemia-reperfusion injury of the liver in rats. Shock. 2013;39:304–10.
145. Zhu L, Zhang J, Zhou J, Lu Y, Huang S, Xiao R, Yu X, Zeng X, Liu B, Liu F, Sun M, Dai M, Hao Q, Li J, Wang T, Li T, Lu Q. Mitochondrial transplantation attenuates hypoxic pulmonary hypertension. Oncotarget. 2016; doi:10.18632/oncotarget.10596.
146. Hayakawa K, Esposito E, Wang X, Terasaki Y, Liu Y, Xing C, Ji X, Lo EH. Transfer of mitochondria from astrocytes to neurons after stroke. Nature. 2016;535:551–5.

Chapter 32
Mitochondria-Targeted Antioxidants for the Treatment of Cardiovascular Disorders

Hyoung Kyu Kim and Jin Han

Oxidative Stress and Antioxidant Systems

Diatomic oxygen (O_2) is essential for the survival of most species that live in aerobic environments. Biologic oxidations inevitably produce metabolites known as reactive oxygen species (ROS), examples of which include superoxide radicals ($O_2^{\bullet-}$) and hydrogen peroxide (H_2O_2). Mammalian tissues and cells produce ROS at various sites including mitochondrial electron-transfer complexes (ETCs), the endoplasmic reticulum, and peroxisomes [1]. The mitochondrial ETCs are the sites where oxidative phosphorylation occurs, producing large amounts of the energy-containing molecule adenosine triphosphate (ATP). Consequently, the mitochondrial ETCs are also the sites that produce the most ROS. ROS play roles as signaling molecules in various biological processes including cell differentiation, cell growth, anti-inflammation, and intercellular signaling under physiological conditions. Imbalanced ROS levels cause the detrimental oxidation of various biomolecules including protein, lipid, and DNA, resulting in dysfunction from the cellular level to the organ level, which is known as oxidative stress [1]. Oxidative stress is a

H.K. Kim
National Research Laboratory for Mitochondrial Signaling, Department of Physiology, Department of Health Sciences and Technology, BK21 plus Project Team, College of Medicine, Cardiovascular and Metabolic Disease Center, Inje University, Bokji-ro 75, Busanjin-gu, Busan 47392, Korea

Department of Integrated Biomedical Science, College of Medicine, Inje University, Busan, Korea

J. Han (✉)
National Research Laboratory for Mitochondrial Signaling, Department of Physiology, Department of Health Sciences and Technology, BK21 plus Project Team, College of Medicine, Cardiovascular and Metabolic Disease Center, Inje University, Bokji-ro 75, Busanjin-gu, Busan 47392, Korea
e-mail: phyhanj@inje.ac.kr

G. Santulli (ed.), *Mitochondrial Dynamics in Cardiovascular Medicine*, Advances in Experimental Medicine and Biology 982,
DOI 10.1007/978-3-319-55330-6_32

fundamental factor in a variety of diseases including inflammatory diseases, neurodegenerative diseases, metabolic syndromes, many cancers, and cardiovascular diseases (CVDs) [1–6].

The human heart is one of the most energy-dependent organs, producing and spending approximately 30 kg ATP/day to continuously pump blood throughout the body [7]. Mitochondria make up more than 30% of its mass [8]. Mitochondrial ETCs I and III have long been known as major sites of ROS generation. A recent study demonstrated that complexes II and IV also produce ROS in disease conditions [9]. Cardiac myocytes can easily be exposed to mitochondrial oxidative stress, resulting in irreversible oxidative damage and cell death. Normal heart function cannot be maintained in the presence of widespread apoptotic or necrotic cardiac-cell death and decreased mitochondrial function in surviving cardiomyocytes, resulting in heart failure and death as a final consequence.

To prevent irreversible oxidative damage, human cells have an effective primary antioxidant system consisting of endogenous antioxidant enzymes such as manganese or copper and zinc superoxide dismutase (Enzyme Commission Number, EC 1.15.1.1), catalase (EC 1.11.1.6), glutathione reductase (EC 1.8.1.7), and multiple isoforms of glutathione peroxidase (GPx 1–8, EC 1.11.1.9) and peroxiredoxin (Prx 1–6, EC 1.11.1.15) [1, 6]. As a second line of defense, the human body has endogenous, non-enzymatic antioxidant systems, which include thiols (glutathione, lipoic acid, and N-acetyl cysteine), L-arginine, NADPH and NADH, ubiquinone (coenzyme Q10), melatonin, uric acid, bilirubin, and metal-binding proteins (ferritin, myoglobin, transferrin, albumin, ceruloplasmin, and metallothionein) [6]. In addition, exogenous antioxidants provided by food or supplements can complement the action of the endogenous antioxidants or can scavenge ROS directly. The exogenous antioxidants include some vitamins and minerals, carotenoids, polyphenols, flavonoids, and omega-3 and omega-6 fatty acids [6] (Fig. 32.1).

Antioxidant Therapy

Various antioxidant therapies have been tested to treat a variety of CVDs [10–16]. Despite some preclinical successes, most global antioxidants failed to improve, or even worsened, the targeted disease states in clinical trials [14–18] (Table 32.1). Several reasons were suggested to explain those results [17, 18].

First, the selected antioxidants were chosen not for their efficiency but instead for their availability. For example, some clinical studies used vitamins E or C, because those vitamins are relatively easy to obtain commercially. Second, there is a lack of biochemical markers to directly evaluate the antioxidant potential of the drugs. Third, the relatively low doses and short durations of the therapies were not enough to effectively treat the tissue damage caused by more than 40 years of oxidative stress in adult patients. Furthermore, the follow-up periods might not have been long enough to realize the benefits of the therapies. Fourth, the trials might not have included patients whose disease was actually related to an antioxidant deficiency. If

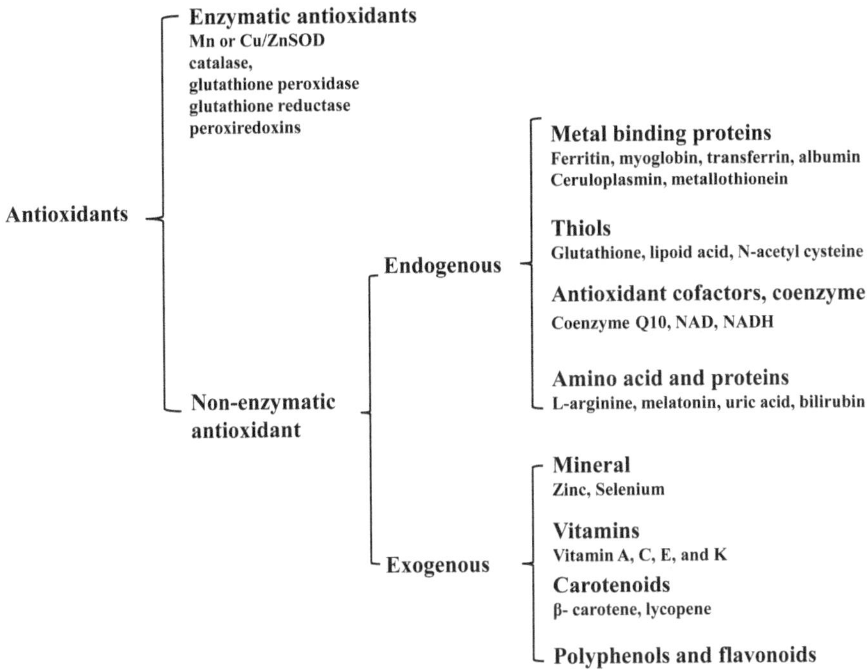

Fig. 32.1 Antioxidant systems

a patient's antioxidant levels are normal or at least sufficient to balance ROS production and removal, then antioxidant therapy would not likely provide any benefit. Finally, the antioxidants tested might not have been able to efficiently reach the mitochondria, where the major ROS generation and oxidative stress take place [32].

Despite the failures of antioxidant therapies in previous clinical trials, oxidative stress remains a focus of research on many diseases. Clinical researchers and pharmaceutical companies therefore continue to develop new strategies to provide effective antioxidant therapy.

Mitochondria-Targeted Antioxidants

In light of the unmet need for effective antioxidant therapy, mitochondria-targeted antioxidants that are selectively taken into the mitochondria in vivo have been developed over the last decade. There is accumulating evidence that the targeted antioxidant therapies are better than their nonspecific predecessors at protecting the heart against oxidative stress and damage [33, 34]. Antioxidants that selectively accumulate in the mitochondria have had beneficial effects in animal models of diseases involving oxidative damage in the mitochondria [34].

Table 32.1 Selected clinical trials in which antioxidants had no effect or had negative effects on events related to cardiovascular disorders

Study name (year) (reference)	No of patients	Subjects	Agent and dose	Study duration (years)	Outcome
Vitamin E					
ATBC (1994) [19]	29,133	Smokers with no medical problems	50 mg vitamin E	6.1	Increase in hemorrhagic stroke
GISSI (1999) [20]	11,324	Post-myocardial infarction adults	300 mg (synthetic) vitamin E	3.5	No beneficial effect
HOPE (2000) [21]	9,541	Patients with high CVD risk	400 IU (natural) vitamin E	4.5	No beneficial effect
PPP (2001) [22]	4,495	Patients with high CVD risk	300 mg (synthetic) vitamin E	3.6	No beneficial effect
MICRO-HOPE (2002) [23]	3,654	Patients with diabetes	400 IU (natural) vitamin E	4.5	No beneficial effect
VEAPS (2002) [24]	353	Patients with elevated LDL-C	400 IU dl-α-tocopherol	3	No beneficial effect
β-carotene					
ATBC (1994) [19]	29,133	Smokers with no medical problems	20 mg β-carotene	6.1	Increase in overall mortality and mortality due to ischemic heart disease, hemorrhagic stroke, and ischemic stroke
ATBC (1998) [25]	27,271	Smokers with no history of myocardial infarction	20 mg β-carotene	6.1	No beneficial effect
SCPS (1996) [26]	1,805	Patients with skin cancer	50 mg β-carotene	8.2	No beneficial effect
PHS (1996) [27]	22,071	Healthy individuals	50 mg β-carotene	12	No beneficial effect

(continued)

Table 32.1 (continued)

Antioxidant cocktails					
CARET (1996) [28]	4,060	Asbestos workers	30 mg β-carotene, 2,500 IU retinol	5.5	Increase in all-cause mortality
	14,254	Current/former smokers			No beneficial effect
ATBC (1998) [25]	27,271	Smokers with no history of myocardial infarction	50 mg vitamin E, 20 mg β-carotene	6.1	No beneficial effect
HATS (2001) [29]	160	Patients high CVD	800 IU vitamin E (as d-α-tocopherol), 1,000 mg vitamin C, 25 mg (natural) β-carotene, 100 μg selenium, simvastatin (10 mg/day)+ niacin (250–1,000 mg twice/day)	3.5	Simvastatin/niacin induced atheroregression; antioxidant cocktail resulted in stenosis progression, cardiovascular disease-related mortality, nonfatal infarction, or revascularization
HPS (2002) [30]	20,536	Patients with high CVD risk	600 mg vitamin E, 250 mg vitamin C, 20 mg β-carotene	5	No beneficial effect
WAVE (2002) [31]	423	Postmenopausal women with CVD	800 IU vitamin E, 1,000 mg vitamin C plus HRT	2.8	All-cause mortality (hazard ratio) was higher in the treatment group vs. that in the placebo group

Adapted from Kris-Etheron's review [17]
MI myocardial infarction, *HRT* hormone replacement therapy, *CVD* cardiovascular disease

There are two major issues in the development of successful mitochondria-targeted antioxidant therapies. The first is the selection of optimal antioxidants with high ROS-scavenging potential. The second is the establishment of methods to target specific tissues and mitochondria. Here, we discuss the current status of the mitochondria-targeted antioxidants (Table 32.2) [35].

Table 32.2 Mitochondria-targeted (or membrane permeable) antioxidants

Agent	Drug property	Related disease	Clinical status
MitoQ	Coenzyme Q10 derivative	NASH	Phase II (NASH)
	+TPP	Parkinson's disease	Phase II
		Aging related dysfunction	Phase I
MitoVit E (TPPB)	Vitamin E derivative	Friederich Ataxia, alcohol syndrome, ischemic neural injury, ocular disease	Animal study only
	+ TPP		
MitoPBN	Nitrone radical trap alpha-phenyl-tert-butylnitrone + TPP	Neuroprotection	Not yet tested
LPBNAH			
MitoPeroxidase (Ebselen analog) 1 and 2	Peroxidase derivative (ebselen) + TPP	Radiomitigative effect	Animal study only
SkQ	Plastoquinone + TPP	Aging induced dry eye	Phase II
		IR heart injury, hypertrophy	Animal study only
		Myocardial fibrosis	
MitoGSH	Glutathione + choline ester	Cardiac oxidative stress	Not yet tested
MitoNAC	N-acetyl-l-cysteine +choline ester,	Not yet tested	Not yet tested
SS (Szeto-Schiller) peptides (Bendavia)	Peptide antioxidants targeted to the inner mitochondrial membrane	IR heart injury,	Phase II (STEMI)
		Mitochondrial myopathy	Phase II
		Hypertensive cardiomyopathy	Animal study
Edaravone	Synthetic antioxidant	Brian IR injury	Used in Japan
		Cardiac IR injury,	Phase IV (AMI)
		Diabetic cardiomyopathy, cardiac stem cell therapy	Animal study
Idebenone	Synthetic antioxidant	Friedereich's Ataxia-CVD	Phase III
		MELAS	Phase II
		Duchenne muscular dystrophy	Phase III
		LHON	Drug approved
NecroX	Synthetic antioxidant, mitochondria Ca^{2+} uniporter blocker	IR heart injury	Phase II (STEMI)
		Myocardial fibrosis	Animal study
		Live injury	
		Allergic inflammation	
		IR kidney injury	
		Atherosclerosis	

(continued)

Table 32.2 (continued)

Agent	Drug property	Related disease	Clinical status
Astaxanthin	Dietary carotenoid	Hyperlipidemia	Phase IV
		Atherosclerosis, coronary heart disease and ischemic brain damage, age-related macular degeneration, acute pain, inflammation, cancer, and cardiovascular diseases	Animal study

I/R ischemia/reperfusion, *RNS* reactive nitrogen species, *ROS* reactive oxygen species, *STEMI* ST-segment elevation in myocardial infarction, *AMI* acute myocardial infarction, *NASH* nonalcoholic steatohepatitis, *MELAS* Mitochondrial Encephalopathy Lactic Acidosis & Stroke-like Episodes, *LHON* Leber's hereditary optic neuropathy

MitoQ

MitoQ (or mitoquinone mesylate) is one of the most well-known mitochondria-targeted synthetic antioxidants. MitoQ was developed in the late 1990s to penetrate the mitochondrial membrane and accumulate within the mitochondria to increase the local antioxidant capacity [36]. They used antioxidative ubiquinone, a mimetic structure of coenzyme Q10, to incorporate the lipophilic cation triphenylphosphonium (TPP), which made it easy for the positively charged molecule to target the negatively charged mitochondrial membrane and subsequently accumulate within the mitochondrial compartment (Fig. 32.2a). Within the mitochondria, ETC II, also known as succinate dehydrogenase, converts the ubiquinone of MitoQ to the active antioxidant ubiquinol, which then scavenges excess ROS in the mitochondrial matrix. After reducing the ROS, the ubiquinol is oxidized to ubiquinone and then recycled by ETC II [37] (Fig. 32.2b).

MitoQ was tested in various animal models of diseases including Alzheimer Disease [38], Parkinson's Disease [39], type I diabetes [40], fatty-liver disease [41, 42], metabolic syndrome [43, 44], and alcohol-induced steatohepatitis [42]. The ability of MitoQ to preserve organs for transplantation was also tested [45]. MitoQ showed remarkable therapeutic potential against a wide range of CVDs including ischemic heart disease [46, 47], hypertrophy [48, 49], hypertension [49], and vascular endothelial dysfunction [50]. It also demonstrated cardioprotection against endotoxin-induced heart failure [51], doxorubicin [52], and cocaine-induced cardiotoxicity [53].

Those findings strongly support MitoQ as a promising clinical therapy for CVDs. Antipodean Pharmaceuticals Inc. completed a successful Phase I clinical trial that demonstrated the safety of oral MitoQ tablets in healthy individuals. Two Phase II clinical trials were subsequently conducted in patients with Parkinson's disease (clinicaltrials.gov as NCT00329056) and hepatitis (clinicaltrials.gov as NCT00433108), respectively. In the former, patients were treated with a daily oral dose of 40 or 80 mg MitoQ for 1 year and compared with patients that received a

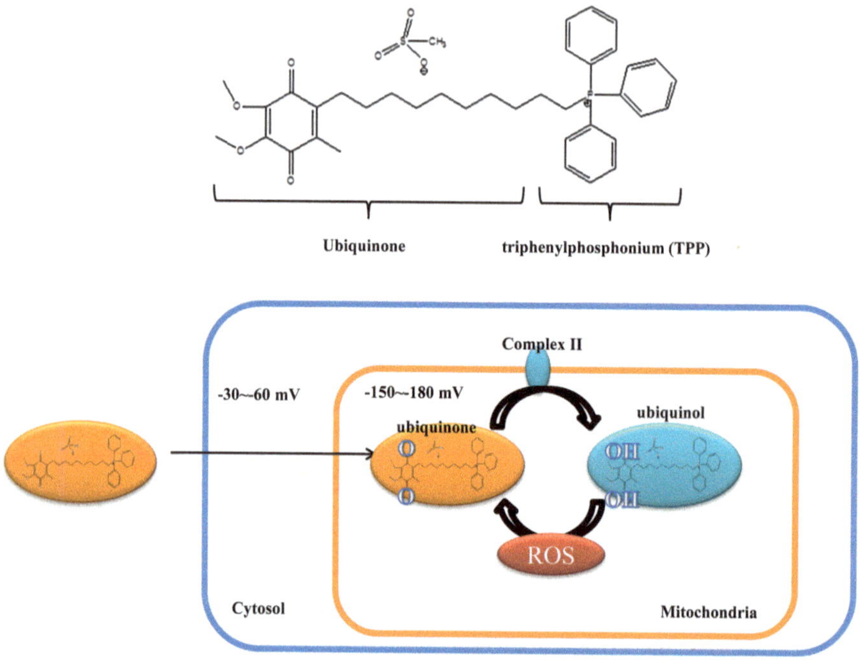

Fig. 32.2 Structure and mechanism of action of MitoQ

placebo. The trial was verified in July 2010. The results were negative, showing no benefit of MitoQ for patients with Parkinson's disease. The negative result might be explained by the relatively short period of the treatment, which was not sufficient to reverse the prolonged neuronal damage suffered by the patients prior to their diagnosis. The study did, however, provide a year's worth of safety data [54]. In the other Phase II clinical trial of MitoQ, 36 patients with chronic hepatitis C infection who were unresponsive to antiviral treatments were treated with MitoQ (40 or 80 mg) over 28 days to prevent liver inflammation, which was determined by serum alanine transaminase (ALT), aminotransferases, and viral RNA. The trial was verified in August 2008. The patients that were treated with MitoQ showed lower serum ALT and aminotransferase levels, demonstrating the therapeutic potential of MitoQ in chronic liver diseases [55].

The two clinical trials showed that it was safe to target human mitochondria with antioxidants over a prolonged period and that the efficacy of the treatment is dependent on the time course of the disease. In 2015, a new clinical trial to study the effects of MitoQ to improve physiological functions (vascular, motor, and cognitive) in middle-aged and older adults was registered on clinicaltrials.gov (NCT02597023) and began recruiting participants.

Mitochondria-Targeted Vitamin E

Together with the successful development of MitoQ, the same group produced a mitochondria-targeted vitamin E (MitoVit E) by binding the active phenolic moiety of vitamin E (α-tocopherol) with a lipophilic TPP cation [56] (Fig. 32.3). The major limitation of vitamin E, or N-acetyl cysteine, is its global distribution in cell, which renders it ineffective for the removal of mitochondrial ROS. MitoVit E showed much greater accumulation in the mitochondria with 5,000–6,000: 1 (accumulated in mitochondria: remained outside) in the incubated mitochondria with 1–20 μM MitoVit E. The accumulated MitoVit E protected the mitochondria against oxidative damage [56]. Although MitoVit E showed therapeutic potential in various models of diseases including Friedreich ataxia (FRDA) [57], aortic endothelial dysfunction [58], fetal alcohol syndrome [59], ischemic neuronal injury [60], obesity and liver disease [61], cadmium toxicity [62], and ocular diseases [63, 64], its therapeutic potential in CVDs has not yet been tested.

MitoPBN

MitoPBN is a derivative form of α-phenyl-*N*-*tert*-butylnitrone (PBN) (Fig. 32.3). Unlike other mitochondria-targeted antioxidants, MitoPBN specifically reacts with carbon-centered radicals but not with superoxide [65]. It was developed to counter the effects of ROS on the activity of uncoupling proteins, which readily accumulate at 2.2–4.0 mM concentrations in mitochondria [65, 66]. The therapeutic effects of MitoPBN have not yet been tested in any disease condition. Another derivative of PBN, *N*-[4-(octa-*O*-acetyllactobionamidomethylene) benzylidene]-*N*-[1,1-dimethyl-2-(*N*-octanoyl) amido]-ethylamine *N*-oxide (LPBNAH), has shown antioxidative and neuroprotective activity in preclinical experiments [67]. The cardioprotective effects of mitochondria-targeted PBN derivatives need to be investigated further.

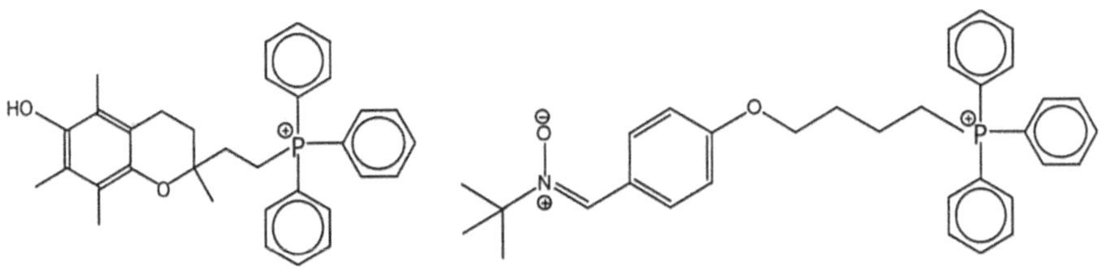

MitoVit E **Mito PBN**

Fig. 32.3 Structure of MitoVit E and Mito PBN

MitoPeroxidase (Ebselen Analog)

MitoPeroxidase is a conjugated compound consisting of the peroxidase mimetic ebselen and a TPP cation [68]. Ebselen, first synthesized in 1924, effectively degrades peroxides and also shows antioxidant activity. The conjugation of ebselen and TPP only slightly enhanced the mitochondrial uptake of ebselen, because the ebselen moiety covalently binds to proteins. MitoPeroxidase enters the mitochondria, and mitochondrial GSH or thioredoxin converts it to effective antioxidants that can remove peroxides and prevent oxidative stress [68]. Recently, Stoyanovsky et al. developed MitoPeroxidase-2 [69], a modified version of MitoPeroxidase (Fig. 32.4). Whereas MitoPeroxidase showed cytotoxicity in mouse embryonic cells at concentrations of 10–20 μM, MitoPeroxidase-2 showed comparatively lower cytotoxicity at a concentration of ~40 μM. In a test of the therapeutic potential of MitoPeroxidase-2 against ionizing radiation-induced cell death, MitoPeroxidase-2 showed radiomitigative properties in radiation-exposed mice [69]. The application of the MitoPeroxidases has so far been limited, so more clinical investigations are required to test their therapeutic potential in CVDs.

SkQ

SkQ is an organic, cationic plastoquinone derivative containing positively charged phosphonium or rhodamine moieties (Fig. 32.5) [70]. SkQ was developed by Antonenko et al. to overcome the unexpected negative effect of MitoQ, which

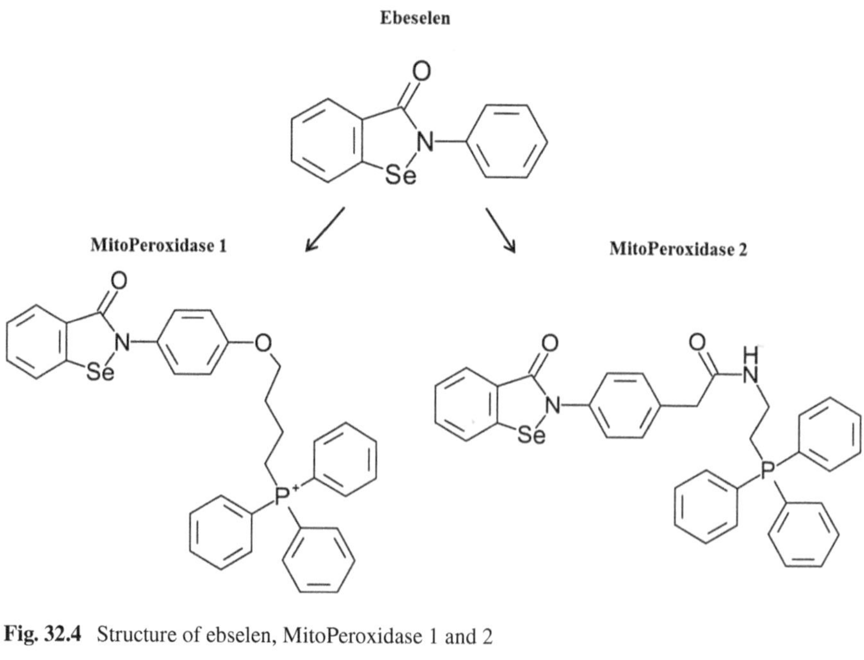

Fig. 32.4 Structure of ebselen, MitoPeroxidase 1 and 2

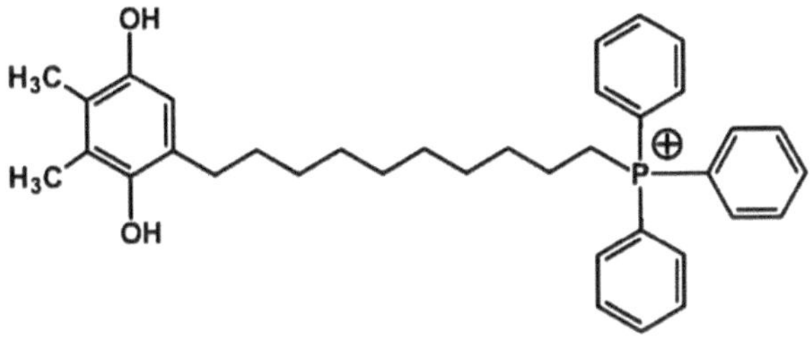

Fig. 32.5 Structure of SkQ

sometimes acts as a pro-oxidant, depending on the treatment dose [37]. For that purpose, the ubiquinone in MitoQ was replaced with plastoquinone, which has better antioxidant capacity. SkQ strongly targets the mitochondria and has remarkably enhanced antioxidant potential, which inhibited H_2O_2-induced apoptosis at very low (picomolar) concentrations in cell culture [70]. A nanomolar oral dose of SkQ1 (0.02 nmol/day for 3 weeks) abolished ROS-induced or Ischemia/Reperfusion (IR)-induced cardiac arrhythmia in ex vivo rat hearts. Higher doses of SkQ1 (125–250 nmol/day for 3 weeks) decreased MI in an in vivo IR model, lowering plasma levels of cardiac damage markers including lactate dehydrogenase and creatin kinase [71]. SkQ1 treatment also effectively attenuated age-associated myocardial diseases, cardiac hypertrophy, and myocardial fibrosis in a mouse model of aging [72].

SkQ has been used to treat various age-related pathologies [73]. A successful Phase II clinical trial (NCT02121301) showed that SkQ1 is a safe and effective treatment for dry-eye syndrome [74]. Preclinical and clinical studies of SkQ to treat CVDs have been limited. Because SkQ showed low toxicity and high antioxidant potential at low doses, it would likely be worthwhile to test the therapeutic potential of SkQ in cardiovascular disorders.

MitoGSH and MitoNAC

MitoGSH and MitoNAC are mitochondria-targeted forms of gluthione choline ester and *N*-acetyl-l-cysteine choline ester, respectively (Fig. 32.6). Both compounds were developed by the Shey-Shing Sheu group [33]. The hydrophilic forms of GSH and NAC were designed to easily enter and accumulate in the mitochondria. MitoGSH protected mitochondria from H_2O_2-induced oxidative stress in neonatal rat ventricular myocytes. The treatment of those cells with 50 μM MitoGSH preserved the mitochondrial membrane potential against H_2O_2-induced depolarization [33]. There have since been no follow-up studies of MitoGSH and MitoNAC.

Fig. 32.6 Structure of MitoGSH and MitoNAC [33]

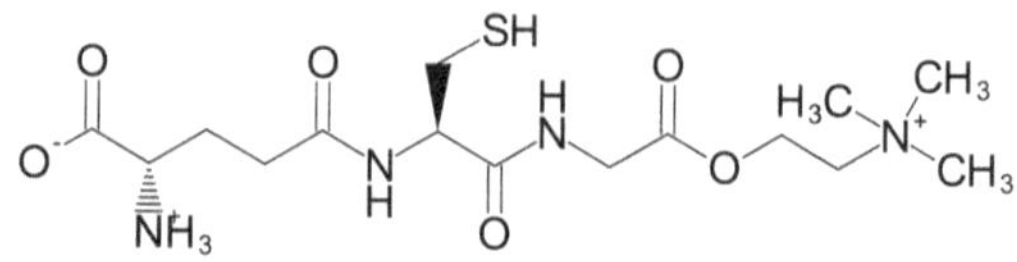

MitoGSH

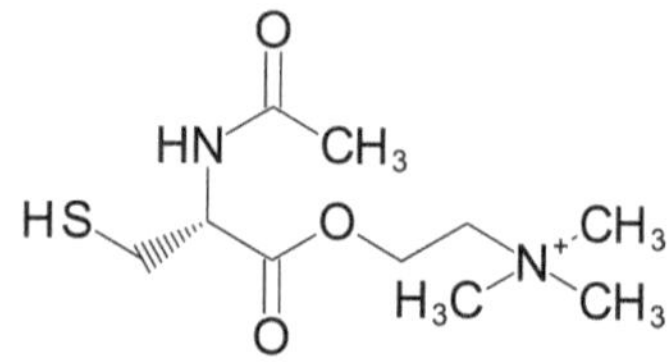

MitoNAC

SS Peptides

SS (Szeto-Schiller) peptides (SS-01, 02, 31, and 20) are short-sequence, antioxidant tetrapeptides that were designed by Szeto and Schiller to target the inner mitochondrial membrane (IMM) [75]. Because mitochondrial ETCs are located in the IMM, targeting the IMM could be a brilliant strategy to reduce mitochondrial ROS levels. The SS peptides possess alternating basic and aromatic residues, providing them with ROS scavenging ability (see [75, 76] for detailed sequences). The cationic SS peptides interact with the negative charge of the IMM potential to enter the mitochondria, reaching concentrations in the IMM that are up to 1,000-fold higher than those in the cytoplasm.

SS-02 and SS-31 prevented tBHP-induced ROS overproduction and sequential membrane-potential depolarization and MPTP opening in neuronal cells [77]. SS-31 pretreatment in a mouse model of islet transplantation prevented mitochondrial depolarization, reduced apoptosis, increased cell yield, and improved post-transplantation function in the islet cells, suggesting that SS-31 might have clinical value in organ transplantation [78]. In other animal models, SS-31 treatment protected kidneys from IR injury [79, 80], unilateral ureteral obstruction [81], and diabetic nephropathy [82] by reducing mitochondrial damage and preserving mitochondrial energy metabolism. SS-31 protected neurons from damage in models of cognitive deficits [83, 84] and protected or recovered mitochondrial function in skeletal muscle damaged by aging [85] and burn trauma [86, 87].

Both SS-02 and SS-31 showed significant cardioprotective effects in ex vivo animal models of IR [75, 88]. Treatment with SS-02 (100 μM) or SS-31 (1 nM) during the reperfusion period significantly improved the contractility of IR-injured hearts. Cho et al. tested the cardioprotective effects of both peptides in an in vivo rat model of left anterior descending artery ligation and found that they significantly reduced IR-induced arrhythmia, lipid peroxidation, and infarct size [89]. In a study of hypertensive cardiomyopathy, 4 weeks of SS-31 treatment attenuated angiotensin II (Ang)-induced cardiomyopathy without changing blood pressure [90]; however, treatment with SS-20, a non-antioxidant peptide, did not protect the heart at all. Those results suggest that the cardioprotective effects of SS-02 and SS-31 are facilitated by the mitochondria-targeted antioxidative properties of the peptides.

Bendavia (or Elamipretide), a drug form of SS-31, was developed by Stealth BioTherapeutics and applied in clinical trials to treat heart failure. The PREVIEW study – a Phase I trial (NCT02388464) to study the safety, tolerability, and pharmacokinetics of escalating doses of Bendavia in patients with heart failure – was completed in September 2015. According to Stealth BioTherapeutics, the Bendavia treatments demonstrated acceptable safety and tolerability and improved heart function in the patients (report not yet published).

The Phase II EMBRACE trial (NCT01572909) [91] enrolled 300 patients with ST-segment elevation myocardial infarction (STEMI). The patients received intravenous infusion of either Bendavia (0.05 mg/kg/h) or a placebo (60 mL/h). The primary endpoint was infarct size. Secondary endpoints included physiological and clinical outcomes [91]. The EMBRACE study was completed in November 2015, and its results are yet to be released.

In addition to the studies of CVDs, the Phase II MMPOWER trial (NCT02367014) was conducted to evaluate the safety and efficacy of Bendavia in 36 patients with mitochondrial myopathy associated with genetically confirmed mitochondrial disease. The patients received Bendavia infusions over a period of 5 days. The study was completed in May 2016. According to Dr. Amel Karaa, the trial investigator, Bendavia passed its safety and tolerability endpoints and enhanced the exercise capacity of the patients (data unpublished).

Edaravone

Edaravone (also known as MCI-186 or Radicut) is a scavenger of free radicals that has been used to treat acute ischemic stroke (Fig. 32.7). There is substantial clinical and experimental evidence demonstrating the neurovascular-protective effects of edaravone (for detail information see the review article [92]). The cardioprotective effects of edaravone have been studied in various conditions including diabetic cardiomyopathy [93], cardiac stem-cell therapy [94], and isoproterenol-induced cardiac toxicity [95]. It is unclear whether edaravone specifically targets the mitochondria or accumulates in any particular mitochondrial compartment. Edaravone did, however, demonstrate remarkable mitochondria-protective potential in a model of myocardial infraction [96].

In a Phase IV clinical trial (NCT00265239) examining the effects of edaravone in patients with acute myocardial infarction, edaravone prevented myocardial injury following IR [97]. Intravenous infusion of edaravone (30 mg) before myocardial reperfusion reduced myocardial infarction and enhanced ejection fraction at early periods of reperfusion [97].

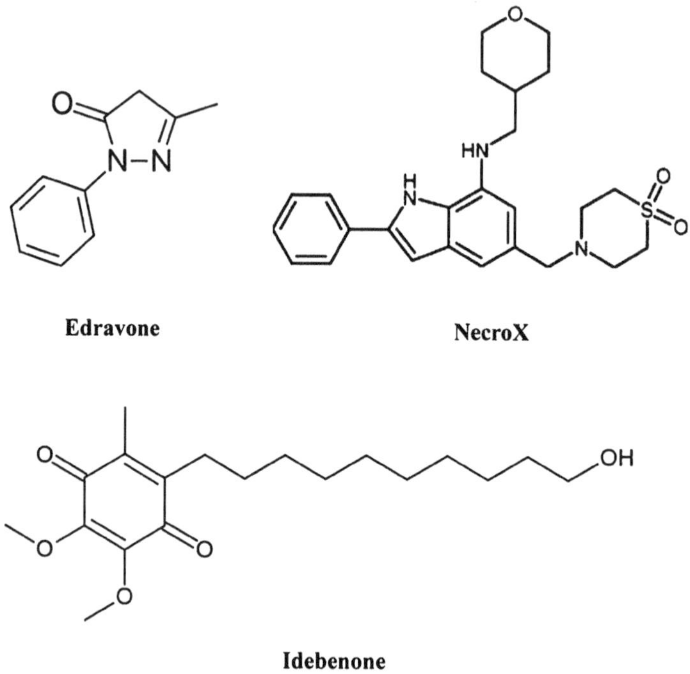

Fig. 32.7 Structure of edravone, NecroX and idebenone

Idebenone

Idebenone (also known as CV-2619) is a CoQ10 analog developed in the 1980s by Takeda Pharmaceuticals (Fig. 32.7) [98]. Idebenone has properties similar to those of CoQ10, a mitochondrial coenzyme that acts as an antioxidant and electron carrier [99, 100]. Idebenone differs structurally from CoQ10 in that it has a shorter lipophilic tail, which reduces its lipophilicity. The extreme lipophilicity of CoQ10 limits its localization and action to membranes, whereas idebenone localizes and functions in membranes, the cytoplasm, and the mitochondrial matrix [98]. In the mitochondria, idebenone plays multiple roles including those of an electron carrier in the ETC, a complex I inhibitor, a complex II-III activator, and an antioxidant (for the detailed mechanism, see the review [98]).

Because of its unique properties related to mitochondrial regulation, pharmaceutical companies developed idebenone to treat mitochondrial dysfunction in diseases including Alzheimer disease [101], FRDA (Phase II, NCT00229632; Phase III, NCT01303406) [102, 103], Duchenne muscular dystrophy (Phase III, NCT01027884) [104, 105], mitochondrial encephalopathy lactic acidosis and stroke-like episodes (MELAS; Phase II, NCT00887562) and Leber's hereditary optic neuropathy (LHON; Phase II, NCT00747487) [106]. Based on the results of clinical trials, the European Medicines Agency approved idebenone (Raxone) for the treatment of LHON, and the drug has been available on the European market since 2015 [106].

The therapeutic potential of idebenone in cardiac hypertrophy or cardiomyopathy was investigated in patients with FRDA [102, 103, 107–110]. The results suggested that idebenone has positive effects on cardiac hypertrophy and cardiomyopathy; however, the efficacy of idebenone is still debated with regards to the conditions of the studies [103]. The cardioprotective effect of idebenone needs to be investigated further in the broad range of heart diseases.

NecroX (LC28-0126)

NecroX (C25H31N3O3S · 2CH4O3S) is a novel mitochondria-targeted antioxidant developed by LG life science (Fig. 32.7) [111]. NecroX shows strong antioxidant effects in in vitro models of oxidative stresses induced by tBHP, doxorubicin, CCl4, or hypoxic injury. NecroX prevented oxidative stress-induced mitochondrial dysfunction [112, 113] and showed therapeutic potential in animal models of diseases including hepatic IR injury [114], myocardial fibrosis [115], myocardial IR injury [113], drug-induced liver injury [116], allergic inflammation [117], renal IR injury [118], and atherosclerosis [119, 120].

A Phase I clinical trial of NecroX (named LC28-0126 at clinicaltrials.gov, NCT01737424) to investigate the drug's safety, tolerability, and pharmacokinetics in healthy males was finished successfully. A Phase II trial (NCT02070471) to evaluate the efficacy, safety and pharmacokinetics of NecroX prior to percutaneous coronary intervention in patients with STEMI was completed in September 2015. An additional Phase II trial (NCT02770664) to further study NecroX under those conditions has been registered and is recruiting patients.

Astaxanthin

Astaxanthin is a membrane-permeable dietary carotenoid found in various marine organisms (shrimp, lobster, crab, and salmon), algae, fungi, and plants (Fig. 32.8) [35, 121]. The antioxidant potential of astaxanthin is 100–1,000 times greater than those of other organic antioxidants including vitamins C and E, CoQ10, and polyphenols [35]. The mitochondrial localization of astaxanthin was tested in blastocysts [122]. The antioxidant and mitochondria-protective effects of astaxanthin were demonstrated in human mesangial cells in a model of diabetic nephropathy [123], in neuronal cells [124, 125], in blastocysts [122], in vitamin E-deficient rats [126], in a model of Parkinson's disease [127, 128], in a model of liver IR injury

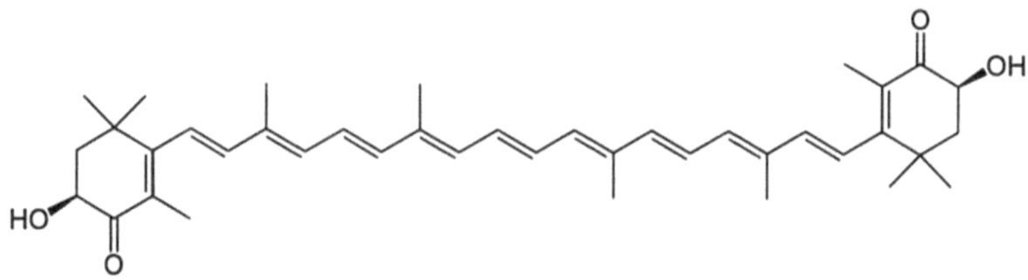

Fig. 32.8 Structure of astaxanthin

[129], and in models of CVDs including various animal models of IR injury and hypertensive rats [121, 130–134].

From 2000 to 2010, various clinical trials investigated the therapeutic potential of astaxanthin in both healthy and disease conditions. In healthy individuals, astaxanthin showed dose safety, antioxidant effects, reduced oxidation of lipids, and anti-inflammation effects, which might help to reduce the risk of CVDs (for detailed information about the clinical trials, see the review of Fassett and Coombes [121]).

In recent human trials, the effects of pure or antioxidant-containing medical-food formulations of astaxanthin on the cardiovascular system were tested. In a Phase IV clinical trial (NCT01490229) entitled “Ezetimibe Versus Nutraceuticals in Statin-intolerant Patients (ECLIPSE)”, an astaxanthin-containing nutraceutical improved the lipid profiles of statin-intolerant patients with coronary heart disease without producing any adverse side effects [135]. Similarly, 8 weeks of treatment with an astaxanthin-containing medical food (PDL-0101) significantly reduced triacylglycerol levels in patients with hyperlipidemia (Pilot study, NCT02582424) [136]. In contrast, 12 months of treatment with astaxanthin alone showed no beneficial effects in recipients of renal transplants (ACTRN12608000159358) [137].

Albeit there are not yet any direct studies of the effects of astaxanthin in cardiovascular patients, the demonstrated safety, antioxidant potential, and beneficial metabolic effects of astaxanthin suggest that it could have therapeutic potential in CVDs.

Next-Generation Mitochondria-Targeted Antioxidants

Recently, several novel mitochondria-targeted antioxidants were developed that can be considered next-generation drug candidates [138–141].

Mitochonic Acid 5 (MA-5) is a new indole derivative synthesized by Suzuki et al. [138]. MA-5 enhanced ATP production and reduced mitochondrial ROS levels without affecting the activity of the mitochondrial OXPHOS complex [139]. MA-5 can bind to mitofilin, an IMM protein. In mice with dysfunctional mitochondria and

IR injury, MA-5 improved the function of cardiac and renal mitochondria [139]. The mitochondria-targeted clinical potential of MA-5 was tested in fibroblasts of patients with different mitochondrial diseases including Leigh syndrome, MELAS, LHON, and Kearns-Sayre syndrome. MA-5 prolonged cell survival and increased the ATP production in all the fibroblasts with failed mitochondrial OXPHOS [138]. Those results suggest that MA-5 could have therapeutic value for patients with genetic mitochondrial diseases and mitochondria-associated cardiac and renal diseases because of its ATP compensation and antioxidant effects.

Mitochondrial CPP 1 (mtCPP-1), a novel mitochondria-targeting peptide, is a modified version of SS-31 with enhanced mitochondrial targeting and antioxidant capacity [140]. One of a series of developed peptides (mtCPP-1~9), mtCPP-1 showed greater mitochondria-targeting efficiency (1.8-fold) and antioxidant potential (twofold) than an equivalent concentration of SS-31. Remarkably, mtCCP-1 did not affect the mitochondrial membrane potential and ATP level, which were reduced by SS-31 treatment [140]. Those advantages make mtCCP-1 more suitable for clinical application and overcome some of the shortcomings of previous mitochondria-targeting peptides.

Kyotorphin-nitroxide hybrid molecules are novel synthetic lipophilic molecules that conjugate nitronyl nitroxide to Kyotrophin (KTP; l-Tyr-l-Arg) [141]. Among the ten such compounds synthesized, compounds 9i and 9j showed strong antioxidant effects and protected mitochondria against IR injury in vitro [141]. Investigation of the exact working mechanism and follow-up studies are needed to further develop these compounds as therapeutic candidates.

Conclusion

The integrity of mitochondrial function is essential for cardiac homeostasis. Excess mitochondrial ROS in cardiomyocytes leads inevitably to oxidative stress, damaging the mitochondria, causing heart function to deteriorate. Antioxidant therapy was suggested as a promising therapeutic strategy for a variety of cardiovascular disorders, but a number of critical obstacles blocked the successful connection between scientific benches and clinical beds. Although only a few clinical studies showed benefits of mitochondria-targeted antioxidants in the treatment of cardiovascular disorders, properly timed treatments with optimal mitochondria-targeting antioxidants are likely to achieve breakthrough therapeutic results in the near future. In addition to the development of optimal mitochondria-targeted antioxidants, techniques to selectively deliver the drugs to cardiac tissues (e.g. ventricle, atrium, and coronary arteries or nodes) and to damaged mitochondria are also necessary for successful clinical treatment.

Acknowledgements This study was supported by a grant from the Priority Research Centers Program through the National Research Foundation of Korea (NRF), Funded by the Ministry of Education, Science and Technology (2010-0020224, 2015R1A2A1A13001900 and 2015R1D1A1A01057937), Republic Korea.

References

1. Kim HK, Nilius B, Kim N, Ko KS, Rhee BD, Han J. Cardiac response to oxidative stress induced by mitochondrial dysfunction. Rev Physiol Biochem Pharmacol. 2016;170:101–27.
2. Mittal M, Siddiqui MR, Tran K, Reddy SP, Malik AB. Reactive oxygen species in inflammation and tissue injury. Antioxid Redox Signal. 2014;20:1126–67.
3. Kim GH, Kim JE, Rhie SJ, Yoon S. The role of oxidative stress in neurodegenerative diseases. Exp Neurobiol. 2015;24:325–40.
4. Ando K, Fujita T. Metabolic syndrome and oxidative stress. Free Radic Biol Med. 2009;47:213–8.
5. Sosa V, Moline T, Somoza R, Paciucci R, Kondoh H, Me LL. Oxidative stress and cancer: an overview. Ageing Res Rev. 2013;12:376–90.
6. Xie W, Santulli G, Reiken SR, Yuan Q, Osborne BW, Chen BX, Marks AR. Mitochondrial oxidative stress promotes atrial fibrillation. Sci Rep. 2015;5:11427.
7. Santulli G, Xie W, Reiken SR, Marks AR. Mitochondrial calcium overload is a key determinant in heart failure. Proc Natl Acad Sci U S A. 2015;112:11389–94.
8. Page E, McCallister LP. Quantitative electron microscopic description of heart muscle cells. Application to normal, hypertrophied and thyroxin-stimulated hearts. Am J Cardiol. 1973;31:172–81.
9. Chen YR, Zweier JL. Cardiac mitochondria and reactive oxygen species generation. Circ Res. 2014;114:524–37.
10. Firuzi O, Miri R, Tavakkoli M, Saso L. Antioxidant therapy: current status and future prospects. Curr Med Chem. 2011;18:3871–88.
11. Mitra S, Deshmukh A, Sachdeva R, Lu J, Mehta JL. Oxidized low-density lipoprotein and atherosclerosis implications in antioxidant therapy. Am J Med Sci. 2011;342:135–42.
12. Vaziri ND. Roles of oxidative stress and antioxidant therapy in chronic kidney disease and hypertension. Curr Opin Nephrol Hypertens. 2004;13:93–9.
13. Hinerfeld D, Traini MD, Weinberger RP, Cochran B, Doctrow SR, Harry J, Melov S. Endogenous mitochondrial oxidative stress: neurodegeneration, proteomic analysis, specific respiratory chain defects, and efficacious antioxidant therapy in superoxide dismutase 2 null mice. J Neurochem. 2004;88:657–67.
14. Mecocci P, Polidori MC. Antioxidant clinical trials in mild cognitive impairment and Alzheimer's disease. Biochim Biophys Acta. 1822;2012:631–8.
15. Galasko DR, Peskind E, Clark CM, Quinn JF, Ringman JM, Jicha GA, Cotman C, Cottrell B, Montine TJ, Thomas RG, Aisen P. Antioxidants for Alzheimer disease: a randomized clinical trial with cerebrospinal fluid biomarker measures. Arch Neurol. 2012;69:836–41.
16. Greenberg ER, Baron JA, Tosteson TD, Freeman Jr DH, Beck GJ, Bond JH, Colacchio TA, Coller JA, Frankl HD, Haile RW, et al. A clinical trial of antioxidant vitamins to prevent colorectal adenoma. Polyp Prevention Study Group. N Engl J Med. 1994;331:141–7.
17. Kris-Etherton PM, Lichtenstein AH, Howard BV, Steinberg D, Witztum JL. Antioxidant vitamin supplements and cardiovascular disease. Circulation. 2004;110:637–41.
18. Steinhubl SR. Why have antioxidants failed in clinical trials? Am J Cardiol. 2008;101:14D–9D.
19. The effect of vitamin E and beta carotene on the incidence of lung cancer and other cancers in male smokers. The Alpha-Tocopherol, Beta Carotene Cancer Prevention Study Group. N Engl J Med. 1994;330:1029–35.
20. Dietary supplementation with n-3 polyunsaturated fatty acids and vitamin E after myocardial infarction: results of the GISSI-Prevenzione trial. Gruppo Italiano per lo Studio della Sopravvivenza nell'Infarto miocardico. Lancet. 1999;354:447–55.
21. Yusuf S, Dagenais G, Pogue J, Bosch J, Sleight P. Vitamin E supplementation and cardiovascular events in high-risk patients. The Heart Outcomes Prevention Evaluation Study Investigators. N Engl J Med. 2000;342:154–60.

22. de Gaetano G. Low-dose aspirin and vitamin E in people at cardiovascular risk: a randomised trial in general practice. Collaborative Group of the Primary Prevention Project. Lancet. 2001;357:89–95.
23. Lonn E, Yusuf S, Hoogwerf B, Pogue J, Yi Q, Zinman B, Bosch J, Dagenais G, Mann JF, Gerstein HC. Effects of vitamin E on cardiovascular and microvascular outcomes in high-risk patients with diabetes: results of the HOPE study and MICRO-HOPE substudy. Diabetes Care. 2002;25:1919–27.
24. Hodis HN, Mack WJ, LaBree L, Mahrer PR, Sevanian A, Liu CR, Liu CH, Hwang J, Selzer RH, Azen SP. Alpha-tocopherol supplementation in healthy individuals reduces low-density lipoprotein oxidation but not atherosclerosis: the Vitamin E Atherosclerosis Prevention Study (VEAPS). Circulation. 2002;106:1453–9.
25. Virtamo J, Rapola JM, Ripatti S, Heinonen OP, Taylor PR, Albanes D, Huttunen JK. Effect of vitamin E and beta carotene on the incidence of primary nonfatal myocardial infarction and fatal coronary heart disease. Arch Intern Med. 1998;158:668–75.
26. Greenberg ER, Baron JA, Karagas MR, Stukel TA, Nierenberg DW, Stevens MM, Mandel JS, Haile RW. Mortality associated with low plasma concentration of beta carotene and the effect of oral supplementation. JAMA. 1996;275:699–703.
27. Hennekens CH, Buring JE, Manson JE, Stampfer M, Rosner B, Cook NR, Belanger C, LaMotte F, Gaziano JM, Ridker PM, Willett W, Peto R. Lack of effect of long-term supplementation with beta carotene on the incidence of malignant neoplasms and cardiovascular disease. N Engl J Med. 1996;334:1145–9.
28. Omenn GS, Goodman GE, Thornquist MD, Balmes J, Cullen MR, Glass A, Keogh JP, Meyskens FL, Valanis B, Williams JH, Barnhart S, Hammar S. Effects of a combination of beta carotene and vitamin A on lung cancer and cardiovascular disease. N Engl J Med. 1996;334:1150–5.
29. Brown BG, Zhao XQ, Chait A, Fisher LD, Cheung MC, Morse JS, Dowdy AA, Marino EK, Bolson EL, Alaupovic P, Frohlich J, Albers JJ. Simvastatin and niacin, antioxidant vitamins, or the combination for the prevention of coronary disease. N Engl J Med. 2001;345:1583–92.
30. MRC/BHF Heart Protection Study of antioxidant vitamin supplementation in 20,536 high-risk individuals: a randomised placebo-controlled trial. Lancet. 2002;360:23–33.
31. Waters DD, Alderman EL, Hsia J, Howard BV, Cobb FR, Rogers WJ, Ouyang P, Thompson P, Tardif JC, Higginson L, Bittner V, Steffes M, Gordon DJ, Proschan M, Younes N, Verter JI. Effects of hormone replacement therapy and antioxidant vitamin supplements on coronary atherosclerosis in postmenopausal women: a randomized controlled trial. JAMA. 2002;288:2432–40.
32. Murphy MP, Smith RA. Targeting antioxidants to mitochondria by conjugation to lipophilic cations. Annu Rev Pharmacol Toxicol. 2007;47:629–56.
33. Sheu S-S, Nauduri D, Anders MW. Targeting antioxidants to mitochondria: a new therapeutic direction. Biochim Biophys Acta (BBA) – Mol Basis Dis. 2006;1762:256–65.
34. Smith RA, Murphy MP. Mitochondria-targeted antioxidants as therapies. Discov Med. 2011;11:106–14.
35. Zhang ZW, Xu XC, Liu T, Yuan S. Mitochondrion-permeable antioxidants to treat ROS-burst-mediated acute diseases. Oxidative Med Cell Longev. 2016;2016:6859523.
36. Kelso GF, Porteous CM, Coulter CV, Hughes G, Porteous WK, Ledgerwood EC, Smith RA, Murphy MP. Selective targeting of a redox-active ubiquinone to mitochondria within cells: antioxidant and antiapoptotic properties. J Biol Chem. 2001;276:4588–96.
37. James AM, Cocheme HM, Smith RA, Murphy MP. Interactions of mitochondria-targeted and untargeted ubiquinones with the mitochondrial respiratory chain and reactive oxygen species. Implications for the use of exogenous ubiquinones as therapies and experimental tools. J Biol Chem. 2005;280:21295–312.
38. McManus MJ, Murphy MP, Franklin JL. The mitochondria-targeted antioxidant MitoQ prevents loss of spatial memory retention and early neuropathology in a transgenic mouse model of Alzheimer's disease. J Neurosci. 2011;31:15703–15.

39. Ghosh A, Chandran K, Kalivendi SV, Joseph J, Antholine WE, Hillard CJ, Kanthasamy A, Kalyanaraman B. Neuroprotection by a mitochondria-targeted drug in a Parkinson's disease model. Free Radic Biol Med. 2010;49:1674–84.
40. Chacko BK, Reily C, Srivastava A, Johnson MS, Ye Y, Ulasova E, Agarwal A, Zinn KR, Murphy MP, Kalyanaraman B, Darley-Usmar V. Prevention of diabetic nephropathy in Ins2(+/)(−)(AkitaJ) mice by the mitochondria-targeted therapy MitoQ. Biochem J. 2010;432:9–19.
41. von Montfort C, Matias N, Fernandez A, Fucho R, Conde de la Rosa L, Martinez-Chantar ML, Mato JM, Machida K, Tsukamoto H, Murphy MP, Mansouri A, Kaplowitz N, Garcia-Ruiz C, Fernandez-Checa JC. Mitochondrial GSH determines the toxic or therapeutic potential of superoxide scavenging in steatohepatitis. J Hepatol. 2012;57:852–9.
42. Chacko BK, Srivastava A, Johnson MS, Benavides GA, Chang MJ, Ye Y, Jhala N, Murphy MP, Kalyanaraman B, Darley-Usmar VM. Mitochondria-targeted ubiquinone (MitoQ) decreases ethanol-dependent micro and macro hepatosteatosis. Hepatology. 2011;54:153–63.
43. Pung YF, Rocic P, Murphy MP, Smith RA, Hafemeister J, Ohanyan V, Guarini G, Yin L, Chilian WM. Resolution of mitochondrial oxidative stress rescues coronary collateral growth in Zucker obese fatty rats. Arterioscler Thromb Vasc Biol. 2012;32:325–34.
44. Mercer JR, Yu E, Figg N, Cheng KK, Prime TA, Griffin JL, Masoodi M, Vidal-Puig A, Murphy MP, Bennett MR. The mitochondria-targeted antioxidant MitoQ decreases features of the metabolic syndrome in ATM+/−/ApoE−/− mice. Free Radic Biol Med. 2012;52:841–9.
45. Mitchell T, Rotaru D, Saba H, Smith RA, Murphy MP, MacMillan-Crow LA. The mitochondria-targeted antioxidant mitoquinone protects against cold storage injury of renal tubular cells and rat kidneys. J Pharmacol Exp Ther. 2011;336:682–92.
46. Adlam VJ, Harrison JC, Porteous CM, James AM, Smith RA, Murphy MP, Sammut IA. Targeting an antioxidant to mitochondria decreases cardiac ischemia-reperfusion injury. FASEB J. 2005;19:1088–95.
47. Dare AJ, Logan A, Prime TA, Rogatti S, Goddard M, Bolton EM, Bradley JA, Pettigrew GJ, Murphy MP, Saeb-Parsy K. The mitochondria-targeted anti-oxidant MitoQ decreases ischemia-reperfusion injury in a murine syngeneic heart transplant model. J Heart Lung Transplant. 2015;34:1471–80.
48. McLachlan J, Beattie E, Murphy MP, Koh-Tan CH, Olson E, Beattie W, Dominiczak AF, Nicklin SA, Graham D. Combined therapeutic benefit of mitochondria-targeted antioxidant, MitoQ10, and angiotensin receptor blocker, losartan, on cardiovascular function. J Hypertens. 2014;32:555–64.
49. Graham D, Huynh NN, Hamilton CA, Beattie E, Smith RA, Cocheme HM, Murphy MP, Dominiczak AF. Mitochondria-targeted antioxidant MitoQ10 improves endothelial function and attenuates cardiac hypertrophy. Hypertension. 2009;54:322–8.
50. Gioscia-Ryan RA, LaRocca TJ, Sindler AL, Zigler MC, Murphy MP, Seals DR. Mitochondria-targeted antioxidant (MitoQ) ameliorates age-related arterial endothelial dysfunction in mice. J Physiol. 2014;592:2549–61.
51. Supinski GS, Murphy MP, Callahan LA. MitoQ administration prevents endotoxin-induced cardiac dysfunction. Am J Physiol Regul Integr Comp Physiol. 2009;297:R1095–102.
52. Chandran K, Aggarwal D, Migrino RQ, Joseph J, McAllister D, Konorev EA, Antholine WE, Zielonka J, Srinivasan S, Avadhani NG, Kalyanaraman B. Doxorubicin inactivates myocardial cytochrome c oxidase in rats: cardioprotection by Mito-Q. Biophys J. 2009;96:1388–98.
53. Vergeade A, Mulder P, Vendeville-Dehaudt C, Estour F, Fortin D, Ventura-Clapier R, Thuillez C, Monteil C. Mitochondrial impairment contributes to cocaine-induced cardiac dysfunction: prevention by the targeted antioxidant MitoQ. Free Radic Biol Med. 2010;49:748–56.
54. Snow BJ, Rolfe FL, Lockhart MM, Frampton CM, O'Sullivan JD, Fung V, Smith RA, Murphy MP, Taylor KM. A double-blind, placebo-controlled study to assess the mitochondria-

targeted antioxidant MitoQ as a disease-modifying therapy in Parkinson's disease. Mov Disord. 2010;25:1670–4.
55. Gane EJ, Weilert F, Orr DW, Keogh GF, Gibson M, Lockhart MM, Frampton CM, Taylor KM, Smith RA, Murphy MP. The mitochondria-targeted anti-oxidant mitoquinone decreases liver damage in a phase II study of hepatitis C patients. Liver Int. 2010;30:1019–26.
56. Smith RA, Porteous CM, Coulter CV, Murphy MP. Selective targeting of an antioxidant to mitochondria. Eur J Biochem. 1999;263:709–16.
57. Jauslin ML, Meier T, Smith RA, Murphy MP. Mitochondria-targeted antioxidants protect Friedreich Ataxia fibroblasts from endogenous oxidative stress more effectively than untargeted antioxidants. FASEB J. 2003;17:1972–4.
58. Dhanasekaran A, Kotamraju S, Kalivendi SV, Matsunaga T, Shang T, Keszler A, Joseph J, Kalyanaraman B. Supplementation of endothelial cells with mitochondria-targeted antioxidants inhibit peroxide-induced mitochondrial iron uptake, oxidative damage, and apoptosis. J Biol Chem. 2004;279:37575–87.
59. Siler-Marsiglio KI, Pan Q, Paiva M, Madorsky I, Khurana NC, Heaton MB. Mitochondrially targeted vitamin E and vitamin E mitigate ethanol-mediated effects on cerebellar granule cell antioxidant defense systems. Brain Res. 2005;1052:202–11.
60. Covey MV, Murphy MP, Hobbs CE, Smith RA, Oorschot DE. Effect of the mitochondrial antioxidant, Mito Vitamin E, on hypoxic-ischemic striatal injury in neonatal rats: a dose-response and stereological study. Exp Neurol. 2006;199:513–9.
61. Mao G, Kraus GA, Kim I, Spurlock ME, Bailey TB, Zhang Q, Beitz DC. A mitochondria-targeted vitamin E derivative decreases hepatic oxidative stress and inhibits fat deposition in mice. J Nutr. 2010;140:1425–31.
62. Kurochkin IO, Etzkorn M, Buchwalter D, Leamy L, Sokolova IM. Top-down control analysis of the cadmium effects on molluscan mitochondria and the mechanisms of cadmium-induced mitochondrial dysfunction. Am J Physiol Regul Integr Comp Physiol. 2011;300:R21–31.
63. Babizhayev MA, Yegorov YE. Reactive oxygen species and the aging eye: specific role of metabolically active mitochondria in maintaining lens function and in the initiation of the oxidation-induced maturity onset cataract – a novel platform of mitochondria-targeted antioxidants with broad therapeutic potential for redox regulation and detoxification of oxidants in eye diseases. Am J Ther. 2016;23:e98–117.
64. Babizhayev MA. Generation of reactive oxygen species in the anterior eye segment. Synergistic codrugs of N-acetylcarnosine lubricant eye drops and mitochondria-targeted antioxidant act as a powerful therapeutic platform for the treatment of cataracts and primary open-angle glaucoma. BBA Clin. 2016;6:49–68.
65. Murphy MP, Echtay KS, Blaikie FH, Asin-Cayuela J, Cocheme HM, Green K, Buckingham JA, Taylor ER, Hurrell F, Hughes G, Miwa S, Cooper CE, Svistunenko DA, Smith RA, Brand MD. Superoxide activates uncoupling proteins by generating carbon-centered radicals and initiating lipid peroxidation: studies using a mitochondria-targeted spin trap derived from alpha-phenyl-N-tert-butylnitrone. J Biol Chem. 2003;278:48534–45.
66. Reddy PH. Mitochondrial oxidative damage in aging and Alzheimer's disease: implications for mitochondrially targeted antioxidant therapeutics. J Biomed Biotechnol. 2006;2006:31372.
67. Poeggeler B, Durand G, Polidori A, Pappolla MA, Vega-Naredo I, Coto-Montes A, Boker J, Hardeland R, Pucci B. Mitochondrial medicine: neuroprotection and life extension by the new amphiphilic nitrone LPBNAH acting as a highly potent antioxidant agent. J Neurochem. 2005;95:962–73.
68. Filipovska A, Kelso GF, Brown SE, Beer SM, Smith RA, Murphy MP. Synthesis and characterization of a triphenylphosphonium-conjugated peroxidase mimetic. Insights into the interaction of ebselen with mitochondria. J Biol Chem. 2005;280:24113–26.
69. Stoyanovsky DA, Jiang J, Murphy MP, Epperly M, Zhang X, Li S, Greenberger J, Kagan V, Bayir H. Design and synthesis of a mitochondria-targeted mimic of glutathione peroxidase, MitoEbselen-2, as a radiation mitigator. ACS Med Chem Lett. 2014;5:1304–7.
70. Antonenko YN, Avetisyan AV, Bakeeva LE, Chernyak BV, Chertkov VA, Domnina LV, Ivanova OY, Izyumov DS, Khailova LS, Klishin SS, Korshunova GA, Lyamzaev KG, Muntyan MS, Nepryakhina OK, Pashkovskaya AA, Pletjushkina OY, Pustovidko AV,

Roginsky VA, Rokitskaya TI, Ruuge EK, Saprunova VB, Severina II, Simonyan RA, Skulachev IV, Skulachev MV, Sumbatyan NV, Sviryaeva IV, Tashlitsky VN, Vassiliev JM, Vyssokikh MY, Yaguzhinsky LS, Zamyatnin Jr AA, Skulachev VP. Mitochondria-targeted plastoquinone derivatives as tools to interrupt execution of the aging program. 1. Cationic plastoquinone derivatives: synthesis and in vitro studies. Biochemistry (Mosc). 2008;73:1273–87.

71. Bakeeva LE, Barskov IV, Egorov MV, Isaev NK, Kapelko VI, Kazachenko AV, Kirpatovsky VI, Kozlovsky SV, Lakomkin VL, Levina SB, Pisarenko OI, Plotnikov EY, Saprunova VB, Serebryakova LI, Skulachev MV, Stelmashook EV, Studneva IM, Tskitishvili OV, Vasilyeva AK, Victorov IV, Zorov DB, Skulachev VP. Mitochondria-targeted plastoquinone derivatives as tools to interrupt execution of the aging program. 2. Treatment of some ROS- and age-related diseases (heart arrhythmia, heart infarctions, kidney ischemia, and stroke). Biochemistry (Mosc). 2008;73:1288–99.
72. Manskikh VN, Gancharova OS, Nikiforova AI, Krasilshchikova MS, Shabalina IG, Egorov MV, Karger EM, Milanovsky GE, Galkin II, Skulachev VP, Zinovkin RA. Age-associated murine cardiac lesions are attenuated by the mitochondria-targeted antioxidant SkQ1. Histol Histopathol. 2015;30:353–60.
73. Skulachev MV, Antonenko YN, Anisimov VN, Chernyak BV, Cherepanov DA, Chistyakov VA, Egorov MV, Kolosova NG, Korshunova GA, Lyamzaev KG, Plotnikov EY, Roginsky VA, Savchenko AY, Severina II, Severin FF, Shkurat TP, Tashlitsky VN, Shidlovsky KM, Vyssokikh MY, Zamyatnin Jr AA, Zorov DB, Skulachev VP. Mitochondrial-targeted plastoquinone derivatives. Effect on senescence and acute age-related pathologies. Curr Drug Targets. 2011;12:800–26.
74. Petrov A, Perekhvatova N, Skulachev M, Stein L, Ousler G. SkQ1 ophthalmic solution for dry eye treatment: results of a phase 2 safety and efficacy clinical study in the environment and during challenge in the controlled adverse environment model. Adv Ther. 2016;33:96–115.
75. Zhao K, Zhao GM, Wu D, Soong Y, Birk AV, Schiller PW, Szeto HH. Cell-permeable peptide antioxidants targeted to inner mitochondrial membrane inhibit mitochondrial swelling, oxidative cell death, and reperfusion injury. J Biol Chem. 2004;279:34682–90.
76. Szeto HH. Cell-permeable, mitochondrial-targeted, peptide antioxidants. AAPS J. 2006;8:E277–83.
77. Zhao K, Luo G, Giannelli S, Szeto HH. Mitochondria-targeted peptide prevents mitochondrial depolarization and apoptosis induced by tert-butyl hydroperoxide in neuronal cell lines. Biochem Pharmacol. 2005;70:1796–806.
78. Thomas DA, Stauffer C, Zhao K, Yang H, Sharma VK, Szeto HH, Suthanthiran M. Mitochondrial targeting with antioxidant peptide SS-31 prevents mitochondrial depolarization, reduces islet cell apoptosis, increases islet cell yield, and improves posttransplantation function. J Am Soc Nephrol. 2007;18:213–22.
79. Szeto HH, Liu S, Soong Y, Wu D, Darrah SF, Cheng FY, Zhao Z, Ganger M, Tow CY, Seshan SV. Mitochondria-targeted peptide accelerates ATP recovery and reduces ischemic kidney injury. J Am Soc Nephrol. 2011;22:1041–52.
80. Liu S, Soong Y, Seshan SV, Szeto HH. Novel cardiolipin therapeutic protects endothelial mitochondria during renal ischemia and mitigates microvascular rarefaction, inflammation, and fibrosis. Am J Physiol Renal Physiol. 2014;306:F970–80.
81. Mizuguchi Y, Chen J, Seshan SV, Poppas DP, Szeto HH, Felsen D. A novel cell-permeable antioxidant peptide decreases renal tubular apoptosis and damage in unilateral ureteral obstruction. Am J Physiol Renal Physiol. 2008;295:F1545–53.
82. Hou Y, Li S, Wu M, Wei J, Ren Y, Du C, Wu H, Han C, Duan H, Shi Y. Mitochondria-targeted peptide SS-31 attenuates renal injury via an antioxidant effect in diabetic nephropathy. Am J Physiol Renal Physiol. 2016;310:F547–59.
83. Wu J, Zhang M, Li H, Sun X, Hao S, Ji M, Yang J, Li K. BDNF pathway is involved in the protective effects of SS-31 on isoflurane-induced cognitive deficits in aging mice. Behav Brain Res. 2016;305:115–21.

84. Wu J, Zhang M, Hao S, Jia M, Ji M, Qiu L, Sun X, Yang J, Li K. Mitochondria-targeted peptide reverses mitochondrial dysfunction and cognitive deficits in sepsis-associated encephalopathy. Mol Neurobiol. 2015;52:783–91.
85. Siegel MP, Kruse SE, Percival JM, Goh J, White CC, Hopkins HC, Kavanagh TJ, Szeto HH, Rabinovitch PS, Marcinek DJ. Mitochondrial-targeted peptide rapidly improves mitochondrial energetics and skeletal muscle performance in aged mice. Aging Cell. 2013;12:763–71.
86. Righi V, Constantinou C, Mintzopoulos D, Khan N, Mupparaju SP, Rahme LG, Swartz HM, Szeto HH, Tompkins RG, Tzika AA. Mitochondria-targeted antioxidant promotes recovery of skeletal muscle mitochondrial function after burn trauma assessed by in vivo 31P nuclear magnetic resonance and electron paramagnetic resonance spectroscopy. FASEB J. 2013;27:2521–30.
87. Lee HY, Kaneki M, Andreas J, Tompkins RG, Martyn JA. Novel mitochondria-targeted antioxidant peptide ameliorates burn-induced apoptosis and endoplasmic reticulum stress in the skeletal muscle of mice. Shock. 2011;36:580–5.
88. Wu D, Soong Y, Zhao GM, Szeto HH. A highly potent peptide analgesic that protects against ischemia-reperfusion-induced myocardial stunning. Am J Physiol Heart Circ Physiol. 2002;283:H783–91.
89. Cho J, Won K, Wu D, Soong Y, Liu S, Szeto HH, Hong MK. Potent mitochondria-targeted peptides reduce myocardial infarction in rats. Coron Artery Dis. 2007;18:215–20.
90. Dai DF, Chen T, Szeto H, Nieves-Cintron M, Kutyavin V, Santana LF, Rabinovitch PS. Mitochondrial targeted antioxidant Peptide ameliorates hypertensive cardiomyopathy. J Am Coll Cardiol. 2011;58:73–82.
91. Chakrabarti AK, Feeney K, Abueg C, Brown DA, Czyz E, Tendera M, Janosi A, Giugliano RP, Kloner RA, Weaver WD, Bode C, Godlewski J, Merkely B, Gibson CM. Rationale and design of the EMBRACE STEMI study: a phase 2a, randomized, double-blind, placebo-controlled trial to evaluate the safety, tolerability and efficacy of intravenous Bendavia on reperfusion injury in patients treated with standard therapy including primary percutaneous coronary intervention and stenting for ST-segment elevation myocardial infarction. Am Heart J. 2013;165:509–14. e7
92. Kikuchi K, Tancharoen S, Takeshige N, Yoshitomi M, Morioka M, Murai Y, Tanaka E. The efficacy of edaravone (radicut), a free radical scavenger, for cardiovascular disease. Int J Mol Sci. 2013;14:13909–30.
93. Ji L, Liu Y, Zhang Y, Chang W, Gong J, Wei S, Li X, Qin L. The antioxidant edaravone prevents cardiac dysfunction by suppressing oxidative stress in type 1 diabetic rats and in high-glucose-induced injured H9c2 cardiomyoblasts. Can J Physiol Pharmacol. 2016;94:996–1006.
94. Zhang GW, Gu TX, Sun XJ, Wang C, Qi X, Wang XB, Li-Ling J. Edaravone promotes activation of resident cardiac stem cells by transplanted mesenchymal stem cells in a rat myocardial infarction model. J Thorac Cardiovasc Surg. 2016;152:570–82.
95. Hassan MQ, Akhtar MS, Akhtar M, Ali J, Haque SE, Najmi AK. Edaravone, a potent free radical scavenger and a calcium channel blocker attenuate isoproterenol induced myocardial infarction by suppressing oxidative stress, apoptotic signaling and ultrastructural damage. Ther Adv Cardiovasc Dis. 2016;10:214–23.
96. Rajesh KG, Sasaguri S, Suzuki R, Maeda H. Antioxidant MCI-186 inhibits mitochondrial permeability transition pore and upregulates Bcl-2 expression. Am J Physiol Heart Circ Physiol. 2003;285:H2171–8.
97. Tsujita K, Shimomura H, Kawano H, Hokamaki J, Fukuda M, Yamashita T, Hida S, Nakamura Y, Nagayoshi Y, Sakamoto T, Yoshimura M, Arai H, Ogawa H. Effects of edaravone on reperfusion injury in patients with acute myocardial infarction. Am J Cardiol. 2004;94:481–4.
98. Gueven N, Woolley K, Smith J. Border between natural product and drug: comparison of the related benzoquinones idebenone and coenzyme Q10. Redox Biol. 2015;4:289–95.

99. Suno M, Nagaoka A. Inhibition of lipid peroxidation by a novel compound (CV-2619) in brain mitochondria and mode of action of the inhibition. Biochem Biophys Res Commun. 1984;125:1046–52.
100. Sugiyama Y, Fujita T. Stimulation of the respiratory and phosphorylating activities in rat brain mitochondria by idebenone (CV-2619), a new agent improving cerebral metabolism. FEBS Lett. 1985;184:48–51.
101. Thal LJ, Grundman M, Berg J, Ernstrom K, Margolin R, Pfeiffer E, Weiner MF, Zamrini E, Thomas RG. Idebenone treatment fails to slow cognitive decline in Alzheimer's disease. Neurology. 2003;61:1498–502.
102. Lagedrost SJ, Sutton MS, Cohen MS, Satou GM, Kaufman BD, Perlman SL, Rummey C, Meier T, Lynch DR. Idebenone in Friedreich ataxia cardiomyopathy-results from a 6-month phase III study (IONIA). Am Heart J. 2011;161:639–45. e1
103. Kearney M, Orrell RW, Fahey M, Brassington R, Pandolfo M. Pharmacological treatments for Friedreich ataxia. Cochrane Database Syst Rev. 2016;8:CD007791.
104. Buyse GM, Voit T, Schara U, Straathof CS, D'Angelo MG, Bernert G, Cuisset JM, Finkel RS, Goemans N, McDonald CM, Rummey C, Meier T. Efficacy of idebenone on respiratory function in patients with Duchenne muscular dystrophy not using glucocorticoids (DELOS): a double-blind randomised placebo-controlled phase 3 trial. Lancet. 2015;385:1748–57.
105. McDonald CM, Meier T, Voit T, Schara U, Straathof CS, D'Angelo MG, Bernert G, Cuisset JM, Finkel RS, Goemans N, Rummey C, Leinonen M, Spagnolo P, Buyse GM. Idebenone reduces respiratory complications in patients with Duchenne muscular dystrophy. Neuromuscul Disord. 2016;26:473–80.
106. Gueven N. Idebenone for Leber's hereditary optic neuropathy. Drugs Today (Barc). 2016;52:173–81.
107. Rustin P, Rotig A, Munnich A, Sidi D. Heart hypertrophy and function are improved by idebenone in Friedreich's ataxia. Free Radic Res. 2002;36:467–9.
108. Mariotti C, Solari A, Torta D, Marano L, Fiorentini C, Di Donato S. Idebenone treatment in Friedreich patients: one-year-long randomized placebo-controlled trial. Neurology. 2003;60:1676–9.
109. Velasco-Sanchez D, Aracil A, Montero R, Mas A, Jimenez L, O'Callaghan M, Tondo M, Capdevila A, Blanch J, Artuch R, Pineda M. Combined therapy with idebenone and deferiprone in patients with Friedreich's ataxia. Cerebellum. 2011;10:1–8.
110. Arpa J, Sanz-Gallego I, Rodriguez-de-Rivera FJ, Dominguez-Melcon FJ, Prefasi D, Oliva-Navarro J, Moreno-Yanguela M. Triple therapy with deferiprone, idebenone and riboflavin in Friedreich's ataxia – open-label trial. Acta Neurol Scand. 2014;129:32–40.
111. Kim HJ, Koo SY, Ahn BH, Park O, Park DH, Seo DO, Won JH, Yim HJ, Kwak HS, Park HS, Chung CW, Oh YL, Kim SH. NecroX as a novel class of mitochondrial reactive oxygen species and ONOO(-) scavenger. Arch Pharm Res. 2010;33:1813–23.
112. Thu VT, Kim HK, Long le T, Nyamaa B, Song IS, Thuy TT, Huy NQ, Marquez J, Kim SH, Kim N, Ko KS, Rhee BD, Han J. NecroX-5 protects mitochondrial oxidative phosphorylation capacity and preserves PGC1alpha expression levels during hypoxia/reoxygenation injury. Korean J Physiol Pharmacol. 2016;20:201–11.
113. Thu VT, Kim HK, Long le T, Lee SR, Hanh TM, Ko TH, Heo HJ, Kim N, Kim SH, Ko KS, Rhee BD, Han J. NecroX-5 prevents hypoxia/reoxygenation injury by inhibiting the mitochondrial calcium uniporter. Cardiovasc Res. 2012;94:342–50.
114. Choi JM, Park KM, Kim SH, Hwang DW, Chon SH, Lee JH, Lee SY, Lee YJ. Effect of necrosis modulator necrox-7 on hepatic ischemia-reperfusion injury in beagle dogs. Transplant Proc. 2010;42:3414–21.
115. Lee DW, Lee TK, Cho IS, Park HE, Jin S, Cho HJ, Kim SH, Oh S, Kim HS. Creation of myocardial fibrosis by transplantation of fibroblasts primed with survival factors. Am J Physiol Heart Circ Physiol. 2011;301:H1004–14.
116. Park JH, Seo KS, Tadi S, Ahn BH, Lee JU, Heo JY, Han J, Song MS, Kim SH, Yim YH, Choi HS, Shong M, Kweon G. An indole derivative protects against acetaminophen-induced liver

injury by directly binding to N-acetyl-p-benzoquinone imine in mice. Antioxid Redox Signal. 2013;18:1713–22.
117. Kim SR, Kim DI, Kim SH, Lee H, Lee KS, Cho SH, Lee YC. NLRP3 inflammasome activation by mitochondrial ROS in bronchial epithelial cells is required for allergic inflammation. Cell Death Dis. 2014;5:e1498.
118. Jin SA, Kim SK, Seo HJ, Jeong JY, Ahn KT, Kim JH, Choi DE, Park JH, Lee JH, Choi SW, Seong IW, Kim SH, Suh KS, Jeong JO. Beneficial effects of necrosis modulator, indole derivative NecroX-7, on renal ischemia-reperfusion injury in rats. Transplant Proc. 2016;48:199–204.
119. Biswas SK. NecroX-7 may appear as a new molecule to stabilize atherosclerotic plaques. Atherosclerosis. 2016;252:190–1.
120. Grootaert MO, Schrijvers DM, Van Spaendonk H, Breynaert A, Hermans N, Van Hoof VO, Takahashi N, Vandenabeele P, Kim SH, De Meyer GR, Martinet W. NecroX-7 reduces necrotic core formation in atherosclerotic plaques of Apoe knockout mice. Atherosclerosis. 2016;252:166–74.
121. Fassett RG, Coombes JS. Astaxanthin: a potential therapeutic agent in cardiovascular disease. Mar Drugs. 2011;9:447–65.
122. Kuroki T, Ikeda S, Okada T, Maoka T, Kitamura A, Sugimoto M, Kume S. Astaxanthin ameliorates heat stress-induced impairment of blastocyst development in vitro: – astaxanthin colocalization with and action on mitochondria. J Assist Reprod Genet. 2013;30:623–31.
123. Manabe E, Handa O, Naito Y, Mizushima K, Akagiri S, Adachi S, Takagi T, Kokura S, Maoka T, Yoshikawa T. Astaxanthin protects mesangial cells from hyperglycemia-induced oxidative signaling. J Cell Biochem. 2008;103:1925–37.
124. Liu X, Osawa T. Astaxanthin protects neuronal cells against oxidative damage and is a potent candidate for brain food. Forum Nutr. 2009;61:129–35.
125. Liu X, Shibata T, Hisaka S, Osawa T. Astaxanthin inhibits reactive oxygen species-mediated cellular toxicity in dopaminergic SH-SY5Y cells via mitochondria-targeted protective mechanism. Brain Res. 2009;1254:18–27.
126. Kurashige M, Okimasu E, Inoue M, Utsumi K. Inhibition of oxidative injury of biological membranes by astaxanthin. Physiol Chem Phys Med NMR. 1990;22:27–38.
127. Lee DH, Kim CS, Lee YJ. Astaxanthin protects against MPTP/MPP+-induced mitochondrial dysfunction and ROS production in vivo and in vitro. Food Chem Toxicol. 2011;49:271–80.
128. Gaki GS, Papavassiliou AG. Oxidative stress-induced signaling pathways implicated in the pathogenesis of Parkinson's disease. Neruomol Med. 2014;16:217–30.
129. Curek GD, Cort A, Yucel G, Demir N, Ozturk S, Elpek GO, Savas B, Aslan M. Effect of astaxanthin on hepatocellular injury following ischemia/reperfusion. Toxicology. 2010;267:147–53.
130. Nakao R, Nelson OL, Park JS, Mathison BD, Thompson PA, Chew BP. Effect of astaxanthin supplementation on inflammation and cardiac function in BALB/c mice. Anticancer Res. 2010;30:2721–5.
131. Gross GJ, Lockwood SF. Cardioprotection and myocardial salvage by a disodium disuccinate astaxanthin derivative (Cardax). Life Sci. 2004;75:215–24.
132. Gross GJ, Hazen SL, Lockwood SF. Seven day oral supplementation with Cardax (disodium disuccinate astaxanthin) provides significant cardioprotection and reduces oxidative stress in rats. Mol Cell Biochem. 2006;283:23–30.
133. Hussein G, Nakamura M, Zhao Q, Iguchi T, Goto H, Sankawa U, Watanabe H. Antihypertensive and neuroprotective effects of astaxanthin in experimental animals. Biol Pharm Bull. 2005;28:47–52.
134. Gross GJ, Lockwood SF. Acute and chronic administration of disodium disuccinate astaxanthin (Cardax) produces marked cardioprotection in dog hearts. Mol Cell Biochem. 2005;272:221–7.
135. Marazzi G, Pelliccia F, Campolongo G, Quattrino S, Cacciotti L, Volterrani M, Gaudio C, Rosano G. Usefulness of nutraceuticals (armolipid plus) versus ezetimibe and combination in

statin-intolerant patients with dyslipidemia with coronary heart disease. Am J Cardiol. 2015;116:1798–801.

136. Maki KC, Geohas JG, Dicklin MR, Huebner M, Udani JK. Safety and lipid-altering efficacy of a new omega-3 fatty acid and antioxidant-containing medical food in men and women with elevated triacylglycerols. Prostaglandins Leukot Essent Fat Acids. 2015;99:41–6.
137. Coombes JS, Sharman JE, Fassett RG. Astaxanthin has no effect on arterial stiffness, oxidative stress, or inflammation in renal transplant recipients: a randomized controlled trial (the XANTHIN trial). Am J Clin Nutr. 2016;103:283–9.
138. Suzuki T, Yamaguchi H, Kikusato M, Matsuhashi T, Matsuo A, Sato T, Oba Y, Watanabe S, Minaki D, Saigusa D, Shimbo H, Mori N, Mishima E, Shima H, Akiyama Y, Takeuchi Y, Yuri A, Kikuchi K, Toyohara T, Suzuki C, Kohzuki M, Anzai J, Mano N, Kure S, Yanagisawa T, Tomioka Y, Toyomizu M, Ito S, Osaka H, Hayashi K, Abe T. Mitochonic acid 5 (MA-5), a derivative of the plant hormone indole-3-acetic acid, improves survival of fibroblasts from patients with mitochondrial diseases. Tohoku J Exp Med. 2015;236:225–32.
139. Suzuki T, Yamaguchi H, Kikusato M, Hashizume O, Nagatoishi S, Matsuo A, Sato T, Kudo T, Matsuhashi T, Murayama K, Ohba Y, Watanabe S, Kanno S, Minaki D, Saigusa D, Shinbo H, Mori N, Yuri A, Yokoro M, Mishima E, Shima H, Akiyama Y, Takeuchi Y, Kikuchi K, Toyohara T, Suzuki C, Ichimura T, Anzai J, Kohzuki M, Mano N, Kure S, Yanagisawa T, Tomioka Y, Toyomizu M, Tsumoto K, Nakada K, Bonventre JV, Ito S, Osaka H, Hayashi K, Abe T. Mitochonic acid 5 binds mitochondria and ameliorates renal tubular and cardiac myocyte mamage. J Am Soc Nephrol. 2016;27:1925–32.
140. Cerrato CP, Pirisinu M, Vlachos EN, Langel U. Novel cell-penetrating peptide targeting mitochondria. FASEB J. 2015;29:4589–99.
141. Bi W, Bi Y, Gao X, Yan X, Zhang Y, Xue P, Bammert CE, Legalley TD, Michael Gibson K, Bi L, Wang JX. Anti-inflammatory, analgesic and antioxidant activities of novel kyotorphin-nitroxide hybrid molecules. Bioorg Med Chem Lett. 2016;26:2005–13.

Kolibri 360 GmbH
Ferdinand-Jühlke-Straße 7
99095 Erfurt, Deutschland
produktsicherheit@kolibri360.de